The Essential Guide to Image Processing

Second Edition

The Essential Guide to Image Processing

Second Edition

EDITOR

Al Bovik

Department of Electrical and Computer Engineering
The University of Texas at Austin
Austin, Texas

AMSTERDAM • BOSTON • HEIDELBERG • LONDON
NEW YORK • OXFORD • PARIS • SAN DIEGO
SAN FRANCISCO • SINGAPORE • SYDNEY • TOKYO

Academic Press is an imprint of Elsevier

ELSEVIER

Academic Press is an imprint of Elsevier
30 Corporate Drive, Suite 400, Burlington, MA 01803, USA
525 B Street, Suite 1900, San Diego, California 92101-4495, USA
84 Theobald's Road, London WC1X 8RR, UK

Library of Congress Cataloging-in-Publication Data
Application submitted

British Library Cataloguing-in-Publication Data
A catalogue record for this book is available from the British Library.

ISBN: 978-0-12-374457-9

For information on all Academic Press publications
visit our Web site at *www.elsevierdirect.com*

Typeset by: diacriTech, India

Printed in the United States of America
09 10 11 12   9 8 7 6 5 4 3 2 1

Contents

Preface

The visual experience is the principal way that humans sense and communicate with their world. We are visual beings and images are being made increasing available to us in electronic digital format via digital cameras, the internet, and hand-held devices with large-format screens. With much of the technology being introduced to the consumer marketplace being rather new, digital image processing remains a "hot" topic and promises to be one for a very long time. Of course, digital image processing has been around for quite awhile, and indeed, methods pervade nearly every branch of science and engineering. One only has to view the latest space telescope images or read about the newest medical image modality to be aware of this.

With this introduction, welcome to *The Essential Guide to Image Processing*! The reader will find that this *Guide* covers introductory, intermediate and advanced topics of digital image processing, and is intended to be highly accessible for those entering the field or wishing to learn about the topic for the first time. As such, the *Guide* can be effectively used as a classroom textbook. Since many intermediate and advanced topics are also covered, the *Guide* is a useful reference for the practicing image processing engineer, scientist, or researcher. As a learning tool, the *Guide* offers easy-to-read material at different levels of presentation, including introductory and tutorial chapters on the most basic image processing techniques. Further, there is included a chapter that explains digital image processing software that is included on a CD with the book. This software is part of the award-winning *SIVA* educational courseware that has been under development at The University of Texas for more than a decade, and which has been adopted for use by more than 400 educational, industry, and research institutions around the world. Image processing educators are invited these user-friendly and intuitive live image processing demonstrations into their teaching curriculum.

The *Guide* contains 27 chapters, beginning with an introduction and a description of the educational software that is included with the book. This is followed by tutorial chapters on the basic methods of gray-level and binary image processing, and on the essential tools of image Fourier analysis and linear convolution systems. The next series of chapters describes tools and concepts necessary to more advanced image processing algorithms, including wavelets, color, and statistical and noise models of images. Methods for improving the appearance of images follow, including enhancement, denoising and restoration (deblurring). The important topic of image compression follows, including chapters on lossless compression, the JPEG and JPEG-2000 standards, and wavelet image compression. Image analysis chapters follow, including two chapters on edge detection and one on the important topic of image quality assessment. Finally, the *Guide* concludes with six exciting chapters dealing explaining image processing applications on such diverse topics as image watermarking, fingerprint recognition, digital microscopy, face recognition, and digital tomography. These have been selected for their timely interest, as well as their illustrative power of how image processing and analysis can be effectively applied to problems of significant practical interest.

The *Guide* then concludes with a chapter pointing towards the topic of digital *video* processing, which deals with visual signals that vary over time. These very broad and more advanced field is covered in a companion volume suitably entitled *The Essential Guide to Video Processing*. The topics covered in the two companion *Guides* are, of course closely related, and it may interest the reader that earlier editions of most of this material appeared in a highly popular but gigantic volume known as *The Handbook of Image and Video Processing*. While this previous book was very well-received, its sheer size made it highly un-portable (but a fantastic doorstop). For this newer rendition, in addition to updating the content, I made the decision to divide the material into two distinct books, separating the material into coverage of still images and moving images (video). I am sure that you will find the resulting volumes to be information-rich as well as highly accessible.

As Editor and Co-Author of *The Essential Guide to Image Processing*, I would thank the many co-authors who have contributed such wonderful work to this *Guide*. They are all models of professionalism, responsiveness, and patience with respect to my cheerleading and cajoling. The group effort that created this book is much larger, deeper, and of higher quality than I think that any individual could have created. Each and every chapter in this *Guide* has been written by a carefully selected distinguished specialist, ensuring that the greatest depth of understanding be communicated to the reader. I have also taken the time to read each and every word of every chapter, and have provided extensive feedback to the chapter authors in seeking to perfect the book. Owing primarily to their efforts, I feel certain that this *Guide* will prove to be an essential and indispensable resource for years to come.

I would also like to thank the staff at Elsevier—the Senior Commissioning Editor, Tim Pitts, for his continuous stream of ideas and encouragement, and for keeping after me to do this project; Melanie Benson for her tireless efforts and incredible organization and accuracy in making the book happen; Eric DeCicco, the graphic artist for his efforts on the wonderful cover design, and Greg Dezarn-O'Hare for his flawless typesetting.

National Instruments, Inc., has been a tremendous support over the years in helping me develop courseware for image processing classes at The University of Texas at Austin, and has been especially generous with their engineer's time. I particularly thank NI engineers George Panayi, Frank Baumgartner, Nate Holmes, Carleton Heard, Matthew Slaughter, and Nathan McKimpson for helping to develop and perfect the many Labview demos that have been used for many years and are now available on the CD-ROM attached to this book.

Al Bovik
Austin, Texas
April, 2009

About the Author

Al Bovik currently holds the Curry/Cullen Trust Endowed Chair Professorship in the Department of Electrical and Computer Engineering at The University of Texas at Austin, where he is the Director of the Laboratory for Image and Video Engineering (LIVE). He has published over 500 technical articles and six books in the general area of image and video processing and holds two US patents.

Dr. Bovik has received a number of major awards from the IEEE Signal Processing Society, including the Education Award (2007); the Technical Achievement Award (2005), the Distinguished Lecturer Award (2000); and the Meritorious Service Award (1998). He is also a recipient of the IEEE Third Millennium Medal (2000), and has won two journal paper awards from the Pattern Recognition Society (1988 and 1993). He is a Fellow of the IEEE, a Fellow of the Optical Society of America, and a Fellow of the Society of Photo-Optical and Instrumentation Engineers. Dr. Bovik has served Editor-in-Chief of the *IEEE Transactions on Image Processing* (1996–2002) and created and served as the first General Chairman of the *IEEE International Conference on Image Processing*, which was held in Austin, Texas, in 1994.

Introduction to Digital Image Processing

1

Alan C. Bovik

The University of Texas at Austin

We are in the middle of an exciting period of time in the field of image processing. Indeed, scarcely a week passes where we do not hear an announcement of some new technological breakthrough in the areas of digital computation and telecommunication. Particularly exciting has been the participation of the general public in these developments, as affordable computers and the incredible explosion of the World Wide Web have brought a flood of instant information into a large and increasing percentage of homes and businesses. Indeed, the advent of broadband wireless devices is bringing these technologies into the pocket and purse. Most of this information is designed for *visual* consumption in the form of text, graphics, and pictures, or integrated *multimedia* presentations. *Digital images* are pictures that have been converted into a computer-readable binary format consisting of logical 0s and 1s. Usually, by an image we mean a still picture that does not change with time, whereas a video evolves with time and generally contains moving and/or changing objects. This *Guide* deals primarily with still images, while a second (companion) volume deals with moving images, or videos. Digital images are usually obtained by converting continuous signals into digital format, although "direct digital" systems are becoming more prevalent. Likewise, digital images are viewed using diverse display media, included digital printers, computer monitors, and digital projection devices. The frequency with which information is transmitted, stored, processed, and displayed in a digital visual format is increasing rapidly, and as such, the design of engineering methods for efficiently transmitting, maintaining, and even improving the visual integrity of this information is of heightened interest.

One aspect of image processing that makes it such an interesting topic of study is the amazing diversity of applications that make use of image processing or analysis techniques. Virtually every branch of science has subdisciplines that use recording devices or sensors to collect image data from the universe around us, as depicted in Fig. 1.1. This data is often multidimensional and can be arranged in a format that is suitable for human viewing. Viewable datasets like this can be regarded as images and processed using established techniques for image processing, even if the information has not been derived from visible light sources.

1

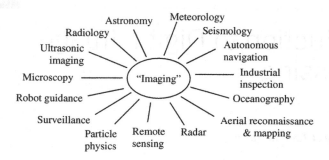

FIGURE 1.1

Part of the universe of image processing applications.

1.1 TYPES OF IMAGES

Another rich aspect of digital imaging is the diversity of image types that arise, and which can derive from nearly every type of radiation. Indeed, some of the most exciting developments in medical imaging have arisen from new sensors that record image data from previously little used sources of radiation, such as PET (positron emission tomography) and MRI (magnetic resonance imaging), or that sense radiation in new ways, as in CAT (computer-aided tomography), where X-ray data is collected from multiple angles to form a rich aggregate image.

There is an amazing availability of radiation to be sensed, recorded as images, and viewed, analyzed, transmitted, or stored. In our daily experience, we think of "what we see" as being "what is there," but in truth, our eyes record very little of the information that is available at any given moment. As with any sensor, the human eye has a limited bandwidth. The band of electromagnetic (EM) radiation that we are able to see, or "visible light," is quite small, as can be seen from the plot of the EM band in Fig. 1.2. Note that the horizontal axis is logarithmic! At any given moment, we see very little of the available radiation that is going on around us, although certainly enough to get around. From an evolutionary perspective, the band of EM wavelengths that the human eye perceives is perhaps optimal, since the volume of data is reduced and the data that is used is highly reliable and abundantly available (the sun emits strongly in the visible bands, and the earth's atmosphere is also largely transparent in the visible wavelengths). Nevertheless, radiation from other bands can be quite useful as we attempt to glean the fullest possible amount of information from the world around us. Indeed, certain branches of science sense and record images from nearly all of the EM spectrum, and use the information to give a better picture of physical reality. For example, astronomers are often identified according to the type of data that they specialize in, e.g., radio astronomers and X-ray astronomers. Non-EM radiation is also useful for imaging. Some good examples are the high-frequency sound waves (ultrasound) that are used to create images of the human body, and the low-frequency sound waves that are used by prospecting companies to create images of the earth's subsurface.

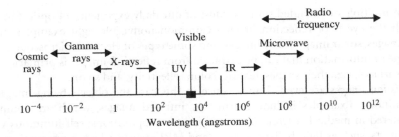

FIGURE 1.2

The electromagnetic spectrum.

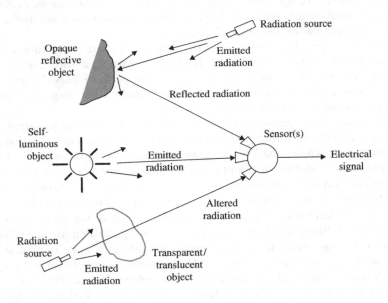

FIGURE 1.3

Recording the various types of interaction of radiation with matter.

One commonality that can be made regarding nearly all images is that radiation is emitted from some source, then interacts with some material, then is sensed and ultimately transduced into an electrical signal which may then be digitized. The resulting images can then be used to extract information about the radiation source and/or about the objects with which the radiation interacts.

We may loosely classify images according to the way in which the interaction occurs, understanding that the division is sometimes unclear, and that images may be of multiple types. Figure 1.3 depicts these various image types.

Reflection images sense radiation that has been reflected from the surfaces of objects. The radiation itself may be ambient or artificial, and it may be from a localized source or from multiple or extended sources. Most of our daily experience of optical imaging through the eye is of reflection images. Common nonvisible light examples include radar images, sonar images, laser images, and some types of electron microscope images. The type of information that can be extracted from reflection images is primarily about object surfaces, viz., their shapes, texture, color, reflectivity, and so on.

Emission images are even simpler, since in this case the objects being imaged are self-luminous. Examples include thermal or infrared images, which are commonly encountered in medical, astronomical, and military applications; self-luminous visible light objects, such as light bulbs and stars; and MRI images, which sense particle emissions. In images of this type, the information to be had is often primarily internal to the object; the image may reveal how the object creates radiation and thence something of the internal structure of the object being imaged. However, it may also be external; for example, a thermal camera can be used in low-light situations to produce useful images of a scene containing warm objects, such as people.

Finally, *absorption images* yield information about the internal structure of objects. In this case, the radiation passes through objects and is partially absorbed or attenuated by the material composing them. The degree of absorption dictates the level of the sensed radiation in the recorded image. Examples include X-ray images, transmission microscopic images, and certain types of sonic images.

Of course, the above classification is informal, and a given image may contain objects, which interacted with radiation in different ways. More important is to realize that images come from many different radiation sources and objects, and that the purpose of imaging is usually to extract information about either the source and/or the objects, by sensing the reflected/transmitted radiation and examining the way in which it has interacted with the objects, which can reveal physical information about both source and objects.

Figure 1.4 depicts some representative examples of each of the above categories of images. Figures 1.4(a) and 1.4(b) depict reflection images arising in the visible light band and in the microwave band, respectively. The former is quite recognizable; the latter is a synthetic aperture radar image of DFW airport. Figures 1.4(c) and 1.4(d) are emission images and depict, respectively, a forward-looking infrared (FLIR) image and a visible light image of the globular star cluster Omega Centauri. Perhaps the reader can guess the type of object that is of interest in Fig. 1.4(c). The object in Fig. 1.4(d), which consists of over a million stars, is visible with the unaided eye at lower northern latitudes. Lastly, Figs. 1.4(e) and 1.4(f), which are absorption images, are of a digital (radiographic) mammogram and a conventional light micrograph, respectively.

1.2 SCALE OF IMAGES

Examining Fig. 1.4 reveals another image diversity: *scale.* In our daily experience, we ordinarily encounter and visualize objects that are within 3 or 4 orders of magnitude of 1 m. However, devices for image magnification and amplification have made it possible to extend the realm of "vision" into the cosmos, where it has become possible to image structures extending over as much as 10^{30} m, and into the microcosmos, where it has

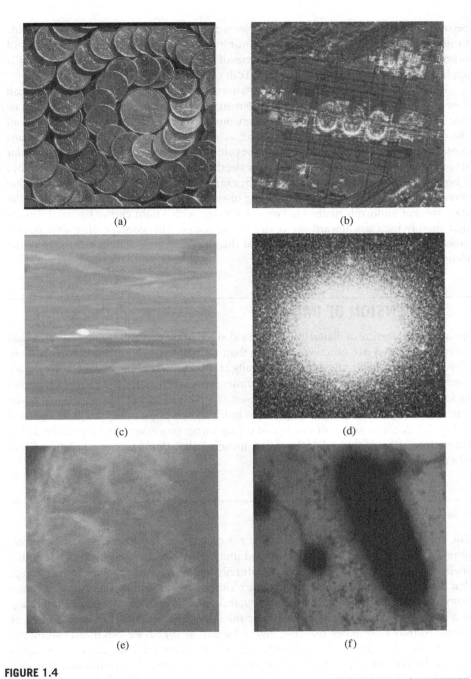

FIGURE 1.4

Examples of reflection (a), (b), emission (c), (d), and absorption (e), (f) image types.

become possible to acquire images of objects as small as 10^{-10} m. Hence we are able to image from the grandest scale to the minutest scales, over a range of 40 orders of magnitude, and as we will find, the techniques of image and video processing are generally applicable to images taken at any of these scales.

Scale has another important interpretation, in the sense that any given image can contain objects that exist at scales different from other objects in the same image, or that even exist at multiple scales simultaneously. In fact, this is the rule rather than the exception. For example, in Fig. 1.4(a), at a small scale of observation, the image contains the bas-relief patterns cast onto the coins. At a slightly larger scale, strong circular structures arose. However, at a yet larger scale, the coins can be seen to be organized into a highly coherent spiral pattern. Similarly, examination of Fig. 1.4(d) at a small scale reveals small bright objects corresponding to stars; at a larger scale, it is found that the stars are non uniformly distributed over the image, with a tight cluster having a density that sharply increases toward the center of the image. This concept of multiscale is a powerful one, and is the basis for many of the algorithms that will be described in the chapters of this *Guide*.

1.3 DIMENSION OF IMAGES

An important feature of digital images and video is that they are *multidimensional signals,* meaning that they are functions of more than a single variable. In the classic study of *digital signal processing,* the signals are usually 1D functions of time. Images, however, are functions of two and perhaps three space dimensions, whereas digital video as a function includes a third (or fourth) time dimension as well. The dimension of a signal is the number of coordinates that are required to index a given point in the image, as depicted in Fig. 1.5. A consequence of this is that digital image processing, and especially digital video processing, is quite data-intensive, meaning that significant computational and storage resources are often required.

1.4 DIGITIZATION OF IMAGES

The environment around us exists, at any reasonable scale of observation, in a space/time continuum. Likewise, the signals and images that are abundantly available in the environment (before being sensed) are naturally *analog*. By analog we mean two things: that the signal exists on a continuous (space/time) domain, and that it also takes values from a continuum of possibilities. However, this *Guide* is about processing *digital* image and video signals, which means that once the image/video signal is sensed, it must be converted into a computer-readable, digital format. By digital we also mean two things: that the signal is defined on a discrete (space/time) domain, and that it takes values from a discrete set of possibilities. Before digital processing can commence, a process of *analog-to-digital conversion* (A/D conversion) must occur. A/D conversion consists of two distinct subprocesses: *sampling* and *quantization*.

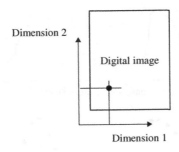

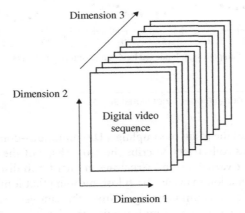

FIGURE 1.5

The dimensionality of images and video.

1.5 SAMPLED IMAGES

Sampling is the process of converting a continuous-space (or continuous-space/time) signal into a discrete-space (or discrete-space/time) signal. The sampling of continuous signals is a rich topic that is effectively approached using the tools of linear systems theory. The mathematics of sampling, along with practical implementations is addressed elsewhere in this *Guide*. In this introductory chapter, however, it is worth giving the reader a feel for the process of sampling and the need to sample a signal sufficiently densely. For a continuous signal of given space/time dimensions, there are mathematical reasons why there is a lower bound on the space/time sampling frequency (which determines the minimum possible number of samples) required to retain the information in the signal. However, image processing is a visual discipline, and it is more fundamental to realize that what is usually important is that the process of sampling does not lose *visual information*. Simply stated, the sampled image/video signal must "look good," meaning that it does not suffer too much from a loss of visual resolution or from artifacts that can arise from the process of sampling.

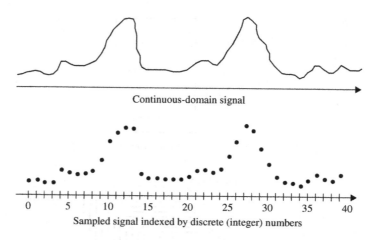

Continuous-domain signal

Sampled signal indexed by discrete (integer) numbers

FIGURE 1.6

Sampling a continuous-domain one-dimensional signal.

Figure 1.6 illustrates the result of sampling a 1D continuous-domain signal. It is easy to see that the samples collectively describe the gross shape of the original signal very nicely, but that smaller variations and structures are harder to discern or may be lost. Mathematically, information may have been lost, meaning that it might not be possible to reconstruct the original continuous signal from the samples (as determined by the Sampling Theorem, see Chapter 5). Supposing that the signal is part of an image, e.g., is a single scan-line of an image displayed on a monitor, then the visual quality may or may not be reduced in the sampled version. Of course, the concept of visual quality varies from person-to-person, and it also depends on the conditions under which the image is viewed, such as the viewing distance.

Note that in Fig. 1.6 the samples are indexed by integer numbers. In fact, the sampled signal can be viewed as a vector of numbers. If the signal is finite in extent, then the signal vector can be stored and digitally processed as an array, hence the integer indexing becomes quite natural and useful. Likewise, image signals that are space/time sampled are generally indexed by integers along each sampled dimension, allowing them to be easily processed as multidimensional arrays of numbers. As shown in Fig. 1.7, a sampled image is an array of sampled image values that are usually arranged in a row-column format. Each of the indexed array elements is often called a *picture element,* or *pixel* for short. The term *pel* has also been used, but has faded in usage probably since it is less descriptive and not as catchy. The number of rows and columns in a sampled image is also often selected to be a power of 2, since it simplifies computer addressing of the samples, and also since certain algorithms, such as discrete Fourier transforms, are particularly efficient when operating on signals that have dimensions that are powers of 2. Images are nearly always rectangular (hence indexed on a Cartesian grid) and are often square, although the horizontal dimensional is often longer, especially in video signals, where an aspect ratio of 4:3 is common.

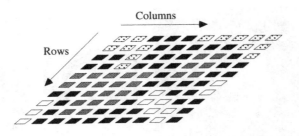

FIGURE 1.7

Depiction of a very small (10 × 10) piece of an image array.

As mentioned earlier, the effects of insufficient sampling ("undersampling") can be visually obvious. Figure 1.8 shows two very illustrative examples of image sampling. The two images, which we will call "mandrill" and "fingerprint," both contain a significant amount of interesting visual detail that substantially defines the content of the images. Each image is shown at three different sampling densities: 256×256 (or $2^8 \times 2^8 = 65,536$ samples), 128×128 (or $2^7 \times 2^7 = 16,384$ samples), and 64×64 (or $2^6 \times 2^6 = 4,096$ samples). Of course, in both cases, all three scales of images are digital, and so there is potential loss of information relative to the original analog image. However, the perceptual quality of the images can easily be seen to degrade rather rapidly; note the whiskers on the mandrill's face, which lose all coherency in the 64×64 image. The 64×64 fingerprint is very interesting since the pattern has completely changed! It almost appears as a different fingerprint. This results from an undersampling effect known as *aliasing*, where image frequencies appear that have no physical meaning (in this case, creating a false pattern). Aliasing, and its mathematical interpretation, will be discussed further in Chapter 2 in the context of the Sampling Theorem.

1.6 QUANTIZED IMAGES

The other part of image digitization is *quantization*. The values that a (single-valued) image takes are usually *intensities* since they are a record of the intensity of the signal incident on the sensor, e.g., the photon count or the amplitude of a measured wave function. Intensity is a positive quantity. If the image is represented visually using shades of gray (like a black-and-white photograph), then the pixel values are referred to as *gray levels*. Of course, broadly speaking, an image may be multivalued at each pixel (such as a color image), or an image may have negative pixel values, in which case, it is not an intensity function. In any case, the image values must be quantized for digital processing.

Quantization is the process of converting a *continuous-valued image* that has a continuous range (set of values that it can take) into a *discrete-valued image* that has a discrete range. This is ordinarily done by a process of rounding, truncation, or some

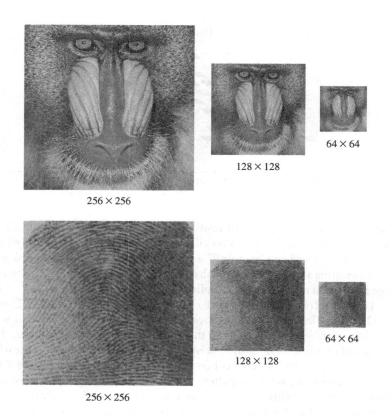

256 × 256

128 × 128

64 × 64

256 × 256

128 × 128

64 × 64

FIGURE 1.8

Examples of the visual effect of different image sampling densities.

other irreversible, nonlinear process of information destruction. Quantization is a necessary precursor to digital processing, since the image intensities must be represented with a finite precision (limited by wordlength) in any digital processor.

When the gray level of an image pixel is quantized, it is assigned to be one of a finite set of numbers which is the *gray level range.* Once the discrete set of values defining the gray-level range is known or decided, then a simple and efficient method of quantization is simply to round the image pixel values to the respective nearest members of the intensity range. These rounded values can be any numbers, but for conceptual convenience and ease of digital formatting, they are then usually mapped by a linear transformation into a finite set of non-negative integers $\{0, \ldots, K - 1\}$, where K is a power of two: $K = 2^B$. Hence the number of allowable gray levels is K, and the number of bits allocated to each pixel's gray level is B. Usually $1 \cdot B \cdot 8$ with $B = 1$ (for binary images) and $B = 8$ (where each gray level conveniently occupies a byte) are the most common bit depths (see Fig. 1.9). Multivalued images, such as color images, require quantization of the components either

a pixel

8-bit representation

FIGURE 1.9

Illustration of 8-bit representation of a quantized pixel.

individually or collectively ("vector quantization"); for example, a three-component color image is frequently represented with 24 bits per pixel of color precision.

Unlike sampling, quantization is a difficult topic to analyze since it is nonlinear. Moreover, most theoretical treatments of signal processing assume that the signals under study are *not* quantized, since it tends to greatly complicate the analysis. On the other hand, quantization is an essential ingredient of any (lossy) signal compression algorithm, where the goal can be thought of as finding an optimal quantization strategy that simultaneously minimizes the volume of data contained in the signal, while disturbing the fidelity of the signal as little as possible. With simple quantization, such as gray level rounding, the main concern is that the pixel intensities or gray levels must be quantized with sufficient precision that excessive information is not lost. Unlike sampling, there is no simple mathematical measurement of information loss from quantization. However, while the effects of quantization are difficult to express mathematically, the effects are visually obvious.

Each of the images depicted in Figs. 1.4 and 1.8 is represented with 8 bits of gray level resolution—meaning that bits less significant than the 8th bit have been rounded or truncated. This number of bits is quite common for two reasons: first, using more bits will generally *not* improve the visual appearance of the image—the adapted human eye usually is unable to see improvements beyond 6 bits (although the total range that can be seen under different conditions can exceed 10 bits)—hence using more bits would be of no use. Secondly, each pixel is then conveniently represented by a byte. There are exceptions: in certain scientific or medical applications, 12, 16, or even more bits may be retained for more exhaustive examination by human or by machine.

Figures 1.10 and 1.11 depict two images at various levels of gray level resolution. Reduced resolution (from 8 bits) was obtained by simply truncating the appropriate number of less significant bits from each pixel's gray level. Figure 1.10 depicts the 256×256 digital image "fingerprint" represented at 4, 2, and 1 bits of gray level resolution. At 4 bits, the fingerprint is nearly indistinguishable from the 8-bit representation of Fig 1.8. At 2 bits, the image has lost a significant amount of information, making the print difficult to read. At 1 bit, the *binary* image that results is likewise hard to read. In practice, binarization of fingerprints is often used to make the print more distinctive. Using simple truncation-quantization, most of the print is lost since it was inked insufficiently on the left, and excessively on the right. Generally, bit truncation is a poor method for creating a binary image from a gray level image. See Chapter 2 for better methods of image binarization.

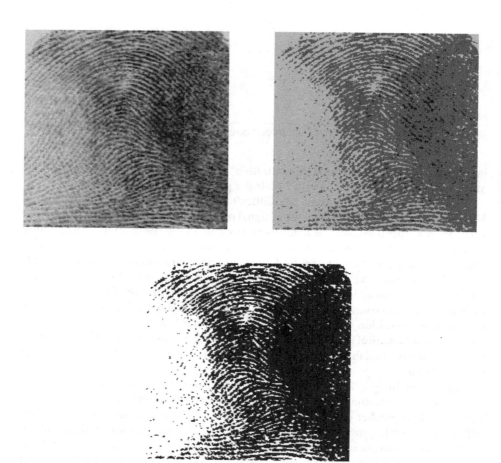

FIGURE 1.10

Quantization of the 256 × 256 image "fingerprint." Clockwise from upper left: 4, 2, and 1 bit(s) per pixel.

Figure 1.11 shows another example of gray level quantization. The image "eggs" is quantized at 8, 4, 2, and 1 bit(s) of gray level resolution. At 8 bits, the image is very agreeable. At 4 bits, the eggs take on the appearance of being striped or painted like Easter eggs. This effect is known as "false contouring," and results when inadequate grayscale resolution is used to represent smoothly varying regions of an image. In such places, the effects of a (quantized) gray level can be visually exaggerated, leading to an appearance of false structures. At 2 bits and 1 bit, significant information has been lost from the image, making it difficult to recognize.

A quantized image can be thought of as a stacked set of single-bit images (known as "bit planes") corresponding to the gray level resolution depths. The most significant

FIGURE 1.11

Quantization of the 256 × 256 image "eggs." Clockwise from upper left: 8, 4, 2, and 1 bit(s) per pixel.

bits of every pixel comprise the top bit plane and so on. Figure 1.12 depicts a 10 × 10 digital image as a stack of B bit planes. Special-purpose image processing algorithms are occasionally applied to the individual bit planes.

1.7 COLOR IMAGES

Of course, the visual experience of the normal human eye is not limited to grayscales—*color* is an extremely important aspect of images. It is also an important aspect of digital images. In a very general sense, color conveys a variety of rich information that describes

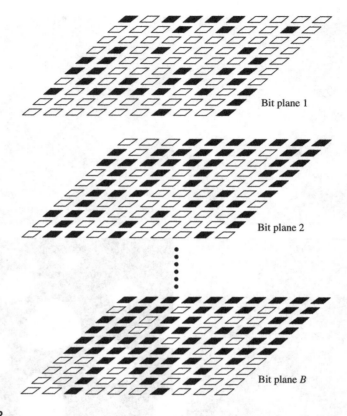

FIGURE 1.12

Depiction of a small (10 × 10) digital image as a stack of bit planes ranging from most significant (top) to least significant (bottom).

the *quality* of objects, and as such, it has much to do with visual *impression*. For example, it is known that different colors have the potential to evoke different emotional responses. The perception of color is allowed by the color-sensitive neurons known as *cones* that are located in the retina of the eye. The cones are responsive to normal light levels and are distributed with greatest density near the center of the retina, known as the *fovea* (along the direct line of sight). The *rods* are neurons that are sensitive at low-light levels and are not capable of distinguishing color wavelengths. They are distributed with greatest density around the periphery of the fovea, with very low density near the line-of-sight. Indeed, this may be observed by observing a dim point target (such as a star) under dark conditions. If the gaze is shifted slightly off-center, then the dim object suddenly becomes easier to see.

In the normal human eye, colors are sensed as near-linear combinations of long, medium, and short wavelengths, which roughly correspond to the three *primary colors* that are used in standard video camera systems: Red (R), Green (G), and Blue (B). The way in which visible light wavelengths map to RGB camera color coordinates is a complicated topic, although standard tables have been devised based on extensive experiments. A number of other color coordinate systems are also used in image processing, printing, and display systems, such as the YIQ (luminance, in-phase chromatic, quadratic chromatic) color coordinate system. Loosely speaking, the YIQ coordinate system attempts to separate the perceived image *brightness* (luminance) from the chromatic components of the image via an invertible linear transformation:

$$\begin{bmatrix} Y \\ I \\ Q \end{bmatrix} = \begin{bmatrix} 0.299 & 0.587 & 0.114 \\ 0.596 & -0.275 & -0.321 \\ 0.212 & -0.523 & 0.311 \end{bmatrix} \begin{bmatrix} R \\ G \\ B \end{bmatrix}. \tag{1.1}$$

The RGB system is used by color cameras and video display systems, while the YIQ is the standard color representation used in broadcast television. Both representations are used in practical image and video processing systems along with several other representations.

Most of the theory and algorithms for digital image and video processing has been developed for single-valued, monochromatic (gray level), or intensity-only images, whereas color images are vector-valued signals. Indeed, many of the approaches described in this *Guide* are developed for single-valued images. However, these techniques are often applied (sub-optimally) to color image data by regarding each color component as a separate image to be processed and recombining the results afterwards. As seen in Fig. 1.13, the R, G, and B components contain a considerable amount of overlapping information. Each of them is a valid image in the same sense as the image seen through colored spectacles and can be processed as such. Conversely, however, if the color components are collectively available, then vector image processing algorithms can often be designed that achieve optimal results by taking this information into account. For example, a vector-based image enhancement algorithm applied to the "cherries" image in Fig. 1.13 might adapt by giving less importance to enhancing the Blue component, since the image signal is weaker in that band.

Chrominance is usually associated with slower amplitude variations than is luminance, since it usually is associated with fewer image details or rapid changes in value. The human eye has a greater spatial bandwidth allocated for luminance perception than for chromatic perception. This is exploited by compression algorithms that use alternative color representations, such as YIQ, and store, transmit, or process the chromatic components using a lower bandwidth (fewer bits) than the luminance component. Image and video compression algorithms achieve increased efficiencies through this strategy.

1.8 SIZE OF IMAGE DATA

The amount of data in visual signals is usually quite large and increases geometrically with the dimensionality of the data. This impacts nearly every aspect of image and

FIGURE 1.13

Color image "cherries" (top left) and (clockwise) its Red, Green, and Blue components.

video processing; data volume is a major issue in the processing, storage, transmission, and display of image and video information. The storage required for a single monochromatic digital still image that has (row $\times$ column) dimensions $N \times M$ and B bits of gray level resolution is NMB bits. For the purpose of discussion, we will assume that the image is square ($N = M$), although images of any aspect ratio are common. Most commonly, $B = 8$ (1 byte/pixel) unless the image is binary or is special-purpose. If the image is vector-valued, e.g., color, then the data volume is multiplied by the vector dimension. Digital images that are delivered by commercially available image digitizers are typically of approximate size 512×512 pixels, which is large enough to fill much of a monitor screen. Images both larger (ranging up to 4096×4096 or

TABLE 1.1   Data volume requirements for digital still images of various sizes, bit depths, and vector dimension.

Spatial dimensions	Pixel resolution (bits)	Image type	Data volume (bytes)
128 × 128	1	Monochromatic	2,048
256 × 256	1	Monochromatic	8,192
512 × 512	1	Monochromatic	32,768
1,024 × 1,024	1	Monochromatic	131,072
128 × 128	8	Monochromatic	16,384
256 × 256	8	Monochromatic	65,536
512 × 512	8	Monochromatic	262,144
1,024 × 1,024	8	Monochromatic	1,048,576
128 × 128	3	Trichromatic	6,144
256 × 256	3	Trichromatic	24,576
512 × 512	3	Trichromatic	98,304
1,024 × 1,024	3	Trichromatic	393,216
128 × 128	24	Trichromatic	49,152
256 × 256	24	Trichromatic	196,608
512 × 512	24	Trichromatic	786,432
1,024 × 1,024	24	Trichromatic	3,145,728

more) and smaller (as small as 16 × 16) are commonly encountered. Table 1.1 depicts the required storage for a variety of image resolution parameters, assuming that there has been no compression of the data. Of course, the spatial extent (area) of the image exerts the greatest effect on the data volume. A single 512 × 512 × 8 color image requires nearly a megabyte of digital storage space, which only a few years ago, was a lot. More recently, even large images are suitable for viewing and manipulation on home personal computers, although somewhat inconvenient for transmission over existing telephone networks.

1.9   OBJECTIVES OF THIS *GUIDE*

The goals of this *Guide* are ambitious, since it is intended to reach a broad audience that is interested in a wide variety of image and video processing applications. Moreover, it is intended to be accessible to readers who have a diverse background and who represent a wide spectrum of levels of preparation and engineering/computer education. However, a *Guide* format is ideally suited for this multiuser purpose, since it allows for a presentation that adapts to the reader's needs. In the early part of the *Guide*, we present very basic material that is easily accessible even for novices to the image processing field. These chapters are also useful for review, for basic reference, and as support for latter chapters. In every major section of the *Guide*, basic introductory material is presented as well as more advanced chapters that take the reader deeper into the subject.

Unlike textbooks on image processing, this *Guide* is, therefore, not geared toward a specified level of presentation, nor does it uniformly assume a specific educational background. There is material that is available for the beginning image processing user, as well as for the expert. The *Guide* is also unlike a textbook in that it is not limited to a specific point of view given by a single author. Instead, leaders from image and video processing education, industry, and research have been called upon to explain the topical material from their own daily experience. By calling upon most of the leading experts in the field, we have been able to provide a complete coverage of the image and video processing area without sacrificing any level of understanding of any particular area.

Because of its broad spectrum of coverage, we expect that the *Essential Guide to Image Processing* and its companion, the *Essential Guide to Video Processing*, will serve as excellent textbooks as well as references. It has been our objective to keep the students, needs in mind, and we feel that the material contained herein is appropriate to be used for classroom presentations ranging from the introductory undergraduate level, to the upper-division undergraduate, and to the graduate level. Although the *Guide* does not include "problems in the back," this is not a drawback since the many examples provided in every chapter are sufficient to give the student a deep understanding of the functions of the various image processing algorithms. This field is very much a visual science, and the principles underlying it are best taught via visual examples. Of course, we also foresee the *Guide* as providing easy reference, background, and guidance for image processing professionals working in industry and research.

Our specific objectives are to:

- provide the practicing engineer and the student with a highly accessible resource for learning and using image processing algorithms and theory;

- provide the essential understanding of the various image processing standards that exist or are emerging, and that are driving today's explosive industry;

- provide an understanding of what images are, how they are modeled, and give an introduction to how they are perceived;

- provide the necessary practical background to allow the engineer student to acquire and process his/her own digital image data;

- provide a diverse set of example applications, as separate complete chapters, that are explained in sufficient depth to serve as extensible models to the reader's own potential applications.

The *Guide* succeeds in achieving these goals, primarily because of the many years of broad educational and practical experience that the many contributing authors bring to bear in explaining the topics contained herein.

1.10 ORGANIZATION OF THE *GUIDE*

It is our intention that this *Guide* be adopted by both researchers and educators in the image processing field. In an effort to make the material more easily accessible and immediately usable, we have provided a CD-ROM with the *Guide*, which contains image processing demonstration programs written in the LabVIEW language. The overall suite of algorithms is part of the SIVA (Signal, Image and Video Audiovisual) Demonstration Gallery provided by the Laboratory for Image and Video Engineering at The University of Texas at Austin, which can be found at http://live.ece.utexas.edu/class/siva/ and which is broadly described in [1]. The SIVA systems are currently being used by more than 400 institutions from more than 50 countries around the world. Chapter 2 is devoted to a more detailed description of the image processing programs available on the disk, how to use them, and how to learn from them.

Since this *Guide* is emphatically about *processing* images and video, the next chapter is immediately devoted to basic algorithms for image processing, instead of surveying methods and devices for image acquisition at the outset, as many textbooks do. Chapter 3 lays out basic methods for gray level image processing, which includes point operations, the image histogram, and simple image algebra. The methods described there stand alone as algorithms that can be applied to most images but they also set the stage and the notation for the more involved methods discussed in later chapters. Chapter 4 describes basic methods for image binarization and binary image processing with emphasis on morphological binary image processing. The algorithms described there are among the most widely used in applications, especially in the biomedical area. Chapter 5 explains the basics of Fourier transform and frequency-domain analysis, including discretization of the Fourier transform and discrete convolution. Special emphasis is laid on explaining frequency-domain concepts through visual examples. Fourier image analysis provides a unique opportunity for visualizing the meaning of frequencies as components of signals. This approach reveals insights which are difficult to capture in 1D, graphical discussions.

More advanced, yet basic topics and image processing tools are covered in the next few chapters, which may be thought of as a core reference section of the *Guide* that supports the entire presentation. Chapter 6 introduces the reader to multiscale decompositions of images and wavelets, which are now standard tools for the analysis of images over multiple scales or over space and frequency simultaneously. Chapter 7 describes basic statistical image noise models that are encountered in a wide diversity of applications. Dealing with noise is an essential part of most image processing tasks. Chapter 8 describes color image models and color processing. Since color is a very important attribute of images from a perceptual perspective, it is important to understand the details and intricacies of color processing. Chapter 9 explains statistical models of natural images. Images are quite diverse and complex yet can be shown to broadly obey statistical laws that prove useful in the design of algorithms.

The following chapters deal with methods for correcting distortions or uncertainties in images. Quite frequently, the visual data that is acquired has been in some way corrupted. Acknowledging this and developing algorithms for dealing with it is especially critical since the human capacity for detecting errors, degradations, and delays in digitally-delivered visual data is quite high. Image signals are derived from imperfect sensors, and the processes of digitally converting and transmitting these signals are subject to errors. There are many types of errors that can occur in image data, including, for example, blur from motion or defocus; noise that is added as part of a sensing or transmission process; bit, pixel, or frame loss as the data is copied or read; or artifacts that are introduced by an image compression algorithm. Chapter 10 describes methods for reducing image noise artifacts using linear systems techniques. The tools of linear systems theory are quite powerful and deep and admit optimal techniques. However, they are also quite limited by the constraint of linearity, which can make it quite difficult to separate signal from noise. Thus, the next three chapters broadly describe the three most popular and complementary nonlinear approaches to image noise reduction. The aim is to remove noise while retaining the perceptual fidelity of the visual information; these are often conflicting goals. Chapter 11 describes powerful wavelet-domain algorithms for image denoising, while Chapter 12 describes highly nonlinear methods based on robust statistical methods. Chapter 13 is devoted to methods that shape the image signal to smooth it using the principles of mathematical morphology. Finally, Chapter 14 deals with the more difficult problem of image restoration, where the image is presumed to have been possibly distorted by a linear transformation (typically a blur function, such as defocus, motion blur, or atmospheric distortion) and more than likely, by noise as well. The goal is to remove the distortion and attenuate the noise, while again preserving the perceptual fidelity of the information contained within. Again, it is found that a balanced attack on conflicting requirements is required in solving these difficult, ill-posed problems.

As described earlier in this introductory chapter, image information is highly data-intensive. The next few chapters describe methods for compressing images. Chapter 16 describes the basics of lossless image compression, where the data is compressed to occupy a smaller storage or bandwidth capacity, yet nothing is lost when the image is decompressed. Chapters 17 and 18 describe lossy compression algorithms, where data is thrown away, but in such a way that the visual loss of the decompressed images is minimized. Chapter 17 describes the existing JPEG standards (JPEG and JPEG2000) which include both lossy and lossless modes. Although these standards are quite complex, they are described in detail to allow for the practical design of systems that accept and transmit JPEG datasets. The more recent JPEG2000 standard is based on a subband (wavelet) decomposition of the image. Chapter 18 goes deeper into the topic of wavelet-based image compression, since these methods have been shown to provide the best performance to date in terms of compression efficiency versus visual quality.

The *Guide* next turns to basic methods for the fascinating topic of image analysis. Not all images are intended for direct human visual consumption. Instead, in many situations it is of interest to automate the process of repetitively interpreting the content of multiple images through the use of an *image analysis algorithm*. For example, it may be desired to *classify* parts of images as being of some type, or it may be desired to *detect* or *recognize* objects contained in the images. Chapter 19 describes the basic methods for detecting *edges* in images. The goal is to find the boundaries of regions, viz., sudden changes in image intensities, rather than finding (segmenting out) and classifying regions directly. The approach taken depends on the application. Chapter 20 describes more advanced approaches to edge detection based on the principles of anisotropic diffusion. These methods provide stronger performance in terms of edge detection ability and noise suppression, but at an increased computational expense. Chapter 21 deals with methods for assessing the *quality* of images. This topic is quite important, since quality must be assessed relative to human subjective impressions of quality. Verifying the efficacy of image quality assessment algorithms requires that they be correlated against the result of large, statistically significant human studies, where volunteers are asked to give their impression of the quality of a large number of images that have been distorted by various processes.

Chapter 22 describes methods for securing image information through the process of watermarking. This process is important since in the age of the internet and other broadcast digital transmission media, digital images are shared and used by the general population. It is important to be able to protect copyrighted images.

Next, the *Guide* includes five chapters (Chapters 23–27) on a diverse set of image processing and analysis applications that are quite representative of the universe of applications that exist. Several of the chapters have *analysis, classification,* or *recognition* as a main goal, but reaching these goals inevitably requires the use of a broad spectrum of image processing subalgorithms for enhancement, restoration, detection, motion, and so on. The work that is reported in these chapters is likely to have significant impact on science, industry, and even on daily life. It is hoped that the reader is able to translate the lessons learned in these chapters, and in the preceding chapters, into their own research or product development work in image processing. For the student, it is hoped that s/he now possesses the required reference material that will allow her/him to acquire the basic knowledge to be able to begin a research or development career in this fast-moving and rapidly growing field.

For those looking to extend their knowledge beyond still image processing to video processing, Chapter 28 points the way with some introductory and transitional comments. However, for an in-depth discussion of digital video processing, the reader is encouraged to consult the companion volume, the *Essential Guide to Video Processing*.

REFERENCE

[1] U. Rajashekar, G. Panayi, F. P. Baumgartner, and A. C. Bovik. The SIVA demonstration gallery for signal, image, and video processing education. *IEEE Trans. Educ.*, 45(4):323–335, November 2002.

The SIVA Image Processing Demos

2

Umesh Rajashekar[1], Al Bovik[2], and Dinesh Nair[3]

[1] *New York University;* [2] *The University of Texas at Austin;* [3] *National Instruments*

2.1 INTRODUCTION

Given the availability of inexpensive digital cameras and the ease of sharing digital photos on Web sites dedicated to amateur photography and social networking, it will come as no surprise that a majority of computer users have performed some form of image processing. Irrespective of their familiarity with the theory of image processing, most people have used image editing software such as Adobe Photoshop, GIMP, Picasa, ImageMagick, or iPhoto to perform simple image processing tasks, such as resizing a large image for emailing, or adjusting the brightness and contrast of a photograph. The fact that "to Photoshop" is being used as a verb in everyday parlance speaks of the popularity of image processing among the masses.

As one peruses the wide spectrum of topics and applications discussed in *The Essential Guide to Image Processing*, it becomes obvious that the field of digital image processing (DIP) is highly interdisciplinary and draws upon a great variety of areas such as mathematics, computer graphics, computer vision, visual psychophysics, optics, and computer science. DIP is a subject that lends itself to a rigorous, analytical treatment and which, depending on how it is presented, is often perceived as being rather theoretical. Although many of these mathematical topics may be unfamiliar (and often superfluous) to a majority of the general image processing audience, we believe it is possible to present the theoretical aspects of image processing as an intuitive and exciting "visual" experience. Surely, the cliché "A picture is worth a thousand words" applies very effectively to the teaching of image processing.

In this chapter, we explain and make available a popular courseware for image processing education known as SIVA—The Signal, Image, and Video Audiovisualization—gallery [1]. This SIVA gallery was developed in the Laboratory for Image and Video Engineering (LIVE) at the University of Texas (UT) at Austin with the purpose of making DIP "accessible" to an audience with a wide range of academic backgrounds, while offering a highly visual and interactive experience. The image and video processing section of the SIVA gallery consists of a suite of special-purpose LabVIEW-based programs (known as

23

Virtual Instruments or VIs). Equipped with informative visualization and a user-friendly interface, these VIs were carefully designed to facilitate a gentle introduction to the fascinating concepts in image and video processing. At UT-Austin, SIVA has been used (for more than 10 years) in an undergraduate image and video processing course as an in-class demonstration tool to illustrate the concepts and algorithms of image processing. The demos have also been seamlessly integrated into the class notes to provide contextual illustrations of the principles being discussed. Thus, they play a dual role: as in-class live demos of image processing algorithms in action, and as online resources for the students to test the image processing concepts on their own. Toward this end, the SIVA demos are much more than simple image processing subroutines. They are user-friendly programs with attractive graphical user interfaces, with button- and slider-enabled selection of the various parameters that control the algorithms, and with before-and-after image windows that show the visual results of the image processing algorithms (and intermediate results as well).

Stand-alone implementations of the SIVA image processing demos, which do not require the user to own a copy of LabVIEW, are provided on the CD that accompanies this *Guide*. SIVA is also available for free download from the Web site mentioned in [2]. The reader is encouraged to experiment with these demos as they read the chapters in this *Guide*. Since the *Guide* contains a very large number of topics, only a subset has associated demonstration programs. Moreover, by necessity, the demos are aligned more with the simpler concepts in the *Guide*, rather than the more complex methods described later, which involve suites of combined image processing algorithms to accomplish tasks.

To make things even easier, the demos are accompanied by a comprehensive set of help files that describe the various controls, and that highlight some illustrative examples and instructive parameter settings. A demo can be activated by clicking the rightward pointing arrow in the top menu bar. Help for the demo can be activated by clicking the "?" button and moving the cursor over the icon that is located immediately to the right of the "?" button. In addition, when the cursor is placed over any other button/control, the help window automatically updates to describe the function of that button/control. We are confident that the user will find this visual, hands-on, interactive introduction to image processing to be a fun, enjoyable, and illuminating experience. In the rest of the chapter, we will describe the software framework used by the SIVA demonstration gallery (Section 2.2), illustrate some of the image processing demos in SIVA (Section 2.3), and direct the reader to other popular tools for image and video processing education (Section 2.4).

2.2 LabVIEW FOR IMAGE PROCESSING

National Instrument's LabVIEW [3] (Laboratory Virtual Instrument Engineering Workbench) is a graphical development environment used for creating flexible and scalable design, control, and test applications. LabVIEW is used worldwide in both industry and academia for applications in a variety of fields: automotive, communications, aerospace, semiconductor, electronic design and production, process control, biomedical, and many more. Applications cover all phases of product development from research to test, manufacturing, and service.

LabVIEW uses a dataflow programming model that frees you from the sequential architecture of text-based programming, where instructions determine the order of program execution. You program LabVIEW using a graphical programming language, G, that uses icons instead of lines of text to create applications. The graphical code is highly intuitive for engineers and scientists familiar with block diagrams and flowcharts. The flow of data through the nodes (icons) in the program determines the execution order of the functions, allowing you to easily create programs that execute multiple operations in parallel. The parallel nature of LabVIEW also makes multitasking and multithreading simple to implement.

LabVIEW includes hundreds of powerful graphical and textual measurement analysis, mathematics, signal and image processing functions that seamlessly integrate with LabVIEW data acquisition, instrument control, and presentation capabilities. With LabVIEW, you can build simulations with interactive user interfaces; interface with real-world signals; analyze data for meaningful information; and share results through intuitive displays, reports, and the Web.

Additionally, LabVIEW can be used to program a real-time operating system, field-programmable gate arrays, handheld devices, such as PDAs, touch screen computers, DSPs, and 32-bit embedded microprocessors.

2.2.1 The LabVIEW Development Environment

In LabVIEW, you build a user interface by using a set of tools and objects. The user interface is known as the front panel. You then add code using graphical representations of functions to control the front panel objects. This graphical source code is also known as G code or block diagram code. The block diagram contains this code. In some ways, the block diagram resembles a flowchart.

LabVIEW programs are called virtual instruments, or VIs, because their appearance and operation imitate physical instruments, such as oscilloscopes and multimeters. Every VI uses functions that manipulate input from the user interface or other sources and display that information or move it to other files or other computers.

A VI contains the following three components:

- **Front panel**—serves as the user interface. The front panel contains the user interface control inputs, such as knobs, sliders, and push buttons, and output indicators to produce items such as charts, graphs, and image displays. Inputs can be fed into the system using the mouse or the keyboard. A typical front panel is shown in Fig. 2.1(a).

- **Block diagram**—contains the graphical source code that defines the functionality of the VI. The blocks are interconnected, using wires to indicate the dataflow. Front panel indicators pass data from the user to their corresponding terminals on

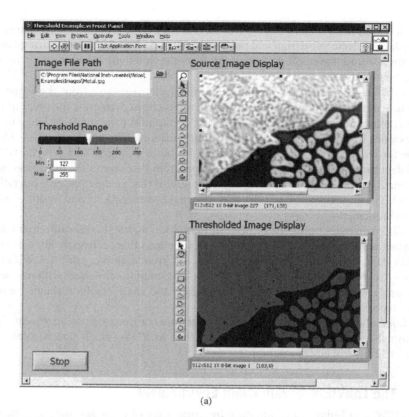

(a)

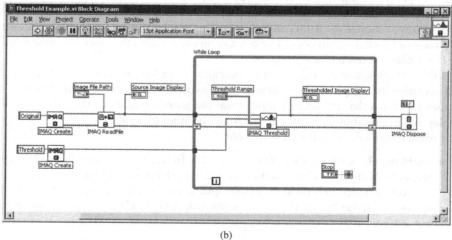

(b)

FIGURE 2.1

Typical development environment in LabVIEW. (a) Front panel; (b) Block diagram.

the block diagram. The results of the operation are then passed back to the front panel indicators. A typical block diagram is shown in Fig. 2.1(b). Within the block diagram, you have access to a full-featured graphical programming language that includes all the standard features of a general-purpose programming environment, such as data structures, looping structures, event handling, and object-oriented programming.

- **Icon and connector pane**—identifies the interface to the VI so that you can use the VI in another VI. A VI within another VI is called a sub-VI. Sub-VIs are analogous to subroutines in conventional programming languages. A sub-VI is a virtual instrument and can be run as a program, with the front panel serving as a user interface, or, when dropped as a node onto the block diagram, the front panel defines the inputs and outputs for the given node through the connector pane. This allows you to easily test each sub-VI before being embedded as a subroutine into a larger program.

LabVIEW also includes debugging tools that allow you to watch data move through a program and see precisely which data passes from one function to another along the wires, a process known as execution highlighting. This differs from text-based languages, which require you to step from function to function to trace your program execution. An excellent introduction to LabVIEW is provided in [4, 5].

2.2.2 Image Processing and Machine Vision in LabVIEW

LabVIEW is widely used for programming scientific imaging and machine vision applications because engineers and scientists find that they can accomplish more in a shorter period of time by working with flowcharts and block diagrams instead of text-based function calls. The NI Vision Development Module [6] is a software package for engineers and scientists who are developing machine vision and scientific imaging applications. The development module includes NI Vision for LabVIEW—a library of over 400 functions for image processing and machine vision and NI Vision Assistant—an interactive environment for quick prototyping of vision applications without programming. The development module also includes NI Vision Acquisition—software with support for thousands of cameras including IEEE 1394 and GigE Vision cameras.

2.2.2.1 *NI Vision*

NI Vision is the image processing toolkit, or library, that adds high-level machine vision and image processing to the LabVIEW environment. NI Vision includes an extensive set of MMX-optimized functions for the following machine vision tasks:

- Grayscale, color, and binary image display
- Image processing—including statistics, filtering, and geometric transforms
- Pattern matching and geometric matching

- Particle analysis
- Gauging
- Measurement
- Object classification
- Optical character recognition
- 1D and 2D barcode reading.

NI Vision VIs are divided into three categories: Vision Utilities, Image Processing, and Machine Vision.

Vision Utilities VIs   Allow you to create and manipulate images to suit the needs of your application. This category includes VIs for image management and manipulation, file management, calibration, and region of interest (ROI) selection.
You can use these VIs to:

- create and dispose of images, set and read attributes of an image, and copy one image to another;

- read, write, and retrieve image file information. The file formats NI Vision supports are BMP, TIFF, JPEG, PNG, AIPD (internal file format), and AVI (for multiple images);

- display an image, get and set ROIs, manipulate the floating ROI tools window, configure an ROI constructor window, and set up and use an image browser;

- modify specific areas of an image. Use these VIs to read and set pixel values in an image, read and set values along a row or column in an image, and fill the pixels in an image with a particular value;

- overlay figures, text, and bitmaps onto an image without destroying the image data. Use these VIs to overlay the results of your inspection application onto the images you inspected;

- spatially calibrate an image. Spatial calibration converts pixel coordinates to real-world coordinates while compensating for potential perspective errors or nonlinear distortions in your imaging system;

- manipulate the colors and color planes of an image. Use these VIs to extract different color planes from an image, replace the planes of a color image with new data, convert a color image into a 2D array and back, read and set pixel values in a color image, and convert pixel values from one color space to another.

Image Processing VIs   Allow you to analyze, filter, and process images according to the needs of your application. This category includes VIs for analysis, grayscale and binary image processing, color processing, frequency processing, filtering, morphology, and operations.

You can use these VIs to:

- transform images using predefined or custom lookup tables, change the contrast information in an image, invert the values in an image, and segment the image;

- filter images to enhance the information in the image. Use these VIs to smooth your image, remove noise, and find edges in the image. You can use a predefined filter kernel or create custom filter kernels;

- perform basic morphological operations, such as dilation and erosion, on grayscale and binary images. Other VIs improve the quality of binary images by filling holes in particles, removing particles that touch the border of an image, removing noisy particles, and removing unwanted particles based on different characteristics of the particle;

- compute the histogram information and grayscale statistics of an image, retrieve pixel information and statistics along any 1D profile in an image, and detect and measure particles in binary images;

- perform basic processing on color images; compute the histogram of a color image; apply lookup tables to color images; change the brightness, contrast, and gamma information associated with a color image; and threshold a color image;

- perform arithmetic and bit-wise operations in NI Vision; add, subtract, multiply, and divide an image with other images or constants or apply logical operations and make pixel comparisons between an image and other images or a constant;

- perform frequency processing and other tasks on images; convert an image from the spatial domain to the frequency domain using a 2D Fast Fourier Transform (FFT) and convert an image from the frequency domain to the spatial domain using the inverse FFT. These VIs also extract the magnitude, phase, real, and imaginary planes of the complex image.

Machine Vision VIs   Can be used to perform common machine vision inspection tasks, including checking for the presence or absence of parts in an image and measuring the dimensions of parts to see if they meet specifications.

You can use these VIs to:

- measure the intensity of a pixel on a point or the intensity statistics of pixels along a line or in a rectangular region of an image;

- measure distances in an image, such as the minimum and maximum horizontal separation between two vertically oriented edges or the minimum or maximum vertical separation between two horizontally oriented edges;

- locate patterns and subimages in an image. These VIs allow you to perform color and grayscale pattern matching as well as shape matching;

- derive results from the coordinates of points returned by image analysis and machine vision algorithms; fit lines, circles, and ellipses to a set of points in the image; compute the area of a polygon represented by a set of points; measure distances between points; and find angles between lines represented by points;

- compare images to a golden template reference image;

- classify unknown objects by comparing significant features to a set of features that conceptually represent classes of known objects;

- read text and/or characters in an image;

- develop applications that require reading from seven-segment displays, meters or gauges, or 1D barcodes.

2.2.2.2 *NI Vision Assistant*

NI Vision Assistant is a tool for prototyping and testing image processing applications. You can create custom algorithms with the Vision Assistant scripting feature, which records every step of your processing algorithm. After completing the algorithm, you can test it on other images to check its reliability. Vision Assistant uses the NI Vision library but can be used independently of LabVIEW. In addition to being a tool for prototyping vision systems, you can use Vision Assistant to learn how different image processing functions perform.

The Vision Assistant interface makes prototyping your application easy and efficient because of features such as a reference window that displays your original image, a script window that stores your image processing steps, and a processing window that reflects changes to your images as you apply new parameters (Fig. 2.2). The result of prototyping an application in Vision Assistant is usually a script of exactly which steps are necessary to properly analyze the image. For example, as shown in Fig. 2.2, the prototype of bracket inspection application to determine if it meets specifications has basically five steps: find the hole at one end of the bracket using pattern matching, find the hole at the other end of the bracket using pattern matching, find the center of the bracket using edge detection, and measure the distance and angle between the holes from the center of the bracket.

Once you have developed a script that correctly analyzes your images, you can use Vision Assistant to tell you the time it takes to run the script. This information is extremely valuable if your inspection has to finish in a certain amount of time. As shown in Fig. 2.3, the bracket inspection takes 10.58 ms to complete.

After prototyping and testing, Vision Assistant automatically generates a block diagram in LabVIEW.

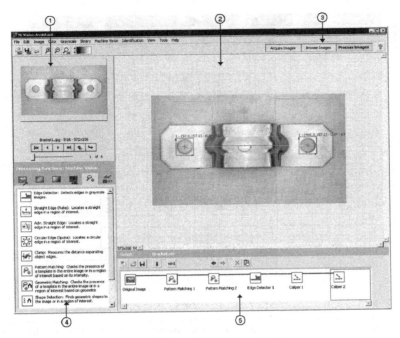

1 Reference window   4 Processing functions palette
2 Processing window   5 Script window
3 Navigation buttons

FIGURE 2.2

NI Vision Assistant, part of the NI Vision Development Module, prototypes vision applications, benchmarks inspections and generates ready-to-run LabVIEW Code.

2.3 EXAMPLES FROM THE SIVA IMAGE PROCESSING DEMOS

The SIVA gallery includes demos for 1D signals, image, and video processing. In this chapter, we focus only on the image processing demos. The image processing gallery of SIVA contains over 40 VIs (Table 2.1) that can be used to visualize many of the image processing concepts described in this book. In this section, we illustrate a few of these demos to familiarize the reader with SIVA's simple, intuitive interface and show the results of processing images using the VIs.

- **Image Quantization and Sampling:** Quantization and sampling are fundamental operations performed by any digital image acquisition device. Many people are familiar with the process of resizing a digital image to a smaller size (for the purpose of emailing photos or uploading them to social networking or photography Web sites). While a thorough mathematical analysis of these operations is rather

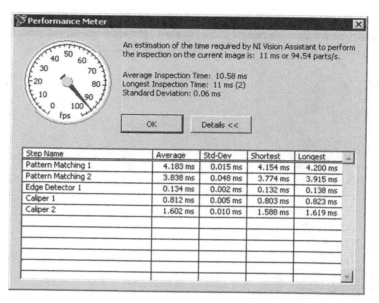

FIGURE 2.3

The Performance Meter inside NI Vision Assistant allows you to benchmark your application and help identify bottlenecks and optimize your vision code.

(a)   (b)   (c)

FIGURE 2.4

Grayscale quantization. (a) Front panel; (b) Original "Eggs" (8 bits per pixel); (c) Quantized "Eggs" (4 bits per pixel).

involved and difficult to interpret, it is nevertheless very easy to visually appreciate the effects and artifacts introduced by these processes using the VIs provided in the SIVA gallery. Figure 2.4, for example, illustrates the "false contouring" effect of grayscale quantization. While discussing the process of sampling any signal, students are introduced to the importance of "Nyquist sampling" and warned of "aliasing" or "false frequency" artifacts introduced by this process. The VI shown in

TABLE 2.1   A list of image and video processing demos available in the SIVA gallery.

Basics of Image Processing:
  Image quantization
  Image sampling
  Image histogram

Binary Image Processing:
  Image thresholding
  Image complementation
  Binary morphological filters
  Image skeletonization

Linear Point Operations:
  Full-scale contrast stretch
  Histogram shaping
  Image differencing
  Image interpolation

Discrete Fourier Analysis:
  Digital 2D sinusoids
  Discrete Fourier transform (DFT)
  DFTs of important 2D functions
  Masked DFTs
  Directional DFTs

Linear Filtering:
  Low, high, and bandpass filters
  Ideal lowpass filtering
  Gaussian filtering
  Noise models
  Image deblurring
  Inverse filter
  Wiener filter

Nonlinear Filtering:
  Median filtering
  Gray level morphological filters
  Trimmed mean filters
  Peak and valley detection
  Homomorphic filters

Digital Image Coding & Compression:
  Block truncation image coding
  Entropy reduction via DPCM
  JPEG coding

Edge Detection:
  Gradient-based edge detection
  Laplacian-of-Gaussian
  Canny edge detection
  Double thresholding
  Contour thresholding
  Anisotropic diffusion

Digital Video Processing:
  Motion compensation
  Optical flow calculation
  Block motion estimation

Other Applications:
  Hough transform
  Template matching
  Image quality using structural similarity

Fig. 2.5 demonstrates these artifacts caused by sampling. The patterns in the scarf, the books in the bookshelf, and the chair in the background of the "Barbara" image clearly change their orientation in the sampled images.

- **Binary Image Processing:** Binary images have only two possible "gray levels" and are therefore represented using only 1 bit per pixel. Besides the simple VIs used for thresholding grayscale images to binary images, SIVA has a demo that demonstrates the effects of various morphological operations on binary images, such as Median, Dilation, Erosion, Open, Close, Open-Clos, Clos-Open, and other

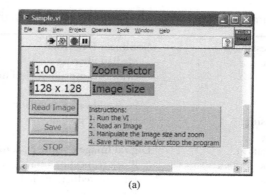

(a)

(b)

(c)

(d)

FIGURE 2.5

Effects of sampling. (a) Front panel; (b) Original "Barbara" image (256 × 256); (c) "Barbara" subsampled to 128 × 128; (d) Image c resized to 256 × 256 to show details.

binary operations including skeletonization. The user has the option to vary the shape and the size of the structuring element. The interface for the Morphology VI along with a binary image processed using the Erode, CLOS, and Majority operations is shown in Fig. 2.6.

- **Linear Point Operations and their Effects on Histograms:** Irrespective of their familiarity with the theory of DIP, most computer and digital camera users are familiar, if not proficient, with some form of an image editing software, such as Adobe Photoshop, Gimp, Picasa, or iPhoto. One of the frequently performed operations (on-camera or using software packages) is that of changing the brightness and/or contrast of an underexposed or overexposed photograph. To illustrate how these operations affect the histogram of the image, a VI in SIVA provides the user with controls to perform linear point operations, such as adding an offset,

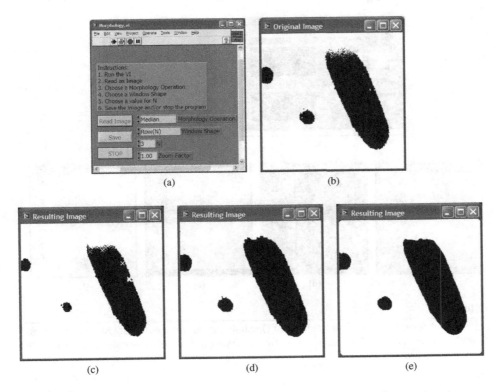

(a)

(b)

(c)

(d)

(e)

FIGURE 2.6

Binary morphological operations. (a) Front panel; (b) Original image; (c) Erosion using X-shaped window; (d) CLOS operation using square window; (e) Median (majority) operation using square window.

scaling the pixel values by scalar multiplication, and performing full-scale contrast stretch. Figure 2.7 shows a simple example where the histogram of the input image is either shifted to the right (increasing brightness), compressed while retaining shape, flipped to create an image negative, or stretched to fill the range (corresponding to full-scale contrast stretch). Advanced VIs allow the user to change the shape of the input histogram—an operation that is useful in cases where full-scale contrast stretch fails.

- **Discrete Fourier Transform:** Most of introductory DIP is based on the theory of linear systems. Therefore, a lucid understanding of frequency analysis techniques such as the Discrete Fourier Transform (DFT) is important to appreciate more advanced topics such as image filtering and spectral theory. SIVA has many VIs that provide an intuitive understanding of the DFT by first introducing the concept of spatial frequency using images of 2D digital sinusoidal gratings. The DFT VI can be used to compute and display the magnitude and the phase of the DFT for gray level images. Masking sections of the DFT using zero-one masks

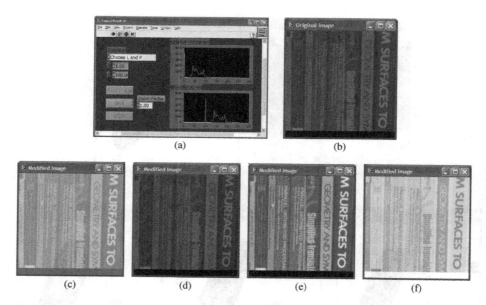

FIGURE 2.7

Linear point operations. (a) Front panel; (b) Original "Books" image; (c) Brightness enhanced by adding a constant; (d) Contrast reduced by multiplying by 0.9; (e) Full-scale contrast stretch; (f) Image negative.

of different shapes and then performing inverse DFT is a very intuitive way of understanding the granularity and directionality of the DFT (see Chapter 5 of this book). To demonstrate the directionality of the DFT, the VI shown in Fig. 2.8 was implemented. As shown on the front panel, the input parameters, Theta 1 and Theta 2, are used to control the angle of the wedge-like zero-one mask in Fig. 2.8(d). It is instructive to note that zeroing out some of the oriented components in the DFT results in the disappearance of one of the tripod legs in the "Cameraman" image in Fig. 2.8(e).

- **Linear and Nonlinear Image Filtering:** SIVA includes several demos to illustrate the use of linear and nonlinear filters for image enhancement and restoration. Low-pass filters for noise smoothing and inverse, pseudo inverse, and Wiener filters for deconvolving images that have been blurred are examples of some demos for linear image enhancement. SIVA also includes demos to illustrate the power of nonlinear filters over their linear counterparts. Figure 2.9, for example, demonstrates the result of filtering a noisy image corrupted with "salt and pepper noise" with a linear filter (average) and with a nonlinear (median) filter.

- **Image Compression:** Given the ease of capturing and publishing digital images on the Internet, it is no surprise most people are familiar with the terminology of compressed image formats such as JPEG. SIVA incorporates demos that highlight

(a)   (b)

(c)   (d)   (e)

FIGURE 2.8

Directionality of the Fourier Transform. (a) Front panel; (b) Original "Cameraman;" (c) DFT magnitude; (d) Masked DFT magnitude; (e) Reconstructed image.

fundamental ideas of image compression, such as the ability to reduce the entropy of an image using pulse code modulation. The gallery also contains a VI to illustrate block truncation coding (BTC)—a very simple yet powerful image compression scheme. As shown in the front panel in Fig. 2.10, the user can select the number of bits, B1, used to represent the mean of each block in BTC and the number of bits, B2, for the block variance. The compression ratio is computed and displayed on the front panel in the CR indicator in Fig. 2.10.

- **Hough Transform:** The Hough transform is useful for detecting straight lines in images. The transform operates on the edge map of an image. It uses an "accumulator" matrix to keep a count of the number of pixels that lie on a straight line of a certain parametric form, say, $y = mx + c$, where (x, y) are the coordinates of an edge location, m is the slope of the line, and c is the y-intercept. (In practice, a polar form of the straight line is used). In the above example, the accumulator matrix is 2D, with the two dimensions being the slope and the intercept. Each entry in the matrix corresponds to the number of pixels in the edge map that satisfy that particular equation of the line. The slope and intercept corresponding to the largest

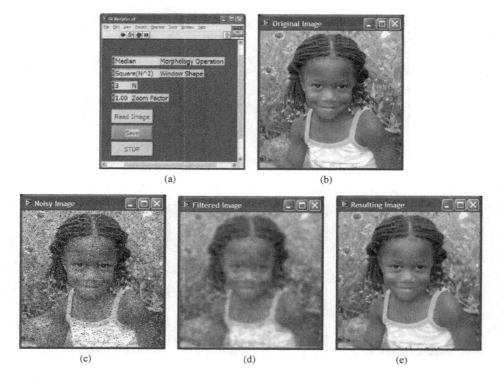

(a) (b)

(c) (d) (e)

FIGURE 2.9

Linear and nonlines image denoising. (a) Front panel; (b) Original "Mercy"; (c) Image corrupted by salt and pepper noise; (d) Denoised by blurring with a Gaussian filter; (e) Denoised using median filter.

entry in the matrix, therefore, correspond to the strongest straight line in the image. Figure 2.11 shows the result of applying the Hough transform in the SIVA gallery on the edges detected in the "Tower" image. As seen from Fig. 2.11(d), the simple algorithm presented above will be unable to distinguish partial line segments from a single straight line.

We have illustrated only a few VIs to whet the reader's appetite. As listed in Table 2.1, SIVA has many other advanced VIs that include many linear and nonlinear fileters for image enhancement, other lossy and lossless image compression schemes, and a large number of edge detectors for image feature analysis. The reader is encouraged to try out these demos at their leisure.

2.4 CONCLUSIONS

The SIVA gallery for image processing demos presented in this chapter was originally developed at UT-Austin to make the subject more accessible to students who came

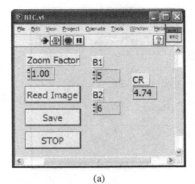

(a)

(b)

(c)

(d)

FIGURE 2.10

Block truncation coding. (a) Front panel; (b) original "Dhivya" image; (c) 5 bits for mean and 0 bits for variance (compression ratio = 6.1:1); (d) 5 bits for mean and 6 bits for variance (compression ratio = 4.74:1).

from varied academic disciplines, such as astronomy, math, genetics, remote sensing, video communications, and biomedicine, to name a few. In addition to the SIVA gallery presented here, there are several other excellent tools for image processing education [7], a few of which are listed below:

- IPLab [8]—A java-based plug-in to the popular ImageJ software from the Swiss Federal Institute of Technology, Lausanne, Switzerland.

- ALMOT 2D DSP and 2D J-DSP [9]—Java-based education tools from Arizona State University, USA.

- VcDemo [10]—A Microsoft Windows-based interactive video and image compression tool from Delft University of Technology, The Netherlands.

Since its release in November 2002, the SIVA demonstration gallery has been gaining in popularity and is currently being widely used by instructors in many educational institutions over the world for teaching their signal, image and video processing courses,

(a)

(b)

(c)

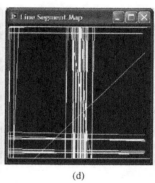

(d)

FIGURE 2.11

Hough transform. (a) Front panel; (b) original "Tower" image; (c) edge map; (d) lines detected by Hough Transform.

and by many individuals in industry for testing their image processing algorithms. To date, there are over 450 institutional users from 54 countries using SIVA. As mentioned earlier, the entire image processing gallery of SIVA is included in the CD that accompanies this book as a stand-alone version that does not need the user to own a copy of LabVIEW. All VIs may also be downloaded directly for free from the Web site mentioned in [2]. We hope that the intuition provided by the demos will make the reader's experience with image processing more enjoyable. Perhaps, the reader's newly found image processing lingo will compel them to mention how they "Designed a pseudo-inverse filter for deblurring" in lieu of "I photoshopped this image to make it sharp."

ACKNOWLEDGMENTS

Most of the image processing demos in the SIVA gallery were developed by National Instruments engineer and LIVE student George Panayi as a part of his M.S. Thesis [11].

The demos were upgraded to be compatible with LabVIEW 7.0 by National Instrument engineer and LIVE student Frank Baumgartner, who also implemented several video processing demos. The authors of this chapter would also like to thank National Instruments engineers Nate Holmes, Matthew Slaughter, Carleton Heard, and Nathan McKimpson for their invaluable help in upgrading the demos to be compatible with the latest version of LabVIEW, for creating a stand-alone version of the SIVA demos, and for their excellent effort in improving the uniformity of presentation and final debugging of the demos for this book. Finally, Umesh Rajashekar would also like to thank Dinesh Nair, Mark Walters, and Eric Luther at National Instruments for their timely assistance in providing him with the latest release of LabVIEW and LabVIEW Vision.

REFERENCES

[1] U. Rajashekar, G. C. Panayi, F. P. Baumgartner, and A. C. Bovik. The SIVA demonstration gallery for signal, image, and video processing education. *IEEE Trans. Educ.*, 45:323–335, 2002.

[2] U. Rajashekar, G. C. Panayi, F. P. Baumgartner, and A. C. Bovik. *SIVA – Signal, Image and Video Audio Visualizations.* The Univeristy of Texas at Austin, Austin, TX, 1999. http://live.ece.utexas.edu/class/siva.

[3] National Instruments. *LabVIEW Home Page.* http://www.ni.com/labview.

[4] R. H. Bishop. *LabVIEW 8 Student Edition. s.l.* National Instruments Inc, Austin, TX, 2006.

[5] J. Travis and J. Kring. *LabVIEW for Everyone: Graphical Programming Made Easy and Fun,* 3rd ed. Upper Saddle River, NJ, Prentice Hall PTR, 2006.

[6] National Instruments. *NI Vision Home Page.* http://www.ni.com/vision.

[7] U. Rajashekar, A. C. Bovik, L. Karam, R. L. Lagendijk, D. Sage, and M. Unser. Image processing education. In A. C. Bovik, editor, *The Handbook of Image and Video Processing,* 2nd ed., pages 73–95. Academic Press, New York, NY, 2005.

[8] D. Sage and M. Unser. Teaching image-processing programming in Java. *IEEE Signal Process. Mag.*, 20:43–52, 2003.

[9] A. Spanias. *JAVA Digital Signal Processing Editor.* Arizona State University. http://jdsp.asu.edu/jdsp.html.

[10] R. L. Lagendijk. *VcDemo Software.* Delft University of Technology, The Netherlands. http://www-ict.ewi.tudelft.nl/vcdemo.

[11] G. C. Panayi. *Implementation of Digital Image Processing Functions Using LabVIEW.* M.S. Thesis, Dept. of Electrical and Computer Engineering, The University of Texas at Austin, Austin, TX, 1999.

Basic Gray Level Image Processing

3

Alan C. Bovik

The University of Texas at Austin

3.1 INTRODUCTION

This chapter, and the two that follow, describe the most commonly used and most basic tools for digital image processing. For many simple image analysis tasks, such as contrast enhancement, noise removal, object location, and frequency analysis, much of the necessary collection of instruments can be found in Chapters 3–5. Moreover, these chapters supply the basic groundwork that is needed for the more extensive developments that are given in the subsequent chapters of the *Guide*.

In the current chapter, we study basic gray level digital image processing operations. The types of operations studied fall into three classes.

The first are *point operations*, or image processing operations, that are applied to individual pixels only. Thus, interactions and dependencies between neighboring pixels are not considered, nor are operations that consider multiple pixels simultaneously to determine an output. Since spatial information, such as a pixel's location and the values of its neighbors, are not considered, point operations are defined as functions of pixel intensity only. The basic tool for understanding, analyzing, and designing image point operations is the *image histogram*, which will be introduced below.

The second class includes *arithmetic operations* between images of the same spatial dimensions. These are also point operations in the sense that spatial information is not considered, although information is shared between images on a pointwise basis. Generally, these have special purposes, e.g., for noise reduction and change or motion detection.

The third class of operations are *geometric image operations*. These are complementary to point operations in the sense that they are not defined as functions of image intensity. Instead, they are functions of spatial position only. Operations of this type change the appearance of images by changing the coordinates of the intensities. This can be as simple as image translation or rotation, or may include more complex operations that distort or bend an image, or "morph" a video sequence. Since our goal, however, is to concentrate on digital image processing of real-world images, rather than the production of special effects, only the most basic geometric transformations will be considered. More complex and time-varying geometric effects are more properly considered within the science of *computer graphics.*

3.2 NOTATION

Point operations, algebraic operations, and geometric operations are easily defined on images of any dimensionality, including digital video data. For simplicity of presentation, we will restrict our discussion to 2D images only. The extensions to three or higher dimensions are not difficult, especially in this case of point operations, which are independent of dimensionality. In fact, spatial/temporal information is not considered in their definition or application.

We will also only consider monochromatic images, since extensions to color or other multispectral images is either trivial, in that the same operations are applied identically to each band (e.g., R, G, B), or they are defined as more complex color space operations, which goes beyond what we want to cover in this basic chapter.

Suppose then that the single-valued image $f(\mathbf{n})$ to be considered is defined on a two-dimensional discrete-space coordinate system $\mathbf{n} = (n_1, n_2)$ or $\mathbf{n} = (m, n)$. The image is assumed to be of finite support, with image domain $[0, M - 1] \times [0, N - 1]$. Hence the nonzero image data can be contained in a matrix or array of dimensions $M \times N$ (rows, columns). This *discrete-space* image will have originated by sampling a continuous image $f(x, y)$. Furthermore, the image $f(\mathbf{n})$ is assumed to be *quantized* to k levels $\{0, \ldots, K - 1\}$, hence each pixel value takes one of these integer values. For simplicity, we will refer to these values as *gray levels*, reflecting the way in which monochromatic images are usually displayed. Since $f(\mathbf{n})$ is both discrete-spaced and quantized, it is *digital.*

3.3 IMAGE HISTOGRAM

The basic tool that is used in designing point operations on digital images (and many other operations as well) is the *image histogram*. The histogram H_f of the digital image f is a plot or graph of the *frequency of occurrence* of each gray level in f. Hence, H_f is a one-dimensional function with domain $\{0, \ldots, K - 1\}$ and possible range extending from 0 to the number of pixels in the image, MN.

The histogram is given explicitly by

$$H_f(k) = J \tag{3.1}$$

if f contains exactly J occurrences of gray level k, for each $k = 0, \ldots, K - 1$. Thus, an algorithm to compute the image histogram involves a simple counting of gray levels, which can be accomplished even as the image is scanned. Every image processing development environment and software library contains basic histogram computation, manipulation, and display routines.

Since the histogram represents a reduction of dimensionality relative to the original image f, information is lost—the image f cannot be deduced from the histogram H_f except in trivial cases (when the image is constant-valued). In fact, the number of images that share the same arbitrary histogram H_f is astronomical. Given an image f with a particular histogram H_f, every image that is a spatial shuffling of the gray levels of f has the same histogram H_f.

The histogram H_f contains no spatial information about f—it describes the frequency of the gray levels in f and nothing more. However, this information is still very rich, and many useful image processing operations can be derived from the image histogram. Indeed, a simple visual display of H_f reveals much about the image. By examining the appearance of a histogram, it is possible to ascertain whether the gray levels are distributed primarily at lower (darker) gray levels, or vice versa. Although this can be ascertained to some degree by visual examination of the image itself, the human eye has a tremendous ability to adapt to overall changes in luminance, which may obscure shifts in the gray level distribution. The histogram supplies an absolute method of determining an image's gray level distribution.

For example, the *average optical density*, or AOD, is the basic measure of an image's overall average brightness or gray level. It can be computed directly from the image:

$$\text{AOD}(f) = \frac{1}{NM} \sum_{n_1=0}^{N-1} \sum_{n_2=0}^{M-1} f(n_1, n_2) \tag{3.2}$$

or it can be computed from the image histogram:

$$\text{AOD}(f) = \frac{1}{NM} \sum_{k=0}^{K-1} k H_f(k). \tag{3.3}$$

The AOD is a useful and simple meter for estimating the center of an image's gray level distribution. A target value for the AOD might be specified when designing a point operation to change the overall gray level distribution of an image.

Figure 3.1 depicts two hypothetical image histograms. The one on the left has a heavier distribution of gray levels close to zero (and a low AOD), while the one on the right is skewed toward the right (a high AOD). Since image gray levels are usually displayed with lower numbers indicating darker pixels, the image on the left corresponds to a predominantly dark image. This may occur if the image f was originally underexposed

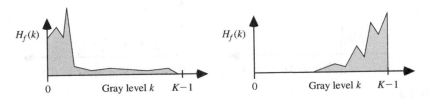

FIGURE 3.1

Histograms of images with gray level distribution skewed towards darker (left) and brighter (right) gray levels. It is possible that these images are underexposed and overexposed, respectively.

prior to digitization, or if it was taken under poor lighting levels, or perhaps the process of digitization was performed improperly. A skewed histogram often indicates a problem in gray level allocation. The image on the right may have been overexposed or taken in very bright light.

Figure 3.2 depicts the 256×256 ($M = N = 256$) gray level digital image "students" with grayscale range $\{0, \ldots, 255\}$ and its computed histogram. Although the image contains a broad distribution of gray levels, the histogram is heavily skewed toward the dark end, and the image appears to be poorly exposed. It is of interest to consider techniques that attempt to "equalize" this distribution of gray levels. One of the important applications of image point operations is to correct for poor exposures like the one in Fig. 3.2. Of course, there may be limitations on the effectiveness of any attempt to recover an image from poor exposure since information may be lost. For example, in Fig. 3.2, the gray levels saturate at the low end of the scale, making it difficult or impossible to distinguish features at low brightness levels.

More generally, an image may have a histogram that reveals a poor usage of the available grayscale range. An image with a compact histogram, as depicted in Fig. 3.3,

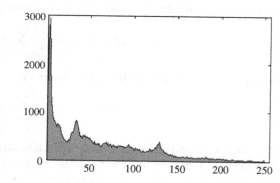

FIGURE 3.2

The digital image "students" (left) and its histogram (right). The gray levels of this image are skewed towards the left, and the image appears slightly underexposed.

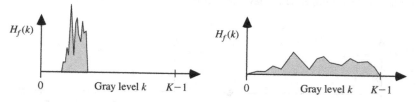

FIGURE 3.3

Histograms of images that make poor (left) and good (right) use of the available grayscale range. A compressed histogram often indicates an image with a poor visual contrast. A well-distributed histogram often has a higher contrast and better visibility of detail.

 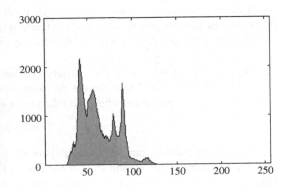

FIGURE 3.4

Digital image "books" (left) and its histogram (right). The image makes poor use of the available grayscale range.

will often have a poor visual contrast or a "washed-out" appearance. If the grayscale range is filled out, also depicted in Fig. 3.3, then the image tends to have a higher contrast and a more distinctive appearance. As will be shown, there are specific point operations that effectively expand the grayscale distribution of an image.

Figure 3.4 depicts the 256 × 256 gray level image "books" and its histogram. The histogram clearly reveals that nearly all of the gray levels that occur in the image fall within a small range of grayscales, and the image is of correspondingly poor contrast.

It is possible that an image may be taken under correct lighting and exposure conditions, but that there is still a skewing of the gray level distribution toward one end of the grayscale or that the histogram is unusually compressed. An example would be an image of the night sky, which is dark nearly everywhere. In such a case, the appearance of the image may be normal but the histogram will be very skewed. In some situations, it may still be of interest to attempt to enhance or reveal otherwise difficult-to-see details in the image by application of an appropriate point operation.

3.4 LINEAR POINT OPERATIONS ON IMAGES

A *point operation* on a digital image $f(\mathbf{n})$ is a function h of a single variable applied identically to every pixel in the image, thus creating a new, modified image $g(\mathbf{n})$. Hence at each coordinate $\mathbf{n}$,

$$g(\mathbf{n}) = h[f(\mathbf{n})]. \tag{3.4}$$

The form of the function h is determined by the task at hand. However, since each output $g(\mathbf{n})$ is a function of a single pixel value only, the effects that can be obtained by a point operation are somewhat limited. Specifically, no spatial information is utilized in (3.4), and there is no change made in the spatial relationships between pixels in the transformed image. Thus, point operations do not affect the spatial positions of objects in an image, nor their shapes. Instead, each pixel value or gray level is increased or decreased (or unchanged) according to the relation in (3.4). Therefore, a point operation h does change the gray level distribution or histogram of an image, and hence the overall appearance of the image.

Of course, there is an unlimited variety of possible effects that can be produced by selection of the function h that defines the point operation (3.4). Of these, the simplest are the *linear point operations*, where h is taken to be a simple linear function of gray level:

$$g(\mathbf{n}) = Pf(\mathbf{n}) + L. \tag{3.5}$$

Linear point operations can be viewed as providing a gray level additive offset L and a gray level multiplicative scaling P of the image f. Offset and scaling provide different effects, and so we will consider them separately before examining the overall linear point operation (3.5).

The *saturation* conditions $|g(\mathbf{n})| < 0$ and $|g(\mathbf{n})| > K - 1$ are to be avoided if possible, since the gray levels are then not properly defined, which can lead to severe errors in processing or display of the result. The designer needs to be aware of this so steps can be taken to ensure that the image is not distorted by values falling outside the range. If a specific wordlength has been allocated to represent the gray level, then saturation may result in an overflow or underflow condition, leading to very large errors. A simple way to handle this is to simply clip those values falling outside of the allowable grayscale range to the endpoint values. Hence, if $|g(\mathbf{n}_0)| < 0$ at some coordinate $\mathbf{n}_0$, then set $|g(\mathbf{n}_0)| = 0$ instead. Likewise, if $|g(\mathbf{n}_0)| > K - 1$, then fix $|g(\mathbf{n}_0)| = K - 1$. Of course, the result is no longer strictly a linear point operation. Care must be taken since information is lost in the clipping operation, and the image may appear artificially flat in some areas if whole regions become clipped.

3.4.1 Additive Image Offset

Suppose $P = 1$ and L is an integer satisfying $|L| \leq K - 1$. An *additive image offset* has the form

$$g(\mathbf{n}) = f(\mathbf{n}) + L. \tag{3.6}$$

Here we have prescribed a range of values that L can take. We have taken L to be an integer, since we are assuming that images are quantized into integers in the range $\{0, \ldots, K - 1\}$. We have also assumed that $|L|$ falls in this range, since otherwise, all of the values of $g(\mathbf{n})$ will fall outside the allowable grayscale range.

In (3.6), if $L > 0$, then $g(\mathbf{n})$ will be a brightened version of the image $f(\mathbf{n})$. Since spatial relationships between pixels are unaffected, the appearance of the image will otherwise be essentially the same. Likewise, if $L < 0$, then $g(\mathbf{n})$ will be a dimmed version of the $f(\mathbf{n})$. The histograms of the two images have a simple relationship:

$$H_g(k) = H_f(k - L). \tag{3.7}$$

Thus, an offset L corresponds to a shift of the histogram by amount L to the left or to the right, as depicted in Fig. 3.5.

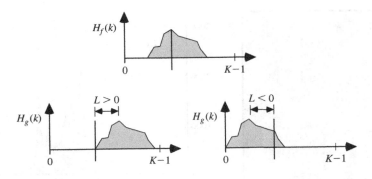

FIGURE 3.5

Effect of additive offset on the image histogram. Top: original image histogram; bottom: positive (left) and negative (right) offsets shift the histogram to the right and to the left, respectively.

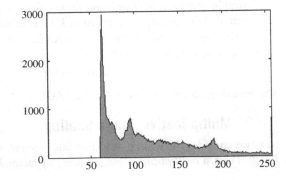

FIGURE 3.6

Left: Additive offset of the image "students" in Fig. 3.2 by amount 60. Observe the clipping spike in the histogram to the right at gray level 255.

Figures 3.6 and 3.7 show the result of applying an additive offset to the images "students" and "books" in Figs. 3.2 and 3.4, respectively. In both cases, the overall visibility of the images has been somewhat increased, but there has not been an improvement in the contrast. Hence, while each image as a whole is easier to see, the details in the image are no more visible than they were in the original. Figure 3.6 is a good example of saturation; a large number of gray levels were clipped at the high end (gray level 255). In this case, clipping did not result in much loss of information.

Additive image offsets can be used to calibrate images to a given average brightness level. For example, suppose we desire to compare multiple images $f_1, f_2, \ldots, f_n$ of the same scene, taken at different times. These might be surveillance images taken of a secure area that experiences changes in overall ambient illumination. These variations could occur because the area is exposed to daylight.

 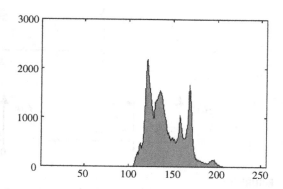

FIGURE 3.7

Left: Additive offset of the image "books" in Fig. 3.4 by amount 80.

A simple approach to counteract these effects is to equalize the AODs of the images. A reasonable AOD is the grayscale center $K/2$, although other values may be used depending on the application. Letting $L_m = \text{AOD}(f_m)$, for $m = 1, \ldots, n$, the "AOD-equalized" images $g_1, g_2, \ldots, g_n$ are given by

$$g_m(\mathbf{n}) = f_m(\mathbf{n}) - L_m + K/2. \tag{3.8}$$

The resulting images then have identical AOD $K/2$.

3.4.2 Multiplicative Image Scaling

Next we consider the scaling aspect of linear point operations. Suppose that $L = 0$ and $P > 0$. Then, a *multiplicative image scaling* by factor P is given by

$$g(\mathbf{n}) = Pf(\mathbf{n}). \tag{3.9}$$

Here P is assumed positive since $g(\mathbf{n})$ must be positive. Note that we have not constrained P to be an integer, since this would usually leave few useful values of P; for example, even taking $P = 2$ will severely saturate most images. If an integer result is required, then a practical definition for the output is to *round* the result in (3.9):

$$g(\mathbf{n}) = \text{INT}[Pf(n) + 0.5], \tag{3.10}$$

where $\text{INT}[R]$ denotes the nearest integer that is less than or equal to R.

The effect that multiplicative scaling has on an image depends largely on whether P is larger or smaller than one. If $P > 1$, then the gray levels of g will cover a broader range than those of f. Conversely, if $P < 1$, then g will have a narrower gray level distribution than f. In terms of the image histogram,

$$H_g\{\text{INT}[Pk + 0.5]\} = H_f(k). \tag{3.11}$$

Hence multiplicative scaling by a factor P either stretches or compresses the image histogram. Note that for quantized images, it is not proper to assume that (3.11) implies $H_g(k) = H_f(k/P)$ since the argument of $H_f(k/P)$ may not be an integer.

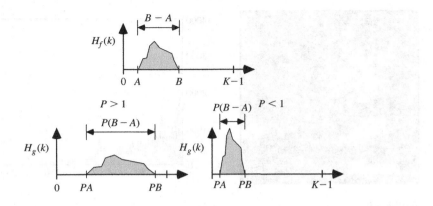

FIGURE 3.8

Effects of multiplicative image scaling on the histogram. If $P > 1$, the histogram is expanded, leading to more complete use of the grayscale range. If $P < 1$, the histogram is contracted, leading to possible information loss and (usually) a less striking image.

Figure 3.8 depicts the effect of multiplicative scaling on a hypothetical histogram. For $P > 1$, the histogram is expanded (and hence, saturation is quite possible), while for $P < 1$, the histogram is contracted. If the histogram is contracted, then multiple gray levels in f may map to single gray levels in g since the number of gray levels is finite. This implies a possible loss of information. If the histogram is expanded, then spaces may appear between the histogram bins where gray levels are not being mapped. This, however, does not represent a loss of information and usually will not lead to visual information loss.

As a rule of thumb, histogram expansion often leads to a more distinctive image that makes better use of the grayscale range, provided that saturation effects are not visually noticeable. Histogram contraction usually leads to the opposite: an image with reduced visibility of detail that is less striking. However, these are only rules of thumb, and there are exceptions. An image may have a grayscale spread that is too extensive, and may benefit from scaling with $P < 1$.

Figure 3.9 shows the image "students" following a multiplicative scaling with $P = 0.75$, resulting in compression of the histogram. The resulting image is darker and less contrasted. Figure 3.10 shows the image "books" following scaling with $P = 2$. In this case, the resulting image is much brighter and has a better visual resolution of gray levels. Note that most of the high end of the grayscale range is now used, although the low end is not.

3.4.3 Image Negative

The first example of a linear point operation that uses both scaling and offset is the *image negative*, which is given by $P = -1$ and $L = K - 1$. Hence

$$g(\mathbf{n}) = -f(\mathbf{n}) + (K - 1) \tag{3.12}$$

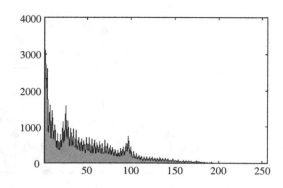

FIGURE 3.9

Histogram compression by multiplicative image scaling with $P = 0.75$. The resulting image is less distinctive. Note also the regularly-spaced tall spikes in the histogram; these are gray levels that are being "stacked," resulting in a loss of information, since they can no longer be distinguished.

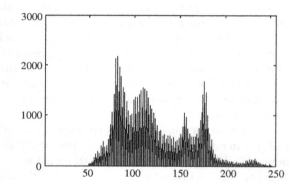

FIGURE 3.10

Histogram expansion by multiplicative image scaling with $P = 2.0$. The resulting image is much more visually appealing. Note the regularly-spaced gaps in the histogram that appear when the discrete histogram values are spread out. This does not imply a loss of information or visual fidelity.

and

$$H_g(k) = H_f(K - 1 - k). \tag{3.13}$$

Scaling by $P = -1$ reverses (flips) the histogram; the additive offset $L = K - 1$ is required so that all values of the result are positive and fall in the allowable grayscale range. This operation creates a *digital negative image*, unless the image is already a negative,

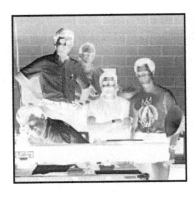
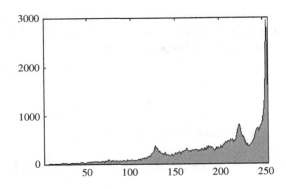

FIGURE 3.11

Example of image negative with resulting reversed histogram.

in which case a positive is created. It should be mentioned that unless the digital negative (3.12) is being computed, $P > 0$ in nearly every application of linear point operations.

An important application of (3.12) occurs when a negative is scanned (digitized), and it is desired to view the positive image. Figure 3.11 depicts the negative image associated with "students." Sometimes, the negative image is viewed intentionally, when the positive image itself is very dark. A common example of this is for the examination of telescopic images of star fields and faint galaxies. In the negative image, faint bright objects appear as dark objects against a bright background, which can be easier to see.

3.4.4 Full-Scale Histogram Stretch

We have already mentioned that an image that has a broadly distributed histogram tends to be more visually distinctive. The *full-scale histogram stretch*, which is also often called a *contrast stretch*, is a simple linear point operation that expands the image histogram to fill the entire available grayscale range. This is such a desirable operation that the full-scale histogram stretch is easily the most common linear point operation. Every image processing programming environment and library contains it as a basic tool. Many image display routines incorporate it as a basic feature. Indeed, commercially-available digital video cameras for home and professional use generally apply a full-scale histogram stretch to the acquired image before being stored in camera memory. It is called automatic gain control on these devices.

The definition of the multiplicative scaling and additive offset factors in the full-scale histogram stretch depend on the image f. Suppose that f has a compressed histogram with maximum gray level value B and minimum value A, as shown in Fig. 3.8 (top):

$$A = \min_{\mathbf{n}}\{f(\mathbf{n})\} \quad \text{and} \quad B = \max_{\mathbf{n}}\{f(\mathbf{n})\}. \tag{3.14}$$

The goal is to find a linear point operation of the form (3.5) that maps gray levels A and B in the original image to gray levels 0 and $K-1$ in the transformed image. This can be expressed in two linear equations:

$$PA + L = 0 \tag{3.15}$$

and

$$PB + L = K - 1 \tag{3.16}$$

in the two unknowns (P, L), with solutions

$$P = \left(\frac{K-1}{B-A}\right) \tag{3.17}$$

and

$$L = -A\left(\frac{K-1}{B-A}\right). \tag{3.18}$$

Hence, the overall full-scale histogram stretch is given by

$$g(\mathbf{n}) = \text{FSHS}(f) = \left(\frac{K-1}{B-A}\right)[f(\mathbf{n}) - A]. \tag{3.19}$$

We make the shorthand notation FSHS, since (3.19) will prove to be commonly useful as an addendum to other algorithms. The operation in (3.19) can produce dramatic improvements in the visual quality of an image suffering from a poor (narrow) grayscale distribution. Figure 3.12 shows the result of applying the full-scale histogram stretch to the image "books." The contrast and visibility of the image was, as expected, greatly improved. The accompanying histogram, which now fills the available range, also shows the characteristic gaps of an expanded discrete histogram.

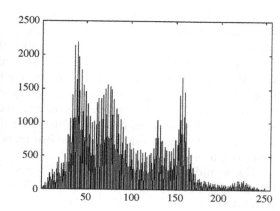

FIGURE 3.12

Full-scale histogram stretch of image "books."

If the image f already has a broad gray level range, then the histogram stretch may produce little or no effect. For example, the image "students" (Fig. 3.2) has grayscales covering the entire available range, as seen in the histogram accompanying the image. Therefore, (3.19) has no effect on "students." This is unfortunate, since we have already commented that "students" might benefit from a histogram manipulation that would redistribute the gray level densities. Such a transformation would need to nonlinearly reallocate the image's gray level values. Such nonlinear point operations are described next.

3.5 NONLINEAR POINT OPERATIONS ON IMAGES

We now consider *nonlinear point* operations of the form

$$g(\mathbf{n}) = h[f(\mathbf{n})], \tag{3.20}$$

where the function h is nonlinear. Obviously, this encompasses a wide range of possibilities. However, there are only a few functions h that are used with any great degree of regularity. Some of these are functional tools that are used as part of larger, multistep algorithms, such as absolute value, square, and square-root functions. One such simple nonlinear function that is very commonly used is the logarithmic point operation, which we describe in detail.

3.5.1 Logarithmic Point Operations

Assuming that the image $f(\mathbf{n})$ is positive-valued, the *logarithmic point operation* is defined by a composition of two operations: a point logarithmic operation, followed by a full-scale histogram stretch:

$$g(\mathbf{n}) = \text{FSHS}\{\log[1 + f(\mathbf{n})]\}. \tag{3.21}$$

Adding unity to the image avoids the possibility of taking the logarithm of zero. The logarithm itself acts to nonlinearly compress the gray level range. All of the gray levels are compressed to the range $[0, \log(K)]$. However, larger (brighter) gray levels are compressed much more severely than are smaller gray levels. The subsequent FSHS operation then acts to linearly expand the log-compressed gray levels to fill the grayscale range. In the transformed image, dim objects in the original are now allocated a much larger percentage of the grayscale range, hence improving their visibility.

The logarithmic point operation is an excellent choice for improving the appearance of the image "students," as shown in Fig. 3.13. The original image (Fig. 3.2) was not a candidate for FSHS because of its broad histogram. The appearance of the original suffers because many of the important features of the image are obscured by darkness. The histogram is significantly spread at these low brightness levels, as can be seen by comparing to Fig. 3.2, and also by the gaps that appear in the low end of the histogram. This does not occur at brighter gray levels.

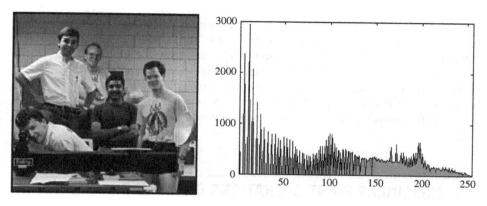

FIGURE 3.13

Logarithmic grayscale range compression followed by FSHS applied to image "students."

Certain applications quite commonly use logarithmic point operations. For example, in astronomical imaging, a relatively few bright pixels (stars and bright galaxies, etc.) tend to dominate the visual perception of the image, while much of the interesting information lies at low bright levels (e.g., large, faint nebulae). By compressing the bright intensities much more heavily, then applying FSHS, the faint, interesting details visually emerge.

Later, in Chapter 5, the Fourier transforms of images will be studied. The Fourier transform magnitudes, which are of the same dimensionalities as images, will be displayed as intensity arrays for visual consumption. However, the Fourier transforms of most images are dominated visually by the Fourier coefficients of a relatively few low frequencies, so the coefficients of important high frequencies are usually difficult or impossible to see. However, a point logarithmic operation usually suffices to ameliorate this problem, and so image Fourier transforms are usually displayed following application of (3.21), both in this *Guide* and elsewhere.

3.5.2 Histogram Equalization

One of the most important nonlinear point operations is *histogram equalization*, also called *histogram flattening*. The idea behind it extends that of FSHS: not only should an image fill the available grayscale range but also it should be uniformly distributed over that range. Hence an idealized goal is a flat histogram. Although care must be taken in applying a powerful nonlinear transformation that actually changes the shape of the image histogram, rather than just stretching it, there are good mathematical reasons for regarding a flat histogram as a desirable goal. In a certain sense,[1] an image with a perfectly flat histogram contains the largest possible amount of *information* or complexity.

---

[1] In the sense of maximum entropy.

In order to explain histogram equalization, it will be necessary to make some refined definitions of the image histogram. For an image containing MN pixels, the *normalized image histogram* is given by

$$p_f(k) = \frac{1}{MN} H_f(k) \tag{3.22}$$

for $k = 0, \ldots, K - 1$. This function has the property that

$$\sum_{k=0}^{K-1} p_f(k) = 1. \tag{3.23}$$

The normalized histogram $p_f(k)$ has a valid interpretation as the empirical probability density (mass function) of the gray level values of image f. In other words, if a pixel coordinate $\mathbf{n}$ is chosen at random, then $p_f(k)$ is the probability that $f(\mathbf{n}) = k : p_f(k) = \Pr\{f(\mathbf{n}) = k\}$.

We also define the *cumulative normalized image histogram* to be

$$P_f(r) = \sum_{k=0}^{r} p_f(k); \quad r = 0, \ldots, K - 1. \tag{3.24}$$

The function $P_f(r)$ is an empirical probability distribution function, hence it is a non-decreasing function, and also $P_f(K - 1) = 1$. It has the probabilistic interpretation that for a randomly selected image coordinate $\mathbf{n}$, $P_f(r) = \Pr\{f(\mathbf{n}) \le r\}$. From (3.24), it is also true that

$$p_f(k) = P_f(k) - P_f(k - 1); k = 0, \ldots, K - 1 \tag{3.25}$$

so $P_f(k)$ and $p_f(k)$ can be obtained from each other. Both are complete descriptions of the gray level distribution of the image f.

In order to understand the process of *digital histogram equalization*, we first explain the process supposing that the normalized and cumulative histograms are functions of continuous variables. We will then formulate the digital case of an approximation of the continuous process. Hence suppose that $p_f(x)$ and $P_f(x)$ are functions of a continuous variable x. They may be regarded as image probability density function (pdf) and cumulative distribution function (cdf), with relationship $p_f(x) = dP_f(x)/dx$. We will also assume that P_f^{-1} exists. Since P_f is nondecreasing, this is either true or P_f^{-1} can be defined by a convention. In this hypothetical continuous case, we claim that the image

$$FSHS(g), \tag{3.26}$$

where

$$g = P_f(f) \tag{3.27}$$

has a uniform (flat) histogram. In (3.27), $P_f(f)$ denotes that P_f is applied on a pixelwise basis to f:

$$g(\mathbf{n}) = P_f[f(\mathbf{n})] \tag{3.28}$$

for all **n**. Since P_f is a continuous function, (3.26)–(3.28) represents a smooth mapping of the histogram of image f to an image with a smooth histogram. At first, (3.27) may seem confusing since the function P_f that is computed from f is then applied to f. To see that a flat histogram is obtained, we use the probabilistic interpretation of the histogram. The cumulative histogram of the resulting image g is:

$$P_g(x) = \Pr\{g \le x\} = \Pr\{P_f(f) \le x\}$$
$$= \Pr\{f \le P_f^{-1}(x)\} = P_f\{P_f^{-1}(x)\} = x \tag{3.29}$$

for $0 \le x \le 1$. Finally, the normalized histogram of g is

$$p_g(x) = dP_g(x)/dx = 1 \tag{3.30}$$

for $0 \le x \le 1$. Since $p_g(x)$ is defined only for $0 \le x \le 1$, FSHS in (3.26) is required to stretch the flattened histogram to fill the grayscale range.

To flatten the histogram of a digital image f, first compute the discrete cumulative normalized histogram $P_f(k)$, apply (3.28) at each **n**, then (3.26) to the result. However, while an image with a perfectly flat histogram is the result in the ideal continuous case outlined above, in the digital case, the output histogram is only approximately flat, or more accurately flatter than the input histogram. This follows since (3.26)–(3.28) collectively is a point operation on the image f, so every occurrence of gray level k maps to $P_f(k)$ in g. Hence, histogram bins are never reduced in amplitude by (3.26)–(3.28), although they may increase if multiple gray levels map to the same value (thus destroying information). Hence, the histogram cannot be truly equalized by this procedure.

Figures 3.14 and 3.15 show histogram equalization applied to our ongoing example images "students" and "books," respectively. Both images are much more striking and viewable than the original. As can be seen, the resulting histograms are not really flat; it is "flatter" in the sense that the histograms are spread as much as possible. However, the heights of peaks are *not* reduced. As is often the case with expansive point operations,

 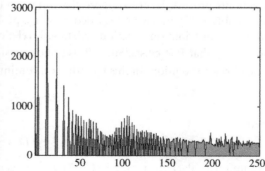

FIGURE 3.14

Histogram equalization applied to the image "students."

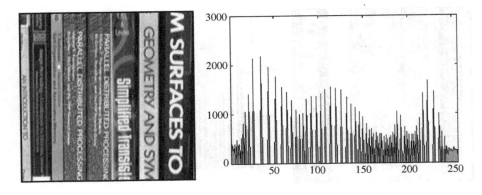

FIGURE 3.15

Histogram equalization applied to the image "books."

gaps or spaces appear in the output histogram. These are not a problem unless the gaps become large and some of the histogram bins become isolated. This amounts to an excess of quantization in that range of gray levels, which may result in false contouring (Chapter 1).

3.5.3 Histogram Shaping

In some applications, it is desired to transform the image into one that has a histogram of a specific shape. The process of histogram shaping generalizes histogram equalization, which is the special case where the target shape is flat. Histogram shaping can be applied when multiple images of the same scene, taken under mildly different lighting conditions, are to be compared. This extends the idea of AOD-equalization described earlier in this chapter. By shaping the histograms to match, the comparison may exclude minor lighting effects. Alternately, it may be that the histogram of one image is shaped to match that of another, again usually for the purpose of comparison. Or it might simply be that a certain histogram shape, such as a Gaussian, produces visually agreeable results for a certain class of images.

Histogram shaping is also accomplished by a nonlinear point operation defined in terms of the empirical image probabilities or histogram functions. Again, exact results are obtained in the hypothetical continuous-scale case. Suppose that the target (continuous) cumulative histogram function is $Q(x)$, and that Q^{-1} exists. Then let

$$g = Q^{-1}[P_f(f)], \tag{3.31}$$

where both functions in the composition are applied on a pixelwise basis. The cumulative histogram of g is then:

$$P_g(x) = \Pr\{g \leq x\} = \Pr\{Q^{-1}[P_f(f)] \leq x\}$$
$$= \Pr\{P_f(f) \leq Q(x)\} = \Pr\{f \leq P_f^{-1}[Q(x)]\}$$
$$= P_f\{P_f^{-1}[Q(x)]\} = Q(x), \tag{3.32}$$

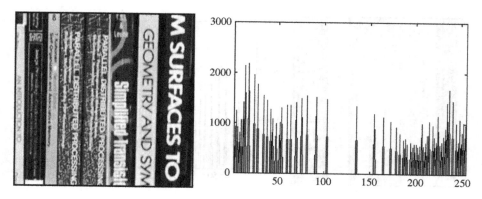

FIGURE 3.16

Histogram of the image "books" shaped to match a "V."

as desired. Note that FSHS is not required in this instance. Of course, (3.32) can only be approximated when the image f is digital. In such cases, the specified target cumulative histogram function $Q(k)$ is discrete, and some convention for defining Q^{-1} should be adopted, particularly if Q is computed from a target image and is unknown in advance. One common convention is to define

$$Q^{-1}(k) = \min_s \{s : Q(s) \geq k\}. \tag{3.33}$$

As an example, Fig. 3.16 depicts the result of shaping the histogram of "books" to match the shape of an inverted "V" centered at the middle gray level and extending across the entire grayscale. Again, a perfect "V" is not produced, although an image of very high contrast is still produced. Instead, the histogram shape that results is a crude approximation to the target.

3.6 ARITHMETIC OPERATIONS BETWEEN IMAGES

We now consider *arithmetic operations* defined on multiple images. The basic operations are pointwise image addition/subtraction and pointwise image multiplication/division. Since digital images are defined as arrays of numbers, these operations need to be defined carefully.

Suppose we have n $N \times M$ images $f_1, f_2, \ldots, f_n$. It is important they are of the same dimensions since we will be defining operations between corresponding array elements (having the same indices).

The *sum* of n images is given by

$$f_1 + f_2 + \cdots + f_n = \sum_{m=1}^{n} f_m \tag{3.34}$$

while for any two images f_r and f_s the *image difference* is

$$f_r - f_s. \tag{3.35}$$

The *pointwise product* of the $n\ N \times M$ images $f_1, \ldots, f_n$ is denoted by

$$f_1 \otimes f_2 \otimes \ldots \otimes f_n = \prod_{m=1}^{n} f_m, \tag{3.36}$$

where in (3.36) we do *not* infer that the matrix product is being taken. Instead, the product is defined on a pointwise basis. Hence $g = f_1 \otimes f_2 \otimes \ldots \otimes f_n$ if and only if

$$g(\mathbf{n}) = f_1(\mathbf{n}) f_2(\mathbf{n}) \ldots f_n(\mathbf{n}) \tag{3.37}$$

for every $\mathbf{n}$. In order to clarify the distinction between matrix product and pointwise array product, we introduce the special notation "$\otimes$" to denote the pointwise product. Given two images f_r and f_s the *pointwise image quotient* is denoted

$$g = f_r \Delta f_s \tag{3.38}$$

if for every $\mathbf{n}$ it is true that $f_s(\mathbf{n}) \neq 0$ and

$$g(\mathbf{n}) = f_r(\mathbf{n})/f_s(\mathbf{n}). \tag{3.39}$$

The pointwise matrix product and quotient are mainly useful when manipulating Fourier transforms of images, as will be seen in Chapter 5. However, the pointwise image sum and difference, despite their simplicity, have important applications that we will examine next.

3.6.1 Image Averaging for Noise Reduction

Images that occur in practical applications invariably suffer from random degradations that are collectively referred to as *noise*. These degradations arise from numerous sources, including radiation scatter from the surface before the image is sensed; electrical noise in the sensor or camera; channel noise as the image is transmitted over a communication channel; bit errors after the image is digitized, and so on. A good review of various image noise models is given in Chapter 7 of this *Guide*.

The most common generic noise model is *additive noise*, where a noisy observed image is taken to be the sum of an original, uncorrupted image g and a noise image q:

$$f = g + q, \tag{3.40}$$

where q is a 2D $N \times M$ random matrix, with elements $q(\mathbf{n})$ that are random variables. Chapter 7 develops the requisite mathematics for understanding random quantities and provides the basis for noise filtering. In this basic chapter, we will not require this more advanced development. Instead, we make the simple assumption that the noise is *zero mean.* If the noise is zero mean, then the average (or sample mean) of n independently occurring noise matrices $q_1, q_2, \ldots, q_n$ tends toward zero as n grows large:[2]

$$\left(\frac{1}{n}\right) \sum_{m=1}^{n} q_m \approx \mathbf{0}, \tag{3.41}$$

where $\mathbf{0}$ denotes the $N \times M$ matrix of zeros.

Now suppose that we are able to obtain n images $f_1, f_2, \ldots, f_n$ of the same scene. The images are assumed to be noisy versions of an original image g, where the noise is zero-mean and additive:

$$f_m = g + q_m \tag{3.42}$$

for $m = 1, \ldots, n$. Hence, the images are assumed either to be taken in rapid succession, so that there is no motion between frames, or under conditions where there is no motion in the scene. In this way only the noise contribution varies from image to image.

By averaging the multiple noisy images (3.42):

$$\left(\frac{1}{n}\right) \sum_{m=1}^{n} f_m = \left(\frac{1}{n}\right) \sum_{m=1}^{n} (g + q_m) = \left(\frac{1}{n}\right) \sum_{m=1}^{n} g + \left(\frac{1}{n}\right) \sum_{m=1}^{n} q_m$$

$$= g + \left(\frac{1}{n}\right) \sum_{m=1}^{n} q_m$$

$$\approx g \tag{3.43}$$

using (3.41). If a large enough number of frames are averaged together, then the resulting image should be nearly noise-free, and hence should approximate the original image. The amount of noise reduction can be quite significant; one can expect a reduction in the noise variance by a factor n. Of course, this is subject to inaccuracies in the model, e.g., if there is any change in the scene itself, or if there are any dependencies between the noise images (e.g., in an extreme case, the noise images might be identical), then the reduction in the noise will be limited.

Figure 3.17 depicts the process of noise reduction by frame averaging in an actual example of confocal microscope imaging. The image(s) are of Macroalga Valonia microphysa, imaged with a laser scanning confocal microscope (LSCM). The dark ring is chlorophyll fluorescing under Ar laser excitation. As can be seen, in this case the process of image averaging is quite effective in reducing the apparent noise content and in improving the visual resolution of the object being imaged.

---

[2]More accurately, the noise must be assumed *mean-ergodic*, which means that the sample mean approaches the statistical mean over large sample sizes. This assumption is usually quite reasonable.

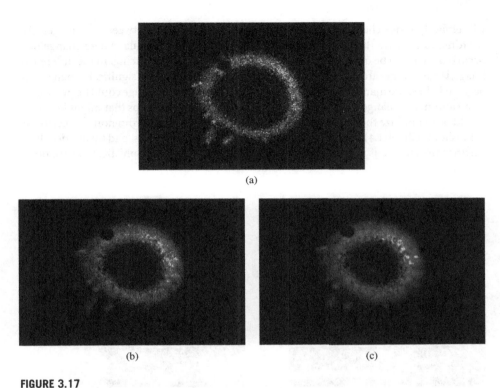

(a)

(b)   (c)

FIGURE 3.17

Example of image averaging for noise reduction. (a) single noisy image; (b) average of 4 frames; (c) average of 16 frames (courtesy of Chris Neils).

3.6.2 Image Differencing for Change Detection

Often it is of interest to detect changes that occur in images taken of the same scene but at different times. If the time instants are closely placed, e.g., adjacent frames in a video sequence, then the goal of change detection amounts to image motion detection. There are many applications of motion detection and analysis. For example, in video compression algorithms, compression performance is improved by exploiting redundancies that are tracked along the motion trajectories of image objects that are in motion. Detected motion is also useful for tracking targets, for recognizing objects by their motion, and for computing three-dimensional scene information from 2D motion.

If the time separation between frames is not small, then change detection can involve the discovery of gross scene changes. This can be useful for security or surveillance cameras, or in automated visual inspection systems, for example. In either case, the basic technique for change detection is the *image difference*. Suppose that f_1 and f_2 are images to be compared. Then the absolute difference image

$$g = |f_1 - f_2| \tag{3.44}$$

will embody those changes or differences that have occurred between the images. At coordinates **n** where there has been little change, $g(\mathbf{n})$ will be small. Where change has occurred, $g(\mathbf{n})$ can be quite large. Figure 3.18 depicts image differencing. In the difference image, large changes are displayed as brighter intensity values. Since significant change has occurred, there are many bright intensity values. This difference image could be processed by an automatic change detection algorithm. A simple series of steps that might be taken would be to binarize the difference image, thus separating change from nonchange, using a threshold (Chapter 4), counting the number of high-change pixels, and finally, deciding whether the change is significant enough to take some action. Sophisticated variations

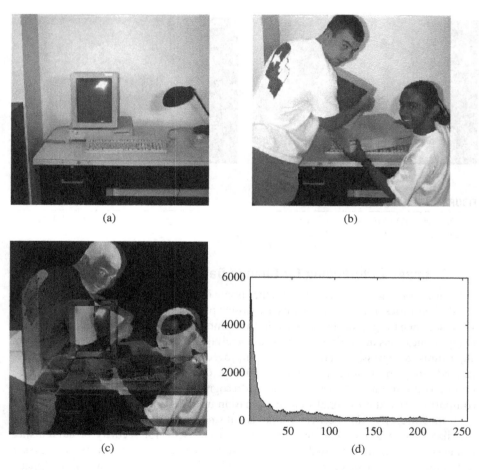

(a)

(b)

(c)

(d)

FIGURE 3.18

Image differencing example. (a) Original placid scene; (b) a theft is occurring! (c) the difference image with brighter points signifying larger changes; (d) the histogram of (c).

of this theme are currently in practical use. The histogram in Fig. 3.18(d) is instructive, since it is characteristic of differenced images; many zero or small gray level changes occur, with the incidence of larger changes falling off rapidly.

3.7 GEOMETRIC IMAGE OPERATIONS

We conclude this chapter with a brief discussion of *geometric image operations*. Geometric image operations are, in a sense, the opposite of point operations: they modify the spatial positions and spatial relationships of pixels, but they do not modify gray level values. Generally, these operations can be quite complex and computationally intensive, especially when applied to video sequences. However, the more complex geometric operations are not much used in engineering image processing, although they are heavily used in the computer graphics field. The reason for this is that image processing is primarily concerned with correcting or improving images of the real world, hence complex geometric operations, which distort images, are less frequently used. Computer graphics, however, is primarily concerned with creating images of an unreal world, or at least a visually modified reality, and subsequently geometric distortions are commonly used in that discipline.

A geometric image operation generally requires two steps: First, a spatial mapping of the coordinates of an original image f to define a new image g:

$$g(\mathbf{n}) = f(\mathbf{n}') = f[\mathbf{a}(\mathbf{n})]. \tag{3.45}$$

Thus, geometric image operations are defined as functions of position rather than intensity. The 2D, two-valued mapping function $\mathbf{a}(\mathbf{n}) = [a_1(n_1, n_2), a_2(n_1, n_2)]$ is usually defined to be continuous and smoothly changing, but the coordinates $\mathbf{a}(\mathbf{n})$ that are delivered are not generally integers. For example, if $\mathbf{a}(\mathbf{n}) = (n_1/3, n_2/4)$, then $g(\mathbf{n}) = f(n_1/3, n_2/4)$, which is not defined for most values of (n_1, n_2). The question then is, which value(s) of f are used to define $g(\mathbf{n})$, when the mapping does not fall on the standard discrete lattice?

This implies the need for the second operation: *interpolation* of noninteger coordinates $a_1(n_1, n_2)$ and $a_2(n_1, n_2)$ to integer values, so that g can be expressed in a standard row-column format. There are many possible approaches for accomplishing interpolation; we will look at two of the simplest: *nearest neighbor interpolation* and *bilinear interpolation*. The first of these is too simplistic for many tasks, while the second is effective for most.

3.7.1 Nearest Neighbor Interpolation

Here, the geometrically transformed coordinates are mapped to the nearest integer coordinates of f:

$$g(\mathbf{n}) = f\{\text{INT}[a_1(n_1, n_2) + 0.5], \text{INT}[a_2(n_1, n_2) + 0.5]\}, \tag{3.46}$$

where INT[R] denotes the nearest integer that is less than or equal to R. Hence, the coordinates are *rounded* prior to assigning them to g. This certainly solves the problem of finding integer coordinates of the input image, but it is quite simplistic, and, in practice, it may deliver less than impressive results. For example, several coordinates to be mapped may round to the same values, creating a block of pixels in the output image of the same value. This may give an impression of "blocking," or of structure that is not physically meaningful. The effect is particularly noticeable along sudden changes in intensity, or "edges," which may appear jagged following nearest neighbor interpolation.

3.7.2 Bilinear Interpolation

Bilinear interpolation produces a smoother interpolation than does the nearest neighbor approach. Given four neighboring image coordinates $f(n_{10}, n_{20}), f(n_{11}, n_{21}), f(n_{12}, n_{22})$, and $f(n_{13}, n_{23})$ (these can be the four nearest neighbors of $f[\mathbf{a(n)}]$), then the geometrically transformed image $g(n_1, n_2)$ is computed as

$$g(n_1, n_2) = A_0 + A_1 n_1 + A_2 n_2 + A_3 n_1 n_2, \tag{3.47}$$

which is a bilinear function in the coordinates (n_1, n_2). The bilinear weights A_0, A_1, A_2, and A_3 are found by solving

$$\begin{bmatrix} A_0 \\ A_1 \\ A_2 \\ A_3 \end{bmatrix} = \begin{bmatrix} 1 & n_{10} & n_{20} & n_{10}n_{20} \\ 1 & n_{11} & n_{21} & n_{11}n_{21} \\ 1 & n_{12} & n_{22} & n_{12}n_{22} \\ 1 & n_{13} & n_{23} & n_{13}n_{23} \end{bmatrix}^{-1} \begin{bmatrix} f(n_{10}, n_{20}) \\ f(n_{11}, n_{21}) \\ f(n_{12}, n_{22}) \\ f(n_{13}, n_{23}) \end{bmatrix}. \tag{3.48}$$

Thus, $g(n_1, n_2)$ is defined to be a linear combination of the gray levels of its four nearest neighbors. The linear combination defined by (3.48) is in fact the value assigned to $g(n_1, n_2)$ when the best (least squares) planar fit is made to these four neighbors. This process of optimal averaging produces a visually smoother result.

Regardless of the interpolation approach that is used, it is possible that the mapping coordinates $a_1(n_1, n_2), a_2(n_1, n_2)$ do not fall within the pixel ranges

$$0 \le a_1(n_1, n_2) \le M - 1$$

and/or $\tag{3.49}$

$$0 \le a_2(n_1, n_2) \le N - 1,$$

in which case it is not possible to define the geometrically transformed image at these coordinates. Usually a nominal value is assigned, such as $g(\mathbf{n}) = 0$, at these locations.

3.7.3 Image Translation

The most basic geometric transformation is the *image translation*, where (b_1, b_2) are integer constants. In this case $g(n_1, n_2) = f(n_1 - b_1, n_2 - b_2)$, which is a simple shift or translation of g by an amount b_1 in the vertical (row) direction and an amount b_2 in the horizontal direction. This operation is used in image display systems, when it is desired to move an image about, and it is also used in algorithms, such as image convolution (Chapter 5), where images are shifted relative to a reference. Since integer shifts can be defined in either direction, there is no need for the interpolation step.

3.7.4 Image Rotation

Rotation of the image g by an angle θ relative to the horizontal (n_1) axis is accomplished by the following transformations:

$$a_1(n_1, n_2) = n_1 \cos\theta - n_2 \sin\theta$$

and   (3.50)

$$a_2(n_1, n_2) = n_1 \sin\theta + n_2 \cos\theta.$$

The simplest cases are $\theta = 90°$, where $[a_1(n_1, n_2), a_2(n_1, n_2)] = (-n_2, n_1)$; $\theta = 180°$, where $[a_1(n_1, n_2), a_2(n_1, n_2)] = (-n_1, -n_2)$; and $\theta = -90°$, where $[a_1(n_1, n_2), a_2(n_1, n_2)] = (n_2, -n_1)$. Since the rotation point is not defined here as the center of the image, the arguments (3.50) may fall outside of the image domain. This may be ameliorated by applying an image translation either before or after the rotation to obtain coordinate values in the nominal range.

3.7.5 Image Zoom

The *image zoom* either magnifies or minifies the input image according to the mapping functions

$$a_1(n_1, n_2) = n_1/c \quad \text{and} \quad a_2(n_1, n_2) = n_2/d,$$ (3.51)

where $c \geq 1$ and $d \geq 1$ to achieve magnification, and $c < 1$ and $d < 1$ to achieve minification. If applied to the entire image, then the image size is also changed by a factor $c(d)$ along the vertical (horizontal) direction. If only a small part of an image is to be zoomed, then a translation may be made to the corner of that region, the zoom applied, and then the image cropped.

The image zoom is a good example of a geometric operation for which the type of interpolation is important, particularly at high magnifications. With nearest neighbor interpolation, many values in the zoomed image may be assigned the same grayscale, resulting in a severe "blotching" or "blocking" effect. The bilinear interpolation usually supplies a much more viable alternative.

Figure 3.19 depicts a 4x zoom operation applied to the image in Fig. 3.13 (logarithmically transformed "students"). The image was first zoomed, creating a much larger image (16 times as many pixels). The image was then translated to a point of interest (selected, e.g., by a mouse), then was cropped to size 256×256 pixels around this point. Both nearest neighbor and bilinear interpolation were applied for the purpose of comparison. Both provide a nice "close-up" of the original, making the faces much more identifiable. However, the bilinear result is much smoother and does not contain the blocking artifacts that can make recognition of the image difficult.

(a)   (b)

FIGURE 3.19

Example of (4x) image zoom followed by interpolation. (a) Nearest-neighbor interpolation; (b) bilinear interpolation.

It is important to understand that image zoom followed by interpolation does not inject *any* new information into the image, although the magnified image may appear easier to see and interpret. The image zoom is only an interpolation of known information.

Basic Binary Image Processing

4

Alan C. Bovik

The University of Texas at Austin

4.1 INTRODUCTION

In this second chapter on basic methods, we explain and demonstrate fundamental tools for the processing of *binary* digital images. Binary image processing is of special interest, since an image in binary format can be processed using very fast logical (Boolean) operators. Often a binary image has been obtained by abstracting essential information from a gray level image, such as object location, object boundaries, or the presence or absence of some image property.

As seen in the previous two chapters, a digital image is an array of numbers or sampled image intensities. Each gray level is quantized or assigned one of a finite set of numbers represented by B bits. In a binary image, only one bit is assigned to each pixel: $B = 1$ implying two possible gray level values, 0 and 1. These two values are usually interpreted as Boolean, hence each pixel can take on the logical values '0' or '1', or equivalently, "true" or "false." For example, these values might indicate the absence or presence of some image property in an associated gray level image of the same size, where '1' at a given coordinate indicates the presence of the property at that coordinate in the gray level image and '0' otherwise. This image property is quite commonly a sufficiently high or low intensity (brightness), although more abstract properties, such as the presence or absence of certain objects, or smoothness/nonsmoothness, might be indicated.

Since most image display systems and software assume images of eight or more bits per pixel, the question arises as to how binary images are displayed. Usually they are displayed using the two extreme gray tones, black and white, which are ordinarily represented by 0 and 255, respectively, in a grayscale display environment, as depicted in Fig. 4.1. There is no established convention for the Boolean values that are assigned to "black" and to "white." In this chapter, we will uniformly use '1' to represent "black" (displayed as gray level 0) and '0' to represent "white" (displayed as gray level 255). However, the assignments are quite commonly reversed, and it is important to note that the Boolean values '0' and '1' have no physical significance other than what the user assigns to them.

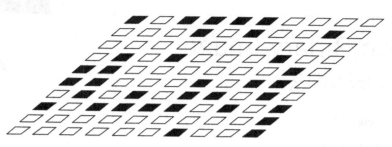

FIGURE 4.1

A 10×10 binary image.

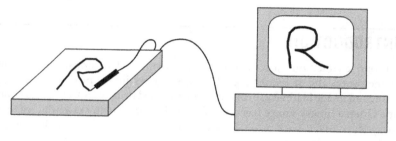

FIGURE 4.2

Simple binary image device.

Binary images arise in a number of ways. Usually they are created from gray level images for simplified processing or for printing. However, certain types of sensors directly deliver a binary image output. Such devices are usually associated with printed, handwritten, or line drawing images, with the input signal being entered by hand on a pressure sensitive tablet, a resistive pad, or a light pen.

In such a device, the (binary) image is first initialized prior to image acquisition:

$$g(\mathbf{n}) = \text{'}0\text{'} \tag{4.1}$$

at all coordinates $\mathbf{n}$. When pressure, a change of resistance, or light is sensed at some image coordinate $\mathbf{n}_0$, then the image is assigned the value '1':

$$g(\mathbf{n}_0) = \text{'}1\text{'}. \tag{4.2}$$

This continues until the user completes the drawing, as depicted in Fig. 4.2. These simple devices are quite useful for entering engineering drawings, handprinted characters, or other binary graphics in a binary image format.

4.2 IMAGE THRESHOLDING

Usually a binary image is obtained from a gray level image by some process of information abstraction. The advantage of the B-fold reduction in the required image storage space is offset by what can be a significant loss of information in the resulting binary image. However, if the process is accomplished with care, then a simple abstraction of information can be obtained that can enhance subsequent processing, analysis, or interpretation of the image.

The simplest such abstraction is the process of *image thresholding*, which can be thought of as an extreme form of gray level quantization. Suppose that a gray level image f can take K possible gray levels $0, 1, 2, \ldots, K-1$. Define an integer threshold, T, that lies in the grayscale range: $T \in \{0, 1, 2, \ldots, K-1\}$. The process of thresholding is a process of simple comparison: each pixel value in f is compared to T. Based on this comparison, a binary decision is made that defines the value of the corresponding pixel in an output binary image g:

$$g(\mathbf{n}) = \begin{cases} \text{`0'} & \text{if } f(\mathbf{n}) \geq T \\ \text{`1'} & \text{if } f(\mathbf{n}) < T. \end{cases} \qquad (4.3)$$

Of course, the threshold T that is used is of critical importance, since it controls the particular abstraction of information that is obtained. Indeed, different thresholds can produce different valuable abstractions of the image. Other thresholds may produce little valuable information at all. It is instructive to observe the result of thresholding an image at many different levels in sequence. Figure 4.3 depicts the image "mandrill" (Fig. 1.8 of Chapter 1) thresholded at four different levels. Each produces different information, or in the case of Figs. 4.3(a) and 4.3(d), very little useful information. Among these, Fig. 4.3(c) probably contains the most visual information, although it is far from ideal. The four threshold values $(50, 100, 150, 200)$ were chosen without using any visual criterion.

As will be seen, image thresholding can often produce a binary image result that is quite useful for simplified processing, interpretation, or display. However, some gray level images do not lead to any interesting binary result regardless of the chosen threshold T.

Several questions arise: given a gray level image, how does one decide whether binarization of the image by gray level thresholding will produce a useful result? Can this be decided automatically by a computer algorithm? Assuming that thresholding is likely to be successful, how does one decide on a threshold level T? These are apparently simple questions pertaining to a very simple operation. However, the answers to these questions turn out to be quite difficult to answer in the general case. In other cases, the answer is simpler. In all cases, however, the basic tool for understanding the process of image thresholding is the image histogram, which was defined and studied in Chapter 3.

Thresholding is most commonly and effectively applied to images that can be characterized as having *bimodal histograms*. Figure 4.4 depicts two hypothetical image histograms. The one on the left has two clear modes; the one at the right either has a single mode or two heavily-overlapping, poorly separated modes.

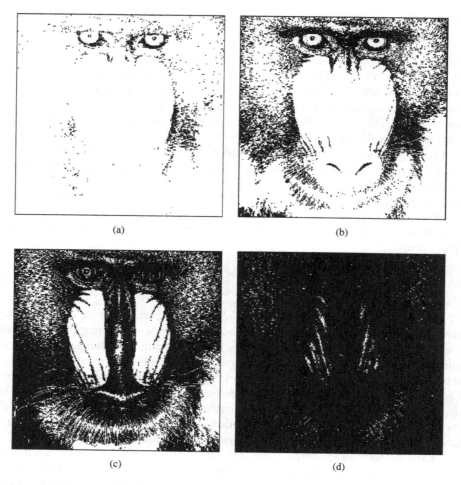

FIGURE 4.3

Image "mandrill" thresholded at gray levels (a) 50; (b) 100; (c) 150; and (d) 200.

Bimodal histograms are often (but not always!) associated with images that contain objects and background having significantly different average brightness. This may imply bright objects on a dark background, or dark objects on a bright background. The goal, in many applications, is to separate the objects from the background, and to label them as object or as background. If the image histogram contains well-separated modes associated with object and with background, then thresholding can be the means for achieving this separation. Practical examples of gray level images with well-separated bimodal histograms are not hard to find. For example, an image of machine-printed type (like that being currently read), or of handprinted characters, will have a very distinctive separation between object and background. Examples abound in biomedical applications, where it

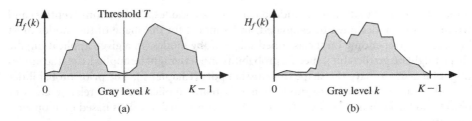

FIGURE 4.4

Hypothetical histograms. (a) Well-separated modes; (b) poorly separated or indistinct modes.

is often possible to control the lighting of objects and background. Standard bright-field microscope images of single or multiple cells (micrographs) typically contain bright objects against a darker background. In many industry applications, it is also possible to control the relative brightness of objects of interest and the backgrounds they are set against. For example, machine parts that are being imaged (perhaps in an automated inspection application) may be placed on a mechanical conveyor that has substantially different reflectance properties than the objects.

Given an image with a bimodal histogram, a general strategy for thresholding is to place the threshold T between the image modes, as depicted in Fig. 4.4(a). Many "optimal" strategies have been suggested for deciding the exact placement of the threshold between the peaks. Most of these are based on an assumed statistical model for the histogram, and by posing the decision of labeling a given pixel as "object" versus "background" as a statistical inference problem. In the simplest version, two hypotheses are posed:

H_0: The pixel belongs to gray level Population 0

H_1: The pixel belongs to gray level Population 1

where pixels from Populations 0 and 1 have conditional probability density functions (pdfs) $p_f(a|H_0)$ and $p_f(a|H_1)$, respectively, under the two hypotheses. If it is also known (or estimated) that H_0 is true with probability p_0 and that H_1 is true with probability p_1 ($p_0 + p_1 = 1$), then the decision may be cast as a likelihood ratio test. If an observed pixel has gray level $f(\mathbf{n}) = k$, then the decision may be rendered according to

$$\frac{p_f(k|H_1)}{p_f(k|H_0)} \mathop{\gtrless}_{H_0}^{H_1} \frac{p_0}{p_1}. \tag{4.4}$$

The decision whether to assign logical '0' or '1' to a pixel can thus be regarded as applying a simple statistical test to each pixel. In (4.4), the conditional pdfs may be taken as the modes of a bimodal histogram. Algorithmically, this means that they must be fit to the histogram using some criterion, such as least-squares. This is usually quite difficult, since it must be decided that there are indeed two separate modes, the locations (centers) and widths of the modes must be estimated, and a model for the shape of the modes must be assumed. Depending on the assumed shape of the modes (in a given application, the shape might be predictable), specific probability models might be applied, e.g., the modes might be taken to have the shape of Gaussian pdfs (Chapter 7). The prior probabilities p_0 and p_1 are often easier to model, since in many applications the relative areas of object and background can be estimated or given reasonable values based on empirical observations.

A likelihood ratio test such as (4.4) will place the image threshold T somewhere between the two modes of the image histogram. Unfortunately, any simple statistical model of the image does not account for such important factors as object/background continuity, visual appearance to a human observer, nonuniform illumination or surface reflectance effects, and so on. Hence, with rare exceptions, a statistical approach such as (4.4) will not produce as good a result as would a human decision-maker making a manual threshold selection.

Placing the threshold T between two obvious modes of a histogram may yield acceptable results, as depicted in Fig. 4.4(a). The problem is significantly complicated, however, if the image contains multiple distinct modes or if the image is nonmodal or level. Multimodal histograms can occur when the image contains multiple objects of different average brightness on a uniform background. In such cases, simple thresholding will exclude some objects (Fig. 4.5). Nonmodal or flat histograms usually imply more complex images, containing significant gray level variation, detail, nonuniform lighting or reflection, etc. (Fig. 4.5). Such images are often not amenable to a simple thresholding process, especially if the goal is to achieve figure-ground separation. However, all of these comments are, at best, rules of thumb. An image with a bimodal histogram might not yield good results when thresholded at any level, while an image with a perfectly flat histogram might yield an ideal result. It is a good mental exercise to consider when these latter cases might occur.

Figures 4.6–4.8 show several images, their histograms, and the thresholded image results. In Fig. 4.6, a good threshold level for the micrograph of the cellular specimens was taken to be $T = 180$. This falls between the two large modes of the histogram (there are many smaller modes) and was deemed to be visually optimal by one user. In the

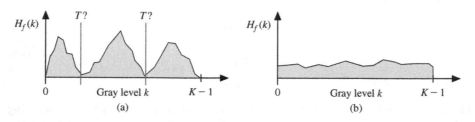

FIGURE 4.5

Hypothetical histograms. (a) Multimodal histogram, showing difficulty of threshold selection; (b) Non-modal histogram, for which threshold selection is quite difficult or impossible.

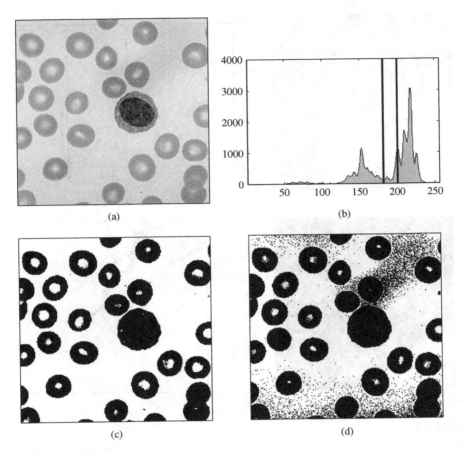

FIGURE 4.6

Binarization of "micrograph." (a) Original; (b) histogram showing two threshold locations (180 and 200); (c) and (d) resulting binarized images.

binarized image, the individual cells are not perfectly separated from the background. The reason for this is that the illuminated cells have nonuniform brightness profiles, being much brighter toward the centers. Taking the threshold higher ($T = 200$), however, does not lead to improved results, since the bright background then begins to fall below threshold.

Figure 4.7 depicts a *negative* (for better visualization) of a digitized mammogram. Mammography is the key diagnostic tool for the detection of breast cancer, and in the future, digital tools for mammographic imaging and analysis. The image again shows two strong modes, with several smaller modes. The first threshold chosen ($T = 190$) was selected at the minimum point between the large modes. The resulting binary image has the nice result of separating the region of the breast from the background.

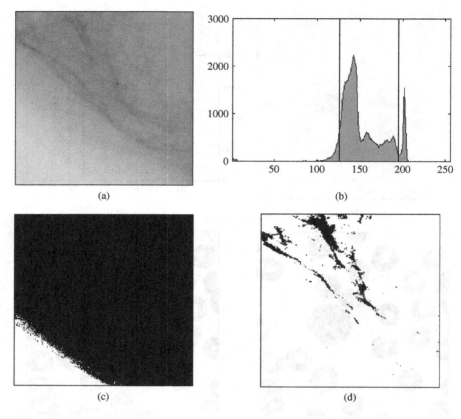

FIGURE 4.7

Binarization of "mammogram." (a) Original negative mammogram; (b) histogram showing two threshold locations (190 and 125); (c) and (d) resulting binarized images.

However, radiologists are often interested in the detailed structure of the breast and in the brightest (darkest in the negative) areas which might indicate tumors or microcalcifications. Figure 4.7(d) shows the result of thresholding at the lower level of 125 (higher level in the positive image), successfully isolating much of the interesting structure.

Generally the best binarization results via thresholding are obtained by direct human operator intervention. Indeed, most general-purpose image processing environments have thresholding routines that allow user interaction. However, even with a human picking a visually "optimal" value of T, thresholding rarely gives "perfect" results. There is nearly always some misclassification of object as background, and vice versa. For example in the image "micrograph," no value of T is able to successfully extract the objects from the background; instead, most of the objects have "holes" in them, and there is a sprinkling of black pixels in the background as well.

Because of these limitations of the thresholding process, it is usually necessary to apply some kind of *region correction* algorithms to the binarized image. The goal of such algorithms is to correct the misclassification errors that occur. This requires identifying misclassified background points as object points, and vice versa. These operations are usually applied directly to the binary images, although it is possible to augment the process by also incorporating information from the original grayscale image. Much of the remainder of this chapter will be devoted to algorithms for region correction of thresholded binary images.

4.3 REGION LABELING

A simple but powerful tool for identifying and labeling the various objects in a binary image is a process called *region labeling, blob coloring,* or *connected component identification.* It is useful since once they are individually labeled, the objects can be separately manipulated, displayed, or modified. For example, the term "blob coloring" refers to the possibility of displaying each object with a different identifying color, once labeled.

Region labeling seeks to identify connected groups of pixels in a binary image f that all have the same binary value. The simplest such algorithm accomplishes this by scanning the entire image (left-to-right, top-to-bottom), searching for occurrences of pixels of the same binary value and connected along the horizontal or vertical directions. The algorithm can be made slightly more complex by also searching for diagonal connections, but this is usually unnecessary. A record of connected pixel groups is maintained in a separate label array r having the same dimensions as f, as the image is scanned. The following algorithm steps explain the process, where the region labels used are positive integers.

4.3.1 Region Labeling Algorithm

1. Given an $N \times M$ binary image f, initialize an associated $N \times M$ region label array: $r(\mathbf{n}) = $ '0' for all $\mathbf{n}$. Also initialize a region number counter: $k = 1$.

   Then, scanning the image from left-to-right and top-to-bottom, for every $\mathbf{n}$ do the following:

2. If $f(\mathbf{n}) = $ '0' then do nothing.

3. If $f(\mathbf{n}) = $ '1' and also $f(\mathbf{n} - (1,0)) = f(\mathbf{n} - (0,1)) = $ '0' (as depicted in Fig. 4.8(a)), then set $r(\mathbf{n}) = $ '0' and $k = k + 1$. In this case, the left and upper neighbors of $f(\mathbf{n})$ do not belong to objects.

4. If $f(\mathbf{n}) = $ '1', $f(\mathbf{n} - (1,0)) = $ '1', and $f(\mathbf{n} - (0,1)) = $ '0' (Fig. 4.8(b)), then set $r(\mathbf{n}) = r(\mathbf{n} - (1,0))$. In this case, the upper neighbor $f(\mathbf{n} - (1,0))$ belongs to the same object as $f(\mathbf{n})$.

5. If $f(\mathbf{n}) = $ '1', $f(\mathbf{n} - (1,0)) = $ '0', and $f(\mathbf{n} - (0,1)) = $ '1' (Fig. 4.8(c)), then set $r(\mathbf{n}) = r(\mathbf{n} - (0,1))$. In this case, the left neighbor $f(\mathbf{n} - (0,1))$ belongs to the same object as $f(\mathbf{n})$.

FIGURE 4.8

Pixel neighbor relationships used in a region labeling algorithm. In each of (a)–(d), $f(\mathbf{n})$ is the lower right pixel.

6. If $f(\mathbf{n}) = \text{`1'}$ and $f(\mathbf{n} - (1,0)) = f(\mathbf{n} - (0,1)) = \text{`1'}$ (Fig. 4.8(d)), then set $r(\mathbf{n}) = r(\mathbf{n} - (0,1))$. If $r(\mathbf{n} - (0,1)) \neq r(\mathbf{n} - (1,0))$, then record the labels $r(\mathbf{n} - (0,1))$ and $r(\mathbf{n} - (1,0))$ as equivalent. In this case, both the left and upper neighbors belong to the same object as $f(\mathbf{n})$, although they may have been labeled differently.

A simple application of region labeling is the measurement of object area. This can be accomplished by defining a vector $\mathbf{c}$ with elements $c(k)$ that are the pixel area (pixel count) of region k.

4.3.2 Region Counting Algorithm

Initialize $\mathbf{c} = 0$. For every $\mathbf{n}$ do the following:

1. If $f(\mathbf{n}) = \text{`0,'}$ then do nothing.

2. If $f(\mathbf{n}) = \text{`1,'}$ then $c[r(\mathbf{n})] = c[r(\mathbf{n})] + 1$.

Another simple but powerful application of region labeling is the removal of minor regions or objects from a binary image. The way in which this is done depends on the application. It may be desired that only a single object should remain (generally the largest object), or it may be desired that any object with a pixel area less than some minimum value should be deleted. A variation is that the minimum value is computed as a percentage of the largest object in the image. The following algorithm depicts the second possibility.

4.3.3 Minor Region Removal Algorithm

Assume a minimum allowable object size of S pixels. For every $\mathbf{n}$ do the following:

1. If $f(\mathbf{n}) = \text{`0,'}$ then do nothing.

2. If $f(\mathbf{n}) = \text{`1'}$ and $c[r(\mathbf{n})] < S$, then set $g(\mathbf{n}) = \text{`0.'}$

Of course, all of the above algorithms can be operated in reverse polarity, by interchanging '0' for '1' and '1' for '0' everywhere.

An important application of region labeling/region counting/minor region removal is in the correction of thresholded binary images. Application of a binarizing threshold to a gray level image inevitably produces an imperfect binary image, with such errors as extraneous objects or holes or holes in objects. These can arise from noise, unexpected

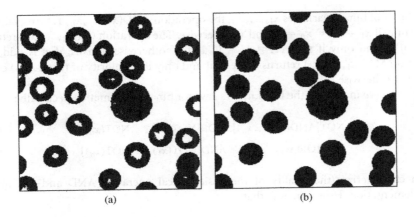

(a)   (b)

FIGURE 4.9

Result of applying the region labeling/counting/removal algorithms to (a) the binarized image in Fig. 4.6(c); (b) and then to the image in (b), but in polarity-reversed mode.

objects (such as dust on a lens), and general nonuniformities in the surface reflectances and illuminations of the objects and background.

Figure 4.9 depicts the result of sequentially applying the region labeling/region counting/minor region removal algorithms to the binarized "micrograph" image in Fig. 4.6(c). The series of algorithms was first applied to the image in Fig. 4.6(c) as above to remove extraneous small black objects, using a size threshold of 500 pixels as shown in Fig. 4.9(a). It was then applied again to this modified image, but in polarity reversed mode, to remove the many object holes, this time using a threshold of 1000 pixels. The result shown in Fig. 4.9(b) is a dramatic improvement over the original binarized result, given that the goal was to achieve a clean separation of the objects in the image from the background.

4.4 BINARY IMAGE MORPHOLOGY

We next turn to a much broader and more powerful class of binary image processing operations that collectively fall under the name *binary image morphology*. These are closely related to (in fact, are the same in a mathematical sense) the gray level morphological operations described in Chapter 13. As the name indicates, these operators modify the *shapes* of the objects in an image.

4.4.1 Logical Operations

The morphological operators are defined in terms of simple logical operations on local groups of pixels. The logical operators that are used are the simple NOT, AND, OR, and MAJ (majority) operators. Given a binary variable x, NOT(x) is its logical complement.

Given a set of binary variables $x_1, \ldots, x_n$, the operation $\text{AND}(x_1, \ldots, x_n)$ returns value '1' if and only if $x_1 = \ldots = x_n =$ '1' and '0' otherwise. The operation $\text{OR}(x_1, \ldots, x_n)$ returns value '0' if and only if $x_1 = \ldots = x_n =$ '0' and '1' otherwise. Finally, if n is odd, the operation $\text{MAJ}(x_1, \ldots, x_n)$ returns value '1' if and only if a majority of $(x_1, \ldots, x_n)$ equal '1' and '0' otherwise.

We observe in passing the DeMorgan's laws for binary arithmetic, specifically:

$$\text{NOT}[\text{AND}(x_1, \ldots, x_n)] = \text{OR}[\text{NOT}(x_1), \ldots, \text{NOT}(x_n)] \tag{4.5}$$

$$\text{NOT}[\text{OR}(x_1, \ldots, x_n)] = \text{AND}[\text{NOT}(x_1), \ldots, \text{NOT}(x_n)], \tag{4.6}$$

which characterizes the duality of the basic logical operators AND and OR under complementation. However, note that

$$\text{NOT}[\text{MAJ}(x_1, \ldots, x_n)] = \text{MAJ}[\text{NOT}(x_1), \ldots, \text{NOT}(x_n)] \tag{4.7}$$

hence MAJ is its own dual under complementation.

4.4.2 Windows

As mentioned, morphological operators change the shapes of objects using *local logical operations*. Since they are local operators, a formal methodology must be defined for making the operations occur on a local basis. The mechanism for doing this is the *window*.

A window defines a geometric rule according to which gray levels are collected from the vicinity of a given pixel coordinate. It is called a window since it is often visualized as a moving collection of empty pixels that is passed over the image. A morphological operation is (conceptually) defined by moving a window over the binary image to be modified, in such a way that it is eventually centered over every image pixel, where a local logical operation is performed. Usually this is done row-by-row, column-by-column, although it can be accomplished at every pixel simultaneously if a massively parallel-processing computer is used.

Usually a window is defined to have an approximate circular shape (a digital circle cannot be exactly realized) since it is desired that the window, and hence, the morphological operator, be rotation-invariant. This means that if an object in the image is rotated through some angle, then the response of the morphological operator will be unchanged other than also being rotated. While rotational symmetry cannot be exactly obtained, symmetry across two axes can be obtained, guaranteeing that the response be at least reflection-invariant. Window size also significantly effects the results, as will be seen.

A formal definition of windowing is needed in order to define the various morphological operators. A window **B** is a set of $2P + 1$ coordinate shifts $b_i = (n_i, m_i)$ centered around $(0, 0)$:

$$\mathbf{B} = \{b_1, \ldots, b_{2P+1}\} = \{(n_1, m_1), \ldots, (n_{2P+1}, m_{2P+1})\}.$$

Some examples of common 1D (row and column) windows are

$$\mathbf{B} = \mathrm{ROW}[2P + 1] = \{(0, m); m = -P, \dots, P\} \qquad (4.8)$$

$$\mathbf{B} = \mathrm{COL}[2P + 1] = \{(n, 0); n = -P, \dots, P\} \qquad (4.9)$$

and some common 2D windows are

$$\mathbf{B} = \mathrm{SQUARE}[(2P + 1)^2] = \{(n, m); n, m = -P, \dots, P\} \qquad (4.10)$$

$$\mathbf{B} = \mathrm{CROSS}[4P + 1] = \mathrm{ROW}(2P + 1) \cup \mathrm{COL}(2P + 1) \qquad (4.11)$$

with obvious shape-descriptive names. In each of (4.8)–(4.11), the quantity in brackets is the number of coordinate shifts in the window, hence also the number of local gray levels that will be collected by the window at each image coordinate. Note that the windows (4.8)–(4.11) are each defined with an odd number $2P + 1$ coordinate shifts. This is because the operators are symmetrical: pixels are collected in pairs from opposite sides of the center pixel or $(0, 0)$ coordinate shift, plus the $(0, 0)$ coordinate shift is always included. Examples of each of the windows (4.8)–(4.11) are shown in Fig. 4.10.

The example window shapes in (4.8)–(4.11) and in Fig. 4.10 are by no means the only possibilities, but they are (by far) the most common implementations because of the simple row-column indexing of the coordinate shifts.

The action of gray level collection by a moving window creates the *windowed set*. Given a binary image f and a window $\mathbf{B}$, the windowed set at image coordinate $\mathbf{n}$ is given by

$$\mathbf{B}f(\mathbf{n}) = \{f(\mathbf{n} - \mathbf{m}); \mathbf{m} \in \mathbf{B}\}, \qquad (4.12)$$

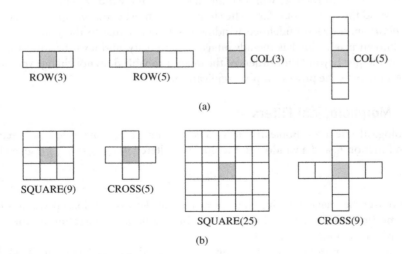

(a)

(b)

FIGURE 4.10

Examples of windows. The window is centered over the shaded pixel. (a) One-dimensional windows ROW($2P + 1$) and COL($2P + 1$) for $P = 1, 2$; (b) Two-dimensional windows SQUARE $[(2P + 1)^2]$ and CROSS$[4P + 1]$ for $P = 1, 2$.

which, conceptually, is the set of image pixels covered by **B** when it is centered at coordinate **n**. Examples of windowed sets associated with some of the windows in (4.8)–(4.11) and Fig. 4.10 are:

$$\mathbf{B} = \text{ROW}(3): \qquad \mathbf{B}f(n_1, n_2) = \{f(n_1, n_2 - 1), f(n_1, n_2), f(n_1, n_2 + 1)\} \qquad (4.13)$$

$$\mathbf{B} = \text{COL}(3): \qquad \mathbf{B}f(n_1, n_2) = \{f(n_1 - 1, n_2), f(n_1, n_2), f(n_1 + 1, n_2)\} \qquad (4.14)$$

$$\mathbf{B} = \text{SQUARE}(9): \quad \mathbf{B}f(n_1, n_2) = \{f(n_1 - 1, n_2 - 1), f(n_1 - 1, n_2),$$
$$f(n_1 - 1, n_2 + 1), f(n_1, n_2 - 1), f(n_1, n_2),$$
$$f(n_1, n_2 + 1), f(n_1 + 1, n_2 - 1), \qquad (4.15)$$
$$f(n_1 + 1, n_2), f(n_1 + 1, n_2 + 1)\}$$

$$\mathbf{B} = \text{CROSS}(5): \qquad \mathbf{B}f(n_1, n_2) = \{f(n_1 - 1, n_2), f(n_1, n_2 - 1),$$
$$f(n_1, n_2), f(n_1, n_2 + 1), \qquad (4.16)$$
$$f(n_1 + 1, n_2)\}$$

where the elements of (4.13)–(4.16) have been arranged to show the geometry of the windowed sets when centered over coordinate $\mathbf{n} = (n_1, n_2)$. Conceptually, the window may be thought of as capturing a series of miniature images as it is passed over the image, row-by-row, column-by-column.

One last note regarding windows involves the definition of the windowed set when the window is centered near the boundary edge of the image. In this case, some of the elements of the windowed set will be undefined, since the window will overlap "empty space" beyond the image boundary. The simplest and most common approach is to use *pixel replication*: set each undefined windowed set value equal to the gray level of the nearest known pixel. This has the advantage of simplicity, and also the intuitive value that the world just beyond the borders of the image probably does not change very much. Figure 4.11 depicts the process of pixel replication.

4.4.3 Morphological Filters

Morphological filters are Boolean filters. Given an image f, a many-to-one binary or Boolean function h, and a window **B**, the Boolean-filtered image $g = h(f)$ is given by

$$g(\mathbf{n}) = h[\mathbf{B}f(\mathbf{n})] \qquad (4.17)$$

at every **n** over the image domain. Thus, at each **n**, the filter collects local pixels according to a geometrical rule into a windowed set, performs a Boolean operation on them, and returns the single Boolean result $g(\mathbf{n})$.

The most common Boolean operations that are used are AND, OR, and MAJ. They are used to create the following simple, yet powerful *morphological filters*. These filters act on the objects in the image by shaping them: expanding or shrinking them, smoothing them, and eliminating too-small features.

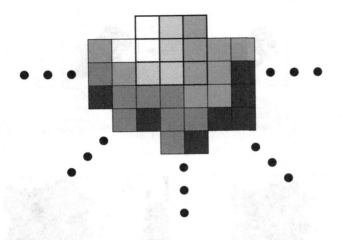

FIGURE 4.11

Depiction of pixel replication for a window centered near the (top) image boundary.

The binary *dilation filter* is defined by

$$g(\mathbf{n}) = \text{OR}[\mathbf{B}f(\mathbf{n})] \tag{4.18}$$

and is denoted $g = dilate(f, \mathbf{B})$. The binary *erosion filter* is defined by

$$g(\mathbf{n}) = \text{AND}[\mathbf{B}f(\mathbf{n})] \tag{4.19}$$

and is denoted $g = erode(f, \mathbf{B})$. Finally, the binary *majority filter* is defined by

$$g(\mathbf{n}) = \text{MAJ}[\mathbf{B}f(\mathbf{n})] \tag{4.20}$$

and is denoted $g = majority(f, \mathbf{B})$. Next we explain the response behavior of these filters.

The *dilate* filter expands the size of the foreground, object, or '1'-valued regions in the binary image f. Here the '1'-valued pixels are assumed to be black because of the convention we have assumed, but this is not necessary. The process of dilation also smoothes the boundaries of objects, removing gaps or bays of too-narrow width, and also removing object holes of too-small size. Generally a hole or gap will be filled if the dilation window cannot fit into it. These actions are depicted in Fig. 4.12, while Fig. 4.13 shows the result of dilating an actual binary image. Note that dilation using $\mathbf{B} = \text{SQUARE}(9)$ removed most of the small holes and gaps, while using $\mathbf{B} = \text{SQUARE}(25)$ removed nearly all of them. It is also interesting to observe that dilation with the larger window nearly completed a bridge between two of the large masses. Dilation with CROSS(9) highlights an interesting effect: individual, isolated '1'-valued or BLACK pixels were dilated into larger objects having the same shape as the window. This can also be seen with the results using the SQUARE windows. This effect underlines the importance of using symmetric

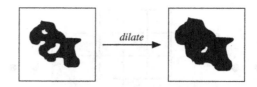

FIGURE 4.12

Illustration of dilation of a binary '1'-valued object. The smallest hole and gap were filled.

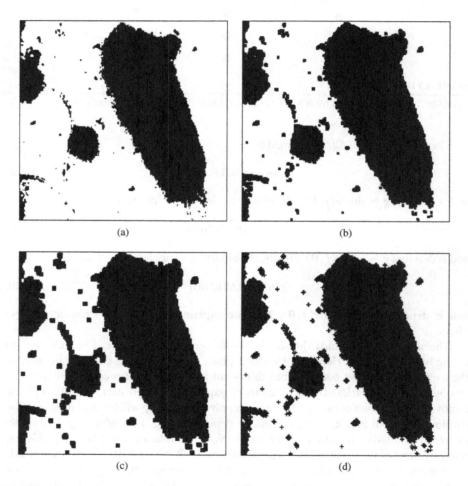

(a)

(b)

(c)

(d)

FIGURE 4.13

Dilation of a binary image. (a) Binarized image "cells." *Dilate* with: (b) **B** = SQUARE(9); (c) **B** = SQUARE(25); (d) **B** = CROSS(9).

windows, preferably with near rotational symmetry, since then smoother results are obtained.

The *erode* filter shrinks the size of the foreground, object, or '1'-valued regions in the binary image f. Alternately, it expands the size of the background or '0'-valued regions. The process of erosion smoothes the boundaries of objects, but in a different way than dilation: it removes peninsulas or fingers of too-narrow width, and also it removes '1'-valued objects of too-small size. Generally an isolated object will be eliminated if the dilation window cannot fit into it. The effects of *erode* are depicted in Fig. 4.14.

Figure 4.15 shows the result of applying the *erode* filter to the binary image "cell." Erosion using $\mathbf{B} = \text{SQUARE}(9)$ removed many of the small objects and fingers, while using $\mathbf{B} = \text{SQUARE}(25)$ removed most of them. As an example of intense smoothing, $\mathbf{B} = \text{SQUARE}(81)$ (a 9×9 square window) was also applied. Erosion with $\text{CROSS}(9)$ again produced a good result, except at a few isolated points where isolated '0'-valued or WHITE pixels were expanded into larger '+'-shaped objects.

An important property of the *erode* and *dilate* filters is the relationship that exists between them. In fact, in reality they are the same operation, in the dual (complementary) sense. Indeed, given a binary image f and an arbitrary window $\mathbf{B}$, it is true that

$$dilate(f, \mathbf{B}) = \text{NOT}\{erode[\text{NOT}(f), \mathbf{B}]\} \tag{4.21}$$

$$erode(f, \mathbf{B}) = \text{NOT}\{dilate[\text{NOT}(f), \mathbf{B}]\}. \tag{4.22}$$

Equations (4.21) and (4.22) are a simple consequence of the DeMorgan's laws (4.5) and (4.6). A correct interpretation of this is that erosion of the '1'-valued or BLACK regions of an image is the same as dilation of the '0'-valued or WHITE regions—and vice versa.

An important and common misconception must be mentioned. *Erode* and *dilate* shrink and expand the sizes of '1'-valued objects in a binary image. However, they are *not* inverse operations of one another. Dilating an eroded image (or eroding a dilated image) very rarely yields the original image. In particular, dilation cannot recreate peninsulas, fingers, or small objects that have been eliminated by erosion. Likewise, erosion cannot unfill holes filled by dilation or recreate gaps or bays filled by dilation. Even without these effects, erosion generally will not exactly recreate the same shapes that have been modified by dilation, and vice versa.

Before discussing the third common Boolean filter, the majority, we will consider further the idea of sequentially applying *erode* and *dilate* filters to an image. One reason

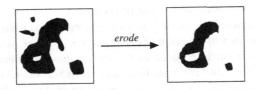

FIGURE 4.14

Illustration of erosion of a binary '1'-valued object. The smallest objects and peninsula were eliminated.

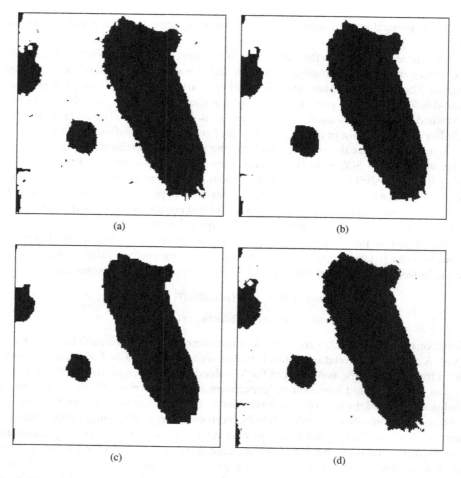

FIGURE 4.15

Erosion of the binary image "cells." *Erode* with: (a) **B** = SQUARE(9); (b) **B** = SQUARE(25); (c) **B** = SQUARE(81); (d) **B** = CROSS(9).

for doing this is that the *erode* and *dilate* filters have the effect of changing the sizes of objects, as well as smoothing them. For some objects this is desirable, e.g., when an extraneous object is shrunk to the point of disappearing; however, often it is undesirable, since it may be desired to further process or analyze the image. For example, it may be of interest to label the objects and compute their sizes, as in Section 4.3 of this chapter.

Although *erode* and *dilate* are not inverse operations of one another, they are approximate inverses in the sense that if they are performed in sequence on the same image with the same window **B**, then object and holes that are not eliminated will be returned to their approximate sizes. We thus define the size-preserving smoothing morphological operators termed *open filter* and *close filter* as follows:

$$open(f, \mathbf{B}) = dilate[erode(f, \mathbf{B}), \mathbf{B}] \tag{4.23}$$

$$close(f, \mathbf{B}) = erode[dilate(f, \mathbf{B}), \mathbf{B}]. \tag{4.24}$$

Hence the opening (closing) of image f is the erosion (dilation) with window $\mathbf{B}$ followed by dilation (erosion) with window $\mathbf{B}$. The morphological filters *open* and *close* have the same smoothing properties as *erode* and *dilate*, respectively, but they do not generally effect the sizes of sufficiently large objects much (other than pixel loss from pruned holes, gaps or bays, or pixel gain from eliminated peninsulas).

Figure 4.16 depicts the results of applying the *open* and *close* operations to the binary image "cell," using the windows $\mathbf{B}$ = SQUARE(25) and $\mathbf{B}$ = SQUARE(81). Large windows were used to illustrate the powerful smoothing effect of these morphological smoothers. As can be seen, the *open* filters did an excellent job of eliminating what might be referred to as "black noise"—the extraneous '1'-valued objects and other features, leaving smooth, connected, and appropriately-sized large objects. By comparison, the *close* filters smoothed the image intensely as well, but without removing the undesirable "black noise." In this particular example, the result of *open* is probably preferable to that of *close*, since the extraneous BLACK structures present more of a problem in the image.

It is important to understand that the *open* and *close* filters are *unidirectional* or *biased* filters in the sense that they remove one type of "noise" (either extraneous WHITE or BLACK features), but not both. Hence *open* and *close* are somewhat special-purpose binary image smoothers that are used when too-small BLACK and WHITE objects (respectively) are to be removed.

It is worth noting that the *close* and *open* filters are again in fact, the same filters, in the dual sense. Given a binary image f and an arbitrary window $\mathbf{B}$:

$$close(f, \mathbf{B}) = NOT\{open[NOT(f), \mathbf{B}]\} \tag{4.25}$$

$$open(f, \mathbf{B}) = NOT\{close[NOT(f), \mathbf{B}]\}. \tag{4.26}$$

In most binary smoothing applications, it is desired to create an unbiased smoothing of the image. This can be accomplished by a further concatenation of filtering operations, applying *open* and *close* operations in sequence on the same image with the same window $\mathbf{B}$. The resulting images will then be smoothed *bidirectionally*. We thus define the unbiased smoothing morphological operators *close-open filter* and *open-close filter*, as follows:

$$close\text{-}open(f, \mathbf{B}) = close[open(f, \mathbf{B}), \mathbf{B}] \tag{4.27}$$

$$open\text{-}close(f, \mathbf{B}) = open[close(f, \mathbf{B}), \mathbf{B}]. \tag{4.28}$$

Hence the *close-open* (*open-close*) of image f is the *open* (*close*) of f with window $\mathbf{B}$ followed by the *close* (*open*) of the result with window $\mathbf{B}$. The morphological filters

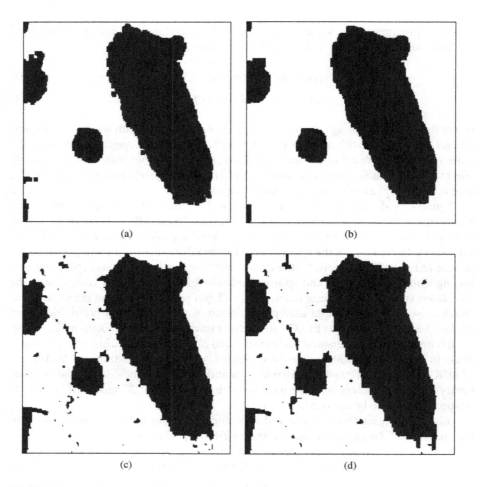

(a) (b)

(c) (d)

FIGURE 4.16

Open and *close* filtering of the binary image "cells." *Open* with: (a) **B** = SQUARE(25); (b) **B** = SQUARE(81); *Close* with: (c) **B** = SQUARE(25); (d) **B** = SQUARE(81).

close-open and *open-close* in (4.27) and (4.28) are general-purpose, bi-directional, size-preserving smoothers. Of course, they may each be interpreted as a sequence of four basic morphological operations (erosions and dilations).

The *close-open* and *open-close* filters are quite similar but are not mathematically identical. Both remove too-small structures without affecting the size much. Both are powerful shape smoothers. However, differences between the processing results can be easily seen. These mainly manifest as a function of the first operation performed in the processing sequence. One notable difference between *close-open* and *open-close* is that *close-open* often links together neighboring holes (since *erode* is the first step), while

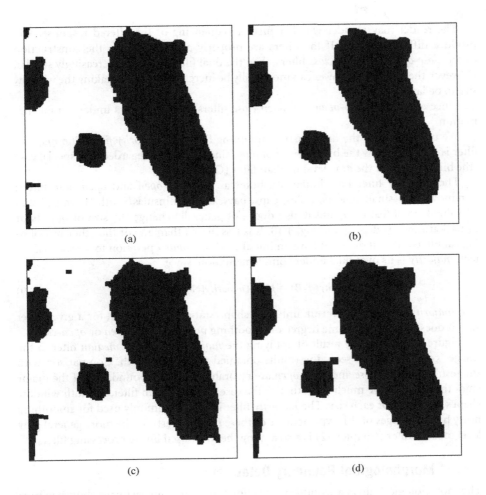

(a)   (b)

(c)   (d)

FIGURE 4.17

Close-open and *open-close* filtering of the binary image "cells." *Close-open* with: (a) **B** = SQUARE(25); (b) **B** = SQUARE(81); *Open-close* with: (c) **B** = SQUARE(25); (d) **B** = SQUARE(81).

open-close often links neighboring objects together (since *dilate* is the first step). The differences are usually somewhat subtle, yet often visible upon close inspection.

Figure 4.17 shows the result of applying the *close-open* and the *open-close* filters to the ongoing binary image example. As can be seen, the results (for **B** fixed) are very similar, although the *close-open* filtered results are somewhat cleaner, as expected. There are also only small differences between the results obtained using the medium and larger windows because of the intense smoothing that is occurring. To fully appreciate the power of these smoothers, it is worth comparing to the original binarized image "cells" in Fig. 4.13(a).

The reader may wonder whether further sequencing of the filtered responses will produce different results. If the filters are properly alternated as in the construction of the *close-open* and *open-close* filters, then the dual filters become increasingly similar. However, the smoothing power can most easily be increased by simply taking the window size to be larger.

Once again, the *close-open* and *open-close* filters are dual filters under complementation.

We now return to the final binary smoothing filter, the *majority filter*. The *majority* filter is also known as the binary *median filter*, since it may be regarded as a special case (the binary case) of the gray level median filter (Chapter 12).

The *majority* filter has similar attributes as the *close-open* and *open-close* filters: it removes too-small objects, holes, gaps, bays, and peninsulas (both '1'-valued and '0'-valued small features), and it also does not generally change the size of objects or of background, as depicted in Fig. 4.18. It is less biased than any of the other morphological filters, since it does not have an initial *erode* or *dilate* operation to set the bias. In fact, *majority* is its own dual under complementation, since

$$majority(f, \mathbf{B}) = \text{NOT}\{majority[\text{NOT}(f), \mathbf{B}]\}. \tag{4.29}$$

The *majority* filter is a powerful, unbiased shape smoother. However, for a given filter size, it does not have the same degree of smoothing power as *close-open* or *open-close*.

Figure 4.19 shows the result of applying the *majority* or *binary median* filter to the image "cell." As can be seen, the results obtained are very smooth. Comparison with the results of *open-close* and *close-open* are favorable, since the boundaries of the major smoothed objects are much smoother in the case of the median filter, for both window shapes used and for each size. The *majority* filter is quite commonly used for smoothing noisy binary images of this type because of these nice properties. The more general gray level median filter (Chapter 12) is also among the most used image processing filters.

4.4.4 Morphological Boundary Detection

The morphological filters are quite effective for smoothing binary images but they have other important applications as well. One such application is *boundary detection*, which is the binary case of the more general edge detectors studied in Chapters 19 and 20.

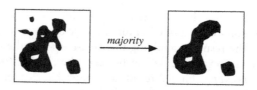

FIGURE 4.18

Effect of *majority* filtering. The smallest holes, gaps, fingers, and extraneous objects are eliminated.

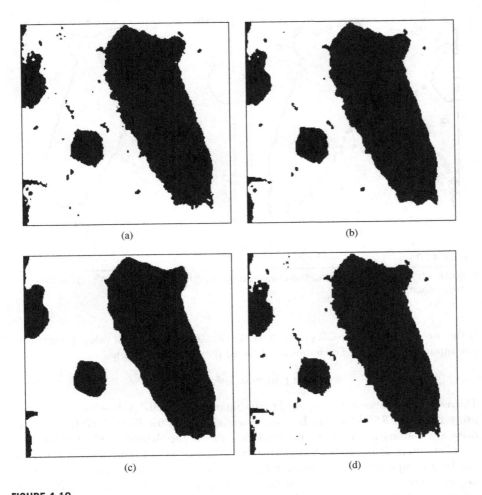

FIGURE 4.19

Majority or *median* filtering of the binary image "cells." *Majority* with: (a) **B** = SQUARE(9); (b) **B** = SQUARE(25); *Majority* with (c) **B** = SQUARE(81); (d) **B** = CROSS(9).

At first glance, boundary detection may seem trivial, since the boundary points can be simply defined as the transitions from '1' to '0' (and vice versa). However, when there is noise present, boundary detection becomes quite sensitive to small noise artifacts, leading to many useless detected edges. Another approach which allows for smoothing of the object boundaries involves the use of morphological operators.

The "difference" between a binary image and a dilated (or eroded) version of it is one effective way of detecting the object boundaries. Usually it is best that the window **B** that is used be small, so that the difference between image and dilation is not too large (leading to thick, ambiguous detected edges). A simple and effective "difference" measure

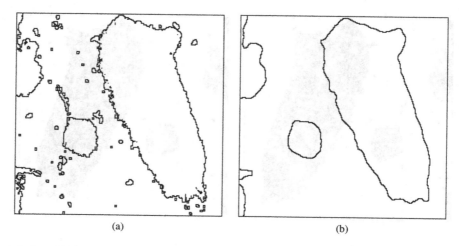

(a)   (b)

FIGURE 4.20

Object boundary detection. Application of *boundary*(f, **B**) to (a) the image "cells"; (b) the *majority-filtered* image in Fig. 4.19(c).

is the two-input exclusive-OR operator XOR. The XOR takes logical value '1' only if its two inputs are different. The boundary detector then becomes simply:

$$boundary(f, \mathbf{B}) = \text{XOR}[f, dilate(f, \mathbf{B})]. \qquad (4.30)$$

The result of this operation as applied to the binary image "cells" is shown in Fig. 4.20(a) using $\mathbf{B} = \text{SQUARE}(9)$. As can be seen, essentially all of the BLACK/WHITE transitions are marked as boundary points. Often this is the desired result. However, in other instances, it is desired to detect only the major object boundary points. This can be accomplished by first smoothing the image with a *close-open, open-close,* or *majority* filter. The result of this smoothed boundary detection process is shown in Fig. 4.20(b). In this case, the result is much cleaner, as only the major boundary points are discovered.

4.5 BINARY IMAGE REPRESENTATION AND COMPRESSION

In several later chapters, methods for compressing gray level images are studied in detail. Compressed images are representations that require less storage than the nominal storage. This is generally accomplished by coding of the data based on measured statistics, rearrangement of the data to exploit patterns and redundancies in the data, and (in the case of lossy compression) quantization of information. The goal is that the image, when decompressed, either looks very much like the original despite a loss of some information (lossy compression), or is not different from the original (lossless compression).

Methods for lossless compression of images are discussed in Chapter 16. Those methods can generally be adapted to both gray level and binary images. Here, we will look at two methods for lossless binary image representation that exploit an assumed structure for the images. In both methods the image data is represented in a new format that exploits the structure. The first method is *run-length coding*, which is so-called because it seeks to exploit the redundancy of long run-lengths or runs of constant value '1' or '0' in the binary data. It is thus appropriate for the coding/compression of binary images containing large areas of constant value '1' and '0.' The second method, *chain coding*, is appropriate for binary images containing binary contours, such as the boundary images shown in Fig. 4.20. Chain coding achieves compression by exploiting this assumption. The chain code is also an information-rich, highly manipulable representation that can be used for shape analysis.

4.5.1 Run-Length Coding

The number of bits required to naively store a $N \times M$ binary image is NM. This can be significantly reduced if it is known that the binary image is smooth in the sense that it is composed primarily of large areas of constant '1' and/or '0' value.

The basic method of run-length coding is quite simple. Assume that the binary image f is to be stored or transmitted on a row-by-row basis. Then for each image row numbered m, the following algorithm steps are used:

1. Store the first pixel value ('0' or '1') in row m in a 1-bit buffer as a reference;

2. Set the run counter $c = 1$;

3. For each pixel in the row:

   - Examine the next pixel to the right;

   - If it is the same as the current pixel, set $c = c + 1$;

   - If different from the current pixel, store c in a buffer of length b and set $c = 1$;

   - Continue until end of row is reached.

Thus, each run-length is stored using b bits. This requires that an overall buffer with segments of lengths b be reserved to store the run-lengths. Run-length coding yields excellent lossless compressions, provided that the image contains lots of constant runs. Caution is necessary, since if the image contains only very short runs, then run-length coding can actually increase the required storage.

Figure 4.21 depicts two hypothetical image rows. In each case, the first symbol stored in a 1-bit buffer will be logical '1.' The run-length code for Fig. 4.21(a) would be '1', 7, 5, 8, 3, 1.... with symbols after the '1' stored using b bits. The first five runs in this sequence

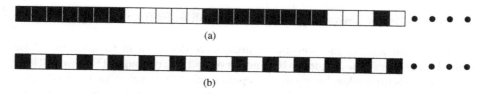

FIGURE 4.21

Example rows of a binary image, depicting (a) reasonable and (b) unreasonable scenarios for run-length coding.

have average length $24/5 = 4.8$, hence if $b \leq 4$, then compression will occur. Of course, the compression can be much higher, since there may be runs of lengths in the dozens or hundreds, leading to very high compressions.

In Fig. 4.21(b), however, in this worst-case example, the storage actually increases b-fold! Hence, care is needed when applying this method. The apparent rule, if it can be applied *a priori*, is that the average run-length L of the image should satisfy $L > b$ if compression is to occur. In fact, the compression ratio will be approximately L/b.

Run-length coding is also used in other scenarios than binary image coding. It can also be adapted to situations where there are run-lengths of any value. For example, in the JPEG lossy image compression standard for gray level images (see Chapter 17), a form of run-length coding is used to code runs of zero-valued frequency-domain coefficients. This run-length coding is an important factor in the good compression performance of JPEG. A more abstract form of run-length coding is also responsible for some of the excellent compression performance of recently developed wavelet image compression algorithms (Chapters 17 and 18).

4.5.2 Chain Coding

Chain coding is an efficient representation of binary images composed of contours. We will refer to these as "contour images." We assume that contour images are composed only of single-pixel width, connected contours (straight or curved). These arise from processes of edge detection or boundary detection, such as the morphological boundary detection method just described above, or the results of some of the edge detectors described in Chapters 19 and 20 when applied to grayscale images.

The basic idea of chain coding is to code contour directions instead of naïve bit-by-bit binary image coding or even coordinate representations of the contours. Chain coding is based on identifying and storing the directions from each pixel to its neighbor pixel on each contour. Before defining this process, it is necessary to clarify the various types of neighbors that are associated with a given pixel in a binary image. Figure 4.22 depicts two neighborhood systems around a pixel (shaded). To the left are depicted the *4-neighbors* of the pixel, which are connected along the horizontal and vertical directions. The set of 4-neighbors of a pixel located at coordinate $\mathbf{n}$ will be denoted $N_4(\mathbf{n})$. To the right

FIGURE 4.22

Depiction of the 4-neighbors and the 8-neighbors of a pixel (shaded).

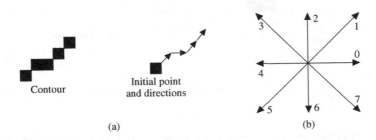

(a)   (b)

FIGURE 4.23

Representation of a binary contour by direction codes. (a) A connected contour can be represented exactly by an initial point and the subsequent directions; (b) only 8 direction codes are
required.

are the *8-neighbors* of the shaded pixel in the center of the grouping. These include the
pixels connected along the diagonal directions. The set of 8-neighbors of a pixel located
at coordinate $\mathbf{n}$ will be denoted $N_8(\mathbf{n})$.

If the initial coordinate $\mathbf{n}_0$ of an 8-connected contour is known, then the rest of the
contour can be represented without loss of information by the directions along which the
contour propagates, as depicted in Fig. 4.23(a). The initial coordinate can be an endpoint,
if the contour is open, or an arbitrary point, if the contour is closed. The contour can be
reconstructed from the directions, if the initial coordinate is known. Since there are only
eight directions that are possible, then a simple 8-neighbor direction code may be used.
The integers $\{0, \ldots, 7\}$ suffice for this, as shown in Fig. 4.23(b).

Of course, the direction codes $0, 1, 2, 3, 4, 5, 6, 7$ can be represented by their 3-bit binary
equivalents: $000, 001, 010, 011, 100, 101, 110, 111$. Hence, each point on the contour *after*
the initial point can be coded by three bits. The initial point of each contour requires
$\lceil \log_2(MN) \rceil$ bits, where $\lceil \cdot \rceil$ denotes the ceiling function: $\lceil x \rceil$ = the smallest integer that
is greater than or equal to x. For long contours, storage of the initial coordinates is
incidental.

Figure 4.24 shows an example of chain coding of a short contour. After the initial
coordinate $\mathbf{n}_0 = (n_0, m_0)$ is stored, the chain code for the remainder of the contour is: $1, 0, 1, 1, 1, 1, 3, 3, 3, 4, 4, 5, 4$ in integer format, or $001, 000, 001, 001, 001, 001, 011,$
$011, 011, 100, 100, 101, 100$ in binary format.

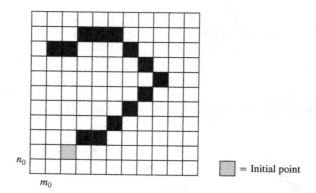

FIGURE 4.24

Depiction of chain coding.

Chain coding is an efficient representation. For example, if the image dimensions are $N = M = 512$, then representing the contour by storing the coordinates of each contour point requires six times as much storage as the chain code.

Basic Tools for Image Fourier Analysis

Alan C. Bovik

The University of Texas at Austin

5.1 INTRODUCTION

In this third chapter on basic methods, the basic mathematical and algorithmic tools for the frequency domain analysis of digital images are explained. Also, 2D discrete-space convolution is introduced. Convolution is the basis for linear filtering, which plays a central role in many places in this *Guide*. An understanding of frequency domain and linear filtering concepts is essential to be able to comprehend such significant topics as image and video enhancement, restoration, compression, segmentation, and wavelet-based methods. Exploring these ideas in a 2D setting has the advantage that frequency domain concepts and transforms can be visualized as images, often enhancing the accessibility of ideas.

5.2 DISCRETE-SPACE SINUSOIDS

Before defining any frequency-based transforms, first we shall explore the concept of *image frequency*, or more generally, of *2D frequency*. Many readers may have a basic background in the frequency domain analysis of 1D signals and systems. The basic theories in two dimensions are founded on the same principles. However, there are some extensions. For example, a 2D frequency component, or sinusoidal function, is characterized not only by its location (phase shift) and its frequency of oscillation but also by its direction of oscillation.

Sinusoidal functions will play an essential role in all of the developments in this chapter. A *2D discrete-space sinusoid* is a function of the form

$$\sin[2\pi(Um + Vn)]. \tag{5.1}$$

Unlike a 1D sinusoid, the function (5.1) has two frequencies, U and V (with units of cycles/pixel) which represent the frequency of oscillation along the vertical (m) and

horizontal (n) spatial image dimensions. Generally, a 2D sinusoid oscillates (is non constant) along every direction except for the direction orthogonal to the direction of fastest
oscillation. The frequency of this fastest oscillation is the *radial frequency*:

$$\Omega = \sqrt{U^2 + V^2}, \tag{5.2}$$

which has the same units as U and V, and the direction of this fastest oscillation is the
angle:

$$\theta = \tan^{-1}\left(\frac{V}{U}\right) \tag{5.3}$$

with units of radians. Associated with (5.1) is the complex exponential function

$$\exp[j2\pi(Um + Vn)] = \cos[2\pi(Um + Vn)] + j\sin[2\pi(Um + Vn)], \tag{5.4}$$

where $j = \sqrt{-1}$ is the pure imaginary number.

In general, sinusoidal functions can be defined on discrete integer grids, hence (5.1)
and (5.4) hold for all integers — $< m$, $n <$. However, sinusoidal functions of infinite
duration are not encountered in practice, although they are useful for image modeling
and in certain image decompositions that we will explore.

In practice, discrete-space images are confined to finite $M \times N$ sampling grids, and
we will also find it convenient to utilize *finite-extent* ($M \times N$) *2D discrete-space sinusoids*
which are defined only for integers

$$0 \le m \le M - 1, \quad 0 \le n \le N - 1, \tag{5.5}$$

and undefined elsewhere. A sinusoidal function that is confined to the domain (5.5) can
be contained within an image matrix of dimension $M \times N$, and is thus easily manipulated
digitally.

In the case of finite sinusoids defined on finite grids (5.5) it will often be convenient
to use the scaled frequencies

$$(u, v) = (MU, NV), \tag{5.6}$$

which have the visually intuitive units of cycles/image. With this, the 2D sinusoid (5.1)
defined on finite grid (5.5) can be re-expressed as:

$$\sin\left[2\pi\left(\frac{u}{M}m + \frac{v}{N}n\right)\right] \tag{5.7}$$

with similar redefinition of the complex exponential (5.4).

Figure 5.1 depicts several discrete-space sinusoids of dimensions 256×256 displayed
as intensity images after linear mapping the grayscale of each to the range $0 - 255$. Because
of the nonlinear response of the eye, the functions in Fig. 5.1 look somewhat more
like square waves than smoothly-varying sinusoids, particularly at higher frequencies.
However, if any of the images in Fig. 5.1 is sampled along a straight line of arbitrary
orientation, the result is an ideal (sampled) sinusoid.

A peculiarity of discrete-space (or discrete-time) sinusoids is that they have a maximum possible *physical* frequency at which they can oscillate. Although the frequency
variables (u, v) or (U, V) may be taken arbitrarily large, these large values do not

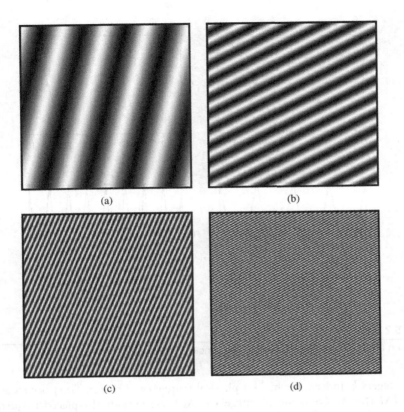

(a)   (b)

(c)   (d)

FIGURE 5.1

Examples of finite 2D discrete-space sinusoidal functions. The scaled frequencies (5.6) measured in cycles/image are (a) $u = 1$, $v = 4$; (b) $u = 10$, $v = 5$; (c) $u = 15$, $v = 35$; and (d) $u = 65$, $v = 35$.

correspond to arbitrarily large physical oscillation frequencies. The ramifications of this are quite deep and significant and relate to the restrictions placed on sampling of continuous-space images (the Sampling Theorem) and the Nyquist frequency.

As an example of this principle we will study a 1D example of a discrete sinusoid. Consider the finite cosine function $\cos\left[2\pi\left(\frac{u}{M}m + \frac{v}{N}n\right)\right] = \cos\left(2\pi\frac{u}{16}m\right)$ which results by taking $M = N = 16$, and $v = 0$. This is a cosine wave propagating in the m-direction only (all columns are the same) at frequency u (cycles/image).

Figure 5.2 depicts the 1D cosine for various values of u. As can be seen, the physical oscillation frequency increases until $u = 8$; for incrementally larger values of u, however, the physical frequency diminishes. In fact, the function is period-16 in the frequency index u:

$$\cos\left(2\pi\frac{u}{16}m\right) = \cos\left[2\pi\frac{(u+16k)}{16}m\right] \tag{5.8}$$

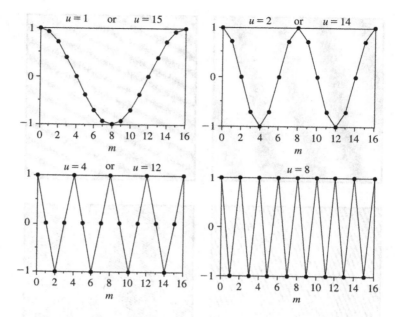

FIGURE 5.2

Illustration of physical versus numerical frequencies of discrete-space sinusoids.

for all integers k. Indeed, the highest physical frequency of $\cos\left(2\pi \frac{u}{M} m\right)$ occurs at $u = M/2 + kM$ (for M even) for all integers k. At these periodically-placed frequencies, (5.8) is equal to $(-1)^m$; the fastest discrete-index oscillation is the alternating signal. This observation will be important next as we define the various frequency domain image transforms.

5.3 DISCRETE-SPACE FOURIER TRANSFORM

The *discrete-space Fourier transform* (DSFT) of a given discrete-space image f is given by

$$F(U, V) = \sum_{m=-\infty}^{\infty} \sum_{n=-\infty}^{\infty} f(m, n) e^{-j2\pi(Um + Vn)} \tag{5.9}$$

with *inverse discrete-space Fourier transform* (IDSFT):

$$f(m, n) = \int_{-0.5}^{0.5} \int_{-0.5}^{0.5} F(U, V) e^{j2\pi(Um + Vn)} \, dU \, dV. \tag{5.10}$$

When (5.9) and (5.10) hold, we will often make the notation $f \overset{\mathfrak{D}}{\leftrightarrow} F$ and say that f, F form a *DSFT pair*. The units of the frequencies (U, V) in (5.9) and (5.10) are cycles/pixel. It should be noted that, unlike continuous Fourier transforms, the DSFT is asymmetrical in that the forward transform F is continuous in the frequency variables (U, V), while the image or inverse transform is discrete. Thus, the DSFT is defined as a summation, while the IDSFT is defined as an integral.

There are several ways of interpreting the DSFT (5.9) and (5.10). The most usual mathematical interpretation of (5.10) is as a decomposition of $f(m, n)$ into orthonormal complex exponential basis functions $e^{j2\pi(Um+Vn)}$ that satisfy

$$\int_{-0.5}^{0.5}\int_{-0.5}^{0.5} e^{j2\pi(Um+Vn)} e^{-j2\pi(Up+Vq)} dU\, dV = \begin{cases} 1; & m = p \quad \text{and} \quad n = q \\ 0; & \text{else} \end{cases}. \tag{5.11}$$

Another (somewhat less precise) interpretation is the engineering concept of the transformation, without loss, of space domain image information into frequency domain image information. Representing the image information in the frequency domain has significant conceptual and algorithmic advantages, as will be seen. A third interpretation is a physical one, where the image is viewed as the result of a sophisticated constructive-destructive interference wave pattern. By assigning each of the infinite number of complex exponential wave functions $e^{j2\pi(Um+Vn)}$ with the appropriate complex weights $F(U, V)$, the intricate structure of any discrete-space image can be recreated exactly as an interference-sum.

The DSFT possesses a number of important properties that will be useful in defining applications. In the following, assume that $f \overset{\Im}{\leftrightarrow} F$, $g \overset{\Im}{\leftrightarrow} G$, and $h \overset{\Im}{\leftrightarrow} H$.

5.3.1 Linearity of DSFT

Given images f, g and arbitrary complex constants a, b, the following holds:

$$af + bg \overset{\Im}{\leftrightarrow} aF + bG. \tag{5.12}$$

This property of linearity follows directly from (5.9), and can be extended to a weighted sum of any countable number of images. It is fundamental to many of the properties of, and operations involving, the DSFT.

5.3.2 Inversion of DSFT

The 2D function $F(U, V)$ uniquely satisfies the relationships (5.9) and (5.10). That the inversion holds can be easily shown by substituting (5.9) into (5.10), reversing the order of sum and integral, and then applying (5.11).

5.3.3 Magnitude and Phase of DSFT

The DSFT F of an image f is generally complex-valued. As such it can be written in the form

$$F(U, V) = R(U, V) + jI(U, V), \tag{5.13}$$

where

$$R(U,V) = \sum_{m=-\infty}^{\infty} \sum_{n=-\infty}^{\infty} f(m,n)\cos[2\pi(Um + Vn)] \qquad (5.14)$$

and

$$I(U,V) = -\sum_{m=-\infty}^{\infty} \sum_{n=-\infty}^{\infty} f(m,n)\sin[2\pi(Um + Vn)] \qquad (5.15)$$

are the real and imaginary parts of $F(U,V)$, respectively.

The DSFT can also be written in the often-convenient phasor form

$$F(U,V) = |F(U,V)|e^{j\angle F(U,V)}, \qquad (5.16)$$

where the *magnitude spectrum* of image f is

$$|F(U,V)| = \sqrt{R^2(U,V) + I^2(U,V)} \qquad (5.17)$$

$$= \sqrt{F(U,V)F^*(U,V)}, \qquad (5.18)$$

where '$*$' denotes the complex conjugation. The *phase spectrum* of image f is

$$\angle F(U,V) = \tan^{-1}\left[\frac{I(U,V)}{R(U,V)}\right]. \qquad (5.19)$$

5.3.4 Symmetry of DSFT

If the image f is real, which is usually the case, then the DSFT is *conjugate symmetric*:

$$F(U,V) = F^*(-U,-V), \qquad (5.20)$$

which means that the DSFT is completely specified by its values over any half-plane. Hence, if f is real, the DSFT is redundant. From (5.20), it follows that the magnitude spectrum is *even symmetric*:

$$|F(U,V)| = |F(-U,-V)|, \qquad (5.21)$$

while the phase spectrum is *odd symmetric*:

$$\angle F(U,V) = -\angle F(-U,-V). \qquad (5.22)$$

5.3.5 Translation of DSFT

Multiplying (or modulating) the discrete-space image $f(m,n)$ by a 2D complex exponential wave function $\exp[j2\pi(U_0 m + V_0 n)]$ results in a translation of the DSFT:

$$f(m,n)\exp[j2\pi(U_0 m + V_0 n)] \overset{\Im}{\leftrightarrow} F(U-U_0, V-V_0). \qquad (5.23)$$

Likewise, translating the image f by amounts m_0 and n_0 produces a modulated DSFT:

$$f(m - m_0, n - n_0) \overset{\Im}{\leftrightarrow} F(U,V)\exp[-j2\pi(Um_0 + Vn_0)]. \qquad (5.24)$$

5.3.6 Convolution and the DSFT

Given two images or 2D functions f and h, their *2D discrete-space linear convolution* is given by

$$g(m,n) = f(m,n) * h(m,n) = h(m,n) * f(m,n) = \sum_{p=-\infty}^{\infty} \sum_{q=-\infty}^{\infty} f(p,q)h(m-p,n-q). \quad (5.25)$$

The linear convolution expresses the result of passing an image signal f through a 2D linear convolution system h (or vice versa). The commutativity of the convolution is easily seen by making a substitution of variables in the double sum in (5.25).

If g, f, and h satisfy the spatial convolution relationship (5.25), then their DSFT's satisfy

$$G(U,V) = F(U,V)H(U,V), \quad (5.26)$$

hence convolution in the space domain corresponds directly to multiplication in the spatial frequency domain. This important property is significant both conceptually, as a simple and direct means for effecting the frequency content of an image, and computationally, since the linear convolution has such a simple expression in the frequency domain.

The 2D DSFT is the basic mathematical tool for analyzing the frequency domain content of 2D discrete-space images. However, it has a major drawback for digital image processing applications: the DSFT $F(U,V)$ of a discrete-space image $f(m,n)$ is continuous in the frequency coordinates (U,V); there are an uncountably infinite number of values to compute. As such, discrete (digital) processing or display in the frequency domain is not possible using the DSFT unless it is modified in some way. Fortunately, this is possible when the image f is of finite dimensions. In fact, by sampling the DSFT in the frequency domain we are able to create a computable Fourier-domain transform.

5.4 2D DISCRETE FOURIER TRANSFORM (DFT)

Now we restrict our attention to the practical case of discrete-space images that are of finite extent. Hence assume that image $f(m,n)$ can be expressed as a matrix $\mathbf{f} = [f(m,n), 0 \le m \le M-1, 0 \le n \le N-1]$. As we will show, a finite-extent image matrix $\mathbf{f}$ can be represented exactly as a *finite* weighted sum of 2D frequency components, instead of an infinite number. This leads to computable and numerically manipulable frequency domain representations. Before showing how this is done, we shall introduce a special notation for the complex exponential that will simplify much of the ensuing development.

We will use

$$W_K = \exp\left[-j\frac{2\pi}{K}\right] \quad (5.27)$$

as a shorthand for the basic complex exponential, where K is the dimension along one of the image axes ($K = N$ or $K = M$). The notation (5.27) makes it possible to index the various elementary frequency components at arbitrary spatial and frequency coordinates by simple exponentiation:

$$W_M^{um} W_N^{vn} = \cos\left[2\pi\left(\frac{u}{M}m + \frac{v}{N}n\right)\right] - j\sin\left[2\pi\left(\frac{u}{M}m + \frac{v}{N}n\right)\right]. \qquad (5.28)$$

This process of space and frequency indexing by exponentiation greatly simplifies the manipulation of frequency components and the definition of the discrete Fourier transform (DFT). Indeed, it is possible to develop frequency domain concepts and frequency transforms without the use of complex numbers (and in fact some of these, such as the discrete cosine transform, or DCT, are widely used, especially in image compression— See Chapters 16 and 17 of this *Guide*).

For the purpose of analysis and basic theory, it is much simpler to use W_M^{um} and W_N^{vn} to represent finite-extent (of dimensions M and N) frequency components oscillating at u (cycles/image) and v (cycles/image) in the m- and n-directions, respectively. Clearly,

$$\left| W_M^{um} W_N^{vn} \right| = 1 \qquad (5.29)$$

and

$$\angle W_M^{um} W_N^{vn} = -2\pi\left(\frac{u}{M}m + \frac{v}{N}n\right). \qquad (5.30)$$

Observe that the minimum physical frequency of W_M^{um} periodically occurs at the indices $u = kM$ for all integers k:

$$W_M^{kMm} = 1 \qquad (5.31)$$

for any integer m; the minimum oscillation is no oscillation. If M is even, the maximum physical frequency periodically occurs at the indices $u = kM + M/2$:

$$W_M^{(kM+M/2)m} = 1 \cdot e^{-j\pi m} = (-1)^m, \qquad (5.32)$$

which is the discrete period-2 (alternating) function, the highest possible discrete oscillation frequency.

The *2D* DFT of the finite-extent ($M \times N$) image **f** is given by

$$\tilde{F}(u,v) = \sum_{m=0}^{M-1}\sum_{n=0}^{N-1} f(m,n) W_M^{um} W_N^{vn} \qquad (5.33)$$

for integer frequencies $0 \le u \le M-1, 0 \le v \le N-1$. Hence the DFT is also of finite-extent $M \times N$, and can be expressed as a (generally complex-valued) matrix $\tilde{\mathbf{F}} = [\tilde{F}(u,v); 0 \le u \le M-1, 0 \le v \le N-1]$. It has a unique *inverse discrete Fourier transform*, or *IDFT*:

$$f(m,n) = \frac{1}{MN}\sum_{u=0}^{M-1}\sum_{v=0}^{N-1} \tilde{F}(u,v) W_M^{-um} W_N^{-vn} \qquad (5.34)$$

for $0 \le m \le M - 1, 0 \le n \le N - 1$. When (5.33) and (5.34) hold, it is often denoted $\mathbf{f} \overset{DFT}{\leftrightarrow} \tilde{\mathbf{F}}$ and we say that $\mathbf{f}, \tilde{\mathbf{F}}$ form a *DFT pair*.

A number of observations regarding the DFT and its relationship to the DSFT are necessary. First, the DFT and IDFT are symmetrical, since both forward and inverse transforms are defined as sums. In fact they have the same form, except for the polarity of the exponents and a scaling factor. Secondly, both forward and inverse transforms are finite sums; both $\tilde{\mathbf{F}}$ and $\mathbf{f}$ can be represented uniquely as *finite* weighted sums of *finite-extent* complex exponentials with integer-indexed frequencies. Thus, for example, any 256×256 digital image can be expressed as the weighted sum of $256^2 = 65{,}536$ complex exponential (sinusoid) functions including those with real parts shown in Fig. 5.1. Note that the frequencies (u, v) are scaled so that their units are in cycles/image, as in (5.6) and Fig. 5.1.

Most importantly, the DFT has a direct relationship to the DSFT. In fact, the DFT of an $M \times N$ image $\mathbf{f}$ is a uniformly *sampled* version of the DSFT of $\mathbf{f}$:

$$\tilde{F}(u,v) = F(U,V)\Big|_{U=\frac{u}{M}, V=\frac{v}{N}} \tag{5.35}$$

for integer frequency indices $0 \le u \le M - 1, 0 \le v \le N - 1$. Since $\mathbf{f}$ is of finite extent, and contains MN elements, the DFT $\tilde{\mathbf{F}}$ is conservative in that it also requires only MN elements to contain complete information about $\mathbf{f}$ (to be exactly invertible). Also, since $\tilde{\mathbf{F}}$ is simply evenly-spaced samples of F, many of the properties of the DSFT translate directly with little or no modification to the DFT.

5.4.1 Linearity and Invertibility of DFT

The DFT is linear in the sense of (5.12). It is uniquely invertible, as can be established by substituting (5.33) into (5.34), reversing the order of summation, and using the fact that the discrete complex exponentials are also orthogonal

$$\sum_{u=0}^{M-1} \sum_{v=0}^{N-1} \left(W_M^{um} W_M^{-up} \right) \left(W_N^{vn} W_N^{-vq} \right) = \begin{cases} MN; & m = p \text{ and } n = q \\ 0; & \text{else} \end{cases}. \tag{5.36}$$

The DFT matrix $\tilde{\mathbf{F}}$ is generally complex, hence it has an associated magnitude spectrum matrix, denoted

$$|\tilde{\mathbf{F}}| = [|\tilde{F}(u,v)|; \quad 0 \le u \le M - 1, 0 \le v \le N - 1] \tag{5.37}$$

and phase spectrum matrix denoted

$$\angle\tilde{\mathbf{F}} = [\angle\tilde{F}(u,v); \quad 0 \le u \le M - 1, 0 \le v \le N - 1]. \tag{5.38}$$

The elements of $|\tilde{\mathbf{F}}|$ and $\angle\tilde{\mathbf{F}}$ are computed in the same way as the DSFT magnitude and phase (5.16)–(5.19).

5.4.2 Symmetry of DFT

Like the DSFT, if **f** is real-valued, then the DFT matrix is conjugate symmetric, but in the matrix sense:

$$\tilde{F}(u,v) = \tilde{F}^*(M-u, N-v) \tag{5.39}$$

for $0 \leq u \leq M-1, 0 \leq v \leq N-1$. This follows easily by substitution of the reversed and translated frequency indices $(M-u, N-v)$ into the forward DFT equation (5.33). An apparent repercussion of (5.39) is that the DFT $\tilde{F}$ matrix is redundant, and hence can represent the $M \times N$ image with only about $MN/2$ DFT coefficients. This mystery is resolved by realizing that $\tilde{F}$ is complex-valued, hence it requires twice the storage for real and imaginary components. If **f** is not real-valued, then (5.39) does not hold.

Of course, (5.39) implies symmetries of the magnitude and phase spectra:

$$|\tilde{F}(u,v)| = |\tilde{F}(M-u, N-v)| \tag{5.40}$$

and

$$\angle\tilde{F}(u,v) = -\angle\tilde{F}(M-u, N-v) \tag{5.41}$$

for $0 \leq u \leq M-1, 0 \leq v \leq N-1$.

5.4.3 Periodicity of DFT

Another property of the DSFT that carries over to the DFT is frequency periodicity. Recall that the DSFT $F(U,V)$ has unit period in U and V. The DFT matrix $\tilde{F}$ was defined to be of finite-extent $M \times N$. However, the forward DFT equation (5.33) admits the possibility of evaluating $\tilde{F}(u,v)$ outside of the range $0 \leq u \leq M-1, 0 \leq v \leq N-1$. It turns out that $\tilde{F}(u,v)$ is period-M and period-N along the u and v dimensions, respectively. For any integers k, l:

$$\tilde{F}(u+kM, v+lN) = \tilde{F}(u,v) \tag{5.42}$$

for every $0 \leq u \leq M-1, 0 \leq v \leq N-1$. This follows easily by substitution of the periodically extended frequency indices $(u+kM, v+lN)$ into the forward DFT equation (5.33). The interpretation (5.42) of the DFT is called the *periodic extension* of the DFT. It is defined for all integer frequencies u, v.

Although many properties of the DFT are the same, or similar to those of the DSFT, certain important properties are different. These effects arise from sampling the DSFT to create the DFT.

5.4.4 Image Periodicity Implied by DFT

A seemingly innocuous, yet extremely important consequence of sampling the DSFT is that the resulting DFT equations imply that the image **f** is itself periodic. In fact, the IDFT equation (5.34) implies that for any integers k, l:

$$f(m+kM, n+lN) = f(m,n) \tag{5.43}$$

for every $0 \leq m \leq M - 1, 0 \leq n \leq N - 1$. This follows easily by substitution of the periodically extended space indices $(m + kM, n + lN)$ into the inverse DFT equation (5.34).

Clearly, finite-extent digital images arise from imaging the real world through finite field of view (FOV) devices, such as cameras, and outside that FOV, the world does not repeat itself periodically, *ad infinitum*. The implied periodicity of f is purely a synthetic effect that derives from sampling the DSFT. Nevertheless, it is of paramount importance, since any algorithm that is developed, and that uses the DFT, will operate as though the DFT-transformed image were spatially periodic in the sense (5.43). One important property and application of the DFT that is effected by this spatial periodicity is the frequency-domain convolution property.

5.4.5 Cyclic Convolution Property of the DFT

One of the most significant properties of the DSFT is the linear convolution property (5.25) and (5.26), which says that space domain convolution corresponds to frequency domain multiplication:

$$f * h \overset{\Im}{\leftrightarrow} FH. \tag{5.44}$$

This useful property makes it possible to analyze and design linear convolution-based systems in the frequency domain. Unfortunately, property (5.44) does not hold for the DFT; a product of DFT's does not correspond (inverse transform) to the linear convolution of the original DFT-transformed functions or images. However, it does correspond to another type of convolution, variously known as *cyclic convolution, circular convolution,* or *wraparound convolution.*

We will demonstrate the form of the cyclic convolution by deriving it. Consider the two $M \times N$ image functions $\mathbf{f} \overset{\text{DFT}}{\leftrightarrow} \tilde{\mathbf{F}}$ and $\mathbf{h} \overset{\text{DFT}}{\leftrightarrow} \tilde{\mathbf{H}}$. Define the *pointwise* matrix product[1]

$$\tilde{\mathbf{G}} = \tilde{\mathbf{F}} \otimes \tilde{\mathbf{H}} \tag{5.45}$$

according to

$$\tilde{G}(u,v) = \tilde{F}(u,v)\tilde{H}(u,v) \tag{5.46}$$

for $0 \leq u \leq M - 1, 0 \leq v \leq N - 1$. Thus we are interested in the form of **g**. For each $0 \leq m \leq M - 1, 0 \leq n \leq N - 1$, we have

$$g(m,n) = \frac{1}{MN} \sum_{u=0}^{M-1} \sum_{v=0}^{N-1} \tilde{G}(u,v) \, W_M^{-um} W_N^{-vn}$$

$$= \frac{1}{MN} \sum_{u=0}^{M-1} \sum_{v=0}^{N-1} \tilde{F}(u,v)\tilde{H}(u,v) \, W_M^{-um} W_N^{-vn}$$

---

[1] As opposed to the standard matrix product.

$$= \frac{1}{MN} \sum_{u=0}^{M-1} \sum_{v=0}^{N-1} \left\{ \sum_{p=0}^{M-1} \sum_{q=0}^{N-1} f(p,q) W_M^{up} W_N^{vq} \right\}$$

$$\times \left\{ \sum_{r=0}^{M-1} \sum_{s=0}^{N-1} h(r,s) W_M^{ur} W_N^{vs} \right\} W_M^{-um} W_N^{-vn} \qquad (5.47)$$

by substitution of the definitions of $\tilde{F}(u,v)$ and $\tilde{H}(u,v)$. Rearranging the order of the summations to collect all of the complex exponentials inside the innermost summation reveals that

$$g(m,n) = \frac{1}{MN} \sum_{p=0}^{M-1} \sum_{q=0}^{N-1} f(p,q) \sum_{r=0}^{M-1} \sum_{s=0}^{N-1} h(r,s) \sum_{u=0}^{M-1} \sum_{v=0}^{N-1} W_M^{u(p+r-m)} W_N^{v(q+s-n)}. \qquad (5.48)$$

Now, from (5.36), the innermost summation

$$\sum_{u=0}^{M-1} \sum_{v=0}^{N-1} W_M^{u(p+r-m)} W_N^{v(q+s-n)} = \begin{cases} MN; & r = m-p \quad \text{and} \quad s = n-q \\ 0; & \text{else} \end{cases} \qquad (5.49)$$

hence

$$g(m,n) = \sum_{p=0}^{M-1} \sum_{q=0}^{N-1} f(p,q) h\left[(m-p)_M, (n-q)_N\right] \qquad (5.50)$$

$$= f(m,n) \circledast h(m,n) = h(m,n) \circledast f(m,n), \qquad (5.51)$$

where $(x)_N = x \bmod N$ and the symbol '$\circledast$' denotes the 2D cyclic convolution.[2] The final step of obtaining (5.50) from (5.49) follows since the argument of the shifted and twice-reversed (along each axis) function $h(m-p, n-q)$ finds no meaning whenever $(m-p) \notin \{0,\ldots,M-1\}$ or $(n-q) \notin \{0,\ldots,N-1\}$, since h is undefined outside of those coordinates. However, because the DFT was used to compute $g(m,n)$, then the periodic extension of $h(m-p, n-q)$ is implied, which can be expressed as $h\left[(m-p)_M, (n-q)_N\right]$. Hence (5.50) follows. That '$\circledast$' is commutative and easily established by a substitution of variables in (5.50). It can also be seen that cyclic convolution is a form of linear convolution, but with one (either, but not both) of the two functions being periodically extended. Hence

$$f(m,n) \circledast h(m,n) = f(m,n) * h[(m)_M, (n)_N] = f[(m)_M, (n)_N] * h(m,n). \qquad (5.52)$$

This *cyclic convolution property* of the DFT is unfortunate, since in the majority of applications it is not desired to compute the cyclic convolution of two image functions. Instead, what is frequently desired is the linear convolution of two functions, as in the case of linear filtering. In both linear and cyclic convolution, the two functions are superimposed, with one function reversed along both axes and shifted to the point at which

---

[2]Modular arithmetic is remaindering. Hence $(x)_N$ is the integer remainder of (x/N).

the convolution is being computed. The product of the functions is computed at every point of overlap, with the sum of products being the convolution. In the case of the cyclic convolution, one (not both) of the functions is periodically extended, hence the overlap is much larger and wraps around the image boundaries. This produces a significant error with respect to the correct linear convolution result. This error is called *spatial aliasing*, since the wraparound error contributes false information to the convolution sum.

Figure 5.3 depicts the linear and cyclic convolutions of two hypothetical $M \times N$ images f and h at a point (m_0, n_0). From the figure, it can be seen that the wraparound

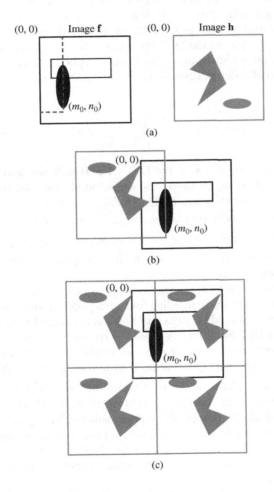

FIGURE 5.3

Convolution of two images. (a) Images f and h; (b) linear convolution result at (m_0, n_0) is computed as the sum of products where f and h overlap; (c) cyclic convolution result at (m_0, n_0) is computed as the sum of products where f and the periodically extended h overlap.

error can overwhelm the linear convolution contribution. Note in Fig. 5.3(b) that although the linear convolution sum (5.25) extends over the indices $0 \leq m \leq M - 1$ and $0 \leq n \leq N - 1$, the overlap is restricted to the indices.

5.4.6 Linear Convolution Using the DFT

Fortunately, it turns out that it is possible to compute the linear convolution of two arbitrary finite-extent 2D discrete-space functions or images using the DFT. The process requires modifying the functions to be convolved prior to taking the product of their DFT's. The modification acts to cancel the effects of spatial aliasing.

Suppose more generally that f and h are two arbitrary finite-extent images of dimensions $M \times N$ and $P \times Q$, respectively. We are interested in computing the linear convolution $g = f * h$ using the DFT. We assume the general case where the images f, h do not have the same dimensions, since in most applications an image is convolved with a filter function of different (usually much smaller) extent.

Clearly,

$$g(m, n) = f(m, n) * h(m, n) = \sum_{p=0}^{M-1} \sum_{q=0}^{N-1} f(p, q) h(m - p, n - q). \qquad (5.53)$$

Inverting the pointwise products of the DFT's $\tilde{\mathbf{F}} \otimes \tilde{\mathbf{H}}$ will not lead to (5.53), since wraparound error will occur. To cancel the wraparound error, the functions $\mathbf{f}$ and $\mathbf{h}$ are modified by increasing their size by *zero-padding* them. Zero-padding means that the arrays $\mathbf{f}$ and $\mathbf{h}$ are expanded into larger arrays, denoted $\hat{\mathbf{f}}$ and $\hat{\mathbf{h}}$, by filling the empty spaces with zeroes. To compute the linear convolution, the pointwise product $\tilde{\hat{\mathbf{G}}} = \tilde{\hat{\mathbf{F}}} \otimes \tilde{\hat{\mathbf{H}}}$ of the DFTs of the zero-padded functions $\hat{\mathbf{f}}$ and $\hat{\mathbf{h}}$ is computed. The inverse DFT $\hat{\mathbf{g}}$ of $\tilde{\hat{\mathbf{G}}}$ then contains the correct linear convolution result.

The question remains as to how many zeroes are used to pad the functions $\mathbf{f}$ and $\mathbf{h}$. The answer to this lies in understanding how zero-padding works and how large the linear convolution result should be. Zero-padding acts to cancel the spatial aliasing error (wraparound) of the DFT by supplying zeroes where the wraparound products occur. Hence the wraparound products are all zero and contribute nothing to the convolution sum. This leaves only the linear convolution contribution to the result. To understand how many zeroes are needed, it must be realized that the resulting product DFT $\tilde{\hat{\mathbf{G}}}$ corresponds to a periodic function $\hat{\mathbf{g}}$. If the horizontal/vertical periods are too small (not enough zero-padding), the periodic replicas will overlap (spatial aliasing). If the periods are just large enough, then the periodic replicas will be contiguous instead of overlapping, hence spatial aliasing will be canceled. Padding with more zeroes than this results in excess computation. Figure 5.4 depicts the successful result of zero-padding to eliminate wraparound error.

The correct period lengths are equal to the lengths of the correct linear convolution result. The linear convolution result of two arbitrary $M \times N$ and $P \times Q$ image functions will generally be $(M + P - 1) \times (N + Q - 1)$, hence we would like the DFT $\tilde{\hat{\mathbf{G}}}$ to have

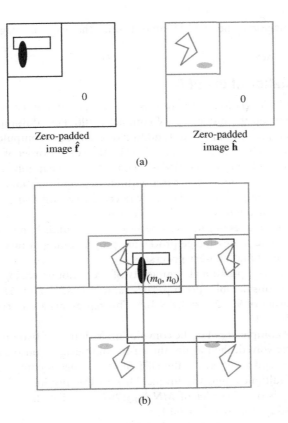

FIGURE 5.4

Linear convolution of the same two images as Fig. 5.2 by zero-padding and cyclic convolution (via the DFT). (a) Zero-padded images $\hat{f}$ and $\hat{h}$; (b) cyclic convolution at (m_0, n_0) computed as the sum of products where $\hat{f}$ and the periodically extended $\hat{h}$ overlap. These products are zero except over the range $0 \le p \le m_0$ and $0 \le q \le n_0$.

these dimensions. Therefore, the $M \times N$ function f and the $P \times Q$ function h must both be zero-padded to size $(M + P - 1) \times (N + Q - 1)$. This yields the correct linear convolution result:

$$\hat{g} = \hat{f} \circledast \hat{h} = f * h. \tag{5.54}$$

In most cases, linear convolution is performed between an image and a filter function much smaller than the image: $M \gg P$ and $N \gg Q$. In such cases the result is not much larger than the image, and often only the $M \times N$ portion indexed $0 \le m \le M - 1, 0 \le n \le N - 1$ is retained. The reason behind this is, firstly, it may be desirable to retain images of size MN only, and secondly, the linear convolution result beyond the borders of the original image may be of little interest, since the original image was zero there anyway.

5.4.7 Computation of the DFT

Inspection of the DFT relation (5.33) reveals that computation of each of the MN DFT coefficients requires on the order of MN complex multiplies/additions. Hence, on the order of M^2N^2 complex, multiplies and additions are needed to compute the overall DFT of an $M \times N$ image $\mathbf{f}$. For example, if $M = N = 512$, then on the order of $2^{36} = 6.9 \times 10^{10}$ complex multiplies/additions are needed, which is a very large number. Of course, these numbers assume a naïve implementation without any optimization. Fortunately, fast algorithms for DFT computation, collectively referred to as *fast fourier transform* (FFT) algorithms, have been intensively studied for many years. We will not delve into the design of these, since it goes beyond what we want to accomplish in this *Guide* and also since they are available in any image processing programming library or development environment and most math library programs.

The FFT offers a computational complexity of order not exceeding $MN \log_2(MN)$, which represents a considerable speedup. For example, if $M = N = 512$, then the complexity is on the order of $9 \times 2^{19} = 4.7 \times 10^6$. This represents a very common speedup of more than 14,500:1 !

Analysis of the complexity of cyclic convolution is similar. If two images of the same size $M \times N$ are convolved, then again, the naïve complexity is on the order of M^2N^2 complex multiplies and additions. If the DFT of each image is computed, the resulting DFTs pointwise multiplied, and the inverse DFT of this product calculated, then the overall complexity is on the order of $MN \log_2(2M^3N^3)$. For the common case $M = N = 512$, the speedup still exceeds 4700:1.

If linear convolution is computed via the DFT, the computation is increased somewhat since the images are increased in size by zero-padding. Hence the speedup of DFT-based linear convolution is somewhat reduced (although in a fixed hardware realization, the known existence of these zeroes can be used to effect a speedup). However, if the functions being linearly convolved are both not small, then the DFT approach will always be faster. If one of the functions is very small, say covering fewer than 32 samples (such as a small linear filter template), then it is possible that direct space domain computation of the linear convolution may be faster than DFT-based computation. However, there is no strict rule of thumb to determine this lower cutoff size, since it depends on the filter shape, the algorithms used to compute DFTs and convolutions, any special-purpose hardware, and so on.

5.4.8 Displaying the DFT

It is often of interest to visualize the DFT of an image. This is possible since the DFT is a sampled function of finite (periodic) extent. Displaying one period of the DFT of image $\mathbf{f}$ reveals a picture of the frequency content of the image. Since the DFT is complex, one can display either the magnitude spectrum $|\tilde{\mathbf{F}}|$ or the phase spectrum $\angle\, \tilde{\mathbf{F}}$ as a single 2D intensity image.

However, the phase spectrum $\angle\tilde{\mathbf{F}}$ is usually not visually revealing when displayed. Generally it appears quite random, and so usually the magnitude spectrum $|\tilde{\mathbf{F}}|$ only is absorbed visually. This is not intended to imply that image phase information is not important; in fact, it is exquisitely important, since it determines the relative shifts of the component complex exponential functions that make up the DFT decomposition. Modifying or ignoring image phase will destroy the delicate constructive-destructive interference pattern of the sinusoids that make up the image.

As briefly noted in Chapter 3, displays of the Fourier transform magnitude will tend to be visually dominated by the low-frequency and zero-frequency coefficients, often to such an extent that the DFT magnitude appears as a single spot. This is highly undesirable, since most of the interesting information usually occurs at frequencies away from the lowest frequencies. An effective way to bring out the higher frequency coefficients for visual display is via a point logarithmic operation: instead of displaying $|\tilde{\mathbf{F}}|$, display

$$\log_2[1 + |\tilde{F}(u,v)|] \qquad (5.55)$$

for $0 \le u \le M - 1, 0 \le v \le N - 1$. This has the effect of compressing all of the DFT magnitudes, but larger magnitudes much more so. Of course, since all of the logarithmic magnitudes will be quite small, a full-scale histogram stretch should then be applied to fill the grayscale range.

Another consideration when displaying the DFT of a discrete-space image is illustrated in Fig. 5.5. In the DFT formulation, a single $M \times N$ period of the DFT is sufficient to represent the image information, and also for display. However, the DFT matrix is even symmetric across both diagonals. More importantly, the center of symmetry occurs in the image center, where the high-frequency coefficients are clustered near $(u,v) = (M/2, N/2)$. This is contrary to conventional intuition, since in most engineering applications Fourier transform magnitudes are displayed with zero and low-frequency coefficients at the center. This is particularly true of 1D continuous Fourier transform magnitudes, which are plotted as graphs with the zero frequency at the origin. This is also visually convenient, since the dominant lower frequency coefficients then are clustered together at the center, instead of being scattered about the display.

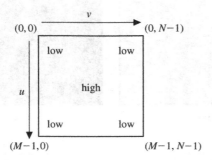

FIGURE 5.5

Distribution of high- and low-frequency DFT coefficients.

A natural way of remedying this is to instead display the shifted DFT magnitude

$$|\tilde{F}(u - M/2, v - N/2)| \tag{5.56}$$

for $0 \le u \le M - 1, 0 \le v \le N - 1$. This can be accomplished in a simple way by taking the DFT of

$$(-1)^{m+n} f(m,n) \overset{\text{DFT}}{\leftrightarrow} \tilde{F}(u - M/2, v - N/2). \tag{5.57}$$

Relation (5.57) follows since $(-1)^{m+n} = e^{j\pi(m+n)}$, hence from (5.23) the DSFT is shifted by amount ½ cycles/pixel along both dimensions; since the DFT uses the scaled frequencies (5.6), the DFT is shifted by $M/2$ and $N/2$ cycles/image in the u- and v- directions, respectively.

Figure 5.6 illustrates the display of the DFT of the "fingerprint" image, which is Fig. 1.8 of Chapter 1. As can be seen, the DFT phase is visually unrevealing, while

(a)   (b)

(c)   (d)

FIGURE 5.6

Display of DFT of image "fingerprint" from Chapter 1 (a) DFT magnitude (logarithmically compressed and histogram stretched); (b) DFT phase; (c) centered DFT (logarithmically compressed and histogram stretched); (d) centered DFT (without logarithmic compression).

the DFT magnitude is most visually revealing when it is centered and logarithmically compressed.

5.5 UNDERSTANDING IMAGE FREQUENCIES AND THE DFT

It is sometimes easy to lose track of the meaning of the DFT and of the frequency content of an image in all of the (necessary!) mathematics. When using the DFT, it is important to remember that the DFT is a detailed map of the frequency content of the image, which can be visually digested as well as digitally processed. It is a useful exercise to examine the DFT of images, particularly the DFT magnitudes, since it reveals much about the distribution and meaning of image frequencies. It is also useful to consider what happens when the image frequencies are modified in certain simple ways, since this both reveals further insights into spatial frequencies, and it also moves toward understanding how image frequencies can be systematically modified to produce useful results.

In the following we will present and discuss a number of interesting digital images along with their DFT magnitudes represented as intensity images. When examining these, recall that bright regions in the DFT magnitude "image" correspond to frequencies that have large magnitudes in the real image. Also, in all cases, the DFT magnitudes have been logarithmically compressed and centered via (5.55) and (5.57), respectively, for improved visual interpretation.

Most engineers and scientists are introduced to Fourier-domain concepts in a 1D setting. One-dimensional signal frequencies have a single attribute, that of being either "high" or "low" frequency. Two-dimensional (and higher dimensional) signal frequencies have richer descriptions characterized by both magnitude and direction,[3] which lend themselves well to visualization. We will seek intuition into these attributes as we separately consider the *granularity* of image frequencies, corresponding to radial frequency (5.2), and the *orientation* of image frequencies, corresponding to frequency angle (5.3).

5.5.1 Frequency Granularity

The granularity of an image frequency refers to its radial frequency. "Granularity" describes the appearance of an image that is strongly characterized by the radial frequency portrait of the DFT. An abundance of large coefficients near the DFT origin corresponds to the existence of large, smooth, image components, often of smooth image surfaces or background. Note that nearly every image will have a significant peak at the DFT origin (unless it is very dark), since from (5.33) it is the summed intensity of the image (integrated optical density):

$$\tilde{F}(0,0) = \sum_{m=0}^{M-1} \sum_{n=0}^{N-1} f(m,n). \tag{5.58}$$

---

[3]Strictly speaking, 1D frequencies can be positive- or negative-going. This polarity may be regarded as a directional attribute, although without much meaning for real-valued 1D signals.

The image "fingerprint" (Fig. 1.8 of Chapter 1) with DFT magnitude shown in Fig. 5.6 (c) is an excellent example of image granularity. The image contains relatively little low frequency or very high frequency energy, but does contain an abundance of mid-frequency energy as can be seen in the symmetrically placed half arcs above and below the frequency origin. The "fingerprint" image is a good example of an image that is primarily bandpass.

Figure 5.7 depicts image "peppers" and its DFT magnitude. The image contains primarily smooth intensity surfaces separated by abrupt intensity changes. The smooth surfaces contribute to the heavy distribution of low-frequency DFT coefficients, while the intensity transitions ("edges") contribute a noticeable amount of mid-to-higher frequencies over a broad range of orientations.

Finally, in Fig. 5.8, "cane" depicts an image of a repetitive weave pattern that exhibits a number of repetitive peaks in the DFT magnitude image. These are *harmonics* that naturally appear in signals (such as music signals) or images that contain periodic or nearly-periodic structures.

As an experiment toward understanding frequency content, suppose that we define several zero-one image frequency masks, as depicted in Fig. 5.9.

Masking (multiplying) the DFT $\tilde{F}$ of an image f with each of these will produce, following an inverse DFT, a resulting image containing only low, mid, or high frequencies. In the following, we show examples of this operation. The astute reader may have observed that the zero-one frequency masks, which are defined in the DFT domain, may be regarded as DFTs with IDFTs defined in the space domain. Since we are taking the products of functions in the DFT domain, it has the interpretation of cyclic convolution (5.46)–(5.51) in the space domain. Therefore, the following examples should not be thought of as lowpass, bandpass, or highpass linear filtering operations in the proper sense. Instead, these are instructive examples where image frequencies are being directly removed. The approach is not a substitute for a proper linear filtering of the image using a space domain filter that has been DFT-transformed with proper zero-padding. In particular, the

FIGURE 5.7

Image "peppers" (left) and DFT magnitude (right).

 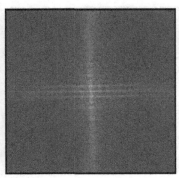

FIGURE 5.8

Image "cane" (left) and DFT magnitude (right).

Low-frequency mask   Mid-frequency mask   High-frequency mask

FIGURE 5.9

Image radial frequency masks. Black pixels take value '1,' white pixels take value '0.'

naïve demonstration here does dictate how the frequencies between the DFT frequencies (frequency samples) are effected, as a properly designed linear filter does.

In all of the examples, the image DFT was computed, multiplied by a zero-one frequency mask, and inverse DFT-ed. Finally, a full-scale histogram stretch was applied to map the result to the gray level range $(0, 255)$, since otherwise, the resulting image is not guaranteed to be positive.

In the first example, shown in Fig. 5.10, the image "fingerprint" is shown following treatment with the low-frequency mask and the mid-frequency mask. The low-frequency result looks much more blurred, and there is an apparent loss of information. However, the mid-frequency result seems to enhance and isolate much of the interesting ridge information about the fingerprint.

In the second example (Fig. 5.10), the image "peppers" was treated with the mid-frequency DFT mask and the high-frequency DFT mask. The mid-frequency image is visually quite interesting since it is apparent that the sharp intensity changes were

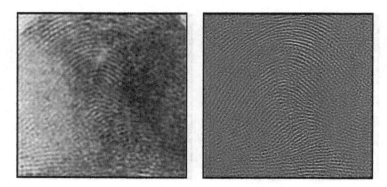

FIGURE 5.10

Image "fingerprint" processed with the (left) low-frequency DFT mask and the (right) mid-frequency DFT mask.

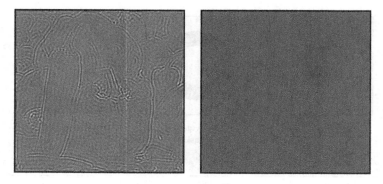

FIGURE 5.11

Image "peppers" processed with the (left) mid-frequency DFT mask and the (right) high-frequency DFT mask.

significantly enhanced. A similar effect was produced with the higher frequency mask, but with greater emphasis on sharp details.

5.5.2 Frequency Orientation

The orientation of an image frequency refers to its angle. The term "orientation" applied to an image or image component describes those aspects of the image that contribute to an appearance that is strongly characterized by the frequency orientation portrait of the DFT. If the DFT is brighter along a specific orientation, then the image contains highly oriented components along that direction.

The image "fingerprint" (with DFT magnitude in Fig. 5.6(c)) is also an excellent example of image orientation. The DFT contains significant mid-frequency energy between the approximate orientations $45° - 135°$ from the horizontal axis. This corresponds perfectly to the orientations of the ridge patterns in the fingerprint image.

Figure 5.12 shows the image "planks," which contains a strong directional component. This manifests as a very strong extended peak extending from lower left to upper right in the DFT magnitude. Figure 5.13 ("escher") exhibits several such extended peaks, corresponding to strongly oriented structures in the horizontal and slightly off-diagonal directions.

Again, an instructive experiment can be developed by defining zero-one image frequency masks, this time tuned to different orientation frequency bands instead of radial frequency bands. Several such oriented frequency masks are depicted in Fig. 5.14.

As a first example, the DFT of the image "planks" was modified by two orientation masks. In Fig. 5.15 (left), an orientation mask that allows the frequencies in the range $40°$ to $50°$ only (as well as the symmetrically placed frequencies $220°$ to $230°$) was applied. This was designed to capture the bright ridge of DFT coefficients easily seen in Fig. 5.12. As can be seen, the strong oriented information describing the cracks in the planks and some of the oriented grain is all that remains. Possibly, this information could be used by some automated process. Then, in Fig. 5.15 (right), the frequencies in the much larger ranges $50°$ to $220°$ (and $-130°$ to $40°$) were admitted. These are the complementary frequencies to the first range chosen, and they contain all the other information other than the strongly oriented component. As can be seen, this residual image contains little oriented structure.

As another example, the DFT of the image "escher" was also modified by two orientation masks. In Fig. 5.16 (left), an orientation mask that allows the frequencies in the range $-25°$ to $25°$ (and $155°$ to $205°$) only was applied. This captured the strong horizontal frequency ridge in the image, corresponding primarily to the strong vertical (building) structures. Then, in Fig. 5.16 (right), frequencies in the vertically-oriented ranges $45°$ to $135°$ (and $225°$ to $315°$) were admitted. This time completely different

FIGURE 5.12

Image "planks" (left) and DFT magnitude (right).

FIGURE 5.13

Image "escher" (left) and DFT magnitude (right).

FIGURE 5.14

Examples of image frequency orientation masks.

FIGURE 5.15

Image "planks" processed with oriented DFT masks that allow frequencies in the range (measured from the horizontal axis): (left) 40° to 50° (and 220° to 230°), and (right) 50° to 220° (and −130° to 40°).

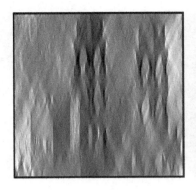

FIGURE 5.16

Image "escher" processed with oriented DFT masks that allow frequencies in the range (measured from the horizontal axis): (left) −25° to 25° (and 155° to 205°) and (right) 45° to 135° (and 225° to 315°).

structures were highlighted, including the diagonal waterways, the background steps, and the paddlewheel.

5.6 RELATED TOPICS IN THIS *GUIDE*

The Fourier transform is one of the most basic tools for image processing, or for that matter, the processing of any kind of signal. It appears throughout this *Guide* in various contexts, since linear filtering and enhancement (Chapters 10 and 11), restoration (Chapter 14), and reconstruction (Chapter 25) all depend on these concepts, as do concepts and applications of wavelet-based image processing (Chapters 6 and 11) which extend the ideas of Fourier techniques in very powerful ways. Extended frequency domain concepts are also heavily utilized in Chapters 16 and 17 (image compression) of the *Guide*, although the transforms used differ somewhat from the DFT.

FIGURE 5.16

RELATED TOPICS IN THIS BOOK

Multiscale Image Decompositions and Wavelets

6

Pierre Moulin

University of Illinois at Urbana-Champaign

6.1 OVERVIEW

The concept of scale, or resolution of an image, is very intuitive. A person observing a scene perceives the objects in that scene at a certain level of resolution that depends on the distance to these objects. For instance, walking toward a distant building, she would first perceive a rough outline of the building. The main entrance becomes visible only in relative proximity to the building. Finally, the door bell is visible only in the entrance area. As this example illustrates, the notions of resolution and scale loosely correspond to the size of the details that can be perceived by the observer. It is of course possible to formalize these intuitive concepts, and indeed signal processing theory gives them a more precise meaning.

These concepts are particularly useful in image and video processing and in computer vision. A variety of digital image processing algorithms decompose the image being analyzed into several components, each of which captures information present at a given scale. While our main purpose is to introduce the reader to the basic concepts of multiresolution image decompositions and wavelets, applications will also be briefly discussed throughout this chapter. The reader is referred to other chapters of this *Guide* for more details.

Throughout, we assume that the images to be analyzed are rectangular with $N \times M$ pixels. While there exists several types of multiscale image decompositions, we consider three main methods [1–6]:

1. In a **Gaussian pyramid** representation of an image (Fig. 6.1(a)), the original image appears at the bottom of a pyramidal stack of images. This image is then lowpass filtered and subsampled by a factor of two in each coordinate. The resulting $N/2 \times M/2$ image appears at the second level of the pyramid. This procedure can be iterated several times. Here resolution can be measured by the size of the

123

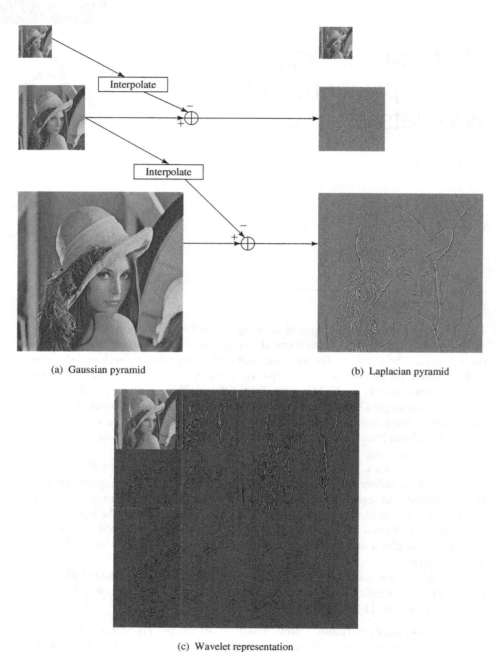

(a) Gaussian pyramid

(b) Laplacian pyramid

(c) Wavelet representation

FIGURE 6.1

Three multiscale image representations applied to *Lena*: (a) Gaussian pyramid; (b) Laplacian pyramid; (c) Wavelet representation.

image at any given level of the pyramid. The pyramid in Fig. 6.1(a) has three resolution levels, or scales. In the original application of this method to computer vision, the lowpass filter used was often a Gaussian filter,[1] hence the terminology *Gaussian pyramid*. We shall use this terminology even when a lowpass filter is not a Gaussian filter. Another possible terminology in that case is simply *lowpass pyramid*. Note that the total number of pixels in a pyramid representation is $NM + NM/4 + NM/16 + \cdots \approx \frac{4}{3}NM$. This is said to be an *overcomplete* representation of the original image, due to the increase in the number of pixels.

2. The **Laplacian pyramid** representation of the image is closely related to the Gaussian pyramid, but here the difference between approximations at two successive scales is computed and displayed for different scales, see Fig. 6.1(b). The precise meaning of the *interpolate* operation in the figure will be given in Section 6.2.1. The displayed images represent details of the image that are significant at each scale. An equivalent way to obtain the image at a given scale is to apply the difference between two Gaussian filters to the original image. This is analogous to filtering the image using a Laplacian filter, a technique commonly employed for edge detection (see Chapter 4). Laplacian filters are bandpass, hence the name Laplacian pyramid, also termed *bandpass pyramid*.

3. In a **wavelet decomposition**, the image is decomposed into a set of subimages (or subbands) which also represent details at different scales (Fig. 6.1(c)). Unlike pyramid representations, the subimages also represent details with different spatial orientations (such as edges with horizontal, vertical, and diagonal orientations). The number of pixels in a wavelet decomposition is only NM. As we shall soon see, the signal processing operations involved here are more sophisticated than those for pyramid image representations.

The pyramid and wavelet decompositions are presented in more detail in Sections 6.2 and 6.3, respectively. The basic concepts underlying these techniques are applicable to other multiscale decomposition methods, some of which are listed in Section 6.4.

Hierarchical image representations such as those in Fig. 6.1 are useful in many applications. In particular, they lend themselves to effective designs of reduced–complexity algorithms for texture analysis and segmentation, edge detection, image analysis, motion analysis, and image understanding in computer vision. Moreover, the Laplacian pyramid and wavelet image representations are *sparse* in the sense that most detail images contain few significant pixels (little significant detail). This sparsity property is very useful in image compression, as bits are allocated only to the few significant pixels; in image recognition, because the search for significant image features is facilitated; and in the restoration of images corrupted by noise, as images and noise possess rather distinct properties in the wavelet domain. The recent JPEG 2000 international standard for image compression is based on wavelets [7], unlike its predecessor JPEG which was based on the discrete cosine transform [8].

---

[1] This design was motivated by analogies to the Human Visual System, see Section 6.3.6.

6.2 PYRAMID REPRESENTATIONS

In this section, we shall explain how the Gaussian and Laplacian pyramid representations in Fig. 6.1 can be obtained from a few basic signal processing operations. To this end, we first describe these operations in Section 6.2.1 for the case of 1D signals. The extension to 2D signals is presented in Sections 6.2.2 and 6.2.3 for Gaussian and Laplacian pyramids, respectively.

6.2.1 Decimation and Interpolation

Consider the problem of decimating a 1D signal by a factor of two, namely, reducing the sample rate by a factor of two. This operation generally entails some loss of information, so it is desired that the decimated signal retain as much fidelity as possible to the original. The basic operations involved in decimation are lowpass filtering (using a digital antialiasing filter) and subsampling, as shown in Fig. 6.2. The impulse response of the lowpass filter is denoted by $h(n)$, and its discrete-time Fourier transform [9] by $H(e^{j\omega})$. The relationship between input $x(n)$ and output $y(n)$ of the filter is the convolution equation

$$y(n) = x(n) * h(n) = \sum_k h(k)x(n-k).$$

The downsampler discards every other sample of its input $y(n)$. Its output is given by

$$z(n) = y(2n).$$

Combining these two operations, we obtain

$$z(n) = \sum_k h(k)x(2n-k). \tag{6.1}$$

Downsampling usually implies a loss of information, as the original signal $x(n)$ cannot be exactly reconstructed from its decimated version $z(n)$. The traditional solution for reducing this information loss consists in using an "ideal" digital antialiasing filter $h(n)$ with cutoff frequency $\omega_c = \pi/2$ [9].[2] However, such "ideal" filters have infinite length.

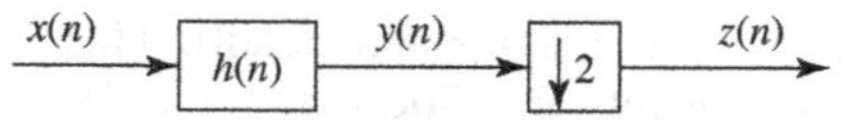

FIGURE 6.2

Decimation of a signal by a factor of two, obtained by cascade of a lowpass filter $h(n)$ and a subsampler ↓ 2.

---

[2]The paper [10] derives the filter that actually minimizes this information loss in the mean-square sense, under some assumptions on the input signal.

In image processing, short finite impulse response (FIR) filters are preferred for obvious computational reasons. Furthermore, approximations to the "ideal" filters above have an oscillating impulse response, which unfortunately results in visually annoying ringing artifacts in the vicinity of edges. The FIR filters typically used in image processing are symmetric, with length between 3 and 20 taps. Two common examples are the 3-tap FIR filter $h(n) = \left(\frac{1}{4}, \frac{1}{2}, \frac{1}{4}\right)$, and the length $-(2L+1)$ truncated Gaussian, $h(n) = Ce^{-n^2/(2\sigma^2)}, |n| \leq L$, where $C = 1/\sum_{|n| \leq L} e^{-n^2/(2\sigma^2)}$. The coefficients of both filters add up to one: $\sum_n h(n) = 1$, which implies that the DC response of these filters is unity.

Another common image processing operation is *interpolation*, which increases the sample rate of a signal. Signal processing theory tells us that interpolation may be performed by cascading two basic signal processing operations: upsampling and lowpass filtering, see Fig. 6.3. The upsampler inserts a zero between every other sample of the signal $x(n)$:

$$y(n) = \begin{cases} x(n/2) & : n \text{ even} \\ 0 & : n \text{ odd} \end{cases}$$

The upsampled signal is then filtered using a lowpass filter $h(n)$. The interpolated signal is given by $z(n) = h(n) * y(n)$ or, in terms of the original signal $x(n)$,

$$z(n) = \sum_k h(k)x(n - 2k). \tag{6.2}$$

The so-called ideal interpolation filters have infinite length. Again, in practice, short FIR filters are used.

6.2.2 Gaussian Pyramid

The construction of a Gaussian pyramid involves 2D lowpass filtering and subsampling operations. The 2D filters used in image processing practice are *separable*, which means that they can be implemented as the cascade of 1D filters operating along image rows and columns. This is a convenient choice in many respects, and the 2D decimation scheme is then separable as well. Specifically, 2D decimation is implemented by applying 1D decimation to each row of the image (using Eq. 6.1) followed by 1D decimation to each column of the resulting image (using Eq. 6.1 again). The same result would be obtained by first processing columns and then rows. Likewise, 2D interpolation is obtained by first applying Eq. 6.2 to each row of the image, and then again to each column of the resulting image, or vice versa.

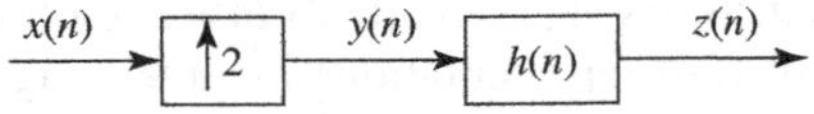

FIGURE 6.3

Interpolation of a signal by a factor of two, obtained by cascade of an upsampler ↑ 2 and a lowpass filter $h(n)$.

This technique was used at each stage of the Gaussian pyramid decomposition in Fig. 6.1(a). The lowpass filter used for both horizontal and vertical filtering was the 3-tap filter $h(n) = \left(\frac{1}{4}, \frac{1}{2}, \frac{1}{4}\right)$.

Gaussian pyramids have found applications to certain types of image storage problems. Suppose for instance that remote users access a common image database (say an Internet site) but have different requirements with respect to image resolution. The representation of image data in the form of an image pyramid would allow each user to directly retrieve the image data at the desired resolution. While this storage technique entails a certain amount of redundancy, the desired image data are available directly and are in a form that does not require further processing. Another application of Gaussian pyramids is in motion estimation for video [1, 2]: in a first step, coarse motion estimates are computed based on low-resolution image data, and in subsequent steps, these initial estimates are refined based on higher resolution image data. The advantages of this multiresolution, coarse-to-fine, approach to motion estimation are a significant reduction in algorithmic complexity (as the crucial steps are performed on reduced-size images) and the generally good quality of motion estimates, as the initial estimates are presumed to be relatively close to the ideal solution. Another closely related application that benefits from a multiscale approach is pattern matching [1].

6.2.3 Laplacian Pyramid

We define a *detail image* as the difference between an image and its approximation at the next coarser scale. The Gaussian pyramid generates images at multiple scales, but these images have different sizes. In order to compute the difference between a $N \times M$ image and its approximation at resolution $N/2 \times M/2$, one should interpolate the smaller image to the $N \times M$ resolution level before performing the subtraction. This operation was used to generate the Laplacian pyramid in Fig. 6.1(b). The interpolation filter used was the 3-tap filter $h(n) = \left(\frac{1}{2}, 1, \frac{1}{2}\right)$.

As illustrated in Fig. 6.1(b), the Laplacian representation is *sparse* in the sense that most pixel values are zero or near zero. The significant pixels in the detail images correspond to edges and textured areas such as *Lena*'s hair. Just like the Gaussian pyramid representation, the Laplacian representation is also *overcomplete*, as the number of pixels is greater (by a factor $\approx 33\%$) than in the original image representation.

Laplacian pyramid representations have found numerous applications in image processing, and in particular texture analysis and segmentation [1]. Indeed, different textures often present very different spectral characteristics which can be analyzed at appropriate levels of the Laplacian pyramid. For instance, a nearly uniform region such as the surface of a lake contributes mostly to the coarse-level image, while a textured region like grass often contributes significantly to other resolution levels. Some of the earlier applications of Laplacian representations include image compression [11, 12], but the emergence of wavelet compression techniques has made this approach somewhat less attractive. However, a Laplacian-type compression technique was adopted in the hierarchical mode of the lossy JPEG image compression standard [8], also see Chapter 5.

6.3 WAVELET REPRESENTATIONS

While the sparsity of the Laplacian representation is useful in many applications, over-completeness is a serious disadvantage in applications such as compression. The wavelet transform offers both the advantages of a sparse image representation and a complete representation. The development of this transform and its theory has had a profound impact on a variety of applications. In this section, we first describe the basic tools needed to construct the wavelet representation of an image. We begin with filter banks, which are elementary building blocks in the construction of wavelets. We then show how filter banks can be cascaded to compute a wavelet decomposition. We then introduce wavelet bases, a concept that provides additional insight into the choice of filter banks. We conclude with a discussion of the relation of wavelet representations to the human visual system and a brief overview of some applications.

6.3.1 Filter Banks

Figure 6.4(a) depicts an *analysis filter bank*, with one input $x(n)$ and two outputs $x_0(n)$ and $x_1(n)$. The input signal $x(n)$ is processed through two paths. In the upper path, $x(n)$ is passed through a lowpass filter $H_0(e^{j\omega})$ and decimated by a factor of two. In the lower path, $x(n)$ is passed through a highpass filter $H_1(e^{j\omega})$ and also decimated by a factor of two. For convenience, we make the following assumptions. First, the number N of available samples of $x(n)$ is even. Second, the filters perform a circular convolution (see Chapter 5), which is equivalent to assuming that $x(n)$ is a periodic signal. Under these assumptions, the output of each path is periodic with period equal to $N/2$ samples. Hence the analysis filter bank can be thought of as a *transform* that maps the original set $\{x(n)\}$ of N samples into a new set $\{x_0(n), x_1(n)\}$ of N samples.

Figure 6.4(b) shows a *synthesis filter bank*. Here there are two inputs $y_0(n)$ and $y_1(n)$, and one single output $y(n)$. The input signal $y_0(n)$ (respectively $y_1(n)$) is upsampled by a factor of two and filtered using a lowpass filter $G_0(e^{j\omega})$ (respectively highpass filter $G_1(e^{j\omega})$). The output $y(n)$ is obtained by summing the two filtered signals. We assume that the input signals $y_0(n)$ and $y_1(n)$ are periodic with period $N/2$. This implies that

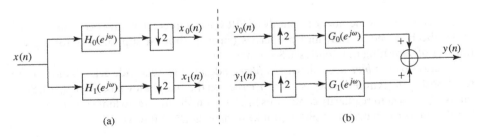

(a)   (b)

FIGURE 6.4

(a) Analysis filter bank, with lowpass filter $H_0(e^{j\omega})$ and highpass filter $H_1(e^{j\omega})$; (b) Synthesis filter bank, with lowpass filter $G_0(e^{j\omega})$ and highpass filter $G_1(e^{j\omega})$.

the output $y(n)$ is periodic with period equal to N. So the synthesis filter bank can also be thought of as a transform that maps the original set of N samples $\{y_0(n), y_1(n)\}$ into a new set of N samples $\{y(n)\}$.

What happens when the output $x_0(n), x_1(n)$ of an analysis filter bank is applied to the input of a synthesis filter bank? As it turns out, under some specific conditions on the four filters $H_0(e^{j\omega}), H_1(e^{j\omega}), G_0(e^{j\omega})$, and $G_1(e^{j\omega})$, the output $y(n)$ of the resulting *analysis/synthesis* system is *identical* (possibly up to a constant delay) to its input $x(n)$. This condition is known as *perfect reconstruction*. It holds, for instance, for the following trivial set of 1-tap filters: $h_0(n)$ and $g_1(n)$ are unit impulses, and $h_1(n)$ and $g_0(n)$ are unit delays. In this case, the reader can verify that $y(n) = x(n-1)$. In this simple example, all four filters are allpass. It is, however, not obvious to design more useful sets of FIR filters that also satisfy the perfect reconstruction condition. A general methodology for doing so was discovered in the mid-1980s. We refer the reader to [4, 5] for more details.

Under some additional conditions on the filters, the transforms associated with both the analysis and the synthesis filter banks are orthonormal. Orthonormality implies that the energy of the samples is preserved under the transformation. If these conditions are met, the filters possess the following remarkable properties: the synthesis filters are a time-reversed version of the analysis filters, and the highpass filters are modulated versions of the lowpass filters, namely $g_0(n) = (-1)^n h_1(n)$, $g_1(n) = (-1)^{n+1} h_0(n)$, and $h_1(n) = (-1)^{-n} h_0(K - n)$, where K is an integer delay. Such filters are often known as quadrature mirror filters (QMF), or conjugate quadrature filters (CQF), or power-complementary filters [5], because both lowpass (respectively highpass) filters have the same frequency response, and the frequency responses of the lowpass and highpass filters are related by the power-complementary property $|H_0(e^{j\omega})|^2 + |H_1(e^{j\omega})|^2 = 2$, valid at all frequencies. The filter $h_0(n)$ is viewed as a prototype filter, because it automatically determines the other three filters.

Finally, if the prototype lowpass filter $H_0(e^{j\omega})$ has a zero at frequency $\omega = \pi$, the filters are said to be *regular filters*, or *wavelet filters*. The meaning of this terminology will become apparent in Section 6.3.4. Figure 6.5 shows the frequency responses of the four filters generated from a famous 4-tap filter designed by Daubechies [4, p. 195]:

$$h_0(n) = \frac{1}{4\sqrt{2}}(1 + \sqrt{3}, 3 + \sqrt{3}, 3 - \sqrt{3}, 1 - \sqrt{3}).$$

This filter is the first member of a family of FIR wavelet filters that have been constructed by Daubechies and possess nice properties (such as shortest support size for a given number of vanishing moments, see Section 6.3.4).

There also exist *biorthogonal wavelet filters*, a design that sets aside degrees of freedom for choosing the synthesis lowpass filter $h_1(n)$ given the analysis lowpass filter $h_0(n)$. Such filters are subject to regularity conditions [4]. The transforms are no longer orthonormal, but the filters can have linear phase (unlike nontrivial QMF filters).

6.3.2 Wavelet Decomposition

An analysis filter bank decomposes 1D signals into lowpass and highpass components. One can perform a similar decomposition on images by first applying 1D filtering along

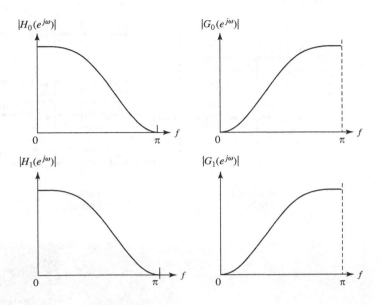

FIGURE 6.5

Magnitude frequency response of the four subband filters for a QMF filter bank generated from the prototype Daubechies' 4-tap lowpass filter.

rows of the image and then along columns, or vice versa [13]. This operation is illustrated in Fig. 6.6(a). The same filters $H_0(e^{j\omega})$ and $H_1(e^{j\omega})$ are used for horizontal and vertical filtering. The output of the analysis system is a set of four $N/2 \times M/2$ subimages: the so-called LL (low low), LH (low high), HL (high high), and HH (high high) *subbands*, which correspond to different spatial frequency bands in the image. The decomposition of *Lena* into four such subbands is shown in Fig. 6.6(b). Observe that the LL subband is a coarse (low resolution) version of the original image, and that the HL, LH, and HH subbands, respectively, contain details with vertical, horizontal, and diagonal orientations. The total number of pixels in the four subbands is equal to the original number of pixels, NM.

In order to perform the wavelet decomposition of an image, one recursively applies the scheme of Fig. 6.6(a) to the LL subband. Each stage of this recursion produces a coarser version of the image as well as three new detail images at that particular scale. Figure 6.7 shows the cascaded filter banks that implement this wavelet decomposition, and Fig. 6.1(c) shows a 3-stage wavelet decomposition of *Lena*. There are seven subbands, each corresponding to a different set of scales and orientations (different spatial frequency bands).

Both the Laplacian decomposition in Fig. 6.1(b) and the wavelet decomposition in Fig. 6.1(c) provide a coarse version of the image as well as details at different scales, but the wavelet representation is complete and provides information about image components at different spatial orientations.

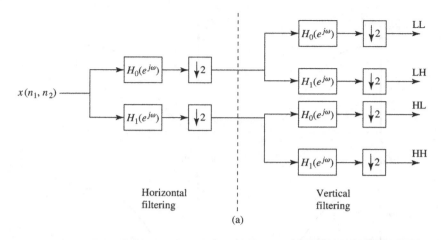

Horizontal filtering   Vertical filtering

(a)

(b)

FIGURE 6.6

Decomposition of $N \times M$ image into four $N/2 \times M/2$ subbands: (a) basic scheme; (b) application to *Lena*, using Daubechies' 4-tap wavelet filters.

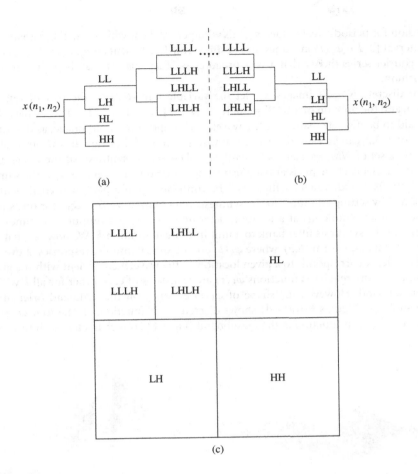

FIGURE 6.7

Implementation of wavelet image decomposition using cascaded filter banks: (a) wavelet decomposition of input image $x(n_1, n_2)$; (b) reconstruction of $x(n_1, n_2)$ from its wavelet coefficients; (c) nomenclature of subbands for a 3-level decomposition.

6.3.3 Discrete Wavelet Bases

So far we have described the mechanics of the wavelet decomposition in Fig. 6.7, but we have yet to explain what wavelets are and how they relate to the decomposition in Fig. 6.7. In order to do so, we first introduce discrete wavelet bases. Consider the following representation of a signal $x(t)$ defined over some (discrete or continuous) domain $\mathcal{T}$:

$$x(t) = \sum_k a_k \varphi_k(t), \quad t \in \mathcal{T}. \tag{6.3}$$

Here $\varphi_k(t)$ are termed *basis functions* and a_k are the coefficients of the signal $x(t)$ in the *basis* $\mathcal{B} = \{\varphi_k(t)\}$. A familiar example of such signal representations is the Fourier series expansion for periodic real-valued signals with period T, in which case the domain $\mathcal{T}$ is the interval $[0, T)$, $\varphi_k(t)$ are sines and cosines, and k represents frequency. It is known from Fourier series theory that a very broad class of signals $x(t)$ can be represented in this fashion.

For discrete $N \times M$ images, we let the variable t in (6.3) be the pair of integers (n_1, n_2), and the domain of x be $\mathcal{T} = \{0, 1, \ldots, N-1\} \times \{0, 1, \ldots, M-1\}$. The basis $\mathcal{B}$ is then said to be discrete. Note that the wavelet decomposition of an image, as described in Section 6.3.2, can be viewed as a linear transformation of the original NM pixel values $x(t)$ into a set of NM wavelet coefficients a_k. Likewise, the synthesis of the image $x(t)$ from its wavelet coefficients is also a linear transformation, and hence $x(t)$ is the sum of contributions of individual coefficients. The contribution of a particular coefficient a_k is obtained by setting all inputs to the synthesis filter bank to zero, except for one single sample with amplitude a_k, at a location determined by k. The output is a_k times the response of the synthesis filter bank to a unit impulse at location k. We now see that the signal $x(t)$ takes the form (6.3), where $\varphi_k(t)$ are the spatial impulse responses above.

The index k corresponds to a given location of the wavelet coefficient within a given subband. The discrete basis functions $\varphi_k(t)$ are translates of each other for all k within a given subband. However, the shape of $\varphi_k(t)$ depends on the scale and orientation of the subband. Figures 6.8(a)–(d) shows discrete basis functions in the four coarsest subbands. The basis function in the LL subband (Fig. 6.8(a)) is characterized by a strong

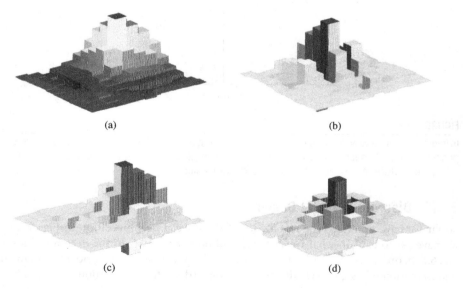

(a)

(b)

(c)

(d)

FIGURE 6.8

Discrete basis functions for image representation: (a) discrete scaling function from LLLL subband; (b)–(d) discrete wavelets from LHLL, LLLH, and LHLH subbands. These basis functions are generated from Daubechies' 4-tap filter.

central bump, while the basis functions in the other three subbands (detail images) have zero mean. Notice that the basis functions in the HL and LH subbands are related through a simple 90-degree rotation. The orientation of these basis functions make them suitable to represent patterns with the same orientation. For reasons that will become apparent in the next section, the basis functions in the low subband are called *discrete scaling functions*, while those in the other subbands are called *discrete wavelets*. The size of the support set of the basis functions is determined by the length of the wavelet filter, and essentially quadruples from one scale to the next.

6.3.4 Continuous Wavelet Bases

Basis functions corresponding to different subbands with the *same orientation* have a similar shape. This is illustrated in Fig. 6.9 which shows basis functions corresponding to two subbands with vertical orientation (Figs. 6.9(a)–(c)). The shape of the basis functions converges to a limit (Fig. 6.9(d)) as the scale becomes coarser. This phenomenon is due to the regularity of the wavelet filters used (Section 6.3.1). One of the remarkable results of Daubechies' wavelet theory [4] is that, under regularity conditions, the shape of the impulse responses corresponding to subbands with the same orientation does converge to a limit shape at coarse scales. Essentially the basis functions come in four shapes, which are displayed in Figs. 6.10(a)–(d). The limit shapes corresponding to the vertical, horizontal, and diagonal orientations are called *wavelets*. The limit shape corresponding to the coarse scale is called a *scaling function*. The three wavelets and the scaling function depend on

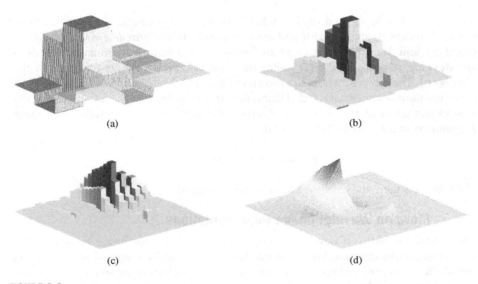

(a)   (b)

(c)   (d)

FIGURE 6.9

Discrete wavelets with vertical orientation at three consecutive scales: (a) in HL band; (b) in LHLL band; (c) in LLHLLL band; (d) Continuous wavelet is obtained as a limit of (normalized) discrete wavelets as scale becomes coarser.

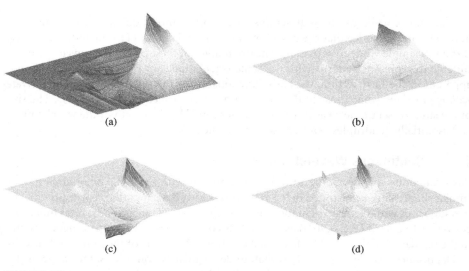

(a)

(b)

(c)

(d)

FIGURE 6.10

Basis functions for image representation: (a) scaling function; (b)–(d) wavelets with horizontal, vertical, and diagonal orientations. These four functions are tensor products of the 1D scaling function and wavelet in Fig. 6.11. The horizontal wavelet has been rotated by 180 degrees so that its negative part is visible on the display.

the wavelet filter $h_0(n)$ used (in Fig. 6.8, Daubechies' 4-tap filter). The four functions in Figs. 6.10(a)–(d) are separable and are respectively of the form $\phi(x)\phi(y)$, $\phi(x)\psi(y)$, $\psi(x)\phi(y)$, and $\psi(x)\psi(y)$. Here (x, y) are horizontal and vertical coordinates, and $\phi(x)$ and $\psi(x)$ are, respectively, the 1D scaling function and the 1D wavelet generated by the filter $h_0(n)$. These two functions are shown in Fig. 6.11, respectively. While the aspect of these functions is somewhat rough, Daubechies' theory shows that the smoothness of the wavelet increases with the number K of zeroes of $H_0(e^{j\omega})$ at $\omega = \pi$. In this case, the first K moments of the wavelet $\psi(x)$ are zero:

$$\int x^k \psi(x)\,dx = 0, \quad 0 \le k < K.$$

The wavelet is then said to possess K vanishing moments.

6.3.5 More on Wavelet Image Representations

The connection between wavelet decompositions and bases for image representation shows that images are sparse linear combinations of elementary images (discrete wavelets and scaling functions) and provides valuable insights for selecting the wavelet filter. Some wavelets are better able to compactly represent certain types of images than others. For instance, images with sharp edges would benefit from the use of short wavelet filters, due to the spatial localization of such edges. Conversely, images with mostly smooth areas would benefit from the use of longer wavelet filters with several vanishing moments, as

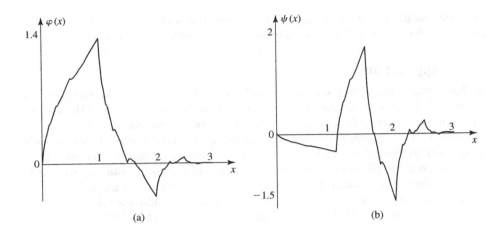

FIGURE 6.11

(a) 1D scaling function and (b) 1D wavelet generated from Daubechies' D4 filter.

such filters generate smooth wavelets. See [14] for a performance comparison of wavelet filters in image compression.

6.3.6 Relation to Human Visual System

Experimental studies of the human visual system (HVS) have shown that the eye's sensitivity to a visual stimulus strongly depends upon the spatial frequency contents of this stimulus. Similar observations have been made about other mammals. Simplified linear models have been developed in the psychophysics community to explain these experimental findings. For instance, the modulation transfer function describes the sensitivity of the HVS to spatial frequency. Additionally, several experimental studies have shown that images sensed by the eye are decomposed into bandpass channels as they move toward and through the visual cortex of the brain [15]. The bandpass components correspond to different scales and spatial orientations. Figure 6.5 in [16] shows the spatial impulse response and spatial frequency response corresponding to a channel at a particular scale and orientation. While the Laplacian representation provides a decomposition based on scale (rather than orientation), the wavelet transform has a limited ability to distinguish between patterns at different orientations, as each scale is comprised of three channels which are respectively associated with the horizontal, vertical, and diagonal orientations. This may not be not sufficient to capture the complexity of early stages of visual information processing, but the approximation is useful. Note there exist linear multiscale representations that more closely approximate the response of the HVS. One of them is the *Gabor transform*, for which the basis functions are Gaussian functions modulated by sine waves [17]. Another one is the *cortical transform* developed by Watson [18]. However, as discussed by Mallat [19], the goal of multiscale image processing and computer vision is not to design a transform that mimics the HVS. Rather, the analogy to the HVS motivates the use of multiscale image decompositions as a front end to complex image processing algorithms, as Nature already contains successful examples of such a design.

6.3.7 Applications

We have already mentioned several applications in which a wavelet decomposition is useful. This is particularly true of applications where the completeness of the wavelet representation is desirable. One such application is image and video compression, see Chapters 3 and 5. Another one is image denoising, as several powerful methods rely on the formulation of statistical models in an orthonormal transform domain [20]. There exist other applications in which wavelets present a plausible (but not necessarily superior) alternative to other multiscale decomposition techniques. Examples include texture analysis and segmentation [3, 21, 22], recognition of handwritten characters [23], inverse image halftoning [24], and biomedical image reconstruction [25].

6.4 OTHER MULTISCALE DECOMPOSITIONS

For completeness, we also mention two useful extensions of the methods covered in this chapter.

6.4.1 Undecimated Wavelet Transform

The wavelet transform is not invariant to shifts of the input image, in the sense that an image and its translate will in general produce different wavelet coefficients. This is a disadvantage in applications such as edge detection, pattern matching, and image recognition in general. The lack of translation invariance can be avoided if the outputs of the filter banks are not decimated. The undecimated wavelet transform then produces a set of bandpass images which have the same size as the original dataset ($N \times M$).

6.4.2 Wavelet Packets

Although the wavelet transform often provides a sparse representation of images, the spatial frequency characteristics of some images may not be best suited for a wavelet representation. Such is the case of fingerprint images, as ridge patterns constitute relatively narrowband bandpass components of the image. An even sparser representation of such images can be obtained by recursively splitting the appropriate subbands (instead of systematically splitting the low-frequency band as in a wavelet decomposition). This scheme is simply termed *subband decomposition*. This approach was already developed in signal processing during the 1970s [5]. In the early 1990s, Coifman and Wickerhauser developed an ingenious algorithm for finding the subband decomposition that gives the sparsest representation of the input signal (or image) in a certain sense [26]. The idea has been extended to find the best subband decomposition for compression of a given image [27].

6.4.3 Geometric Wavelets

One of the main strengths of 1D wavelets is their ability to represent abrupt transitions in a signal. This property does not extend straightforwardly to higher dimensions. In particular, the extension of wavelets to two dimensions, using tensor-product constructions, has two shortcomings: (1) limited ability to represent patterns at arbitrary orientations and (2) limited ability to represent image edges. For instance, the tensor-product construction is suitable for capturing the discontinuity across an edge, but is ineffective for exploiting the smoothness along the edge direction. To represent a simple, straight edge, one needs many wavelets.

To remedy this problem, several researchers have recently developed improved 2D multiresolution representations. The idea was pioneered by Candès and Donoho [28]. They introduced the *ridgelet transform*, which decomposes images as a superposition of ridgelets such as the one shown in Fig. 6.12. A ridgelet is parameterized by three parameters: resolution, angle, and location. Ridgelets are also known as geometric wavelets, a growing family which includes exotically named functions such as curvelets, bandelets, and contourlets. Signal processing algorithms for discrete images and applications to denoising and compression have been developed by Starck *et al.* [29], Do and Vetterli [30, 31], and Le Pennec and Mallat [32]. Remarkable results have been obtained by exploiting the sparse representation of object contours offered by geometric wavelets.

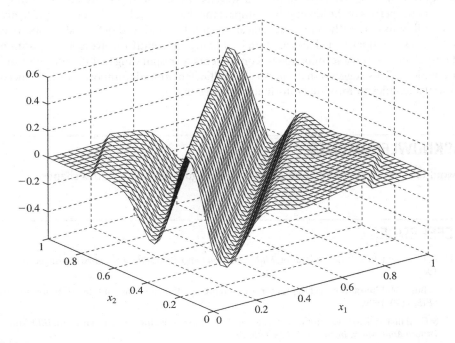

FIGURE 6.12

Ridgelet. (courtesy of M. Do).

6.5 CONCLUSION

We have introduced basic concepts of multiscale image decompositions and wavelets. We have focused on three main techniques: Gaussian pyramids, Laplacian pyramids, and wavelets. The Gaussian pyramid provides a representation of the same image at multiple scales, using simple lowpass filtering and decimation techniques. The Laplacian pyramid provides a coarse representation of the image as well as a set of detail images (bandpass components) at different scales. Both the Gaussian and the Laplacian representations are overcomplete, in the sense that the total number of pixels is approximately 33% higher than in the original image.

Wavelet decompositions are a more recent addition to the arsenal of multiscale signal processing techniques. Unlike the Gaussian and Laplacian pyramids, they provide a complete image representation and perform a decomposition according to both scale and orientation. They are implemented using cascaded filter banks in which the lowpass and highpass filters satisfy certain specific constraints. While classical signal processing concepts provide an operational understanding of such systems, there exist remarkable connections with work in applied mathematics (by Daubechies, Mallat, Meyer and others) and in psychophysics, which provide a deeper understanding of wavelet decompositions and their role in vision. From a mathematical standpoint, wavelet decompositions are equivalent to signal expansions in a wavelet basis. The regularity and vanishing-moment properties of the lowpass filter impact the shape of the basis functions and hence their ability to efficiently represent typical images. From a psychophysical perspective, early stages of human visual information processing apparently involve a decomposition of retinal images into a set of bandpass components corresponding to different scales and orientations. This suggests that multiscale/multiorientation decompositions are indeed natural and efficient for visual information processing.

ACKNOWLEDGMENTS

I would like to thank Juan Liu for generating the figures and plots in this chapter.

REFERENCES

[1] A. Rosenfeld. In A. Rosenfeld, editor, *Multiresolution Image Processing and Analysis,* Springer-Verlag, 1984.

[2] P. Burt. Multiresolution techniques for image representation, analysis, and 'smart' transmission. *SPIE,* 1199, 1989.

[3] S. G. Mallat. A theory for multiresolution signal decomposition: the wavelet transform. *IEEE Trans. Pattern Anal. Mach. Intell.,* 11(7):674–693, 1989.

[4] I. Daubechies. *Ten Lectures on Wavelets, CBMS-NSF Regional Conference Series in Applied Mathematics,* Vol. 61. SIAM, Philadelphia, PA, 1992.

[5] M. Vetterli and J. Kovačević. *Wavelets and Subband Coding*. Prentice-Hall, Englewood Cliffs, NJ, 1995.

[6] S. G. Mallat. *A Wavelet Tour of Signal Processing*. Academic Press, San Diego, CA, 1998.

[7] D. S. Taubman and M. W. Marcellin. *JPEG 2000: Image Compression Fundamentals, Standards and Practice*. Kluwer, Norwell, MA, 2001.

[8] W. B. Pennebaker and J. L. Mitchell. *JPEG: Still Image Data Compression Standard*. Van Nostrand Reinhold, 1993.

[9] J. Proakis and Manolakis. *Digital Signal Processing: Principles, Algorithms, and Applications*, 3rd ed. Prentice-Hall, 1996.

[10] M. K. Tsatsanis and G. B. Giannakis. Principal component filter banks for optimal multiresolution analysis. *IEEE Trans. Signal Process.*, 43(8):1766–1777, 1995.

[11] P. Burt and A. H. Adelson. The Laplacian pyramid as a compact image code. *IEEE Trans. Commun.*, 31:532–540, 1983.

[12] M. Vetterli and K. M. Uz. Multiresolution coding techniques for digital video: a review. *Multidimensional Syst. Signal Process.*, Special Issue on Multidimensional Proc. of Video Signals, 3:161–187, 1992.

[13] M. Vetterli. Multi-dimensional sub-band coding: some theory and algorithms. *Signal Processing*, 6(2):97–112, 1984.

[14] J. D. Villasenor, B. Belzer, and J. Liao. Wavelet filter comparison for image compression. *IEEE Trans. Image Process.*, 4(8):1053–1060, 1995.

[15] F. W. Campbell and J. G. Robson. Application of Fourier analysis to cortical cells. *J. Physiol.*, 197:551–566, 1968.

[16] M. Webster and R. De Valois. Relationship between spatial-frequency and orientation tuning of striate-cortex cells. *J. Opt. Soc. Am. A*, 2(7):1124–1132, 1985.

[17] J. G. Daugmann. Two-dimensional spectral analysis of cortical receptive field profile. *Vision Res.*, 20:847–856, 1980.

[18] A. B. Watson. The cortex transform: rapid computation of simulated neural images. *Comput. Vis. Graph. Image Process.*, 39:311–327, 1987.

[19] S. G. Mallat. Multifrequency channel decompositions of images and wavelet models. *IEEE Trans. Acoust.*, 37(12):2091–2110, 1989.

[20] P. Moulin and J. Liu. Analysis of multiresolution image denoising schemes using generalized-Gaussian and complexity priors. In *IEEE Trans. Inf. Theory*, Special Issue on Multiscale Analysis, 1999.

[21] M. Unser. Texture classification and segmentation using wavelet frames. *IEEE Trans. Image Process.*, 4(11):1549–1560, 1995.

[22] R. Porter and N. Canagarajah. A robust automatic clustering scheme for image segmentation using wavelets. *IEEE Trans. Image Process.*, 5(4):662–665, 1996.

[23] Y. Qi and B. R. Hunt. A multiresolution approach to computer verification of handwritten signatures. *IEEE Trans. Image Process.*, 4(6):870–874, 1995.

[24] J. Luo, R. de Queiroz, and Z. Fan. A robust technique for image descreening based on the wavelet transform. *IEEE Trans. Signal Process.*, 46(4):1179–1184, 1998.

[25] A. H. Delaney and Y. Bresler. Multiresolution tomographic reconstruction using wavelets. *IEEE Trans. Image Process.*, 4(6):799–813, 1995.

[26] R. R. Coifman and M. V. Wickerhauser. Entropy-based algorithms for best basis selection. *IEEE Trans. Inf. Theory*, Special Issue on Wavelet Tranforms and Multiresolution Signal Analysis, 38(2):713–718, 1992.

[27] K. Ramchandran and M. Vetterli. Best wavelet packet bases in a rate-distortion sense. *IEEE Trans. Image Process.*, 2:160–175, 1993.

[28] E. J. Candès. Ridgelets: theory and applications. *Ph.D. Thesis*, Department of Statistics, Stanford University, 1998.

[29] J.-L. Starck, E. J. Candès, and D. L. Donoho. The curvelet transform for image denoising. *IEEE Trans. Image Process.*, 11(6):670–684, 2002.

[30] M. N. Do and M. Vetterli. The finite ridgelet transform for image representation. *IEEE Trans. Image Process.*, 12(1):16–28, 2003.

[31] M. N. Do and M. Vetterli. Contourlets. In G. V. Welland, editor, *Beyond Wavelets*, Academic Press, New York, 2003.

[32] E. Le Pennec and S. G. Mallat. Sparse geometrical image approximation with bandelets. In *IEEE Trans. Image Process.*, 2005.

Image Noise Models

Charles Boncelet

University of Delaware

7

7.1 SUMMARY

This chapter reviews some of the more commonly used *image noise models*. Some of these are naturally occurring, e.g., Gaussian noise, some sensor induced, e.g., photon counting noise and speckle, and some result from various processing, e.g., quantization and transmission.

7.2 PRELIMINARIES

7.2.1 What is Noise?

Just what is noise, anyway? Somewhat imprecisely we will define *noise* as an unwanted component of the image. Noise occurs in images for many reasons. Gaussian noise is a part of almost any signal. For example, the familiar white noise on a weak television station is well modeled as Gaussian. Since image sensors must count photons—especially in low-light situations—and the number of photons counted is a random quantity, images often have photon counting noise. The grain noise in photographic films is sometimes modeled as Gaussian and sometimes as Poisson. Many images are corrupted by salt and pepper noise, as if someone had sprinkled black and white dots on the image. Other noises include quantization noise and speckle in coherent light situations.

Let $f(\cdot)$ denote an image. We will decompose the image into a desired component, $g(\cdot)$, and a noise component, $q(\cdot)$. The most common decomposition is *additive*:

$$f(\cdot) = g(\cdot) + q(\cdot). \tag{7.1}$$

For instance, Gaussian noise is usually considered to be an additive component.

The second most common decomposition is *multiplicative*:

$$f(\cdot) = g(\cdot)q(\cdot). \tag{7.2}$$

An example of a noise often modeled as multiplicative is speckle.

143

Note, the multiplicative model can be transformed into the additive model by taking logarithms and the additive model into the multiplicative one by exponentiation. For instance, (7.1) becomes

$$e^{\mathbf{f}} = e^{g+\mathbf{q}} = e^g e^{\mathbf{q}}. \tag{7.3}$$

Similarly, (7.2) becomes

$$\log \mathbf{f} = \log(g\mathbf{q}) = \log g + \log \mathbf{q}. \tag{7.4}$$

If the two models can be transformed into one another, what is the point? Why do we bother? The answer is that we are looking for *simple* models that properly describe the behavior of the system. The additive model, (7.1), is most appropriate when the noise *in that model* is independent of f. There are many applications of the additive model. Thermal noise, photographic noise, and quantization noise, for instance, obey the additive model well.

The multiplicative model is most appropriate when the noise in that model is independent of f. One common situation where the multiplicative model is used is for speckle in coherent imagery.

Finally, there are important situations when neither the additive nor the multiplicative model fits the noise well. Poisson counting noise and salt and pepper noise fit neither model well.

The questions about noise models one might ask include: What are the properties of $\mathbf{q}(\cdot)$? Is $\mathbf{q}$ related to g or are they independent? Can $\mathbf{q}(\cdot)$ be eliminated or at least, mitigated? As we will see in this chapter and in others, it is only occasionally true that $\mathbf{q}(\cdot)$ will be independent of $g(\cdot)$. Furthermore, it is usually impossible to remove all the effects of the noise.

Figure 7.1 is a picture of the San Francisco, CA, skyline. It will be used throughout this chapter to illustrate the effects of various noises. The image is 432×512, 8 bits per pixel, grayscale. The largest value (the whitest pixel) is 220 and the minimum value is 32. This image is relatively noise free with sharp edges and clear details.

7.2.2 Notions of Probability

The various noises considered in this chapter are random in nature. Their exact values are random variables whose values are best described using probabilistic notions. In this section, we will review some of the basic ideas of probability. A fuller treatment can be found in many texts on probability and randomness, including Feller [1], Billingsley [2], and Woodroofe [3].

Let $\mathbf{a} \in R^n$ be a n-dimensional *random vector* and $a \in R^n$ be a point. Then the *distribution function* of $\mathbf{a}$ (also known as the cumulative distribution function) will be denoted as $P_{\mathbf{a}}(a) = \Pr[\mathbf{a} \le a]$ and the corresponding *density function*, $p_{\mathbf{a}}(a) = dP_{\mathbf{a}}(a)/da$. Probabilities of events will be denoted as $\Pr[A]$.

The *expected value* of a function, $\psi(\mathbf{a})$ is

$$E[\psi(\mathbf{a})] = \int_{-\infty}^{\infty} \psi(a) p_{\mathbf{a}}(a)\, da. \tag{7.5}$$

FIGURE 7.1

Original picture of San Francisco skyline.

Note that for discrete distributions the integral is replaced by the corresponding sum:

$$E[\psi(\mathbf{a})] = \sum_k \psi(a_k)\Pr[\mathbf{a} = a_k]. \tag{7.6}$$

The *mean* is $\mu_{\mathbf{a}} = E[\mathbf{a}]$ (i.e., $\psi(\mathbf{a}) = \mathbf{a}$), the *variance* of a single random variable is $\sigma_{\mathbf{a}}^2 = E\big[(\mathbf{a} - \mu_{\mathbf{a}})^2\big]$, and the *covariance matrix* of a random vector is $\Sigma_{\mathbf{a}} = E\big[(\mathbf{a} - \mu_{\mathbf{a}})(\mathbf{a} - \mu_{\mathbf{a}})^T\big]$.

Related to the covariance matrix is the *correlation matrix*,

$$\mathbf{R_a} = E\big[\mathbf{a}\mathbf{a}^T\big]. \tag{7.7}$$

The various moments are related by the well-known relation, $\Sigma = \mathbf{R} - \mu\mu^T$.

The *characteristic function*, $\Phi_{\mathbf{a}}(u) = E\big[\exp(ju\mathbf{a})\big]$, has two main uses in analyzing probabilistic systems: calculating moments and calculating the properties of sums of independent random variables. For calculating moments, consider the power series of $\exp(ju\mathbf{a})$:

$$e^{ju\mathbf{a}} = 1 + ju\mathbf{a} + \frac{(ju\mathbf{a})^2}{2!} + \frac{(ju\mathbf{a})^3}{3!} + \cdots \tag{7.8}$$

After taking expected values,

$$E\big[e^{ju\mathbf{a}}\big] = 1 + juE[\mathbf{a}] + \frac{(ju)^2 E[\mathbf{a}^2]}{2!} + \frac{(ju)^3 E[\mathbf{a}^3]}{3!} + \cdots, \tag{7.9}$$

One can isolate the kth moment by taking k derivatives with respect to u and then setting $u = 0$:

$$E\left[\mathbf{a}^k\right] = \frac{1}{j^k} \frac{d^k E\left[e^{jua}\right]}{d^k u}\Bigg|_{u=0}. \tag{7.10}$$

Consider two independent random variables, $\mathbf{a}$ and $\mathbf{b}$, and their sum $\mathbf{c}$. Then,

$$\Phi_{\mathbf{c}}(u) = E\left[e^{ju(\mathbf{c})}\right] \tag{7.11}$$

$$= E\left[e^{ju(\mathbf{a}+\mathbf{b})}\right] \tag{7.12}$$

$$= E\left[e^{ju\mathbf{a}} e^{ju\mathbf{b}}\right] \tag{7.13}$$

$$= E\left[e^{ju\mathbf{a}}\right] E\left[e^{ju\mathbf{b}}\right] \tag{7.14}$$

$$= \Phi_{\mathbf{a}}(u)\Phi_{\mathbf{b}}(u), \tag{7.15}$$

where (7.14) used the independence of $\mathbf{a}$ and $\mathbf{b}$. Since the characteristic function is the (complex conjugate of the) Fourier transform of the density, the density of $\mathbf{c}$ is easily calculated by taking an inverse Fourier transform of $\Phi_{\mathbf{c}}(u)$.

7.3 ELEMENTS OF ESTIMATION THEORY

As we said in the introduction, noise is generally an *unwanted* component in an image. In this section, we review some of the techniques to eliminate—or at least minimize—the noise.

The basic estimation problem is to find a good estimate of the noise-free image, g, given the noisy image, $\mathbf{f}$. Some authors refer to this as an *estimation problem*, while others say it is a *filtering* problem. Let the estimate be denoted $\hat{\mathbf{g}} = \hat{\mathbf{g}}(\mathbf{f})$. The most common performance criterion is the mean squared error (MSE):

$$\text{MSE}(g,\hat{\mathbf{g}}) = E\left[(g - \hat{\mathbf{g}})^2\right]. \tag{7.16}$$

The estimator that minimizes the MSE is called the *minimum mean squared error estimator* (MMSE). Many authors prefer to measure the performance in a positive way using the *peak signal-to-noise ratio* (PSNR) measured in dB:

$$\text{PSNR} = 10\log_{10}\left(\frac{\text{MAX}^2}{\text{MSE}}\right), \tag{7.17}$$

where MAX is the maximum pixel value, e.g., 255 for 8 bit images.

While the MSE is the most common error criterion, it is by no means the only one. Many researchers argue that MSE results are not well correlated with the human visual system. For instance, the mean absolute error (MAE) is often used in motion compensation in video compression. Nevertheless, MSE has the advantages of easy tractability and intuitive appeal since MSE can be interpreted as "noise power." Estimators can be classified in many different ways. The primary division we will consider here is linear versus nonlinear estimators.

The *linear estimators* form estimates by taking linear combinations of the sample values. For example, consider a small region of an image modeled as a constant value plus additive noise:

$$\mathbf{f}(x,y) = \mu + \mathbf{q}(x,y). \tag{7.18}$$

A linear estimate of μ is

$$\hat{\mu} = \sum_{x,y} \alpha(x,y)\mathbf{f}(x,y) \tag{7.19}$$

$$= \mu \sum_{x,y} \alpha(x,y) + \sum_{x,y} \alpha(x,y)\mathbf{q}(x,y). \tag{7.20}$$

An estimator is called *unbiased* if $E[\mu - \hat{\mu}] = 0$. In this case, assuming $E[\mathbf{q}] = 0$, unbiasedness requires $\sum_{x,y} \alpha(x,y) = 1$. If the $\mathbf{q}(x,y)$ are independent and identically distributed (i.i.d.), meaning that the random variables are independent and each has the same distribution function, then the MMSE for this example is the sample mean:

$$\hat{\mu} = \frac{1}{M} \sum_{(x,y)} \mathbf{f}(x,y), \tag{7.21}$$

where M is the number of samples averaged over.

Linear estimators in image filtering get more complicated primarily for two reasons: Firstly, the noise may not be i.i.d., and secondly and more commonly, the noise-free image is not well modeled as a constant. If the noise-free image is Gaussian and the noise is Gaussian, then the optimal estimator is the well-known Weiner filter [4].

In many image filtering applications, linear filters do not perform well. Images are not well modeled as Gaussian, and linear filters are not optimal. In particular, images have small details and sharp edges. These are blurred by linear filters. It is often true that the filtered image is more objectionable than the original. The blurriness is worse than the noise.

Largely because of the blurring problems of linear filters, nonlinear filters have been widely studied in image filtering. While there are many classes of nonlinear filters, we will concentrate on the class based on order statistics. Many of these filters were invented to solve image processing problems.

Order statistics are the result of sorting the observations from smallest to largest. Consider an image *window* (a small piece of an image) centered on the pixel to be estimated. Some windows are square, some are "x" shaped, some are "+" shaped, and some more oddly shaped. The choice of a window size and shape is usually up to the practitioner. Let the samples in the window be denoted simply as f_i for $i = 1, \ldots, N$. The order statistics are denoted $f_{(i)}$ for $i = 1, \ldots, N$ and obey the ordering $f_{(1)} \leq f_{(2)} \leq \cdots \leq f_{(N)}$.

The simplest order statistic-based estimator is the sample median, $f_{((N+1)/2)}$. For example, if $N = 9$, the median is $f_{(5)}$. The median has some interesting properties. Its value is one of the samples. The median tends to blur images much less than the mean. The median can pass an edge without any blurring at all.

Some other order statistic estimators are the following:

Linear Combinations of Order Statistics $\hat{\mu} = \sum_{i=1}^{N} \alpha_i f_{(i)}$. The α_i determine the behavior of the filter. In some cases, the coefficients can be determined optimally, see Lloyd [5] and Bovik *et al.* [6].

Weighted Medians and the LUM Filter Another way to weight the samples is to repeat certain samples more than once before the data is sorted. The most common situation is to repeat the center sample more than once. The center weighted median does "less filtering" than the ordinary median and is suitable when the noise is not too severe. (See Salt and Pepper Noise below.)

The LUM filter [7] is a rearrangement of the center weighted median. It has the advantages of being easy to understand and extensible to image sharpening applications.

Iterated and Recursive Forms The various filtering operations can be combined or iterated upon. One might first filter horizontally, then vertically. One might compute the outputs of three or more filters and then use "majority rule" techniques to choose between them.

To analyze or optimally design order statistics filters, we need descriptions of the probability distributions of the order statistics. Initially, we will assume the f_i are i.i.d. Then the $\Pr[f_{(i)} \leq x]$ equals the probability that at least i of the f_i are less than or equal to x. Thus,

$$\Pr[f_{(i)} \leq x] = \sum_{k=i}^{N} \binom{N}{k} (P_f(x))^k (1 - P_f(x))^{N-k}. \tag{7.22}$$

We see immediately that the order statistic probabilities are related to the binomial distribution.

Unfortunately (7.22) does not hold when the observations are not i.i.d. In the special case when the observations are independent (or Markov), but not identically distributed, there are simple recursive formulas to calculate the probabilities [8, 9]. For example, even if the additive noise in (7.1) is i.i.d, the image may not be constant throughout the window. One may be interested in how much blurring of an edge is done by a particular order statistics filter.

7.4 TYPES OF NOISE AND WHERE THEY MIGHT OCCUR

In this section, we present some of the more common image noise models and show sample images illustrating the various degradations.

7.4.1 Gaussian Noise

Probably the most frequently occurring noise is additive Gaussian noise. It is widely used to model thermal noise and, under some often reasonable conditions, is the limiting behavior of other noises, e.g., photon counting noise and film grain noise. Gaussian noise is used in many places in this *Guide*.

The density function of univariate Gaussian noise, $\mathbf{q}$, with mean μ and variance σ^2 is

$$p_{\mathbf{q}}(x) = (2\pi\sigma^2)^{-1/2} e^{-(x-\mu)^2/2\sigma^2} \tag{7.23}$$

for $-\infty < x < \infty$. Notice that the support, which is the range of values of x where the probability density is nonzero, is infinite in both the positive and negative directions. But, if we regard an image as an intensity map, then the values must be nonnegative. In other words, the noise cannot be strictly Gaussian. If it were, there would be some nonzero probability of having negative values. In practice, however, the range of values of the Gaussian noise is limited to about $\pm 3\sigma$, and the Gaussian density is a useful and accurate model for many processes. If necessary, the noise values can be truncated to keep $f > 0$.

In situations where $\mathbf{a}$ is a random vector, the multivariate Gaussian density becomes

$$p_{\mathbf{a}}(a) = (2\pi)^{-n/2} |\Sigma|^{-1/2} e^{-(a-\mu)^T \Sigma^{-1}(a-\mu)/2}, \tag{7.24}$$

where $\mu = E[\mathbf{a}]$ is the mean vector and $\Sigma = E\left[(\mathbf{a}-\mu)(\mathbf{a}-\mu)^T\right]$ is the covariance matrix. We will use the notation $\mathbf{a} \sim N(\mu, \Sigma)$ to denote that $\mathbf{a}$ is Gaussian (also known as *Normal*) with mean μ and covariance Σ.

The Gaussian characteristic function is also Gaussian in shape:

$$\Phi_{\mathbf{a}}(u) = e^{u^T \mu - u^T \Sigma u/2}. \tag{7.25}$$

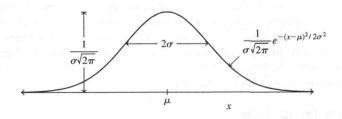

FIGURE 7.2

The Gaussian density.

The Gaussian distribution has many convenient mathematical properties—and some not so convenient ones. Certainly the least convenient property of the Gaussian distribution is that the cumulative distribution function cannot be expressed in closed form using elementary functions. However, it is tabulated numerically. See almost any text on probability, e.g., [10].

Linear operations on Gaussian random variables yield Gaussian random variables. Let $\mathbf{a}$ be $N(\mu, \Sigma)$ and $\mathbf{b} = G\mathbf{a} + h$. Then a straightforward calculation of $\Phi_{\mathbf{b}}(u)$ yields

$$\Phi_{\mathbf{b}}(u) = e^{ju^T(G\mu+h) - u^T G\Sigma G^T u/2}, \tag{7.26}$$

which is the characteristic function of a Gaussian random variable with mean, $G\mu + h$, and covariance, $G\Sigma G^T$.

Perhaps the most significant property of the Gaussian distribution is called the Central Limit Theorem, which states that the distribution of a sum of a large number of independent, small random variables has a Gaussian distribution. Note the individual random variables do not need to have a Gaussian distribution themselves, nor do they even need to have the same distribution. For a detailed development, see, e.g., Feller [1] or Billingsley [2]. A few comments are in order:

- There must be a large number of random variables that contribute to the sum. For instance, thermal noise is the result of the thermal vibrations of an astronomically large number of tiny electrons.

- The individual random variables in the sum must be independent, or nearly so.

- Each term in the sum must be small compared to the sum.

As one example, thermal noise results from the vibrations of a very large number of electrons, the vibration of any one electron is independent of that of another, and no one electron contributes significantly more than the others. Thus, all three conditions are satisfied and the noise is well modeled as Gaussian. Similarly, binomial probabilities approach the Gaussian. A binomial random variable is the sum of N independent Bernoulli (0 or 1) random variables. As N gets large, the distribution of the sum approaches a Gaussian distribution.

In Fig. 7.3 we see the effect of a small amount of Gaussian noise ($\sigma = 10$). Notice the "fuzziness" overall. It is often counterproductive to try to use signal processing techniques to remove this level of noise—the filtered image is usually visually less pleasing than the original noisy one (although sometimes the image is filtered to reduce the noise, then sharpened to eliminate the blurriness introduced by the noise reducing filter).

In Fig. 7.4, the noise has been increased by a factor of 3 ($\sigma = 30$). The degradation is much more objectionable. Various filtering techniques can improve the quality, though usually at the expense of some loss of sharpness.

7.4.2 Heavy Tailed Noise

In many situations, the conditions of the Central Limit Theorem are almost, but not quite, true. There may not be a large enough number of terms in the sum, or the terms

FIGURE 7.3

San Francisco corrupted by additive Gaussian noise with standard deviation equal to 10.

FIGURE 7.4

San Francisco corrupted by additive Gaussian noise with standard deviation equal to 30.

may not be sufficiently independent, or a small number of the terms may contribute a disproportionate amount to the sum. In these cases, the noise may only be approximately Gaussian. One should be careful. Even when the center of the density is approximately Gaussian, the tails may not be.

The *tails* of a distribution are the areas of the density corresponding to large x, i.e., as $|x| \to \infty$. A particularly interesting case is when the noise has *heavy tails*. "Heavy tails" means that for large values of x, the density, $p_a(x)$, approaches 0 more slowly than the Gaussian. For example, for large values of x, the Gaussian density goes to 0 as $\exp(-x^2/2\sigma^2)$; the Laplacian density (also known as the double exponential density) goes to 0 as $\exp(-\lambda|x|)$. The Laplacian density is said to have heavy tails.

In Table 7.1, we present the tail probabilities, $\Pr[|\mathbf{x}| > x_0]$, for the "standard" Gaussian and Laplacian ($\mu = 0, \sigma = 1$, and $\lambda = 1$). Note the probability of exceeding 1 is approximately the same for both distributions, while the probability of exceeding 3 is about 20 times greater for the double exponential than for the Gaussian.

An interesting example of heavy tailed noise that should be familiar is static on a weak, broadcast AM radio station during a lightning storm. Most of the time, the

TABLE 7.1   Comparison of tail probabilities for the Gaussian and Laplacian distributions. Specifically, the values of $\Pr[|\mathbf{x}| > x_0]$ are listed for both distributions (with $\sigma = 1$ and $\lambda = 1$)

x_0	Gaussian	Laplacian
1	0.32	0.37
2	0.046	0.14
3	0.0027	0.05

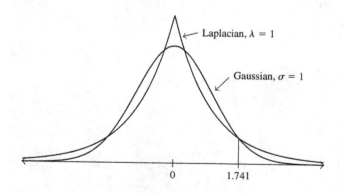

FIGURE 7.5

Comparison of the Laplacian ($\lambda = 1$) and Gaussian ($\sigma = 1$) densities, both with $\mu = 0$. Note, for deviations larger than 1.741, the Laplacian density is larger than the Gaussian.

conditions of the central limit theorem are well satisfied and the noise is Gaussian. Occasionally, however, there may be a lightning bolt. The lightning bolt overwhelms the tiny electrons and dominates the sum. During the time period of the lightning bolt, the noise is non-Gaussian and has much heavier tails than the Gaussian.

Some of the heavy tailed models that arise in image processing include the following:

7.4.2.1 *Laplacian or Double Exponential*

$$p_a(x) = \frac{\lambda}{2} e^{-\lambda |x - \mu|} \tag{7.27}$$

The mean is μ and the variance is $2/\lambda^2$. The Laplacian is interesting in that the best estimate of μ is the median, not the mean, of the observations. Not truly "noise," the prediction error in many image compression algorithms is modeled as Laplacian. More simply, the difference between successive pixels is modeled as Laplacian.

7.4.2.2 *Negative Exponential*

$$p_a(x) = \lambda e^{-\lambda x} \tag{7.28}$$

for $x > 0$. The mean is $1/\lambda > 0$ and variance, $1/\lambda^2$. The negative exponential is used to model speckle, for example, in SAR systems.

7.4.2.3 *Alpha-Stable*

In this class, appropriately normalized sums of independent and identically distributed random variables have the same distribution as the individual random variables. We have already seen that sums of Gaussian random variables are Gaussian, so the Gaussian is in the class of alpha-stable distributions. In general, these distributions have characteristic functions that look like $\exp(-|u|^\alpha)$ for $0 < \alpha \leq 2$. Unfortunately, except for the Gaussian ($\alpha = 2$) and the Cauchy ($\alpha = 1$), it is not possible to write the density functions of these distributions in closed form.

As $\alpha \rightarrow 0$, these distributions have very heavy tails.

7.4.2.4 *Gaussian Mixture Models*

$$p_a(x) = (1 - \alpha)p_0(x) + \alpha p_1(x), \tag{7.29}$$

where $p_0(x)$ and $p_1(x)$ are Gaussian densities with differing means, μ_0 and μ_1, or variances, σ_0^2 and σ_1^2. In modeling heavy tailed distributions, it is often true that α is small, say $\alpha = 0.05$, $\mu_0 = \mu_1$, and $\sigma_1^2 \gg \sigma_0^2$.

In the "static in the AM radio" example above, at any given time, α would be the probability of a lightning strike, σ_0^2 the average variance of the thermal noise, and σ_1^2 the variance of the lightning induced signal.

Sometimes this model is generalized further and $p_1(x)$ is allowed to be non-Gaussian (and sometimes completely arbitrary). See Huber [11].

7.4.2.5 *Generalized Gaussian*

$$p_{\mathbf{a}}(x) = A e^{-\beta|x-\mu|^{\alpha}}, \tag{7.30}$$

where μ is the mean and A, β, and α are constants. α determines the shape of the density: $\alpha = 2$ corresponds to the Gaussian and $\alpha = 1$ to the double exponential. Intermediate values of α correspond to densities that have tails in between the Gaussian and double exponential. Values of $\alpha < 1$ give even heavier tailed distributions.

The constants, A and β, can be related to α and the standard deviation, σ, as follows:

$$\beta = \frac{1}{\sigma}\left(\frac{\Gamma(3/\alpha)}{\Gamma(1/\alpha)}\right)^{0.5} \tag{7.31}$$

$$A = \frac{\beta\alpha}{2\Gamma(1/\alpha)}. \tag{7.32}$$

The generalized Gaussian has the advantage of being able to fit a large variety of (symmetric) noises by appropriate choice of the three parameters, μ, σ, and α [12].

One should be careful to use estimators that behave well in heavy tailed noise. The sample mean, optimal for a constant signal in additive Gaussian noise, can perform quite poorly in heavy tailed noise. Better choices are those estimators designed to be robust against the occasional outlier [11]. For instance, the median is only slightly worse than the mean in Gaussian noise, but can be much better in heavy tailed noise.

7.4.3 Salt and Pepper Noise

Salt and pepper noise refers to a wide variety of processes that result in the same basic image degradation: only a few pixels are noisy, but they are *very* noisy. The effect is similar to sprinkling white and black dots—salt and pepper—on the image.

One example where salt and pepper noise arises is in transmitting images over noisy digital links. Let each pixel be quantized to B bits in the usual fashion. The value of the pixel can be written as $X = \sum_{i=0}^{B-1} b_i 2^i$. Assume the channel is a binary symmetric one with a crossover probability of ϵ. Then each bit is flipped with probability ϵ. Call the received value, Y. Then, assuming the bit flips are independent,

$$\Pr\left[|X - Y| = 2^i\right] = \epsilon(1-\epsilon)^{B-1} \tag{7.33}$$

for $i = 0, 1, \ldots, B-1$. The MSE due to the most significant bit is $\epsilon 4^{B-1}$ compared to $\epsilon(4^{B-1} - 1)/3$ for all the other bits combined. In other words, the contribution to the MSE from the most significant bit is approximately three times that of all the other bits. The pixels whose most significant bits are changed will likely appear as black or white dots.

Salt and pepper noise is an example of (very) heavy tailed noise. A simple model is the following: Let $f(x,y)$ be the original image and $\mathbf{q}(x,y)$ be the image after it has been

FIGURE 7.6

San Francisco corrupted by salt and pepper noise with a probability of occurrence of 0.05.

altered by salt and pepper noise.

$$\Pr[\mathbf{q} = f] = 1 - \alpha \tag{7.34}$$

$$\Pr[\mathbf{q} = \text{MAX}] = \alpha/2 \tag{7.35}$$

$$\Pr[\mathbf{q} = \text{MIN}] = \alpha/2, \tag{7.36}$$

where MAX and MIN are the maximum and minimum image values, respectively. For 8 bit images, MIN = 0 and MAX = 255. The idea is that with probability $1 - \alpha$ the pixels are unaltered; with probability α the pixels are changed to the largest or smallest values. The altered pixels look like black and white dots sprinkled over the image.

Figure 7.6 shows the effect of salt and pepper noise. Approximately 5% of the pixels have been set to black or white (95% are unchanged). Notice the sprinkling of the black and white dots. Salt and pepper noise is easily removed with various order statistic filters, especially the center weighted median and the LUM filter [13].

7.4.4 Quantization and Uniform Noise

Quantization noise results when a continuous random variable is converted to a discrete one or when a discrete random variable is converted to one with fewer levels. In images, quantization noise often occurs in the acquisition process. The image may be continuous initially, but to be processed it must be converted to a digital representation.

As we shall see, quantization noise is usually modeled as uniform. Various researchers use uniform noise to model other impairments, e.g., dither signals. Uniform noise is the opposite of the heavy tailed noise discussed above. Its tails are very light (zero!).

Let $\mathbf{b} = Q(\mathbf{a}) = \mathbf{a} + \mathbf{q}$, where $-\Delta/2 \leq \mathbf{q} \leq \Delta/2$ is the quantization noise and $\mathbf{b}$ is a discrete random variable usually represented with β bits. In the case where the number of quantization levels is large (so Δ is small), $\mathbf{q}$ is usually modeled as being uniform between $-\Delta/2$ and $\Delta/2$ and independent of $\mathbf{a}$. The mean and variance of $\mathbf{q}$ are

$$E[\mathbf{q}] = \frac{1}{\Delta} \int_{-\Delta/2}^{\Delta/2} s \, ds = 0 \qquad (7.37)$$

and

$$E\left[(\mathbf{q} - E[\mathbf{q}])^2\right] = \frac{1}{\Delta} \int_{-\Delta/2}^{\Delta/2} s^2 \, ds = \Delta^2/12. \qquad (7.38)$$

Since $\Delta \sim 2^{-\beta}$, $\sigma_\nu^2 \sim 2^{2\beta}$, the signal-to-noise ratio increases by 6 dB for each additional bit in the quantizer.

When the number of quantization levels is small, the quantization noise becomes signal dependent. In an image of the noise, signal features can be discerned. Also, the noise is correlated on a pixel by pixel basis and not uniformly distributed.

The general appearance of an image with too few quantization levels may be described as "scalloped." Fine graduations in intensities are lost. There are large areas of constant color separated by clear boundaries. The effect is similar to transforming a smooth ramp into a set of discrete steps.

In Fig. 7.7, the San Francisco image has been quantized to only 4 bits. Note the clear "stair-stepping" in the sky. The previously smooth gradations have been replaced by large constant regions separated by noticeable discontinuities.

7.4.5 Photon Counting Noise

Fundamentally, most image acquisition devices are photon counters. Let $\mathbf{a}$ denote the number of photons counted at some location (a pixel) in an image. Then, the distribution of $\mathbf{a}$ is usually modeled as Poisson with parameter λ. This noise is also called *Poisson noise* or Poisson counting noise.

$$P(\mathbf{a} = k) = \frac{e^{-\lambda} \lambda^k}{k!} \qquad (7.39)$$

for $k = 0, 1, 2, \ldots$

The Poisson distribution is one for which calculating moments by using the characteristic function is much easier than by the usual sum.

FIGURE 7.7

San Francisco quantized to 4 bits.

$$\Phi(u) = \sum_{k=0}^{\infty} \frac{e^{juk} e^{-\lambda} \lambda^k}{k!} \tag{7.40}$$

$$= e^{-\lambda} \sum_{k=0}^{\infty} \frac{(\lambda e^{ju})^k}{k!} \tag{7.41}$$

$$= e^{-\lambda} e^{\lambda e^{ju}} \tag{7.42}$$

$$= e^{\lambda(e^{ju}-1)}. \tag{7.43}$$

While this characteristic function does not *look* simple, it does yield the moments:

$$E[\mathbf{a}] = \frac{1}{j} \frac{d}{du} e^{\lambda(e^{ju}-1)} \bigg|_{u=0} \tag{7.44}$$

$$= \frac{1}{j} \lambda j e^{ju} e^{\lambda(e^{ju}-1)} \bigg|_{u=0} \tag{7.45}$$

$$= \lambda. \tag{7.46}$$

Similarly, $E[\mathbf{a}^2] = \lambda + \lambda^2$ and $\sigma^2 = (\lambda + \lambda^2) - \lambda^2 = \lambda$. We see one of the most interesting properties of the Poisson distribution, that the variance is equal to the expected value.

When λ is large, the central limit theorem can be invoked and the Poisson distribution is well approximated by the Gaussian with mean and variance both equal to λ.

Consider two different regions of an image, one brighter than the other. The brighter one has a higher λ and therefore a higher noise variance.

As another example of Poisson counting noise, consider the following:

Example: Effect of Shutter Speed on Image Quality   Consider two pictures of the same scene, one taken with a shutter speed of 1 unit time and the other with $\Delta > 1$ unit of time. Assume that an area of an image emits photons at the rate λ per unit time. The first camera measures a random number of photons, whose expected value is λ and whose variance is also λ. The second, however, has an expected value and variance equal to $\lambda\Delta$. When time averaged (divided by Δ), the second now has an expected value of λ and a variance of $\lambda/\Delta < \lambda$. Thus, we are led to the intuitive conclusion: all other things being equal, slower shutter speeds yield better pictures.

For example, astro-photographers traditionally used long exposures to average over a long enough time to get good photographs of faint celestial objects. Today's astronomers use CCD arrays and average many short exposure photographs, but the principal is the same.

Figure 7.8 shows the image with Poisson noise. It was constructed by taking each pixel value in the original image and generating a Poisson random variable with λ equal to that value. Careful examination reveals that the white areas are noisier than the dark areas. Also, compare this image with Fig. 7.3 which shows Gaussian noise of almost the same power.

FIGURE 7.8

San Francisco corrupted by Poisson noise.

7.4.6 **Photographic Grain Noise**

Photographic grain noise is a characteristic of photographic films. It limits the effective magnification one can obtain from a photograph. A simple model of the photography process is as follows:

A photographic film is made up from millions of tiny *grains*. When light strikes the film, some of the grains absorb the photons and some do not. The ones that do change their appearance by becoming metallic silver. In the developing process, the unchanged grains are washed away.

We will make two simplifying assumptions: (1) the grains are uniform in size and character and (2) the probability that a grain changes is proportional to the number of photons incident upon it. Both assumptions can be relaxed, but the basic answer is the same. In addition, we will assume the grains are independent of each other.

Slow film has a large number of small fine grains, while fast film has a smaller number of larger grains. The small grains give slow film a better, less grainy picture; the large grains in fast film cause a grainier picture.

In a given area, A, assume there are L grains, with the probability of each grain changing, p, proportionate to the number of incident photons. Then the number of grains that change, $\mathbf{N}$, is binomial

$$\Pr[\mathbf{N} = k] = \binom{L}{k} p^k (1 - p)^{L-k}. \tag{7.47}$$

Since L is large, when p small but $\lambda = Lp = E[\mathbf{N}]$ moderate, this probability is well approximated by a Poisson distribution

$$\Pr[\mathbf{N} = k] = \frac{e^{-\lambda} \lambda^k}{k!} \tag{7.48}$$

and by a Gaussian when p is larger:

$$\Pr[k \leq \mathbf{N} < k + \Delta_k]$$

$$= \Pr\left[\frac{k - Lp}{\sqrt{Lp(1-p)}} \leq \frac{\mathbf{N} - Lp}{\sqrt{Lp(1-p)}} \leq \frac{k + \Delta_k - Lp}{\sqrt{Lp(1-p)}} \right] \tag{7.49}$$

$$\approx e^{-0.5\left(\frac{k-Lp}{Lp(1-p)} \right)^2} \Delta_k \tag{7.50}$$

The probability interval on the right-hand side of (7.49) is exactly the same as that on the left except that it has been normalized by subtracting the mean and dividing by the standard deviation. (7.50) results from (7.49) by applying the central limit theorem. In other words, the distribution of grains that change is approximately Gaussian with mean Lp and variance $Lp(1 - p)$. This variance is maximized when $p = 0.5$. Sometimes, however, it is sufficiently accurate to ignore this variation and model grain noise as additive Gaussian with a constant noise power.

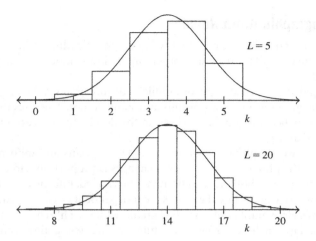

FIGURE 7.9

Illustration of the Gaussian approximation to the binomial. In both figures, $p = 0.7$ and the Gaussians have the same means and variances as the binomials. Even for L as small as 5, the Gaussian reasonably approximates the binomial PMF. For $L = 20$, the approximation is very good.

7.5 CCD IMAGING

In the past 20 years or so, CCD (charge-coupled devices) imaging has replaced photographic film as the dominant imaging form. First CCDs appeared in scientific applications, such as astronomical imaging and microscopy. Recently, CCD digital cameras and videos have become widely used consumer items. In this section, we analyze the various noise sources affecting CCD imagery.

CCD arrays work on the photoelectric principle (first discovered by Hertz and explained by Einstein, for which he was awarded the Nobel prize). Incident photons are absorbed, causing electrons to be elevated into a high energy state. These electrons are captured in a well. After some time, the electrons are counted by a "read out" device.

The number of electrons counted, N, can be written as

$$N = N_I + N_{th} + N_{ro}, \tag{7.51}$$

where N_I is the number of electrons due to the image, N_{th} the number due to thermal noise, and N_{ro} the number due to read out effects.

N_I is Poisson, with the expected value $E[N_I] = \lambda$ proportional to the incident image intensity. The variance of N_I is also λ, thus the standard deviation is $\sqrt{\lambda}$. The signal-to-noise ratio (neglecting the other noises) is $\lambda/\sqrt{\lambda} = \sqrt{\lambda}$. The only way to increase the signal-to-noise ratio is to increase the number of electrons recorded. Sometimes the image intensity can be increased (e.g., a photographer's flash), the aperture increased (e.g., a large telescope), or the exposure time increased. However, CCD arrays saturate: only a finite number of electrons can be captured. The effect of long exposures is achieved by averaging many short exposure images.

Even without incident photons, some electrons obtain enough energy to get captured. This is due to thermal effects and is called thermal noise or *dark current*. The amount of thermal noise is proportional to the temperature, T, and the exposure time. N_{th} is modeled as Gaussian.

The read out process introduces its own uncertainties and can inject electrons into the count. Read out noise is a function of the read out process and is independent of the image and the exposure time. Like image noise, N_{ro} is modeled as Poisson noise.

There are two different regimes in which CCD imaging is used: low light and high light levels. In low light, the number of image electrons is small. In this regime, thermal noise and read out noise are both significant and can dominate the process. For instance, much scientific and astronomical imaging is in low light. Two important steps are taken to reduce the effects of thermal and read out noise. The first is obvious: since thermal noise increases with temperature, the CCD is cooled as much as practicable. Often liquid nitrogen is used to lower the temperature.

The second is to estimate the means of the two noises and subtract them from measured image. Since the two noises arise from different effects, the means are measured separately. The mean of the thermal noise is measured by averaging several images taken with the shutter closed, but with the same shutter speed and temperature. The mean of the read out noise is estimated by taking the median of several (e.g., 9) images taken with the shutter closed and a zero exposure time (so that any signal measured is due to read out effects).

In high light levels, the image noise dominates and thermal and read out noises can be ignored. This is the regime in which consumer imaging devices are normally used. For large values of N_I, the Poisson distribution is well modeled as Gaussian. Thus the overall noise looks Gaussian, but the signal-to-noise ratio is higher in bright regions than in dark regions.

7.6 SPECKLE

In this section, we discuss two kinds of *speckle*, a curious distortion in images created by coherent light or by atmospheric effects. Technically not noise in the same sense as other noise sources considered so far, speckle is noise-like in many of its characteristics.

7.6.1 Speckle in Coherent Light Imaging

Speckle is one of the more complex image noise models. It is signal dependent, non-Gaussian, and spatially dependent. Much of this discussion is taken from [14, 15]. We will first discuss the origins of speckle, then derive the first-order density of speckle, and conclude this section with a discussion of the second-order properties of speckle.

In coherent light imaging, an object is illuminated by a coherent source, usually a laser or a radar transmitter. For the remainder of this discussion, we will consider the illuminant to be a light source, e.g., a laser, but the principles apply to radar imaging as well.

When coherent light strikes a surface, it is reflected back. Due to the microscopic variations in the surface roughness within one pixel, the received signal is subjected to random variations in phase and amplitude. Some of these variations in phase add constructively, resulting in strong intensities, and others add deconstructively, resulting in low intensities. This variation is called *speckle*.

Of crucial importance in the understanding of speckle is the point spread function of the optical system. There are three regimes:

- The point spread function is so narrow that the individual variations in surface roughness can be resolved. The reflections off the surface are random (if, indeed, we can model the surface roughness as random in this regime), but we cannot appeal to the central limit theorem to argue that the reflected signal amplitudes are Gaussian. Since this case is uncommon in most applications, we will ignore it.

- The point spread function is broad compared to the feature size of the surface roughness, but small compared to the features of interest in the image. This is a common case and leads to the conclusion, presented below, that the noise is exponentially distributed and uncorrelated on the scale of the features in the image. Also, in this situation, the noise is often modeled as multiplicative.

- The point spread function is broad compared to both the feature size of the object and the feature size of the surface roughness. Here, the speckle is correlated and its size distribution is interesting and is determined by the point spread function.

The development will proceed in two parts. Firstly, we will derive the first-order probability density of speckle and, secondly, we will discuss the correlation properties of speckle.

In any given macroscopic area, there are many microscopic variations in the surface roughness. Rather than trying to characterize the surface, we will content ourselves with finding a statistical description of the speckle.

We will make the (standard) assumptions that the surface is very rough on the scale of the optical wavelengths. This roughness means that each microscopic reflector in the surface is at a random height (distance from the observer) and a random orientation with respect to the incoming polarization field. These random reflectors introduce random changes in the reflected signal's amplitude, phase, and polarization. Further, we assume these variations at any given point are independent from each other and independent from the changes at any other point.

These assumptions amount to assuming that the system cannot resolve the variations in roughness. This is generally true in optical systems, but may not be so in some radar applications.

The above assumptions on the physics of the situation can be translated to statistical equivalents: the amplitude of the reflected signal at any point, (x,y), is multiplied by a random amplitude, denoted $\mathbf{a}(x,y)$, and the polarization, $\phi(x,y)$, is uniformly distributed between 0 and 2π.

Let $\mathbf{u}(x,y)$ be the complex phasor of the incident wave at a point (x,y), $\mathbf{v}(x,y)$ be the reflected signal, and $\mathbf{w}(x,y)$ be the received phasor. From the above assumptions,

$$\mathbf{v}(x,y) = \mathbf{u}(x,y)\mathbf{a}(x,y)e^{j\phi(x,y)} \tag{7.52}$$

and, letting $h(\cdot,\cdot)$ denote the 2D point spread function of the optical system,

$$\mathbf{w}(x,y) = h(x,y) * \mathbf{v}(x,y). \tag{7.53}$$

One can convert the phasors to rectangular coordinates:

$$\mathbf{v}(x,y) = \mathbf{v}_R(x,y) + j\mathbf{v}_I(x,y) \tag{7.54}$$

and

$$\mathbf{w}(x,y) = \mathbf{w}_R(x,y) + j\mathbf{w}_I(x,y). \tag{7.55}$$

Since the change in polarization is uniform between 0 and 2π, $\mathbf{v}_R(x,y)$ and $\mathbf{v}_I(x,y)$ are statistically independent. Similarly, $\mathbf{w}_R(x,y)$ and $\mathbf{w}_I(x,y)$ are statistically independent. Thus,

$$\mathbf{w}_R(x,y) = \int_{-\infty}^{\infty} \int_{-\infty}^{\infty} h(\alpha,\beta)v_R(x-\alpha, y-\beta)\, d\alpha\, d\beta \tag{7.56}$$

and similarly for $\mathbf{w}_I(x,y)$.

The integral in (7.56) is basically a sum over many tiny increments in x and y. By assumption, the increments are independent of one another. Thus, we can appeal to the central limit theorem and conclude that the distributions of $\mathbf{w}_R(x,y)$ and $\mathbf{w}_I(x,y)$ are each Gaussian with mean 0 and variance σ^2. Note, this conclusion does not depend on the details of the roughness, as long as the surface is rough on the scale of the wavelength of the incident light and the optical system cannot resolve the individual components of the surface.

The measured intensity, $\mathbf{f}(x,y)$, is the squared magnitude of the received phasors:

$$\mathbf{f}(x,y) = \mathbf{w}_R(x,y)^2 + \mathbf{w}_I(x,y)^2. \tag{7.57}$$

The distribution of $\mathbf{f}$ can be found by integrating the joint density of $\mathbf{w}_R$ and $\mathbf{w}_I$ over a circle of radius $f^{0.5}$:

$$\Pr[\mathbf{f}(x,y) \le f] = \int_0^{2\pi} \int_0^{f^{0.5}} \frac{1}{2\pi\sigma^2} e^{-\rho/2\sigma^2} \rho\, d\rho\, d\phi \tag{7.58}$$

$$= 1 - e^{-f/2\sigma^2}. \tag{7.59}$$

The corresponding density is $p_{\mathbf{f}}(f)$:

$$p_{\mathbf{f}}(f) = \begin{cases} \frac{1}{g}e^{-f/g} & f \geq 0 \\ 0 & f < 0, \end{cases} \qquad (7.60)$$

where we have taken the liberty to introduce the mean intensity, $g = g(x,y) = 2\sigma^2(x,y)$. A little rearrangement can put this into a multiplicative noise model:

$$\mathbf{f}(x,y) = g(x,y)\mathbf{q}, \qquad (7.61)$$

where $\mathbf{q}$ has a exponential density

$$p_{\mathbf{q}}(x) = \begin{cases} e^{-x} & x \geq 0 \\ 0 & x < 0. \end{cases} \qquad (7.62)$$

The mean of $\mathbf{q}$ is 1 and the variance is 1.

The exponential density is much heavier tailed than the Gaussian density, meaning that much greater excursions from the mean occur. In particular, the standard deviation of $\mathbf{f}$ equals $E[\mathbf{f}]$, i.e., the typical deviation in the reflected intensity is equal to the typical intensity. It is this large variation that causes speckle to be so objectionable to human observers.

It is sometimes possible to obtain multiple images of the same scene with independent realizations of the speckle pattern, i.e., the speckle in any one image is independent of the speckle in the others. For instance, there may be multiple lasers illuminating the same object from different angles or with different optical frequencies. One means of speckle reduction is to average these images:

$$\hat{\mathbf{f}}(x,y) = \frac{1}{M}\sum_{i=1}^{M}\mathbf{f}_i(x,y) \qquad (7.63)$$

$$= g(x,y)\frac{\sum_{i=1}^{M}\mathbf{q}_i(x,y)}{M}. \qquad (7.64)$$

Now, the average of the negative exponentials has mean 1 (the same as each individual negative exponential) and variance $1/M$. Thus, the average of the speckle images has a mean equal to $g(x,y)$ and variance $g^2(x,y)/M$.

Figure 7.10 shows an uncorrelated speckle image of San Francisco. Notice how severely degraded this image is. Careful examination will show that the light areas are noisier than the dark areas. This image was created by generating an "image" of exponential variates and multiplying each by the corresponding pixel value. Intensity values beyond 255 were truncated to 255.

The correlation structure of speckle is largely determined by the width of the point spread function. As above the real and imaginary components (or, equivalently, the X and Y components) of the reflected wave are independent Gaussian. These components ($\mathbf{w_R}$ and $\mathbf{w_I}$ above) are individually filtered by the point spread function of the imaging

FIGURE 7.10

San Francisco with uncorrelated speckle.

system. The intensity image is formed by taking the complex magnitude of the resulting filtered components.

Figure 7.11 shows a correlated speckle image of San Francisco. The image was created by filtering $\mathbf{w_R}$ and $\mathbf{w_I}$ with a 2D square filter of size 5×5. This size filter is too big for the fine details in the original image, but is convenient to illustrate the correlated speckle. As above, intensity values beyond 255 were truncated to 255. Notice the correlated structure to the "speckles." The image has a pebbly appearance.

We will conclude this discussion with a quote from Goodman [16]:

> *The general conclusions to be drawn from these arguments are that, in any speckle pattern, large-scale-size fluctuations are the most populous, and no scale sizes are present beyond a certain small-size cutoff. The distribution of scale sizes in between these limits depends on the autocorrelation function of the object geometry, or on the autocorrelation function of the pupil function of the imaging system in the imaging geometry.*

7.6.2 Atmospheric Speckle

The twinkling of stars is similar in cause to speckle in coherent light, but has important differences. Averaging multiple frames of independent coherent imaging speckle results in an image estimate whose mean equals the underlying image and whose variance is reduced by the number of frames averaged over. However, averaging multiple images of twinkling stars results in a blurry image of the star.

FIGURE 7.11

San Francisco with correlated speckle.

From the earth, stars (except the Sun!) are point sources. Their light is spatially coherent and planar when it reaches the atmosphere. Due to thermal and other variations, the diffusive properties of the atmosphere changes in an irregular way. This causes the index of refraction to change randomly. The star appears to twinkle. If one averages multiple images of the star, one obtains a blurry image.

Until recently, the preferred way to eliminate atmospheric-induced speckle (the "twinkling") was to move the observer to a location outside the atmosphere, i.e., in space. In recent years, new techniques to estimate and track the fluctuations in atmospheric conditions have allowed astronomers to take excellent pictures from the earth. One class is called "speckle interferometry" [17]. It uses multiple short duration (typically less than 1 second each) images and a nearby star to estimate the random speckle pattern. Once estimated, the speckle pattern can be removed, leaving the unblurred image.

7.7 CONCLUSIONS

In this chapter, we have tried to summarize the various image noise models and give some recommendations for minimizing the noise effects. Any such summary is, by necessity, limited. We do, of course, apologize to any authors whose work we may have omitted.

For further information, the interested reader is urged to consult the references for this and other chapters.

REFERENCES

[1] W. Feller. *An Introduction to Probability Theory and its Applications*. J. Wiley & Sons, New York, 1968.

[2] P. Billingsley. *Probability and Measure*. J. Wiley & Sons, New York, 1979.

[3] M. Woodroofe. *Probability with Applications*. McGraw-Hill, New York, 1975.

[4] C. Helstrom. *Probability and Stochastic Processes for Engineers*. Macmillan, New York, 1991.

[5] E. H. Lloyd. Least-squares estimations of location and scale parameters using order statistics. *Biometrika*, 39:88–95, 1952.

[6] A. C. Bovik, T. S. Huang, and D. C. Munson, Jr. A generalization of median filtering using linear combinations of order statistics. *IEEE Trans. Acoust.*, ASSP-31(6):1342–1350, 1983.

[7] R. C. Hardie and C. G. Boncelet, Jr. LUM filters: a class of order statistic based filters for smoothing and sharpening. *IEEE Trans. Signal Process.*, 41(3):1061–1076, 1993.

[8] C. G. Boncelet, Jr. Algorithms to compute order statistic distributions. *SIAM J. Sci. Stat. Comput.*, 8(5):868–876, 1987.

[9] C. G. Boncelet, Jr. Order statistic distributions with multiple windows. *IEEE Trans. Inf. Theory*, IT-37(2):436–442, 1991.

[10] P. Peebles. *Probability, Random Variables, and Random Signal Principles*. McGraw Hill, New York, 1993.

[11] P. J. Huber. *Robust Statistics*. J. Wiley & Sons, New York, 1981.

[12] J. H. Miller and J. B. Thomas. Detectors for discrete-time signals in non-Gaussian noise. *IEEE Trans. Inf. Theory*, IT-18(2):241–250, 1972.

[13] J. Astola and P. Kuosmanen. *Fundamentals of Nonlinear Digital Filtering*. CRC Press, Boca Raton, FL, 1997.

[14] D. Kuan, A. Sawchuk, T. Strand, and P. Chavel. Adaptive restoration of images with speckle. *IEEE Trans. Acoust.*, ASSP-35(3):373–383, 1987.

[15] J. Goodman. *Statistical Optics*. Wiley-Interscience, New York, 1985.

[16] J. Goodman. Some fundamental properties of speckle. *J. Opt. Soc. Am.*, 66:1145–1150, 1976.

[17] A. Labeyrie. Attainment of diffraction limited resolution in large telescopes by fourier analysis speckle patterns in star images. *Astron. Astrophys.*, VI:85–87, 1970.

Color and Multispectral Image Representation and Display

8

H. J. Trussell

North Carolina State University

8.1 INTRODUCTION

One of the most fundamental aspects of image processing is the representation of the image. The basic concept that a digital image is a matrix of numbers is reinforced by virtually all forms of image display. It is another matter to interpret how that value is related to the physical scene or object that is represented by the recorded image and how closely displayed results represent the data obtained from digital processing. It is these relationships to which this chapter is addressed.

Images are the result of a spatial distribution of radiant energy. The most common images are 2D color images seen on television. Other everyday images include photographs, magazine and newspaper pictures, computer monitors and motion pictures. Most of these images represent realistic or abstract versions of the real world. Medical and satellite images form classes of images where there is no equivalent scene in the physical world. Because of the limited space in this chapter, we will concentrate on the pictorial images.

The representation of an image goes beyond the mere designation of independent and dependent variables. In that limited case, an image is described by a function

$$f(x,y,\lambda,t), \tag{8.1}$$

where x,y are spatial coordinates (angular coordinates can also be used), λ indicates the wavelength of the radiation, and t represents time. It is noted that images are inherently 2D spatial distributions. Higher dimensional functions can be represented by a straightforward extension. Such applications include medical CT and MRI, as well as seismic surveys. For this chapter, we will concentrate on the spatial and wavelength variables associated with still images. The temporal coordinate will be left for another chapter.

In addition to the stored numerical values in a discrete coordinate system, the representation of multidimensional information includes the relationship between the samples and the real world. This relationship is important in the determination of appropriate sampling and subsequent display of the image.

Before presenting the fundamentals of image presentation, it is necessary to define our notation and to review the prerequisite knowledge that is required to understand the following material. A review of rules for the display of images and functions is presented in Section 8.2, followed by a review of mathematical preliminaries in Section 8.3. Section 8.4 will cover the physical basis for multidimensional imaging. The foundations of colorimetry are reviewed in Section 8.5. This material is required to lay a foundation for a discussion of color sampling. Section 8.6 describes multidimensional sampling with concentration on sampling color spectral signals. We will discuss the fundamental differences between sampling the wavelength and spatial dimensions of the multidimensional signal. Finally, Section 8.7 contains a mathematical description of the display of multidimensional data. This area is often neglected by many texts. The section will emphasize the requirements for displaying data in a fashion that is both accurate and effective. The final section briefly considers future needs in this basic area.

8.2 PRELIMINARY NOTES ON DISPLAY OF IMAGES

One difference between 1D and 2D functions is the way they are displayed. One-dimensional functions are easily displayed in a graph where the scaling is obvious. The observer will need to examine the numbers which label the axes to determine the scale of the graph and get a mental picture of the function. With 2D scalar-valued functions the display becomes more complicated. The accurate display of vector-valued 2D functions, e.g., color images, will be discussed after covering the necessary material on sampling and colorimetery.

2D functions can be displayed in several different ways. The most common are supported by MATLAB [1]. The three most common are the isometric plot, the grayscale plot, and the contour plot. The user should choose the right display for the information to be conveyed. Let us consider each of the three display modalities. As simple example, consider the 2D Gaussian functional form

$$f(m, n) = \mathrm{sinc}\left(\frac{m^2}{a^2} + \frac{n^2}{b^2}\right),$$

where, for the following plots, $a = 1$ and $b = 2$.

The isometric or surface plots give the appearance of a 3D drawing. The surface can be represented as a wire mesh or as a shaded solid, as in Fig. 8.1. In both cases, portions of the function will be obscured by other portions, for example, one cannot see through the main lobe. This representation is reasonable for observing the behavior of mathematical functions, such as, point spread functions, or filters in the space or frequency domains. An advantage of the surface plot is that it gives a good indication of the values of the

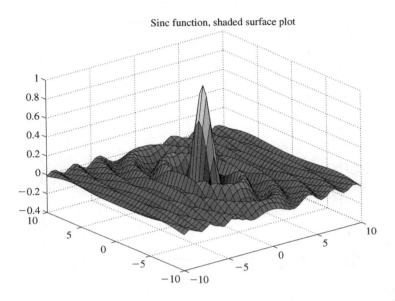

FIGURE 8.1

Shaded surface plot.

function since a scale is readily displayed on the axes. It is rarely effective for the display of images.

Contour plots are analogous to the contour or topographic maps used to describe geographical locations. The sinc function is shown using this method in Fig. 8.2. All points which have a specific value are connected to form a continuous line. For a continuous function the lines must form closed loops. This type of plot is useful in locating the position of maxima or minima in images or 2D functions. It is used primarily in spectrum analysis and pattern recognition applications. It is difficult to read values from the contour plot and takes some effort to determine whether the functional trend is up or down. The filled contour plot, available in MATLAB, helps in this last task.

Most monochrome images are displayed using the grayscale plot where the value of a pixel is represented by it relative lightness. Since in most cases high values are displayed as light and low values are displayed as dark, it is easy to determine functional trends. It is almost impossible to determine exact values. For images, which are nonnegative functions, the display is natural; but for functions, which have negative values, it can be quite artificial.

In order to use this type of display with functions, the representation must be scaled to fit in the range of displayable gray levels. This is most often done using a min/max scaling, where the function is linearly mapped such that the minimum value appears as black and the maximum value appears as white. This method was used for the sinc function shown in Fig. 8.3. For the display of functions, the min/max scaling can be effective to indicate trends in the behavior. Scaling for images is another matter.

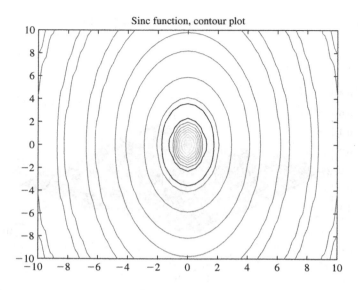

FIGURE 8.2

Contour plot.

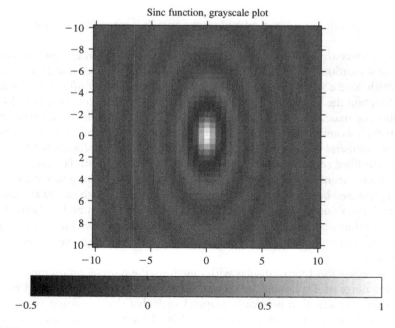

FIGURE 8.3

Grayscale plot.

Let us consider a monochrome image which has been digitized by some device, e.g., a scanner or camera. Without knowing the physical process of digitization, it is impossible to determine the best way to display the image. The proper display of images requires calibration of both the input and output devices. For now, it is reasonable to give some general rules about the display of monochrome images.

1. For the comparison of a sequence of images, it is *imperative* that all images be displayed using the same scaling. It is hard to emphasize this rule sufficiently and hard to count all the misleading results that have occurred when it has been ignored. The most common violation of this rule occurs when comparing an original and processed image. The user scales both images independently using min/max scaling. In many cases, the scaling can produce significant enhancement of low-contrast images which can be mistaken for improvements produced by an algorithm under investigation. For example, consider an algorithm designed to reduce noise, with the noisy image modeled by

$$\mathbf{g = f + n}.$$

   Since the noise is both positive and negative, the noisy image, $\mathbf{g}$, has a larger range than the clean image, $\mathbf{f}$. Almost any noise reduction method will reduce the range of the processed image, thus, the output image undergoes additional contrast enhancement if min/max scaling is used. The result is greater apparent dynamic range and a better looking image.

   There are several ways to implement this rule. The most appropriate way will depend on the application. The scaling may be done using the min/max of the collection of all images to be compared. In some cases, it is appropriate to truncate values at the limits of the display, rather than force the entire range into the range of the display. This is particularly true of images containing a few outliers. It may be advantageous to reduce the region of the image to a particular region of interest which will usually reduce the range to be reproduced.

2. Display a step-wedge, a strip of sequential gray levels from minimum to maximum values, with the image to show how the image gray levels are mapped to brightness or density. This allows some idea of the quantitative values associated with the pixels. This is routinely done on images which are used for analysis, such as the digital photographs from space probes.

3. Use a graytone mapping which allows a wide range of gray levels to be visually distinguished. In software such as MATLAB, the user can control the mapping between the continuous values of the image and the values sent to the display device. For example, consider the CRT monitor as the output device. The visual tonal qualities of the output depend on many factors including the brightness and contrast setting of the monitor, the specific phosphors used in the monitor, the linearity of the electron guns, and the ambient lighting. It is recommended that adjustments be made so that a user is able to distinguish all levels of a step-wedge of about 32 levels.

Most displays have problems with gray levels at the ends of the range being indistinguishable. This can be overcome by proper adjustment of the contrast and gain controls and an appropriate mapping from image values to display values. For hardcopy devices, the medium should be taken into account. For example, changes in paper type or manufacturer can result in significant tonal variations.

8.3 NOTATION AND PREREQUISITE KNOWLEDGE

In most cases, the multidimensional process can be represented as a straightforward extension of 1D processes. Thus, it is reasonable to mention the 1D operations which are prerequisite to the chapter and will form the basis of the multidimensional processes.

8.3.1 Practical Sampling

Mathematically, ideal sampling is usually represented with the use of a *generalized function*, the Dirac delta function, $\delta(t)$ [2]. The entire sampled sequence can be represented using the *comb* function

$$\text{comb}(t) = \sum_{n=-\infty}^{\infty} \delta(t - n), \tag{8.2}$$

where the sampling interval is unity. The sampled signal is obtained by multiplication

$$s_d(t) = s(t)\text{comb}(t) = s(t) \sum_{n=-\infty}^{\infty} \delta(t - n) = \sum_{n=-\infty}^{\infty} s(t)\delta(t - n). \tag{8.3}$$

It is common to use the notation of $\{s(n)\}$ or $s(n)$ to represent the collection of samples in discrete space. The arguments n and t will serve to distinguish the discrete or continuous space.

Practical imaging devices, such as video cameras, CCD arrays, and scanners, must use a finite aperture for sampling. The *comb* function cannot be realized by actual devices. The finite aperture is required to obtain a finite amount of energy from the scene. The engineering tradeoff is that large apertures receive more light and thus will have higher SNR's than smaller apertures; while smaller apertures have higher spatial resolution than larger ones. This is true for apertures larger than the order of the wavelength of light. At that point diffraction limits the resolution.

The aperture may cause the light intensity to vary over the finite region of integration. For a single sample of a 1D signal at time, nT, the sample value can be obtained by

$$s(n) = \int_{(n-1)T}^{nT} s(t)a(nT - t)dt, \tag{8.4}$$

where $a(t)$ represents the impulse response (or light variation) of the aperture. This is simple convolution. The sampling of the signal can be represented by

$$s(n) = [s(t) * a(t)]\text{comb}(t/T), \tag{8.5}$$

where * represents convolution. This model is reasonably accurate for spatial sampling of most cameras and scanning systems.

The sampling model can be generalized to include the case where each sample is obtained with a different aperture. For this case, the samples which need not be equally spaced, are given by

$$s(n) = \int_l^u s(t)a_n(t)\,dt, \tag{8.6}$$

where the limits of integration correspond to the region of support for the aperture. While there may be cases where this form is used in spatial sampling, its main use is in sampling the wavelength dimension of the image signals. That topic will be covered later. The generalized signal reconstruction equation has the form

$$s(t) = \sum_{n=-\infty}^{\infty} s(n)g_n(t), \tag{8.7}$$

where the collection of functions, $\{g_n(t)\}$, provide the interpolation from discrete to continuous space. The exact form of $\{g_n(t)\}$ depends on the form of $\{a_n(t)\}$.

8.3.2 One-Dimensional Discrete System Representation

Linear operations on signals and images can be represented as simple matrix multiplications. The internal form of the matrix may be complicated, but the conceptual manipulation of images is very easy. Let us consider the representation of a one-dimensional convolution before going on to multidimensions. Consider the linear, time-invariant system

$$g(t) = \int_{-\infty}^{\infty} h(u)s(t - u)\,du.$$

The discrete approximation to continuous convolution is given by

$$g(n) = \sum_{k=0}^{L-1} h(k)s(n - k), \tag{8.8}$$

where the indices n and k represent sampling of the analog signals, e.g., $s(n) = s(n\Delta T)$. Since it is assumed that the signals under investigation have finite support, the summation is over a finite number of terms. If $s(n)$ has M nonzero samples and $h(n)$ has L nonzero samples, then $g(n)$ can have at most $N = M + L - 1$ nonzero samples. It is assumed that the reader is familiar with what conditions are necessary so that we can represent the analog system by discrete approximation. Using the definition of the signal as a vector, $s = [s(0), s(1), \ldots s(M - 1)]$, the summation of Eq. (8.8) can be written

$$\mathbf{g} = \mathbf{Hs}, \tag{8.9}$$

where the vectors **s** and **g** are of length M and N, respectively, and the $N \times M$ matrix **H** is defined by

$$
\mathbf{H} = \begin{bmatrix}
h_0 & 0 & 0 & \cdots & 0 & 0 & 0 \\
h_1 & h_0 & 0 & \cdots & 0 & 0 & 0 \\
h_2 & h_1 & h_0 & \cdots & 0 & 0 & 0 \\
\vdots & \vdots & \vdots & \vdots & \vdots & \vdots & \vdots \\
h_{L-1} & h_{L-2} & h_{L-3} & \cdots & 0 & 0 & 0 \\
0 & h_{L-1} & h_{L-2} & \cdots & 0 & 0 & 0 \\
\vdots & \vdots & \vdots & \vdots & \vdots & \vdots & \vdots \\
0 & 0 & 0 & \cdots & h_0 & 0 & 0 \\
0 & 0 & 0 & \cdots & h_1 & h_0 & 0 \\
0 & 0 & 0 & \cdots & h_2 & h_1 & h_0 \\
0 & 0 & 0 & \cdots & h_3 & h_2 & h_1 \\
\vdots & \vdots & \vdots & \vdots & \vdots & \vdots & \vdots \\
0 & 0 & 0 & \cdots & 0 & h_{L-1} & h_{L-2} \\
0 & 0 & 0 & \cdots & 0 & 0 & h_{L-1}
\end{bmatrix}.
$$

It is often desirable to work with square matrices. In this case, the input vector can be padded with zeros to the same size as **g** and the matrix **H** modified to produce an $N \times N$ Toeplitz form

$$
\mathbf{H}_t = \begin{bmatrix}
h_0 & 0 & 0 & \cdots & 0 & 0 & 0 & \cdots & 0 & 0 & 0 \\
h_1 & h_0 & 0 & \cdots & 0 & 0 & 0 & \cdots & 0 & 0 & 0 \\
h_2 & h_1 & h_0 & \cdots & 0 & 0 & 0 & \cdots & 0 & 0 & 0 \\
\vdots & \vdots & \vdots & \vdots & \vdots & \vdots & \vdots & \vdots & \vdots & \vdots & \vdots \\
h_{L-1} & h_{L-2} & h_{L-3} & \cdots & h_0 & 0 & 0 & \cdots & 0 & 0 & 0 \\
0 & h_{L-1} & h_{L-2} & \cdots & h_1 & h_0 & 0 & \cdots & 0 & 0 & 0 \\
\vdots & \vdots & \vdots & \vdots & \vdots & \vdots & \vdots & \vdots & \vdots & \vdots & \vdots \\
0 & 0 & 0 & \cdots & h_k & h_{k-1} & h_{k-2} & \cdots & 0 & 0 & 0 \\
0 & 0 & 0 & \cdots & h_{k+1} & h_k & h_{k-1} & \cdots & 0 & 0 & 0 \\
0 & 0 & 0 & \cdots & h_{k+2} & h_{k+1} & h_k & \cdots & 0 & 0 & 0 \\
\vdots & \vdots & \vdots & \vdots & \vdots & \vdots & \vdots & \vdots & \vdots & \vdots & \vdots \\
0 & 0 & 0 & \cdots & 0 & h_{L-1} & h_{L-2} & \cdots & h_1 & h_0 & 0 \\
0 & 0 & 0 & \cdots & 0 & 0 & h_{L-1} & \cdots & h_2 & h_1 & h_0
\end{bmatrix}.
$$

The output can now be written as

$$
\mathbf{g} = \mathbf{H}_t \mathbf{s}_0,
$$

where $\mathbf{s}_0 = [s(0), s(1), \ldots s(M-1), 0, \ldots 0]^T$.

It is often useful, because of the efficiency of the FFT, to approximate the Toeplitz form by a circulant form

$$
\mathbf{H}_c =
\begin{bmatrix}
h_0 & 0 & 0 & \cdots & 0 & h_{L-1} & h_{L-2} & \cdots & h_3 & h_2 & h_1 \\
h_1 & h_0 & 0 & \cdots & 0 & 0 & 0 & \cdots & h_4 & h_3 & h_2 \\
h_2 & h_1 & h_0 & \cdots & 0 & 0 & 0 & \cdots & h_5 & h_4 & h_3 \\
\vdots & \vdots & \vdots & \vdots & \vdots & \vdots & \vdots & \vdots & \vdots & \vdots & \vdots \\
h_{L-1} & h_{L-2} & h_{L-3} & \cdots & 0 & 0 & 0 & \cdots & 0 & 0 & 0 \\
0 & h_{L-1} & h_{L-2} & \cdots & 0 & 0 & 0 & \cdots & 0 & 0 & 0 \\
\vdots & \vdots & \vdots & \vdots & \vdots & \vdots & \vdots & \vdots & \vdots & \vdots & \vdots \\
0 & 0 & 0 & \cdots & h_k & h_{k-1} & h_{k-2} & \cdots & 0 & 0 & 0 \\
0 & 0 & 0 & \cdots & h_{k+1} & h_k & h_{k-1} & \cdots & 0 & 0 & 0 \\
0 & 0 & 0 & \cdots & h_{k+2} & h_{k+1} & h_k & \cdots & 0 & 0 & 0 \\
\vdots & \vdots & \vdots & \vdots & \vdots & \vdots & \vdots & \vdots & \vdots & \vdots & \vdots \\
0 & 0 & 0 & \cdots & 0 & h_{L-1} & h_{L-2} & \cdots & h_1 & h_0 & 0 \\
0 & 0 & 0 & \cdots & 0 & 0 & h_{L-1} & \cdots & h_2 & h_1 & h_0
\end{bmatrix}.
$$

The approximation of a Toeplitz matrix by a circulant gets better as the dimension of the matrix increases. Consider the matrix norm

$$
||\mathbf{H}||^2 = \frac{1}{N^2} \sum_{k=1}^{N} \sum_{l=1}^{N} h_{kl}^2,
$$

then $||\mathbf{H}_t - \mathbf{H}_c|| \to 0$ as $N \to \infty$. This approximation works well with impulse responses of short duration and autocorrelation matrices with small correlation distances.

8.3.3 Multidimensional System Representation

The images of interest are described by two spatial coordinates and a wavelength coordinate, $f(x, y, \lambda)$. This continuous image will be sampled in each dimension. The result is a function defined on a discrete coordinate system, $f(m, n, l)$. This would usually require a 3D matrix. However, to allow the use of standard matrix algebra, it is common to use stacked notation [3]. Each band, defined by wavelength λ_l or simply l, of the image is a $P \times P$ image. Without loss of generality, we will assume a square image for notational simplicity. This image can be represented as a $P^2 \times 1$ vector. The Q bands of the image can be stacked in a like manner forming a $QP^2 \times 1$ vector.

Optical blurring is modeled as convolution of the spatial image. Each wavelength of the image may be blurred by a slightly different point spread function (PSF). This is represented by

$$
\mathbf{g}_{(QP^2 \times 1)} = \mathbf{H}_{(QP^2 \times QP^2)} \mathbf{f}_{(QP^2 \times 1)}, \tag{8.10}
$$

where the matrix $\mathbf{H}$ has a block form

$$\mathbf{H} = \begin{bmatrix} \mathbf{H}_{1,1} & \mathbf{H}_{1,2} & \cdots & \mathbf{H}_{1,Q} \\ \mathbf{H}_{2,1} & \mathbf{H}_{2,2} & \cdots & \mathbf{H}_{2,Q} \\ \vdots & \vdots & \cdots & \vdots \\ \mathbf{H}_{Q,1} & \mathbf{H}_{Q,2} & \cdots & \mathbf{H}_{Q,Q} \end{bmatrix}. \tag{8.11}$$

The submatrix $\mathbf{H}_{i,j}$ is of dimension $P^2 \times P^2$ and represents the contribution of the jth band of the input to the ith band of the output. Since an optical system does not modify the frequency of an optical signal, $\mathbf{H}$ will be block diagonal. There are cases, e.g., imaging using color filter arrays, where the diagonal assumption does not hold.

In many cases, multidimensional processing is a straightforward extension of 1D processing. The use of matrix notation permits the use of simple linear algebra to derive many results that are valid in any dimension. Problems arise primarily during the implementation of the algorithms when simplifying assumptions are usually made. Some of the similarities and differences are listed below.

8.3.3.1 *Similarities*

1. Derivatives and Taylor expansions are extensions of 1D

2. Fourier transforms are straightforward extension of 1D

3. Linear systems theory is the same

4. Sampling theory is straightforward extension of 1D

5. Separable 2D signals are treated as 1D signals

8.3.3.2 *Differences*

1. Continuity and derivatives have directional definitions

2. 2D signals are usually not causal; causality is not intuitive

3. 2D polynomials cannot always be factored; this limits use of rational polynomial models

4. More variation in 2D sampling, hexagonal lattices are common in nature, random sampling makes interpolation much more difficult

5. Periodic functions may have a wide variety of 2D periods

6. 2D regions of support are more variable, the boundaries of objects are often irregular instead of rectangular or elliptical

7. 2D systems can be mixed IIR and FIR, causal and noncausal

8. Algebraic representation using stacked notation for 2D signals is more difficult to manipulate and understand

Algebraic representation using stacked notation for 2D signals is more difficult to manipulate and understand than in 1D. An example of this is illustrated by considering the autocorrelation of multiband images which are used in multispectral restoration methods. This is easily written in terms of the matrix notation reviewed earlier:

$$\mathbf{R}_{ff} = E\{\mathbf{f}\mathbf{f}^T\},$$

where $\mathbf{f}$ is a $QP^2 \times 1$ vector. In order to compute estimates we must be able to manipulate this matrix. While the $QP^2 \times QP^2$ matrix is easily manipulated symbolically, direct computation with the matrix is not practical for realistic values of P and Q, e.g., $Q = 3$ and $P = 256$. For practical computation, the matrix form is simplified by using various assumptions, such as separability, circularity, and independence of bands. These assumptions result in block properties of the matrix which reduces the dimension of the computation. A good example is shown in the multidimensional restoration problem [4].

8.4   ANALOG IMAGES AS PHYSICAL FUNCTIONS

The image which exists in the analog world is a spatio-temporal distribution of radiant energy. As was mentioned earlier, this chapter will not discuss the temporal dimension but concentrate on the spatial and wavelength aspects of the image. The function is represented by $f(x, y, \lambda)$. While it is often overlooked by students eager to process their first image, it is fundamental to define what the value of the function represents. Since we are dealing with radiant energy, the value of the function represents energy flux, exactly like electromagnetic theory. The units will be energy per unit area (or angle) per unit time per unit wavelength. From the imaging point of view, the function is described by the spatial energy distribution at the sensor. It does not matter whether the object in the image emits light or reflects light.

To obtain a sample of the analog image we must integrate over space, time and wavelength to obtain a finite amount of energy. Since we have eliminated time from the description, we can have watts per unit area per unit wavelength. To obtain overall lightness, the wavelength dimension is integrated out using the luminous efficiency function discussed in the following section on colorimetry. The common units of light intensity are lux (lumens/m^2) or footcandles. See [5] for an exact definition of radiometric quantities. A table of typical light levels is given in Table 8.1. The most common instrument for measuring light intensity is the light meter used in professional and amateur photography.

In order to sample an image correctly, we must be able to characterize its energy distribution in each of the dimensions. There is little that can be said about the spatial distribution of energy. From experience, we know that images vary greatly in spatial content. Objects in an image usually may appear at any spatial location and at any orientation. This implies that there is no reason to apply varying sample spacing over the spatial range of an image. In the cases of some very restricted ensembles of images, variable spatial sampling has been used to advantage. Since these examples are quite rare, they will not be discussed here.

TABLE 8.1   Qualitative description of luminance levels.

Description	Lux (Cd/m^2)	Footcandles
Moonless night	$\sim 10^{-6}$	$\sim 10^{-7}$
Full moon night	$\sim 10^{-3}$	$\sim 10^{-4}$
Restaurant	~ 100	~ 9
Office	~ 350	~ 33
Overcast day	$\sim 5,000$	~ 465
Sunny day	$\sim 200,000$	$\sim 18,600$

Spatial sampling is done using a regular grid. The grid is most often rectilinear but hexagonal sampling has been thoroughly investigated [6]. Hexagonal sampling is used for efficiency when the images have a natural circular region of support or circular symmetry. All the mathematical operations, such as Fourier transforms and convoutions, exist for hexagonal grids. It is noted that the reasons for uniform sampling of the temporal dimension follow the same arguments.

The distribution of energy in the wavelength dimension is not as straightforward to characterize. In addition, we are often not interested in reconstructing the radiant spectral distribution as we are for the spatial distribution. We are interested in constructing an image which appears to the human observer to be the same colors as the original image. In this sense, we are actually using color aliasing to our advantage. Because of this aspect of color imaging, we need to characterize the color vision system of the eye in order to determine proper sampling of the wavelength dimension.

8.5 COLORIMETRY

To understand the fundamental difference in the wavelength domain, it is necessary to describe some of the fundamentals of color vision and color measurement. What is presented here is only a brief description that will allow us to proceed with the description of the sampling and mathematical representation of color images. A more complete description of the human color visual system can be found in [7, 8].

The retina contains two types of light sensors, rods and cones. The rods are used for monochrome vision at low light levels; the cones are used for color vision at higher light levels. There are three types of cones. Each type is maximally sensitive to a different part of the spectrum. They are often referred to as long, medium, and short wavelength regions. A common description refers to them as red, green, and blue cones, although their maximal sensitivity is in the yellow, green, and blue regions of the spectrum. Recall that the visible spectrum extends from about 400 nm (blue) to about 700 nm (red). Cones sensitivites are related to the absorption sensitivity of the pigments in the cones. The absorption sensitivity of the different cones has been measured by several methods. An example of the curves is shown in Fig. 8.4. Long before the technology was

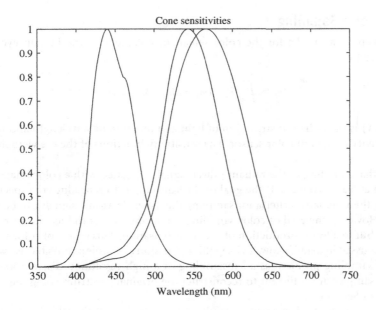

FIGURE 8.4

Cone sensitivities.

available to measure the curves directly, they were estimated from a clever color-matching experiment. A description of this experiment which is still used today can be found in [5, 7].

Grassmann formulated a set of laws for additive color mixture in 1853 [5, 9, 10]. Additive in this sense refers to the addition of two or more radiant sources of light. In addition, Grassmann conjectured that any additive color mixture could be matched by the proper amounts of three primary stimuli. Considering what was known about the physiology of the eye at that time, these laws represent considerable insight. It should be noted that these "laws" are not physically exact but represent a good approximation under a wide range of visibility conditions. There is current research in the vision and color science community on the refinements and reformulations of the laws.

Grassmann's laws are essentially unchanged as printed in recent texts on color science [5]. With our current understanding of the physiology of the eye and a basic background in linear algebra, Grassmann's laws can be stated more concisely. Furthermore, extensions of the laws and additional properties are easily derived using the mathematics of matrix theory. There have been several papers which have taken a linear systems approach to describing color spaces as defined by a standard human observer [11–14]. This section will briefly summarize these results and relate them to simple signal processing concepts. For the purposes of this work, it is sufficient to note that the spectral responses of the three types of sensors are sufficiently different so as to define a 3D vector space.

8.5.1 Color Sampling

The mathematical model for the color sensor of a camera or the human eye can be represented by

$$v_k = \int_{-\infty}^{\infty} r_a(\lambda) m_k(\lambda) d\lambda, \quad k = 1,2,3 \tag{8.12}$$

where $r_a(\lambda)$ is the radiant distribution of light as a function of wavelength and $m_k(\lambda)$ is the sensitivity of the kth color sensor. The sensitivity functions of the eye were shown in Fig. 8.4.

Note that sampling of the radiant power signal associated with a color image can be viewed in at least two ways. If the goal of the sampling is to reproduce the spectral distribution, then the same criteria for sampling the usual electronic signals can be directly applied. However, the goal of color sampling is not often to reproduce the spectral distribution but to allow reproduction of the color sensation. This aspect of color sampling will be discussed in detail below. To keep this discussion as simple as possible, we will treat the color sampling problem as a subsampling of a high-resolution discrete space, that is, the N samples are sufficient to reconstruct the original spectrum using the uniform sampling of Section 8.3.

It has been assumed in most research and standard work that thevisual frequency spectrum can be sampled finely enough to allow the accurate use of numerical approximation of integration. A common sample spacing is 10 nm over the range 400–700 nm, although ranges as wide as 360–780 nm have been used. This is used for many color tables and lower priced instrumentation. Precision color instrumentation produces data at 2 nm intervals. Finer sampling is required for some illuminants with line emitters. Reflective surfaces are usually smoothly varying and can be accurately sampled more coarsely. Sampling of color signals is discussed in Section 8.6 and in detail in [15].

Proper sampling follows the same bandwidth restrictions that govern all digital signal processing. Following the assumption that the spectrum can be adequately sampled, the space of all possible visible spectra lies in an N-dimensional vector space, where $N = 31$ is the range if 400–700 nm is used. The spectral response of each of the eye's sensors can be sampled as well, giving three linearly independent N-vectors which define the visual subspace.

Under the assumption of proper sampling, the integral of Eq. (8.12) can be well approximated by a summation

$$v_k = \sum_{n=L}^{U} r_a(n\Delta\lambda) s_k(n\Delta\lambda), \tag{8.13}$$

where $\Delta\lambda$ represents the sampling interval and the summation limits are determined by the region of support of the sensitivity of the eye. The above equations can be generalized to represent any color sensor by replacing $s_k(\cdot)$ with $m_k(\cdot)$. This discrete form is easily represented in matrix/vector notation. This will be done in the following sections.

8.5.2 Discrete Representation of Color-Matching

The response of the eye can be represented by a matrix, $S = [s_1, s_2, s_3]$, where the N-vectors, s_i, represent the response of the ith type sensor (cone). Any visible spectrum can be represented by an N-vector, f. The response of the sensors to the input spectrum is a 3-vector, t, obtained by

$$t = S^T f. \tag{8.14}$$

Two visible spectra are said to have the same color if they appear the same to the human observer. In our linear model, this means that if f and g are two N-vectors representing different spectral distributions, they are equivalent colors if

$$S^T f = S^T g. \tag{8.15}$$

It is clear that there may be many different spectra that appear to be the same color to the observer. Two spectra that appear the same are called metamers. Metamerism (meh-tam-er-ism) is one of the greatest and most fascinating problems in color science. It is basically color "aliasing" and can be described by the generalized sampling described earlier.

It is difficult to find the matrix, S, that defines the response of the eye. However, there is a conceptually simple experiment which is used to define the human visual space defined by S. A detailed discussion of this experiment is given in [5, 7]. Consider the set of monochromatic spectra e_i, for $i = 1, 2, \ldots N$. The N-vectors, e_i, have a one in the ith position and zeros elsewhere. The goal of the experiment is to match each of the monochromatic spectra with a linear combination of primary spectra. Construct three lighting sources that are linearly independent in N-space. Let the matrix $P = [p_1, p_2, p_3]$ represent the spectral content of these primaries. The phosphors of a color television are a common example, Fig. 8.5.

An experiment is conducted where a subject is shown one of the monochromactic spectra, e_i, on one half of a visual field. On the other half of the visual field appears a linear combination of the primary sources. The subject attempts to visually match an input monochromatic spectrum by adjusting the relative intensities of the primary sources. Physically, it may be impossible to match the input spectrum by adjusting the intensities of the primaries. When this happens, the subject is allowed to change the field of one of the primaries so that it falls on the same field as the monochromatic spectrum. This is mathematically equivalent to subtracting that amount of primary from the primary field. Denoting the relative intensities of the primaries by the 3 vector $a_i = [a_{i1}, a_{i2}, a_{i3}]^T$, the match is written mathematically as

$$S^T e_i = S^T P a_i. \tag{8.16}$$

Combining the results of all N monochromatic spectra, Eq. (8.5) can be written

$$S^T I = S^T = S^T P A^T, \tag{8.17}$$

where $I = [e_1, e_2, \ldots, e_N]$ is the $N \times N$ identity matrix.

Note that because the primaries, P, are not metameric, the product matrix is nonsingular, i.e., $(S^T P)^{-1}$ exists. The Human Visual Subspace (HVSS) in the N-dimensional

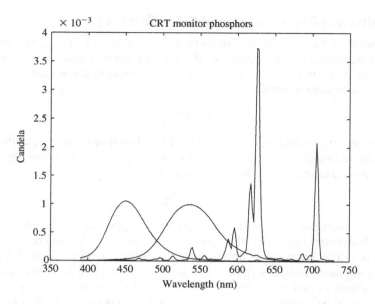

FIGURE 8.5

CRT monitor phosphors.

vector space is defined by the column vectors of **S**; however, this space can be equally well defined by any nonsingular transformation of those basis vectors. The matrix,

$$\mathbf{A} = \mathbf{S}(\mathbf{P}^T\mathbf{S})^{-1} \tag{8.18}$$

is one such transformation. The columns of the matrix **A** are called the color-matching functions associated with the primaries **P**.

To avoid the problem of negative values which cannot be realized with transmission or reflective filters, the CIE developed a standard transformation of the color-matching functions which have no negative values. This set of color-matching functions is known as the *standard observer* or the CIE XYZ color-matching functions. These functions are shown in Fig. 8.6. For the remainder of this chapter, the matrix, **A**, can be thought of as this standard set of functions.

8.5.3 Properties of Color-Matching Functions

Having defined the HVSS, it is worthwhile examining some of the common properties of this space. Because of the relatively simple mathematical definition of color-matching given in the last section, the standard properties enumerated by Grassmann are easily derived by simple matrix manipulations [14]. These properties play an important part in color sampling and display.

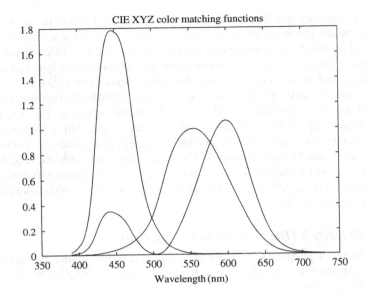

FIGURE 8.6

CIE XYZ color-matching functions.

8.5.3.1 *Property 1 (Dependence of Color on A)*

Two visual spectra, $\mathbf{f}$ and $\mathbf{g}$, appear the same if and only if $\mathbf{A}^T\mathbf{f} = \mathbf{A}^T\mathbf{g}$. Writing this mathematically, $\mathbf{S}^T\mathbf{f} = \mathbf{S}^T\mathbf{g}$ if $\mathbf{A}^T\mathbf{f} = \mathbf{A}^T\mathbf{g}$. Metamerism is color aliasing. Two signals $\mathbf{f}$ and $\mathbf{g}$ are sampled by the cones or equivalently by the color-matching functions and produce the same tristimulus values.

The importance of this property is that any linear transformation of the sensitivities of the eye or the CIE color-matching functions can be used to determine a color match. This gives more latitude in choosing color filters for cameras and scanners as well as for color measurement equipment. It is this property that is the basis for the design of optimal color scanning filters [16, 17].

A note on terminology is appropriate here. When the color-matching matrix is the CIE standard [5], the elements of the 3-vector defined by $\mathbf{t} = \mathbf{A}^T\mathbf{f}$ are called tristimulus values and usually denoted by X, Y, Z; i.e., $\mathbf{t}^T = [X, Y, Z]$. The chromaticity of a spectrum is obtained by normalizing the tristimulus values,

$$x = X/(X + Y + Z)$$

$$y = Y/(X + Y + Z)$$

$$z = Z/(X + Y + Z).$$

Since the chromaticity coordinates have been normalized, any two of them are sufficient to characterize the chromaticity of a spectrum. The x and y terms are the standard for describing chromaticity. It is noted that the convention of using different variables for the elements of the tristimulus vector may make mental conversion between the vector space notation and notation in common color science texts more difficult.

The CIE has chosen the $\mathbf{a}_2$ sensitivity vector to correspond to the luminance efficiency function of the eye. This function, shown as the middle curve in Fig. 8.6, gives the relative sensitivity of the eye to the energy at each wavelength. The Y tristimulus value is called luminance and indicates the perceived brightness of a radiant spectrum. It is this value that is used to calculate the effective light output of light bulbs in lumens. The chromaticities x and y indicate the hue and saturation of the color. Often the color is described in terms of $[x, y, Y]$ because of the ease of interpretation. Other color coordinate systems will be discussed later.

8.5.3.2 *Property 2 (Transformation of Primaries)*

If a different set of primary sources, $\mathbf{Q}$, are used in the color-matching experiment, a different set of color-matching functions, $\mathbf{B}$, are obtained. The relation between the two color-matching matrices is given by

$$\mathbf{B}^T = (\mathbf{A}^T\mathbf{Q})^{-1}\mathbf{A}^T. \tag{8.19}$$

The more common interpretation of the matrix $\mathbf{A}^T\mathbf{Q}$ is obtained by a direct examination. The jth column of $\mathbf{Q}$, denoted $\mathbf{q}_j$, is the spectral distribution of the jth primary of the new set. The element $[\mathbf{A}^T\mathbf{Q}]_{i,j}$ is the amount of the primary $\mathbf{p}_i$ required to match primary $\mathbf{q}_j$. It is noted that the above form of the change of primaries is restricted to those that can be adequately represented under the assumed sampling discussed previously. In the case that one of the new primaries is a Dirac delta function located between sample frequencies, the transformation $\mathbf{A}^T\mathbf{Q}$ must be found by interpolation. The CIE RGB color-matching functions are defined by the monochromatic lines at 700 nm, 546.1 nm, and 435.8 nm, shown in Fig. 8.7. The negative portions of these functions are particularly important since it implies that all color-matching functions associated with realizable primaries have negative portions.

One of the uses of this property is in determining the filters for color television cameras. The color-matching functions associated with the primaries used in a television monitor are the ideal filters. The tristimulus values obtained by such filters would directly give the values to drive the color guns. The NTSC standard $[R, G, B]$ are related to these color-matching functions. For coding purposes and efficient use of bandwidth, the RGB values are transformed to YIQ values, where Y is the CIE Y (luminance) and, I and Q carry the hue and saturation information. The transformation is a 3×3 matrix multiplication [3] (see Property 3).

Unfortunately, since the TV primaries are realizable, the color-matching functions which correspond to them are not. This means that the filters which are used in TV cameras are only an approximation to the ideal filters. These filters are usually obtained by simply clipping the part of the ideal filter which falls below zero. This introduces an error which cannot be corrected by any postprocessing.

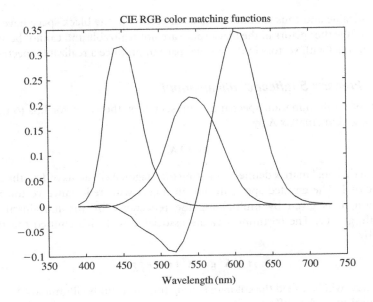

FIGURE 8.7

CIE XYZ color-matching functions.

8.5.3.3 *Property 3 (Transformation of Color Vectors)*

If **c** and **d** are the color vectors in 3-space associated with the visible spectrum, **f**, under the primaries **P** and **Q**, respectively, then

$$\mathbf{d} = (\mathbf{A}^T\mathbf{Q})^{-1}\mathbf{c}, \tag{8.20}$$

where **A** is the color-matching function matrix associated with primaries **P**. This states that a 3×3 transformation is all that is required to go from one color space to another.

8.5.3.4 *Property 4 (Metamers and the Human Visual Subspace)*

The N-dimensional spectral space can be decomposed into a 3D subspace known as the HVSS and an N-3D subspace known as the black space. All metamers of a particular visible spectrum, **f**, are given by

$$\mathbf{x} = \mathbf{P}_v\mathbf{f} + \mathbf{P}_b\mathbf{g}, \tag{8.21}$$

where $\mathbf{P}_v = \mathbf{A}(\mathbf{A}^T\mathbf{A})^{-1}\mathbf{A}^T$ is the orthogonal projection operator to the visual space, $\mathbf{P}_b = \left[\mathbf{I} - \mathbf{A}(\mathbf{A}^T\mathbf{A})^{-1}\mathbf{A}^T\right]$ is the orthogonal projection operator to the black space, and **g** is any vector in N-space.

It should be noted that humans cannot see (or detect) all possible spectra in the visual space. Since it is a vector space, there exist elements with negative values. These elements are not realizable and thus cannot be seen. All vectors in the black space have negative elements. While the vectors in the black space are not realizable and cannot be seen, they can be combined with vectors in the visible space to produce a realizable spectrum.

8.5.3.5 *Property 5 (Effect of Illumination)*

The effect of an illumination spectrum, represented by the N-vector $\mathbf{l}$, is to transform the color-matching matrix $\mathbf{A}$ by

$$\mathbf{A}_l = \mathbf{LA}, \tag{8.22}$$

where $\mathbf{L}$ is a diagonal matrix defined by setting the diagonal elements of $\mathbf{L}$ to the elements of the vector $\mathbf{l}$. The emitted spectrum for an object with reflectance vector, $\mathbf{r}$, under illumination, $\mathbf{l}$, is given by multiplying the reflectance by the illuminant at each wavelength, $\mathbf{g} = \mathbf{Lr}$. The tristimulus values associated with this emitted spectrum are obtained by

$$\mathbf{t} = \mathbf{A}^T\mathbf{g} = \mathbf{A}^T\mathbf{Lr} = \mathbf{A}_l^T\mathbf{r}. \tag{8.23}$$

The matrix $\mathbf{A}_l$ will be called the color-matching functions under illuminant $\mathbf{l}$.

Metamerism under different illuminants is one of the greatest problems in color science. A common imaging example occurs in making a digital copy of an original color image, e.g., a color copier. The user will compare the copy to the original under the light in the vicinity of the copier. The copier might be tuned to produce good matches under the fluorescent lights of a typical office but may produce copies that no longer match the original when viewed under the incandescent lights of another office or viewed near a window which allows a strong daylight component.

A typical mismatch can be expressed mathematically by relations

$$\mathbf{A}^T\mathbf{L}_f\mathbf{r}_1 = \mathbf{A}^T\mathbf{L}_f\mathbf{r}_2, \tag{8.24}$$

$$\mathbf{A}^T\mathbf{L}_d\mathbf{r}_1 \neq \mathbf{A}^T\mathbf{L}_d\mathbf{r}_2, \tag{8.25}$$

where $\mathbf{L}_f$ and $\mathbf{L}_d$ are diagonal matrices representing standard fluorescent and daylight spectra, respectively, and $\mathbf{r}_1$ and $\mathbf{r}_2$ represent the reflectance spectra of the original and the copy, respectively. The ideal images would have $\mathbf{r}_2$ matching $\mathbf{r}_1$ under all illuminations which would imply they are equal. This is virtually impossible since the two images are made with different colorants.

If the appearance of the image under a particular illuminant is to be recorded, then the scanner must have sensitivities that are within a linear transformation of the color-matching functions under that illuminant. In this case, the scanner consists of an illumination source, a set of filters, and a detector. The product of the three must duplicate the desired color-matching functions

$$\mathbf{A}_l = \mathbf{LA} = \mathbf{L}_s\mathbf{DM}, \tag{8.26}$$

where $\mathbf{L}_s$ is a diagonal matrix defined by the scanner illuminant, $\mathbf{D}$ is the diagonal matrix defined by the spectral sensitivity of the detector, and $\mathbf{M}$ is the $N \times 3$ matrix defined by the transmission characteristics of the scanning filters. In some modern scanners, three colored lamps are used instead of a single lamp and three filters. In this case, the $\mathbf{L}_s$ and $\mathbf{M}$ matrices can be combined.

In most applications, the scanner illumination is a high-intensity source so as to minimize scanning time. The detector is usually a standard CCD array or photomultiplier tube. The design problem is to create a filter set $\mathbf{M}$ which brings the product in Eq. (8.26) to within a linear transformation of $\mathbf{A}_l$. Since creating a perfect match with real materials is a problem, it is of interest to measure the goodness of approximations to a set of scanning filters which can be used to design optimal realizable filter sets [16, 17].

8.5.4 Notes on Sampling for Color Aliasing

Sampling of the radiant power signal associated with a color image can be viewed in at least two ways. If the goal of the sampling is to reproduce the spectral distribution, then the same criteria for sampling the usual electronic signals can be directly applied. However, the goal of color sampling is not often to reproduce the spectral distribution but to allow reproduction of the color sensation. To illustrate this problem, let us consider the case of a television system. The goal is to sample the continuous color spectrum in such a way that the color sensation of the spectrum can be reproduced by the monitor.

A scene is captured with a television camera. We will consider only the color aspects of the signal, i.e., a single pixel. The camera uses three sensors with sensitivities $\mathbf{M}$ to sample the radiant spectrum. The measurements are given by

$$\mathbf{v} = \mathbf{M}^T \mathbf{r}, \tag{8.27}$$

where $\mathbf{r}$ is a high-resolution sampled representation of the radiant spectrum and $\mathbf{M} = [\mathbf{m}_1, \mathbf{m}_2, \mathbf{m}_3]$ represent the high-resolution sensitivities of the camera. The matrix $\mathbf{M}$ includes the effects of the filters, detectors, and optics.

These values are used to reproduce colors at the television receiver. Let us consider the reproduction of color at the receiver by a linear combination of the radiant spectra of the three phosphors on the screen, denoted $\mathbf{P} = [\mathbf{p}_1, \mathbf{p}_2, \mathbf{p}_3]$, where $\mathbf{p}_k$ represent the spectra of the red, green, and blue phosphors. We will also assume that the driving signals, or control values, for the phosphors are linear combinations of the values measured by the camera, $\mathbf{c} = \mathbf{B}\mathbf{v}$. The reproduced spectrum is $\hat{\mathbf{r}} = \mathbf{P}\mathbf{c}$.

The appearance of the radiant spectra is determined by the response of the human eye

$$\mathbf{t} = \mathbf{S}^T \mathbf{r}, \tag{8.28}$$

where $\mathbf{S}$ is defined by Eq. (8.14). The tristimulus values of the spectrum reproduced by the TV are obtained by

$$\hat{\mathbf{t}} = \mathbf{S}^T \hat{\mathbf{r}} = \mathbf{S}^T \mathbf{P}\mathbf{B}\mathbf{M}^T \mathbf{r}. \tag{8.29}$$

If the sampling is done correctly, the tristimulus values can be computed, that is, $\mathbf{B}$ can be chosen so that $\mathbf{t} = \hat{\mathbf{t}}$. Since the three primaries are not metameric and the eye's sensitivities are linearly independent, $(\mathbf{S}^T\mathbf{P})^{-1}$ exists and from the equality we have

$$(\mathbf{S}^T\mathbf{P})^{-1}\mathbf{S}^T = \mathbf{B}\mathbf{M}^T, \tag{8.30}$$

since equality of tristimulus values holds for all $\mathbf{r}$. This means that the color spectrum is sampled properly if the sensitivities of the camera are within a linear transformation of the sensitivities of the eye, or equivalently the color-matching functions.

Considering the case where the number of sensors Q in the camera or any color measuring device is larger than three, the condition is that the sensitivities of the eye must be a linear combination of the sampling device sensitivities. In this case,

$$(\mathbf{S}^T\mathbf{P})^{-1}\mathbf{S}^T = \mathbf{B}_{3\times Q}\mathbf{M}^T_{Q\times N}. \tag{8.31}$$

There are still only three types of cones which are described by $\mathbf{S}$. However, the increase in the number of basis functions used in the measuring device allows more freedom to the designer of the instrument. From the vector space viewpoint, the sampling is correct if the 3D vector space defined by the cone sensitivity functions lies within the N-dimensional vector space defined by the device sensitivity functions.

Let us now consider the sampling of reflective spectra. Since color is measured for radiant spectra, a reflective object must be illuminated to be seen. The resulting radiant spectra is the product of the illuminant and the reflection of the object

$$\mathbf{r} = \mathbf{L}\mathbf{r}_0, \tag{8.32}$$

where $\mathbf{L}$ is a diagonal matrix containing the high-resolution sampled radiant spectrum of the illuminant and the elements of the reflectance of the object are constrained, $0 \leq \mathbf{r}_0(k) \leq 1$.

To consider the restrictions required for sampling a reflective object, we must account for two illuminants: the illumination under which the object is to be viewed and the illumination under which the measurements are made. The equations for computing the tristimulus values of reflective objects under the viewing illuminant $\mathbf{L}_v$ are given by

$$\mathbf{t} = \mathbf{A}^T\mathbf{L}_v\mathbf{r}_0, \tag{8.33}$$

where we have used the CIE color-matching functions instead of the sensitivities of the eye (Property 1). The equation for estimating the tristimulus values from the sampled data is given by

$$\hat{\mathbf{t}} = \mathbf{B}\mathbf{M}^T\mathbf{L}_d\mathbf{r}_0, \tag{8.34}$$

where $\mathbf{L}_d$ is a matrix containing the illuminant spectrum of the device. The sampling is proper if there exists a $\mathbf{B}$ such that

$$\mathbf{B}\mathbf{M}^T\mathbf{L}_d = \mathbf{A}^T\mathbf{L}_v. \tag{8.35}$$

It is noted that in practical applications the device illuminant usually placed severe limitations on the problem of approximating the color-matching functions under the viewing illuminant. In most applications the scanner illumination is a high-intensity source so as to minimize scanning time. The detector is usually a standard CCD array or photomultiplier tube. The design problem is to create a filter set **M** which brings the product of the filters, detectors, and optics to within a linear transformation of $\mathbf{A}_l$. Since creating a perfect match with real materials is a problem, it is of interest to measure the goodness of approximations to a set of scanning filters which can be used to design optimal realizable filter sets [16, 17].

8.5.5 A Note on the Nonlinearity of the Eye

It is noted here that most physical models of the eye include some type of nonlinearity in the sensing process. This nonlinearity is often modeled as a logarithm; in any case, it is always assumed to be monotonic within the intensity range of interest. The nonlinear function, $\mathbf{v} = V(\mathbf{c})$, transforms the 3-vector in an element-independent manner; that is,

$$[v_1, v_2, v_3]^T = [V(c_1), V(c_2), V(c_3)]^T. \tag{8.36}$$

Since equality is required for a color match by Eq. (8.2), the function $V(\cdot)$ does not affect our definition of equivalent colors. Mathematically,

$$V(\mathbf{S}^T\mathbf{f}) = V(\mathbf{S}^T\mathbf{g}) \tag{8.37}$$

is true if, and only if, $\mathbf{S}^T\mathbf{f} = \mathbf{S}^T\mathbf{g}$. This nonlinearity does have a definite effect on the relative sensitivity in the color-matching process and is one of the causes of much searching for the "uniform color space" discussed next.

8.5.6 Uniform Color Spaces

It has been mentioned that the psychovisual system is known to be nonlinear. The problem of color matching can be treated by linear systems theory since the receptors behave in a linear mode and exact equality is the goal. In practice, it is seldom that an engineer can produce an exact match to any specification. The nonlinearities of the visual system play a critical role in the determination of a color-sensitivity function. Color vision is too complex to be modeled by a simple function. A measure of sensitivity that is consistent with the observations of arbitrary scenes are well beyond present capability. However, much work has been done to determine human color sensitivity in matching two color fields which subtend only a small portion of the visual field.

   Some of the first controlled experiments in color sensitivity were done by MacAdam [18]. The observer viewed a disk made of two hemispheres of different colors on a neutral background. One color was fixed; the other could be adjusted by the user. Since MacAdam's pioneering work there have been many additional studies of color sensitivity. Most of these have measured the variability in three dimensions which yields sensitivity ellipsoids in tristimulus space. The work by Wyszecki and Felder [19] is of particular interest as it shows the variation between observers and between a single observer at different times. The large variation of the sizes and orientation of the ellipsoids indicates that mean square error in tristimulus space is a very poor measure of color error. A common method of treating the nonuniform error problem is to transform the space into one where the euclidean distance is more closely correlated with perceptual error. The CIE recommended two transformations in 1976 in an attempt to standardize measures in the industry.

Neither of the CIE standards exactly achieves the goal of a uniform color space. Given the variability of the data, it is unreasonable to expect that such a space could be found. The transformations do reduce the variations in the sensitivity ellipses by a large degree. They have another major feature in common: the measures are made relative to a reference white point. By using the reference point the transformations attempt to account for the adpative characteristics of the visual system. The CIELab (see-lab) space is defined by

$$L^* = 116 \left(\frac{Y}{Y_n} \right)^{\frac{1}{3}} - 16 \tag{8.38}$$

$$a^* = 500 \left[\left(\frac{X}{X_n} \right)^{\frac{1}{3}} - \left(\frac{Y}{Y_n} \right)^{\frac{1}{3}} \right] \tag{8.39}$$

$$b^* = 200 \left[\left(\frac{Y}{Y_n} \right)^{\frac{1}{3}} - \left(\frac{Z}{Z_n} \right)^{\frac{1}{3}} \right] \tag{8.40}$$

for $\frac{X}{X_n}, \frac{Y}{Y_n}, \frac{Z}{Z_n} > 0.01$. The values X_n, Y_n, Z_n are the tristimulus values of the reference white under the reference illumination, and X, Y, Z are the tristimulus values which are to be mapped to the Lab color space. The restriction that the normalized values be greater than 0.01 is an attempt to account for the fact that at low illumination the cones become less sensitive and the rods (monochrome receptors) become active. A linear model is used at low light levels. The exact form of the linear portion of CIELab and the definition of the CIELuv (see-luv) transformation can be found in [3, 5].

A more recent modification of the CIELab space was created in 1994, appropriately called CIELab94, [20]. This modification addresses some of the shortcomings of the 1931 and 1976 versions. However, it is significantly more complex and costly to compute. A major difference is the inclusion of weighting factors in the summation of square errors, instead of using a strict Euclidean distance in the space.

The color error between two colors c_1 and c_2 is measured in terms of

$$\Delta E_{ab} = [(L_1^* - L_2^*)^2 + (a_1^* - a_2^*)^2 + (b_1^* - b_2^*)^2]^{1/2}, \tag{8.41}$$

where $c_i = [L_i^*, a_i^*, b_i^*]$. A useful rule of thumb is that two colors cannot be distinguished in a scene if their ΔE_{ab} value is less than 3. The ΔE_{ab} threshold is much lower in the experimental setting than in pictorial scenes. It is noted that the sensitivities discussed above are for flat fields. The sensitivity to modulated color is a much more difficult problem.

8.6 SAMPLING OF COLOR SIGNALS AND SENSORS

It has been assumed in most of this chapter that the color signals of interest can be sampled sufficiently well to permit accurate computation using discrete arithmetic. It is appropriate to consider this assumption quantitatively. From the previous sections, it is seen that there are three basic types of color signals to consider: reflectances, illuminants, and sensors. Reflectances usually characterize everyday objects but occasionally man-made items with special properties such as filters and gratings are of interest. Illuminants vary a great deal from natural daylight or moonlight to special lamps used in imaging equipment. The sensors most often used in color evaluation are those of the human eye. However, because of their use in scanners and cameras, CCD's and photomultiplier tubes are of great interest.

The most important sensor characteristics are the cone sensitivities of the eye or equivalently, the color-matching functions, e.g., Fig. 8.6. It is easily seen that the functions in Figs. 8.4, 8.6, and 8.7 are very smooth functions and have limited bandwidths. A note on bandwidth is appropriate here. The functions represent continuous functions with finite support. Because of the finite support constraint, they cannot be bandlimited. However, they are clearly smooth and have very low power outside of a very small frequency band. Using 2 nm representations of the functions, the power spectra of these signals are shown in Fig. 8.8. The spectra represent the Welch estimate where the data is first windowed, then the magnitude of the DFT is computed [2]. It is seen that 10 nm sampling produces very small aliasing error.

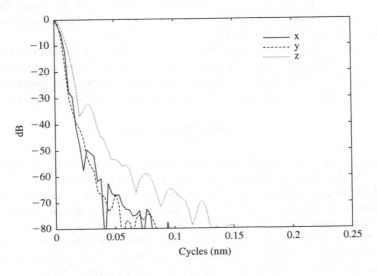

FIGURE 8.8

Power spectrum of CIE XYZ color-matching functions.

In the context of cameras and scanners, the actual photo-electric sensor should be considered. Fortunately, most sensors have very smooth sensitivity curves which have bandwidths comparable to those of the color-matching functions. See any handbook of CCD sensors or photomultiplier tubes. Reducing the variety of sensors to be studied can also be justified by the fact that filters can be designed to compensate for the characteristics of the sensor and bring the combination within a linear combination of the color-matching functions.

The function $r(\lambda)$, which is sampled to give the vector $\mathbf{r}$ used in the Colorimetry section, can represent either reflectance or transmission. Desktop scanners usually work with reflective media. There are, however, several film scanners on the market which are used in this type of environment. The larger dynamic range of the photographic media implies a larger bandwidth. Fortunately, there is not a large difference over the range of everyday objects and images. Several ensembles were used for a study in an attempt to include the range of spectra encountered by image scanners and color measurement instrumentation [21]. The results showed again that 10 nm sampling was sufficient [15].

There are three major types of viewing illuminants of interest for imaging: daylight, incandescent, and fluorescent. There are many more types of illuminants used for scanners and measurement instruments. The properties of the three viewing illuminants can be used as a guideline for sampling and signal processing which involves other types. It has been shown that the illuminant is the determining factor for the choice of sampling interval in the wavelength domain [15].

Incandescent lamps and natural daylight can be modeled as filtered blackbody radiators. The wavelength spectra are relatively smooth and have relatively small bandwidths. As with previous color signals they are adequately sampled at 10 nm. Office lighting is dominated by fluorescent lamps. Typical wavelength spectra and their frequency power spectra are shown in Figs. 8.9 and 8.10.

It is with the fluorescent lamps that the 2 nm sampling becomes suspect. The peaks that are seen in the wavelength spectra are characteristic of mercury and are delta function signals at 404.7 nm, 435.8 nm, 546.1 nm, and 578.4 nm. The flourescent lamp can be modeled as the sum of a smoothly varying signal and a delta function series:

$$l(\lambda) = l_d(\lambda) + \sum_{k=1}^{q} \alpha_k \delta(\lambda - \lambda_k), \tag{8.42}$$

where α_k represents the strength of the spectral line at wavelength λ_k. The wavelength spectra of the phosphors is relatively smooth as seen from Fig. 8.9.

It is clear that the fluorescent signals are not bandlimited in the sense used previously. The amount of power outside of the band is a function of the positions and strengths of the line spectra. Since the lines occur at known wavelengths, it remains only to estimate their power. This can be done by signal restoration methods which can use the information about this specific signal. Using such methods, the frequency spectrum of the lamp may be estimated by combining the frequency spectra of its components

$$L(\omega) = L_d(\omega) + \sum_{k=1}^{q} \alpha_k e^{j\omega(\lambda_0 - \lambda_k)}, \tag{8.43}$$

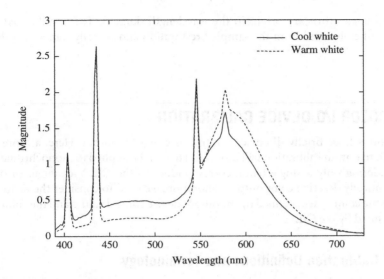

FIGURE 8.9

Cool white fluorescent and warm white fluorescent.

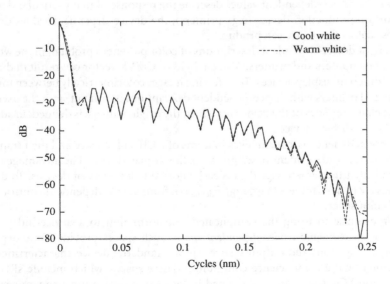

FIGURE 8.10

Power spectra of cool white fluorescent and warm white fluorescent.

where λ_0 is an arbitrary origin in the wavelength domain. The bandlimited spectra $L_d(\omega)$ can be obtained from the sampled restoration and is easily represented by 2 nm sampling.

8.7 COLOR I/O DEVICE CALIBRATION

In Section 8.2, we briefly discussed control of grayscale output. Here, a more formal approach to output calibration will be given. This can be applied to monochrome images by considering only a single band, corresponding to the CIE Y channel. In order to mathematically describe color output calibration, we need to consider the relationships between the color spaces defined by the output device control values and the colorimetric space defined by the CIE.

8.7.1 Calibration Definitions and Terminology

A *device-independent color space* is defined as any space that has a one-to-one mapping onto the CIE XYZ color space. Examples of CIE device-independent color spaces include XYZ, Lab, Luv, and Yxy. Current image format standards, such as JPEG, support the description of color in Lab. By definition, a *device-dependent color space* cannot have a one-to-one mapping onto the CIE XYZ color space. In the case of a recording device (e.g., scanners), the device-dependent values describe the response of that particular device to color. For a reproduction device (e.g., printers), the device-dependent values describe only those colors the device can produce.

The use of device-dependent descriptions of color presents a problem in the world of networked computers and printers. A single RGB or CMYK vector can result in different colors on different display devices. Transferring images colorimetrically between multiple monitors and printers with device-dependent descriptions is difficult since the user must know the characteristics of the device for which the original image is defined, in addition to those of the display device.

It is more efficient to define images in terms of a CIE color space and then transform this data to device-dependent descriptors for the display device. The advantage of this approach is that the same image data is easily ported to a variety of devices. To do this, it is necessary to determine a mapping, $\mathcal{F}_{\text{device}}(\cdot)$, from device-dependent control values to a CIE color space.

A compromise to using the complicated transformation to a device-independent space is to use a *pseudo-device-dependent space*. Such spaces provide some degree of matching across input and output devices since "standard" device characteristics have been defined by the color science community. These spaces, which include sRGB and Kodak's PhotoYCC space, are well defined in terms of a device-independent space. As such, a device manufacturer can design an input or output device such that when given sRGB values the proper device-independent color value is displayed. However, there do exist limitations with this approach such as nonuniformity and limited gamut.

Modern printers and display devices are limited in the colors they can produce. This limited set of colors is defined as the *gamut* of the device. If Ω_{cie} is the range of values in the selected CIE color space and Ω_{print} is the range of the device control values then the set

$$G = \{\, \mathbf{t} \in \Omega_{cie} \mid \text{there exists } \mathbf{c} \in \Omega_{print} \text{ where } \mathcal{F}_{device}(\mathbf{c}) = \mathbf{t} \,\}$$

defines the gamut of the color output device. For colors in the gamut, there will exist a mapping between the device-dependent control values and the CIE XYZ color space. Colors which are in the complement, G^c, cannot be reproduced and must be *gamut-mapped* to a color which is within G. The gamut mapping algorithm $\mathcal{D}$ is a mapping from Ω_{cie} to G, that is $\mathcal{D}(\mathbf{t}) \in G \; \forall \mathbf{t} \in \Omega_{cie}$. A more detailed discussion of gamut mapping is found in [22].

The mappings $\mathcal{F}_{device}$, $\mathcal{F}_{device}^{-1}$, and $\mathcal{D}$ make up what is defined as a *device profile*. These mappings describe how to transform between a CIE color space and the device control values. The International Color Commission (ICC) has suggested a standard format for describing a profile. This standard profile can be based on a physical model (common for monitors) or a look-up-table (LUT) (common for printers and scanners) [23]. In the next sections, we will mathematically discuss the problem of creating a profile.

8.7.2 CRT Calibration

A monitor is often used to provide a preview for the printing process, as well as comparison of image processing methods. Monitor calibration is almost always based on a physical model of the device [24–26]. A typical model is

$$r' = (r - r_0)/(r_{max} - r_0)^{\gamma_r},$$

$$g' = (g - g_0)/(g_{max} - g_0)^{\gamma_g},$$

$$b' = (b - b_0)/(b_{max} - b_0)^{\gamma_b},$$

$$\mathbf{t} = \mathbf{H}[r', g', b']^T,$$

where $\mathbf{t}$ is the CIE value produced by driving the monitor with control value $\mathbf{c} = [r, g, b]^T$. The value of the tristimulus vector is obtained using a colorimeter or spectrophometer.

Creating a profile for a monitor involves the determination of these parameters where r_{max}, g_{max}, and b_{max} are the maximum values of the control values (e.g., 255). To determine the parameters, a series of color patches is displayed on the CRT and measured with a colorimeter which will provide pairs of CIE values $\{\mathbf{t}_k\}$ and control values $\{\mathbf{c}_k\}$, $k = 1, \ldots, M$.

Values for γ_r, γ_g, γ_b, r_0, g_0, and b_0 are determined such that the elements of $[r', g', b']$ are linear with respect to the elements of XYZ and scaled between the range $[0,1]$.

The matrix $\mathbf{H}$ is then determined from the tristimulus values of the CRT phosphors at maximum luminance. Specifically the mapping is given by

$$\begin{bmatrix} X \\ Y \\ Z \end{bmatrix} = \begin{bmatrix} X_{Rmax} & X_{Rmax} & X_{Rmax} \\ Y_{Gmax} & Y_{Gmax} & Y_{Gmax} \\ Z_{Bmax} & Z_{Bmax} & Z_{Bmax} \end{bmatrix} \begin{bmatrix} r' \\ b' \\ g' \end{bmatrix},$$

where $[X_{Rmax}\, Y_{Rmax}\, Z_{Rmax}]^T$ is the CIE XYZ tristimulus value of the red phosphor for control value $\mathbf{c} = [r_{max}, 0, 0]^T$.

This standard model is often used to provide an approximation to the mapping $\mathcal{F}_{monitor}(\mathbf{c}) = \mathbf{t}$. Problems such as spatial variation of the screen or electron gun dependence are typically ignored. A LUT can also be used for the monitor profile in a manner similar to that described below for scanner calibration.

8.7.3 Scanners and Cameras

Mathematically, the recording process of a scanner or camera can be expressed as

$$\mathbf{z}_i = \mathcal{H}(\mathbf{M}^T \mathbf{r}_i),$$

where the matrix $\mathbf{M}$ contains the spectral sensitivity (including the scanner illuminant) of the three (or more) bands of the device, $\mathbf{r}_i$ is the spectral reflectance at spatial point i, $\mathcal{H}$ models any nonlinearities in the scanner (invertible in the range of interest), and $\mathbf{z}_i$ is the vector of recorded values.

We define *colorimetric recording* as the process of recording an image such that the CIE values of the image can be recovered from the recorded data. This reflects the requirements of ideal sampling in Section 8.5.4. Given such a scanner, the calibration problem is to determine the continuous mapping $\mathcal{F}_{scan}$ which will transform the recorded values to a CIE color space:

$$\mathbf{t} = \mathbf{A}^T \mathbf{Lr} = \mathcal{F}_{scan}(\mathbf{z}) \quad \text{for all } \mathbf{r} \in \Omega_r.$$

Unfortunately, most scanners and especially desktop scanners are not colorimetric. This is caused by physical limitations on the scanner illuminants and filters which prevent them from being within a linear transformation of the CIE color-matching functions. Work related to designing optimal approximations is found in [27, 28].

For the noncolorimetric scanner, there will exist spectral reflectances which look different to the standard human observer but when scanned produce the same recorded values. These colors are defined as being metameric to the scanner. This cannot be corrected by any transformation $\mathcal{F}_{scan}$.

Fortunately, there will always (except for degenerate cases) exist a set of reflectance spectra over which a transformation from scan values to CIE XYZ values will exist. Such a set can be expressed mathematically as

$$B_{scan} = \{ \mathbf{r} \in \Omega_r \mid \mathcal{F}_{scan}(\mathcal{H}(\mathbf{Mr})) = \mathbf{A}^T \mathbf{Lr} \},$$

where $\mathcal{F}_{scan}$ is the transformation from scanned values to colorimetric descriptors for the set of reflectance spectra in B_{scan}. This is a restriction to a set of reflectance spectra over which the continuous mapping $\mathcal{F}_{scan}$ exists.

Look-up tables, neural nets, nonlinear and linear models for $\mathcal{F}_{scan}$ have been used to calibrate color scanners [29–33]. In all of these approaches, the first step is to select a collection of color patches which span the colors of interest. These colors should not be metameric to the scanner or to the standard observer under the viewing illuminant. This constraint assures a one-to-one mapping between the scan values and the device-independent values across these samples. In practice, this constraint is easily obtained. The reflectance spectra of these M_q color patches will be denoted by $\{\mathbf{q}\}_k$ for $1 \leq k \leq M_q$.

These patches are measured using a spectrophotometer or a colorimeter which will provide the device-independent values

$$\{\mathbf{t}_k = \mathbf{A}^T \mathbf{q}_k\} \quad \text{for} \quad 1 \leq k \leq M_q.$$

Without loss of generality, $\{\mathbf{t}_k\}$ could represent any colorimetric or device-independent values, e.g., CIELAB, CIELUV, in which case $\{\mathbf{t}_k = \mathcal{L}(\mathbf{A}^T \mathbf{q}_k)\}$ where $\mathcal{L}(\cdot)$ is the transformation from CIEXYZ to the appropriate color space. The patches are also measured with the scanner to be calibrated providing $\{\mathbf{z}_k = \mathcal{H}(\mathbf{M}^T \mathbf{q}_k)\}$ for $1 \leq k \leq M_q$. Mathematically, the calibration problem is: find a transformation $\mathcal{F}_{scan}$ where

$$\mathcal{F}_{scan} = \arg\left(\min_{\mathcal{F}} \sum_{i=1}^{M_q} ||\mathcal{F}(\mathbf{z}_i) - \mathbf{t}_i||^2\right)$$

and $||.||^2$ is the error metric in the CIE color space. In practice, it may be necessary and desirable to incorporate constraints on $\mathcal{F}_{scan}$ [22].

8.7.4 Printers

Printer calibration is difficult due to the nonlinearity of the printing process and the wide variety of methods used for color printing (e.g., lithography, inkjet, dye sublimation, etc.). Thus, printing devices are often calibrated with an LUT with the continuum of values found by interpolating between points in the LUT [29, 34].

To produce a profile of a printer, a subset of values spanning the space of allowable control values, $\mathbf{c}_k$ for $1 \leq k \leq M_p$, for the printer is first selected. These values produce a set of reflectance spectra which are denoted by $\mathbf{p}_k$ for $1 \leq k \leq M_p$.

The patches $\mathbf{p}_k$ are measured using a colorimetric device which provides the values

$$\{\mathbf{t}_k = \mathbf{A}^T \mathbf{p}_k\} \quad \text{for} \quad 1 \leq k \leq M_p.$$

The problem is then to determine a mapping $\mathcal{F}_{print}$ which is the solution to the optimization problem

$$\mathcal{F}_{print} = \arg\left(\min_{\mathcal{F}} \sum_{i=1}^{M_p} ||\mathcal{F}(\mathbf{c}_i) - \mathbf{t}_i||^2\right),$$

where as in the scanner calibration problem, there may be constraints which $\mathcal{F}_{print}$ must satisfy.

8.7.5 Calibration Example

Before presenting an example of the need for calibrated scanners and displays, it is necessary to state some problems with the display to be used, i.e., the color printed page. Currently, printers and publishers do not use the CIE values for printing but judge the quality of their prints by subjective methods. Thus, it is impossible to numerically specify the image values to the publisher of this book. We have to rely on the experience of the company to produce images which faithfully reproduce those given to them. Every effort has been made to reproduce the images as accurately as possible. The tiff image format allows the specification of CIE values and the images defined by those values can be found on the ftp site, ftp.ncsu.edu in directory pub/hjt/calibration. Even in the tiff format, problems arise because of quantization to 8 bits.

The original color Lena image is available in many places as an RGB image. The problem is that there is no standard to which the RGB channels refer. The image is usually printed to an RGB device (one that takes RGB values as input) with no transformation. An example of this is shown in Fig. 8.11. This image compares well with current printed versions of this image, e.g., those shown in papers in the special issue on color image processing of the IEEE Transactions on Image Processing [35]. However, the displayed image does not compare favorably with the original. An original copy of the image was obtained and scanned using a calibrated scanner and then printed using a calibrated printer. The result, shown in Fig. 8.12, does compare well with the original. Even with the display problem mentioned above, it is clear that the images are sufficiently different to

FIGURE 8.11

Original Lena.

FIGURE 8.12

Calibrated Lena.

FIGURE 8.13

New scan of Lena.

make the point that calibration is necessary for accurate comparisons of any processing method that uses color images. To complete the comparison, the RGB image that was used to create the corrected image shown in Fig. 8.12 was also printed directly on the RGB printer. The result shown in Fig. 8.13 further demonstrates the need for calibration. A complete discussion of this calibration experiment is found in [22].

8.8 SUMMARY AND FUTURE OUTLOOK

The major portion of the chapter emphasized the problems and differences in treating the color dimension of image data. Understanding of the basics of uniform sampling is required to proceed to the problems of sampling the color component. The phenomenon of aliasing is generalized to color sampling by noting that the goal of most color sampling is to reproduce the sensation of color and not the actual color spectrum. The calibration of recording and display devices is required for accurate representation of images. The proper recording and display outlined in Section 8.7 cannot be overemphasized.

While the fundamentals of image recording and display are well understood by experts in that area, they are not well appreciated by the general image processing community. It is hoped that future work will help widen the understanding of this aspect of image processing. At present, it is fairly difficult to calibrate color image I/O devices. The interface between the devices and the interpretation of the data is still problematic. Future work can make it easier for the average user to obtain, process and display accurate color images.

ACKNOWLEDGMENT

The author would like to acknowledge Michael Vrhel for his contribution to the section on color calibration. Most of the material in that section was the result of a joint paper with him [22].

REFERENCES

[1] MATLAB. *High Performance Numeric Computation and Visualization Software*. The Mathworks Inc., Natick, MA.

[2] A. V. Oppenheim and R. W. Schafer. *Discrete-Time Signal Processing*. Prentice-Hall, Upper Saddle River, NJ, 1989.

[3] A. K. Jain. *Fundamentals of Digital Image Processing*. Prentice-Hall, Englewood Cliffs, NJ, 1989.

[4] N. P. Galatsanos and R. T. Chin. Digital restoration of multichannel images. *IEEE Trans. Acoust.*, ASSP-37(3):415–421, 1989.

[5] G. Wyszecki and W. S. Stiles. *Color Science: Concepts and Methods, Quantitative Data and Formulae*, 2nd ed. John Wiley and Sons, New York, 1982.

[6] D. E. Dudgeon and R. M. Mersereau. *Multidimensional Digital Signal Processing*. Prentice-Hall, Upper Saddle River, NJ, 1984.

[7] B. A. Wandell. *Foundations of Vision*. Sinauer Assoc. Inc., Sunderland, MA, 1995.

[8] H. B. Barlow and J. D. Mollon. *The Senses*. Cambridge University Press, Cambridge, UK, 1982.

[9] H. Grassmann. Zur therorie der farbenmischung. *Ann. Phys.*, 89:69–84, 1853.

[10] H. Grassmann. On the theory of compound colours. *Philos. Mag.*, 7(4):254–264, 1854.

[11] B. K. P. Horn. Exact reproduction of colored images. *Comput. Vision Graph. Image Process.*, 26:135–167, 1984.

[12] B. A. Wandell. The synthesis and analysis of color images. *IEEE Trans. Pattern. Anal. Mach. Intell.*, PAMI-9(1):2–13, 1987.

[13] J. B. Cohen and W. E. Kappauf. Metameric color stimuli, fundamental metamers, and Wyszecki's metameric blacks. *Am. J. Psychol.*, 95(4):537–564, 1982.

[14] H. J. Trussell. Application of set theoretic methods to color systems. *Color Res. Appl.*, 16(1):31–41, 1991.

[15] H. J. Trussell and M. S. Kulkarni. Sampling and processing of color signals. *IEEE Trans. Image Process.*, 5(4):677–681, 1996.

[16] P. L. Vora and H. J. Trussell. Measure of goodness of a set of colour scanning filters. *J. Opt. Soc. Am.*, 10(7):1499–1508, 1993.

[17] M. J. Vrhel and H. J. Trussell. Optimal color filters in the presence of noise. *IEEE Trans. Image Process.*, 4(6):814–823, 1995.

[18] D. L. MacAdam. Visual sensitivities to color differences in daylight. *J. Opt. Soc. Am.*, 32(5):247–274, 1942.

[19] G. Wyszecki and G. H. Felder. New color matching ellipses. *J. Opt. Soc. Am.*, 62:1501–1513, 1971.

[20] CIE. *Industrial Colour Difference Evaluation.* Technical Report 116–1995, CIE, 1995.

[21] M. J. Vrhel, R. Gershon, and L. S. Iwan. Measurement and analysis of object reflectance spectra. *Color Res. Appl.*, 19:4–9, 1994.

[22] M. J. Vrhel and H. J. Trussell. Color device calibration: a mathematical formulation. *IEEE Trans. Image Process.*, 1999.

[23] International Color Consortium. *Int. Color Consort. Profile Format Ver. 3.4*, available at http://color.org/.

[24] W. B. Cowan. An inexpensive scheme for calibration of a color monitor in terms of standard CIE coordinates. *Comput. Graph.*, 17:315–321, 1983.

[25] R. S. Berns, R. J. Motta, and M. E. Grozynski. CRT colorimetry. Part I: theory and *practice. Color Res. Appl.*, 18:5–39, 1988.

[26] R. S. Berns, R. J. Motta, and M. E. Grozynski. CRT colorimetry. Part II: metrology. *Color Res. Appl.*, 18:315–325, 1988.

[27] P. L. Vora and H. J. Trussell. Mathematical methods for the design of color scanning filters. *IEEE Trans. Image Process.*, IP-6(2):312–320, 1997.

[28] G. Sharma, H. J. Trussell, and M. J. Vrhel. Optimal nonnegative color scanning filters. *IEEE Trans. Image Process.*, 7(1):129–133, 1998.

[29] P. C. Hung. Colorimetric calibration in electronic imaging devices using a look-up table model and interpolations. *J. Electron. Imaging*, 2:53–61, 1993.

[30] H. R. Kang and P. G. Anderson. Neural network applications to the color scanner and printer calibrations. *J. Electron. Imaging*, 1:125–134, 1992.

[31] H. Haneishi, T. Hirao, A. Shimazu, and Y. Mikaye. Colorimetric precision in scanner calibration using matrices. In *Proc. Third IS&T/SID Color Imaging Conference: Color Science, Systems and Applications*, 106–108, 1995.

[32] H. R. Kang. Color scanner calibration. *J. Imaging Sci. Technol.*, 36:162–170, 1992.

[33] M. J. Vrhel and H. J. Trussell. Color scanner calibration via neural networks. In *Proc. Conf. on Acoust., Speech and Signal Process.*, Phoenix, AZ, March 15–19, 1999.

[34] J. Z. Chang, J. P. Allebach, and C. A. Bouman. Sequential linear interpolation of multidimensional functions. *IEEE Trans. Image Process.*, 6(9):1231–1245, 1997.

[35] *IEEE Trans. Image Process.*, 6(7): 1997.

Capturing Visual Image Properties with Probabilistic Models

9

Eero P. Simoncelli

New York University

The set of all possible visual images is enormous, but not all of these are equally likely to be encountered by your eye or a camera. This nonuniform distribution over the image space is believed to be exploited by biological visual systems, and can be used as an advantage in most applications in image processing and machine vision. For example, loosely speaking, when one observes a visual image that has been corrupted by some sort of noise, the process of estimating the original source image may be viewed as one of looking for the highest probability image that is "close to" the noisy observation. Image compression amounts to using a larger proportion of the available bits to encode those regions of the image space that are more likely. And problems such as resolution enhancement or image synthesis involve selecting (sampling) a high-probability image, subject to some set of constraints. Specific examples of these applications can be found in many chapters throughout this *Guide.*

In order to develop a probability model for visual images, we first must decide which images to model. In a practical sense, this means we must (a) decide on imaging conditions, such as the field of view, resolution, sensor or postprocessing nonlinearities and (b) decide what kind of scenes, under what kind of lighting, are to be captured in the images. It may seem odd, if one has not encountered such models, to imagine that all images are drawn from a single universal probability run. In particular, the features and properties in any given image are often specialized. For example, outdoor nature scenes contain structures that are quite different from city streets, which in turn are nothing like human faces. There are two means by which this dilemma is resolved. First, the statistical properties that we will examine are basic enough that they are relevant for essentially *all* visual scenes. Second, we will use parametric models, in which a set of hyperparameters (possibly random variables themselves) govern the detailed behavior of the model, and thus allow a certain degree of adaptability of the model to different types of source material.

205

In this chapter, we will describe an empirical methodology for building and testing probability models for discretized (pixelated) images. Currently available digital cameras record such images, typically containing millions of pixels. Naively, one could imagine examining a large set of such images to try to determine how they are distributed. But a moment's thought leads one to realize the hopelessness of the endeavor. The amount of data needed to estimate a probability distribution from samples grows exponentially in D, the dimensionality of the space (in this case, the number of pixels). This is known as the "curse of dimensionality." For example, if we wanted to build a histogram for images with one million pixels, and each pixel value was partitioned into just two possibilites (low or high), we would need $2^{1,000,000}$ bins, which greatly exceeds estimates of the number of atoms in the universe!

Thus, in order to make progress on image modeling, it is *essential* that we reduce the dimensionality of the space. Two types of simplifying assumptions can help in this regard. The first, known as a Markov assumption, is that the probability density of a pixel, when conditioned on a set of pixels in a small spatial neighborhood, is independent of the pixels outside of the neighborhood. A second type of simplification comes from imposing symmetries or invariances on the probability structure. The most common of these is that of translation-invariance (i.e., sometimes called homogeneity, or strict-sense stationarity): the probability density of pixels in a neighborhood does not depend on the absolute location of that neighborhood within the image. This seems intuitively sensible, given that a lateral or vertical translation of the camera leads (approximately) to translation of the image intensities across the pixel array. Note that translation-invariance is not well defined at the boundaries, and as is often the case in image processing, these locations must be handled specially.

Another common assumption is scale-invariance: resizing the image does not alter the probability structure. This may also be loosely justified by noting that adjusting the focal length (zoom) of a camera lens approximates (apart from perspective distortions) image resizing. As with translation-invariance, scale-invariance will clearly fail to hold at certain "boundaries." Specifically, scale-invariance must fail for discretized images at fine scales approaching the size of the pixels. And similarly, it will also fail for finite-size images at coarse scales approaching the size of the entire image.

With these sort of simplifying structural assumptions in place, we can return to the problem of developing a probability model. In recent years, researchers from image processing, computer vision, physics, psychology, applied math, and statistics have proposed a wide variety of different types of models. In this chapter, I will review the most basic statistical properties of photographic images and describe several models that have been developed to incorporate these properties. I will give some indication of how these models have been validated by examining how well they fit the data. In order to keep the discussion focused, I will limit the discussion to discretized *grayscale* photographic images. Many of the principles are easily extended to color photographs [1, 2], or temporal image sequences (movies) [3], as well as more specialized image classes such as portraits, landscapes, or textures. In addition, the general concepts are often applicable to nonvisual imaging devices, such as medical images, infrared images, radar and other types of range images, or astronomical images.

9.1 THE GAUSSIAN MODEL

The classical model of image statistics was developed by television engineers in the 1950s (see [4] for a review), who were interested in optimal signal representation and transmission. The most basic motivation for these models comes from the observation that pixels at nearby locations tend to have similar intensity values. This is easily confirmed by measurements like those shown in Fig. 9.1(a). Each scatterplot shows values of a pair of pixels[1] with a different relative horizontal displacement. Implicit in these measurements is the assumption of homogeneity mentioned in the introduction: the distributions are assumed to be independent of the absolute location within the image.

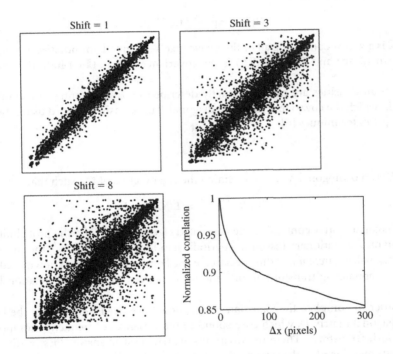

FIGURE 9.1

(a) Scatterplots comparing values of pairs of pixels at three different spatial displacements, averaged over five example images; (b) Autocorrelation function. Photographs are of New York City street scenes, taken with a Canon 10D digital camera in RAW mode (these are the sensor measurements which are approximately proportional to light intensity). The scatterplots and correlations were computed on the logs of these sensor intensity values [4].

---

[1]Pixel values recorded by digital cameras are generally nonlinearly related to the light intensity that fell on the sensor. Here, we used linear measurements in a single image of a New York City street scene, as recorded by the CMOS sensor, and took the log of these.

The most striking behavior observed in the plots is that the pixel values are highly correlated: when one is large, the other tends to also be large. This correlation weakens with the distance between pixels. This behavior is summarized in Fig. 9.1(b), which shows the image autocorrelation (pixel correlation as a function of separation).

The correlation statistics of Fig. 9.1 place a strong constraint on the structure of images, but they do not provide a full probability model. Specifically, there are many probability densities that would share the same correlation (or equivalently, covariance) structure. How should we choose a model from amongst this set? One natural criterion is to select a density that has maximal entropy, subject to the covariance constraint [5]. Solving for this density turns out to be relatively straighforward, and the result is a multidimensional Gaussian:

$$\mathcal{P}(\vec{x}) \propto \exp(-\vec{x}^T \mathbf{C_x}^{-1} \vec{x}/2), \tag{9.1}$$

where $\vec{x}$ is a vector containing all of the image pixels (assumed, for notational simplicity, to be zero-mean) and $\mathbf{C_x} \equiv \mathbb{E}(\vec{x}\vec{x}^T)$ is the covariance matrix ($\mathbb{E}(\cdot)$ indicates expected value).

Gaussian densities are more succinctly described by transforming to a coordinate system in which the covariance matrix is diagonal. This is easily achieved using standard linear algebra techniques [6]:

$$\vec{y} = E^T \vec{x},$$

where E is an orthogonal matrix containing the eigenvectors of $\mathbf{C_x}$, such that

$$\mathbf{C_x} = EDE^T, \qquad \Rightarrow E^T \mathbf{C_x} E = D. \tag{9.2}$$

D is a diagonal matrix containing the associated eigenvalues. When the probability distribution on $\vec{x}$ is stationary (assuming periodic handling of boundaries), the covariance matrix, $\mathbf{C_x}$, will be *circulant*. In this special case, the Fourier transform is known in advance to be a diagonalizing transformation,[2] and is guaranteed to satisfy the relationship of Eq. (9.2).

In order to complete the Gaussian image model, we need only specify the entries of the diagonal matrix D, which correspond to the variances of frequency components in the Fourier transform. There are two means of arriving at an answer. First, setting aside the caveats mentioned in the introduction, we can assume that image statistics are scale-invariant. Specifically, suppose that the second-order (covariance) statistical properties of the image are invariant to resizing of the image. We can express scale-invariance in the frequency domain as:

$$\mathbb{E}\left(|F(s\vec{\omega})|^2\right) = h(s)\mathbb{E}\left(|F(\vec{\omega})|^2\right), \qquad \forall \vec{\omega}, s$$

---

[2]More generally, the Fourier transform diagonalizes any matrix that represents a translation-invariant (i.e., convolution) operation.

where $F(\vec{\omega})$ indicates the (2D) Fourier transform of the image. That is, rescaling the frequency axis does not change the shape of the function; it merely multiplies the spectrum by a constant. The only functions that satisfy this identity are power laws:

$$\mathbb{E}\left(|F(\vec{\omega})|^2\right) = \frac{A}{|\vec{\omega}|^\gamma},$$

where the exponent γ controls the rate at which the spectrum falls. Thus, the dual assumptions of translation- and scale-invariance constrains the covariance structure of images to a model with two parameters!

Alternatively, the form of the power spectrum may be estimated empirically [e.g., 7–11]. For many "typical" images, it turns out to be quite well approximated by a power law, consistent with the scale-invariance assumption. In these empirical measurements, the value of the exponent is typically near two. Examples of power spectral estimates for several example images are shown in Fig. 9.2. It has also been demonstrated that scale-invariance holds for statistics other than the power spectrum [e.g., 10, 12].

The spectral model is the classic model of image processing. In addition to accounting for spectra of typical image data, the simplicity of the Gaussian form leads to direct solutions for image compression and denoising that may be found in nearly every textbook on signal or image processing. As an example, consider the problem of removing additive Gaussian white noise from an image, $\vec{x}$. The degradation process is described

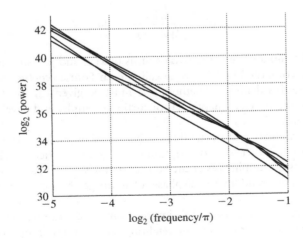

FIGURE 9.2

Power spectral estimates for five example images (see Fig. 9.1 for image description), as a function of spatial frequency, averaged over orientation. These are well described by power law functions with an exponent, γ, slightly larger than 2.0.

by the conditional density of the observed (noisy) image, $\vec{y}$, given the original (clean) image $\vec{x}$:

$$\mathcal{P}(\vec{y}|\vec{x}) \propto \exp(-||\vec{y} - \vec{x}||^2/2\sigma_n^2),$$

where σ_n^2 is the variance of the noise. Using Bayes' rule, we can reverse the conditioning by multiplying by the prior probability density on $\vec{x}$:

$$\mathcal{P}(\vec{x}|\vec{y}) \propto \exp(-||\vec{y} - \vec{x}||^2/2\sigma_n^2) \cdot \mathcal{P}(\vec{x}).$$

An estimate $\hat{x}$ for $\vec{x}$ may now be obtained from this posterior density. One can, for example, choose the $\vec{x}$ that maximizes the probability (the *maximum a posteriori* or MAP estimate), or the mean of the density (the *minimum mean squared error* (MMSE) or *Bayes Least Squares* (BLS estimate). If we assume that the prior density is Gaussian, then the posterior density will also be Gaussian, and the maximum and the mean will then be identical:

$$\hat{x}(\vec{y}) = \mathbf{C_x}(\mathbf{C_x} + \mathbf{I}\sigma_n^2)^{-1}\vec{y},$$

where $\mathbf{I}$ is an identity matrix. Note that this solution is linear in the observed (noisy) image $\vec{y}$.

This linear estimator is particularly simple when both the noise and signal covariance matrices are diagonalized. As mentioned previously, under the spectral model , the signal covariance matrix may be diagonlized by transforming to the Fourier domain, where the estimator may be written as:

$$\hat{F}(\vec{\omega}) = \frac{A/|\vec{\omega}|^\gamma}{A|\vec{\omega}|^\gamma + \sigma_n^2} \cdot G(\vec{\omega}),$$

where $\hat{F}(\vec{\omega})$ and $G(\vec{\omega})$ are the Fourier transforms of $\hat{x}(\vec{y})$ and $\vec{y}$, respectively. Thus, the estimate may be computed by linearly rescaling each Fourier coefficient individually. In order to apply this denoising method, one must be given (or must estimate) the parameters A, γ, and σ_n (see Chapter 11 for further examples and development of the denoising problem).

Despite the simplicity and tractability of the Gaussian model, it is easy to see that the model provides a rather weak description of images. In particular, while the model strongly constrains the amplitudes of the Fourier coefficients, it places no constraint on their *phases*. When one randomizes the phases of an image, the appearance is completely destroyed [13].

As a direct test, one can draw sample images from the distribution by simply generating white noise in the Fourier domain, weighting each sample appropriately by $1/|\vec{\omega}|^\gamma$, and then inverting the transform to generate an image. The fact that this experiment invariably produces images of clouds (an example is shown in Fig. 9.3) implies that a Gaussian model is insufficient to capture the structure of features that are found in photographic images.

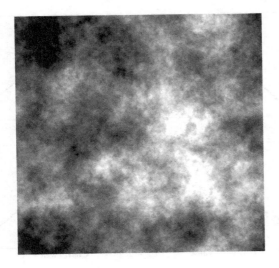

FIGURE 9.3

Example image randomly drawn from the Gaussian spectral model, with $\gamma = 2.0$.

9.2 THE WAVELET MARGINAL MODEL

For decades, the inadequacy of the Gaussian model was apparent. But direct improvement, through introduction of constraints on the Fourier phases, turned out to be quite difficult. Relationships between phase components are not easily measured, in part because of the difficulty of working with joint statistics of circular variables, and in part because the dependencies between phases of different frequencies do not seem to be well captured by a model that is localized in frequency. A breakthrough occurred in the 1980s, when a number of authors began to describe more direct indications of non-Gaussian behaviors in images. Specifically, a multidimensional Gaussian statistical model has the property that all conditional or marginal densities must also be Gaussian. But these authors noted that histograms of bandpass-filtered natural images were highly non-Gaussian [8, 14–17]. Specifically, their marginals tend to be much more sharply peaked at zero, with more extensive tails, when compared with a Gaussian of the same variance. As an example, Fig. 9.4 shows histograms of three images, filtered with a Gabor function (a Gaussian-windowed sinuosoidal grating). The intuitive reason for this behavior is that images typically contain smooth regions, punctuated by localized "features" such as lines, edges, or corners. The smooth regions lead to small filter responses that generate the sharp peak at zero, and the localized features produce large-amplitude responses that generate the extensive tails.

   This basic behavior holds for essentially any zero-mean local filter, whether it is nondirectional (center-surround), or oriented, but some filters lead to responses that are

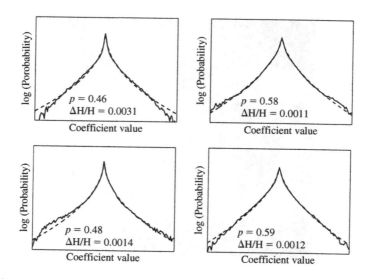

FIGURE 9.4

Log histograms of bandpass (Gabor) filter responses for four example images (see Fig. 9.1 for image description). For each histogram, tails are truncated so as to show 99.8% of the distribution. Also shown (dashed lines) are fitted generalized Gaussian densities, as specified by Eq. (9.3). Text indicates the maximum-likelihood value of p of the fitted model density, and the relative entropy (Kullback-Leibler divergence) of the model and histogram, as a fraction of the total entropy of the histogram.

more non-Gaussian than others. By the mid-1990s, a number of authors had developed methods of optimizing a basis of filters in order to maximize the non-Gaussianity of the responses [e.g., 18, 19]. Often these methods operate by optimizing a higher-order statistic such as kurtosis (the fourth moment divided by the squared variance). The resulting basis sets contain oriented filters of different sizes with frequency bandwidths of roughly one octave. Figure 9.5 shows an example basis set, obtained by optimizing kurtosis of the marginal responses to an ensemble of 12×12 pixel blocks drawn from a large ensemble of natural images. In parallel with these statistical developments, authors from a variety of communities were developing multiscale orthonormal bases for signal and image analysis, now generically known as "wavelets" (see Chapter 6 in this *Guide*). These provide a good approximation to optimized bases such as that shown in Fig. 9.5.

Once we have transformed the image to a multiscale representation, what statistical model can we use to characterize the coefficients? The statistical motivation for the choice of basis came from the shape of the marginals, and thus it would seem natural to assume that the coefficients within a subband are independent and identically distributed. With this assumption, the model is completely determined by the marginal statistics of the coefficients, which can be examined empirically as in the examples of Fig. 9.4. For natural images, these histograms are surprisingly well described by a two-parameter

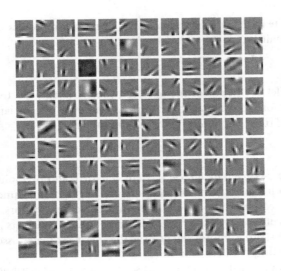

FIGURE 9.5

Example basis functions derived by optimizing a marginal kurtosis criterion [see 22].

generalized Gaussian (also known as a *stretched*, or *generalized* exponential) distribution [e.g., 16, 20, 21]:

$$\mathcal{P}_c(c;s,p) = \frac{\exp(-|c/s|^p)}{Z(s,p)}, \tag{9.3}$$

where the normalization constant is $Z(s,p) = 2\frac{s}{p}\Gamma(\frac{1}{p})$. An exponent of $p = 2$ corresponds to a Gaussian density, and $p = 1$ corresponds to the Laplacian density. In general, smaller values of p lead to a density that is both more concentrated at zero and has more expansive tails. Each of the histograms in Fig. 9.4 is plotted with a dashed curve corresponding to the best fitting instance of this density function, with the parameters $\{s,p\}$ estimated by maximizing the probability of the data under the model. The density model fits the histograms remarkably well, as indicated numerically by the relative entropy measures given below each plot. We have observed that values of the exponent p typically lie in the range $[0.4, 0.8]$. The factor s varies monotonically with the scale of the basis functions, with correspondingly higher variance for coarser-scale components.

This wavelet marginal model is significantly more powerful than the classical Gaussian (spectral) model. For example, when applied to the problem of compression, the entropy of the distributions described above is significantly less than that of a Gaussian with the same variance, and this leads directly to gains in coding efficiency. In denoising, the use of this model as a prior density for images yields to significant improvements over the Gaussian model [e.g., 20, 21, 23–25]. Consider again the problem of removing additive Gaussian white noise from an image. If the wavelet transform is orthogonal, then the noise remains white in the wavelet domain. The degradation process may be described in the wavelet domain as:

$$\mathcal{P}(d|c) \propto \exp(-(d-c)^2/2\sigma_n^2),$$

where d is a wavelet coefficient of the observed (noisy) image, c is the corresponding wavelet coefficient of the original (clean) image, and σ_n^2 is the variance of the noise. Again, using Bayes' rule, we can reverse the conditioning:

$$\mathcal{P}(c|d) \propto \exp(-(d-c)^2/2\sigma_n^2) \cdot \mathcal{P}(c),$$

where the prior on c is given by Eq. (9.3). Here, the MAP and BLS solutions cannot, in general, be written in closed form, and they are unlikely to be the same. But numerical solutions are fairly easy to compute, resulting in nonlinear estimators, in which small-amplitude coefficients are suppressed and large-amplitude coefficients preserved. These estimates show substantial improvement over the linear estimates associated with the Gaussian model of the previous section.

Despite these successes, it is again easy to see that important attributes of images are not captured by wavelet marginal models. When the wavelet transform is orthonormal, we can easily draw statistical samples from the model. Figure 9.6 shows the result of drawing the coefficients of a wavelet representation independently from generalized Gaussian densities. The density parameters for each subband were chosen as those that best fit an example photographic image. Although it has more structure than an image of white noise, and perhaps more than the image drawn from the spectral model (Fig. 9.3), the result still does not look very much like a photographic image!

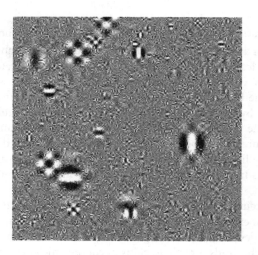

FIGURE 9.6

A sample image drawn from the wavelet marginal model, with subband density parameters chosen to fit the image of Fig. 9.7.

The wavelet marginal model may be improved by extending it to an *overcomplete* wavelet basis. In particular, Zhu *et al.* have shown that large numbers of marginals are sufficient to uniquely constrain a high-dimensional probability density [26] (this is a variant of the Fourier projection-slice theorem used for tomographic reconstruction). Marginal models have been shown to produce better denoising results when the multiscale representation is overcomplete [20, 27–30]. Similar benefits have been obtained for texture representation and synthesis [26, 31]. The drawback of these models is that the joint statistical properties are defined *implicitly* through the marginal statistics. They are thus difficult to study directly, or to utilize in deriving optimal solutions for image processing applications. In the next section, we consider the more direct development of joint statistical descriptions.

9.3 WAVELET LOCAL CONTEXTUAL MODELS

The primary reason for the poor appearance of the image in Fig. 9.6 is that the coefficients of the wavelet transform are not independent. Empirically, the coefficients of orthonormal wavelet decompositions of visual images are found to be moderately well decorrelated (i.e., their covariance is near zero). But this is only a statement about their *second-order* dependence, and one can easily see that there are important higher order dependencies. Figure 9.7 shows the amplitudes (absolute values) of coefficients in a four-level separable orthonormal wavelet decomposition. First, we can see that individual subbands are not homogeneous: Some regions have large-amplitude coefficients, while other regions are relatively low in amplitude. The variability of the local amplitude is characteristic of most photographic images: the large-magnitude coefficients tend to occur near each other within subbands, and also occur at the same relative spatial locations in subbands at adjacent scales and orientations.

The intuitive reason for the clustering of large-amplitude coefficients is that typical localized and isolated image features are represented in the wavelet domain via the superposition of a group of basis functions at different positions, orientations, and scales. The signs and relative magnitudes of the coefficients associated with these basis functions will depend on the precise location, orientation, and scale of the underlying feature. The magnitudes will also scale with the contrast of the structure. Thus, measurement of a large coefficient at one scale means that large coefficients at adjacent scales are more likely.

This clustering property was exploited in a heuristic but highly effective manner in the Embedded Zerotree Wavelet (EZW) image coder [32], and has been used in some fashion in nearly all image compression systems since. A more explicit description had been first developed for denoising, when Lee [33] suggested a two-step procedure, in which the local signal variance is first estimated from a neighborhood of observed pixels, after which the pixels in the neighborhood are denoised using a standard linear least squares method. Although it was done in the pixel domain, this chapter introduced the idea that variance is a local property that should be estimated *adaptively*, as compared

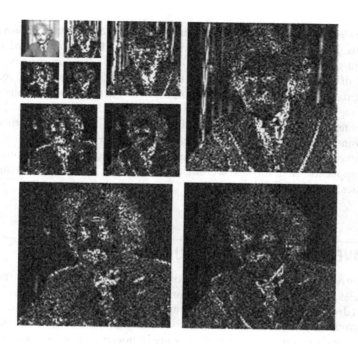

FIGURE 9.7

Amplitudes of multiscale wavelet coefficients for an image of Albert Einstein. Each subimage shows coefficient amplitudes of a subband obtained by convolution with a filter of a different scale and orientation, and subsampled by an appropriate factor. Coefficients that are spatially near each other within a band tend to have similar amplitudes. In addition, coefficients at different orientations or scales but in nearby (relative) spatial positions tend to have similar amplitudes.

with the classical Gaussian model in which one assumes a fixed global variance. It was not until the 1990s that a number of authors began to apply this concept to denoising in the wavelet domain, estimating the variance of clusters of wavelet coefficients at nearby positions, scales, and/or orientations, and then using these estimated variances in order to denoise the cluster [20, 34–39].

The locally-adaptive variance principle is powerful, but does not constitute a full probability model. As in the previous sections, we can develop a more explicit model by directly examining the statistics of the coefficients. The top row of Fig. 9.8 shows joint histograms of several different pairs of wavelet coefficients. As with the marginals, we assume homogeneity in order to consider the joint histogram of this pair of coefficients, gathered over the spatial extent of the image, as representative of the underlying density. Coefficients that come from adjacent basis functions are seen to produce contours that are nearly circular, whereas the others are clearly extended along the axes.

The joint histograms shown in the first row of Fig. 9.8 do not make explicit the issue of whether the coefficients are independent. In order to make this more explicit, the bottom row shows *conditional* histograms of the same data. Let x_2 correspond to the

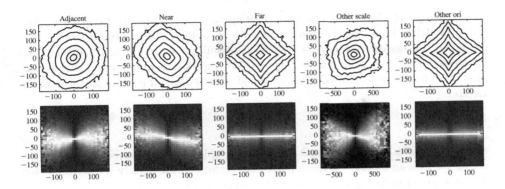

FIGURE 9.8

Empirical joint distributions of wavelet coefficients associated with different pairs of basis functions, for a single image of a New York City street scene (see Fig. 9.1 for image description). The top row shows joint distributions as contour plots, with lines drawn at equal intervals of log probability. The three leftmost examples correspond to pairs of basis functions at the same scale and orientation, but separated by different spatial offsets. The next corresponds to a pair at adjacent scales (but the same orientation, and nearly the same position), and the rightmost corresponds to a pair at orthogonal orientations (but the same scale and nearly the same position). The bottom row shows corresponding conditional distributions: brightness corresponds to frequency of occurance, except that each column has been independently rescaled to fill the full range of intensities.

density coefficient (vertical axis), and x_1 the conditioning coefficient (horizontal axis). The histograms illustrate several important aspects of the relationship between the two coefficients. First, the expected value of x_2 is approximately zero for all values of x_1, indicating that they are nearly decorrelated (to second order). Second, the variance of the conditional histogram of x_2 clearly depends on the value of x_1, and the strength of this dependency depends on the particular pair of coefficients being considered. Thus, although x_2 and x_1 are uncorrelated, they still exhibit statistical dependence!

The form of the histograms shown in Fig. 9.8 is surprisingly robust across a wide range of images. Furthermore, the qualitative form of these statistical relationships also holds for pairs of coefficients at adjacent spatial locations and adjacent orientations. As one considers coefficients that are more distant (either in spatial position or in scale), the dependency becomes weaker, suggesting that a Markov assumption might be appropriate.

Essentially all of the statistical properties we have described thus far—the circular (or elliptical) contours, the dependency between local coefficient amplitudes, as well as the heavy-tailed marginals—can be modeled using a random field with a spatially fluctuating variance. These kinds of models have been found useful in the speech-processing community [40]. A related set of models, known as autoregressive conditional heteroskedastic (ARCH) models [e.g., 41], have proven useful for many real signals that suffer from abrupt fluctuations, followed by relative "calm" periods (stock market prices, for example). Finally, physicists studying properties of turbulence have noted similar behaviors [e.g., 42].

An example of a local density with fluctuating variance, one that has found particular use in modeling local clusters (neighborhoods) of multiscale image coefficients, is the product of a Gaussian vector and a hidden scalar multiplier. More formally, this model, known as a *Gaussian scale mixture* [43] (GSM), expresses a random vector $\vec{x}$ as the product of a zero-mean Gaussian vector $\vec{u}$ and an independent positive scalar random variable $\sqrt{z}$:

$$\vec{x} \sim \sqrt{z}\vec{u}, \tag{9.4}$$

where $\sim$ indicates equality in distribution. The variable z is known as the *multiplier*. The vector $\vec{x}$ is thus an infinite mixture of Gaussian vectors, whose density is determined by the covariance matrix $\mathbf{C_u}$ of vector $\vec{u}$ and the mixing density, $p_z(z)$:

$$
\begin{aligned}
p_{\vec{x}}(\vec{x}) &= \int p(\vec{x}|z)\, p_z(z)\, dz \\
&= \int \frac{\exp\left(-\vec{x}^T(z\mathbf{C_u})^{-1}\vec{x}/2\right)}{(2\pi)^{N/2}|z\mathbf{C_u}|^{1/2}}\, p_z(z)\, dz,
\end{aligned} \tag{9.5}
$$

where N is the dimensionality of $\vec{x}$ and $\vec{u}$ (in our case, the size of the neighborhood). Notice that since the level surfaces (contours of constant probability) for $P_{\vec{u}}(\vec{u})$ are ellipses determined by the covariance matrix $\mathbf{C_u}$, and the density of $\vec{x}$ is constructed as a mixture of scaled versions of the density of $\vec{u}$, then $P_{\vec{x}}(\vec{x})$ will *also* exhibit the same elliptical level surfaces. In particular, if $\vec{u}$ is spherically symmetric ($\mathbf{C_u}$ is a multiple of the identity), then $\vec{x}$ will also be spherically symmetric. Figure 9.9 demonstrates that this model can capture the strongly kurtotic behavior of the marginal densities of natural image wavelet coefficients, as well as the correlation in their local amplitudes.

A number of recent image models describe the wavelet coefficients within each local neighborhood using a Gaussian mixture model [e.g., 37, 38, 44–48]. Sampling from these models is difficult, since the local description is typically used for *overlapping* neighborhoods, and thus one cannot simply draw independent samples from the model (see [48] for an example). The underlying Gaussian structure of the model allows it to be adapted for problems such as denoising. The resulting estimator is more complex than that described for the Gaussian or wavelet marginal models, but performance is significantly better.

As with the models of the previous two sections, there are indications that the GSM model is insufficient to fully capture the structure of typical visual images. To demonstrate this, we note that normalizing each coefficient by (the square root of) its estimated variance should produce a field of Gaussian white noise [4, 49]. Figure 9.10 illustrates this process, showing an example wavelet subband, the estimated variance field, and the normalized coefficients. But note that there are two important types of structure that remain. First, although the normalized coefficients are certainly closer to a homogeneous field, the *signs* of the coefficients still exhibit important structure. Second, the variance field itself is far from homogeneous, with most of the significant values concentrated on one-dimensional contours. Some of these attributes can be captured by measuring joint statistics of phase and amplitude, as has been demonstrated in texture modeling [50].

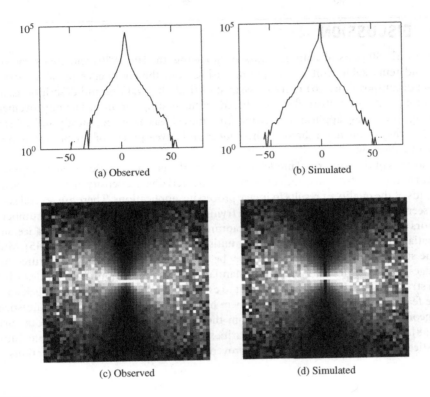

(a) Observed

(b) Simulated

(c) Observed

(d) Simulated

FIGURE 9.9

Comparison of statistics of coefficients from an example image subband (left panels) with those generated by simulation of a local GSM model (right panels). Model parameters (covariance matrix and the multiplier prior density) are estimated by maximizing the likelihood of the subband coefficients (see [47]). (a,b) Log of marginal histograms. (c,d) Conditional histograms of two spatially adjacent coefficients. Pixel intensity corresponds to frequency of occurance, except that each column has been independently rescaled to fill the full range of intensities.

Original coefficients   Estimated $\sqrt{z}$ field   Normalized coefficients

FIGURE 9.10

Example wavelet subband, square root of the variance field, and normalized subband.

9.4 DISCUSSION

After nearly 50 years of Fourier/Gaussian modeling, the late 1980s and 1990s saw sudden and remarkable shift in viewpoint, arising from the confluence of (a) multiscale image decompositions, (b) non-Gaussian statistical observations and descriptions, and (c) locally-adaptive statistical models based on fluctuating variance. The improvements in image processing applications arising from these ideas have been steady and substantial. But the complete synthesis of these ideas and development of further refinements are still underway.

Variants of the contextual models described in the previous section seem to represent the current state-of-the-art, both in terms of characterizing the density of coefficients, and in terms of the quality of results in image processing applications. There are several issues that seem to be of primary importance in trying to extend such models. First, a number of authors are developing models that can capture the regularities in the local variance, such as spatial random fields [48, 51–53], and multiscale tree-structured models [38, 45]. Much of the structure in the variance field may be attributed to discontinuous features such as edges, lines, or corners. There is substantial literature in computer vision describing such structures, but it has proven difficult to establish models that are both explicit about these features and yet flexible. Finally, there have been several recent studies investigating geometric regularities that arise from the continuity of contours and boundaries [54–58]. These and other image regularities will surely be incorporated into future statistical models, leading to further improvements in image processing applications.

REFERENCES

[1] G. Buchsbaum and A. Gottschalk. Trichromacy, opponent color coding, and optimum colour information transmission in the retina. *Proc. R. Soc. Lond., B, Biol. Sci.,* 220:89–113, 1983.

[2] D. L. Ruderman, T. W. Cronin, and C.-C. Chiao. Statistics of cone responses to natural images: implications for visual coding. *J. Opt. Soc. Am. A,* 15(8):2036–2045, 1998.

[3] D. W. Dong and J. J. Atick. Statistics of natural time-varying images. *Network Comp. Neural,* 6:345–358, 1995.

[4] D. L. Ruderman. The statistics of natural images. *Network Comp. Neural,* 5:517–548, 1996.

[5] E. T. Jaynes. Where do we stand on maximum entropy? In R. D. Levine and M. Tribus, editors, *The Maximal Entropy Formalism.* MIT Press, Cambridge, MA, 1978.

[6] G. Strang. *Linear Algebra and its Applications.* Academic Press, Orlando, FL, 1980.

[7] N. G. Deriugin. The power spectrum and the correlation function of the television signal. *Telecomm.,* 1(7):1–12, 1956.

[8] D. J. Field. Relations between the statistics of natural images and the response properties of cortical cells. *J. Opt. Soc. Am. A,* 4(12):2379–2394, 1987.

[9] D. J. Tolhurst, Y. Tadmor, and T. Chao. Amplitude spectra of natural images. *Ophthalmic Physiol. Opt.,* 12:229–232, 1992.

[10] D. L. Ruderman and W. Bialek. Statistics of natural images: scaling in the woods. *Phys. Rev. Lett.*, 73(6):814–817, 1994.

[11] A. van der Schaaf and J. H. van Hateren. Modelling the power spectra of natural images: statistics and information. *Vision Res.*, 28(17):2759–2770, 1996.

[12] A. Turiel and N. Parga. The multi-fractal structure of contrast changes in natural images: from sharp edges to textures. *Neural. Comput.*, 12:763–793, 2000.

[13] A. V. Oppenheim and J. S. Lim. The importance of phase in signals. *Proc. IEEE*, 69:529–541, 1981.

[14] P. J. Burt and E. H. Adelson. The Laplacian pyramid as a compact image code. *IEEE Trans. Comm.*, COM-31(4):532–540, 1983.

[15] J. G. Daugman. Complete discrete 2-D Gabor transforms by neural networks for image analysis and compression. *IEEE Trans. Acoust.*, 36(7):1169–1179, 1988.

[16] S. G. Mallat. A theory for multiresolution signal decomposition: the wavelet representation. *IEEE Trans. Pattern Anal. Mach. Intell.*, 11:674–693, 1989.

[17] C. Zetzsche and E. Barth. Fundamental limits of linear filters in the visual processing of two-dimensional signals. *Vision Res.*, 30:1111–1117, 1990.

[18] B. A. Olshausen and D. J. Field. Sparse coding with an overcomplete basis set: a strategy employed by V1? *Vision Res.*, 37:3311–3325, 1997.

[19] A. J. Bell and T. J. Sejnowski. The independent components of natural scenes are edge filters. *Vision Res.*, 37(23):3327–3338, 1997.

[20] E. P. Simoncelli. Bayesian denoising of visual images in the wavelet domain. In P. Müller and B. Vidakovic, editors, *Bayesian Inference in Wavelet Based Models*, Vol. 141, 291–308. Springer-Verlag, New York, Lecture Notes in Statistics, 1999.

[21] P. Moulin and J. Liu. Analysis of multiresolution image denoising schemes using a generalized Gaussian and complexity priors. *IEEE Trans. Inf. Theory*, 45:909–919, 1999.

[22] B. A. Olshausen and D. J. Field. Emergence of simple-cell receptive field properties by learning a sparse code for natural images. *Nature*, 381:607–609, 1996.

[23] E. P. Simoncelli and E. H. Adelson. Noise removal via Bayesian wavelet coring. In *Proc. 3rd IEEE Int. Conf. on Image Process.*, Vol. I, 379–382, IEEE Signal Processing Society, Lausanne, September 16–19, 1996.

[24] H. A. Chipman, E. D. Kolaczyk, and R. M. McCulloch. Adaptive Bayesian wavelet shrinkage. *J Am. Stat. Assoc.*, 92(440):1413–1421, 1997.

[25] F. Abramovich, T. Sapatinas, and B. W. Silverman. Wavelet thresholding via a Bayesian approach. *J. Roy. Stat. Soc. B*, 60:725–749, 1998.

[26] S. C. Zhu, Y. N. Wu, and D. Mumford. FRAME: filters, random fields and maximum entropy – towards a unified theory for texture modeling. *Int. J. Comput. Vis.*, 27(2):1–20, 1998.

[27] R. R. Coifman and D. L. Donoho. Translationinvariant de-noising. In A. Antoniadis and G. Oppenheim, editors, *Wavelets and Statistics*, Springer-Verlag, Lecture notes, San Diego, CA, 1995.

[28] F. Abramovich, T. Sapatinas, and B. W. Silverman. Stochastic expansions in an overcomplete wavelet dictionary. *Probab. Theory Rel.*, 117:133–144, 2000.

[29] X. Li and M. T. Orchard. Spatially adaptive image denoising under overcomplete expansion. In *IEEE Int. Conf. on Image Process.*, Vancouver, September 2000.

[30] M. Raphan and E. P. Simoncelli. Optimal denoising in redundant representations. *IEEE Trans. Image Process.*, 17(8):1342–1352, 2008.

[31] D. Heeger and J. Bergen. Pyramid-based texture analysis/synthesis. In *Proc. ACM SIGGRAPH*, 229–238. Association for Computing Machinery, August 1995.

[32] J. Shapiro. Embedded image coding using zerotrees of wavelet coefficients. *IEEE Trans. Signal Process.*, 41(12):3445–3462, 1993.

[33] J. S. Lee. Digital image enhancement and noise filtering by use of local statistics. *IEEE T. Pattern Anal.*, PAMI-2:165–168, 1980.

[34] M. Malfait and D. Roose. Wavelet-based image denoising using a Markov random field a priori model. *IEEE Trans. Image Process.*, 6:549–565, 1997.

[35] E. P. Simoncelli. Statistical models for images: compression, restoration and synthesis. In *Proc. 31st Asilomar Conf. on Signals, Systems and Computers*, Vol. 1, 673–678, IEEE Computer Society, Pacific Grove, CA, November 2–5, 1997.

[36] S. G. Chang, B. Yu, and M. Vetterli. Spatially adaptive wavelet thresholding with context modeling for image denoising. In *Fifth IEEE Int. Conf. on Image Process.*, IEEE Computer Society, Chicago, October 1998.

[37] M. K. Mihçak, I. Kozintsev, K. Ramchandran, and P. Moulin. Low-complexity image denoising based on statistical modeling of wavelet coefficients. *IEEE Signal Process. Lett.*, 6(12):300–303, 1999.

[38] M. J. Wainwright, E. P. Simoncelli, and A. S. Willsky. Random cascades on wavelet trees and their use in modeling and analyzing natural imagery. *Appl. Comput. Harmonic Anal.*, 11(1):89–123, 2001.

[39] F. Abramovich, T. Besbeas, and T. Sapatinas. Empirical Bayes approach to block wavelet function estimation. *Comput. Stat. Data. Anal.*, 39:435–451, 2002.

[40] H. Brehm and W. Stammler. Description and generation of spherically invariant speech-model signals. *Signal Processing*, 12:119–141, 1987.

[41] T. Bollersley, K. Engle, and D. Nelson. ARCH models. In B. Engle and D. McFadden, editors, *Handbook of Econometrics* IV, North Holland, Amsterdam, 1994.

[42] A. Turiel, G. Mato, N. Parga, and J. P. Nadal. The self-similarity properties of natural images resemble those of turbulent flows. *Phys. Rev. Lett.*, 80:1098–1101, 1998.

[43] D. Andrews and C. Mallows. Scale mixtures of normal distributions. *J. Roy. Stat. Soc.*, 36:99–102, 1974.

[44] M. S. Crouse, R. D. Nowak, and R. G. Baraniuk. Wavelet-based statistical signal processing using hidden Markov models. *IEEE Trans. Signal Process.*, 46:886–902, 1998.

[45] J. Romberg, H. Choi, and R. Baraniuk. Bayesian wavelet domain image modeling using hidden Markov trees. In *Proc. IEEE Int. Conf. on Image Process.*, Kobe, Japan, October 1999.

[46] S. M. LoPresto, K. Ramchandran, and M. T. Orchard. Wavelet image coding based on a new generalized Gaussian mixture model. In *Data Compression Conf.*, Snowbird, Utah, March 1997.

[47] J. Portilla, V. Strela, M. J. Wainwright, and E. P. Simoncelli. Image denoising using a scale mixture of Gaussians in the wavelet domain. *IEEE Trans. Image Process.*, 12(11):1338–1351, 2003.

[48] S. Lyu and E. P. Simoncelli. Modeling multiscale subbands of photographic images with fields of Gaussian scale mixtures. *IEEE Trans. Pattern Anal. Mach. Intell.*, 2008. Accepted for publication, 4/08.

[49] M. J. Wainwright and E. P. Simoncelli. Scale mixtures of Gaussians and the statistics of natural images. In S. A. Solla, T. K. Leen, and K.-R. Müller, editors, *Advances in Neural Information Processing Systems (NIPS*99)*, Vol. 12, 855–861. MIT Press, Cambridge, MA, 2000.

[50] J. Portilla and E. P. Simoncelli. A parametric texture model based on joint statistics of complex wavelet coefficients. *Int. J. Comput. Vis.*, 40(1):49–71, 2000.

[51] A. Hyvärinen and P. Hoyer. Emergence of topography and complex cell properties from natural images using extensions of ICA. In S. A. Solla, T. K. Leen, and K.-R. Müller, editors, *Advances in Neural Information Processing Systems*, Vol. 12, 827–833. MIT Press, Cambridge, MA, 2000.

[52] Y. Karklin and M. S. Lewicki. Learning higher-order structures in natural images. *Network*, 14:483–499, 2003.

[53] A. Hyvärinen, J. Hurri, and J. Väyrynen. Bubbles: a unifying framework for low-level statistical properties of natural image sequences. *J. Opt. Soc. Am. A*, 20(7):2003.

[54] M. Sigman, G. A. Cecchi, C. D. Gilbert, and M. O. Magnasco. On a common circle: natural scenes and Gestalt rules. *Proc. Natl. Acad. Sci.*, 98(4):1935–1940, 2001.

[55] J. H. Elder and R. M. Goldberg. Ecological statistics of gestalt laws for the perceptual organization of contours. *J. Vis.*, 2(4):324–353, 2002. DOI 10:1167/2.4.5.

[56] W. S. Geisler, J. S. Perry, B. J. Super, and D. P. Gallogly. Edge co-occurance in natural images predicts contour grouping performance. *Vision Res.*, 41(6):711–724, 2001.

[57] P. Hoyer and A. Hyvärinen. A multi-layer sparse coding network learns contour coding from natural images. *Vision Res.*, 42(12):1593–1605, 2002.

[58] S.-C. Zhu. Statistical modeling and conceptualization of visual patterns. *IEEE Trans. Pattern Anal. Mach. Intell.*, 25(6):691–712, 2003.

Basic Linear Filtering with Application to Image Enhancement

10

Alan C. Bovik[1] and Scott T. Acton[2]

[1] *The University of Texas at Austin;* [2] *University of Virginia*

10.1 INTRODUCTION

Linear system theory and linear filtering play a central role in digital image processing. Many potent techniques for modifying, improving, or representing digital visual data are expressed in terms of linear systems concepts. Linear filters are used for generic tasks such as image/video contrast improvement, denoising, and sharpening, as well as for more object- or feature-specific tasks such as target matching and feature enhancement.

Much of this *Guide* deals with the application of linear filters to image and video enhancement, restoration, reconstruction, detection, segmentation, compression, and transmission. The goal of this chapter is to introduce some of the basic supporting ideas of linear systems theory as they apply to digital image filtering, and to outline some of the applications. Special emphasis is given to the topic of linear image enhancement.

We will require some basic concepts and definitions in order to proceed. The basic 2D discrete-space signal is the *2D impulse function,* defined by

$$\delta(m - p, n - q) = \begin{cases} 1; & m = p \text{ and } n = q \\ 0; & \text{else} \end{cases}. \tag{10.1}$$

Thus, (10.1) takes unit value at coordinate (p, q) and is everywhere else zero. The function in (10.1) is often termed the *Kronecker delta function* or the *unit sample sequence* [1]. It plays the same role and has the same significance as the so-called *Dirac delta function* of continuous system theory. Specifically, the response of linear systems to (10.1) will be used to characterize the general responses of such systems.

Any discrete-space image f may be expressed in terms of the impulse function (10.1):

$$f(m,n) = \sum_{p=-\infty}^{\infty} \sum_{q=-\infty}^{\infty} f(m-p, n-q)\,\delta(p,q) = \sum_{p=-\infty}^{\infty} \sum_{q=-\infty}^{\infty} f(p,q)\,\delta(m-p, n-q). \quad (10.2)$$

The expression (10.2), called the *sifting property*, has two meaningful interpretations here. First, any discrete-space image can be written as a sum of weighted, shifted unit impulses. Each weighted impulse comprises one of the pixels of the image. Second, the sum in (10.2) is in fact a discrete-space linear convolution. As is apparent, the linear convolution of any image f with the impulse function δ returns the function unchanged.

The impulse function effectively describes certain systems known as *linear space-invariant (LSI) systems*. We explain these terms next.

A 2D system **L** is a process of image transformation, as shown in Fig. 10.1:

We can write

$$g(m,n) = \mathbf{L}[f(m,n)]. \quad (10.3)$$

The system **L** is *linear* if and only if for any two constants a, b and for any $f_1(m,n)$, $f_2(m,n)$ such that

$$g_1(m,n) = \mathbf{L}[f_1(m,n)] \quad \text{and} \quad g_2(m,n) = \mathbf{L}[f_2(m,n)], \quad (10.4)$$

then

$$a \cdot g_1(m,n) + b \cdot g_2(m,n) = \mathbf{L}[a \cdot f_1(m,n) + b \cdot f_2(m,n)] \quad (10.5)$$

for every (m, n). This is often called the *superposition property* of linear systems.

The system **L** is *shift-invariant* if for every $f(m, n)$ such that (10.3) holds, then also

$$g(m-p, n-q) = \mathbf{L}[f(m-p, n-q)] \quad (10.6)$$

for any (p, q). Thus, a spatial shift in the input to **L** produces no change in the output, except for an identical shift.

The rest of this chapter will be devoted to studying systems that are linear and shift-invariant (LSI). In this and other chapters, it will be found that LSI systems can be used for many powerful image and video processing tasks. In yet other chapters, nonlinearity and/or space-variance will be shown to afford certain advantages, particularly in surmounting the inherent limitations of LSI systems.

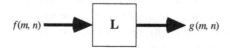

FIGURE 10.1

Two-dimensional input-output system.

10.2 IMPULSE RESPONSE, LINEAR CONVOLUTION, AND FREQUENCY RESPONSE

The *unit impulse response* of a 2D input-output system **L** is

$$\mathbf{L}[\delta(m - p, n - q)] = h(m, n; p, q). \tag{10.7}$$

This is the response of system **L**, at spatial position (m, n), to an impulse located at spatial position (p, q). Generally, the impulse response is a function of these four spatial variables. However, if the system **L** is space-invariant, then if

$$\mathbf{L}[\delta(m, n)] = h(m, n) \tag{10.8}$$

is the response to an impulse applied at the spatial origin, then also

$$\mathbf{L}[\delta(m - p, n - q)] = h(m - p, n - q), \tag{10.9}$$

which means that the response to an impulse applied at any spatial position can be found from the impulse response (10.8).

As already mentioned, the discrete-space impulse response $h(m, n)$ completely characterizes the input-output response of LSI input-output systems. This means that if the impulse response is known, then an expression can be found for the response to any input. The form of the expression is 2D discrete-space linear convolution.

Consider the generic system **L** shown in Fig. 10.1, with input $f(m, n)$ and output $g(m, n)$. Assume that the response is due to the input f only (the system would be at rest without the input). Then from (10.2):

$$g(m, n) = \mathbf{L}[f(m, n)] = \mathbf{L}\left[\sum_{p=-\infty}^{\infty} \sum_{q=-\infty}^{\infty} f(p, q)\delta(m - p, n - q) \right]. \tag{10.10}$$

If the system is known to be linear, then

$$g(m, n) = \sum_{p=-\infty}^{\infty} \sum_{q=-\infty}^{\infty} f(p, q)\mathbf{L}[\delta(m - p, n - q)] \tag{10.11}$$

$$= \sum_{p=-\infty}^{\infty} \sum_{q=-\infty}^{\infty} f(p, q)h(m, n; p, q), \tag{10.12}$$

which is all that generally can be said without further knowledge of the system and the input. If it is known that the system is space-invariant (hence LSI), then (10.12) becomes

$$g(m, n) = \sum_{p=-\infty}^{\infty} \sum_{q=-\infty}^{\infty} f(p, q)h(m - p, n - q) \tag{10.13}$$

$$= f(m, n) * h(m, n), \tag{10.14}$$

which is the 2D discrete-space linear convolution of input f with impulse response h.

The linear convolution expresses the output of a wide variety of electrical and mechanical systems. In continuous systems, the convolution is expressed as an integral. For example, with lumped electrical circuits, the convolution integral is computed in terms of the passive circuit elements (resistors, inductors, capacitors). In optical systems, the integral utilizes the point spread functions of the optics. The operations occur effectively instantaneously, with the computational speed limited only by the speed of the electrons or photons through the system elements.

However, in discrete signal and image processing systems, the discrete convolutions are calculated sums of products. This convolution can be directly evaluated at each coordinate (m, n) by a digital processor, or, as discussed in Chapter 5, it can be computed using the DFT using an FFT algorithm. Of course, if the exact linear convolution is desired, this means that the involved functions must be appropriately *zero-padded* prior to using the DFT, as discussed in Chapter 5. The DFT/FFT approach is usually, but not always faster. If an image is being convolved with a very small spatial filter, then direct computation of (10.14) can be faster.

Suppose that the input to a discrete LSI system with impulse response $h(m, n)$ is a complex exponential function:

$$f(m, n) = e^{2\pi j(Um + Vn)} = \cos[2\pi(Um + Vn)] + j\sin[2\pi(Um + Vn)]. \qquad (10.15)$$

Then the system response is the linear convolution:

$$g(m, n) = \sum_{p=-\infty}^{\infty} \sum_{q=-\infty}^{\infty} h(p, q) f(m - p, n - q) = \sum_{p=-\infty}^{\infty} \sum_{q=-\infty}^{\infty} h(p, q) e^{2\pi j[U(m-p)+V(n-q)]}$$

$$(10.16)$$

$$= e^{2\pi j(Um + Vn)} \sum_{p=-\infty}^{\infty} \sum_{q=-\infty}^{\infty} h(p, q) e^{-2\pi j(Up + Vq)}, \qquad (10.17)$$

which is exactly the input $f(m, n) = e^{2\pi j(Um + Vn)}$ multiplied by a function of (U, V) only:

$$H(U, V) = \sum_{p=-\infty}^{\infty} \sum_{q=-\infty}^{\infty} h(p, q) e^{-2\pi j(Up + Vq)} = |H(U, V)| \cdot e^{j\angle H(U,V)}. \qquad (10.18)$$

The function $H(U, V)$, which is immediately identified as the discrete-space Fourier transform (or DSFT, discussed extensively in Chapter 5) of the system impulse response, is called the *frequency response* of the system.

From (10.17) it may be seen that the response to any complex exponential sinusoid function, with frequencies (U, V), is the same sinusoid, but with its amplitude scaled by the system *magnitude response* $|H(U, V)|$ evaluated at (U, V) and with a shift equal to the system *phase response* $\angle H(U, V)$ at (U, V). The complex sinusoids are the unique functions that have this invariance property in LSI systems.

As mentioned, the impulse response $h(m, n)$ of a LSI system is sufficient to express the response of the system to any input.[1] The frequency response $H(U, V)$ is uniquely

---

[1] Strictly speaking, for any *bounded* input, and provided that the system is stable. In practical image processing systems, the inputs are invariably bounded. Also, almost all image processing filters do not involve feedback, and hence are naturally stable.

obtainable from the impulse response (and vice versa), and so contains sufficient information to compute the response to any input that has a DSFT. In fact, the output can be expressed in terms of the frequency response via $G(U, V) = F(U, V)H(U, V)$ and via the DFT/FFT with appropriate zero-padding. In fact, throughout this chapter and elsewhere, it may be assumed that whenever a DFT is being used to compute linear convolution, the appropriate zero-padding has been applied to avoid the wraparound effect of the cyclic convolution.

Usually, linear image processing filters are characterized in terms of their frequency responses, specifically by their spectrum shaping properties. Coarse descriptions that apply to many 2D image processing filters include lowpass, bandpass, or highpass. In such cases, the frequency response is primarily a function of radial frequency, and may even be circularly symmetric, viz., a function of $U^2 + V^2$ only. In other cases, the filter may be strongly directional or oriented, with response strongly depending on the frequency angle of the input. Of course, the terms lowpass, bandpass, highpass, and oriented are only rough qualitative descriptions of a system frequency response. Each broad class of filters has some generalized applications. For example, lowpass filters strongly attenuate all but the "lower" radial image frequencies (as determined by some bandwidth or cutoff frequency), and so are primarily smoothing filters. They are commonly used to reduce high-frequency noise, or to eliminate all but coarse image features, or to reduce the bandwidth of an image prior to transmission through a low-bandwidth communication channel or before subsampling the image.

A (radial frequency) bandpass filter attenuates all but an intermediate range of "middle" radial frequencies. This is commonly used for the enhancement of certain image features, such as edges (sudden transitions in intensity) or the ridges in a fingerprint. A highpass filter attenuates all but the "higher" radial frequencies, or commonly, significantly amplifies high frequencies without attenuating lower frequencies. This approach is often used for correcting images that are blurred—see Chapter 14.

Oriented filters tend to be more specialized. Such filters attenuate frequencies falling outside of a narrow range of orientations or amplify a narrow range of angular frequencies. For example, it may be desirable to enhance vertical image features as a prelude to detecting vertical structures, such as buildings.

Of course, filters may be a combination of types, such as bandpass and oriented. In fact, such filters are the most common types of basis functions used in the powerful wavelet image decompositions (Chapters 6, 11, 17, 18).

In the remainder of this chapter, we introduce the simple but important application of linear filtering for *linear image enhancement*, which specifically means attempting to smooth image noise while not disturbing the original image structure.[2]

---

[2]The term "image enhancement" has been widely used in the past to describe any operation that improves image quality by some criteria. However, in recent years, the meaning of the term has evolved to denote image-preserving noise smoothing. This primarily serves to distinguish it from similar-sounding terms, such as "image restoration" and "image reconstruction," which also have taken specific meanings.

10.3 LINEAR IMAGE ENHANCEMENT

The term "enhancement" implies a process whereby the visual quality of the image is improved. However, the term "image enhancement" has come to specifically mean a process of smoothing irregularities or noise that has somehow corrupted the image, while modifying the original image information as little as possible. The noise is usually modeled as an additive noise or as a multiplicative noise. We will consider additive noise now. As noted in Chapter 7, multiplicative noise, which is the other common type, can be converted into additive noise in a homomorphic filtering approach.

Before considering methods for image enhancement, we will make a simple model for additive noise. Chapter 7 of this *Guide* greatly elaborates image noise models, which prove particularly useful for studying image enhancement filters that are nonlinear.

We will make the practical assumption that an observed noisy image is of finite extent $M \times N : \mathbf{f} = [f(m, n); 0 \le m \le M - 1, 0 \le n \le N - 1]$. We model $\mathbf{f}$ as a sum of an original image $\mathbf{o}$ and a *noise image* $\mathbf{q}$:

$$\mathbf{f} = \mathbf{o} + \mathbf{q}, \tag{10.19}$$

where $\mathbf{n} = (m, n)$. The additive noise image $\mathbf{q}$ models an undesirable, unpredictable corruption of $\mathbf{o}$. The process $\mathbf{q}$ is called a *2D random process* or a *random field*. Random additive noise can occur as thermal circuit noise, communication channel noise, sensor noise, and so on. Quite commonly, the noise is present in the image signal before it is sampled, so the noise is also sampled coincident with the image.

In (10.19), both the original image and noise image are unknown. The goal of enhancement is to recover an image $\mathbf{g}$ that resembles $\mathbf{o}$ as closely as possible by reducing $\mathbf{q}$. If there is an adequate model for the noise, then the problem of finding $\mathbf{g}$ can be posed as an *image estimation* problem, where $\mathbf{g}$ is found as the solution to a statistical optimization problem. Basic methods for image estimation are also discussed in Chapter 7, and in some of the following chapters on image enhancement using nonlinear filters.

With the tools of Fourier analysis and linear convolution in hand, we will now outline the basic approach of image enhancement by linear filtering. More often than not, the detailed statistics of the noise process $\mathbf{q}$ are unknown. In such cases, a simple linear filter approach can yield acceptable results, if the noise satisfies certain simple assumptions.

We will assume a *zero-mean additive white noise* model. The zero-mean model is used in Chapter 3, in the context of frame averaging. The process $\mathbf{q}$ is *zero-mean* if the average or *sample mean* of R arbitrary noise samples

$$\left(\frac{1}{R}\right) \sum_{r=1}^{R} q(m_r, n_r) \to 0 \tag{10.20}$$

as R grows large (provided that the noise process is mean-ergodic, which means that the sample mean approaches the statistical mean for large samples).

The term *white noise* is an idealized model for noise that has, on the average, a broad spectrum. It is a simplified model for *wideband noise*. More precisely, if $Q(U, V)$ is the DSFT of the noise process $\mathbf{q}$, then Q is also a random process. It is called the *energy spectrum* of the random process **q**. If the noise process is white, then the average squared magnitude of $Q(U, V)$ takes constant over all frequencies in the range $[-\pi, \pi]$. In the ensemble sense, this means that the sample average of the magnitude spectra of R noise images generated from the same source becomes constant for large R:

$$\left(\frac{1}{R}\right) \sum_{r=1}^{R} |Q_r(U, V)| \to \eta \qquad (10.21)$$

for all (U, V) as R grows large. The square η^2 of the constant level is called the *noise power*. Since **q** has finite-extent $M \times N$, it has a DFT $\tilde{Q} = [\tilde{Q}(u, v) : 0 \leq u \leq M - 1, 0 \leq v \leq N - 1]$. On average, the magnitude of the noise DFT $\tilde{Q}$ will also be flat. Of course, it is highly unlikely that a given noise DSFT or DFT will actually have a flat magnitude spectrum. However, it is an effective simplified model for unknown, unpredictable broadband noise.

Images are also generally thought of as relatively broadband signals. Significant visual information may reside at mid-to-high spatial frequencies, since visually significant image details such as edges, lines, and textures typically contain higher frequencies. However, the magnitude spectrum of the image at higher image frequencies is usually relatively low; most of the image power resides in the low frequencies contributed by the dominant luminance effects. Nevertheless, the higher image frequencies are visually significant.

The basic approach to linear image enhancement is lowpass filtering. There are different types of lowpass filters that can be used; several will be studied in the following. For a given filter type, different degrees of smoothing can be obtained by adjusting the filter bandwidth. A narrower bandwidth lowpass filter will reject more of the high-frequency content of white or broadband noise, but it may also degrade the image content by attenuating important high-frequency image details. This is a tradeoff that is difficult to balance.

Next we describe and compare several smoothing lowpass filters that are commonly used for linear image enhancement.

10.3.1 Moving Average Filter

The moving average filter can be described in several equivalent ways. First, using the notion of *windowing* introduced in Chapter 4, the moving average can be defined as an algebraic operation performed on local image neighborhoods according to a geometric rule defined by the window. Given an image **f** to be filtered and a window **B** that collects gray level pixels according to a geometric rule (defined by the window shape), then the moving average-filtered image **g** is given by

$$g(\mathbf{n}) = \text{AVE}[\mathbf{B}f(\mathbf{n})], \qquad (10.22)$$

where the operation AVE computes the sample average of its. Thus, the local average is computed over each local neighborhood of the image, producing a powerful smoothing effect. The windows are usually selected to be symmetric, as with those used for binary morphological image filtering (Chapter 4).

Since the average is a linear operation, it is also true that

$$g(\mathbf{n}) = \text{AVE}[\mathbf{B}o(\mathbf{n})] + \text{AVE}[\mathbf{B}q(\mathbf{n})]. \tag{10.23}$$

Because the noise process $\mathbf{q}$ is assumed to be zero-mean in the sense of (10.20), then the last term in (10.23) will tend to zero as the filter window is increased. Thus, the moving average filter has the desirable effect of reducing zero-mean image noise toward zero. However, the filter also effects the original image information. It is desirable that $\text{AVE}[\mathbf{B}o(\mathbf{n})] \approx o(\mathbf{n})$ at each $\mathbf{n}$, but this will not be the case everywhere in the image if the filter window is too large. The moving average filter, which is lowpass, will blur the image, especially as the window span is increased. Balancing this tradeoff is often a difficult task.

The moving average filter operation (10.22) is actually a linear convolution. In fact, the impulse response of the filter is defined as having value $1/R$ over the span covered by the window when centered at the spatial origin $(0, 0)$, and zero elsewhere, where R is the number of elements in the window.

For example, if the window is SQUARE $[(2P + 1)^2]$, which is the most common configuration (it is defined in Chapter 4), then the average filter impulse response is given by

$$h(m,n) = \begin{cases} 1/(2P+1)^2; & -P \leq m,n \leq P \\ 0 & ; \quad \text{else} \end{cases}. \tag{10.24}$$

The frequency response of the moving average filter (10.24) is:

$$H(U,V) = \frac{\sin[(2P+1)\pi U]}{(2P+1)\sin(\pi U)} \cdot \frac{\sin[(2P+1)\pi V]}{(2P+1)\sin(\pi V)}. \tag{10.25}$$

The *half-peak bandwidth* is often used for image processing filters. The half-peak (or 3 dB) cutoff frequencies occur on the locus of points (U, V) where $|H(U, V)|$ falls to $1/2$. For the filter (10.25), this locus intersects the U-axis and V-axis at the cutoffs $U_{\text{half-peak}}, V_{\text{half-peak}} \approx 0.6/(2P + 1)$ cycles/pixel.

As depicted in Fig. 10.2, the magnitude response $|H(U, V)|$ of the filter (10.25) exhibits considerable *sidelobes*. In fact, the number of sidelobes in the range $[0, \pi]$ is P. As P is increased, the filter bandwidth naturally decreases (more high-frequency attenuation or smoothing), but the overall sidelobe energy does not. The sidelobes are in fact a significant drawback, since there is considerable *noise leakage* at high noise frequencies. These residual noise frequencies remain to degrade the image. Nevertheless, the moving average filter has been commonly used because of its general effectiveness in the sense of (10.21) and because of its simplicity (ease of programming).

The moving average filter can be implemented either as a direct 2D convolution in the space domain, or using DFTs to compute the linear convolution (see Chapter 5).

Since application of the moving average filter balances a tradeoff between noise smoothing and image smoothing, the filter span is usually taken to be an intermediate value. For images of the most common sizes, e.g., 256×256 or 512×512, typical (SQUARE) average filter sizes range from 3×3 to 15×15. The upper end provides significant (and probably excessive) smoothing, since 225 image samples are being averaged

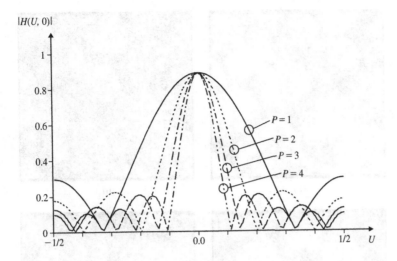

FIGURE 10.2

Plots of $|H(U,V)|$ given in (10.25) along $V = 0$, for $P = 1,2,3,4$. As the filter span is increased, the bandwidth decreases. The number of sidelobes in the range $[0, \pi]$ is P.

to produce each new image value. Of course, if an image suffers from severe noise, then a larger window might be used. A large window might also be acceptable if it is known that the original image is very smooth everywhere.

Figure 10.3 depicts the application of the moving average filter to an image that has had zero-mean white Gaussian noise added to it. In the current context, the distribution (Gaussian) of the noise is not relevant, although the meaning can be found in Chapter 7. The original image is included for comparison. The image was filtered with SQUARE-shaped moving average filters of window sizes 5×5 and 9×9, producing images with significantly different appearances from each other as well as the noisy image. With the 5×5 filter, the noise is inadequately smoothed, yet the image has been blurred noticeably. The result of the 9×9 moving average filter is much smoother, although the noise influence is still visible, with some higher noise frequency components managing to leak through the filter, resulting in a mottled appearance.

10.3.2 Ideal Lowpass Filter

As an alternative to the average filter, a filter may be designed explicitly with no side-lobes by forcing the frequency response to be zero outside of a given radial cutoff frequency Ω_c:

$$H(U,V) = \begin{cases} 1; & \sqrt{U^2 + V^2} \le \Omega_c \\ 0; & \text{else} \end{cases} \qquad (10.26)$$

(a)   (b)

(c)   (d)

FIGURE 10.3

Example of application of moving average filter. (a) Original image "eggs"; (b) image with additive Gaussian white noise; moving average-filtered image using; (c) SQUARE(25) window (5 × 5); and (d) SQUARE(81) window (9 × 9).

or outside of a rectangle defined by cutoff frequencies along the U- and V-axes:

$$H(U, V) = \begin{cases} 1; & |U| \leq U_c \quad \text{and} \quad |V| \leq V_c \\ 0; & \text{else} \end{cases}. \tag{10.27}$$

Such a filter is called an *ideal lowpass filter* (ideal LPF) because of its idealized characteristic. We will study (10.27) rather than (10.26) since it is easier to describe the impulse response of the filter. If the region of frequencies passed by (10.26) is square, then there is little practical difference in the two filters if $U_c = V_c = \Omega_c$.

The impulse response of the ideal lowpass filter (10.26) is given explicitly by

$$h(m, n) = U_c V_c \, \text{sinc} \, (2\pi U_c m) \cdot \text{sinc} \, (2\pi V_c n), \tag{10.28}$$

where $\text{sinc}(x) = \frac{\sin x}{x}$. Despite the seemingly "ideal" nature of this filter, it has some major drawbacks. First, it cannot be implemented exactly as a linear convolution, since the impulse response (10.28) is infinite in extent (it never decays to zero). Therefore, it must be approximated. One way is to simply truncate the impulse response, which in image processing applications is often satisfactory. However, this has the effect of introducing *ripple* near the frequency discontinuity, producing unwanted noise leakage. The introduced ripple is a manifestation of the well-known *Gibbs phenomena* studied in standard signal processing texts [1]. The ripple can be reduced by using a tapered truncation of the impulse response, e.g., by multiplying (10.28) with a Hamming window [1]. If the response is truncated to image size $M \times N$, then the ripple will be restricted to the vicinity of the locus of cutoff frequencies, which may make little difference in the filter performance. Alternately, the ideal LPF can be approximated by a Butterworth filter or other ideal LPF approximating function. The Butterworth filter has frequency response [2]

$$H(U, V) = \frac{1}{1 + \left(\frac{\sqrt{U^2 + V^2}}{\Omega_c} \right)^{2K}} \tag{10.29}$$

and, in principle, can be made to agree with the ideal LPF with arbitrary precision by taking the filter order K large enough. However, (10.29) also has an infinite-extent impulse response with no known closed-form solution. Hence, to be implemented it must also be spatially truncated (approximated), which reduces the approximation effectiveness of the filter [2].

It should be noted that if a filter impulse response is truncated, then it should also be slightly modified by adding a constant level to each coefficient. The constant should be selected such that the filter coefficients sum to unity. This is commonly done since it is generally desirable that the response of the filter to the $(0, 0)$ spatial frequency be unity, and since for any filter

$$H(0,0) = \sum_{p=-\infty}^{\infty} \sum_{q=-\infty}^{\infty} h(p,q). \tag{10.30}$$

The second major drawback of the ideal LPF is the phenomena known as *ringing*. This term arises from the characteristic response of the ideal LPF to highly concentrated bright spots in an image. Such spots are impulse-like, and so the local response has the appearance of the impulse response of the filter. For the circularly-symmetric ideal LPF in (10.26), the response consists of a blurred version of the impulse surrounded by sinc-like spatial sidelobes, which have the appearances of rings surrounding the main lobe.

In practical application, the ringing phenomena create more of a problem because of the *edge response* of the ideal LPF. In the simplistic case, the image consists of a single one-dimensional step edge: $s(m, n) = s(n) = 1$ for $n \geq 0$ and $s(n) = 0$, otherwise. Figure 10.4 depicts the response of the ideal LPF with impulse response (10.28) to the step edge. The step response of the ideal LPF oscillates (rings) because the sinc function oscillates about the zero level. In the convolution sum, the impulse response alternately

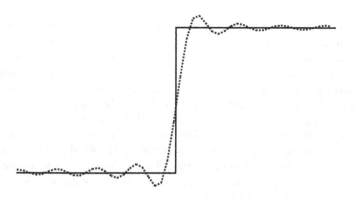

FIGURE 10.4

Depiction of edge ringing. The step edge is shown as a continuous curve, while the linear convolution response of the ideal LPF (10.28) is shown as a dotted curve.

makes positive and negative contribution, creating overshoots and undershoots in the vicinity of the edge profile. Most digital images contain numerous step-like light-to-dark or dark-to-light image transitions; hence, application of the ideal LPF will tend to contribute considerable ringing artifacts to images. Since edges contain much of the significant information about the image, and since the eye tends to be sensitive to ringing artifacts, often the ideal LPF and its derivatives are not a good choice for image smoothing. However, if it is desired to strictly bandlimit the image as closely as possible, then the ideal LPF is a necessary choice.

Once an impulse response for an approximation to the ideal LPF has been decided, then the usual approach to implementation again entails zero-padding both the image and the impulse response, using the periodic extension, taking the product of their DFTs (using an FFT algorithm), and defining the result as the inverse DFT. This was done in the example of Fig. 10.5, which depicts application of the ideal LPF using two cutoff frequencies. This was implemented using a truncated ideal LPF without any special windowing. The dominant characteristic of the filtered images is the ringing, manifested as a strong mottling in both images. A very strong oriented ringing can be easily seen near the upper and lower borders of the image.

10.3.3 Gaussian Filter

As we have seen, *filter sidelobes* in either the space or spatial frequency domains contribute a negative effect to the responses of noise-smoothing linear image enhancement filters. Frequency-domain sidelobes lead to noise leakage, and space-domain sidelobes lead to ringing artifacts. A filter with sidelobes in neither domain is the *Gaussian filter* (see Fig. 10.6), with impulse response

$$h(m, n) = \frac{1}{2\pi\sigma^2} e^{-(m^2 + n^2)/2\sigma^2}. \tag{10.31}$$

(a)   (b)

FIGURE 10.5

Example of application of ideal lowpass filter to noisy image in Fig. 10.3(b). Image is filtered using radial frequency cutoff of (a) 30.72 cycles/image and (b) 17.07 cycles/image. These cutoff frequencies are the same as the half-peak cutoff frequencies used in Fig. 10.3.

(a)   (b)

FIGURE 10.6

Example of application of Gaussian filter to noisy image in Fig. 10.3(b). Image is filtered using radial frequency cutoff of (a) 30.72 cycles/image ($\sigma \approx 1.56$ pixels) and (b) 17.07 cycles/image ($\sigma \approx 2.80$ pixels). These cutoff frequencies are the same as the half-peak cutoff frequencies used in Figs. 10.3 and 10.5.

The impulse response (10.31) is also infinite in extent, but falls off rapidly away from the origin. In this case, the frequency response is closely approximated by

$$H(U, V) \approx e^{-2\pi^2\sigma^2(U^2+V^2)} \quad \text{for} \quad |U|, |V| < 1/2. \tag{10.32}$$

Observe that (10.32) is also a Gaussian function. Neither (10.31) nor (10.32) shows any sidelobes; instead, both impulse and frequency response decay smoothly. The Gaussian filter is noted for the absence of ringing and noise leakage artifacts. The half-peak radial frequency bandwidth of (10.32) is easily found to be

$$\Omega_c = \frac{1}{\pi\sigma}\sqrt{\ln\sqrt{2}} \approx \frac{0.187}{\sigma}. \tag{10.33}$$

If it is possible to decide an appropriate cutoff frequency Ω_c, then the cutoff frequency may be fixed by setting $\sigma = 0.187/\Omega_c$ pixels. The filter may then be implemented by truncating (10.31) using this value of σ, adjusting the coefficients to sum to one, zero-padding both impulse response and image (taking care to use the periodic extension of the impulse response implied by the DFT), multiplying DFTs, and taking the inverse DFT to be the result. The results obtained are much better than those computed using the ideal LPF, and slightly better than those obtained with the moving average filter, because of the reduced noise leakage.

Figure 10.7 shows the result of filtering an image with a Gaussian filter of successively larger σ values. As the value of σ is increased, small-scale structures such as noise and details are reduced to a greater degree. The sequence of images shown in Fig. 10.7(b) is a *Gaussian scale-space*, where each scaled image is calculated by convolving the original image with a Gaussian filter of increasing σ value [3].

The Gaussian scale-space may be thought of as evolving over time t. At time t, the scale-space image $\mathbf{g}_t$ is given by

$$\mathbf{g}_t = \mathbf{h}_\sigma * \mathbf{f}, \tag{10.34}$$

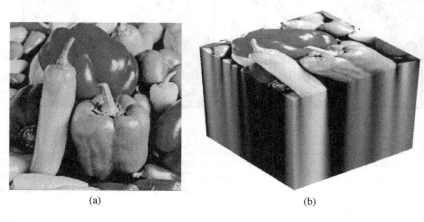

(a)   (b)

FIGURE 10.7

Depiction of scale-space property of Gaussian filter lowpass filter. In (b), the image in (a) is Gaussian-filtered with progressively larger values of σ (narrower bandwidths) producing successively smoother and more diffuse versions of the original. These are "stacked" to produce a data cube with the original image on top to produce the representation shown in (b).

where $\mathbf{h}_\sigma$ is a Gaussian filter with scale factor σ, and $\mathbf{f}$ is the initial image. The time-scale relationship is defined by $\sigma = \sqrt{t}$. As σ is increased, less significant image features and noise begin to disappear, leaving only large-scale image features.

The Gaussian scale-space may also be viewed as the evolving solution of a partial differential equation [3, 4]:

$$\frac{\partial \mathbf{g}_t}{\partial t} = \nabla^2 \mathbf{g}_t, \tag{10.35}$$

where $\nabla^2 \mathbf{g}_t$ is the Laplacian of $\mathbf{g}_t$.

10.4 DISCUSSION

Linear filters are omnipresent in image and video processing. Firmly established in the theory of linear systems, linear filters are the basis of processing signals of arbitrary dimensions. Since the advent of the fast Fourier transform in the 1960s, the linear filter has also been an attractive device in terms of computational expense. However, it must be noted that linear filters are performance-limited for image enhancement applications. From the experiments performed in this chapter, it can be anecdotally observed that the removal of broadband noise from most images via linear filtering is impossible without some degradation (blurring) of the image information content. This limitation is due to the fact that complete frequency separation between signal and broadband noise is rarely viable. Alternative solutions that remedy the deficiencies of linear filtering have been devised, resulting in a variety of powerful nonlinear image enhancement alternatives. These are discussed in Chapters 11–13 of this *Guide*.

REFERENCES

[1] A. V. Oppenheim and R. W. Schafer. *Discrete-Time Signal Processing*. Prentice-Hall, Upper Saddle River, NJ, 1989.

[2] R. C. Gonzalez and R. E. Woods. *Digital Image Processing*. Addison-Wesley, Boston, MA, 1993.

[3] A. P. Witkin. Scale-space filtering. In *Proc. Int. Joint Conf. Artif. Intell.*, 1019–1022, 1983.

[4] J. J. Koenderink. The structure of images. *Biol. Cybern.*, 50:363–370, 1984.

Multiscale Denoising of Photographic Images

11

Umesh Rajashekar and Eero P. Simoncelli

New York University

11.1 INTRODUCTION

Signal acquisition is a noisy business. In photographic images, there is noise within the light intensity signal (e.g., photon noise), and additional noise can arise within the sensor (e.g., thermal noise in a CMOS chip), as well as in subsequent processing (e.g., quantization). Image noise can be quite noticeable, as in images captured by inexpensive cameras built into cellular telephones, or imperceptible, as in images captured by professional digital cameras. Stated simply, the goal of image denoising is to recover the "true" signal (or its best approximation) from these noisy acquired observations. All such methods rely on understanding and exploiting the differences between the properties of signal and noise.

Formally, solutions to the denoising problem rely on three fundamental components: a signal model, a noise model, and finally a measure of signal fidelity (commonly known as the objective function) that is to be minimized. In this chapter, we will describe the basics of image denoising, with an emphasis on signal properties. For noise modeling, we will restrict ourselves to the case in which images are corrupted by additive, white, Gaussian noise—that is, we will assume each pixel is contaminated by adding a sample drawn independently from a Gaussian probability distribution of fixed variance. A variety of other noise models and corruption processes are considered in Chapter 7. Throughout, we will use the well-known mean-squared error (MSE) measure as an objective function.

We develop a sequence of three image denoising methods, motivating each one by observing a particular property of photographic images that emerges when they are decomposed into subbands at different spatial scales. We will examine each of these properties quantitatively by examining statistics across a training set of photographic images and noise samples. And for each property, we will use this quantitative characterization to develop two example denoising functions: a binary threshold function that retains or discards each multiscale coefficient depending on whether it is more likely to be dominated by noise or signal, and a continuous-valued function that multiplies each coefficient by an optimized scalar value. Although these methods are quite simple, they capture many of the concepts that are used in state-of-the-art denoising systems. Toward the end of the chapter, we briefly describe several alternative approaches.

11.2 DISTINGUISHING IMAGES FROM NOISE IN MULTISCALE REPRESENTATIONS

Consider the images in the top row of Fig. 11.3. Your visual system is able to recognize effortlessly that the image in the left column is a photograph while the image in the middle column is filled with noise. How does it do this? We might hypothesize that it simply recognizes the difference in the distributions of pixel values in the two images. But the distribution of pixel values of photographic images is highly inconsistent from image to image, and more importantly, one can easily generate a noise image whose pixel distribution is matched to any given image (by simply spatially scrambling the pixels). So it seems that visual discrimination of photographs and noise cannot be accomplished based on the statistics of individual pixels.

Nevertheless, the joint statistics of pixels reveal striking differences, and these may be exploited to distinguish photographs from noise, and also to restore an image that has been corrupted by noise, a process commonly referred to as *denoising*. Perhaps the most obvious (and historically, the oldest) observation is that spatially proximal pixels of photographs are correlated, whereas the noise pixels are not. Thus, a simple strategy for denoising an image is to separate it into smooth and nonsmooth parts, or equivalently, low-frequency and high-frequency components. This decomposition can then be applied recursively to the lowpass component to generate a multiscale representation, as illustrated in Fig. 11.1. The lower frequency subbands are smoother, and thus can be subsampled to allow a more efficient representation, generally known as a multiscale pyramid [1, 2]. The resulting collection of frequency subbands contains the exact same information as the input image, but, as we shall see, it has been separated in such a way that it is more easily distinguished from noise. A detailed development of multiscale representations can be found in Chapter 6 of this *Guide*.

Transformation of an input image to a multiscale image representation has almost become a *de facto* pre-processing step for a wide variety of image processing and computer vision applications. In this chapter, we will assume a three-step denoising methodology:

1. Compute the multiscale representation of the noisy image.

2. Denoise the noisy coefficients, y, of all bands except the lowpass band using denoising functions $\hat{x}(y)$ to get an estimate, $\hat{x}$, of the true signal coefficient, x.

3. Invert the multiscale representation (i.e., recombine the subbands) to obtain a denoised image.

This sequence is illustrated in Fig. 11.2. Given this general framework, our problem is to determine the form of the denoising functions, $\hat{x}(y)$.

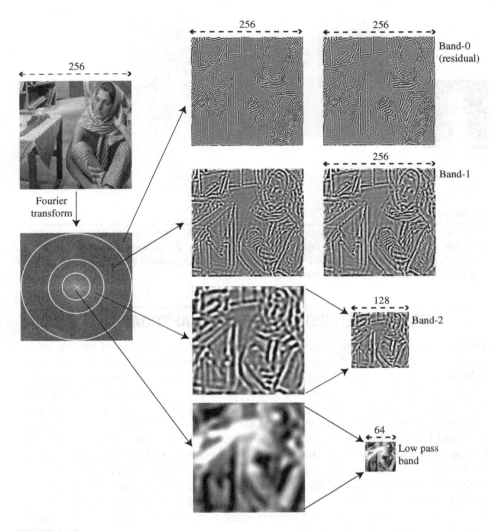

FIGURE 11.1

A graphical depiction of the multiscale image representation used for all examples in this chapter. Left column: An image and its centered Fourier transform. The white circles represent filters used to select bands of spatial frequencies. Middle column: Inverse Fourier transforms of the various spatial frequencies bands selected by the idealized filters in the left column. Each filtered image represents only a subset of the entire frequency space (indicated by the arrows originating from the left column). Depending on their maximum spatial frequency, some of these filtered images can be downsampled in the pixel domain without any loss of information. Right column: Downsampled versions of the filtered images in the middle column. The resulting images form the subbands of a multiscale "pyramid" representation [1, 2]. The original image can be exactly recovered from these subbands by reversing the procedure used to construct the representation.

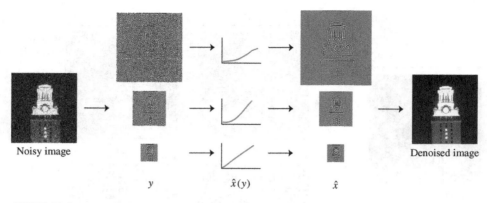

Noisy image · y · $\hat{x}(y)$ · $\hat{x}$ · Denoised image

FIGURE 11.2

Block diagram of multiscale denoising. The noisy photographic image is first decomposed into a multiscale representation. The noisy pyramid coefficients, y, are then denoised using the functions, $\hat{x}(y)$, resulting in denoised coefficients, $\hat{x}$. Finally, the pyramid of denoised coefficients is used to reconstruct the denoised image.

11.3 SUBBAND DENOISING—A GLOBAL APPROACH

We begin by making some observations about the differences between photographic images and random noise. Figure 11.3 shows the multiscale decomposition of an essentially noise-free photograph, random noise, and a noisy image obtained by adding the two. The pixels of the signal (the noise-free photograph) lie in the interval $[0, 255]$. The noise pixels are uncorrelated samples of a Gaussian distribution with zero mean and standard deviation of 60. When we look at the subbands of the noisy image, we notice that band 1 of the noisy image is almost indistinguishable from the corresponding band for the noise image; band 2 of the noisy image is contaminated by noise, but some of the features from the original image remain visible; and band 3 looks nearly identical to the corresponding band of the original image. These observations suggest that, on average, noise coefficients tend to have larger amplitude than signal coefficients in the high-frequency bands (e.g., band 1), whereas signal coefficients tend to be more dominant in the low-frequency bands (e.g., band 3).

11.3.1 Band Thresholding

This observation about the relative strength of signal and noise in different frequency bands leads us to our first denoising technique: we can set each coefficient that lies in a band that is significantly corrupted by noise (e.g., band 1) to zero, and retain the other bands without modification. In other words, we make a binary decision to retain or discard each subband. But how do we decide which bands to keep and which to discard? To address this issue, let us denote the entire band of noise-free image coefficients as a vector, $\vec{x}$, the coefficients of the noise image as $\vec{n}$, and the band of noisy coefficients as $\vec{y} = \vec{x} + \vec{n}$. Then the total squared error incurred if we should decide to retain the noisy

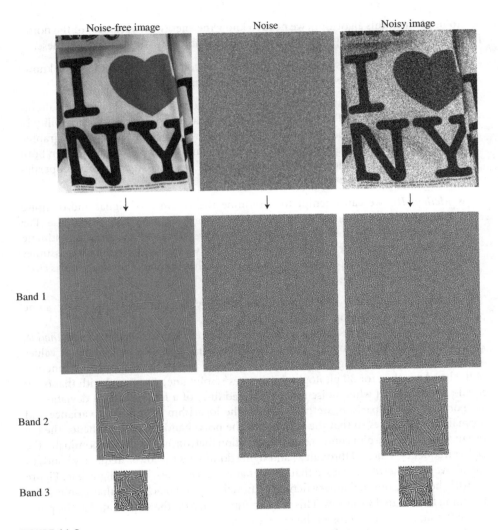

FIGURE 11.3

Multiscale representations of Left: a noise-free photographic image. Middle: a Gaussian white noise image. Right: The noisy image obtained by adding the noise-free image and the white noise.

band is $|\vec{x} - \vec{y}|^2 = |\vec{n}|^2$, and the error incurred if we discard the band is $|\vec{x} - \vec{0}|^2 = |\vec{x}|^2$. Since our objective is to minimize the MSE between the original and denoised coefficients, the optimal decision is to retain the band whenever the signal energy (i.e., the squared norm of the signal vector, $\vec{x}$) is greater than that of the noise (i.e., $|\vec{x}|^2 > |\vec{n}|^2$) and discard it otherwise[1].

---

[1]Minimizing the total energy is equivalent to minimizing the MSE, since the latter is obtained from the former by dividing by the number of elements.

To implement this algorithm, we need to know the energy (or variance) of the noise-free signal, $|\vec{x}|^2$, and noise, $|\vec{n}|^2$. There are several possible ways for us to obtain these.

- *Method I*: we can assume values for either or both, based on some prior knowledge or principles about images or our measurement device.

- *Method II*: we can estimate them in advance from a set of "training" or calibration measurements. For the noise, we might imagine measuring the variability in the pixel values for photographs of a set of known test images. For the photographic images, we could measure the variance of subbands of noise-free images. In both cases, we must assume that our training images have the same variance properties as the images that we will subsequently denoise.

- *Method III*: we can attempt to determine the variance of signal and/or noise from the observed noisy coefficients of the image we are trying to denoise. For example, if we the noise energy is known to have a value of E_n^2, we could estimate the signal energy as $|\vec{x}|^2 = |\vec{y} - \vec{n}|^2 \approx |\vec{y}|^2 - E_n^2$, where the approximation assumes that the noise is independent of the signal, and that the actual noise energy is close to the assumed value: $|\vec{n}|^2 \approx E_n^2$.

These three methods of obtaining parameters may be combined obtaining some parameters with one method and others with another.

For our purposes, we assume that the noise variance is known in advance (*Method I*), and we use *Method II* to obtain estimates of the signal variance by looking at values across a training set of images. Figure 11.4(a) shows a plot of the variance as a function of the band number, for 30 photographic images[2] (solid line) compared with that of 30 equal-sized Gaussian white noise images (dashed line) of a fixed standard deviation of 60. For ease of comparison, we have plotted the logarithm of the band variance and normalized the curves so that the variance of the noise bands is 1.0 (and hence the log variance is zero). The plot confirms our observation that, on average, noise dominates the higher frequency bands (0 through 2) and signal dominates the lower frequency bands (3 and above). Furthermore, we see that the signal variance is nearly a straight line. Figure 11.4(b) shows the optimal binary denoising function (solid black line) that results from assuming these signal variances. This is a step function, with the step located at the point where the signal variance crosses the noise variance.

We can examine the behavior of this method visually, by retaining or discarding the subbands of the pyramid of noisy coefficients according to the optimal rule in Fig. 11.4(b), and then generating a denoised image by inverting the pyramid transformation. Figure 11.8(c) shows the result of applying this denoising technique to the noisy image shown in Fig. 11.8(b). We can see that a substantial amount of the noise has been eliminated, although the denoised image appears somewhat blurred, since the high-frequency bands have been discarded. The performance of this denoising scheme

---

[2]All images in our training set are of New York City street scenes, each of size 1536 × 1024 pixels. The images were acquired using a Canon 30D digital SLR camera.

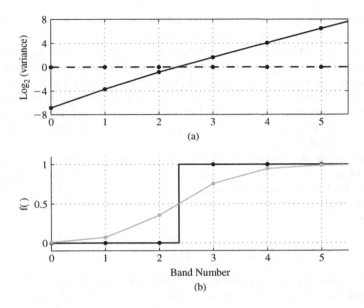

FIGURE 11.4

Band denoising functions. (a) Plot of average log variance of subbands of a multiscale pyramid as a function of the band number averaged over the photographic images in our training set (solid line denoting $\log(|\vec{x}|^2)$) and Gaussian white noise image of standard deviation of 60 (dashed line denoting $\log(|\vec{n}|^2)$). For visualization purposes, the curves have been normalized so that the log of the noise variance was equal to 0.0; (b) Optimal thresholding function (black) and weighting function (gray) as a function of band number.

can be quantified using the mean squared error (MSE), or with the related measure of peak signal-to-noise ratio (PSNR), which is essentially a log-domain version of the MSE. If we define the MSE between two vectors $\vec{x}$ and $\vec{y}$, each of size N, as $\mathrm{MSE}(\vec{x}, \vec{y}) = \frac{1}{N} |\vec{x} - \vec{y}|^2$, then the PSNR (assuming 8-bit images) is defined as $\mathrm{PSNR}(\vec{x}, \vec{y}) = 10 \log_{10} \frac{255^2}{\mathrm{MSE}(\vec{x},\vec{y})}$ and measured in units of decibels (dB). For the current example, the PSNR of the noisy and denoised image were 13.40 dB and 24.45 dB, respectively. Figure 11.9 shows the improvement in PSNR over the noisy image across 5 different images.

11.3.2 Band Weighting

In the previous section, we developed a binary denoising function based on knowledge of the relative strength of signal and noise in each band. In general, we can write the solution for each individual coefficient:

$$\hat{x}(\vec{y}) = f(|\vec{y}|) \cdot y, \tag{11.1}$$

where the binary-valued function, $f(\cdot)$, is written as a function of the energy of the noisy coefficients, $|\vec{y}|$, to allow estimation of signal or noise variance from the observation (as described in *Method III* above). An examination of the pyramid decomposition of the noisy image in Fig. 11.3 suggests that the binary assumption is overly restrictive. Band 1, for example, contains some residual signal that is visible despite the large amount of noise. And band 3 shows some noise in the presence of strong signal coefficients. This observation suggests that instead of the binary retain-or-discard technique, we might obtain better results by allowing $f(\cdot)$ to take on real values that depend on the relative strength of the signal and noise.

But how do we determine the optimal real-valued denoising function $f(\cdot)$? For each band of noisy coefficients $\vec{y}$, we seek a scalar value, a, that minimizes the error $|a\vec{y} - \vec{x}|^2$. To find the optimal value, we can expand the error as $a^2\vec{y}^T\vec{y} - 2a\vec{y}^T\vec{x} + \vec{x}^T\vec{x}$, differentiate it with respect to a, set the result to zero, and solve for a. The optimal value is found to be

$$\hat{a} = \frac{\vec{y}^T\vec{x}}{\vec{y}^T\vec{y}}. \tag{11.2}$$

Using the fact that the noise is uncorrelated with the signal (i.e., $\vec{x}^T\vec{n} \approx 0$), and the definition of the noisy image $\vec{y} = \vec{x} + \vec{n}$, we may express the optimal value as

$$\hat{a} = \frac{|\vec{x}|^2}{|\vec{x}|^2 + |\vec{n}|^2}. \tag{11.3}$$

That is, the optimal scalar multiplier is a value in the range $[0, 1]$, which depends on the relative strength of signal and noise. As described under *Method II* in the previous section, we may estimate this quantity from training examples.

To compute this function $f(\cdot)$, we performed a five-band decomposition of the images and noise in our training set and computed the average values of $|\vec{x}|^2$ and $|\vec{n}|^2$, indicated by the solid and dashed lines in Fig. 11.4(a). The resulting function, is plotted in gray as a function of the band number in Fig. 11.4(b). As expected, bands 0-1, which are dominated by noise, have a weight close to zero; bands 4 and above, which have more signal energy, have a weight close to 1.0; and bands 2-3 are weighted by intermediate values. Since this denoising function includes the binary functions as a special case, the denoising performance cannot be any worse than band thresholding, and will in general be better. To denoise a noisy image, we compute its five-band decomposition, weight each band in accordance to its weight indicated in Fig. 11.4(b) and invert the pyramid to obtain the denoised image. An example of this denoising is shown in Fig. 11.8(d). The PSNR of the noisy and denoised images were 13.40 dB and 25.04 dB—an improvement of more than 11.5 dB! This denoising performance is consistent across images, as shown in Fig. 11.9.

Previously, the value of the optimal scalar was derived using *Method II*. But we can use the fact that $\vec{x} = \vec{y} - \vec{n}$, and the knowledge that noise is uncorrelated with the signal (i.e., $\vec{x}^T\vec{n} \approx 0$), to rewrite Eq. (11.2) as a function of each band as:

$$\hat{a} = f(|\vec{y}|) = \frac{|\vec{y}|^2 - |\vec{n}|^2}{|\vec{y}|^2}. \tag{11.4}$$

If we assume that the noise energy is known, then this formulation is an example of *Method III*, and more generally, we now can rewrite $\hat{x}(\vec{y}) = f(|\vec{y}|) \cdot y$.

The denoising function in Eq. (11.4) is often applied to coefficients in a Fourier transform representation, where it is known as the "Wiener filter". In this case, each Fourier transform coefficient is multiplied by a value that depends on the variances of the signal and noise at each spatial frequency—that is, the power spectra of the signal and noise. The power spectrum of natural images is commonly modeled using a power law, $F(\Omega) = A/\Omega^p$, where Ω is spatial frequency, p is the exponent controlling the falloff of the signal power spectrum (typically near 2), A is a scale factor controlling the overall signal power, is the unique form that is consistent with a process that is both translation- and scale-invariant (see Chapter 9). Note that this model is consistent with the measurements of Fig. 11.4, since the frequency of the subbands grows exponentially with the band number. If, in addition, the noise spectrum is assumed to be flat (as it would be, for example, with Gaussian white noise), then the Wiener filter is simply

$$|H(\Omega)|^2 = \frac{|A/\Omega^p|}{|A/\Omega^p| + \sigma_N^2}, \tag{11.5}$$

where σ_N^2 is the noise variance.

11.4 SUBBAND COEFFICIENT DENOISING—A POINTWISE APPROACH

The general form of denoising in Section 11.3 involved weighting the *entire* band by a single number—0 or 1 for band thresholding, or a scalar between 0 and 1 for band weighting. However, we can observe that in a noisy band such as band 2 in Fig. 11.3, the amplitudes of signal coefficients tend to be either very small, or quite substantial. The simple interpretation is that images have isolated features such as edges that tend to produce large coefficients in a multiscale representation. The noise, on the other hand, is relatively homogeneous.

To verify this observation, we used the 30 images in our training set and 30 Gaussian white noise images (standard deviation of 60) of the same size and computed the distribution of signal and noise coefficients in a band. Figure 11.5 shows the log of the distribution of the magnitude of signal (solid line) and noise coefficients (dashed line) in one band of the multiscale decomposition. We can see that the distribution tails are heavier and the frequency of small values is higher for the signal coefficients, in agreement with our observations above.

From this basic observation, we can see that signal and noise coefficients might be further distinguished based on their magnitudes. This idea has been used for decades in video cassette recorders for removing magnetic tape noise, where it is known as "coring". We capture it using a denoising function of the form:

$$\hat{x}(y) = f(|y|) \cdot y, \tag{11.6}$$

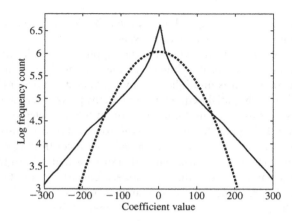

FIGURE 11.5

Log histograms of coefficients of a band in the multiscale pyramid for a photographic image (solid) and Gaussian white noise of standard deviation of 60 (dashed). As expected, the log of the distribution of the Gaussian noise is parabolic.

where $\hat{x}(y)$ is the estimate of a single noisy coefficient y. Note that unlike the denoising scheme in Equation (11.1) the value of the denoising function, $f(\cdot)$, will now be different for each coefficient.

11.4.1 Coefficient Thresholding

Consider first the case where the function $f(\cdot)$ is constrained to be binary, analogous to our previous development of band thresholding. Given a band of noisy coefficients, our goal now is to determine a threshold such that coefficients whose magnitudes are less than this threshold are set to zero, and all coefficients whose magnitudes are greater than or equal to the threshold are retained.

The threshold is again selected so as to minimize the mean squared error.

We determined this threshold empirically using our image training set. We computed the five-band pyramid for the noise-free and noisy images (corrupted by Gaussian noise of standard deviation of 60) to get pairs of noisy coefficients, y, and their corresponding noise-free coefficients, x, for a particular band. Let us now consider an arbitrary threshold value, say T. As in the case of band thresholding, there are two types of error introduced at any threshold level. First, when the magnitude of the observed coefficient, y, is below the threshold and set to zero, we have discarded the signal, x, and hence incur an error of x^2. Second, when the observed coefficient is greater than the threshold, we leave the coefficient (signal and noise) unchanged. The error introduced by passing the noise component is $n^2 = (y - x)^2$. Therefore, given pairs of coefficients, (x_i, y_i), for a subband, the total error at a particular threshold, T, is

$$\sum_{i:|y_i|\le T} x_i^2 + \sum_{i:|y_i|>T} (y_i - x_i)^2.$$

Unlike the band denoising case, the optimal choice of threshold cannot be obtained in closed form. Using the pairs of coefficients obtained from the training set, we searched over the set of threshold values, T, to find the one that gave the smallest total least squared error.

Figure 11.6 shows the optimized threshold functions, $f(\cdot)$, in Eq. (11.6) as solid black lines for three of the five bands that we used in our analysis. For readers who might be more familiar with the input-output form, we also show the denoising functions $\hat{x}(y)$ in Fig. 11.6(b). The resulting plots are intuitive and can be explained as follows. For band 1, we know that all the coefficients are likely to be corrupted heavily by noise. Therefore, the threshold value is so high that essentially all of the coefficients are set to zero. For band 2, the signal-to-noise ratio increases and therefore the threshold values get smaller allowing more of the larger magnitude coefficients to pass unchanged. Finally, once we reach band 3 and above, the signal is so strong compared to noise that the threshold is close to zero, thus allowing all coefficients to be passed without alteration.

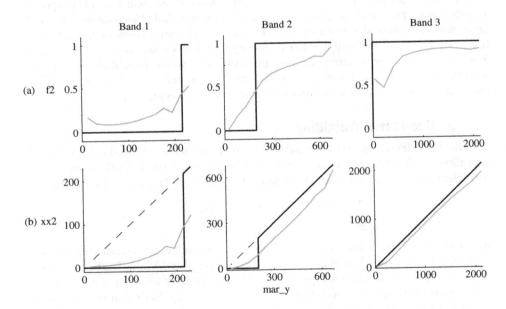

FIGURE 11.6

Coefficient denoising functions for three of the five pyramid bands. (a) Coefficient thresholding (black) and coefficient weighting (gray) functions $f(|y|)$ as a function of $|y|$ (see Eq. (11.6)); (b) Coefficient estimation functions $\hat{x}(y) = f(|y|) \cdot y$. The dashed line depicts the unit slope line. For the sake of uniformity across the various denoising schemes, we show only one half of the denoising curve corresponding to the positive values of the observed noisy coefficient. Jaggedness in the curves occurs at values for which there was insufficient data to obtain a reliable estimate of the function.

To denoise a noisy image, we first decompose it using the multiscale pyramid, and apply an appropriate threshold operation to the coefficients of each band (as plotted in Fig. 11.6). Coefficients whose magnitudes are smaller than the threshold are set to zero, and the rest are left unaltered. The signs of the observed coefficients are retained. Figure 11.8(e) shows the result of this denoising scheme, and additional examples of PSNR improvement are given in Fig. 11.9. We can see that the coefficient-based thresholding has an improvement of roughly 1 dB over band thresholding.

Although this denoising method is more powerful than the whole-band methods described in the previous section, note that it requires more knowledge of the signal and the noise. Specifically, the coefficient threshold values were derived based on knowledge of the distributions of both signal and noise coefficients. The former was obtained from training images, and thus relies on the additional assumption that the image to be denoised has a distribution that is the same as that seen in the training images. The latter was obtained by assuming the noise was white and Gaussian, of known variance. As with the band denoising methods, it is also possible to approximate the optimal denoising function directly from the noisy image data, although this procedure is significantly more complex than the one outlined above. Specifically, Donoho and Johnstone [3] proposed a methodology known as *SUREshrink* for selecting the threshold based on the observed noisy data, and showed it to be optimal for a variety of some classes of regular functions [4]. They also explored another denoising function, known as soft-thresholding, in which a fixed value is subtracted from the coefficients whose magnitudes are greater than the threshold. This function is continuous (as opposed to the hard thresholding function) and has been shown to produce more visually pleasing images.

11.4.2 Coefficient Weighting

As in the band denoising case, a natural extension of the coefficient thresholding method is to allow the function $f(\cdot)$ to take on scalar values between 0.0 and 1.0. Given a noisy coefficient value, y, we are interested in finding the scalar value $f(|y|) = a$ that minimizies

$$\sum_{i:y_i=y} (x_i - f(|y_i|) \cdot y_i)^2 = \sum_{i:y_i=y} (x_i - a \cdot y)^2.$$

We differentiate this equation with respect to a, set the result equal to zero, and solve for a resulting in the optimal estimate $\hat{a} = f(|y|) = (1/y) \cdot (\sum_i x_i / \sum_i 1)$. The best estimate, $\hat{x}(y) = f(|y|) \cdot y$, is therefore simply the conditional mean of all noisy coefficients, x_i, whose noisy coefficients are such that $y_i = y$. In practise, it is likely that no noisy coefficient has a value that is exactly equal to y. Therefore, we bin the coefficients such that $y - \delta \le |y_i| \le y + \delta$, where δ is a small positive value.

The plot of this function $f(|y|)$ as a function of y is shown as a light gray line in Fig. 11.6(a) for three of the five bands that we used in our analysis; the functions for the other bands (4 and above) look identical to band 3. We also show the denoising functions, $\hat{x}(y)$, in Fig. 11.6(b). The reader will notice that, similar to the band weighting functions, these functions are smooth approximations of the hard thresholding functions, whose thresholds always occur when the weighting estimator reaches a value of 0.5.

To denoise a noisy image, we first decompose the image using a five-band multi-scale pyramid. For a given band, we use the smooth function $f(\cdot)$ that was learned in the previous step (for that particular band), and multiply the magnitude of each noisy coefficient, y, by the corresponding value, $f(|y|)$. The sign of the observed coefficients are retained. The modified pyramid is then inverted to result in the denoised image as shown in Fig. 11.8(f). The method outperforms the coefficient thresholding method (since thresholding is again a special case of the scalar-valued denoising function). Improvements in PSNR across five different images are shown in Fig. 11.9.

As in the coefficient thresholding case, this method relies on a fair amount of knowledge about the signal and noise. Although the denoising function can be learned from training images (as was done here), this needs to be done for each band, and for each noise level, and it assumes that the image to be denoised has coefficient distributions similar to those of the training set. An alternative formulation, known as *Bayesian coring* was developed by Simoncelli and Adelson [5], who assumed a generalized Gaussian model (see Chapter 9) for the coefficient distributions. They then fit the parameters of this model adaptively to the noisy image, and then computed the optimal denoising function from the fitted model.

11.5 SUBBAND NEIGHBORHOOD DENOISING—STRIKING A BALANCE

The technique presented in Section 11.3 was global, in that all coefficients in a band were multiplied by the *same* value. The technique in Section 11.4, on the other hand, was completely local: each coefficient was multiplied by a value that depended only on the magnitude of that particular coefficient. Looking again at the bands of the noise-free signal in Fig. 11.3, we can see that a method that treats each coefficient in isolation is not exploiting all of the available information about the signal. Specifically, the large magnitude coefficients tend to be spatially adjacent to other large magnitude coefficients (e.g., because they lie along contours or other spatially localized features). Hence, we should be able to improve the denoising of individual coefficients by incorporating knowledge of neighboring coefficients. In particular, we can use the energy of a small neighborhood around a given coefficient to provide some predictive information about the coefficient being denoised. In the form of our generic equation for denoising, we may write

$$\hat{x}(\bar{y}) = f(|\bar{y}|) \cdot y, \qquad (11.7)$$

where $\bar{y}$ now corresponds to a neighborhood of multiscale coefficients around the coefficient to be denoised, y, and $|\cdot|$ indicates the vector magnitude.

11.5.1 Neighborhood Thresholding

Analogous to previous sections, we first consider a simple form of neighborhood thresholding in which the function, $f(\cdot)$ in Eq. (11.7) is binary. Our methodology for determining the optimal function is identical to the technique previously discussed in Section 11.4.1, with the exception that we are now trying to find a threshold based on the local energy $|\tilde{y}|$ instead of the coefficient magnitude, $|y|$. For this simulation, we used a neighborhood of 5×5 coefficients surrounding the central coefficient.

To find the denoising functions, we begin by computing the five-band pyramid for the noise-free and noisy images in the training set. For a given subband we create triplets of noise-free coefficients, x_i, noisy coefficients, y_i, and the energy, $|\tilde{y}_i|$, of the 5×5 neighborhood around y_i. For a particular threshold value, T, the total error is given by

$$\sum_{i:|\tilde{y}_i| \leq T} x_i^2 + \sum_{i:|\tilde{y}_i| > T} (y_i - x_i)^2.$$

The threshold that provides the smallest error is then selected. A plot of the resulting functions, $f(\cdot)$, is shown by the solid black line in Fig. 11.7. The coefficient estimation functions, $\hat{x}(\tilde{y})$, depend on both $|\tilde{y}|$ and y and not very easy to visualize. The reader should note that the abscissa is now the energy of the neighborhood, and not the amplitude of a coefficient (as in Fig. 11.6(a)).

To denoise a noisy image, we first compute the five-band pyramid decomposition, and for a given band, we first compute the local variance of the noisy coefficient using a 5×5 window, and use this estimate along with the corresponding band thresholding function in Fig. 11.7 to denoise the magnitude of the coefficient. The sign of the noisy coefficient is retained. The pyramid is inverted to obtain the denoised image. The result of denoising a noisy image using this framework is shown in Fig. 11.8(g).

The use of neighborhood (or "contextual") information has permeated many areas of image processing. In denoising, one of the first published methods was a locally adapted version of the Weiner filter by Lee [6], in which the local variance in the pixel domain

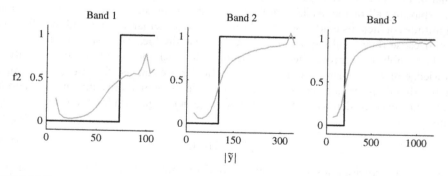

FIGURE 11.7

Neighborhood thresholding (black) and neighborhood weighting (gray) functions $f(|\tilde{y}|)$ as a function of $|\tilde{y}|$ (see Eq. (11.7)) for various bands; Jaggedness in the curves occurs at values for which there was insufficient data to obtain a reliable estimate of the function.

FIGURE 11.8

Example image denoising results. (a) Original image; (b) Noisy image (13.40 dB); (c) Band thresholding (24.45 dB); (d) band weighting (25.04 dB); (e) coefficient thresholding (24.97 dB); (f) coefficient weighting (25.72 dB); (g) neighborhood thresholding (26.24 dB); (h) neighborhood weighting (26.60 dB). All images have been cropped from the original to highlight the details more clearly.

is used to estimate the signal strength, and thus the denoising function. This method is available in MATLAB (through the function *wiener2*). More recently, Chang *et al.* [7] used this idea in a spatially-adaptive thresholding scheme and derived a closed form expression for the threshold. A variation of this implementation known as *NeighShrink* [8] is similar to our implementation, but determines the threshold in closed form based on the observed noisy image, thus obviating the need for training.

11.5.2 Neighborhood Weighting

As in the previous examples, a natural extension of the idea of thresholding a coefficient based on its neighbors is to weight the coefficient by a scalar value that is computed from the neighborhood energy. Once again, our implementation to find these functions is similar to the one presented earlier for the coefficient-weighting in Section 11.4.2. Given the triplets, $(x_i, y_i, |\tilde{y}_i|)$, we now solve for the scalar, $f(|\tilde{y}_i|)$, that minimizes:

$$\sum_{i:|\vec{y}_i|=|\tilde{y}|} (x_i - f(|\tilde{y}_i|) \cdot y_i)^2 .$$

Using the same technique from earlier, the resulting scalar can be shown to be $f(|\tilde{y}_i|) = \sum_i (x_i y_i) / \sum_i (y_i^2)$. The form of the function, $f(\cdot)$, is shown in Fig. 11.7. The coefficient estimation functions, $\hat{x}(\tilde{y})$, depend on both $|\tilde{y}|$ and y and not very easy to visualize.

To denoise an image, we first compute its five-band multiscale decomposition. For a given band, we use a 5×5 kernel to estimate the local energy $|\vec{y}|$ around each coefficient y, and use the denoising functions in Fig. 11.7 to multiply the central coefficient y by $f(|\vec{y}|)$. The pyramid is then inverted to create the denoised image as shown in Fig. 11.8(h). We see in Fig. 11.9 that this method provides consistent PSNR improvement over other schemes.

The use of contextual neighborhoods is found in all of the highest performing recent methods. Miçhak *et al.* [9] exploited the observation that when the central coefficient is divided by the magnitude of its spatial neighbors, the distribution of the multiscale coefficients is approximately Gaussian (see also [10]), and used this to develop a Wiener-like estimate. Of course, the "neighbors" in this formulation need not be restricted to spatially adjacent pixels. Sendur and Selesnick [11] derive a bivariate shrinkage function, where the neighborhood $\vec{y}$ contains the coefficient being denoised, and the coefficient in the same location at the next coarsest scale (the "parent"). The resulting denoising functions are a 2D extension of those shown in Fig. 11.6. Portilla *et al.* [12] present a denoising scheme based on modeling a neighborhood of coefficients as arising from an infinite mixture of Gaussian distributions, known as a "Gaussian scale mixture." The resulting least-squares denoising function uses a more general combination over the neighbors than a simple sum of squares, and this flexibility leads to substantial improvements in denoising performance. The problem of contextual denoising remains an active area of research, with new methods appearing every month.

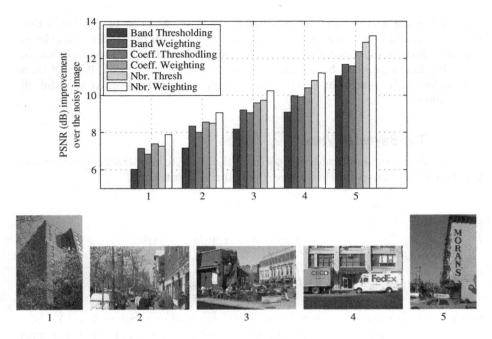

FIGURE 11.9

PSNR improvement (in dB, relative to that of the noisy image). Each group of bars shows the performance of the six denoising schemes for one of the images shown in the bottom row. All denoising schemes used the exact same Gaussian white noise sample of standard deviation 60.

11.6 STATISTICAL MODELING FOR OPTIMAL DENOISING

In order to keep the presentation focused and simple, we have resorted to using a training set of noise-free and noisy coefficients to learn parameters for the denoising function (such as the threshold or weighting values). In particular, given training pairs of noise-free and noisy coefficients, (x_n, y_n), we have solved a regression problem to obtain the parameters of the denoising function: $\hat{\theta} = \mathrm{argmin}_\theta \sum_n (x_n - f(y_n; \theta) \cdot y_n)^2$. This methodology is appealing because it does not depend on models of image or noise, and this directness makes it easy to understand. It can also be useful for image enhancement in practical situations where it might be difficult to model the signal and noise. Recently, such an approach [13] was used to produce denoising results that are comparable to the state-of-the-art. As shown in that work, the data-driven approach can also be used to compensate for other distortions such as blurring.

But there are two clear drawbacks in the regression approach. First, the underlying assumption of such a training scheme is that the ensemble of training images is representative of all images. But some of the photographic image properties we have described, while general, do vary significantly from image to image, and it is thus preferable to adapt the denoising solution to the properties of the specific image being denoised. Second, the simplistic form of training we have described requires that the denoising functions must be separately learned for each noise level. Both of these drawbacks can be somewhat alleviated by considering a more abstract probabilistic formulation.

11.6.1 The Bayesian View

If we consider the noise-free and noisy coefficients, x and y, to be instances of two random variables X and Y, respectively, we may rewrite the MSE criterion

$$\sum_n (x_n - g(y_n))^2 \approx E_{X,Y}(X - g(Y))^2$$

$$= \int dX \int dY \, P(X, Y)(X - g(Y))^2$$

$$= \int dX \underbrace{P(X)}_{\text{Prior}} \int dY \, \underbrace{P(Y|X)}_{\text{Noise model}} \underbrace{(X - g(Y))^2}_{\text{Loss function}}, \qquad (11.8)$$

where $E_{X,Y}(\cdot)$ indicates the expected value, taken over random variables X and Y.

As described earlier in Section 11.4.2, the denoising function, $g(Y)$, that minimizes this expression is the conditional expectation $E(X|Y)$. In the framework described above, we have replaced all our samples (x_n, y_n) by their probability density functions. In general, the prior, $P(X)$, is the model for multiscale coefficients in the ensemble of noise-free images. The conditional density, $P(Y|X)$, is a model for the noise corruption process. Thus, this formulation cleanly separates the description of the noise from the description of the image properties, allowing us to learn the image model, $P(X)$, once and then reuse it for any level or type of noise (since $P(Y|X)$ need not be restricted to additive white Gaussian). The problem of image modeling is an active area of research, and is described in more detail in Chapter 9.

11.6.2 Empirical Bayesian Methods

The Bayesian approach assumes that we know (or have learned from a training set) the densities $P(X)$ and $P(Y|X)$. While the idea of a single prior, $P(X)$, for all images in an ensemble is exciting and motivates much of the work in image modeling, denoising solutions based on this model are unable to adapt to the peculiarities of a particular image. The most successful recent image denoising techniques are based on empirical Bayes methods. The basic idea is to define a parametric prior $P(X; \theta)$ and adjust the parameters, θ, for each image that is to be denoised. This adaptation can be difficult to achieve, since one generally has access only to the noisy data samples, Y, and not the noise-free samples, X. A conceptually simple method is to select the parameters that maximize the probabillity of the noisy, but this utilizes a separate criterion (likelihood) for the parameter estimation and denoising, and can thus lead to suboptimal results.

A more abstract but more consistent method relies on optimizing Stein's unbiased risk estimator (SURE) [14–17].

11.7 CONCLUSIONS

The main objective of this chapter was to lead the reader through a sequence of simple denoising techniques, illustrating how observed properties of noise and image structure can be formalized statistically and used to design and optimize denoising methods. We presented a unified framework for multiscale denoising of the form $\hat{x}(*) = f(\bar{*}) \cdot y$, and developed three different versions, each one using a different definition for $*$. The first was a global model in which entire bands of multiscale coefficients were modified using a common denoising function, while the second was a local technique in which each individual coefficient was modified using a function that depended on its own value. The third approach adopted a compromise between these two extremes, using a function that depended on local neighborhood information to denoise each coefficient. For each of these denoising schemes, we presented two variations: a thresholding operator and a weighting operator.

An important aspect of our examples that we discussed only briefly is the choice of image representation. Our examples were based on an overcomplete multiscale decomposition into octave-width frequency channels. While the development of orthogonal wavelets has had a profound impact on the application of compression, the artifacts that arise from the critical sampling of these decompositions are higly visible and detrimental when they are used for denoising. Since denoising generally less concerned about the economy of representation (and in particular, about the number of coefficients), it makes sense to relax the critical sampling requirement, sampling subbands at rates equal to or higher than their associated Nyquist limits.

In fact, it has been demonstrated repeatedly (e.g., [18]) and recently proven [17] that redundancy in the image representation redundancy in the image representation can lead directly to improved denoising performance. There has also been significant effort in developing multiscale geometric transforms such as ridgelets, curvelets, and wedgelets which aim to provide better signal compaction by representing relevant image features such as edges and contours. And although this chapter has focused on multiscale image denoising, there have also been significant improvements in denoising in the pixel domain [19].

The three primary components of the general statistical formalism of Eq. (11.8) signal model, noise model, and error function—are all active areas of research. As mentioned previously, statistical modeling of images is discussed in Chapter 9. Regarding the noise, we have assumed an additive Gaussian model, but the noise that contaminates real images is often correlated, non-Gaussian, and even signal-dependent. Modeling of image noise is described in Chapter 7. And finally, there is room for improvement in the choice of objective function. Throughout this chapter, we minimized the error in the pyramid domain, but always reported the PSNR results in the image domain. If the multiscale pyramid is orthonormal, minimizing error in the multiscale domain is equivalent to minimizing error in the pixel domain. But in over-complete representations, this is no longer true, and noise that starts out white in the pixel domain is correlated in the pyramid domain. Recent approaches in image denoising attempt to minimize the mean-squared error in the image domain while still operating in an over-complete transform domain [13, 17]. But even if the denoising scheme is designed to minimize PSNR in the pixel domain, it is well known that PSNR does not provide a good description of perceptual image quality (see Chapter 21). An important topic of future research is thus to optimize denoising functions using a perceptual metric for image quality [20].

ACKNOWLEDGMENTS

All photographs used for training and testing were taken by Nicolas Bonnier. We thank Martin Raphan and Siwei Lyu for their comments and suggestions on the presentation of the material in this chapter.

REFERENCES

[1] E. H. Adelson, C. H. Anderson, J. R. Bergen, P. J. Burt, and J. M. Ogden. Pyramid methods in image processing. *RCA Eng.*, 29(6):33–41, 1984.

[2] P. J. Burt and E. H. Adelson. The Laplacian pyramid as a compact image code. *IEEE Trans. Commun.*, COM-31:532–540, 1983.

[3] D. L. Donoho and I. M. Johnstone. Ideal spatial adaptation by wavelet shrinkage. *Biometrika*, 81(3):425–455, 1994.

[4] D. L. Donoho and I. M. Johnstone. Adapting to unknown smoothness via wavelet shrinkage. *J. Am. Stat. Assoc.*, 90(432):1200–1224, 1995.

[5] E. P. Simoncelli and E. H. Adelson. Noise removal via Bayesian wavelet coring. In E. Adelson, editor, *Proceedings of the International Conference on Image Processing*, Vol. 1, 379–382, 1996.

[6] J. S. Lee. Digital image enhancement and noise filtering by use of local statistics. *IEEE. Trans. Pattern. Anal. Mach. Intell.*, PAMI-2:165–168, 1980.

[7] S. G. Chang, B. Yu, and M. Vetterli. Spatially adaptive wavelet thresholding with context modeling for image denoising. *IEEE Trans. Image Process.*, 9(9):1522–1531, 2000.

[8] G. Chen, T. Bui, and A. Krzyzak. Image denoising using neighbouring wavelet coefficients. In T. Bui, editor, *Proceedings of IEEE International Conference on Acoustics, Speech, and Signal Processing (ICASSP '04)*, Vol. 2, ii-917–ii-920, 2004.

[9] M. K. Mihçak, I. Kozintsev, K. Ramchandran, and P. Moulin. Low-complexity image denoising based on statistical modeling of wavelet coefficients. *IEEE Signal Process. Lett.*, 6(12):300–303, 1999.

[10] D. L. Ruderman and W. Bialek. Statistics of natural images: scaling in the woods. *Phys. Rev. Lett.*, 73(6):814–817, 1994.

[11] L. Sendur and I. W. Selesnick. Bivariate shrinkage functions for wavelet-based denoising exploiting interscale dependency. *IEEE Trans. Signal Process.*, 50(11):2744–2756, 2002.

[12] J. Portilla, V. Strela, M. Wainwright, and E. Simoncelli. Image denoising using scale mixtures of Gaussians in the wavelet domain. *IEEE Trans. Image Process.*, 12(11):1338–1351, 2003.

[13] Y. Hel-Or and D. Shaked. A discriminative approach for wavelet denoising. *IEEE Trans. Image Process.*, 17(4):443–457, 2008.

[14] D. L. Donoho. Denoising by soft-thresholding. *IEEE Trans. Inf. Theory*, 43:613–627, 1995.

[15] J. C. Pesquet and D. Leporini. A new wavelet estimator for image denoising. In *6th International Conference on Image Processing and its Applications*, Dublin, Ireland, 249–253, July 1997.

[16] F. Luisier, T. Blu, and M. Unser. A new SURE approach to image denoising: Interscale orthonormal wavelet thresholding. *IEEE Trans. Image Process.*, 16:593–606, 2007.

[17] M. Raphan and E. P. Simoncelli. Optimal denoising in redundant bases. *IEEE Trans. Image Process.*, 17(8), pp. 1342–1352, Aug 2008.

[18] R. R. Coifman and D. L. Donoho. Translation-invariant de-noising. In A. Antoniadis and G. Oppenheim, editors, *Wavelets and Statistics*. Springer-Verlag lecture notes, San Diego, CA, 1995.

[19] K. Dabov, A. Foi, V. Katkovnik, and K. Egiazarian. Image denoising by sparse 3-d transform-domain collaborative filtering. *IEEE Trans. Image Process.*, 16(8):2080–2095, 2007.

[20] S. S. Channappayya, A. C. Bovik, C. Caramanis, and R. W. Heath. Design of linear equalizers optimized for the structural similarity index. *IEEE Trans. Image Process.*, 17:857–872, 2008.

Nonlinear Filtering for Image Analysis and Enhancement

12

Gonzalo R. Arce[1], Jan Bacca[1], and José L. Paredes[2]

[1] *University of Delaware;* [2] *Universidad de Los Andes*

12.1 INTRODUCTION

Digital image enhancement and analysis have played, and will continue to play, an important role in scientific, industrial, and military applications. In addition to these applications, image enhancement and analysis are increasingly being used in consumer electronics. Internet Web users, for instance, rely on built-in image processing protocols such as JPEG and interpolation and in the process have become image processing users equipped with powerful yet inexpensive software such as Photoshop. Users not only retrieve digital images from the Web but are now able to acquire their own by use of digital cameras or through digitization services of standard 35 mm analog film. The end result is that consumers are beginning to use home computers to enhance and manipulate their own digital pictures. Image enhancement refers to processes seeking to improve the visual appearance of an image. As an example, image enhancement might be used to emphasize the edges within the image. This edge-enhanced image would be more visually pleasing to the naked eye, or perhaps could serve as an input to a machine that would detect the edges and perhaps make measurements of shape and size of the detected edges. Image enhancement is important because of its usefulness in virtually all image processing applications.

Image enhancement tools are often classified into (a) point operations and (b) spatial operations. Point operations include contrast stretching, noise clipping, histogram modification, and pseudo-coloring. Point operations are, in general, simple nonlinear operations that are well known in the image processing literature and are covered elsewhere in this *Guide*. Spatial operations used in image processing today are, on the other hand, typically linear operations. The reason for this is that spatial linear operations are simple and easily implemented. Although linear image enhancement tools are often adequate in many applications, significant advantages in image enhancement can be attained if nonlinear techniques are applied [1]. Nonlinear methods effectively preserve edges and details of images while methods using linear operators tend to blur and distort them. Additionally, nonlinear image enhancement tools are less susceptible to noise. Noise is always present due to the physical randomness of image acquisition systems. For example, underexposure and low-light conditions in analog photography conditions lead to images with film-grain noise which, together with the image signal itself, are captured during the digitization process.

This chapter focuses on nonlinear and spatial image enhancement and analysis. The nonlinear tools described in this chapter are easily implemented on currently available computers. Rather than using linear combinations of pixel values within a local window, these tools use the local weighted median (WM). In Section 12.2, the principles of WM are presented. Weighted medians have striking analogies with traditional linear FIR filters, yet their behavior is often markedly different. In Section 12.3, we show how WM filters can be easily used for noise removal. In particular, the center WM filter is described as a tunable filter highly effective in impulsive noise. Section 12.4 focuses on image enlargement, or zooming, using WM filter structures which, unlike standard linear interpolation methods, provide little edge degradation. Section 12.5 describes image sharpening algorithms based on WM filters. These methods offer significant advantages over traditional linear sharpening tools whenever noise is present in the underlying images.

12.2 WEIGHTED MEDIAN SMOOTHERS AND FILTERS

12.2.1 Running Median Smoothers

The running median was first suggested as a nonlinear smoother for time series data by Tukey in 1974 [2]. To define the running median smoother, let $\{x(\cdot)\}$ be a discrete-time sequence. The running median passes a window over the sequence $\{x(\cdot)\}$ that selects, at each instant n, a set of samples to comprise the observation vector $\mathbf{x}(n)$. The observation window is centered at n, resulting in

$$\mathbf{x}(n) = [x(n - N_L), \ldots, x(n), \ldots, x(n + N_R)]^T, \tag{12.1}$$

where N_L and N_R may range in value over the nonnegative integers and $N = N_L + N_R + 1$ is the window size. The median smoother operating on the input sequence $\{x(\cdot)\}$ produces the output sequence $\{y\}$, where at time index n

$$y(n) = \text{MEDIAN}[x(n - N_L), \ldots, x(n), \ldots, x(n + N_R)] \tag{12.2}$$

$$= \text{MEDIAN}[x_1(n), \ldots, x_N(n)], \tag{12.3}$$

where $x_i(n) = x(n - N_L + 1 - i)$ for $i = 1, 2, \ldots, N$. That is, the samples in the observation window are sorted and the middle, or median, value is taken as the output.

If $x_{(1)}, x_{(2)}, \ldots, x_{(N)}$ are the sorted samples in the observation window, the median smoother outputs

$$
y(n) = \begin{cases} x_{\left(\frac{N+1}{2}\right)} & \text{if } N \text{ is odd} \\[2ex] \dfrac{x_{\left(\frac{N}{2}\right)} + x_{\left(\frac{N}{2}+1\right)}}{2} & \text{otherwise.} \end{cases}
\tag{12.4}
$$

In most cases, the window is symmetric about $x(n)$ and $N_L = N_R$.

The input sequence $\{x(\cdot)\}$ may be either finite or infinite in extent. For the finite case, the samples of $\{x(\cdot)\}$ can be indexed as $x(1), x(2), \ldots, x(L)$, where L is the length of the sequence. Due to the symmetric nature of the observation window, the window extends beyond a finite extent input sequence at both the beginning and end. These end effects are generally accounted for by appending N_L samples at the beginning and N_R samples at the end of $\{x(\cdot)\}$. Although the appended samples can be arbitrarily chosen, typically these are selected so that the points appended at the beginning of the sequence have the same value as the first signal point, and the points appended at the end of the sequence all have the value of the last signal point.

To illustrate the appending of an input sequence and the median smoother operation, consider the input signal $\{x(\cdot)\}$ of Fig. 12.1. In this example, $\{x(\cdot)\}$ consists of 20 observations from a 6-level process, $\{x : x(n) \in \{0, 1, \ldots, 5\}, n = 1, 2, \ldots, 20\}$. The figure

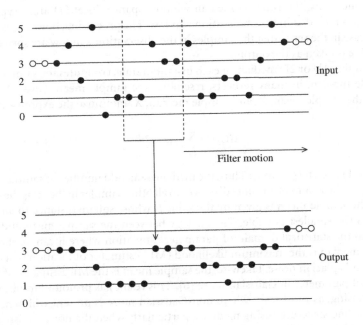

FIGURE 12.1

The operation of the window width 5 median smoother. o: appended points.

shows the input sequence and the resulting output sequence for a window size 5 median smoother. Note that to account for edge effects, two samples have been appended to both the beginning and end of the sequence. The median smoother output at the window location shown in the figure is

$$y(9) = \text{MEDIAN}[x(7), x(8), x(9), x(10), x(11)]$$

$$= \text{MEDIAN}[\,1, 1, 4, 3, 3\,] = 3.$$

Running medians can be extended to a recursive mode by replacing the "causal" input samples in the median smoother by previously derived output samples [3]. The output of the recursive median smoother is given by

$$y(n) = \text{MEDIAN}[y(n - N_L), \ldots, y(n - 1), x(n), \ldots, x(n + N_R)]. \tag{12.5}$$

In recursive median smoothing, the center sample in the observation window is modified before the window is moved to the next position. In this manner, the output at each window location replaces the old input value at the center of the window. With the same amount of operations, recursive median smoothers have better noise attenuation capabilities than their nonrecursive counterparts [4, 5]. Alternatively, recursive median smoothers require smaller window lengths than their nonrecursive counterparts in order to attain a desired level of noise attenuation. Consequently, for the same level of noise attenuation, recursive median smoothers often yield less signal distortion. In image processing applications, the running median window spans a local 2D area. Typically, an $N \times N$ area is included in the observation window. The processing, however, is identical to the 1D case in the sense that the samples in the observation window are sorted and the middle value is taken as the output.

The running 1D or 2D median, at each instant in time, computes the sample median. The sample median, in many respects, resembles the sample mean. Given N samples $x_1, \ldots, x_N$ the sample mean, $\bar{X}$, and sample median, $\tilde{X}$, minimize the expression

$$G(\beta) = \sum_{i=1}^{N} |x_i - \beta|^p \tag{12.6}$$

for $p = 2$ and $p = 1$, respectively. Thus, the median of an odd number of samples emerges as the sample whose sum of absolute distances to all other samples in the set is the smallest. Likewise, the sample mean is given by the value β whose square distance to all samples in the set is the smallest possible. The analogy between the sample mean and median extends into the statistical domain of parameter estimation where it can be shown that the sample median is the maximum likelihood (ML) estimator of location of a constant parameter in Laplacian noise. Likewise, the sample mean is the ML estimator of location of a constant parameter in Gaussian noise [6]. This result has profound implications in signal processing, as most tasks where non-Gaussian noise is present will benefit from signal processing structures using medians, particularly when the noise statistics can be characterized by probability densities having lighter than Gaussian tails (which leads to noise with impulsive characteristics)[7–9].

12.2.2 Weighted Median Smoothers

Although the median is a robust estimator that possesses many optimality properties, the performance of running medians is limited by the fact that it is temporally blind. That is, all observation samples are treated equally regardless of their location within the observation window. Much like weights can be incorporated into the sample mean to form a weighted mean, a WM can be defined as the sample which minimizes the weighted cost function

$$G_p(\beta) = \sum_{i=1}^{N} W_i |x_i - \beta|^p, \tag{12.7}$$

for $p = 1$. For $p = 2$, the cost function (12.7) is quadratic and the value β minimizing it is the normalized weighted mean

$$\hat{\beta} = \arg\min_{\beta} \sum_{i=1}^{N} W_i (x_i - \beta)^2 = \frac{\sum_{i=1}^{N} W_i \cdot x_i}{\sum_{i=1}^{N} W_i} \tag{12.8}$$

with $W_i > 0$. For $p = 1$, $G_1(\beta)$ is piecewise linear and convex for $W_i \geq 0$. The value β minimizing (12.7) is thus guaranteed to be one of the samples $x_1, x_2, \ldots, x_N$ and is referred to as the WM, originally introduced over a hundred years ago by Edgemore [10]. After some algebraic manipulations, it can be shown that the running WM output is computed as

$$y(n) = \text{MEDIAN}[W_1 \diamond x_1(n), W_2 \diamond x_2(n), \ldots, W_N \diamond x_N(n)], \tag{12.9}$$

where $W_i > 0$ and $\diamond$ is the replication operator defined as $W_i \diamond x_i = \overbrace{x_i, x_i, \ldots, x_i}^{W_i \text{ times}}$. Weighted median smoothers were introduced in the signal processing literature by Brownigg in 1984 and have since received considerable attention [11–13]. The WM smoothing operation can be schematically described as in Fig. 12.2.

Weighted Median Smoothing Computation Consider the window size 5 WM smoother defined by the symmetric weight vector $\mathbf{W} = [1, 2, 3, 2, 1]$. For the observation $\mathbf{x}(n) = [12, 6, 4, 1, 9]$, the WM smoother output is found as

$$
\begin{aligned}
y(n) &= \text{MEDIAN}[\, 1 \diamond 12, 2 \diamond 6, 3 \diamond 4, 2 \diamond 1, 1 \diamond 9 \,] \\
&= \text{MEDIAN}[\, 12, 6, 6, 4, 4, 4, 1, 1, 9 \,] \\
&= \text{MEDIAN}[\, 1, 1, 4, 4, \underline{4}, 6, 6, 9, 12 \,] = 4,
\end{aligned} \tag{12.10}
$$

where the median value is underlined in Eq. (12.10). The large weighting on the center input sample results in this sample being taken as the output. As a comparison, the standard median output for the given input is $y(n) = 6$.

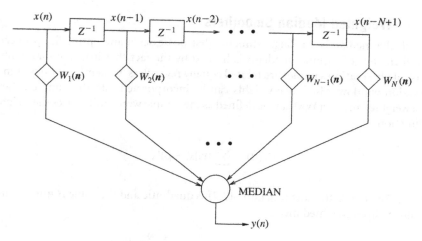

FIGURE 12.2

The weighted median smoothing operation.

Although the smoother weights in the above example are integer-valued, the standard WM smoother definition clearly allows for positive real-valued weights. The WM smoother output for this case is as follows:

1. Calculate the threshold $W_0 = \frac{1}{2}\sum_{i=1}^{N} W_i$.

2. Sort the samples in the observation vector $\mathbf{x}(n)$.

3. Sum the weights corresponding to the sorted samples beginning with the maximum sample and continuing down in order.

4. The output is the sample whose weight causes the sum to become $\geq W_0$.

To illustrate the WM smoother operation for positive real-valued weights, consider the WM smoother defined by $\mathbf{W} = [0.1, 0.1, 0.2, 0.2, 0.1]$. The output for this smoother operating on $\mathbf{x}(n) = [12, 6, 4, 1, 9]$ is found as follows. Summing the weights gives the threshold $W_0 = \frac{1}{2}\sum_{i=1}^{5} W_i = 0.35$. The observation samples, sorted observation samples, their corresponding weight, and the partial sum of weights (from each ordered sample to the maximum) are:

observation samples	12,	6,	4,	1,	9
corresponding weights	0.1,	0.1,	0.2,	0.2,	0.1
sorted observation samples	1,	4,	6,	9,	12
corresponding weights	0.2,	0.2,	0.1,	0.1,	0.1
partial weight sums	0.7,	<u>0.5</u>,	0.3,	0.2,	0.1

(12.11)

Thus, the output is 4 since when starting from the right (maximum sample) and summing the weights, the threshold $W_0 = 0.35$ is not reached until the weight associated with 4 is added.

An interesting characteristic of WM smoothers is that the nature of a WM smoother is not modified if its weights are multiplied by a positive constant. Thus, the same filter characteristics can be synthesized by different sets of weights. Although the WM smoother admits real-valued positive weights, it turns out that any WM smoother based on real-valued positive weights has an equivalent integer-valued weight representation [14]. Consequently, there are only a finite number of WM smoothers for a given window size. The number of WM smoothers, however, grows rapidly with window size [13].

Weighted median smoothers can also operate on a recursive mode. The output of a recursive WM smoother is given by

$$y(n) = \text{MEDIAN}[W_{-N_1} \diamond y(n - N_1), \ldots, W_{-1} \diamond y(n - 1), W_0 \diamond x(n), \ldots,$$
$$W_{N_1} \diamond x(n + N_1)], \tag{12.12}$$

where the weights W_i are as before constrained to be positive-valued. Recursive WM smoothers offer advantages over WM smoothers in the same way that recursive medians have advantages over their nonrecursive counterparts. In fact, recursive WM smoothers can synthesize nonrecursive WM smoothers of much longer window sizes [14].

12.2.2.1 *The Center Weighted Median Smoother*

The weighting mechanism of WM smoothers allows for great flexibility in emphasizing or deemphasizing specific input samples. In most applications, not all samples are equally important. Due to the symmetric nature of the observation window, the sample most correlated with the desired estimate is, in general, the center observation sample. This observation leads to the center weighted median (CWM) smoother, which is a relatively simple subset of the WM smoother that has proven useful in many applications [12].

The CWM smoother is realized by allowing only the center observation sample to be weighted. Thus, the output of the CWM smoother is given by

$$y(n) = \text{MEDIAN}[x_1, \ldots, x_{c-1}, W_c \diamond x_c, x_{c+1}, \ldots, x_N], \tag{12.13}$$

where W_c is an odd positive integer and $c = (N + 1)/2 = N_1 + 1$ is the index of the center sample. When $W_c = 1$, the operator is a median smoother, and for $W_c \geq N$, the CWM reduces to an identity operation.

The effect of varying the center sample weight is perhaps best seen by way of an example. Consider a segment of recorded speech. The voiced waveform "a" noise is shown at the top of Fig. 12.3. This speech signal is taken as the input of a CWM smoother of size 9. The outputs of the CWM, as the weight parameter $W_c = 2w + 1$ for $w = 0, \ldots, 3$, are shown in the figure. Clearly, as W_c is increased less smoothing occurs. This response of the CWM smoother is explained by relating the weight W_c and the CWM smoother output to select order statistics.

The CWM smoother has an intuitive interpretation. It turns out that the output of a CWM smoother is equivalent to computing

$$y(n) = \text{MEDIAN}\left[x_{(k)}, x_c, x_{(N-k+1)}\right], \tag{12.14}$$

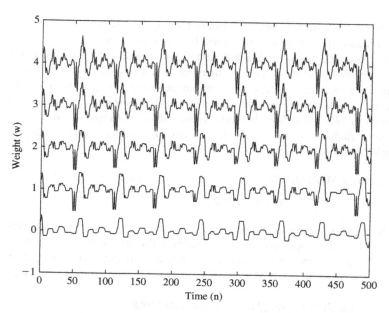

FIGURE 12.3

Effects of increasing the center weight of a CWM smoother of size $N = 9$ operating on the voiced speech "a." The CWM smoother output is shown for $W_c = 2w + 1$, with $w = 0, 1, 2, 3$. Note that for $W_c = 1$ the CWM reduces to median smoothing, and for $W_c = 9$ it becomes the identity operator.

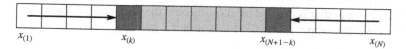

FIGURE 12.4

The center weighted median smoothing operation. The center observation sample is mapped to the order statistic $x_{(k)}$ $(x_{(N+1-k)})$ if the center sample is less (greater) than $x_{(k)}$ $(x_{(N+1-k)})$, and left unaltered otherwise.

where $k = (N + 2 - W_c)/2$ for $1 \le W_c \le N$, and $k = 1$ for $W_c > N$. Since $x(n)$ is the center sample in the observation window, i.e., $x_c = x(n)$, the output of the smoother is identical to the input as long as the $x(n)$ lies in the interval $\left[x_{(k)}, x_{(N+1-k)}\right]$. If the center input sample is greater than $x_{(N+1-k)}$ the smoothing outputs $x_{(N+1-k)}$, guarding against a high rank order (large) aberrant data point being taken as the output. Similarly, the smoother's output is $x_{(k)}$ if the sample $x(n)$ is smaller than this order statistic. This CWM smoother performance characteristic is illustrated in Figs. 12.4 and 12.5. Figure 12.4 shows how the input sample is left unaltered if it is between the trimming statistics $x_{(k)}$ and $x_{(N+1-k)}$ and mapped to one of these statistics if it is outside this range. Figure 12.5

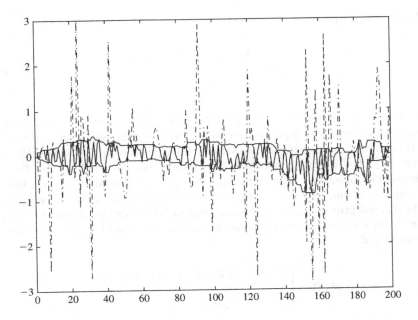

FIGURE 12.5

An example of the CWM smoother operating on a Laplacian distributed sequence with unit variance. Shown are the input $(- \cdot - \cdot -)$ and output (——) sequences as well as the trimming statistics $x_{(k)}$ and $x_{(N+1-k)}$. The window size is 25 and $k = 7$.

shows an example of the CWM smoother operating on a constant-valued sequence in additive Laplacian noise. Along with the input and output, the trimming statistics are shown as an upper and lower bound on the filtered signal. It is easily seen how increasing k will tighten the range in which the input is passed directly to the output.

12.2.2.2 *Permutation Weighted Median Smoothers*

The principle behind the CWM smoother lies in the ability to emphasize, or de-emphasize, the center sample of the window by tuning the center weight, while keeping the weight values of all other samples at unity. In essence, the value given to the center weight indicates the "reliability" of the center sample. If the center sample does not contain an impulse (high reliability), it would be desirable to make the center weight large such that no smoothing takes place (identity filter). On the other hand, if an impulse was present in the center of the window (low reliability), no emphasis should be given to the center sample (impulse), and the center weight should be given the smallest possible weight, i.e., $W_c = 1$, reducing the CWM smoother structure to a simple median. Notably, this adaptation of the center weight can be easily achieved by considering the center sample's rank among all pixels in the window [15, 16]. More precisely, denoting the rank of the center sample of the window at a given location as $R_c(n)$, then the simplest *permutation* WM smoother is defined by the following modification of the

CWM smoothing operation

$$W_c(n) = \begin{cases} N & \text{if } T_L \leq R_c(n) \leq T_U \\ 1 & \text{otherwise,} \end{cases} \qquad (12.15)$$

where N is the window size and $1 \leq T_L \leq T_U \leq N$ are two adjustable threshold parameters that determine the degree of smoothing. Note that the weight in (12.15) is data adaptive and may change between two values with n. The smaller (larger) the threshold parameter T_L (T_U) is set to, the better the detail preservation. Generally, T_L and T_U are set symmetrically around the median. If the underlying noise distribution was not symmetric about the origin, a nonsymmetric assignment of the thresholds would be appropriate.

The data-adaptive structure of the smoother in (12.15) can be extended so that the center weight is not only switched between two possible values but also can take on N different values:

$$W_c(n) = \begin{cases} W_{c(j)}(n) & \text{if } R_c(n) = j, \quad j \in \{1,2,\dots,N\} \\ 0 & \text{otherwise.} \end{cases} \qquad (12.16)$$

Thus, the weight assigned to x_c is drawn from the center weight set $\{W_{c(1)}, W_{c(2)}, \dots, W_{c(N)}\}$. With an increased number of weights, the smoother in (12.16) can perform better although the design of the weights is no longer trivial and optimization algorithms are needed [15, 16]. A further generalization of (12.16) is feasible where weights are given to all samples in the window, but where the value of each weight is data-dependent and determined by the rank of the corresponding sample. In this case, the output of the permutation WM smoother is found as

$$y(n) = \text{MEDIAN}[x_1(n) \diamond W_{1(R_1)}, x_2(n) \diamond W_{1(R_2)}, \dots, x_1(n) \diamond W_{1(R_1)}], \qquad (12.17)$$

where $W_{i(R_i)}$ is the weight assigned to $x_i(n)$ and selected according to the sample's rank R_i. The weight assigned to x_i is drawn from the weight set $\{W_{i(1)}, W_{i(2)}, \dots, W_{i(N)}\}$. Having N weights per sample, a total of N^2 samples need to be stored in the computation of (12.17). In general, optimization algorithms are needed to design the set of weights although in some cases the design is simple, as with the smoother in (12.15). Permutation WM smoothers can provide significant improvement in performance at the higher cost of memory cells [15].

12.2.2.3 *Threshold Decomposition and Stack Smoothers*

An important tool for the analysis and design of WM smoothers is the threshold decomposition property [17]. Given an integer-valued set of samples $x_1, x_2, \dots, x_N$ forming the vector $\mathbf{x} = [x_1, x_2, \dots, x_N]^T$, where $x_i \in \{-M, \dots, -1, 0, \dots, M\}$, the threshold decomposition of $\mathbf{x}$ amounts to decomposing this vector into $2M$ binary vectors $\mathbf{x}^{-M+1}, \ldots, \mathbf{x}^0, \ldots, \mathbf{x}^M$, where the ith element of $\mathbf{x}^m$ is defined by

$$x_i^m = T^m(x_i) = \begin{cases} 1 & \text{if } x_i \geq m, \\ -1 & \text{if } x_i < m, \end{cases} \tag{12.18}$$

where $T^m(\cdot)$ is referred to as the thresholding operator. Using the sign function, the above can be written as $x_i^m = \text{sgn}(x_i - m^-)$, where m^- represents a real number approaching the integer m from the left. Although defined for integer-valued signals, the thresholding operation in (12.18) can be extended to noninteger signals with a finite number of quantization levels. The threshold decomposition of the vector $\mathbf{x} = [0, 0, 2, -2, 1, 1, 0, -1, -1]^T$ with $M = 2$, for instance, leads to the 4 binary vectors

$$\mathbf{x}^2 = [-1, -1, 1, -1, -1, -1, -1, -1, -1]^T$$

$$\mathbf{x}^1 = [-1, -1, 1, -1, 1, 1, -1, -1, -1]^T$$

$$\mathbf{x}^0 = [1, 1, 1, -1, 1, 1, 1, -1, -1]^T \tag{12.19}$$

$$\mathbf{x}^{-1} = [1, 1, 1, -1, 1, 1, 1, 1, 1]^T.$$

Threshold decomposition has several important properties. First, threshold decomposition is reversible. Given a set of thresholded signals, each of the samples in $\mathbf{x}$ can be exactly reconstructed as

$$x_i = \frac{1}{2} \sum_{m=-M+1}^{M} x_i^m. \tag{12.20}$$

Thus, an integer-valued discrete-time signal has a unique threshold signal representation, and vice versa

$$x_i \overset{T.D.}{\longleftrightarrow} \{x_i^m\},$$

where $\overset{T.D.}{\longleftrightarrow}$ denotes the one-to-one mapping provided by the threshold decomposition operation.

The set of threshold decomposed variables obey the following set of partial ordering rules. For all thresholding levels $m > \ell$, it can be shown that $x_i^m \leq x_i^\ell$. In particular, if $x_i^m = 1$, then $x_i^\ell = 1$ for all $\ell < m$. Similarly, if $x_i^\ell = -1$, then $x_i^m = -1$, for all $m > \ell$. The partial order relationships among samples across the various thresholded levels emerge naturally in thresholding and are referred to as the *stacking constraints* [18].

Threshold decomposition is of particular importance in WM smoothing since they are commutable operations. That is, applying a WM smoother to a $2M + 1$ valued signal is equivalent to decomposing the signal to $2M$ binary thresholded signals, processing each binary signal separately with the corresponding WM smoother, and then adding the binary outputs together to obtain the integer-valued output. Thus, the WM smoothing of a set of samples $x_1, x_2, \ldots, x_N$ is related to the set of the thresholded WM smoothed signals as [14, 17]

$$\text{Weighted MEDIAN}(x_1, \ldots, x_N) = \frac{1}{2} \sum_{m=-M+1}^{M} \text{Weighted MEDIAN}(x_1^m, \ldots, x_N^m). \quad (12.21)$$

Since $x_i \stackrel{T.D.}{\longleftrightarrow} \{x_i^m\}$ and Weighted $\text{MEDIAN}(x_i|_{i=1}^N) \stackrel{T.D.}{\longleftrightarrow} \{\text{WeigthedMEDIAN}(x_i^m|_{i=1}^N)\}$, the relationship in (12.21) establishes a *weak* superposition property satisfied by the nonlinear median operator, which is important from the fact that the effects of median smoothing on binary signals are much easier to analyze than that on multilevel signals. In fact, the WM operation on binary samples reduces to a simple Boolean operation. The median of three binary samples x_1, x_2, x_3, for example, is equivalent to: $x_1 x_2 + x_2 x_3 + x_1 x_3$, where the $+$ (OR) and $x_i x_j$ (AND) "Boolean" operators in the $\{-1, 1\}$ domain are defined as

$$x_i + x_j = \max(x_i, x_j)$$
$$x_i x_j = \min(x_i, x_j). \quad (12.22)$$

Note that the operations in (12.22) are also valid for the standard Boolean operations in the $\{0, 1\}$ domain.

The framework of threshold decomposition and Boolean operations has led to the general class of nonlinear smoothers referred to here as stack smoothers [18], whose output is defined by

$$S(x_1, \ldots, x_N) = \frac{1}{2} \sum_{m=-M+1}^{M} f(x_1^m, \ldots, x_N^m), \quad (12.23)$$

where $f(\cdot)$ is a "Boolean" operation satisfying (12.22) and the stacking property. More precisely, if two binary vectors $\mathbf{u} \in \{-1, 1\}^N$ and $\mathbf{v} \in \{-1, 1\}^N$ stack, i.e., $u_i \geq v_i$ for all $i \in \{1, \ldots, N\}$, then their respective outputs stack, $f(\mathbf{u}) \geq f(\mathbf{v})$. A necessary and sufficient condition for a function to possess the stacking property is that it can be expressed as a Boolean function which contains no complements of input variables [19]. Such functions are known as *positive Boolean* functions (PBFs).

Given a PBF $f(x_1^m, \ldots, x_N^m)$ which characterizes a stack smoother, it is possible to find the equivalent smoother in the integer domain by replacing the binary AND and OR Boolean functions acting on the x_i's with *max* and *min* operations acting on the multilevel x_i samples. A more intuitive class of smoothers is obtained, however, if the PBFs are further restricted [14]. When self-duality and separability is imposed, for instance, the equivalent integer domain stack smoothers reduce to the well-known class of WM smoothers with positive weights. For example, if the Boolean function in the stack smoother representation is selected as $f(x_1, x_2, x_3, x_4) = x_1 x_3 x_4 + x_2 x_4 + x_2 x_3 + x_1 x_2$, the equivalent WM smoother takes on the positive weights $(W_1, W_2, W_3, W_4) = (1, 2, 1, 1)$. The procedure of how to obtain the weights W_i from the PBF is described in [14].

12.2.3 Weighted Median Filters

Admitting only positive weights, WM smoothers are severely constrained as they are, in essence, smoothers having "lowpass" type filtering characteristics. A large number of engineering applications require "bandpass" or "highpass" frequency filtering characteristics. Linear FIR equalizers admitting only positive filter weights, for instance, would lead to completely unacceptable results. Thus, it is not surprising that WM smoothers admitting only positive weights lead to unacceptable results in a number of applications.

Much like how the sample mean can be generalized to the rich class of linear FIR filters, there is a logical way to generalize the median to an equivalently rich class of WM filters that admit both positive and negative weights [20]. It turns out that the extension is not only natural, leading to a significantly richer filter class, but it is simple as well. Perhaps the simplest approach to derive the class of WM filters with real-valued weights is by analogy. The sample mean $\bar{\beta} = \text{MEAN}(X_1, X_2, \ldots, X_N)$ can be generalized to the class of linear FIR filters as

$$\beta = \text{MEAN}(W_1 \cdot X_1, W_2 \cdot X_2, \ldots, W_N \cdot X_N), \qquad (12.24)$$

where $X_i \in R$. In order to apply the analogy to the median filter structure (12.24) must be written as

$$\bar{\beta} = \text{MEAN}(|W_1| \cdot \text{sgn}(W_1)X_1, |W_2| \cdot \text{sgn}(W_2)X_2, \ldots, |W_N| \cdot \text{sgn}(W_n)X_N), \qquad (12.25)$$

where the sign of the weight affects the corresponding input sample and the weighting is constrained to be nonnegative. By analogy, the class of WM filters admitting real-valued weights emerges as [20]

$$\tilde{\beta} = \text{MEDIAN}(|W_1| \diamond \text{sgn}(W_1)X_1, |W_2| \diamond \text{sgn}(W_2)X_2, \ldots, |W_N| \diamond \text{sgn}(W_n)X_N), \qquad (12.26)$$

with $W_i \in R$ for $i = 1, 2, \ldots, N$. Again, the weight signs are uncoupled from the weight magnitude values and are merged with the observation samples. The weight magnitudes play the equivalent role of positive weights in the framework of WM smoothers. It is simple to show that the weighted mean (normalized) and the WM operations shown in (12.25) and (12.26), respectively, minimize to

$$G_2(\beta) = \sum_{i=1}^{N} |W_i| (\text{sgn}(W_i)X_i - \beta)^2 \quad \text{and} \quad G_1(\beta) = \sum_{i=1}^{N} |W_i||\text{sgn}(W_i)X_i - \beta|. \qquad (12.27)$$

While $G_2(\beta)$ is a convex continuous function, $G_1(\beta)$ is a convex but piecewise linear function whose minimum point is guaranteed to be one of the "signed" input samples (i.e., $\text{sgn}(W_i) X_i$).

Weighted Median Filter Computation The WM filter output for noninteger weights can be determined as follows [20]:

1. Calculate the threshold $T_0 = \frac{1}{2} \sum_{i=1}^{N} |W_i|$.

2. Sort the "signed" observation samples $\text{sgn}(W_i)X_i$.

3. Sum the magnitude of the weights corresponding to the sorted "signed" samples beginning with the maximum and continuing down in order.

4. The output is the signed sample whose magnitude weight causes the sum to become $\geq T_0$.

The following example illustrates this procedure. Consider the window size 5 WM filter defined by the real-valued weights $[W_1, W_2, W_3, W_4, W_5]^T = [0.1, 0.2, 0.3, -0.2, 0.1]^T$. The output for this filter operating on the observation set $[X_1, X_2, X_3, X_4, X_5]^T = [-2, 2, -1, 3, 6]^T$ is found as follows. Summing the absolute weights gives the threshold $T_0 = \frac{1}{2} \sum_{i=1}^{5} |W_i| = 0.45$. The "signed" observation samples, sorted observation samples, their corresponding weight, and the partial sum of weights (from each ordered sample to the maximum) are:

observation samples	$-2,$	$2,$	$-1,$	$3,$	6
corresponding weights	$0.1,$	$0.2,$	$0.3,$	$-0.2,$	0.1

sorted signed observation samples	$-3,$	$-2,$	$-1,$	$2,$	6
corresponding absolute weights	$0.2,$	$0.1,$	$0.3,$	$0.2,$	0.1
partial weight sums	$0.9,$	$0.7,$	$\underline{0.6},$	$0.3,$	$0.1.$

Thus, the output is -1 since when starting from the right (maximum sample) and summing the weights, the threshold $T_0 = 0.45$ is not reached until the weight associated with -1 is added. The underlined sum value above indicates that this is the first sum which meets or exceeds the threshold.

The effect that negative weights have on the WM operation is similar to the effect that negative weights have on linear FIR filter outputs. Figure 12.6 illustrates this concept where $G_2(\beta)$ and $G_1(\beta)$, the cost functions associated with linear FIR and WM filters, respectively, are plotted as a function of β. Recall that the output of each filter is the value minimizing the cost function. The input samples are again selected as $[X_1, X_2, X_3, X_4, X_5] = [-2, 2, -1, 3, 6]$ and two sets of weights are used. The first set is $[W_1, W_2, W_3, W_4, W_5] = [0.1, 0.2, 0.3, 0.2, 0.1]$, where all the coefficients are positive, and the second set is $[0.1, 0.2, 0.3, -0.2, 0.1]$, where W_4 has been changed, with respect to the first set of weights, from 0.2 to -0.2. Figure 12.6(a) shows the cost functions $G_2(\beta)$ of the linear FIR filter for the two sets of filter weights. Notice that by changing the sign of W_4, we are effectively moving X_4 to its new location $\text{sgn}(W_4)X_4 = -3$. This, in turn, pulls the minimum of the cost function toward the relocated sample $\text{sgn}(W_4)X_4$. Negatively weighting X_4 on $G_1(\beta)$ has a similar effect as shown in Fig. 12.6(b). In this case, the minimum is pulled toward the new location of $\text{sgn}(W_4)X_4$. The minimum, however, occurs at one of the samples $\text{sgn}(W_i)X_i$. More details on WM filtering can be found in [20, 21].

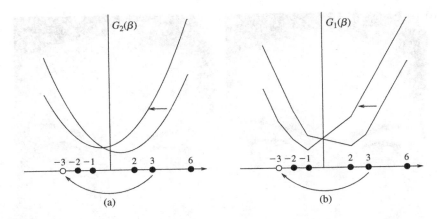

FIGURE 12.6

Effects of negative weighting on the cost functions $G_2(\beta)$ and $G_1(\beta)$. The input samples are $[X_1, X_2, X_3, X_4, X_5]^T = [-2, 2, -1, 3, 6]^T$ which are filtered by the two set of weights $[0.1, 0.2, 0.3, 0.2, 0.1]^T$ and $[0.1, 0.2, 0.3, -0.2, 0.1]^T$, respectively.

12.3 IMAGE NOISE CLEANING

Median smoothers are widely used in image processing to clean images corrupted by noise. Median filters are particularly effective at removing outliers. Often referred to as "salt and pepper" noise, outliers are often present due to bit errors in transmission, or introduced during the signal acquisition stage. Impulsive noise in images can also occur as a result to damage to analog film. Although a WM smoother can be designed to "best" remove the noise, CWM smoothers often provide similar results at a much lower complexity [12]. By simply tuning the center weight, a user can obtain the desired level of smoothing. Of course, as the center weight is decreased to attain the desired level of impulse suppresion, the output image will suffer increased distortion particularly around the image's fine details. Nonetheless, CWM smoothers can be highly effective in removing "salt and pepper" noise while preserving the fine image details. Figures 12.7(a) and (b) depict a noise free grayscale image and the corresponding image with "salt and pepper" noise. Each pixel in the image has a 10 percent probability of being contaminated with an impulse. The impulses occur randomly and were generated by MATLAB's imnoise funtion. Figures 12.7(c) and (d) depict the noisy image processed with a 5×5 window CWM smoother with center weights 15 and 5, respectively. The impulse-rejection and detail-preservation tradeoff in CWM smoothing is clearly illustrated in Figs. 12.7(c) and 12.7(d). A color version of the "portrait" image was also corrupted by "salt and pepper" noise and filtered using CWM independently in each color plane.

At the extreme, for $W_c = 1$, the CWM smoother reduces to the median smoother which is effective at removing impulsive noise. It is, however, unable to preserve the image's fine details [22]. Figure 12.9 shows enlarged sections of the noise-free image

FIGURE 12.7

Impulse noise cleaning with a 5×5 CWM smoother: (a) original grayscale "portrait" image; (b) image with salt and pepper noise; (c) CWM smoother with $W_c = 15$; (d) CWM smoother with $W_c = 5$.

FIGURE 12.8

Impulse noise cleaning with a 5 × 5 CWM smoother: (a) original "portrait" image; (b) image with salt and pepper noise; (c) CWM smoother with $W_c = 16$; (d) CWM smoother with $W_c = 5$.

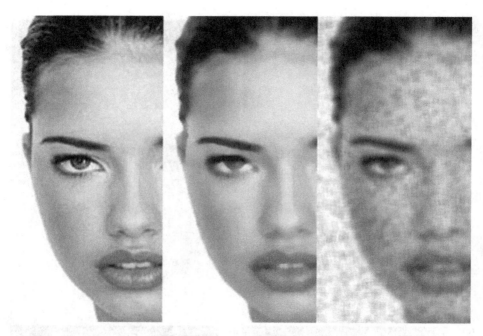

FIGURE 12.9

(Enlarged) Noise-free image (left); 5×5 median smoother output (center); and 5×5 mean smoother (right).

(left), and of the noisy image after the median smoother has been applied (center). Severe blurring is introduced by the median smoother and it is readily apparent in Fig. 12.9. As a reference, the output of a running mean of the same size is also shown in Fig. 12.9 (right). The image is severely degraded as each impulse is smeared to neighboring pixels by the averaging operation.

Figures 12.7 and 12.8 show that CWM smoothers can be effective at removing impulsive noise. If increased detail-preservation is sought and the center weight is increased, CWM smoothers begin to breakdown and impulses appear on the output. One simple way to ameliorate this limitation is to employ a recursive mode of operation. In essence, past inputs are replaced by previous outputs as described in (12.12) with the only difference that only the center sample is weighted. All the other samples in the window are weighted by one. Figure 12.10 shows enlarged sections of the nonrecursive CWM filter (left) and of the corresponding recursive CWM smoother, both with the same center weight ($W_c = 15$). This figure illustrates the increased noise attenuation provided by recursion without the loss of image resolution.

Both recursive and nonrecursive CWM smoothers can produce outputs with disturbing artifacts particularly when the center weights are increased in order to improve

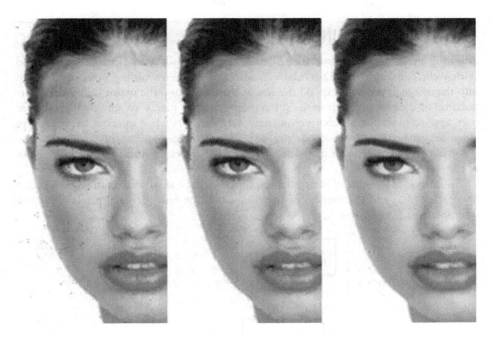

FIGURE 12.10

(Enlarged) CWM smoother output (left); recursive CWM smoother output (center); and permutation CWM smoother output (right). Window size is 5 × 5.

the detail-preservation characteristics of the smoothers. The artifacts are most apparent around the image's edges and details. Edges at the output appear jagged and impulsive noise can break through next to the image detail features. The distinct response of the CWM smoother in different regions of the image is due to the fact that images are nonstationary in nature. Abrupt changes in the image's local mean and texture carry most of the visual information content. CWM smoothers process the entire image with fixed weights and are inherently limited in this sense by their static nature. Although some improvement is attained by introducing recursion or by using more weights in a properly designed WM smoother structure, these approaches are also static and do not properly address the nonstationary nature of images.

Significant improvement in noise attenuation and detail preservation can be attained if permutation WM filter structures are used. Figure 12.10 (right) shows the output of the permutation CWM filter in (12.15) when the "salt and pepper" degraded "portrait" image is inputted. The parameters were given the values $T_L = 6$ and $T_U = 20$. The improvement achieved by switching W_c between just two different values is significant. The impulses are deleted without exception, the details are preserved, and the jagged artifacts typical of CWM smoothers are not present in the output.

12.4 IMAGE ZOOMING

Zooming an image is an important task used in many applications, including the World Wide Web, digital video, DVDs, and scientific imaging. When zooming, pixels are inserted into the image in order to expand the size of the image, and the major task is the interpolation of the new pixels from the surrounding original pixels. Weighted medians have been applied to similar problems requiring interpolation, such as interlace to progressive video conversion for television systems [13]. The advantage of using the WM in interpolation over traditional linear methods is better edge preservation and a less "blocky" look to edges.

To introduce the idea of interpolation, suppose that a small matrix must be zoomed by a factor of 2, and the median of the closest two (or four) original pixels is used to interpolate each new pixel:

$$
\begin{bmatrix} 7 & 8 & 5 \\ 6 & 10 & 9 \end{bmatrix}
\xrightarrow{\boxed{\begin{array}{c} \text{Zero} \\ \text{Interlace} \end{array}}}
\begin{bmatrix} 7 & 0 & 8 & 0 & 5 & 0 \\ 0 & 0 & 0 & 0 & 0 & 0 \\ 6 & 0 & 10 & 0 & 9 & 0 \\ 0 & 0 & 0 & 0 & 0 & 0 \end{bmatrix}
$$

$$
\xrightarrow{\boxed{\begin{array}{c} \text{Median} \\ \text{Interpolation} \end{array}}}
\begin{bmatrix} 7 & 7.5 & 8 & 6.5 & 5 & 5 \\ 6.5 & 7.5 & 9 & 8.5 & 7 & 7 \\ 6 & 8 & 10 & 9.5 & 9 & 9 \\ 6 & 8 & 10 & 9.5 & 9 & 9 \end{bmatrix}.
$$

Zooming commonly requires a change in the image dimensions by a noninteger factor, such as a 50% zoom where the dimensions must be 1.5 times the original. Also, a change in the length-to-width ratio might be needed if the horizontal and vertical zoom factors are different. The simplest way to accomplish zooming of arbitrary scale is to double the size of the original as many times as needed to obtain an image larger than the target size in all dimensions, interpolating new pixels on each expansion. Then the desired image can be attained by subsampling the larger image, or taking pixels at regular intervals from the larger image in order to obtain an image with the correct length and width. The subsampling of images and the possible filtering needed are topics well known in traditional image processing, thus, we will focus on the problem of doubling the size of an image.

A digital image is represented by an array of values, each value defining the color of a pixel of the image. Whether the color is constrained to be a shade of gray, in which case only one value is needed to define the brightness of each pixel, or whether three values are needed to define the red, green, and blue components of each pixel does not affect the definition of the technique of WM interpolation. The only difference between grayscale and color images is that an ordinary WM is used in grayscale images while color requires a vector WM.

To double the size of an image, first an empty array is constructed with twice the number of rows and columns as the original (Fig. 12.11(a)), and the original pixels are placed into alternating rows and columns (the "00" pixels in Fig. 12.11(a)). To interpolate the remaining pixels, the method known as polyphase interpolation is used. In this method, each new pixel with four original pixels at its four corners (the "11" pixels in Fig. 12.11(b)) is interpolated first by using the WM of the four nearest original pixels as the value for that pixel. Since all original pixels are equally trustworthy and the same distance from the pixel being interpolated, a weight of 1 is used for the four nearest original pixels. The resulting array is shown in Fig. 12.11(c). The remaining pixels are determined by taking a WM of the four closest pixels. Thus each of the "01" pixels in Fig. 12.11(c) is interpolated using two original pixels to the left and right and two previously interpolated pixels above and below. Similarly, the "10" pixels are interpolated with original pixels above and below and interpolated pixels ("11" pixels) to the right and left.

Since the "11" pixels were interpolated, they are less reliable than the original pixels and should be given lower weights in determining the "01" and "10" pixels. Therefore, the "11" pixels are given weights of 0.5 in the median to determine the "01" and "10" pixels, while the "00" original pixels have weights of 1 associated with them. The weight of 0.5 is used because it implies that when both "11" pixels have values that are not between the two "00" pixel values then one of the "00" pixels or their average will be used. Thus "11" pixels differing from the "00" pixels do not greatly affect the result of the WM. Only when the "11" pixels lie between the two "00" pixels will they have a direct effect on the interpolation. The choice of 0.5 for the weight is arbitrary, since any weight greater than 0 and less than 1 will produce the same result. When implementing the polyphase method, the "01" and "10" pixels must be treated differently due to the fact that the orientation of the two closest original pixels is different for the two types of pixels. Figure 12.11(d) shows the final result of doubling the size of the original array.

To illustrate the process, consider an expansion of the grayscale image represented by an array of pixels, the pixel in the ith row and jth column having brightness $a_{i,j}$. The array $a_{i,j}$ will be interpolated into the array $x_{i,j}^{pq}$, with p and q taking values 0 or 1 indicating in the same way as above the type of interpolation required:

$$
\begin{bmatrix} a_{1,1} & a_{1,2} & a_{1,3} \\ a_{2,1} & a_{2,2} & a_{2,3} \\ a_{3,1} & a_{3,2} & a_{3,3} \end{bmatrix} \Longrightarrow
\begin{bmatrix}
x_{1,1}^{00} & x_{1,1}^{01} & x_{1,2}^{00} & x_{1,2}^{01} & x_{1,3}^{00} & x_{1,3}^{01} \\
x_{1,1}^{10} & x_{1,1}^{11} & x_{1,2}^{10} & x_{1,2}^{11} & x_{1,3}^{10} & x_{1,3}^{11} \\
x_{2,1}^{00} & x_{2,1}^{01} & x_{2,2}^{00} & x_{2,2}^{01} & x_{2,3}^{00} & x_{2,3}^{01} \\
x_{2,1}^{10} & x_{2,1}^{11} & x_{2,2}^{10} & x_{2,2}^{11} & x_{2,3}^{10} & x_{2,3}^{11} \\
x_{3,1}^{00} & x_{3,1}^{01} & x_{3,2}^{00} & x_{3,2}^{01} & x_{3,3}^{00} & x_{3,3}^{01} \\
x_{3,1}^{10} & x_{3,1}^{11} & x_{3,2}^{10} & x_{3,2}^{11} & x_{3,3}^{10} & x_{3,3}^{11}
\end{bmatrix}.
$$

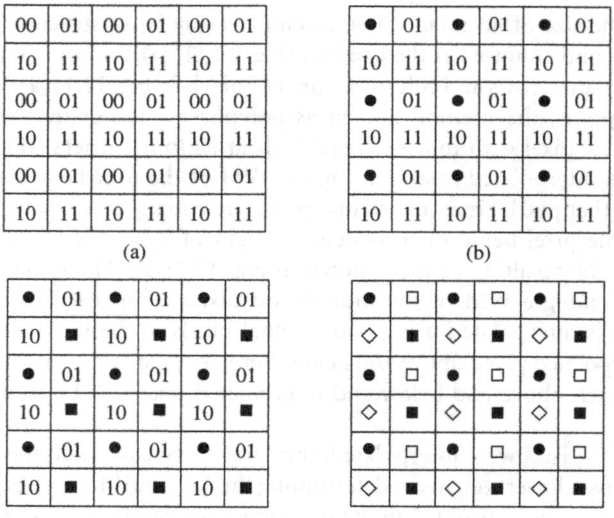

FIGURE 12.11

The steps of polyphase interpolation.

The pixels are interpolated as follows:

$$x_{i,j}^{00} = a_{i,j}$$

$$x_{i,j}^{11} = \text{MEDIAN}[a_{i,j}, a_{i+1,j}, a_{i,j+1}, a_{i+1,j+1}]$$

$$x_{i,j}^{01} = \text{MEDIAN}[a_{i,j}, a_{i,j+1}, 0.5 \diamond x_{i-1,j}^{11}, 0.5 \diamond x_{i+1,j}^{11}]$$

$$x_{i,j}^{10} = \text{MEDIAN}[a_{i,j}, a_{i+1,j}, 0.5 \diamond x_{i,j-1}^{11}, 0.5 \diamond x_{i,j+1}^{11}].$$

An example of median interpolation compared with bilinear interpolation is given in Fig. 12.12. Bilinear interpolation uses the average of the nearest two original pixels to interpolate the "01" and "10" pixels in Fig. 12.11(b) and the average of the nearest four original pixels for the "11" pixels. The edge-preserving advantage of the WM interpolation is readily seen in the figure.

12.5   IMAGE SHARPENING

Human perception is highly sensitive to edges and fine details of an image and since they are composed primarily high-frequency components, the visual quality of an image can be enormously degraded if the high frequencies are attenuated or completely removed.

FIGURE 12.12

Example of zooming. Original is at the top with the area of interest outlined in white. On the lower left is the bilinear interpolation of the area, and on the lower right the weighted median interpolation.

On the other hand, enhancing the high-frequency components of an image leads to an improvement in the visual quality. *Image sharpening refers to any enhancement technique that highlights edges and fine details in an image.* Image sharpening is widely used in printing and photographic industries for increasing the local contrast and sharpening the images. In principle, image sharpening consists of adding to the original image a signal that is proportional to a highpass filtered version of the original image. Figure 12.13 illustrates this procedure often referred to as unsharp masking [23, 24] on a 1D signal. As shown in Fig. 12.13, the original image is first filtered by a highpass filter which extracts the high-frequency components, and then a scaled version of the highpass filter output

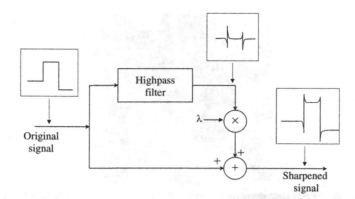

FIGURE 12.13

Image sharpening by high-frequency emphasis.

is added to the original image thus producing a sharpened image of the original. Note that the homogeneous regions of the signal, i.e., where the signal is constant, remain unchanged. The sharpening operation can be represented by

$$s_{i,j} = x_{i,j} + \lambda * \mathcal{F}(x_{i,j}), \tag{12.28}$$

where $x_{i,j}$ is the original pixel value at the coordinate (i, j), $\mathcal{F}(\cdot)$ is the highpass filter, λ is a tuning parameter greater than or equal to zero, and $s_{i,j}$ is the sharpened pixel at the coordinate (i, j). The value taken by λ depends on the grade of sharpness desired. Increasing λ yields a more sharpened image.

If color images are used, $x_{i,j}$, $s_{i,j}$, and λ are three-component vectors, whereas if grayscale images are used, $x_{i,j}$, $s_{i,j}$, and λ are single-component vectors. Thus the process described here can be applied to either grayscale or color images with the only difference that vector-filters have to be used in sharpening color images whereas single-component filters are used with grayscale images.

The key point in the effective sharpening process lies in the choice of the highpass filtering operation. Traditionally, linear filters have been used to implement the highpass filter, however, linear techniques can lead to unacceptable results if the original image is corrupted with noise. A trade-off between noise attenuation and edge highlighting can be obtained if a WM filter with appropriated weights is used. To illustrate this, consider a WM filter applied to a grayscale image where the following filter mask is used

$$W = \frac{1}{3} \begin{bmatrix} -1 & -1 & -1 \\ -1 & 8 & -1 \\ -1 & -1 & -1 \end{bmatrix}. \tag{12.29}$$

Due to the weight coefficients in (12.29), for each position of the moving window, the output is proportional to the difference between the center pixel and the smallest pixel around the center pixel. Thus, the filter output takes relatively large values for prominent edges in an image, and small values in regions that are fairly smooth, being zero only in regions that have constant gray level.

Although this filter can effectively extract the edges contained in a image, the effect that this filtering operation has over negative-slope edges is different from that obtained for positive-slope edges.[1] Since the filter output is proportional to the difference between the center pixel and the smallest pixel around the center, for negative-slope edges, the center pixel takes small values producing small values at the filter output. Moreover, the filter output is zero if the smallest pixel around the center pixel and the center pixel have the same values. This implies that negative-slope edges are not extracted in the same way as positive-slope edges. To overcome this limitation, the basic image sharpening structure shown in Fig. 12.13 must be modified such that positive-slope edges and negative-slope edges are highlighted in the same proportion. A simple way to accomplish that is: (a) extract the positive-slope edges by filtering the original image with the filter mask described above; (b) extract the negative-slope edges by first preprocessing the original image such that the negative-slope edges become positive-slope edges, and then filter the preprocessed image with the filter described above; and (c) combine appropriately the original image, the filtered version of the original image and the filtered version of the preprocessed image to form the sharpened image.

Thus both positive-slope edges and negative-slope edges are equally highlighted. This procedure is illustrated in Fig. 12.14, where the top branch extracts the positive-slope edges and the middle branch extracts the negative-slope edges. In order to understand the effects of edge sharpening, a row of a test image is plotted in Fig. 12.15 together with a row of the sharpened image when only the positive-slope edges are highlighted (Fig. 12.15(a)), only the negative-slope edges are highlighted (Fig. 12.15(b)), and both positive-slope and negative-slope edges are jointly highlighted (Fig. 12.15(c)).

In Fig. 12.14, λ_1 and λ_2 are tuning parameters that control the amount of sharpness desired in the positive-slope direction and in the negative-slope direction, respectively. The values of λ_1 and λ_2 are generally selected to be equal. The output of the pre-filtering operation is defined as

$$x'_{i,j} = M - x_{i,j} \tag{12.30}$$

with M equal to the maximum pixel value of the original image. This pre-filtering operation can be thought of as a flipping and a shifting operation of the values of the original image such that the negative-slope edges are converted to positive-slope edges. Since the original image and the pre-filtered image are filtered by the same WM filter, the positive-slope edges and negative-slope edges are sharpened in the same way.

In Fig. 12.16, the performance of the WM filter image sharpening is compared with that of traditional image sharpening based on linear FIR filters. For the linear sharpener, the scheme shown in Fig. 12.13 was used. The parameter λ was set to 1 for the clean

---

[1] A change from a gray level to a lower gray level is referred to as a negative-slope edge, whereas a change from a gray level to a higher gray level is referred to as a positive-slope edge.

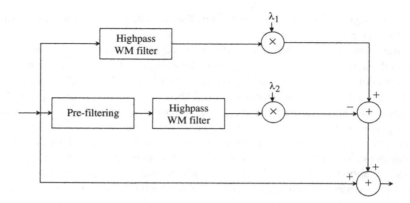

FIGURE 12.14

Image sharpening based on the weighted median filter.

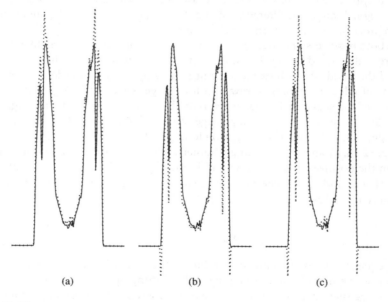

(a) (b) (c)

FIGURE 12.15

Original row of a test image (solid line) and row sharpened (dotted line) with (a) only positive-slope edges; (b) only negative-slope edges; and (c) both positive-slope and negative-slope edges.

image and to 0.75 for the noise image. For the WM sharpener, the scheme of Fig. 12.14 was used with $\lambda_1 = \lambda_2 = 2$ for the clean image, and $\lambda_1 = \lambda_2 = 1.5$ for the noisy image. The filter mask given by (12.29) was used in both linear and median image sharpening. As before each component of the color image was processed separately.

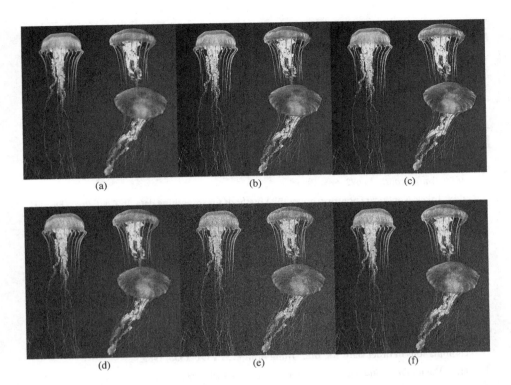

FIGURE 12.16

(a) Original image sharpened with; (b) the FIR-sharpener; and (c) the WM-sharpener; (d) Image with added Gaussian noise sharpened with; (e) the FIR-sharpener; and (f) the WM-sharpener.

12.6 CONCLUSION

The principles behind WM smoothers and WM filters have been presented in this chapter, as well as some of the applications of these nonlinear signal processing structures in image enhancement. It should be apparent to the reader that many similarities exist between linear and median filters. As illustrated in this chapter, there are several applications in image enhancement where WM filters provide significant advantages over traditional image enhancement methods using linear filters. The methods presented here, and other image enhancement methods that can be easily developed using WM filters, are computationally simple and provide significant advantages, and consequently can be used in emerging consumer electronic products, PC and internet imaging tools, medical and biomedical imaging systems, and of course in military applications.

ACKNOWLEDGMENT

This work was supported in part by the National Science Foundation under grant MIP-9530923.

REFERENCES

[1] Y. H. Lee and S. A. Kassam. Generalized median filtering and related nonlinear filtering techniques. *IEEE Trans. Acoust.*, 33:672–683, 1985.

[2] J. W. Tukey. Nonlinear (nonsuperimposable) methods for smoothing data. In *Conf. Rec., (Eascon)*, 1974.

[3] T. A. Nodes and N. C. Gallagher, Jr. Median filters: some modifications and their properties. *IEEE Trans. Acoust.*, 30:739–746, 1982.

[4] G. R. Arce and N. C. Gallagher, Jr. Stochastic analysis of the recursive median filter process. *IEEE Trans. Inf. Theory*, IT-34:669–679, 1988.

[5] G. R. Arce. Statistical threshold decomposition for recursive and nonrecursive median filters. *IEEE Trans. Inf. Theory*, 32:243–253, 1986.

[6] E. L. Lehmann. *Theory of Point Estimation.* J Wiley & Sons, New York, NY, 1983.

[7] A. C. Bovik, T. S. Huang, and J. D.C. Munson. A generalization of median filtering using linear combinations of order statistics. *IEEE Trans. Acoust.*, 31:1342–1350, 1983.

[8] H. A. David. *Order Statistics.* Wiley Interscience, New York, 1981.

[9] B. C. Arnold, N. Balakrishnan, and H. N. Nagaraja. *A First Course in Order Statistics.* John Wiley & Sons, New York, NY, 1992.

[10] F. Y. Edgeworth. A new method of reducing observations relating to several quantities. *Phil. Mag. (Fifth Series)*, 24:222–223, 1887.

[11] D. R. K. Brownrigg. The weighted median filter. *Commun. ACM*, 27:807–818, 1984.

[12] S.-J. Ko and Y. H. Lee. Center weighted median filters and their applications to image enhancement. *Theor. Comput. Sci.*, 38:984–993, 1991.

[13] L. Yin, R. Yang, M. Gabbouj, and Y. Neuvo. Weighted median filters: a tutorial. *IEEE Trans. Circuits Syst. II*, 41:157–192, 1996.

[14] O. Yli-Harja, J. Astola, and Y. Neuvo. Analysis of the properties of median and weighted median filters using threshold logic and stack filter representation. *IEEE Trans. Acoust.*, 39:395–410, 1991.

[15] G. R. Arce, T. A. Hall, and K. E. Barner. Permutation weighted order statistic filters. *IEEE Trans. Image Process.*, 4:1070–1083, 1995.

[16] R. C. Hardie and K. E. Barner. Rank conditioned rank selection filters for signal restoration. *IEEE Trans Image Process.*, 3:192–206, 1994.

[17] J. P. Fitch, E. J. Coyle, and N. C. Gallagher. Median filtering by threshold decomposition. *IEEE Trans. Acoust.*, 32:1183–1188, 1984.

[18] P. D. Wendt, E. J. Coyle, and N. C. Gallagher, Jr. Stack filters. *IEEE Trans. Acoust.*, 34:898–911, 1986.

[19] E. N. Gilbert. Lattice-theoretic properties of frontal switching functions. *J. Math. Phys.*, 33:57–67, 1954.

[20] G. R. Arce. A general weighted median filter structure admitting negative weights. *IEEE Trans. Signal Process.*, 46:3195–3205, 1998.

[21] J. L. Paredes and G. R. Arce. Stack filters, stack smoothers, and mirrored threshold decomposition. *IEEE Trans. Signal Process.*, 47:2757–2767, 1999.

[22] A. C. Bovik. Streaking in median filtered images. *IEEE Trans. Acoust.*, 35:493–503, 1987.

[23] A. K. Jain. *Fundamentals of Digital Image Processing.* Prentice Hall, Upper Saddle River, New Jersey, 1989.

[24] J. S. Lim. *Two-Dimensional Signal and Image Processing.* Prentice Hall, Englewood Cliffs, NJ, 1990.

[25] S. Hoyos, Y. Li, J. Bacca, and G. R. Arce. Weighted median filters admitting complex-valued weights and their optimization. *IEEE Trans. Acoust.*, 52:2776–2787, 2004.

[26] S. Hoyos, J. Bacca, and G. R. Arce. Spectral design of weighted median filters: a general iterative approach. *IEEE Trans. Acoust.*, 53:1045–1056, 2005.

Morphological Filtering

Petros Maragos

National Technical University of Athens

13

13.1 INTRODUCTION

The goals of *image enhancement* include the improvement of the visibility and perceptibility of the various regions into which an image can be partitioned and of the detectability of the image features inside these regions. These goals include tasks such as cleaning the image from various types of noise, enhancing the contrast among adjacent regions or features, simplifying the image via selective smoothing or elimination of features at certain scales, and retaining only features at certain desirable scales. Image enhancement is usually followed by (or is done simultaneously with) *detection of features* such as edges, peaks, and other geometric features, which is of paramount importance in low-level vision. Further, many related vision problems involve the detection of a known template; such problems are usually solved via template matching.

While traditional approaches for solving the above tasks have used mainly tools of linear systems, nowadays a new understanding has matured that linear approaches are not well suited or even fail to solve problems involving geometrical aspects of the image. Thus, there is a need for nonlinear geometric approaches. A powerful nonlinear methodology that can successfully solve the above problems is mathematical morphology.

Mathematical morphology is a set- and lattice-theoretic methodology for image analysis, which aims at quantitatively describing the geometrical structure of image objects. It was initiated [1, 2] in the late 1960s to analyze binary images from geological and biomedical data as well as to formalize and extend earlier or parallel work [3, 4] on binary pattern recognition based on cellular automata and Boolean/threshold logic. In the late 1970s, it was extended to gray-level images [2]. In the mid-1980s, it was brought to the mainstream of image/signal processing and related to other nonlinear filtering approaches [5, 6]. Finally, in the late 1980s and 1990s, it was generalized to arbitrary lattices [7, 8]. The above evolution of ideas has formed what we call nowadays the field of **morphological image processing**, which is a broad and coherent collection of theoretical concepts, nonlinear filters, design methodologies, and applications systems. Its rich theoretical framework, algorithmic efficiency, easy implementability on special hardware, and suitability for many shape-oriented problems have propelled its widespread usage and further advancement by many academic and industry groups working on various problems in image processing, computer vision, and pattern recognition.

This chapter provides a brief introduction to the application of morphological image processing to image enhancement and feature detection. Thus, it discusses four important general problems of low-level (early) vision, progressing from the easiest (or more easily defined) to the more difficult (or harder to define): (i) geometric filtering of binary and gray-level images of the shrink/expand type or of the peak/valley blob removal type; (ii) cleaning noise from the image or improving its contrast; (iii) detecting in the image the presence of known templates; and (iv) detecting the existence and location of geometric features such as edges and peaks whose types are known but not their exact form.

13.2 MORPHOLOGICAL IMAGE OPERATORS

13.2.1 Morphological Filters for Binary Images

Given a sampled[1] binary image signal $f[x]$ with values 1 for the image object and 0 for the background, typical image transformations involving a moving window set $W = \{y_1, y_2, \ldots, y_n\}$ of n sample indexes would be

$$\psi_b(f)[x] = b(f[x - y_1], \ldots, f[x - y_n]), \tag{13.1}$$

where $b(v_1, \ldots, v_n)$ is a Boolean function of n variables. The mapping $f \mapsto \psi_b(f)$ is called a *Boolean filter*. By varying the Boolean function b, a large variety of Boolean filters can be obtained. For example, choosing a Boolean AND for b would *shrink* the input image object, whereas a Boolean OR would *expand* it. Numerous other Boolean filters are possible since there are 2^{2^n} possible Boolean functions of n variables. The main applications of such Boolean image operations have been in biomedical image processing, character recognition, object detection, and general 2D shape analysis [3, 4].

Among the important concepts offered by mathematical morphology was to use *sets* to represent binary images and set operations to represent binary image transformations. Specifically, given a binary image, let the object be represented by the set X and its background by the set complement X^c. The Boolean OR transformation of X by a (window) set B is equivalent to the Minkowski set addition $\oplus$, also called **dilation**, of X by B:

$$X \oplus B \triangleq \{z : (B^s)_{+z} \cap X \neq \varnothing\} = \bigcup_{y \in B} X_{+y}, \tag{13.2}$$

where $X_{+y} \triangleq \{x + y : x \in X\}$ is the *translation* of X along the vector y, and $B^s \triangleq \{x : -x \in B\}$ is the *symmetric* of B with respect to the origin. Likewise, the Boolean AND

---

[1]Signals of a continuous variable $x \in \mathbb{R}^m$ are usually denoted by $f(x)$, whereas for signals with discrete variable $x \in \mathbb{Z}^m$ we write $f[x]$. $\mathbb{R}$ and $\mathbb{Z}$ denote, respectively, the set of reals and integers.

transformation of X by B^s is equivalent to the Minkowski set subtraction $\ominus$, also called **erosion**, of X by B:

$$X \ominus B \triangleq \{z : B_{+z} \subseteq X\} = \bigcap_{y \in B} X_{-y}. \qquad (13.3)$$

Cascading erosion and dilation creates two other operations, the Minkowski **opening** $X \circ B \triangleq (X \ominus B) \oplus B$ and the **closing** $X \bullet B \triangleq (X \oplus B) \ominus B$ of X by B. In applications, B is usually called a *structuring element* and has a simple geometrical shape and a size smaller than the image X. If B has a regular shape, e.g., a small disk, then both opening and closing act as nonlinear filters that smooth the contours of the input image. Namely, if X is viewed as a flat island, the opening suppresses the sharp capes and cuts the narrow isthmuses of X, whereas the closing fills in the thin gulfs and small holes.

There is a *duality* between dilation and erosion since $X \oplus B = (X^c \ominus B^s)^c$; i.e., dilation of an image object by B is equivalent to eroding its background by B^s and complementing the result. A similar duality exists between closing and opening.

13.2.2 Morphological Filters for Gray-level Images

Extending morphological operators from binary to gray-level images can be done by using set representations of signals and transforming these input sets via morphological set operations. Thus, consider an image signal $f(x)$ defined on the continuous or discrete plane $\mathbb{E} = \mathbb{R}^2$ or $\mathbb{Z}^2$ and assuming values in $\overline{\mathbb{R}} = \mathbb{R} \cup \{-\infty, \infty\}$. Thresholding f at all amplitude levels v produces an ensemble of binary images represented by the upper **level sets** (also called *threshold sets*):

$$X_v(f) \triangleq \{x \in \mathbb{E} : f(x) \geq v\}, \quad -\infty < v < +\infty. \qquad (13.4)$$

The image can be exactly reconstructed from all its level sets since

$$f(x) = \sup\{v \in \mathbb{R} : x \in X_v(f)\}, \qquad (13.5)$$

where "sup" denotes supremum.[2] Transforming each level set of the input signal f by a set operator Ψ and viewing the transformed sets as level sets of a new image creates [2, 5] a **flat** image operator ψ, whose output signal is

$$\psi(f)(x) = \sup\{v \in \mathbb{R} : x \in \Psi[X_v(f)]\}. \qquad (13.6)$$

---

[2]Given a set X of real numbers, the *supremum* of X is its lowest upper bound. If X is finite (or infinite but closed from above), its supremum coincides with its maximum.

For example, if Ψ is the set dilation and erosion by B, the above procedure creates the two most elementary morphological image operators: the dilation and erosion of $f(x)$ by a set B:

$$(f \oplus B)(x) \triangleq \bigvee_{y \in B} f(x - y), \tag{13.7}$$

$$(f \ominus B)(x) \triangleq \bigwedge_{y \in B} f(x + y), \tag{13.8}$$

where $\bigvee$ denotes supremum (or maximum for finite B) and $\bigwedge$ denotes infimum (or minimum for finite B). Flat erosion (dilation) of a function f by a small convex set B reduces (increases) the peaks (valleys) and enlarges the minima (maxima) of the function. The flat opening $f \circ B = (f \ominus B) \oplus B$ of f by B smooths the graph of f from below by cutting down its peaks, whereas the closing $f \bullet B = (f \oplus B) \ominus B$ smooths it from above by filling up its valleys.

The most general translation-invariant morphological dilation and erosion of a gray-level image signal $f(x)$ by another signal g are:

$$(f \oplus g)(x) \triangleq \bigvee_{y \in \mathbb{E}} f(x - y) + g(y), \tag{13.9}$$

$$(f \ominus g)(x) \triangleq \bigwedge_{y \in \mathbb{E}} f(x + y) - g(y). \tag{13.10}$$

Note that signal dilation is a nonlinear convolution where the sum-of-products in the standard linear convolution is replaced by a max-of-sums.

13.2.3　Universality of Morphological Operators[3]

Dilations or erosions can be combined in many ways to create more complex morphological operators that can solve a broad variety of problems in image analysis and nonlinear filtering. Their versatility is further strengthened by a theory outlined in [5, 6] that represents a broad class of nonlinear and linear operators as a minimal combination of erosions or dilations. Here we summarize the main results of this theory restricting our discussion only to discrete 2D image signals.

Any *translation-invariant* set operator Ψ is uniquely characterized by its **kernel** $\mathrm{Ker}(\Psi) \triangleq \{X \in \mathbb{Z}^2 : 0 \in \Psi(X)\}$. If Ψ is also *increasing* (i.e., $X \subseteq Y \Longrightarrow \Psi(X) \subseteq \Psi(Y)$), then it can be represented as a union of erosions by all its kernel sets [1]. However, this kernel representation requires an infinite number of erosions. A more efficient (requiring less erosions) representation uses only a substructure of the kernel, its **basis** $\mathrm{Bas}(\Psi)$, defined as the collection of kernel elements that are *minimal* with respect to the partial ordering $\subseteq$. If Ψ is also *upper semicontinuous* (i.e., $\Psi(\bigcap_n X_n) = \bigcap_n \Psi(X_n)$ for any

---

[3]This is a section for mathematically-inclined readers and can be skipped without significant loss of continuity.

decreasing set sequence (X_n)), then Ψ has a nonempty basis and can be represented exactly as a union of erosions by its basis sets:

$$\Psi(X) = \bigcup_{A \in \mathrm{Bas}(\Psi)} X \ominus A. \tag{13.11}$$

The morphological basis representation has also been extended to gray-level signal operators. As a special case, if ϕ is a flat signal operator as in (13.6) that is translation-invariant and commutes with thresholding, then ϕ can be represented as a supremum of erosions by the basis sets of its corresponding set operator Φ:

$$\phi(f) = \bigvee_{A \in \mathrm{Bas}(\Phi)} f \ominus A. \tag{13.12}$$

By duality, there is also an alternative representation where a set operator Ψ satisfying the above three assumptions can be realized exactly as the intersection of dilations by the reflected basis sets of its *dual operator* $\Psi^d(X) \triangleq [\Psi(X^c)]^c$. There is also a similar dual representation of signal operators as an infimum of dilations.

Given the wide applicability of erosions/dilations, their parallellism, and their simple implementations, the morphological representation theory supports a general purpose image processing (software or hardware) module that can perform erosions/dilations, based on which numerous other complex image operations can be built.

13.2.4 Median, Rank, and Stack Filters

Flat erosion and dilation of a discrete image signal $f[x]$ by a finite window $W = \{y_1,\dots,y_n\} \subseteq \mathbb{Z}^2$ is a moving local minimum or maximum. Replacing min/max with a more general rank leads to rank filters. At each location $x \in \mathbb{Z}^2$, sorting the signal values within the reflected and shifted n-point window $(W^s)_{+x}$ in decreasing order and picking the p-th largest value, $p = 1, 2, \dots, n$, yields the output signal from the pth **rank filter**:

$$(f \square_p W)[x] \triangleq p\text{th rank of } (f[x - y_1],\dots,f[x - y_n]). \tag{13.13}$$

For odd n and $p = (n + 1)/2$ we obtain the **median** filter. Rank filters and especially medians have been applied mainly to suppress impulse noise or noise whose probability density has heavier tails than the Gaussian for enhancement of image and other signals, since they can remove this type of noise without blurring edges, as would be the case for linear filtering.

If the input image is binary, the rank filter output is also binary since sorting preserves a signal's range. Rank filtering of binary images involves only counting of points and no sorting. Namely, if the set $S \subseteq \mathbb{Z}^2$ represents an input binary image, the output set produced by the pth **rank set filter** is

$$S \square_p W \triangleq \{x : \mathrm{card}[(W^s)_{+x} \cap S] \geq p\}, \tag{13.14}$$

where $\mathrm{card}(X)$ denotes the cardinality (i.e., number of points) of a set X.

All rank operators *commute with thresholding*; i.e.,

$$X_v[f \square_p W] = [X_v(f)] \square_p W, \quad \forall v, \forall p \tag{13.15}$$

298 CHAPTER 13 Morphological Filtering

where $X_v(f)$ is the level set (binary image) resulting from thresholding f at level v. This property is also shared by all morphological operators that are finite compositions or maxima/minima of flat dilations and erosions by finite structuring elements. All such signal operators ψ that have a corresponding set operator Ψ and commute with thresholding can be alternatively implemented via *threshold superposition* as in (13.6).

Further, since the binary version of all the above discrete translation-invariant finite-window operators can be described by their generating Boolean function as in (13.1), all that is needed in synthesizing their corresponding gray-level image filters is knowledge of this Boolean function. Specifically, let $f_v[x]$ be the binary images represented by the threshold sets $X_v(f)$ of an input gray-level image $f[x]$. Transforming all f_v with an increasing (i.e., containing no complemented variables) Boolean function $b(u_1,\ldots,u_n)$ in place of the set operator Ψ in (13.6) and using threshold superposition creates a class of nonlinear digital filters called **stack filters** [5, 9]:

$$\phi_b(f)[x] \triangleq \sup\{v \in \mathbb{R} : b(f_v[x - y_1],\ldots,f_v[x - y_n]) = 1\}. \quad (13.16)$$

The use of Boolean functions facilitates the design of such discrete flat operators with determinable structural properties. Since each increasing Boolean function can be uniquely represented by an irreducible sum (product) of product (sum) terms, and each product (sum) term corresponds to an erosion (dilation), each stack filter can be represented as a finite maximum (minimum) of flat erosions (dilations) [5]. For example, the window $W = \{-1, 0, 1\}$ and the Boolean function $b_1(u_1, u_2, u_3) = u_1 u_2 + u_2 u_3 + u_1 u_3$ create a stack filter that is identical to the 3-point median by W, which can also be represented as a maximum of three 2-point erosions:

$$\phi_b(f)[x] = \mathrm{median}(f[x - 1], f[x], f[x + 1])$$
$$= \max\big[\min(f[x - 1], f[x]), \min(f[x - 1], f[x + 1]), \min(f[x], f[x + 1])\big]. \quad (13.17)$$

In general, because of their representation via erosions/dilations (which have a geometric interpretation) and Boolean functions (which are related to mathematical logic), stack filters can be analyzed or designed not only in terms of their statistical properties for image denoising but also in terms of their geometric and logic properties for preserving selected image structures.

13.2.5 Algebraic Generalizations of Morphological Operators

A more general formalization [7, 8] of morphological operators views them as operators on complete lattices. A *complete lattice* is a set $\mathcal{L}$ equipped with a partial ordering $\leq$ such that $(\mathcal{L}, \leq)$ has the algebraic structure of a *partially ordered set* where the supremum and infimum of any of its subsets exist in $\mathcal{L}$. For any subset $\mathcal{K} \subseteq \mathcal{L}$, its *supremum* $\bigvee \mathcal{K}$ and *infimum* $\bigwedge \mathcal{K}$ are defined as the lowest (with respect to $\leq$) upper bound and greatest lower bound of $\mathcal{K}$, respectively. The two main examples of complete lattices used in morphological image processing are (i) the space of all binary images represented by subsets of the plane $\mathbb{E}$ where the $\bigvee / \bigwedge$ lattice operations are the set union/intersection, and (ii) the space of all gray-level image signals $f : \mathbb{E} \to \overline{\mathbb{R}}$ where the $\bigvee / \bigwedge$ lattice operations are the supremum/infimum of sets of real numbers. An operator ψ on $\mathcal{L}$ is called **increasing** if it preserves the partial ordering, i.e., $f \le g$ implies $\psi(f) \le \psi(g)$. Increasing operators are of great importance, and among them four fundamental examples are:

$$\delta \text{ is \textbf{dilation}} \iff \delta\left(\bigvee_{i \in I} f_i\right) = \bigvee_{i \in I} \delta(f_i) \tag{13.18}$$

$$\varepsilon \text{ is \textbf{erosion}} \iff \varepsilon\left(\bigwedge_{i \in I} f_i\right) = \bigwedge_{i \in I} \varepsilon(f_i) \tag{13.19}$$

$$\alpha \text{ is \textbf{opening}} \iff \alpha \text{ is increasing, idempotent, and anti-extensive} \tag{13.20}$$

$$\beta \text{ is \textbf{closing}} \iff \beta \text{ is increasing, idempotent, and extensive,} \tag{13.21}$$

where I is an arbitrary index set, *idempotence* of an operator ψ means that $\psi(\psi(f)) = \psi(f)$, and *antiextensivity* and *extensivity* of operators α and β means that $\alpha(f) \le f \le \beta(f)$ for all f.

The above definitions allow broad classes of signal operators to be grouped as lattice dilations, erosions, openings, or closings and their common properties to be studied under the unifying lattice framework. Thus, the translation-invariant Minkowski dilations $\oplus$, erosions $\ominus$, openings $\bigcirc$, and closings $\bullet$ are simple special cases of their lattice counterparts.

In lattice-theoretic morphology, the term *morphological filter* means any increasing and idempotent operator on a lattice of images. However, in this chapter, we shall use the term "morphological operator," which broadly means a morphological signal transformation, interchangeably with the term "morphological filter," in analogy to the terminology "rank or linear filter."

13.3 MORPHOLOGICAL FILTERS FOR IMAGE ENHANCEMENT

Enhancement may be accomplished in various ways including (i) noise suppression, (ii) simplification by retaining only those image components that satisfy certain size or contrast criteria, and (iii) contrast sharpening. The first two cases may also be viewed as examples of "image smoothing."

The simplest morphological image smoother is a Minkowski opening by a disk B. This smooths and simplifies a (binary image) set X by retaining only those parts inside which a translate of B can fit. Namely,

$$X \bigcirc B = \bigcup_{B+z \subseteq X} B_{+z}. \tag{13.22}$$

In the case of gray-level image f, its opening by B performs the above smoothing at all level sets simultaneously. However, this horizontal geometric local and isotropic smoothing performed by the Minkowski disk opening may not be sufficient for several other smoothing tasks that may need directional smoothing, or may need contour preservation based on size or contrast criteria. To deal with these issues, we discuss below several types of morphological filters that are generalized operators in the lattice-theoretic sense and have proven to be very useful for image enhancement.

13.3.1 Noise Suppresion and Image Smoothing

13.3.1.1 Median versus Open-Closing

In their behavior as nonlinear smoothers, as shown in Fig. 13.1, the medians act similarly to an *open-closing* $(f \circ B) \bullet B$ by a convex set B of diameter about half the diameter of the median window [5]. The open-closing has the advantages over the median that it requires less computation and decomposes the noise suppression task into two independent steps, i.e., suppressing positive spikes via the opening and negative spikes via the closing.

The popularity and efficiency of the simple morphological openings and closings to suppress impulse noise is supported by the following theoretical development [10]. Assume a class of sufficiently smooth random input images which is the collection of all subsets of a finite mask W that are open (or closed) with respect to a set B and assign a uniform probability distribution on this collection. Then, a discrete binary input image X is a random realization from this collection; i.e., use ideas from random sets [1, 2] to model X. Further, X is corrupted by a union (or intersection) noise N which is a 2D sequence of i.i.d. binary Bernoulli random variables with probability $p \in (0, 1)$ of occurrence at each pixel. The observed image is the noisy version $Y = X \cup N$ (or $Y = X \cap N$). Then, the maximum-a-posteriori estimate [10] of the original X given the noisy image Y is the opening (or closing) of the observed Y by B.

(a)   (b)   (c)

FIGURE 13.1

(a) Noisy image obtained by corrupting an original with two-level salt-and-pepper noise occuring with probability 0.1 (PSNR = 18.9dB); (b) Open-closing of noisy image by a 2×2-pel square (PSNR = 25.4dB); (c) Median of noisy image by a 3×3-pel square (PSNR = 25.4dB).

13.3.1.2 *Alternating Sequential Filters*

Another useful generalization of openings and closings involves cascading open-closings $\beta_t\alpha_t$ at multiple scales $t = 1,\ldots,r$, where $\alpha_t(f) = f \circ tB$ and $\beta_t(f) = f \bullet tB$. This generates a class of efficient nonlinear smoothing filters,

$$\psi_{\text{asf}}(f) = \beta_r\alpha_r\ldots\beta_2\alpha_2\beta_1\alpha_1(f), \tag{13.23}$$

called **alternating sequential filters (ASF)**, which smooth progressively from the smallest scale possible up to a maximum scale r and have a broad range of applications [7]. Their optimal design is addressed in [11]. Further, the Minkowski open-closings in an ASF can be replaced by other types of lattice open-closings. A simple such generalization is the radial open-closing, discussed next.

13.3.1.3 *Radial Openings*

Consider a 2D image f that contains 1D objects, e.g., lines; then the simple Minkowski opening or closing of f by a disk B will eliminate these 1D objects. Another problem arises when f contains large-scale objects with sharp corners that need to be preserved; in such cases, opening or closing f by a disk B will round these corners. These two problems could be avoided in some cases if we replace the conventional opening with a **radial opening**,

$$\alpha(f) = \bigvee_\theta f \circ L_\theta, \tag{13.24}$$

where the sets L_θ are rotated versions of a line segment L at various angles $\theta \in (0, 2\pi)$. This has the effect of preserving an object in f if this object is left unchanged after the opening by L_θ in at least one of the possible orientations θ (see Fig. 13.2). Dually, in case of dark 1D objects, we can use a radial closing $\beta(f) = \bigwedge_\theta f \bullet L_\theta$.

13.3.2 Connected Filters for Smoothing and Simplification

The **flat zones** of an image signal $f : \mathbb{E} \to \overline{\mathbb{R}}$ are defined as the connected components of the image domain on which f assumes a constant value. A useful class of morphological filters was introduced in [12, 13], which operate by merging flat zones and hence exactly preserving the contours of the image parts remaining in the filter's output. These are called **connected operators**. They cannot create new image structures or new boundaries if they did not exist in the input. Specifically, if $\mathcal{D}$ is a partition of the image domain, let $\mathcal{D}(x)$ denote the (partition member) region that contains the pixel x. Now, given two partitions $\mathcal{D}_1, \mathcal{D}_2$, we say that $\mathcal{D}_1$ is "finer" than $\mathcal{D}_2$ if $\mathcal{D}_1(x) \subseteq \mathcal{D}_2(x)$ for all x. An operator ψ is called *connected* if the flat zone partition of its input f is finer than the flat zone partition of its output $\psi(f)$. Next we discuss two types of connected operators, the area filters and the reconstruction filters.

13.3.2.1 *Area Openings*

There are numerous image enhancement problems where what is needed is suppression of arbitrarily-shaped connected components in the input image whose areas (number

Original image = F Radial opening (F)   Reconstr. opening (rad.open|F)

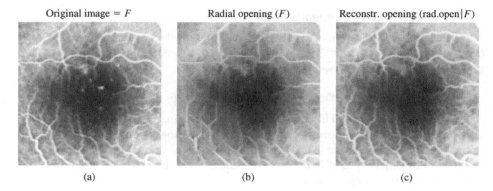

(a)   (b)   (c)

FIGURE 13.2

(a) Original image f of an eye angiogram with microaneurisms; (b) Radial opening $\alpha(f)$ of f as max of four openings by lines oriented at $0°, 45°, 90°, 135°$ of size 20 pixels each; (c) Reconstruction opening $\rho^-(\alpha(f)|f)$ of f using the radial opening as marker.

of pixels) are smaller than a certain threshold n. This can be accomplished by the **area opening** α_n of size n which, for binary images, keeps only the connected components whose area is $\geq n$ and eliminates the rest. Consider an input set $X = \bigsqcup_i X_i$ as a union of disjoint connected components X_i. Then the output from the area opening is

$$\alpha_n(X) = \bigsqcup_{\text{Area}(X_j)\geq n} X_j, \quad X = \bigsqcup_i X_i, \tag{13.25}$$

where $\bigsqcup$ denotes disjoint union. The area opening can be extended to gray-level images f by applying the same binary area opening to all level sets of f and constructing the filtered gray-level image via threshold superposition:

$$\alpha_n(f)(x) = \sup\{v : x \in \alpha_n[X_v(f)]\}. \tag{13.26}$$

Figure 13.3 shows examples of binary and gray area openings. If we apply the above operations to the complements of the level sets of an image, we obtain an *area closing*.

13.3.2.2 *Reconstruction Filters and Levelings*

Consider a *reference* (image) set $X = \bigsqcup_i X_i$ as a union of I disjoint connected components X_i, $i \in I$, and let $M \subseteq X_j$ be a *marker* in some component(s) X_j, indexed by $j \in J \subseteq I$; i.e., M could consist of a single point or some feature sets in X that lie only in the component(s) X_j. Let us define the **reconstruction opening** as the operator:

$$\rho^-(M|X) \triangleq \text{connected components of } X \text{ intersecting } M. \tag{13.27}$$

Its output contains exactly the input component(s) X_j that intersect the marker. It can extract large-scale components of the image from knowledge only of a smaller marker inside them. Note that the reconstruction opening has two inputs. If the marker M is fixed, then the mapping $X \mapsto \rho^-(M|X)$ is a lattice opening since it is increasing,

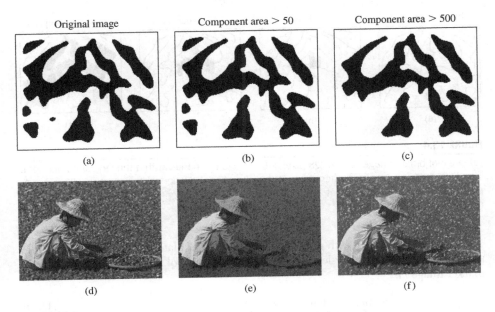

FIGURE 13.3

Top row: (a) Original binary image (192 × 228 pixels); (b) Area opening by keeping connected components with area ≥50; (c) Area opening by keeping components with area ≥500. Bottom row: (d) Gray original image (420 × 300 pixels); (e) Gray area opening by keeping bright components with area ≥500; (f) Gray area closing by keeping dark components with area ≥500.

antiextensive, and idempotent. Its output is called the *morphological reconstruction* of (the component(s) of) X from the marker M. However, if the reference X is fixed, then the mapping $M \mapsto \rho^-(M|X)$ is an idempotent lattice dilation; in this case, the output is called the reconstruction of M under X.

An algorithm to implement the discrete reconstruction opening is based on the *conditional dilation* of M by B within X:

$$\delta_B(M|X) \triangleq (M \oplus B) \cap X, \tag{13.28}$$

where B is the unit-radius discrete disk associated with the selected connectivity of the rectangular grid; i.e., a 5-pixel rhombus or a 9-pixel square depending on whether we have 4- or 8-neighbor connectivity, respectively. By iterating this conditional dilation, we can obtain in the limit the whole marked component(s) X_j, i.e., the *conditional* reconstruction opening,

$$\rho_B^-(M|X) = \lim_{k \to \infty} Y_k, \quad Y_k = \delta_B(Y_{k-1}|X), \ Y_0 = M. \tag{13.29}$$

An example is shown in Fig. 13.4.

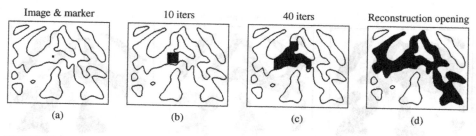

Image & marker	10 iters	40 iters	Reconstruction opening
(a)	(b)	(c)	(d)

FIGURE 13.4

(a) Original binary image (192 × 228 pixels) and a square marker within the largest component. The next three images show iterations of the conditional dilation of the marker with a 3 × 3-pixel square structuring element; (b) 10 iterations; (c) 40 iterations; (d) reconstruction opening, reached after 128 iterations.

Replacing the binary with gray-level images, the set dilation with function dilation, and ∩ with ∧ yields the gray-level reconstruction opening of a gray-level image f from a marker image m:

$$\rho_B^-(m|f) = \lim_{k\to\infty} g_k, \quad g_k = \delta_B(g_{k-1}) \wedge f, \quad g_0 = m \le f. \tag{13.30}$$

This reconstructs the bright components of the reference image f that contains the marker m. For example, as shown in Fig. 13.2, the results of any prior image smoothing, like the radial opening of Fig. 13.2(b), can be treated as a marker which is subsequently reconstructed under the original image as reference to recover exactly those bright image components whose parts have remained after the first operation.

There is a large variety of reconstruction openings depending on the choice of the marker. Two useful cases are (i) *size*-based markers chosen as the Minkowski erosion $m = f \ominus rB$ of the reference image f by a disk of radius r and (ii) *contrast*-based markers chosen as the difference $m(x) = f(x) - h$ of a constant $h > 0$ from the image. In the first case, the reconstruction opening retains only objects whose horizontal size (i.e., diameter of inscribable disk) is not smaller than r. In the second case, only objects whose contrast (i.e., height difference from neighbors) exceeds h will leave a remnant after the reconstruction. In both cases, the marker is a function of the reference signal.

Reconstruction of the dark image components hit by some marker is accomplished by the dual filter, the *reconstruction closing*,

$$\rho_B^+(m|f) = \lim_{k\to\infty} g_k, \quad g_k = \varepsilon_B(g_{k-1}) \vee f, \quad g_0 = m \ge f. \tag{13.31}$$

Examples of gray-level reconstruction filters are shown in Fig. 13.5.

Despite their many applications, reconstruction openings and closings ψ have as a disadvantage the property that they are *not* self-dual operators; hence, they treat the image and its background asymmetrically. A newer operator type that unifies both of them and possesses self-duality is the leveling [14]. Levelings are nonlinear object-oriented filters that simplify a reference image f through a simultaneous use of locally

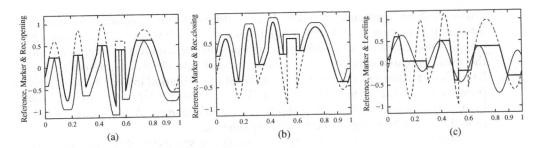

FIGURE 13.5

Reconstruction filters for 1D images. Each figure shows reference signals f (dash), markers (thin solid), and reconstructions (thick solid). (a) Reconstruction opening from marker $= (f \ominus B) - \text{const}$; (b) Reconstruction closing from marker $= (f \oplus B) + \text{const}$; (c) Leveling (self-dual reconstruction) from an arbitrary marker.

expanding and shrinking an initial seed image, called the marker m, and global constraining of the marker evolution by the reference image. Specifically, iterations of the image operator $\lambda(m|f) = (\delta_B(m) \wedge f) \vee \varepsilon_B(m)$, where $\delta_B(\cdot)$ (respectively $\varepsilon_B(\cdot)$) is a dilation (respectively erosion) by the unit-radius discrete disk B of the grid, yield in the limit the **leveling** of f w.r.t. m:

$$\Lambda_B(m|f) = \lim_{k\to\infty} g_k, \quad g_k = (\delta_B(g_{k-1}) \wedge f) \vee \varepsilon_B(g_{k-1}), \quad g_0 = m. \tag{13.32}$$

In contrast to the reconstruction opening (closing) where the marker m is smaller (greater) than f, the marker for a general leveling may have an arbitrary ordering w.r.t. the reference signal (see Fig. 13.5(c)). The leveling reduces to being a reconstruction opening (closing) over regions where the marker is smaller (greater) than the reference image.

If the marker is self-dual, then the leveling is a self-dual filter and hence treats symmetrically the bright and dark objects in the image. Thus, the leveling may be called a *self-dual reconstruction* filter. It simplifies both the original image and its background by completely eliminating smaller objects inside which the marker cannot fit. The reference image plays the role of a *global constraint*.

In general, levelings have many interesting multiscale properties [14]. For example, they preserve the coupling and sense of variation in neighbor image values and do not create any new regional maxima or minima. Also, they are increasing and idempotent filters. They have proven to be very useful for image simplification toward segmentation because they can suppress small-scale noise or small features and keep only large-scale objects with exact preservation of their boundaries.

13.3.3 Contrast Enhancement

Imagine a gray-level image f that has resulted from blurring an original image g by linearly convolving it with a Gaussian function of variance $2t$. This Gaussian blurring can be modeled by running the classic heat diffusion differential equation for the time interval $[0, t]$ starting from the initial condition g at $t = 0$. If we can reverse in time this diffusion process, then we can deblur and sharpen the blurred image. By approximating the spatio-temporal derivatives of the heat equation with differences, we can derive a linear discrete filter that can enhance the contrast of the blurred image f by subtracting from f a discretized version of its Laplacian $\nabla^2 f = \partial^2 f / \partial x^2 + \partial^2 f / \partial y^2$. This is a simple linear deblurring scheme, called *unsharp contrast enhancement*. A conceptually similar procedure is the following nonlinear filtering scheme.

Consider a gray-level image $f[x]$ and a small-size symmetric disk-like structuring element B containing the origin. The following discrete nonlinear filter [15] can enhance the local contrast of f by sharpening its edges:

$$\psi(f)[x] = \begin{cases} (f \oplus B)[x] & \text{if} \quad f[x] \geq ((f \oplus B)[x] + (f \ominus B)[x])/2 \\ (f \ominus B)[x] & \text{if} \quad f[x] < ((f \oplus B)[x] + (f \ominus B)[x])/2. \end{cases} \tag{13.33}$$

At each pixel x, the output value of this filter *toggles* between the value of the dilation of f by B (i.e., the maximum of f inside the moving window B centered) at x and the value of its erosion by B (i.e., the minimum of f within the same window) according to which is closer to the input value $f[x]$. The toggle filter is usually applied not only once but is *iterated*. The more iterations, the more contrast enhancement. Further, the iterations converge to a *limit (fixed point)* [15] reached after a finite number of iterations. Examples are shown in Figs. 13.6 and 13.7.

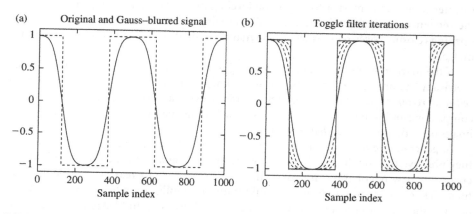

FIGURE 13.6

(a) Original signal (dashed line) $f[x] = \text{sign}(\cos(4\pi x))$, $x \in [0, 1]$, and its blurring (solid line) via convolution with a truncated sampled Gaussian function of $\sigma = 40$; (b) Filtered versions (dashed lines) of the blurred signal in (a) produced by iterating the 1D toggle filter (with $B = \{-1, 0, 1\}$) until convergence to the limit signal (thick solid line) reached at 66 iterations; the displayed filtered signals correspond to iteration indexes that are multiples of 20.

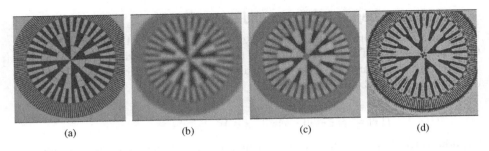

(a)   (b)   (c)   (d)

FIGURE 13.7

(a) Original image f; (b) Blurred image g obtained by an out-of-focus camera digitizing f; (c) Output of the 2D toggle filter acting on g (B was a small symmetric disk-like set); (d) Limit of iterations of the toggle filter on g (reached at 150 iterations).

13.4 MORPHOLOGICAL OPERATORS FOR TEMPLATE MATCHING

13.4.1 Morphological Correlation

Consider two real-valued discrete image signals $f[x]$ and $g[x]$. Assume that g is a signal pattern to be found in f. To find which shifted version of g "best" matches f, a standard approach has been to search for the shift lag y that minimizes the *mean-squared error*, $E_2[y] = \sum_{x \in W}(f[x + y] - g[x])^2$, over some subset W of $\mathbb{Z}^2$. Under certain assumptions, this matching criterion is equivalent to maximizing the *linear cross-correlation* $L_{fg}[y] \triangleq \sum_{x \in W} f[x + y]g[x]$ between f and g.

Although less mathematically tractable than the mean squared error criterion, a statistically more robust criterion is to minimize the *mean absolute error*,

$$E_1[y] = \sum_{x \in W} |f[x + y] - g[x]|.$$

This mean absolute error criterion corresponds to a nonlinear signal correlation used for signal matching; see [6] for a review. Specifically, since $|a - b| = a + b - 2\min(a, b)$, under certain assumptions (e.g., if the error norm and the correlation is normalized by dividing it with the average area under the signals f and g), minimizing $E_1[y]$ is equivalent to maximizing the *morphological* cross-correlation:

$$M_{fg}[y] \triangleq \sum_{x \in W} \min(f[x + y], g[x]). \tag{13.34}$$

It can be shown experimentally and theoretically that the detection of g in f is indicated by a sharper matching peak in $M_{fg}[y]$ than in $L_{fg}[y]$. In addition, the morphological (sum of minima) correlation is faster than the linear (sum of products) correlation. These two advantages of the morphological correlation coupled with the relative robustness of the mean absolute error criterion make it promising for general signal matching.

13.4.2 Binary Object Detection and Rank Filtering

Let us approach the problem of binary image object detection in the presence of noise from the viewpoint of statistical hypothesis testing and rank filtering. Assume that the observed discrete binary image $f[x]$ within a mask W has been generated under one of the following two probabilistic hypotheses:

$$H_0: \quad f[x] = e[x], \ x \in W,$$
$$H_1: \quad f[x] = |g[x-y] - e[x]|, \ x \in W.$$

Hypothesis H_1 (H_0) stands for "object present" ("object not present") at pixel location y. The *object* $g[x]$ is a deterministic binary template. The *noise* $e[x]$ is a stationary binary random field which is a 2D sequence of i.i.d. random variables taking value 1 with probability p and 0 with probability $1 - p$, where $0 < p < 0.5$. The *mask* $W = G_{+y}$ is a finite set of pixels equal to the region G of support of g shifted to location y at which the decision is taken. (For notational simplicity, G is assumed to be symmetric, i.e., $G = G^s$.) The absolute-difference superposition between g and e under H_1 forces f to always have values 0 or 1. Intuitively, such a signal/noise superposition means that the noise e toggles the value of g from 1 to 0 and from 0 to 1 with probability p at each pixel. This noise model can be viewed either as the common binary symmetric channel noise in signal transmission or as a binary version of the salt-and-pepper noise. To decide whether the object g occurs at y, we use a Bayes decision rule that minimizes the total probability of error and hence leads to the *likelihood ratio test*:

$$\frac{Pr(f/H_1)}{Pr(f/H_0)} \underset{\underset{H_0}{<}}{\overset{\overset{H_1}{>}}{}} \frac{Pr(H_0)}{Pr(H_1)}, \tag{13.35}$$

where $Pr(f/H_i)$ are the likelihoods of H_i with respect to the observed image f, and $Pr(H_i)$ are the *a priori* probabilities. This is equivalent to

$$M_{fg}[y] = \sum_{x \in W} \min(f[x], g[x-y]) \underset{\underset{H_0}{<}}{\overset{\overset{H_1}{>}}{}} \theta = \frac{1}{2}\left(\frac{\log[Pr(H_0)/Pr(H_1)]}{\log[(1-p)/p]} + \text{card}(G)\right). \tag{13.36}$$

Thus, the selected statistical criterion and noise model lead to computing the morphological (or equivalently linear) binary correlation between a noisy image and a known image object and comparing it to a threshold for deciding whether the object is present.

Thus, optimum detection in a binary image f of the presence of a binary object g requires comparing the binary correlation between f and g to a threshold θ. This is equivalent[4] to performing a r-th rank filtering on f by a set G equal to the support of

---

[4]An alternative implementation and view of binary rank filtering is via *thresholded convolutions*, where a binary image is linearly convolved with the indicator function of a set G with $n = \text{card}(G)$ pixels, and then the result is thresholded at an integer level r between 1 and n; this yields the output of the r-th rank filter by G acting on the input image.

g, where $1 \leq r \leq \text{card}(G)$ and r is related to θ. Thus, the rank r reflects the area portion of (or a probabilistic confidence score for) the shifted template existing around pixel y. For example, if $Pr(H_0) = Pr(H_1)$, then $r = \theta = \text{card}(G)/2$, and hence the binary median filter by G becomes the optimum detector.

13.4.3 Hit-Miss Filter

The set erosion (13.3) can also be viewed as Boolean template matching since it gives the center points at which the shifted structuring element fits inside the image object. If we now consider a set A probing the image object X and another set B probing the background X^c, the set of points at which the shifted pair (A, B) fits inside the image X is the **hit-miss transformation** of X by (A, B):

$$X \otimes (A, B) \triangleq \{x : A_{+x} \subseteq X, \ B_{+x} \subseteq X^c\}. \tag{13.37}$$

In the discrete case, this can be represented by a Boolean product function whose uncomplemented (complemented) variables correspond to points of A (B). It has been used extensively for binary feature detection [2]. It can actually model all binary template matching schemes in binary pattern recognition that use a pair of a positive and a negative template [3].

    In the presence of noise, the hit-miss filter can be made more robust by replacing the erosions in its definitions with rank filters that do not require an exact fitting of the whole template pair (A, B) inside the image but only a part of it.

13.5 MORPHOLOGICAL OPERATORS FOR FEATURE DETECTION

13.5.1 Edge Detection

By image *edges* we define *abrupt intensity changes* of an image. Intensity changes usually correspond to physical changes in some property of the imaged 3D objects' surfaces (e.g., changes in reflectance, texture, depth or orientation discontinuities, object boundaries) or changes in their illumination. Thus, edge detection is very important for subsequent higher level vision tasks and can lead to some inference about physical properties of the 3D world. Edge types may be classified into three types by approximating their shape with three idealized patterns: lines, steps, and roofs, which correspond, respectively, to the existence of a Dirac impulse in the derivative of order 0, 1, and 2. Next we focus mainly on *step* edges. The problem of edge detection can be separated into three main subproblems:

1. *Smoothing*: image intensities are smoothed via filtering or approximated by smooth analytic functions. The main motivations are to suppress noise and decompose edges at multiple scales.

2. *Differentiation*: amplifies the edges and creates more easily detectable simple geometric patterns.

3. *Decision*: edges are detected as peaks in the magnitude of the first-order derivatives or zero-crossings in the second-order derivatives, both compared with some threshold.

Smoothing and differentiation can be either linear or nonlinear. Further, the differentiation can be either directional or isotropic. Next, after a brief synopsis of the main linear approaches for edge detection, we describe some fully nonlinear ones using morphological gradient-type residuals.

13.5.1.1 *Linear Edge Operators*

In linear edge detection, both smoothing and differentiation are done via linear convolutions. These two stages of smoothing and differentiation can be done in a single stage of convolution with the derivative of the smoothing kernel. Three well-known approaches for edge detection using linear operators in the main stages are the following:

- **Convolution with edge templates:** Historically, the first approach for edge detection, which lasted for about three decades (1950s–1970s), was to use discrete approximations to the image linear partial derivatives, $f_x = \partial f / \partial x$ and $f_y = \partial f / \partial y$, by convolving the digital image f with very small edge-enhancing kernels. Examples include the Prewitt, Sobel and Kirsch edge convolution masks reviewed in [3, 16]. Then these approximations to f_x, f_y were combined nonlinearly to give a gradient magnitude $\|\nabla f\|$ using the ℓ_1, ℓ_2, or ℓ_∞ norm. Finally, peaks in this edge gradient magnitude were detected, via thresholding, for a binary edge decision. Alternatively, edges were identified as zero-crossings in second-order derivatives which were approximated by small convolution masks acting as digital Laplacians. All these above approaches do not perform well because the resulting convolution masks act as poor digital highpass filters that amplify high-frequency noise and do not provide a scale localization/selection.

- **Zero-crossings of Laplacian-of-Gaussian convolution:** Marr and Hildreth [17] developed a theory of edge detection based on evidence from biological vision systems and ideas from signal theory. For image smoothing, they chose linear convolutions with isotropic Gaussian functions $G_\sigma(x, y) = \exp[-(x^2 + y^2)/2\sigma^2]/(2\pi\sigma^2)$ to optimally localize edges both in the space and frequency domains. For differentiation, they chose the Laplacian operator ∇^2 since it is the only isotropic linear second-order differential operator. The combination of Gaussian smoothing and Laplacian can be done using a single convolution with a Laplacian-of-Gaussian (LoG) kernel, which is an approximate bandpass filter that isolates from the original image a scale band on which edges are detected. The scale is determined by σ. Thus, the image edges are defined as the zero-crossings of the image convolution with a LoG kernel. In practice, one does not accept all zero-crossings in the LoG output as edge points but tests whether the slope of the LoG output exceeds a certain threshold.

- **Zero-crossings of directional derivatives of smoothed image:** For detecting edges in 1D signals corrupted by noise, Canny [18] developed an optimal approach where edges were detected as maxima in the output of a linear convolution of the signal with a finite-extent impulse response h. By maximizing the following figures of merit, (i) good detection in terms of robustness to noise, (ii) good edge localization, and (iii) uniqueness of the result in the vicinity of the edge, he found an optimum filter with an impulse response $h(x)$ which can be closely approximated by the derivative of a Gaussian. For 2D images, the Canny edge detector consists of three steps: (1) smooth the image $f(x, y)$ with an isotropic 2D Gaussian G_σ, (2) find the zero-crossings of the second-order directional derivative $\partial^2 f / \partial \eta^2$ of the image in the direction of the gradient $\vec{\eta} = \nabla f / ||\nabla f||$, (3) keep only those zero-crossings and declare them as edge pixels if they belong to connected arcs whose points possess edge strengths that pass a double-threshold hysteresis criterion. Closely related to Canny's edge detector was Haralick's previous work (reviewed in [16]) to regularize the 2D discrete image function by fitting to it bicubic interpolating polynomials, compute the image derivatives from the interpolating polynomial, and find the edges as the zero-crossings of the second directional derivative in the gradient direction. The Haralick-Canny edge detector yields different and usually better edges than the Marr-Hildreth detector.

13.5.1.2 *Morphological Edge Detection*

The **boundary** of a set $X \subseteq \mathbb{R}^m$, $m = 1, 2, \ldots$, is given by

$$\partial X \triangleq \overline{X} \setminus \overset{\circ}{X} = \overline{X} \cap (\overset{\circ}{X})^c, \tag{13.38}$$

where $\overline{X}$ and $\overset{\circ}{X}$ denote the closure and interior of X. Now, if $||x||$ is the Euclidean norm of $x \in \mathbb{R}^m$, B is the unit ball, and $rB = \{x \in \mathbb{R}^m : ||x|| \leq r\}$ is the ball of radius r, then it can be shown that

$$\partial X = \bigcap_{r > 0} (X \oplus rB) \setminus (X \ominus rB). \tag{13.39}$$

Hence, the set difference between erosion and dilation can provide the "edge," i.e., the boundary of a set X.

These ideas can also be extended to signals. Specifically, let us define morphological **sup-derivative** $\mathcal{M}(f)$ of a function $f : \mathbb{R}^m \to \mathbb{R}$ at a point x as

$$\mathcal{M}(f)(x) \triangleq \lim_{r \downarrow 0} \frac{(f \oplus rB)(x) - f(x)}{r} = \lim_{r \downarrow 0} \frac{\bigvee_{||y|| \leq r} f(x + y) - f(x)}{r}. \tag{13.40}$$

By applying $\mathcal{M}$ to $-f$ and using the duality between dilation and erosion, we obtain the *inf-derivative* of f. Suppose now that f is differentiable at $x = (x_1, \ldots, x_m)$ and let its gradient be $\nabla f = \left(\frac{\partial f}{\partial x_1}, \ldots, \frac{\partial f}{\partial x_m} \right)$. Then it can be shown that

$$\mathcal{M}(f)(x) = ||\nabla f(x)||. \tag{13.41}$$

Next, if we take the difference between sup-derivative and inf-derivative when the scale goes to zero, we arrive at an isotropic *second-order morphological derivative*:

$$\mathcal{M}^2(f)(x) \triangleq \lim_{r \downarrow 0} \frac{[(f \oplus rB)(x) - f(x)] - [f(x) - (f \ominus rB)(x)]}{r^2}. \tag{13.42}$$

The peak in the first-order morphological derivative or the zero-crossing in the second-order morphological derivative can detect the location of an edge, in a similar way as the traditional linear derivatives can detect an edge.

By approximating the morphological derivatives with differences, various simple and effective schemes can be developed for extracting edges in *digital images*. For example, for a *binary* discrete image represented as a set X in $\mathbb{Z}^2$, the set difference $(X \oplus B) \setminus (X \ominus B)$ gives the boundary of X. Here B equals the 5-pixel rhombus or 9-pixel square depending on whether we desire 8- or 4-connected image boundaries. An asymmetric treatment between the image foreground and background results if the dilation difference $(X \oplus B) \setminus X$ or the erosion difference $X \setminus (X \ominus B)$ is applied, because they yield a boundary belonging only to X^c or to X, respectively.

Similar ideas apply to gray-level images. Both the *dilation residual* and the *erosion residual*,

$$\text{edge}^{\oplus}(f) \triangleq (f \oplus B) - f, \quad \text{edge}^{\ominus}(f) \triangleq f - (f \ominus B), \tag{13.43}$$

enhance the edges of a gray-level image f. Adding these two operators yields the **discrete morphological gradient**,

$$\text{edge}(f) \triangleq (f \oplus B) - (f \ominus B) = \text{edge}^{\oplus}(f) + \text{edge}^{\ominus}(f), \tag{13.44}$$

that treats more symmetrically the image and its background (see Fig. 13.8).

Threshold analysis can be used to understand the action of the above edge operators. Let the nonnegative discrete-valued image signal $f(x)$ have $L + 1$ possible integer intensity values: $i = 0, 1, \ldots, L$. By thresholding f at all levels, we obtain the *threshold binary images* f_i from which we can resynthesize f via threshold-sum signal superposition:

$$f(x) = \sum_{i=1}^{L} f_i(x), \quad f_i(x) = \begin{cases} 1, & \text{if } f(x) \geq i \\ 0, & \text{if } f(x) < i. \end{cases} \tag{13.45}$$

Since the flat dilation and erosion by a finite B commute with thresholding and f is nonnegative, they obey threshold-sum superposition. Therefore, the dilation-erosion difference operator also obeys threshold-sum superposition:

$$\text{edge}(f) = \sum_{i=1}^{L} \text{edge}(f_i) = \sum_{i=1}^{m} f_i \oplus B - f_i \ominus B. \tag{13.46}$$

This implies that the output of the edge operator acting on the gray-level image f is equal to the sum of the binary signals that are the boundaries of the binary images f (see Fig. 13.8). At each pixel x, the larger the gradient of f, the larger the number of threshold levels i such that $\text{edge}(f_i)(x) = 1$, and hence the larger the value of the gray-level signal $\text{edge}(f)(x)$. Finally, a binarized edge image can be obtained by thresholding $\text{edge}(f)$ or detecting its peaks.

The morphological digital edge operators have been extensively applied to image processing by many researchers. By combining the erosion and dilation differences, various other effective edge operators have also been developed. Examples include 1) the

FIGURE 13.8

(a) Original image f with range in $[0, 255]$; (b) $f \oplus B - f \ominus B$, where B is a 3×3-pixel square; (c) Level set $X = X_i(f)$ of f at level $i = 100$; (d) $X \oplus B \setminus X \ominus B$; (In (c) and (d), black areas represent the sets, while white areas are the complements.)

asymmetric morphological edge-strength operators by Lee *et al.* [19],

$$\min[\text{edge}^{\ominus}(f), \text{edge}^{\oplus}(f)], \quad \max[\text{edge}^{\ominus}(f), \text{edge}^{\oplus}(f)], \tag{13.47}$$

and 2) the edge operator $\text{edge}^{\oplus}(f) - \text{edge}^{\ominus}(f)$ by Vliet *et al.* [20], which behaves as a discrete "*nonlinear Laplacian*,"

$$NL(f) = (f \oplus B) + (f \ominus B) - 2f, \tag{13.48}$$

and at its zero-crossings can yield edge locations. Actually, for a 1D twice differentiable function $f(x)$, it can be shown that if $df(x)/dx \neq 0$ then $\mathcal{M}^2(f)(x) = d^2 f(x)/dx^2$.

For robustness in the presence of noise, these morphological edge operators should be applied after the input image has been smoothed first via either linear or nonlinear filtering. For example, in [19], a small local averaging is used on f before applying the morphological edge-strength operator, resulting in the so-called *min-blur* edge detection operator,

$$\min[f_{av} - f_{av} \ominus B, f_{av} \oplus B - f_{av}], \tag{13.49}$$

with f_{av} being the local average of f, whereas in [21] an opening and closing is used instead of linear preaveraging:

$$\min[f \circ B - f \ominus B, f \oplus B - f \bullet B]. \tag{13.50}$$

Combinations of such smoothings and morphological first or second derivatives have performed better in detecting edges of noisy images. See Fig. 13.9 for an experimental comparison of the LoG and the morphological second derivative in detecting edges.

13.5.2 Peak/Valley Blob Detection

Residuals between openings or closings and the original image offer an intuitively simple and mathematically formal way for peak or valley detection. The general principle for peak detection is to subtract from a signal an opening of it. If the latter is a standard Minkowski opening by a flat compact convex set B, then this yields the peaks of the signal whose base cannot contain B. The morphological peak/valley detectors are simple, efficient, and have some advantages over curvature-based approaches. Their applicability in situations where the peaks or valleys are not clearly separated from their surroundings is further strengthened by generalizing them in the following way. The conventional Minkowski opening in peak detection is replaced by a general lattice opening, usually of the reconstruction type. This generalization allows a more effective estimation of the image background surroundings around the peak and hence a better detection of the peak. Next we discuss peak detectors based on both the standard Minkowski openings as well as on generalized lattice openings like contrast-based reconstructions which can control the peak height.

13.5.2.1 *Top-Hat Transformation*

Subtracting from a signal f its Minkowski opening by a compact convex set B yields an output consisting of the signal peaks whose supports cannot contain B. This is Meyer's *top-hat transformation* [22], implemented by the opening residual,

$$\text{peak}(f) \triangleq f - (f \circ B), \tag{13.51}$$

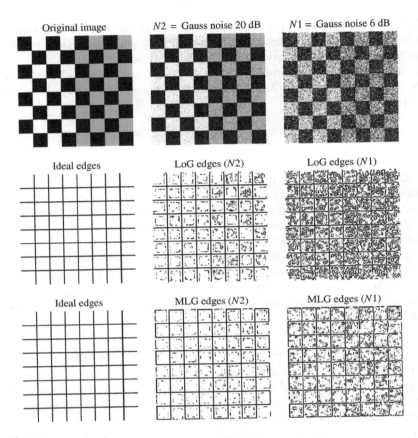

FIGURE 13.9

Top: Test image and two noisy versions with additive Gaussian noise at SNR 20 dB and 6 dB. Middle: Ideal edges and edges from zero-crossings of Laplacian-of-Gaussian of the two noisy images. Bottom: Ideal edges and edges from zero-crossings of 2D morphological second derivative (nonlinear Laplacian) of the two noisy images after some Gaussian presmoothing. In both methods, the edge pixels were the subset of the zero-crossings where the edge strength exceeded some threshold. By using as figure-of-merit the average of the probability of detecting an edge given that it is true and the probability of a true edge given than it is detected, the morphological method scored better by yielding detection probabilities of 0.84 and 0.63 at the noise levels of 20 and 6 dB, respectively, whereas the corresponding probabilities of the LoG method were 0.81 and 0.52.

and henceforth called the **peak** operator. The output peak(f) is always a nonnegative signal, which guarantees that it contains only peaks. Obviously the set B is a very important parameter of the peak operator, because the shape and size of the peak's support obtained by (13.51) are controlled by the shape and size of B. Similarly, to extract the valleys of a signal f, we can apply the closing residual,

$$\text{valley}(f) \triangleq (f \bullet B) - f, \tag{13.52}$$

henceforth called the **valley** operator.

If f is an intensity image, then the opening (or closing) residual is a very useful operator for detecting *blobs*, defined as regions with significantly brighter (or darker) intensities relative to the surroundings. Examples are shown in Fig. 13.10.

If the signal $f(x)$ assumes only the values $0, 1, \ldots, L$ and we consider its threshold binary signals $f_i(x)$ defined in (13.45), then since the opening by $f \bigcirc B$ obeys the threshold-sum superposition,

$$\text{peak}(f) = \sum_{i=1}^{L} \text{peak}(f_i).\qquad(13.53)$$

Thus the peak operator obeys threshold-sum superposition. Hence, its output when operating on a gray-level signal f is the sum of its binary outputs when it operates on all the threshold binary versions of f. Note that, for each binary signal f_i, the binary output peak (f_i) contains only those nonzero parts of f_i inside which no translation of B fits.

The morphological peak and valley operators, in addition to being simple and efficient, avoid several shortcomings of the curvature-based approaches to peak/valley extraction that can be found in earlier computer vision literature. A differential geometry interpretation of the morphological feature detectors was given by Noble [23], who also developed and analyzed simple operators based on residuals from openings and closings to detect corners and junctions.

13.5.2.2 *Dome/Basin Extraction with Reconstruction Opening*

Extracting the peaks of a signal via the simple top-hat operator (13.51) does not constrain the height of the resulting peaks. Specifically, the threshold-sum superposition of the opening difference in (13.53) implies that the peak height at each point is the sum of all binary peak signals at this point. In several applications, however, it is desirable to extract from a signal f peaks that have a maximum height $h > 0$. Such peaks are called *domes* and are defined as follows. Subtracting a contrast height constant h from $f(x)$ yields the smaller signal $g(x) = f(x) - h < f(x)$. Enlarging the maximum peak value of g below

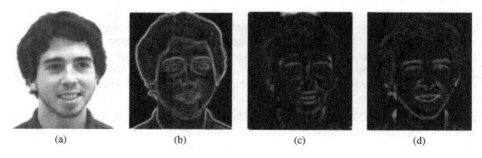

(a)   (b)   (c)   (d)

FIGURE 13.10

Facial image feature extraction. (a) Original image f; (b) Morphological gradient $f \oplus B - f \ominus B$; (c) Peaks: $f - (f \bigcirc 3B)$; (d) Valleys: $(f \bullet 3B) - f$ (B is 21-pixel octagon).

a peak of f by locally dilating g with a symmetric compact and convex set of an ever-increasing diameter and always restricting these dilations to never produce a signal larger than f under this specific peak produces in the limit a signal which consists of valleys interleaved with flat plateaus. This signal is the *reconstruction opening* of g under f, denoted as $\rho^-(g|f)$; namely, f is the reference signal and g is the marker. Subtracting the reconstruction opening from f yields the **domes** of f, defined in [24] as the generalized top-hat:

$$\text{dome}(f) \triangleq f - \rho^-(f - h|f). \tag{13.54}$$

For *discrete-domain* signals f, the above reconstruction opening can be implemented by iterating the conditional dilation as in (13.30). This is a simple but computationally expensive algorithm. More efficient algorithms can be found in [24, 25]. The dome operator extracts peaks whose height cannot exceed h but their supports can be arbitrarily wide. In contrast, the peak operator (using the opening residual) extracts peaks whose supports cannot exceed a set B but their heights are unconstrained.

Similarly, an operator can be defined that extracts signal valleys whose depth cannot exceed a desired maximum h. Such valleys are called **basins** and are defined as the domes of the negated signal. By using the duality between morphological operations, it can be shown that basins of height h can be extracted by subtracting the original image $f(x)$ from its reconstruction closing obtained using as marker the signal $f(x) + h$:

$$\text{basin}(f) \triangleq \text{dome}(-f) = \rho^+(f + h|f) - f. \tag{13.55}$$

Domes and basins have found numerous applications as region-based image features and as markers in image segmentation tasks. Several successful paradigms are discussed in [24–26].

The following example, adapted from [24], illustrates that domes perform better than the classic top-hat in extracting small isolated peaks that indicate pathology points in biomedical images, e.g., detect microaneurisms in eye angiograms without confusing them with the large vessels in the eye image (see Fig. 13.11).

13.6 DESIGN APPROACHES FOR MORPHOLOGICAL FILTERS

Morphological and rank/stack filters are useful for image enhancement and are closely related since they can all be represented as maxima of morphological erosions [5]. Despite the wide application of these nonlinear filters, very few ideas exist for their optimal design. The current four main approaches are as follows: (a) designing morphological filters as a finite union of erosions [27] based on the morphological basis representation theory (outlined in Section 13.2.3); (b) designing stack filters via threshold decomposition and linear programming [9]; (c) designing morphological networks using either voting logic and rank tracing learning or simulated annealing [28]; (d) designing morphological/rank filters via a gradient-based adaptive optimization [29]. Approach (a) is limited to binary increasing filters. Approach (b) is limited to increasing filters processing nonnegative quantized signals. Approach (c) needs a long time to train and convergence is

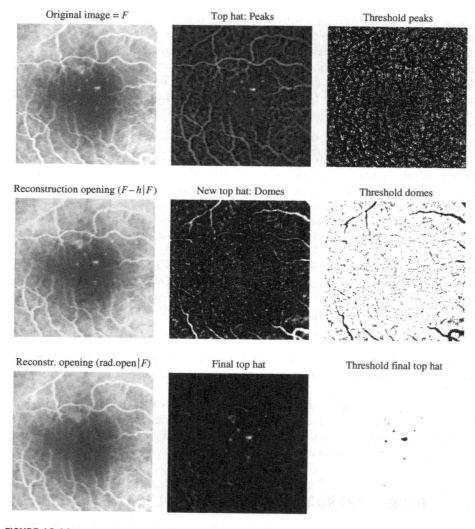

FIGURE 13.11

Top row: Original image F of eye angiogram with microaneurisms, its top hat $F - F \circ B$, where B is a disk of radius 5, and level set of top hat at height $h/2$. Middle row: Reconstruction opening $\rho^-(F - h|F)$, domes $F - \rho^-(F - h|F)$, level set of domes at height $h/2$. Bottom row: New reconstruction opening of F using the radial opening of Fig. 13.2(b) as marker, new domes, and level set detecting microaneurisms.

complex. In contrast, approach (d) is more general since it applies to both increasing and non-increasing filters and to both binary and real-valued signals. The major difficulty involved is that rank functions are *not* differentiable, which imposes a deadlock on how to adapt the coefficients of morphological/rank filters using a gradient-based algorithm.

The methodology described in this section is an extension and improvement to the design methodology (d), leading to a new approach that is simpler, more intuitive, and numerically more robust.

For various signal processing applications, it is sometimes useful to mix in the same system both nonlinear and linear filtering strategies. Thus, hybrid systems, composed of linear and nonlinear (rank-type) sub-systems, have frequently been proposed in the research literature. A typical example is the class of L-filters that are linear combinations of rank filters. Several adaptive algorithms have also been developed for their design, which illustrated the potential of adaptive hybrid filters for image processing applications, especially in the presence of non-Gaussian noise.

Another example of hybrid systems are the **morphological/rank/linear (MRL)** filters [30], which contain as special cases morphological, rank, and linear filters. These MRL filters consist of a linear combination between a morphological/rank filter and a linear finite impulse response filter. Their nonlinear component is based on a rank function, from which the basic morphological operators of erosion and dilation can be obtained as special cases. An efficient method for their adaptive optimal design can be found in [30].

13.7 CONCLUSIONS

In this chapter, we have briefly presented the application of both the standard and some advanced morphological filters to several problems of image enhancement and feature detection. There are several motivations for using morphological filters for such problems. First, it is of paramount importance to preserve, uncover, or detect the geometric structure of image objects. Thus, morphological filters which are more suitable than linear filters for shape analysis, play a major role for geometry-based enhancement and detection. Further, they offer efficient solutions to other nonlinear tasks such as non-Gaussian noise suppression. Although this denoising task can also be accomplished (with similar improvements over linear filters) by the closely related class of median-type and stack filters, the morphological operators provide the additional feature of geometric intuition. Finally, the elementary morphological operators are the building blocks for large classes of nonlinear image processing systems, which include rank and stack filters.

Three important broad research directions in morphological filtering are (1) their optimal design for various advanced image analysis and vision tasks, (2) their scale-space formulation using geometric partial differential equations (PDEs), and (3) their isotropic implementation using numerical algorithms that solve these PDEs. A survey of the last two topics can be found in [31].

REFERENCES

[1] G. Matheron. *Random Sets and Integral Geometry.* John Wiley and Sons, NY, 1975.

[2] J. Serra. *Image Analysis and Mathematical Morphology.* Academic Press, Burlington, MA, 1982.

[3] A. Rosenfeld and A. C. Kak. *Digital Picture Processing*, Vols. 1 & 2. Academic Press, Boston, MA, 1982.

[4] K. Preston, Jr. and M. J. B. Duff. *Modern Cellular Automata*. Plenum Press, NY, 1984.

[5] P. Maragos and R. W. Schafer. Morphological filters. Part I: their set-theoretic analysis and relations to linear shift-invariant filters. Part II: their relations to median, order-statistic, and stack filters. *IEEE Trans. Acoust.*, 35:1153–1184, 1987; *ibid*, 37:597, 1989.

[6] P. Maragos and R. W. Schafer. Morphological systems for multidimensional signal processing. *Proc. IEEE*, 78:690–710, 1990.

[7] J. Serra, editor. *Image Analysis and Mathematical Morphology, Vol. 2: Theoretical Advances*. Academic Press, Burlington, MA, 1988.

[8] H. J. A. M. Heijmans. *Morphological Image Operators*. Academic Press, Boston, MA, 1994.

[9] E. J. Coyle and J. H. Lin. Stack filters and the mean absolute error criterion. *IEEE Trans. Acoust.*, 36:1244–1254, 1988.

[10] N. D. Sidiropoulos, J. S. Baras, and C. A. Berenstein. Optimal filtering of digital binary images corrupted by union/intersection noise. *IEEE Trans. Image Process.*, 3:382–403, 1994.

[11] D. Schonfeld and J. Goutsias. Optimal morphological pattern restoration from noisy binary images. *IEEE Trans. Pattern Anal. Mach. Intell.*, 13:14–29, 1991.

[12] J. Serra and P. Salembier. Connected operators and pyramids. In *Proc. SPIE Vol. 2030, Image Algebra and Mathematical Morphology*, 65–76, 1993.

[13] P. Salembier and J. Serra. Flat zones filtering, connected operators, and filters by reconstruction. *IEEE Trans. Image Process.*, 4:1153–1160, 1995.

[14] F. Meyer and P. Maragos. Nonlinear scale-space representation with morphological levelings. *J. Visual Commun. Image Representation*, 11:245–265, 2000.

[15] H. P. Kramer and J. B. Bruckner. Iterations of a nonlinear transformation for enhancement of digital images. *Pattern Recognit.*, 7:53–58, 1975.

[16] R. M. Haralick and L. G. Shapiro. *Computer and Robot Vision*, Vol. I. Addison-Wesley, Boston, MA, 1992.

[17] D. Marr and E. Hildreth. Theory of edge detection. *Proc. R. Soc. Lond., B, Biol. Sci.*, 207:187–217, 1980.

[18] J. Canny. A computational approach to edge detection. *IEEE Trans. Pattern Anal. Mach. Intell.*, PAMI-8:679–698, 1986.

[19] J. S. J. Lee, R. M. Haralick, and L. G. Shapiro. Morphologic edge detection. *IEEE Trans. Rob. Autom.*, RA-3:142–156, 1987.

[20] L. J. van Vliet, I. T. Young, and G. L. Beckers. A nonlinear Laplace operator as edge detector in noisy images. *Comput. Vis., Graphics, and Image Process.*, 45:167–195, 1989.

[21] R. J. Feehs and G. R. Arce. Multidimensional morphological edge detection. In *Proc. SPIE Vol. 845: Visual Communications and Image Processing II*, 285–292, 1987.

[22] F. Meyer. Contrast feature extraction. In *Proc. 1977 European Symp. on Quantitative Analysis of Microstructures in Materials Science, Biology and Medicine*, France. Published in: *Special Issues of Practical Metallography*, J. L. Chermant, editor, Riederer-Verlag, Stuttgart, 374–380, 1978.

[23] J. A. Noble. Morphological feature detection. In *Proc. Int. Conf. Comput. Vis.*, Tarpon-Springs, FL, 1988.

[24] L. Vincent. Morphological grayscale reconstruction in image analysis: applications and efficient algorithms. *IEEE Trans. Image Process.*, 2:176–201, 1993.

[25] P. Salembier. Region-based filtering of images and video sequences: a morphological view-point. In S. K. Mitra and G. L. Sicuranza, editors, *Nonlinear Image Processing*, Academic Press, Burlington, MA, 2001.

[26] A. Banerji and J. Goutsias. A morphological approach to automatic mine detection problems. *IEEE Trans. Aerosp. Electron Syst.*, 34:1085–1096, 1998.

[27] R. P. Loce and E. R. Dougherty. Facilitation of optimal binary morphological filter design via structuring element libraries and design constraints. *Opt. Eng.*, 31:1008–1025, 1992.

[28] S. S. Wilson. Training structuring elements in morphological networks. In E. R. Dougherty, editor, *Mathematical Morphology in Image Processing*, Marcel Dekker, NY, 1993.

[29] P. Salembier. Adaptive rank order based filters. *Signal Processing*, 27:1–25, 1992.

[30] L. F. C. Pessoa and P. Maragos. MRL-filters: a general class of nonlinear systems and their optimal design for image processing. *IEEE Trans. Image Process.*, 7:966–978, 1998.

[31] P. Maragos. Partial differential equations for morphological scale-spaces and Eikonal applications. In A. C. Bovik, editor, *The Image and Video Processing Handbook*, 2nd ed., 587–612. Elsevier Academic Press, Burlington, MA, 2005.

Basic Methods for Image Restoration and Identification

14

Reginald L. Lagendijk and Jan Biemond

Delft University of Technology, The Netherlands

14.1 INTRODUCTION

Images are produced to record or display useful information. Due to imperfections in the imaging and capturing process, however, the recorded image invariably represents a degraded version of the original scene. The undoing of these imperfections is crucial to many of the subsequent image processing tasks. There exists a wide range of different degradations that need to be taken into account, covering for instance noise, geometrical degradations (pin cushion distortion), illumination and color imperfections (under/overexposure, saturation), and blur. This chapter concentrates on basic methods for removing blur from recorded sampled (spatially discrete) images. There are many excellent overview articles, journal papers, and textbooks on the subject of image restoration and identification. Readers interested in more details than given in this chapter are referred to [1–5].

Blurring is a form of bandwidth reduction of an ideal image owing to the imperfect image formation process. It can be caused by relative motion between the camera and the original scene, or by an optical system that is out of focus. When aerial photographs are produced for remote sensing purposes, blurs are introduced by atmospheric turbulence, aberrations in the optical system, and relative motion between the camera and the ground. Such blurring is not confined to optical images; for example, electron micrographs are corrupted by spherical aberrations of the electron lenses, and CT scans suffer from X-ray scatter.

In addition to these blurring effects, noise always corrupts any recorded image. Noise may be introduced by the medium through which the image is created (random absorption or scatter effects), by the recording medium (sensor noise), by measurement errors due to the limited accuracy of the recording system, and by quantization of the data for digital storage.

323

The field of *image restoration* (sometimes referred to as image deblurring or image deconvolution) is concerned with the reconstruction or estimation of the uncorrupted image from a blurred and noisy one. Essentially, it tries to perform an operation on the image that is the inverse of the imperfections in the image formation system. In the use of image restoration methods, the characteristics of the degrading system and the noise are assumed to be known *a priori*. In practical situations, however, one may not be able to obtain this information directly from the image formation process. The goal of *blur identification* is to estimate the attributes of the imperfect imaging system from the observed degraded image itself prior to the restoration process. The combination of image restoration and blur identification is often referred to as *blind image deconvolution* [4].

Image restoration algorithms distinguish themselves from image *enhancement* methods in that they are based on models for the degrading process and for the ideal image. For those cases where a fairly accurate blur model is available, powerful restoration algorithms can be arrived at. Unfortunately, in numerous practical cases of interest, the modeling of the blur is unfeasible, rendering restoration impossible. The limited validity of blur models is often a factor of disappointment, but one should realize that if none of the blur models described in this chapter are applicable, the corrupted image may well be beyond restoration. Therefore, no matter how powerful blur identification and restoration algorithms are, the objective *when capturing* an image undeniably is to avoid the need for restoring the image.

The image restoration methods that are described in this chapter fall under the class of *linear spatially invariant restoration* filters. We assume that the blurring function acts as a convolution kernel or *point-spread function* $d(n_1, n_2)$ that does not vary spatially. It is also assumed that the statistical properties (mean and correlation function) of the image and noise do not change spatially. Under these conditions the restoration process can be carried out by means of a linear filter of which the point-spread function (PSF) is spatially invariant, i.e., is constant throughout the image. These modeling assumptions can be mathematically formulated as follows. If we denote by $f(n_1, n_2)$ the desired ideal spatially discrete image that does not contain any blur or noise, then the recorded image $g(n_1, n_2)$ is modeled as (see also Fig. 14.1(a)) [6]:

$$g(n_1, n_2) = d(n_1, n_2) * f(n_1, n_2) + w(n_1, n_2)$$

$$= \sum_{k_1=0}^{N-1} \sum_{k_2=0}^{M-1} d(k_1, k_2) f(n_1 - k_1, n_2 - k_2) + w(n_1, n_2). \qquad (14.1)$$

Here $w(n_1, n_2)$ is the noise that corrupts the blurred image. Clearly the objective of image restoration is to make an estimate $f(n_1, n_2)$ of the ideal image, given only the degraded image $g(n_1, n_2)$, the blurring function $d(n_1, n_2)$, and some information about the statistical properties of the ideal image and the noise.

An alternative way of describing (14.1) is through its spectral equivalence. By applying discrete Fourier transforms to (14.1), we obtain the following representation (see also Fig. 14.1(b)):

$$G(u, v) = D(u, v) F(u, v) + W(u, v), \qquad (14.2)$$

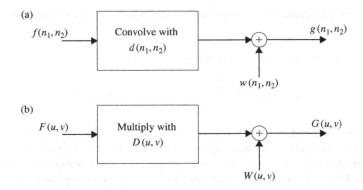

FIGURE 14.1

(a) Image formation model in the spatial domain; (b) Image formation model in the Fourier domain.

where (u, v) are the spatial frequency coordinates and capitals represent Fourier transforms. Either (14.1) or (14.2) can be used for developing restoration algorithms. In practice the spectral representation is more often used since it leads to efficient implementations of restoration filters in the (discrete) Fourier domain.

In (14.1) and (14.2), the noise $w(n_1, n_2)$ is modeled as an additive term. Typically the noise is considered to have a zero-mean and to be white, i.e., spatially uncorrelated. In statistical terms this can be expressed as follows [7]:

$$E[w(n_1, n_2)] \approx \frac{1}{NM} \sum_{k_1=0}^{N-1} \sum_{k_2=0}^{M-1} w(k_1, k_2) = 0 \qquad (14.3a)$$

$$R_w(k_1, k_2) = E[w(n_1, n_2)w(n_1 - k_1, n_2 - k_2)]$$

$$\approx \frac{1}{NM} \sum_{n_1=0}^{N-1} \sum_{n_2=0}^{M-1} w(n_1, n_2)w(n_1 - k_1, n_2 - k_2) = \begin{cases} \sigma_w^2 & \text{if} \quad k_1 = k_2 = 0 \\ 0 & \text{elsewhere} \end{cases}. \qquad (14.3b)$$

Here σ_w^2 is the variance or power of the noise and $E[]$ refers to the expected value operator. The approximate equality indicates that on the average Eq. (14.3) should hold, but that for a given image Eq. (14.3) holds only approximately as a result of replacing the expectation by a pixelwise summation over the image. Sometimes the noise is assumed to have a Gaussian probability density function, but this is not a necessary condition for the restoration algorithms described in this chapter.

In general the noise $w(n_1, n_2)$ may not be independent of the ideal image $f(n_1, n_2)$. This may happen for instance if the image formation process contains nonlinear components, or if the noise is multiplicative instead of additive. Unfortunately, this dependency is often difficult to model or to estimate. Therefore, noise and ideal image are usually assumed to be orthogonal, which is—in this case—equivalent to being uncorrelated because the noise has zero-mean. Expressed in statistical terms, the following condition holds:

$$R_{fw}(k_1, k_2) = E[f(n_1, n_2)w(n_1 - k_1, n_2 - k_2)]$$

$$\approx \frac{1}{NM} \sum_{n_1=0}^{N-1} \sum_{n_2=0}^{M-1} f(n_1, n_2)w(n_1 - k_1, n_2 - k_2) = 0. \qquad (14.4)$$

The above models (14.1)–(14.4) form the foundations for the class of linear spatially invariant image restoration and accompanying blur identification algorithms. In particular these models apply to monochromatic images. For color images, two approaches can be taken. One approach is to extend Eqs. (14.1)–(14.4) to incorporate multiple color components. In many practical cases of interest this is indeed the proper way of modeling the problem of color image restoration since the degradations of the different color components (such as the tri-stimulus signals red-green-blue, luminance-hue-saturation, or luminance-chrominance) are not independent. This leads to a class of algorithms known as "multiframe filters" [3, 8]. A second, more pragmatic, way of dealing with color images is to assume that the noises and blurs in each of the color components are independent. The restoration of the color components can then be carried out independently as well, meaning that each color component is simply regarded as a monochromatic image by itself, forgetting the other color components. Though obviously this model might be in error, acceptable results have been achieved in this way.

The outline of this chapter is as follows. In Section 14.2, we first describe several important models for linear blurs, namely motion blur, out-of-focus blur, and blur due to atmospheric turbulence. In Section 14.3, three classes of restoration algorithms are introduced and described in detail, namely the inverse filter, the Wiener and constrained least-squares filter, and the iterative restoration filters. In Section 14.4, two basic approaches to blur identification will be described briefly.

14.2 BLUR MODELS

The blurring of images is modeled in (14.1) as the convolution of an ideal image with a 2D PSF $d(n_1, n_2)$. The interpretation of (14.1) is that *if* the ideal image $f(n_1, n_2)$ would consist of a single intensity point or point source, this point would be recorded as a spread-out intensity pattern[1] $d(n_1, n_2)$, hence the name *point-spread* function.

It is worth noticing that PSFs in this chapter are not a function of the spatial location under consideration, i.e., they are spatially invariant. Essentially this means that the image is blurred in exactly the *same* way at *every* spatial location. Point-spread functions that do not follow this assumption are, for instance, due to rotational blurs (turning wheels) or local blurs (a person out of focus while the background is in focus). The

---

[1] Ignoring the noise for a moment.

modeling, restoration, and identification of images degraded by spatially varying blurs is outside the scope of this chapter, and is actually still a largely unsolved problem.

In most cases the blurring of images is a spatially continuous process. Since identification and restoration algorithms are always based on spatially discrete images, we present the blur models in their continuous forms, followed by their discrete (sampled) counterparts. We assume that the sampling rate of the images has been chosen high enough to minimize the (aliasing) errors involved in going from the continuous to discrete models.

The spatially continuous PSF $d(x, y)$ of any blur satisfies three constraints, namely:

- $d(x, y)$ takes on nonnegative values only, because of the physics of the underlying image formation process;

- when dealing with real-valued images the PSF $d(x, y)$ is also real-valued;

- the imperfections in the image formation process are modeled as passive operations on the data, i.e., no "energy" is absorbed or generated. Consequently, for spatially continuous blurs the PSF is constrained to satisfy

$$\int_{-\infty}^{\infty} \int_{-\infty}^{\infty} d(x, y) \mathrm{d}x \, \mathrm{d}y = 1, \tag{14.5a}$$

and for spatially discrete blurs:

$$\sum_{n_1=0}^{N-1} \sum_{n_2=0}^{M-1} d(n_1, n_2) = 1. \tag{14.5b}$$

In the following we will present four common PSFs, which are encountered regularly in practical situations of interest.

14.2.1 No Blur

In case the recorded image is imaged perfectly, no blur will be apparent in the discrete image. The spatially continuous PSF can then be modeled as a Dirac delta function:

$$d(x, y) = \delta(x, y) \tag{14.6a}$$

and the spatially discrete PSF as a unit pulse:

$$d(n_1, n_2) = \delta(n_1, n_2) = \begin{cases} 1 & \text{if} \quad n_1 = n_2 = 0 \\ 0 & \text{elsewhere} \end{cases}. \tag{14.6b}$$

Theoretically (14.6a) can never be satisfied. However, as long as the amount of "spreading" in the continuous image is smaller than the sampling grid applied to obtain the discrete image, Eq. (14.6b) will be arrived at.

14.2.2 Linear Motion Blur

Many types of motion blur can be distinguished all of which are due to relative motion between the recording device and the scene. This can be in the form of a translation, a rotation, a sudden change of scale, or some combination of these. Here only the important case of a global translation will be considered.

When the scene to be recorded translates relative to the camera at a constant velocity $v_{relative}$ under an angle of ϕ radians with the horizontal axis during the exposure interval $[0, t_{exposure}]$, the distortion is one-dimensional. Defining the "length of motion" by $L = v_{relative} t_{exposure}$, the PSF is given by

$$d(x,y;L,\phi) = \begin{cases} \dfrac{1}{L} & \text{if } \sqrt{x^2 + y^2} \leq \dfrac{L}{2} \text{ and } \dfrac{x}{y} = -\tan\phi \\ 0 & \text{elsewhere} \end{cases} \tag{14.7a}$$

The discrete version of (14.7a) is not easily captured in a closed form expression in general. For the special case that $\phi = 0$, an appropriate approximation is

$$d(n_1,n_2;L) = \begin{cases} \dfrac{1}{L} & \text{if } n_1 = 0, |n_2| \leq \left\lfloor \dfrac{L-1}{2} \right\rfloor \\ \dfrac{1}{2L}\left\{(L-1) - 2\left\lfloor \dfrac{L-1}{2} \right\rfloor\right\} & \text{if } n_1 = 0, |n_2| = \left\lceil \dfrac{L-1}{2} \right\rceil \\ 0 & \text{elsewhere} \end{cases} \tag{14.7b}$$

Figure 14.2(a) shows the modulus of the Fourier transform of the PSF of motion blur with $L = 7.5$ and $\phi = 0$. This figure illustrates that the blur is effectively a horizontal lowpass filtering operation and that the blur has spectral zeros along characteristic lines. The interline spacing of these characteristic zero-patterns is (for the case that $N = M$) approximately equal to N/L. Figure 14.2(b) shows the modulus of the Fourier transform for the case of $L = 7.5$ and $\phi = \pi/4$.

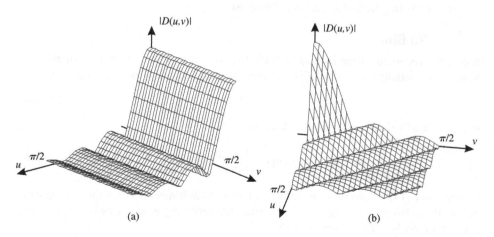

(a)　　　　(b)

FIGURE 14.2

PSF of motion blur in the Fourier domain, showing $|D(u,v)|$, for (a) $L = 7.5$ and $\phi = 0$; (b) $L = 7.5$ and $\phi = \pi/4$.

14.2.3 Uniform Out-of-Focus Blur

When a camera images a 3D scene onto a 2D imaging plane, some parts of the scene are in focus while other parts are not. If the aperture of the camera is circular, the image of any point source is a small disk, known as the circle of confusion (COC). The degree of defocus (diameter of the COC) depends on the focal length and the aperture number of the lens and the distance between camera and object. An accurate model not only describes the diameter of the COC but also the intensity distribution within the COC. However, if the degree of defocusing is large relative to the wavelengths considered, a geometrical approach can be followed resulting in a uniform intensity distribution within the COC. The spatially continuous PSF of this uniform out-of-focus blur with radius R is given by

$$d(x,y;R) = \begin{cases} \dfrac{1}{\pi R^2} & \text{if } \sqrt{x^2 + y^2} \leq R^2 \\ 0 & \text{elsewhere} \end{cases}. \qquad (14.8a)$$

Also for this PSF, the discrete version $d(n_1, n_2)$ is not easily arrived at. A coarse approximation is the following spatially discrete PSF:

$$d(n_1, n_2; R) = \begin{cases} \dfrac{1}{C} & \text{if } \sqrt{n_1^2 + n_2^2} \leq R^2 \\ 0 & \text{elsewhere} \end{cases}, \qquad (14.8b)$$

where C is a constant that must be chosen so that (14.5b) is satisfied. The approximation (14.8b) is incorrect for the fringe elements of the PSF. A more accurate model for the fringe elements would involve the integration of the area covered by the spatially continuous PSF, as illustrated in Fig. 14.3. Figure 14.3(a) shows the fringe elements that need to be

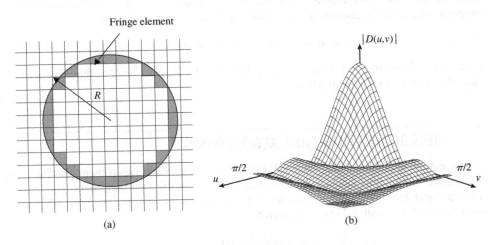

Fringe element

R

$|D(u,v)|$

$\pi/2$

u

$\pi/2$

v

(a)   (b)

FIGURE 14.3

(a) Fringe elements of discrete out-of-focus blur that are calculated by integration; (b) PSF in the Fourier domain, showing $|D(u,v)|$, for $R = 2.5$.

calculated by integration. Figure 14.3(b) shows the modulus of the Fourier transform of the PSF for $R = 2.5$. Again a lowpass behavior can be observed (in this case both horizontally and vertically), as well as a characteristic pattern of spectral zeros.

14.2.4 Atmospheric Turbulence Blur

Atmospheric turbulence is a severe limitation in remote sensing. Although the blur introduced by atmospheric turbulence depends on a variety of factors (such as temperature, wind speed, exposure time), for long-term exposures, the PSF can be described reasonably well by a Gaussian function:

$$d(x, y; \sigma_G) = C \exp\left(-\frac{x^2 + y^2}{2\sigma_G^2} \right). \tag{14.9a}$$

Here σ_G determines the amount of spread of the blur, and the constant C is to be chosen so that (14.5a) is satisfied. Since (14.9a) constitutes a PSF that is separable in a horizontal and a vertical component, the discrete version of (14.9a) is usually obtained by first computing a 1D discrete Gaussian PSF $\tilde{d}(n)$. This 1D PSF is found by a numerical discretization of the continuous PSF. For each PSF element $\tilde{d}(n)$, the 1D continuous PSF is integrated over the area covered by the 1D sampling grid, namely $\left[n - \frac{1}{2}, n + \frac{1}{2} \right]$:

$$\tilde{d}(n; \sigma_G) = C \int_{n - \frac{1}{2}}^{n + \frac{1}{2}} \exp\left(-\frac{x^2}{2\sigma_G^2} \right) dx. \tag{14.9b}$$

Since the spatially continuous PSF does not have a finite support, it has to be truncated properly. The spatially discrete approximation of (14.9a) is then given by

$$d(n_1, n_2; \sigma_G) = \tilde{d}(n_1; \sigma_G) \tilde{d}(n_2; \sigma_G). \tag{14.9c}$$

Figure 14.4 shows this PSF in the spectral domain ($\sigma_G = 1.2$). Observe that Gaussian blurs do not have exact spectral zeros.

14.3 IMAGE RESTORATION ALGORITHMS

In this section, we will assume that the PSF of the blur is satisfactorily known. A number of methods will be introduced for removing the blur from the recorded image $g(n_1, n_2)$ using a linear filter. If the PSF of the linear restoration filter, denoted by $h(n_1, n_2)$, has been designed, the restored image is given by

$$\hat{f}(n_1, n_2) = h(n_1, n_2) * g(n_1, n_2)$$

$$= \sum_{k_1=0}^{N-1} \sum_{k_2=0}^{M-1} h(k_1, k_2) g(n_1 - k_1, n_2 - k_2) \tag{14.10a}$$

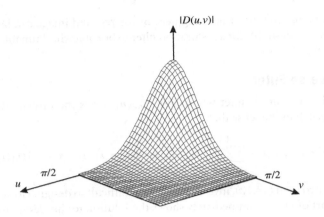

FIGURE 14.4

Gaussian PSF in the Fourier domain ($\sigma_G = 1.2$).

or in the spectral domain by

$$F(u,v) = H(u,v)G(u,v). \qquad (14.10b)$$

The objective of this section is to design appropriate restoration filters $h(n_1, n_2)$ or $H(u,v)$ for use in (14.10).

In image restoration the improvement in quality of the restored image over the recorded blurred one is measured by the signal-to-noise ratio (SNR) improvement. The SNR of the recorded (blurred and noisy) image is defined as follows in decibels:

$$\text{SNR}_g = 10\log_{10}\left(\frac{\text{Variance of the ideal image } f(n_1, n_2)}{\text{Variance of the difference image } g(n_1, n_2) - f(n_1, n_2)}\right) \text{ (dB)}. \qquad (14.11a)$$

The SNR of the restored image is similarly defined as

$$\text{SNR}_{\hat{f}} = 10\log_{10}\left(\frac{\text{Variance of the ideal image } f(n_1, n_2)}{\text{Variance of the difference image } \hat{f}(n_1, n_2) - f(n_1, n_2)}\right) \text{ (dB)}. \qquad (14.11b)$$

Then, the improvement in SNR is given by

$$\Delta\text{SNR} = \text{SNR}_{\hat{f}} - \text{SNR}_g$$

$$= 10\log_{10}\left(\frac{\text{Variance of the difference image } g(n_1, n_2) - f(n_1, n_2)}{\text{Variance of the difference image } \hat{f}(n_1, n_2) - f(n_1, n_2)}\right) \text{ (dB)}. \qquad (14.11c)$$

The improvement in SNR is basically a measure that expresses the reduction of disagreement with the ideal image when comparing the distorted and restored image. Note that all of the above signal-to-noise measures can only be computed in case the ideal image $f(n_1, n_2)$ is available, i.e., in an experimental setup or in a design phase of the restoration algorithm. When applying restoration filters to real images of which the ideal image is not available, often only the visual judgment of the restored image can be relied upon. For this reason it is desirable for a restoration filter to be somewhat "tunable" to the liking of the user.

14.3.1 Inverse Filter

An inverse filter is a linear filter whose PSF $h_{inv}(n_1, n_2)$ is the inverse of the blurring function $d(n_1, n_2)$, in the sense that

$$h_{inv}(n_1, n_2) * d(n_1, n_2) = \sum_{k_1=0}^{N-1} \sum_{k_2=0}^{M-1} h_{inv}(k_1, k_2) d(n_1 - k_1, n_2 - k_2) = \delta(n_1, n_2). \quad (14.12)$$

When formulated as in (14.12), inverse filters seem difficult to design. However, the spectral counterpart of (14.12) immediately shows the solution to this design problem [6]:

$$H_{inv}(u, v) D(u, v) = 1 \Rightarrow H_{inv}(u, v) = \frac{1}{D(u, v)}. \quad (14.13)$$

The advantage of the *inverse filter* is that it requires only the blur PSF as *a priori* knowledge, and that it allows for perfect restoration in the case that noise is absent, as can easily be seen by substituting (14.13) into (14.10b):

$$\hat{F}_{inv}(u, v) = H_{inv}(u, v) G(u, v) = \frac{1}{D(u, v)} (D(u, v) F(u, v) + W(u, v))$$

$$= F(u, v) + \frac{W(u, v)}{D(u, v)}. \quad (14.14)$$

If the noise is absent, the second term in (14.14) disappears so that the restored image is identical to the ideal image. Unfortunately, several problems exist with (14.14). In the first place the inverse filter may not exist because $D(u, v)$ is zero at selected frequencies (u, v). This happens for both the linear motion blur and the out-of-focus blur described in the previous section. Secondly, even if the blurring function's spectral representation $D(u, v)$ does not actually go to zero but becomes small, the second term in (14.14)—known as the inverse filtered noise—will become very large. Inverse filtered images are, therefore, often dominated by excessively amplified noise.[2]

Figure 14.5(a) shows an image degraded by out-of-focus blur ($R = 2.5$) and noise. The inverse filtered version is shown in Fig. 14.5(b), clearly illustrating its uselessness. The Fourier transforms of the restored image and of $H_{inv}(u, v)$ are shown in Fig. 14.5(c) and (d), respectively, demonstrating that indeed the spectral zeros of the PSF cause problems.

14.3.2 Least-Squares Filters

To overcome the noise sensitivity of the inverse filter, a number of restoration filters have been developed that are collectively called least-squares filters. We describe the two most

---

[2]In literature, this effect is commonly referred to as the ill-conditionedness or ill-posedness of the restoration problem.

FIGURE 14.5

(a) Image out-of-focus with $\mathrm{SNR}_g = 10.3\,\mathrm{dB}$ (noise variance $= 0.35$); (b) inverse filtered image; (c) magnitude of the Fourier transform of the restored image. The DC component lies in the center of the image. The oriented white lines are spectral components of the image with large energy; (d) magnitude of the Fourier transform of the inverse filter response.

commonly used filters from this collection, namely the Wiener filter and the constrained least-squares filter.

The Wiener filter is a linear spatially invariant filter of the form (14.10a), in which the PSF $h(n_1, n_2)$ is chosen such that it minimizes the mean-squared error (MSE) between the ideal and the restored image. This criterion attempts to make the difference between the ideal image and the restored one—i.e., the remaining restoration error—as small as possible *on the average*:

$$\text{MSE} = E[(f(n_1,n_2) - \hat{f}(n_1,n_2))^2] \approx \frac{1}{NM} \sum_{n_1=0}^{N-1} \sum_{n_2=0}^{M-1} (f(n_1,n_2) - \hat{f}(n_1,n_2))^2, \quad (14.15)$$

where $\hat{f}(n_1,n_2)$ is given by (14.10a). The solution of this minimization problem is known as the *Wiener filter*, and is easiest defined in the spectral domain:

$$H_{\text{wiener}}(u,v) = \frac{D^*(u,v)}{D^*(u,v)D(u,v) + \dfrac{S_w(u,v)}{S_f(u,v)}}. \quad (14.16)$$

Here $D^*(u,v)$ is the complex conjugate of $D(u,v)$, and $S_f(u,v)$ and $S_w(u,v)$ are the power spectrum of the ideal image and the noise, respectively. The power spectrum is a measure for the average signal power per spatial frequency (u,v) carried by the image. In the noiseless case we have $S_w(u,v) = 0$, so that the Wiener filter approximates the inverse filter:

$$H_{\text{wiener}}(u,v)|_{S_w(u,v)\to 0} = \begin{cases} \dfrac{1}{D(u,v)} & \text{for} \quad D(u,v) \neq 0 \\ 0 & \text{for} \quad D(u,v) = 0 \end{cases}. \quad (14.17)$$

For the more typical situation where the recorded image is noisy, the Wiener filter trades off the restoration by inverse filtering and suppression of noise for those frequencies where $D(u,v) \to 0$. The important factors in this tradeoff are the power spectra of the ideal image and the noise. For spatial frequencies where $S_w(u,v) \ll S_f(u,v)$, the Wiener filter approaches the inverse filter, while for spatial frequencies where $S_w(u,v) \gg S_f(u,v)$ the Wiener filter acts as a frequency rejection filter, i.e., $H_{\text{wiener}}(u,v) \to 0$.

If we assume that the noise is uncorrelated (white noise), its power spectrum is determined by the noise variance only:

$$S_w(u,v) = \sigma_w^2 \quad \text{for all } (u,v). \quad (14.18)$$

Thus, it is sufficient to estimate the noise variance from the recorded image to get an estimate of $S_w(u,v)$. The estimation of the noise variance can also be left to the user of the Wiener filter as if it were a tunable parameter. Small values of σ_w^2 will yield a result close to the inverse filter, while large values will over-smooth the restored image.

The estimation of $S_f(u,v)$ is somewhat more problematic since the ideal image is obviously not available. There are three possible approaches to take. In the first place, one can replace $S_f(u,v)$ by an estimate of the power spectrum of the blurred image and compensate for the variance of the noise σ_w^2:

$$S_f(u,v) \approx S_g(u,v) - \sigma_w^2 \approx \frac{1}{NM} G^*(u,v)G(u,v) - \sigma_w^2. \quad (14.19)$$

The above estimator for the power spectrum $S_g(u,v)$ of $g(n_1,n_2)$ is known as the periodogram. This estimator requires little *a priori* knowledge, but it is known to have several

TABLE 14.1 Prediction coefficients and variance of $v(n_1, n_2)$ for four images, computed in the MSE optimal sense by the Yule-Walker equations.

	$a_{0,1}$	$A_{1,1}$	$a_{1,0}$	σ_v^2
Cameraman	0.709	−0.467	0.739	231.8
Lena	0.511	−0.343	0.812	132.7
Trevor White	0.759	−0.525	0.764	33.0
White noise	−0.008	−0.003	−0.002	5470.1

shortcomings. More elaborate estimators for the power spectrum exist, but these require much more *a priori* knowledge.

A second approach is to estimate the power spectrum $S_f(u, v)$ from a set of representative images. These representative images are to be taken from a collection of images that have a content "similar" to the image that needs to be restored. Of course, one still needs an appropriate estimator to obtain the power spectrum from the set of representative images.

The third and final approach is to use a statistical model for the ideal image. Often these models incorporate parameters that can be tuned to the actual image being used. A widely used image model—not only popular in image restoration but also in image compression—is the following 2D causal autoregressive model [9]:

$$f(n_1, n_2) = a_{0,1}f(n_1, n_2 - 1) + a_{1,1}f(n_1 - 1, n_2 - 1)$$
$$+ a_{1,0}f(n_1 - 1, n_2) + v(n_1, n_2). \tag{14.20a}$$

In this model the intensities at the spatial location (n_1, n_2) are described as the sum of weighted intensities at neighboring spatial locations and a small unpredictable component $v(n_1, n_2)$. The unpredictable component is often modeled as white noise with variance σ_v^2. Table 14.1 gives numerical examples for MSE estimates of the *prediction coefficients* $a_{i,j}$ for some images. For the MSE estimation of these parameters the 2D autocorrelation function has first been estimated, and then used in the Yule-Walker equations [9]. Once the model parameters for (14.20a) have been chosen, the power spectrum can be calculated to be equal to

$$S_f(u, v) = \frac{\sigma_v^2}{\left|1 - a_{0,1}e^{-ju} - a_{1,1}e^{-ju-jv} - a_{1,0}e^{-jv}\right|^2}. \tag{14.20b}$$

The tradeoff between noise smoothing and deblurring that is made by the Wiener filter is illustrated in Fig. 14.6. Going from 14.6(a) to 14.6(c) the variance of the noise in the degraded image, i.e., σ_w^2, has been estimated too large, optimally, and too small, respectively. The visual differences, as well as the differences in improvement in SNR (ΔSNR) are substantial. The power spectrum of the original image has been calculated from the model (14.20a). From the results it is clear that the excessive noise amplification of the earlier example is no longer present because of the masking of the spectral zeros (see Fig. 14.6(d)). Typical artifacts of the Wiener restoration—and actually of most

FIGURE 14.6

(a) Wiener restoration of image in Fig. 14.5(a) with assumed noise variance equal to 35.0 (ΔSNR = 3.7 dB); (b) restoration using the correct noise variance of 0.35 (ΔSNR = 8.8 dB); (c) restoration assuming the noise variance is 0.0035 (ΔSNR = 1.1 dB); (d) Magnitude of the Fourier transform of the restored image in Fig. 14.6(b).

restoration filters—are the residual blur in the image and the "ringing" or "halo" artifacts present near edges in the restored image.

The *constrained least-squares filter* [10] is another approach for overcoming some of the difficulties of the inverse filter (excessive noise amplification) and of the Wiener filter (estimation of the power spectrum of the ideal image), while still retaining the simplicity of a spatially invariant linear filter. If the restoration is a good one, the blurred version of the restored image should be approximately equal to the recorded distorted image. That is

$$d(n_1, n_2) * \hat{f}(n_1, n_2) \approx g(n_1, n_2). \tag{14.21}$$

With the inverse filter the approximation is made exact, which leads to problems because a match is made to noisy data. A more reasonable expectation for the restored image is that it satisfies

$$\left\| g(n_1, n_2) - d(n_1, n_2) * \hat{f}(n_1, n_2) \right\|^2 = \frac{1}{NM} \sum_{k_1=0}^{N-1} \sum_{k_2=0}^{M-1} (g(k_1, k_2) - d(k_1, k_2) * \hat{f}(k_1, k_2))^2 \approx \sigma_w^2. \tag{14.22}$$

There are potentially many solutions that satisfy the above relation. A second criterion must be used to choose among them. A common criterion, acknowledging the fact that the inverse filter tends to amplify the noise $w(n_1, n_2)$, is to select the solution that is as "smooth" as possible. If we let $c(n_1, n_2)$ represent the PSF of a 2D highpass filter, then among the solutions satisfying (14.22) the solution is chosen that minimizes

$$\Omega\left(\hat{f}(n_1, n_2)\right) = \left\| c(n_1, n_2) * \hat{f}(n_1, n_2) \right\|^2 = \frac{1}{NM} \sum_{k_1=0}^{N-1} \sum_{k_2=0}^{M-1} \left(c(k_1, k_2) * \hat{f}(k_1, k_2) \right)^2. \tag{14.23}$$

The interpretation of $\Omega(\hat{f}(n_1, n_2))$ is that it gives a measure for the high-frequency content of the restored image. Minimizing this measure subject to the constraint (14.22) will give a solution that is both within the collection of potential solutions of (14.22) and has as little high-frequency content as possible at the same time. A typical choice for $c(n_1, n_2)$ is the discrete approximation of the second derivative shown in Fig. 14.7, also known as the 2D Laplacian operator.

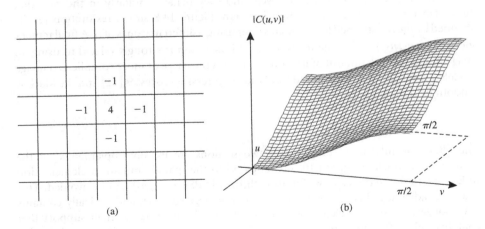

(a)   (b)

FIGURE 14.7

Two-dimensional discrete approximation of the second derivative operation. (a) PSF $c(n_1, n_2)$; (b) spectral representation.

(a)   (b)   (c)

FIGURE 14.8

(a) Constrained least-squares restoration of image in Fig. 14.5(a) with $\alpha = 2 \times 10^{-2}(\Delta\text{SNR} = 1.7\,\text{dB})$; (b) $\alpha = 2 \times 10^{-4}(\Delta\text{SNR} = 6.9\,\text{dB})$; (c) $\alpha = 2 \times 10^{-6}(\Delta\text{SNR} = 0.8\,\text{dB})$.

The solution to the above minimization problem is the constrained least-squares filter $H_{\text{cls}}(u, v)$ that is easiest formulated in the discrete Fourier domain:

$$H_{\text{cls}}(u,v) = \frac{D^*(u,v)}{D^*(u,v)D(u,v) + \alpha C^*(u,v)C(u,v)}. \tag{14.24}$$

Here α is a tuning or *regularization* parameter that should be chosen such that (14.22) is satisfied. Though analytical approaches exist to estimate α [3], the regularization parameter is usually considered user tunable.

It should be noted that although their motivations are quite different, the formulation of the Wiener filter (14.16) and constrained least-squares filter (14.24) are quite similar. Indeed these filters perform equally well, and they behave similarly in the case that the variance of the noise, σ_w^2, approaches zero. Figure 14.8 shows restoration results obtained by the constrained least-squares filter using 3 different values of α. A final remark about $\Omega(\hat{f}(n_1, n_2))$ is that the inclusion of this criterion is strongly related to using an image model. A vast amount of literature exists on the usage of more complicated image models, especially the ones inspired by 2D auto-regressive processes [11] and the Markov random field theory [12].

14.3.3 Iterative Filters

The filters formulated in the previous two sections are usually implemented in the Fourier domain using Eq. (14.10b). Compared to the spatial domain implementation in Eq. (14.10a), the direct convolution with the 2D PSF $h(n_1, n_2)$ can be avoided. This is a great advantage because $h(n_1, n_2)$ has a very large support, and typically contains NM nonzero filter coefficients even if the PSF of the blur has a small support that contains only a few nonzero coefficients. There are, however, two situations in which spatial domain convolutions are preferred over the Fourier domain implementation, namely:

- in situations where the dimensions of the image to be restored are very large;

- in cases where additional knowledge is available about the restored image, especially if this knowledge cannot be cast in the form of Eq. (14.23). An example is the *a priori* knowledge that image intensities are always positive. Both in the Wiener and the constrained least-squares filter the restored image may come out with negative intensities, simply because negative restored signal values are not explicitly prohibited in the design of the restoration filter.

Iterative restoration filters provide a means to handle the above situations elegantly [2, 5, 13]. The basic form of iterative restoration filters is the one that iteratively approaches the solution of the inverse filter, and is given by the following spatial domain iteration:

$$\hat{f}_{i+1}(n_1, n_2) = \hat{f}_i(n_1, n_2) + \beta(g(n_1, n_2) - d(n_1, n_2) * \hat{f}_i(n_1, n_2)). \tag{14.25}$$

Here $\hat{f}_i(n_1, n_2)$ is the restoration result after i iterations. Usually in the first iteration $\hat{f}_0(n_1, n_2)$ is chosen to be identical to zero or identical to $g(n_1, n_2)$. The iteration (14.25) has been independently discovered many times, and is referred to as the van Cittert, Bially, or Landweber iteration. As can be seen from (14.25), during the iterations the blurred version of the current restoration result $\hat{f}_i(n_1, n_2)$ is compared to the recorded image $g(n_1, n_2)$. The difference between the two is scaled and added to the current restoration result to give the next restoration result.

With iterative algorithms, there are two important concerns—does it converge and, if so, to what limiting solution? Analyzing (14.25) shows that convergence occurs if the convergence parameter β satisfies

$$|1 - \beta D(u, v)| < 1 \quad \text{for all } (u, v). \tag{14.26a}$$

Using the fact that $|D(u, v)| \le 1$, this condition simplifies to

$$0 < \beta < 2 \quad \text{and} \quad D(u, v) > 0. \tag{14.26b}$$

If the number of iterations becomes very large, then $\hat{f}_i(n_1, n_2)$ approaches the solution of the inverse filter

$$\lim_{i \to \infty} \hat{f}_i(n_1, n_2) = h_{\text{inv}}(n_1, n_2) * g(n_1, n_2). \tag{14.27}$$

Figure 14.9 shows four restored images obtained by the iteration (14.25). Clearly as the iteration progresses, the restored image is dominated more and more by inverse filtered noise.

The iterative scheme (14.25) has several advantages and disadvantages that we will discuss next. The first advantage is that (14.25) does not require the convolution of images with 2D PSFs containing many coefficients. The only convolution is that of the restored image with the PSF of the blur, which has relatively few coefficients.

The second advantage is that no Fourier transforms are required, making (14.25) applicable to images of arbitrary size. The third advantage is that although the iteration

(a)   (b)

(c)   (d)

FIGURE 14.9

(a) Iterative restoration ($\beta = 1.9$) of the image in Fig. 14.5(a) after 10 iterations (ΔSNR = 1.6 dB); (b) after 100 iterations (ΔSNR = 5.0 dB); (c) after 500 iterations (ΔSNR = 6.6 dB); (d) after 5000 iterations (ΔSNR = −2.6 dB).

produces the inverse filtered image as a result if the iteration is continued indefinitely, the iteration can be terminated whenever an acceptable restoration result has been achieved. Starting off with a blurred image, the iteration progressively deblurs the image. At the same time the noise will be amplified more and more as the iteration continues. It is now usually left to the user to tradeoff the degree of restoration against the noise amplification, and to stop the iteration when an acceptable partially deblurred result has been achieved.

The fourth advantage is that the basic form (14.25) can be extended to include all types of *a priori* knowledge. First all knowledge is formulated in the form of projective operations on the image [14]. After applying a projective operation, the (restored) image satisfies the *a priori* knowledge reflected by that operator. For instance, the fact that image intensities are always positive can be formulated as the following projective operation P:

$$P[\hat{f}(n_1, n_2)] = \begin{cases} \hat{f}(n_1, n_2) & \text{if} \quad \hat{f}(n_1, n_2) \geq 0 \\ 0 & \text{if} \quad \hat{f}(n_1, n_2) < 0 \end{cases}. \tag{14.28}$$

By including this projection P in the iteration, the final image after convergence of the iteration and all of the intermediate images will not contain negative intensities. The resulting iterative restoration algorithm now becomes

$$\hat{f}_{i+1}(n_1, n_2) = P\left[\hat{f}_i(n_1, n_2) + \beta(g(n_1, n_2) - d(n_1, n_2) * \hat{f}_i(n_1, n_2))\right]. \tag{14.29}$$

The requirements on β for convergence as well as the properties of the final image after convergence are difficult to analyze and fall outside the scope of this chapter. Practical values for β are typically around 1. Further, not all projections P can be used in the iteration (14.29), but only *convex* projections. A loose definition of a convex projection is the following. If both images $f^{(1)}(n_1, n_2)$ and $f^{(2)}(n_1, n_2)$ satisfy the *a priori* information described by the projection P, then also the combined image

$$f^{(c)}(n_1, n_2) = \varepsilon f^{(1)}(n_1, n_2) + (1 - \varepsilon)f^{(2)}(n_1, n_2) \tag{14.30}$$

must satisfy this *a priori* information for all values of ε between 0 and 1.

A final advantage of iterative schemes is that they are easily extended for spatially variant restoration, i.e., restoration where either the PSF of the blur or the model of the ideal image (for instance the prediction coefficients in Eq. (14.20)) vary locally [3, 5].

On the negative side, the iterative scheme (14.25) has two disadvantages. First, the second requirement in Eq. (14.26b), namely that $D(u, v) > 0$, is not satisfied by many blurs, like motion blur and out-of-focus blur. This causes (14.25) to diverge for these types of blur. Second, unlike the Wiener and constrained least-squares filter—the basic scheme does not include any knowledge about the spectral behavior of the noise and the ideal image. Both disadvantages can be corrected by modifying the basic iterative scheme as follows:

$$\hat{f}_{i+1}(n_1, n_2) = (\delta(n_1, n_2) - \alpha\beta c(-n_1, -n_2) * c(n_1, n_2)) * \hat{f}_i(n_1, n_2) +$$
$$+ \beta d(-n_1, -n_2) * (g(n_1, n_2) - d(n_1, n_2) * \hat{f}_i(n_1, n_2)). \tag{14.31}$$

Here α and $c(n_1, n_2)$ have the same meaning as in the constrained least-squares filter. Though the convergence requirements are more difficult to analyze, it is no longer necessary for $D(u, v)$ to be positive for all spatial frequencies. If the iteration is continued indefinitely, Eq. (14.31) will produce the constrained least-squares filtered image as a result. In practice the iteration is terminated long before convergence. The precise termination point of the iterative scheme gives the user an additional degree of freedom over the direct implementation of the constrained least-squares filter. It is noteworthy that although (14.31) seems to involve many more convolutions than (14.25), a reorganization of terms is possible revealing that many of those convolutions can be carried out once and offline, and that only one convolution is needed per iteration:

$$\hat{f}_{i+1}(n_1, n_2) = g^d(n_1, n_2) + k(n_1, n_2) * \hat{f}_i(n_1, n_2), \tag{14.32a}$$

where the image $g^d(n_1, n_2)$ and the fixed convolution kernel $k(n_1, n_2)$ are given by

$$g^d(n_1, n_2) = \beta d(-n_1, -n_2) * g(n_1, n_2)$$

$$k(n_1, n_2) = \delta(n_1, n_2) - \alpha\beta c(-n_1, -n_2) * c(n_1, n_2) - \beta d(-n_1, -n_2) * d(n_1, n_2). \tag{14.32b}$$

A second—and very significant—disadvantage of the iterations (14.25) and (14.29)–(14.32) is the slow convergence. Per iteration the restored image $\hat{f}_i(n_1, n_2)$ changes only a little. Many iteration steps are, therefore, required before an acceptable point for termination of the iteration is reached. The reason is that the above iteration is essentially a *steepest descent* optimization algorithm, which is known to be slow in convergence. It is possible to reformulate the iterations in the form of, for instance, a *conjugate gradient* algorithm, which exhibits a much higher convergence rate [5].

14.3.4 Boundary Value Problem

Images are always recorded by sensors of finite spatial extent. Since the convolution of the ideal image with the PSF of the blur extends beyond the borders of the observed degraded image, part of the information that is necessary to restore the border pixels is not available to the restoration process. This problem is known as the *boundary value problem*, and poses a severe problem to restoration filters. Although at first glance the boundary value problem seems to have a negligible effect because it affects only border pixels, this is not true at all. The PSF of the restoration filter has a very large support, typically as large as the image itself. Consequently, the effect of missing information at the borders of the image propagates throughout the image, in this way deteriorating the entire image. Figure 14.10(a) shows an example of a case where the missing information immediately outside the borders of the image is assumed to be equal to the mean value of the image, yielding dominant horizontal oscillation patterns due to the restoration of the horizontal motion blur.

Two solutions to the boundary value problem are used in practice. The choice depends on whether a spatial domain or a Fourier domain restoration filter is used. In a spatial domain filter, missing image information outside the observed image can be estimated by extrapolating the available image data. In the extrapolation, a model for the observed image can be used, such as the one in Eq. (14.20), or more simple procedures can be used such as mirroring the image data with respect to the image border. For instance, image data missing on the left-hand side of the image could be estimated as follows:

$$g(n_1, n_2 - k) = g(n_1, n_2 + k) \quad \text{for } k = 1, 2, 3, \ldots \tag{14.33}$$

When Fourier domain restoration filters are used, such as the ones in (14.16) or (14.24), one should realize that discrete Fourier transforms assume periodicity of the data to be

(a)

(b)

FIGURE 14.10

(a) Restored image illustrating the effect of the boundary value problem. The image was blurred by the motion blur shown in Fig. 14.2(a), and restored using the constrained least-squares filter; (b) preprocessed blurred image at its borders such that the boundary value problem is solved.

transformed. Effectively in 2D Fourier transforms this means that the left- and right-hand sides of the image are implicitly assumed to be connected, as well as the top and bottom parts of the image. A consequence of this property—implicit to discrete Fourier transforms—is that missing image information at the left-hand side of the image will be taken from the right-hand side, and vice versa. Clearly in practice this image data may not correspond to the actual (but missing data) at all. A common way to fix this problem is to interpolate the image data at the borders such that the intensities at the left- and right-hand side as well as the top and bottom of the image transit smoothly. Figure 14.10(b) shows what the blurred image looks like if a border of 5 columns or rows is used for linearly interpolating between the image boundaries. Other forms of interpolation could be used, but in practice mostly linear interpolation suffices. All restored images shown in this chapter have been preprocessed in this way to solve the boundary value problem.

14.4 BLUR IDENTIFICATION ALGORITHMS

In the previous section it was assumed that the PSF $d(n_1, n_2)$ of the blur was known. In many practical cases the actual restoration process has to be preceded by the identification of this PSF. If the camera misadjustment, object distances, object motion, and camera motion are known, we could—in theory—determine the PSF analytically. Such situations are, however, rare. A more common situation is that the blur is estimated from the observed image itself.

The blur identification procedure starts out by choosing a parametric model for the PSF. One category of parametric blur models has been given in Section 14.2. As an example, if the blur were known to be due to motion, the blur identification procedure would estimate the length and direction of the motion.

A second category of parametric blur models describes the PSF $d(n_1, n_2)$ as a (small) set of coefficients within a given finite support. Within this support the value of the PSF coefficients needs to be estimated. For instance, if an initial analysis shows that the blur in the image resembles out-of-focus blur which, however, cannot be described parametrically by Eq. (14.8b), the blur PSF can be modeled as a square matrix of—say— size 3 by 3, or 5 by 5. The blur identification then requires the estimation of 9 or 25 PSF coefficients, respectively. This section describes the basics of the above two categories of blur estimation.

14.4.1 Spectral Blur Estimation

In Figs. 14.2 and 14.3 we have seen that two important classes of blurs, namely motion and out-of-focus blur, have spectral zeros. The structure of the zero-patterns characterizes the type and degree of blur within these two classes. Since the degraded image is described by (14.2), the spectral zeros of the PSF should also be visible in the Fourier transform $G(u, v)$, albeit that the zero-pattern might be slightly masked by the presence of the noise.

Figure 14.11 shows the modulus of the Fourier transform of two images, one subjected to motion blur and one to out-of-focus blur. From these images, the structure and location of the zero-patterns can be estimated. When the pattern contains dominant parallel lines of zeros, an estimate of the length and angle of motion can be made. When dominant

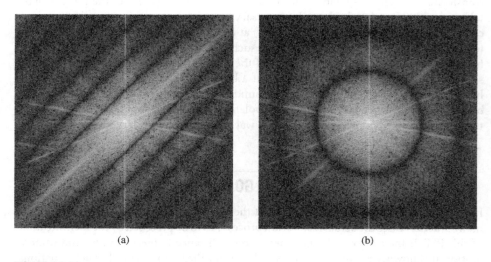

(a)   (b)

FIGURE 14.11

$|G(u, v)|$ of two blurred images.

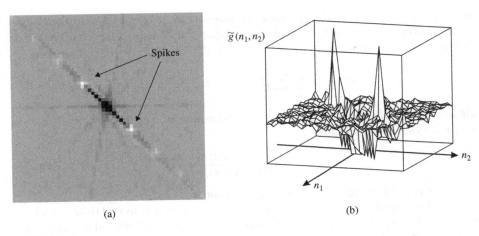

FIGURE 14.12

Cepstrum for motion blur from Fig. 14.2(c). (a) Cepstrum is shown as a 2D image. The spikes appear as bright spots around the center of the image; (b) cepstrum shown as a surface plot.

circular patterns occur, out-of-focus blur can be inferred and the degree of out-of-focus (the parameter R in Eq. (14.8)) can be estimated.

An alternative to the above method for identifying motion blur involves the computation of the 2D cepstrum of $g(n_1, n_2)$. The cepstrum is the inverse Fourier transform of the logarithm of $|G(u, v)|$. Thus

$$\tilde{g}(n_1, n_2) = -\mathsf{F}^{-1}\{\log |G(u, v)|\}, \qquad (14.34)$$

where F^{-1} is the inverse Fourier transform operator. If the noise can be neglected, $\tilde{g}(n_1, n_2)$ has a large spike at a distance L from the origin. Its position indicates the direction and extent of the motion blur. Figure 14.12 illustrates this effect for an image with the motion blur from Fig. 14.2(b).

14.4.2 Maximum Likelihood Blur Estimation

When the PSF does not have characteristic spectral zeros or when a parametric blur model such as motion or out-of-focus blur cannot be assumed, the individual coefficients of the PSF have to be estimated. To this end maximum likelihood estimation procedures for the unknown coefficients have been developed [3, 15, 16, 18]. Maximum likelihood estimation is a well-known technique for parameter estimation in situations where no stochastic knowledge is available about the parameters to be estimated [7].

Most maximum likelihood identification techniques begin by assuming that the ideal image can be described with the 2D auto-regressive model (14.20a). The parameters of this image model—that is, the prediction coefficients $a_{i,j}$ and the variance σ_v^2 of the white noise $v(n_1, n_2)$—are not necessarily assumed to be known.

If we can assume that both the observation noise $w(n_1, n_2)$ and the image model noise $v(n_1, n_2)$ are Gaussian distributed, the log-likelihood function of the observed image, given the image model and blur parameters, can be formulated. Although the log-likelihood function can be formulated in the spatial domain, its spectral version is slightly easier to compute [16]:

$$L(\theta) = -\sum_u \sum_v \left(\log P(u,v) + \frac{|G(u,v)|^2}{P(u,v)} \right), \tag{14.35a}$$

where θ symbolizes the set of parameters to be estimated, i.e., $\theta = \{a_{i,j}, \sigma_v^2, d(n_1, n_2), \sigma_w^2\}$, and $P(u,v)$ is defined as

$$P(u,v) = \sigma_v^2 \frac{|D(u,v)|^2}{|1 - A(u,v)|^2} + \sigma_w^2. \tag{14.35b}$$

Here $A(u,v)$ is the discrete 2D Fourier transform of $a_{i,j}$.

The objective of maximum likelihood blur estimation is now to find those values for the parameters $a_{i,j}, \sigma_v^2, d(n_1, n_2)$, and σ_w^2 that maximize the log-likelihood function $L(\theta)$. From the perspective of parameter estimation, the optimal parameter values best explain the observed degraded image. A careful analysis of (14.35) shows that the maximum likelihood blur estimation problem is closely related to the identification of 2D auto-regressive moving-average (ARMA) stochastic processes [16, 17].

The maximum likelihood estimation approach has several problems that require nontrivial solutions. The differentiation between state-of-the-art blur identification procedures is mostly in the way they handle these problems [4]. In the first place, some constraints must be enforced in order to obtain a unique estimate for the PSF. Typical constraints are:

- the energy conservation principle, as described by Eq. (14.5b);

- symmetry of the PSF of the blur, i.e., $d(-n_1, -n_2) = d(n_1, n_2)$.

Secondly, the log-likelihood function (14.35) is highly nonlinear and has many local maxima. This makes the optimization of (14.35) difficult, no matter what optimization procedure is used. In general, maximum-likelihood blur identification procedures require good initializations of the parameters to be estimated in order to ensure converge to the global optimum. Alternatively, multiscale techniques could be used, but no "ready-to-go" or "best" approach has been agreed upon so far.

Given reasonable initial estimates for θ, various approaches exist for the optimization of $L(\theta)$. They share the property of being iterative. Besides standard gradient-based searches, an attractive alternative exists in the form of the expectation-minimization (EM) algorithm. The EM-algorithm is a general procedure for finding maximum likelihood parameter estimates. When applied to the blur identification procedure, an iterative scheme results that consists of two steps [15, 18] (see Fig. 14.13).

14.4.2.1 *Expectation step*

Given an estimate of the parameters θ, a restored image $\hat{f}_E(n_1, n_2)$ is computed by the Wiener restoration filter (14.16). The power spectrum is computed by (14.20b) using the given image model parameter $a_{i,j}$ and σ_v^2.

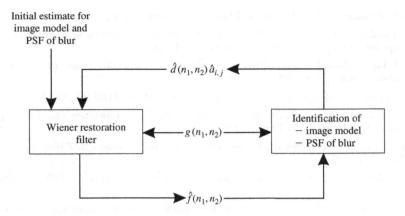

FIGURE 14.13

Maximum-likelihood blur estimation by the EM procedure.

14.4.2.2 *Maximization step*

Given the image restored during the expectation step, a new estimate of θ can be computed. Firstly, from the restored image $\hat{f}_E(n_1, n_2)$ the image model parameters $a_{i,j}$ and σ_v^2 can be estimated directly. Secondly, from the approximate relation

$$g(n_1, n_2) \approx d(n_1, n_2) * \hat{f}_E(n_1, n_2) \tag{14.36}$$

and the constraints imposed on $d(n_1, n_2)$, the coefficients of the PSF can be estimated by standard system identification procedures [5].

By alternating the E-step and the M-step, convergence to a (local) optimum of the log-likelihood function is achieved. A particularly attractive property of this iteration is that although the overall optimization is nonlinear in the parameters θ, the individual steps in the EM-algorithm are entirely linear. Furthermore, as the iteration progresses, intermediate restoration results are obtained that allow for monitoring of the identification process.

In conclusion, we observe that the field of blur identification has been studied and developed significantly less thoroughly than the classical problem of image restoration. Research in image restoration continues with a focus on blur identification using, for example, cumulants and generalized cross-validation [4].

REFERENCES

[1] M. R. Banham and A. K. Katsaggelos. Digital image restoration. *IEEE Signal Process. Mag.*, 14(2): 24–41, 1997.

[2] J. Biemond, R. L. Lagendijk, and R. M. Mersereau. Iterative methods for image deblurring. *Proc. IEEE*, 78(5):856–883, 1990.

[3] A. K. Katsaggelos, editor. *Digital Image Restoration*. Springer Verlag, New York, 1991.

[4] D. Kundur and D. Hatzinakos. Blind image deconvolution: an algorithmic approach to practical image restoration. *IEEE Signal Process. Mag.*, 13(3):43–64, 1996.

[5] R. L. Lagendijk and J. Biemond. *Iterative Identification and Restoration of Images*. Kluwer Academic Publishers, Boston, MA, 1991.

[6] H. C. Andrews and B. R. Hunt. *Digital Image Restoration*. Prentice Hall Inc., New Jersey, 1977.

[7] H. Stark and J. W. Woods. *Probability, Random Processes, and Estimation Theory for Engineers*. Prentice Hall, Upper Saddle River, NJ, 1986.

[8] N. P. Galatsanos and R. Chin. Digital restoration of multichannel images. *IEEE Trans. Signal Process.*, 37:415–421, 1989.

[9] A. K. Jain. Advances in mathematical models for image processing. *Proc. IEEE*, 69(5):502–528, 1981.

[10] B. R. Hunt. The application of constrained least squares estimation to image restoration by digital computer. *IEEE Trans. Comput.*, 2:805–812, 1973.

[11] J. W. Woods and V. K. Ingle. Kalman filtering in two-dimensions – further results. *IEEE Trans. Acoust.*, 29:188–197, 1981.

[12] F. Jeng and J. W. Woods. Compound Gauss-Markov random fields for image estimation. *IEEE Trans. Signal Process.*, 39:683–697, 1991.

[13] A. K. Katsaggelos. Iterative image restoration algorithm. *Opt. Eng.*, 28(7):735–748, 1989.

[14] P. L. Combettes. The foundation of set theoretic estimation. *Proc. IEEE*, 81:182–208, 1993.

[15] R. L. Lagendijk, J. Biemond, and D. E. Boekee. Identification and restoration of noisy blurred images using the expectation-maximization algorithm. *IEEE Trans. Acoust.*, 38:1180–1191, 1990.

[16] R. L. Lagendijk, A. M. Tekalp, and J. Biemond. Maximum likelihood image and blur identification: a unifying approach. *Opt. Eng.*, 29(5):422–435, 1990.

[17] Y. L. You and M. Kaveh. A regularization approach to joint blur identification and image restoration. *IEEE Trans. Image Process.*, 5:416–428, 1996.

[18] A. M. Tekalp, H. Kaufman, and J. W. Woods. Identification of image and blur parameters for the restoration of non-causal blurs. *IEEE Trans. Acoust.*, 34:963–972, 1986.

Iterative Image Restoration

15

Aggelos K. Katsaggelos[1], S. Derin Babacan[1], and Chun-Jen Tsai[2]

[1] *Northwestern University;* [2] *National Chiao Tung University*

15.1 INTRODUCTION

In this chapter we consider a class of iterative image restoration algorithms. Let $\mathbf{g}$ be the observed noisy and blurred image, $\mathbf{D}$ the operator describing the degradation system, $\mathbf{f}$ the input to the system, and $\mathbf{v}$ the noise added to the output image. The input-output relation of the degradation system is then described by [1]

$$\mathbf{g} = \mathbf{Df} + \mathbf{v}. \tag{15.1}$$

The image restoration problem, therefore, to be solved is the inverse problem of recovering $\mathbf{f}$ from knowledge of $\mathbf{g}$, $\mathbf{D}$, and $\mathbf{v}$. If $\mathbf{D}$ is also unknown, then we deal with the blind image restoration problem (semiblind if $\mathbf{D}$ is partially known).

There are numerous imaging applications which are described by (15.1) [1–4]. $\mathbf{D}$, for example, might represent a model of the turbulent atmosphere in astronomical observations with ground-based telescopes, or a model of the degradation introduced by an out-of-focus imaging device. $\mathbf{D}$ might also represent the quantization performed on a signal or a transformation of it, for reducing the number of bits required to represent the signal.

The success in solving any recovery problem depends on the amount of the available prior information. This information refers to properties of the original image, the degradation system (which is in general only partially known), and the noise process. Such prior information can, for example, be represented by the fact that the original image is a sample of a stochastic field, or that the image is "smooth," or that it takes only nonnegative values. Besides defining the amount of prior information, equally critical is the ease of incorporating it into the recovery algorithm.

After the degradation model is established, the next step is the formulation of a solution approach. This might involve the stochastic modeling of the input image (and the noise), the determination of the model parameters, and the formulation of a criterion to be optimized. Alternatively it might involve the formulation of a functional to be optimized subject to constraints imposed by the prior information. In the simplest possible case, the degradation equation defines directly the solution approach. For example, if $\mathbf{D}$ is a square invertible matrix, and the noise is ignored in (15.1), $\mathbf{f} = \mathbf{D}^{-1}\mathbf{g}$ is the desired

349

unique solution. In most cases, however, the solution of (15.1) represents an ill-posed problem [5]. Application of regularization theory transforms it to a well-posed problem which provides meaningful solutions to the original problem.

There are a large number of approaches providing solutions to the image restoration problem. For reviews of such approaches refer, for example, to [2, 4] and references therein. Recent reviews of blind image restoration approaches can be found in [6, 7]. This chapter concentrates on a specific type of iterative algorithm, the *successive approximations* algorithm, and its application to the image restoration problem. The material presented here can be extended in a rather straightforward manner to use other iterative algorithms, such as steepest descent and conjugate gradient methods.

15.2 ITERATIVE RECOVERY ALGORITHMS

Iterative algorithms form an important part of optimization theory and numerical analysis. They date back to Gauss time, but they also represent a topic of active research. A large part of any textbook on optimization theory or numerical analysis deals with iterative optimization techniques or algorithms [8].

Out of all possible iterative recovery algorithms we concentrate on the successive approximations algorithms, which have been successfully applied to the solution of a number of inverse problems ([9] represents a very comprehensive paper on the topic). The basic idea behind such an algorithm is that the solution to the problem of recovering a signal which satisfies certain constraints from its degraded observation can be found by the alternate implementation of the degradation and the constraint operator. Problems reported in [9] which can be solved with such an iterative algorithm are the phase-only recovery problem, the magnitude-only recovery problem, the bandlimited extrapolation problem, the image restoration problem, and the filter design problem [10]. Reviews of iterative restoration algorithms are also presented in [11, 12]. There are a number of advantages associated with iterative restoration algorithms, among which [9, 12]: (i) there is no need to determine or implement the inverse of an operator; (ii) knowledge about the solution can be incorporated into the restoration process in a relatively straightforward manner; (iii) the solution process can be monitored as it progresses; and (iv) the partially restored signal can be utilized in determining unknown parameters pertaining to the solution.

In the following we first present the development and analysis of two simple iterative restoration algorithms. Such algorithms are based on a linear and spatially invariant degradation, when the noise is ignored. Their description is intended to provide a good understanding of the various issues involved in dealing with iterative algorithms. We adopt a "how-to" approach; it is expected that no difficulties will be encountered by anybody wishing to implement the algorithms. We then proceed with the matrix-vector representation of the degradation model and the iterative algorithms. The degradation systems described now are linear but not necessarily spatially invariant. The relation between the matrix-vector and scalar representation of the degradation equation and the iterative solution is also presented. Experimental results demonstrate the capabilities of the algorithms.

15.3 SPATIALLY INVARIANT DEGRADATION

15.3.1 Degradation Model

Let us consider the following degradation model

$$g(n_1, n_2) = d(n_1, n_2) * f(n_1, n_2), \tag{15.2}$$

where $g(n_1, n_2)$ and $f(n_1, n_2)$ represent, respectively, the observed degraded and the original image, $d(n_1, n_2)$ is the impulse response of the degradation system, and $*$ denotes 2D convolution. It is mentioned here that the arrays $d(n_1, n_2)$ and $f(n_1, n_2)$ are appropriately padded with zeros, so that the result of 2D circular convolution equals the result of 2D linear convolution in (15.2) (see Chapter 5). Henceforth, in the following all convolutions involved are circular convolutions and all shifts are circular shifts.

We rewrite (15.2) as follows

$$\Phi(f(n_1, n_2)) = g(n_1, n_2) - d(n_1, n_2) * f(n_1, n_2) = 0. \tag{15.3}$$

The restoration problem, therefore, of finding an estimate of $f(n_1, n_2)$ given $g(n_1, n_2)$ and $d(n_1, n_2)$, becomes the problem of finding a root of $\Phi(f(n_1, n_2)) = 0$.

15.3.2 Basic Iterative Restoration Algorithm

The solution of (15.3) also satisfies the following equation for any value of the parameter β

$$f(n_1, n_2) = f(n_1, n_2) + \beta\Phi(f(n_1, n_2)). \tag{15.4}$$

Equation (15.4) forms the basis of the successive approximations iteration, by interpreting $f(n_1, n_2)$ on the left-hand side as the solution at the current iteration step, and $f(n_1, n_2)$ on the right-hand side as the solution at the previous iteration step. That is, with $f_0(n_1, n_2) = 0$,

$$\begin{aligned} f_{k+1}(n_1, n_2) &= f_k(n_1, n_2) + \beta\Phi(f_k(n_1, n_2)) \\ &= \beta g(n_1, n_2) + (\delta(n_1, n_2) - \beta d(n_1, n_2)) * f_k(n_1, n_2), \end{aligned} \tag{15.5}$$

where $f_k(n_1, n_2)$ denotes the restored image at the k-th iteration step, $\delta(n_1, n_2)$ the discrete delta function, and β the relaxation parameter which controls the convergence, as well as the rate of convergence of the iteration. Iteration (15.5) is the basis of a large number of iterative recovery algorithms, and is therefore analyzed in detail. Perhaps the earliest reference to iteration (15.5) with $\beta = 1$ was by Van Cittert [13] in the 1930s.

15.3.3 **Convergence**

Clearly if a root of $\Phi(f(n_1, n_2))$ exists, this root is a *fixed point* of iteration (15.5), that is, a point for which $f_{k+1}(n_1, n_2) = f_k(n_1, n_2)$. It is not guaranteed, however, that iteration (15.5) will converge, even if (15.3) has one or more solutions. Let us, therefore, examine under what condition (sufficient condition) iteration (15.5) converges. Let us first rewrite it in the discrete frequency domain, by taking the 2D discrete Fourier transform (DFT) of both sides. It then becomes

$$F_{k+1}(u,v) = \beta G(u,v) + (1 - \beta D(u,v))F_k(u,v), \tag{15.6}$$

where $F_k(u,v)$, $G(u,v)$, and $D(u,v)$ represent, respectively, the 2D DFT of $f_k(n_1, n_2)$, $g(n_1, n_2)$, and $d(n_1, n_2)$. We express next $F_k(u,v)$ in terms of $F_0(u,v)$. Clearly

$$F_1(u,v) = \beta G(u,v),$$

$$F_2(u,v) = \beta G(u,v) + (1 - \beta D(u,v))\beta G(u,v)$$

$$= \sum_{\ell=0}^{1}(1 - \beta D(u,v))^{\ell}\beta G(u,v),$$

$$\vdots$$

$$F_k(u,v) = \sum_{\ell=0}^{k-1}(1 - \beta D(u,v))^{\ell}\beta G(u,v)$$

$$= H_k(u,v)G(u,v). \tag{15.7}$$

We, therefore, see that the restoration filter at the k-th iteration step is given by

$$H_k(u,v) = \beta \sum_{\ell=0}^{k-1}(1 - \beta D(u,v))^{\ell}. \tag{15.8}$$

The obvious next question is then under what conditions the series in (15.8) converges and what is this convergence filter equal to. Clearly if

$$|1 - \beta D(u,v)| < 1, \tag{15.9}$$

then

$$\lim_{k \to \infty} H_k(u,v) = \lim_{k \to \infty} \beta \frac{1 - (1 - \beta D(u,v))^k}{1 - (1 - \beta D(u,v))} = \frac{1}{D(u,v)}. \tag{15.10}$$

Notice that (15.9) is not satisfied at the frequencies for which $D(u,v) = 0$. At these frequencies

$$H_k(u,v) = k \cdot \beta, \tag{15.11}$$

and therefore, in the limit $H_k(u,v)$ is not defined. However, since the number of iterations run is always finite, $H_k(u,v)$ is a large but finite number.

Taking a closer look at the sufficient condition for convergence, we see that (15.9) can be rewritten as

$$|1 - \beta \operatorname{Re}\{D(u,v)\} - \beta \operatorname{Im}\{D(u,v)\}|^2 < 1$$
$$\Rightarrow (1 - \beta \operatorname{Re}\{D(u,v)\})^2 + (\beta \operatorname{Im}\{D(u,v)\})^2 < 1. \qquad (15.12)$$

Inequality (15.12) defines the region inside a circle of radius $1/\beta$ centered at $c = (1/\beta, 0)$ in the $(\operatorname{Re}\{D(u,v)\}, \operatorname{Im}\{D(u,v)\})$ domain, as shown in Fig. 15.1. From this figure, it is clear that the left half-plane is not included in the region of convergence. That is, even though by decreasing β the size of the region of convergence increases, if the real part of $D(u,v)$ is negative, the sufficient condition for convergence cannot be satisfied. Therefore, for the class of degradations that this is the case, such as the degradation due to motion, iteration (15.5) is not guaranteed to converge.

The following form of (15.12) results when $\operatorname{Im}\{D(u,v)\} = 0$, which means that $d(n_1, n_2)$ is symmetric:

$$0 < \beta < \frac{2}{D_{\max}(u,v)}, \qquad (15.13)$$

where $D_{\max}(u,v)$ denotes the maximum value of $D(u,v)$ over all frequencies (u,v). If we now also take into account that $d(n_1, n_2)$ is typically normalized, i.e., $\sum_{n_1,n_2} d(n_1, n_2) = 1$, and represents a lowpass degradation, then $D(0,0) = D_{\max}(u,v) = 1$. In this case, (15.12) becomes

$$0 < \beta < 2. \qquad (15.14)$$

From the above analysis, when the sufficient condition for convergence is satisfied, the iteration converges to the original signal. This is also the inverse solution obtained directly

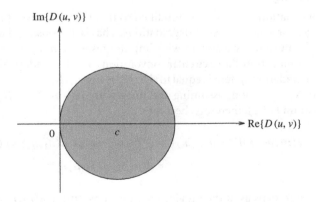

FIGURE 15.1

Geometric interpretation of the sufficient condition for convergence of the basic iteration, where $c = (1/\beta, 0)$.

from the degradation equation. That is, by rewriting (15.2) in the discrete frequency domain

$$G(u,v) = D(u,v) \cdot F(u,v),$$ (15.15)

we obtain,

$$F(u,v) = \begin{cases} \dfrac{G(u,v)}{D(u,v)}, & \text{for } D(u,v) \neq 0 \\ 0, & \text{otherwise,} \end{cases}$$ (15.16)

which represents the *pseudo-inverse* or *generalized inverse* solution.

An important point to be made here is that, unlike the iterative solution, the inverse solution (15.16) can be obtained without imposing any requirements on $D(u,v)$. That is, even if (15.2) or (15.15) has a unique solution, that is, $D(u,v) \neq 0$ for all (u,v), iteration (15.5) may not converge, if the sufficient condition for convergence is not satisfied. It is, therefore, not the appropriate iteration to solve the problem. Iteration (15.5) actually may not offer any advantages over the direct implementation of the inverse filter of (15.16), if no other features of the iterative algorithms are used, as will be explained later. One possible advantage of (15.5) over (15.16) is that the noise amplification in the restored image can be controlled by terminating the iteration before convergence, which represents another form of regularization, as will also be demonstrated experimentally. An iteration which will converge to the inverse solution of (15.2) for any $d(n_1,n_2)$ is described in the next section.

15.3.4 Reblurring

The degradation equation (15.2) can be modified so that the successive approximations iteration converges for a larger class of degradations. That is, the observed data $g(n_1,n_2)$ are first filtered (reblurred) by a system with impulse response $d^*(-n_1,-n_2)$, where $*$ denotes complex conjugation. Since circular convolutions have been adopted, the impulse response of the degradation system is equal to $d^*((N_1 - n_1)_{N_1}, (N_2 - n_2)_{N_2})$, where $(\cdot)_{N_1}$ denotes modulo N_1 operation, assuming the images are of size $N_1 \times N_2$ pixels. The degradation equation (15.2), therefore, becomes

$$\tilde{g}(n_1,n_2) = g(n_1,n_2) * d^*(-n_1,-n_2) = d^*(-n_1,-n_2) * d(n_1,n_2) * f(n_1,n_2)$$
$$= \tilde{d}(n_1,n_2) * f(n_1,n_2).$$ (15.17)

If we follow the same steps as in the previous section substituting $g(n_1,n_2)$ by $\tilde{g}(n_1,n_2)$ and $d(n_1,n_2)$ by $\tilde{d}(n_1,n_2)$ the iteration providing a solution to (15.17) becomes

$$f_{k+1}(n_1,n_2) = \beta d^*(-n_1,-n_2) * g(n_1,n_2) + (\delta(n_1,n_2) - \beta\tilde{d}(n_1,n_2)) * f_k(n_1,n_2),$$ (15.18)

with $f_0(n_1, n_2) = 0$. Following similar steps to the ones shown in the previous section, we find that the restoration filter at the k-th iteration step is now given by

$$H_k(u,v) = \beta \sum_{\ell=0}^{k-1} (1 - \beta |D(u,v)|^2)^\ell D^*(u,v)$$

$$= \beta \frac{1 - (1 - \beta |D(u,v)|^2)^k}{1 - (1 - \beta |D(u,v)|^2)} D^*(u,v). \tag{15.19}$$

Therefore, the sufficient condition for convergence, corresponding to condition (15.9), becomes

$$|1 - \beta |D(u,v)|^2| < 1, \quad \text{or} \quad 0 < \beta < \frac{2}{\max_{u,v} |D(u,v)|^2}. \tag{15.20}$$

In this case

$$\lim_{k \to \infty} H_k(u,v) = \begin{cases} \dfrac{1}{D(u,v)}, & D(u,v) \neq 0 \\[2mm] 0, & \text{otherwise.} \end{cases} \tag{15.21}$$

15.3.5 Experimental Results

In this section the performance of the iterative image restoration algorithms presented so far is demonstrated experimentally. We use a relatively simple degradation model in order to clearly analyze the behavior of the restoration filters. The degradation is due to 1D horizontal motion between the camera and the scene, due, for example, to camera panning or fast object motion. The impulse response of the degradation system is given by

$$d(n_1, n_2) = \begin{cases} \dfrac{1}{L}, & -\dfrac{L-1}{2} \leq n_1 \leq \dfrac{L-1}{2}, & L \text{ odd}, \ n_2 = 0 \\[2mm] \dfrac{1}{L}, & -\dfrac{L}{2} + 1 \leq n_1 \leq \dfrac{L}{2}, & L \text{ even}, \ n_2 = 0 \\[2mm] 0, & \text{otherwise.} \end{cases} \tag{15.22}$$

The blurred signal-to-noise ratio (BSNR) is typically used in the restoration community to measure the degree of the degradation (blur plus additive noise). This figure is given by

$$\text{BSNR} = 10 \log_{10} \frac{\sigma_{\text{Df}}^2}{\sigma_{\text{v}}^2}, \tag{15.23}$$

where σ_{Df}^2 and σ_{v}^2 are, respectively, the variance of the blurred image and the additive noise.

For the purpose of objectively testing the performance of image restoration algorithms, the improvement in SNR (ISNR) is often used. This metric using the restored image at the k-th iteration step is given by

$$\text{ISNR} = 10\log_{10}\frac{||\mathbf{f} - \mathbf{g}||^2}{||\mathbf{f} - \mathbf{f}_k||^2}. \tag{15.24}$$

Obviously, this metric can only be used for simulation cases when the original image is available. While mean-squared error (MSE) metrics such as ISNR do not always reflect the visual quality of the restored image, they serve to provide an objective standard by which to compare different techniques. However, in all cases presented here, it is important to consider the behavior of the various algorithms from the viewpoint of ringing and noise amplification, which can be a key indicator of improvement in quality for subjective comparisons of restoration algorithms.

In Fig. 15.2(a) the image blurred by the 1D motion blur of extent 8 pixels ($L = 8$ in Eq. (15.22)) is shown, along with $|D(u,0)|$, a slice of the magnitude of the 256×256 point DFT of $d(n_1, n_2)$ (notice that all slices of the DFT are the same, independently of

(a)

(b)

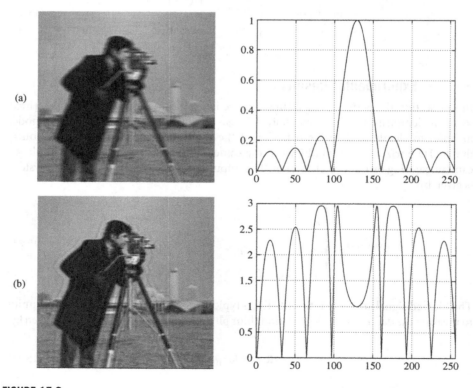

FIGURE 15.2

Continued

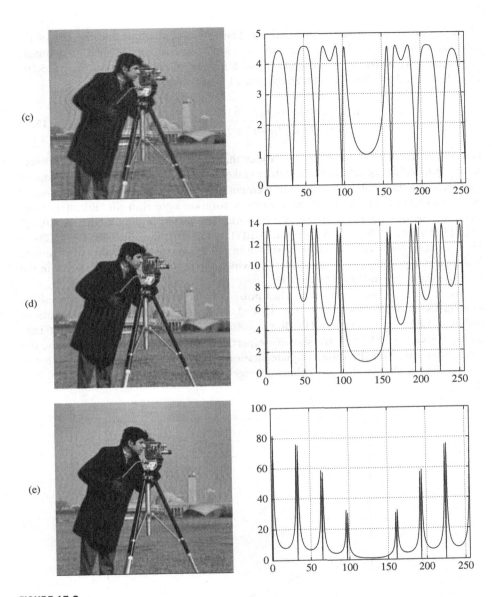

FIGURE 15.2

(a) Blurred image by an 1D motion blur over 8 pixels and the corresponding magnitude of the frequency response of the degradation system; (b)–(d): images restored by iteration (15.18), after 20 iterations (ISNR = 4.03 dB), 50 iterations (ISNR = 6.22 dB) and at convergence after 465 iterations (ISNR = 11.58 dB), and the corresponding magnitude of $H_k(u,0)$ in (15.19); (e) image restored by the direct implementation of the generalized inverse filter in (15.16) (ISNR = 15.50 dB), and the corresponding magnitude of the frequency response of the restoration filter.

the value of v). No noise has been added. The extent of the blur and the size of the DFT were chosen in such a way that exact zeros exist in $D(u,v)$. The next three images represent the restored images using (15.18) with $\beta = 1.0$, along with $|H_k(u,0)|$ in (15.19), after 20, 50, and 465 iterations (at convergence). The criterion

$$\frac{\sum_{n_1,n_2}(f_{k+1}(n_1,n_2) - f_k(n_1,n_2))^2}{\sum_{n_1,n_2}(f_k(n_1,n_2))^2} \leq 10^{-8} \tag{15.25}$$

is used for terminating the iteration. Notice that (15.5) is not guaranteed to converge for this particular degradation since $D(u,v)$ takes negative values. The restored image of Fig. 15.2(e) is the result of the direct implementation of the pseudo-inverse filter, which can be thought of as the result of the iterative restoration algorithm after infinitely many iterations assuming infinite precision arithmetic. The corresponding ISNRs are 4.03 dB (Fig. 15.2(b)), 6.22 dB (Fig. 15.2(c)), 11.58 dB (Fig. 15.2(d)), and 15.50 dB (Fig. 15.2(e)). Finally, the normalized residual error shown in (15.25) versus the number of iterations is shown in Fig. 15.3. The iteration steps at which the restored images are shown in the previous figure are indicated by circles.

We repeat the same experiment when noise is added to the blurred image, resulting in a BSNR of 20 dB, as shown in Fig. 15.4(a). The restored images after 20 iterations (ISNR = 1.83 dB), 50 iterations (ISNR = −0.40 dB), and at convergence after 1376 iterations (ISNR = −9.06 dB) are shown, respectively, in Figs. 15.4(b)–(d). Finally, the restoration based on the direct implementation of the pseudo-inverse filter is shown in Fig. 15.4(e). The iterative algorithm converges slower in this case.

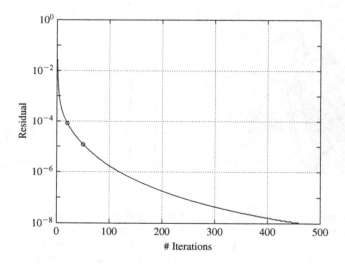

FIGURE 15.3

Normalized residual error as a function of the number of iterations.

FIGURE 15.4

(a) Noisy-blurred image; 1D motion blur over 8 pixels, BSNR = 20 dB; (b)–(d): images restored by iteration (15.18), after 20 iterations (ISNR = 1.83 dB), 50 iterations (ISNR = −0.30 dB), and at convergence after 1376 iterations (ISNR = −9.06 dB); (e) image restored by the direct implementation of the generalized inverse filter in (15.16) (ISNR = −12.09 dB).

What becomes evident from these experiments is that:

- As expected, for the noise-free case, the visual quality as well as the objective quality in terms of ISNR of the restored images increases as the number of iterations increases.

- For the noise-free case the inverse filter outperforms the iterative restoration filter. Based on this experiment there is no reason to implement this particular filter iteratively, except possibly for computational reasons.

- For the noisy-blurred image the noise is amplified and the ISNR decreases as the number of iterations increases. Noise completely dominates the image restored by the pseudo-inverse filter. In this case, the iterative implementation of the restoration filter offers the advantage that the number of iterations can be used to control the amplification of the noise, which represents a form of regularization. The restored image, for example, after 50 iterations (Fig. 15.4(c)) represents a reasonable restoration.

■ The iteratively restored image exhibits noticeable *ringing artifacts*, which will be further analyzed below. Such artifacts can be masked by noise, as demonstrated, for example, with the image in Fig. 15.4(d).

15.3.5.1 *Ringing Artifacts*

Let us compare the magnitudes of the frequency response of the restoration filter after 465 iterations (Fig. 15.2(d)) and the inverse filter (Fig. 15.2(e)). First of all, it is clear that the existence of spectral zeros in $D(u,v)$ does not cause any difficulty in the determination of the restoration filter in both cases, since the restoration filter is also zero at these frequencies. The main difference is that the values of $|H(u,v)|$, the magnitude of the frequency response of the inverse filter, at frequencies close to the zeroes of $D(u,v)$ are considerably larger than the corresponding values of $|H_k(u,v)|$. This is because the values of $H_k(u,v)$ are approximated by a series according to (15.19). The important term in this series is $(1 - \beta|D(u,v)|^2)$, since it determines whether the iteration converges or not (sufficient condition). Clearly this term for values of $D(u,v)$ close to zero is close to one, and therefore, it approaches zero much slower when raised to the power of k, the number of iterations, than the terms for which $D(u,v)$ assumes larger values and the term $(1 - \beta|D(u,v)|^2)$ is close to zero. This means that each frequency component is restored independently and with different convergence rates. Clearly the larger the values of β the faster the convergence.

Let us denote by $h(n_1,n_2)$ the impulse response of the restoration filter and define

$$h_{\text{all}}(n_1,n_2) = d(n_1,n_2) * h(n_1,n_2). \tag{15.26}$$

Ideally, $h_{\text{all}}(n_1,n_2)$ should be equal to an impulse, or its DFT $H_{\text{all}}(u,v)$ should be a constant, that is, the restoration filter is precisely undoing what the degradation system did. Due to the spectral zeros, however, in $D(u,v)$, $H_{\text{all}}(u,v)$ deviates from a constant. For the particular example under consideration, $|H_{\text{all}}(u,0)|$ is shown in Figs. 15.5(a) and 15.5(c), for the inverse filter and the iteratively implemented inverse filter by (15.18), respectively. In Figs. 15.5(b) and 15.5(d) the corresponding impulse responses are shown. Due to the periodic zeros of $D(u,v)$ in this particular case, $h_{\text{all}}(n_1,n_2)$ consists of the sum of an impulse and an impulse train (of period 8 samples). The deviation from a constant or an impulse is greater with the iterative restoration filter than with the direct inverse filter.

Now, in the absence of noise the restored image, $\hat{f}(n_1,n_2)$ is given by

$$\hat{f}(n_1,n_2) = h_{\text{all}}(n_1,n_2) * f(n_1,n_2). \tag{15.27}$$

Clearly due to the shape of $h_{\text{all}}(n_1,n_2)$ shown in Figs. 15.5(b) and 15.5(d) (only $h_{\text{all}}(n_1,0)$ is shown, since it is zero for the rest of the values of n_2), the existence of the periodic train of impulses gives rise to ringing. In the case of the inverse filter (Fig. 15.5(b)) the impulses of the train are small in magnitude and therefore ringing is not visible. In the case of the iterative filter, however, the few impulses close to zero have larger amplitude and therefore ringing is noticeable in this case.

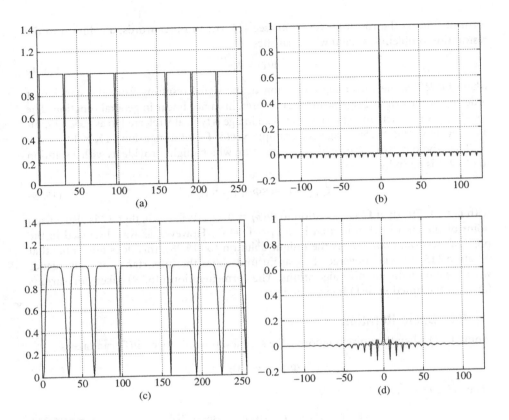

FIGURE 15.5

(a) $|H_{all}(u,0)|$ for direct implementation of the inverse filter; (c) $|H_{all}(u,0)|$ for the iterative implementation of the inverse filter; (b),(d): $h_{all}(n,0)$ corresponding to Figs. 15.5(a) and 15.5(c).

15.4 MATRIX-VECTOR FORMULATION

The presentation so far has followed a rather simple and intuitive path. It hopefully demonstrated some of the issues involved in developing and implementing an iterative algorithm. In this section we present the matrix-vector formulation of the degradation process and the restoration iteration. More general results are therefore obtained, since now the degradation can be spatially varying, while the restoration filter may be spatially varying as well, but even nonlinear. The degradation actually can be nonlinear as well (of course it is not represented by a matrix in this case), but we do not focus on this case, although most of the iterative algorithms discussed below would be applicable.

What became clear from the previous sections is that in applying the successive approximations iteration, the restoration problem to be solved is brought first into the form of finding the root of a function (see (15.3)). In other words, a solution to the restoration problem is sought which satisfies

$$\Phi(\mathbf{f}) = 0, \tag{15.28}$$

where $\mathbf{f} \in \mathcal{R}^N$ is the vector representation of the signal resulting from the stacking or ordering of the original signal, and $\Phi(\mathbf{f})$ represents a nonlinear in general function. The row-by-row from left-to-right stacking of an image is typically referred to as *lexicographic ordering*. For a 256×256 image, for example, vector $\mathbf{f}$ is of dimension 64 K $\times$ 1.

   Then the successive approximations iteration which might provide us with a solution to (15.28) is given by

$$\mathbf{f}_{k+1} = \mathbf{f}_k + \beta\Phi(\mathbf{f}_k) = \Psi(\mathbf{f}_k), \tag{15.29}$$

with $\mathbf{f}_0 = 0$. Clearly if $\mathbf{f}^*$ is a solution to $\Phi(\mathbf{f}) = 0$, i.e., $\Phi(\mathbf{f}^*) = 0$, then $\mathbf{f}^*$ is also a fixed point of the above iteration, that is, $\mathbf{f}_{k+1} = \mathbf{f}_k = \mathbf{f}^*$. However, as was discussed in the previous section, even if $\mathbf{f}^*$ is the unique solution to (15.28), this does not imply that iteration (15.29) will converge. This again underlines the importance of convergence when dealing with iterative algorithms. The form iteration (15.29) taken for various forms of the function $\Phi(\mathbf{f})$ is examined next.

15.4.1 Basic Iteration

From (15.1) when the noise is ignored, the simplest possible form $\Phi(\mathbf{f})$ can take is

$$\Phi(\mathbf{f}) = \mathbf{g} - \mathbf{Df}. \tag{15.30}$$

Then (15.29) becomes

$$\mathbf{f}_{k+1} = \beta\mathbf{g} + (\mathbf{I} - \beta\mathbf{D})\mathbf{f}_k = \beta\mathbf{g} + \mathbf{G}_1\mathbf{f}_k, \tag{15.31}$$

where $\mathbf{I}$ is the identity operator.

15.4.2 Least-Squares Iteration

According to the least-squares approach, a solution to (15.1) is sought by minimizing

$$M(\mathbf{x}) = \|\mathbf{g} - \mathbf{Df}\|^2. \tag{15.32}$$

A necessary condition for $M(\mathbf{f})$ to have a minimum is that its gradient with respect to $\mathbf{f}$ is equal to zero. That is, in this case

$$\Phi(\mathbf{f}) = \nabla_{\mathbf{f}} M(\mathbf{f}) = \mathbf{D}^T(\mathbf{g} - \mathbf{Df}) = 0, \tag{15.33}$$

where T denotes the transpose of a matrix or vector. Application of iteration (15.29) then results in

$$\mathbf{f}_{k+1} = \beta\mathbf{D}^T\mathbf{g} + (\mathbf{I} - \beta\mathbf{D}^T\mathbf{D})\mathbf{f}_k = \beta\mathbf{D}^T\mathbf{g} + \mathbf{G}_2\mathbf{f}_k = \mathbf{T}_2\mathbf{f}_k. \tag{15.34}$$

   The matrix-vector representation of an iteration does not necessarily determine the way the iteration is implemented. In other words, the pointwise version of the iteration may be more efficient from the implementation point of view than the matrix-vector form of the iteration. Now when (15.2) is used to form the matrix-vector equation $\mathbf{g} = \mathbf{Df}$, matrix $\mathbf{D}$ is a block-circulant matrix [1]. A square matrix is circulant when a circular shift of one row produces the next row, and the circular shift of the last row produces the first row. A square matrix is block-circulant when it consists of circular submatrices, which when circularly shifted produce the next row of circulant matrices. This implies that the singular values of $\mathbf{D}$ are the DFT values of $d(n_1, n_2)$, and the eigenvectors are the complex exponential basis functions of the DFT. Iterations (15.31) and (15.34) can, therefore, be written in the discrete frequency domain, and they become identical to iteration (15.6) and the frequency domain version of iteration (15.18), respectively [12].

15.4.3 Constrained Least-Squares Iteration

The image restoration problem is an ill-posed problem, which means that matrix $\mathbf{D}$ is ill-conditioned. A regularization method replaces an ill-posed problem by a well-posed problem, whose solution is an acceptable approximation to the solution of the ill-posed problem [5]. Most regularization approaches transform the original inverse problem into a constrained optimization problem. That is, a functional needs to be optimized with respect to the original image, and possibly other parameters. By using the necessary condition for optimality, the gradient of the functional with respect to the original image is set equal to zero, therefore determining the mathematical form of $\Phi(\mathbf{f})$. The successive approximations iteration becomes in this case a gradient method with a fixed step (determined by β).

As an example, a restored image is sought as the result of the minimization of [14]

$$\|\mathbf{Cf}\|^2 \tag{15.35}$$

subject to the constraint that

$$\|\mathbf{g} - \mathbf{Df}\|^2 \leq \epsilon^2. \tag{15.36}$$

Operator $\mathbf{C}$ is a highpass operator. The meaning then of the minimization of $\|\mathbf{Cf}\|^2$ is to constrain the high-frequency energy of the restored image, therefore requiring that the restored image is smooth. On the other hand, by enforcing inequality (15.36) the fidelity to the data is preserved.

Following the Lagrangian approach which transforms the constrained optimization problem into an unconstrained one, the following functional is minimized:

$$M(\alpha, \mathbf{f}) = \|\mathbf{Df} - \mathbf{g}\|^2 + \alpha \|\mathbf{Cf}\|^2. \tag{15.37}$$

The necessary condition for a minimum is that the gradient of $M(\alpha, \mathbf{f})$ is equal to zero. That is, in this case

$$\Phi(\mathbf{f}) = \nabla_{\mathbf{f}} M(\alpha, \mathbf{f}) = (\mathbf{D}^T \mathbf{D} + \alpha \mathbf{C}^T \mathbf{C})\mathbf{f} - \mathbf{D}^T \mathbf{g} \tag{15.38}$$

is used in iteration (15.29). The determination of the value of the regularization parameter α is a critical issue in regularized restoration, since it controls the tradeoff between fidelity to the data and smoothness of the solution, and therefore the quality of the restored image. A number of approaches for determining its value are presented and compared in [15].

Since the restoration filter resulting from (15.38) is widely used it is worth looking further into its properties. When the degradation matrices $\mathbf{D}$ and $\mathbf{C}$ are block-circulant (15.38) the resulting successive approximations iteration can be written in the discrete frequency domain. The iteration takes the form

$$F_{k+1}(u,v) = \beta D^*(u,v)G(u,v) + (1 - \beta(|D(u,v)|^2 + \alpha|C(u,v)|^2))F_k(u,v), \tag{15.39}$$

where $C(u,v)$ represents the 2D DFT of the impulse response of a highpass filter, such as the 2D Laplacian. Following steps similar to the ones presented in Section 15.3.3, it is straightforward to verify that in this case the restoration filter at the k-th iteration step is given by

$$H_k(u,v) = \beta \sum_{\ell=0}^{k-1}(1 - \beta(|D(u,v)|^2 + \alpha|C(u,v)|^2))^\ell D^*(u,v). \tag{15.40}$$

Clearly if

$$|1 - \beta(|D(u,v)|^2 + \alpha|C(u,v)|^2)| < 1 \tag{15.41}$$

then

$$\lim_{k \to \infty} H_k(u,v) = \lim_{k \to \infty} \beta \frac{1 - (1 - \beta(|D(u,v)|^2 + \alpha|C(u,v)|^2))^k}{1 - (1 - \beta(|D(u,v)|^2 + \alpha|C(u,v)|^2))}$$

$$D^*(u,v) = \frac{D^*(u,v)}{|D(u,v)|^2 + \alpha|C(u,v)|^2}. \tag{15.42}$$

Notice that condition (15.41) is not satisfied at the frequencies for which $H_d(u,v) = |D(u,v)|^2 + \alpha|C(u,v)|^2 = 0$. It is therefore now not the zeros of the degradation matrix which need to be considered, but the zeros of the regularized matrix, with DFT values $H_d(u,v)$. Clearly if $H_d(u,v)$ is zero at certain frequencies, this means that both $D(u,v)$ and $C(u,v)$ are zero at these frequencies. This demonstrates the purpose of regularization, which is to remove the zeros of $D(u,v)$ without altering the rest of its values, or in general to make the matrix $\mathbf{D}^T\mathbf{D} + \alpha\mathbf{C}^T\mathbf{C}$ better conditioned than the matrix $\mathbf{D}^T\mathbf{D}$.

For the frequencies at which $H_d(u,v) = 0$

$$\lim_{k \to \infty} H_k(u,v) = \lim_{k \to \infty} k \cdot \beta \cdot D^*(u,v) = 0, \tag{15.43}$$

since $D^*(u,v) = 0$.

15.4.3.1 *Experimental Results*

The noisy and blurred image of Fig. 15.4(a) (1D motion blur over 8 pixels, BSNR = 20 dB) is now restored using iteration (15.39), with $\alpha = 0.01$, $\beta = 1.0$, and $\mathbf{C}$ the 2D Laplacian operator. It is mentioned here that the regularization parameter is chosen to be equal to $\frac{\sigma_y^2}{\sigma_{Df}^2}$, as determined by a set theoretic restoration approach presented in [16]. The restored images after 20 iterations (ISNR = 2.12 dB), 50 iterations (ISNR = 0.98 dB), and at convergence after 330 iterations (ISNR = −1.01 dB) with the corresponding $|H_k(u,v)|$ in (15.40), are shown respectively in Figs. 15.6(a)–(c). In Fig. 15.6(d) the restored image (ISNR = −1.64 dB) by the direct implementation of the constrained least-squares filter in (15.42) is shown, along with the magnitude of the frequency response of the restoration filter. It is clear now by comparing the restoration filters of Figs. 15.2(d) and 15.6(c) and 15.2(e) and 15.6(d), that the high frequencies have been suppressed, due to regularization, that is the addition in the denominator of the filter of the term $\alpha|C(u,v)|^2$. Due to the iterative approximation of the constrained least-squares filter, however, the two filters shown in Figs. 15.6(c) and 15.6(d) differ primarily in the vicinity of the low-frequency zeros of $D(u,v)$. Ringing is still present, as it can be primarily seen in Figs. 15.6(a) and 15.6(b), although is not as visible in Figs. 15.6(c) and 15.6(d). Due to regularization the results in Figs. 15.6(c) and 15.6(d) are preferred over the corresponding results with no regularization ($\alpha = 0.0$), shown in Figs. 15.4(d) and 15.4(e).

The value of the regularization parameter is very critical for the quality of the restored image. The restored images with three different values of the regularization

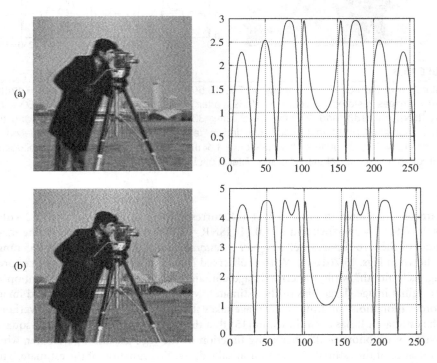

FIGURE 15.6

Continued

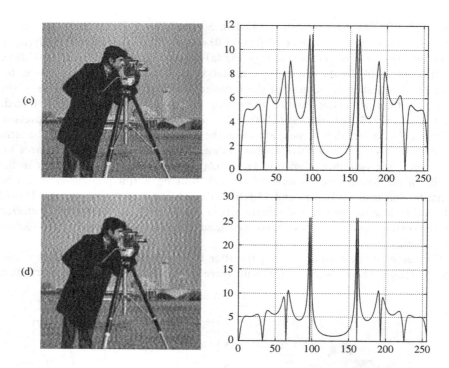

FIGURE 15.6

Restoration of the noisy-blurred image in Fig. 15.5(a) (motion over 8 pixels, $BSNR = 20\,dB$); (a)–(c): images restored by iteration (15.39), after 20 iterations ($ISNR = 2.12\,dB$), 50 iterations ($ISNR = 0.98\,dB$) and at convergence after 330 iterations ($ISNR = -1.01\,dB$), and the corresponding $|H_k(u,0)|$ in (15.40); (d): image restored by the direct implementation of the constrained least-squares filter ($ISNR = -1.64\,dB$), and the corresponding magnitude of the frequency response of the restoration filter (Eq. (15.42)).

parameter are shown in Figs. 15.7(a)–(c), corresponding to $\alpha = 1.0$ ($ISNR = 2.4\,dB$), $\alpha = 0.1$ ($ISNR = 2.96\,dB$), and $\alpha = 0.01$ ($ISNR = -1.80\,dB$). The corresponding magnitudes of the error images, i.e., $|original - restored|$, scaled linearly to the 32–255 range are shown in Figs. 15.7(d)–(f). What is observed is that for large values of α the restored image is "smooth" while the error image contains the high-frequency information of the original image (large bias of the estimate), while as α decreases the restored image becomes more noisy and the error image takes the appearance of noise (large variance of the estimate). It has been shown in [15] that the bias of the constrained least-squares estimate is a monotonically increasing function of the regularization parameter, while the variance of the estimate is a monotonically decreasing function of the estimate. This implies that the MSE of the estimate, the sum of the bias and the variance, has a unique minimum for a specific value of α.

FIGURE 15.7

Direct constrained least-squares restorations of the noisy-blurred image in Fig. 15.5(a) (motion over 8 pixels, BSNR = 20 dB) with α equal to: (a) 1; (b) 0.1; (c) 0.01; (d)–(f): corresponding |original – restored| linearly mapped to the range [32, 255].

15.4.4 Spatially Adaptive Iteration

Spatially adaptive image restoration is the next natural step in improving the quality of the restored images. There are various ways to argue the introduction of spatial adaptivity, the most commonly used ones being the nonhomogeneity or nonstationarity of the image field and the properties of the human visual system. In either case, the functional to be minimized takes the form [11, 12]

$$M(\alpha, \mathbf{f}) = \|\mathbf{Df} - \mathbf{g}\|^2_{\mathbf{W}_1} + \alpha \|\mathbf{Cf}\|^2_{\mathbf{W}_2}, \tag{15.44}$$

in which case

$$\Phi(\mathbf{f}) = \nabla_{\mathbf{f}} M(\alpha, \mathbf{f}) = (\mathbf{D}^T \mathbf{W}_1^T \mathbf{W}_1 \mathbf{D} + \alpha \mathbf{C}^T \mathbf{W}_2^T \mathbf{W}_2 \mathbf{C})\mathbf{f} - \mathbf{D}^T \mathbf{W}_1 \mathbf{g}. \tag{15.45}$$

The choice of the diagonal weighting matrices $\mathbf{W}_1$ and $\mathbf{W}_2$ can be justified in various ways. In [12] both matrices are determined by the diagonal noise visibility matrix $\mathbf{V}$ [17]. That is, $\mathbf{W}_1 = \mathbf{V}^T\mathbf{V}$ and $\mathbf{W}_2 = \mathbf{I} - \mathbf{V}^T\mathbf{V}$. The entries of $\mathbf{V}$ take values between 0 and 1. They are equal to 0 at the edges (noise is not visible), equal to 1 at the flat regions (noise is visible) and take values in between at the regions with moderate spatial activity.

15.4.4.1 *Experimental Results*

The resulting successive approximations iteration from the use of $\Phi(\mathbf{f})$ in (15.45) has been tested with the noisy and blurred image we have been using so far in our experiments, which is shown in Fig. 15.4(a). It should be emphasized here that although matrices $\mathbf{D}$ and $\mathbf{C}$ are block-circulant, the iteration cannot be implemented in the discrete frequency domain, since the weight matrices, $\mathbf{W}_1$ and $\mathbf{W}_2$, are diagonal, but not circulant. Therefore, the iterative algorithm is implemented exclusively in the spatial domain, or by switching between the frequency domain (where the convolutions are implemented) and the spatial domain (where the weighting takes place). Clearly, from an implementation point of view the use of iterative algorithms offers a distinct advantage in this particular case.

The iteratively restored image with $\mathbf{W}_1 = 1 - \mathbf{W}_2$, $\alpha = 0.01$, and $\beta = 0.1$, is shown in Fig. 15.8(a), at convergence after 381 iterations and ISNR $= 0.61$ dB. The entries of the diagonal matrix $\mathbf{W}_2$, denoted by $w_2(i)$, are computed according to

$$w_2(i) = \frac{1}{\theta\sigma^2(i) + 1},$$

(15.46)

where $\sigma^2(i)$ is the local variance at the ordered i-th pixel location, and θ a tuning parameter. The resulting values of $w_2(i)$ are linearly mapped into the $[0, 1]$ range. These weights computed from the degraded image are shown in Fig. 15.8(c), linearly mapped to the $[32, 255]$ range, using a 3×3 window to find the local variance and $\theta = 0.001$. The image restored by the nonadaptive algorithm, that is, $\mathbf{W}_1 = \mathbf{W}_2 = \mathbf{I}$ and the rest of the parameters the same, is shown in Fig. 15.8(b) (ISNR $= -0.20$ dB). The absolute value of the difference between the images linearly mapped in the $[32, 255]$ range is shown in Fig. 15.8(d). It is clear that the two algorithms differ primarily at the vicinity of edges, where the smoothing is downweighted or disabled with the adaptive algorithm. Spatially adaptive algorithms in general can greatly improve the restoration results, since they can adapt to the local characteristics of each image.

15.5 USE OF CONSTRAINTS

Iterative signal restoration algorithms regained popularity in the 1970s due to the realization that improved solutions can be obtained by incorporating prior knowledge about the solution into the restoration process. For example, we may know in advance that $\mathbf{f}$ is bandlimited or spacelimited, or we may know on physical grounds that $\mathbf{f}$ can only have nonnegative values. A convenient way of expressing such prior knowledge is to define a constraint operator $\mathcal{C}$, such that

$$\mathbf{f} = \mathcal{C}\mathbf{f},$$

(15.47)

(a)   (b)

(c)   (d)

FIGURE 15.8

Restoration of the noisy-blurred image in Fig. 15.5(a) (motion over 8 pixels, $\mathrm{BSNR} = 20\,\mathrm{dB}$), using (a) the adaptive algorithm of (15.45); (b) the nonadaptive algorithm of iteration (15.39); (c) values of the weight matrix in Eq. (15.46); (d) amplitude of the difference between images (a) and (b) linearly mapped to the range [32, 255].

if and only if $\mathbf{f}$ satisfies the constraint. In general, $\mathcal{C}$ represents the concatenation of constraint operators. With the use of constraints, iteration (15.29) becomes [9]

$$\mathbf{f}_0 = 0,$$
$$\tilde{\mathbf{f}}_k = \mathcal{C}\mathbf{f}_k,$$
$$\mathbf{f}_{k+1} = \Psi(\tilde{\mathbf{f}}_k). \tag{15.48}$$

As already mentioned, a number of recovery problems, such as the bandlimited extrapolation problem, and the reconstruction from phase or magnitude problem, can be solved with the use of algorithms of the form (15.48), by appropriately describing the distortion and constraint operators [9].

The *contraction mapping theorem* [8] usually serves as a basis for establishing convergence of iterative algorithms. Sufficient conditions for the convergence of the algorithms presented in Section 15.4 are presented in [12]. Such conditions become identical to the ones derived in Section 15.3 when all matrices involved are block-circulant. When constraints are used, the sufficient condition for convergence of the iteration is that at least one of the operators $\mathcal{C}$ and Ψ is contractive while the other is nonexpansive. Usually it is harder to prove convergence and determine the convergence rate of the constrained iterative algorithm, taking also into account that some of the constraint operators are nonlinear, such as the positivity constraint operator.

15.5.1 Experimental Results

We demonstrate the effectiveness of the positivity constraint with the use of a simple example. An 1D impulsive signal is shown in Fig. 15.9(a). Its degraded version by

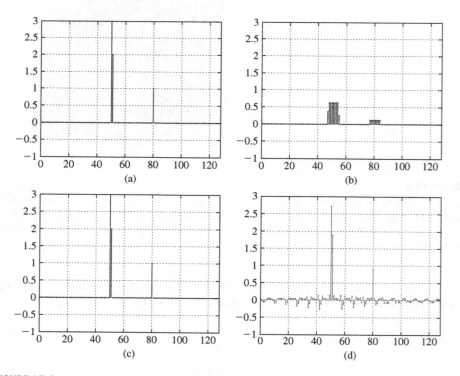

FIGURE 15.9

(a) Original signal; (b) blurred signal by motion blur over 8 samples; signals restored by iteration (15.18); (c) with positivity constraint; (d) without positivity constraint.

a motion blur over 8 samples is shown in Fig. 15.9(b). The blurred signal is restored by iteration (15.18) ($\beta = 1.0$) with the use of the positivity constraint (Fig. 15.9(c), 370 iterations, ISNR = 41.35), and without the use of the positivity constraint (Fig. 15.9(d), 543 iterations, ISNR = 11.05). The application of the positivity constraint, which represents a nonexpansive mapping, simply sets to zero all negative values of the signal. Clearly a considerably better restoration is represented by Fig. 15.9(c).

15.6 ADDITIONAL CONSIDERATIONS

In the previous sections we dealt exclusively with the image restoration problem, as described by Eq. (15.1). As was mentioned in the introduction, there is a plethora of inverse problems, i.e., problems described by Eq. (15.1), for which the iterative algorithms presented so far can be applied. Inverse problems are representative examples of more general recovery problems, i.e., problems for which information that is lost (due, for example, to the imperfections of the imaging system or the transmission medium, or the specific processing the signal is undergoing, such as compression), is attempted to be recovered. A critical step in solving any such problem is the modeling of the signals and systems involved, or in other words, the derivation of the degradation model. After this is accomplished the solution approach needs to be decided (of course these two steps do not need to be independent). In this chapter we dealt primarily with the image restoration problem under a deterministic formulation and a successive approximations based iterative solution approach. In the following four subsections we describe some additional forms the successive approximations iteration can take, a stochastic modeling of the restoration problem which results in successive approximations type of iterations, the blind image deconvolution problem, and finally additional recent image recovery applications.

15.6.1 Other Forms of the Iterative Algorithm

The basic iteration presented in previous sections can be extended in a number of ways. One such way is to utilize the partially restored image at each iteration step in evaluating unknown problem parameters or refining our prior knowledge about the original image. A critical such parameter which directly controls the quality of the restoration results, as was experimentally demonstrated in Fig. 15.8, is the regularization parameter α in Eq. (15.37). As was already mentioned in Section 15.4.3, a number of approaches have appeared in the literature for the evaluation of α [15]. It depends on the value of $\|\mathbf{g} - \mathbf{Df}\|^2$ or its upper bound ϵ in Eq. (15.36), but also on the value of $\|\mathbf{Cf}\|^2$ or an upper bound of it, or in other words on the value of $\mathbf{f}$. This dependency of α on the unknown original image $\mathbf{f}$ is expressed explicitly in [18], by rewriting the functional to be minimized in Eq. (15.37) as

$$M(\alpha(\mathbf{f}),\mathbf{f}) = \|\mathbf{g} - \mathbf{Df}\|^2 + \alpha(\mathbf{f})\|\mathbf{Cf}\|^2. \tag{15.49}$$

The desirable properties of $\alpha(\mathbf{f})$ and various functional forms it can take are investigated in detail in [18]. One such choice is given by

$$\alpha(\mathbf{f}) = \frac{\|\mathbf{g} - \mathbf{D}\mathbf{f}\|^2}{(1/\gamma) - \|\mathbf{C}\mathbf{f}\|^2}, \tag{15.50}$$

with γ constrained so that the denominator in Eq. (15.50) is positive. The successive approximations iteration in this case then becomes

$$\mathbf{f}_{k+1} = \mathbf{f}_k + \beta[\mathbf{D}^T\mathbf{g} - (\mathbf{D}^T\mathbf{D} + \alpha(\mathbf{f}_k)\mathbf{C}^T\mathbf{C})\mathbf{f}_k]. \tag{15.51}$$

Sufficient conditions for the convergence of iteration (15.51) are derived in [18] in terms of the parameter γ, and also conditions which guarantee $M(\alpha(\mathbf{f}),\mathbf{f})$ to be convex (the relaxation parameter β can be set equal to 1 since it can be combined with the parameter γ). Iteration (15.51) represents a major improvement toward the solution of the restoration problem because (i) no prior knowledge, such as knowledge of the noise variance, is required for the determination of the regularization parameter, as instead such information is extracted from the partially restored image; and (ii) the determination of the regularization parameter does not constitute a separate, typically iterative step, as it is performed simultaneously with the restoration of the image. The performance of iteration (15.51) is studied in detail in [18] for various forms of the functional $\alpha(\mathbf{f})$ and various initial conditions.

This framework of extracting information required by the restoration process at each iteration step from the partially restored image has also been applied to the evaluation of the weights $\mathbf{W}_1$ and $\mathbf{W}_2$ in iteration (15.45) [19] and in deriving algorithms which use a different iteration-dependent regularization parameter for each discrete frequency component [20].

Additional extensions of the basic form of the successive approximations algorithm are represented by algorithms with higher rates of convergence [21, 22], algorithms with a relaxation parameter β which depends on the iteration step (steepest descent and conjugate gradient algorithms are examples of this), algorithms which depend on more than one previous restoration steps (multistep algorithms [23]), and algorithms which utilize the number of iterations as a means of regularizing the solution.

15.6.2 Hierarchical Bayesian Image Restoration

In the presentation so far we have assumed that the degradation and the images in Eq. (15.1) are deterministic and the noise only represents a stochastic signal. A different approach towards the derivation of the degradation model and a restoration solution is represented by the Bayesian paradigm. According to it, knowledge about the structural form of the noise and the structural behavior of the reconstructed image is used in forming respectively $p(\mathbf{g}|\mathbf{f},\tau)$ and $p(\mathbf{f}|\delta)$, where $p(\cdot|\cdot)$ denotes a conditional probability density function (pdf). For example, the following conditional pdf is typically used to describe the structural form of the noise:

$$p(\mathbf{g}|\mathbf{f},\tau) = \frac{1}{Z_{\text{noise}}(\tau)}\left\{\exp\left[-\frac{1}{2}\tau\|\mathbf{g} - \mathbf{D}\mathbf{f}\|^2\right]\right\}, \tag{15.52}$$

where $Z_{noise}(\tau) = (2\pi/\tau)^{N/2}$, with N, as mentioned earlier, the dimension of the vectors $\mathbf{f}$ and $\mathbf{g}$. Smoothness constraints on the original image can be incorporated under the form of

$$p(\mathbf{f}|\delta) \propto \delta^{q/2} \exp\left\{-\frac{1}{2}\delta S(\mathbf{f})\right\}, \tag{15.53}$$

where $S(\mathbf{f})$ is a nonnegative quadratic form which usually corresponds to a conditional or simultaneous autoregressive model in the statistical community or to placing constraints on the first or second differences in the engineering community and q is the number of positive eigenvalues of S [24]. A form of $S(\mathbf{f})$ which has been used widely in the engineering community and also in this chapter is

$$S(\mathbf{f}) = \|\mathbf{Cf}\|^2,$$

with $\mathbf{C}$ the Laplacian operator.

The parameters δ and τ are typically referred to as hyperparameters. If they are known, according to the Bayesian paradigm, the image $\mathbf{f}_{(\delta,\tau)}$ is selected as the restored image, defined by

$$\mathbf{f}_{(\delta,\tau)} = \arg\{\max_{\mathbf{f}} p(\mathbf{f}|\delta)p(\mathbf{g}|\mathbf{f},\tau)\} = \arg\{\min_{\mathbf{f}}[\alpha S(\mathbf{f}) + \tau\|\mathbf{g} - \mathbf{Df}\|^2]\}. \tag{15.54}$$

If the hyperparameters are not known then they can be treated as random variables and the hierarchical Bayesian approach can be followed. It consists of two stages. In the first stage the conditional probabilities shown in Eqs. (15.52) and (15.53) are formed. In the second stage the hyperprior $p(\delta,\tau)$ is also formulated, resulting in the distribution $p(\delta,\tau,\mathbf{f},\mathbf{g})$. With the so-called evidence analysis, $p(\delta,\tau,\mathbf{f},\mathbf{g})$ is integrated over $\mathbf{f}$ to give the likelihood $p(\delta,\tau|\mathbf{g})$, which is then maximized over the hyperparameters. Alternatively, with the maximum *a posteriori* (MAP) analysis, $p(\delta,\tau,\mathbf{f},\mathbf{g})$ is integrated over δ and τ to obtain the true likelihood, which is then maximized over $\mathbf{f}$ to obtain the restored image.

Although in some cases it would be possible to establish relationships between the hyperpriors, the following model of the global probability is typically used:

$$p(\delta,\tau,\mathbf{f},\mathbf{g}) = p(\delta)p(\tau)p(\mathbf{f}|\delta)p(\mathbf{g}|\mathbf{f},\tau). \tag{15.55}$$

Flat or noninformative hyperpriors are used for $p(\delta)$ and $p(\tau)$ if no prior knowledge about the hyperpriors exists. If such knowledge exists, as an example, a gamma distribution can be used [24]. As expected, the form of these pdf impacts the subsequent calculations.

Clearly the hierarchical Bayesian analysis offers a methodical procedure to evaluate unknown parameters in the context of solving a recovery problem. A critical step in its application is the determination of $p(\delta)$ and $p(\tau)$ and the above-mentioned integration of $p(\delta,\tau,\mathbf{f},\mathbf{g})$ either over $\mathbf{f}$, or δ and τ. Both flat and gamma hyperpriors $p(\delta)$ and $p(\tau)$ have been considered in [24], utilizing both the evidence and MAP analyses. They resulted in iterative algorithms for the evaluation of δ, τ, and $\mathbf{f}$. The important connection between the hierarchical Bayesian approach and the iterative approach presented in Section 15.6.1 is that iteration (15.51) with $\alpha(\mathbf{f})$ given by Eq. (15.50) or any of the forms proposed in

[18, 20] can now be derived by the hierarchical Bayesian analysis with the appropriate choice of the required hyperpriors and the integration method. It should be made clear that the regularization parameter α is equal to the ratio (τ/δ). A related result has been obtained in [25] by deriving through a Bayesian analysis the same expressions for the iterative evaluation of the weight matrices $\mathbf{W}_1$ and $\mathbf{W}_2$ as in iteration (15.45) and Eq. (15.46). It is therefore significant that there is a precise interpretation of the framework briefly described in the previous section, based on the stochastic modeling of the signals and the unknown parameters.

15.6.3 Blind Deconvolution

Throughout this chapter, a fundamental assumption was that the exact form of the degradation system is known. This assumption is valid in certain applications where the degradation can be modeled accurately using the information about the technical design of the system, or can be obtained through experimental approaches (as was done, for example, with the Hubble Space Telescope). However, in many other applications the exact form of the degradation system may not be known. In such cases, it is also desired for the algorithm to provide an estimate of the unknown degradation system as well as the original image. The problem of estimating the unknown original image $\mathbf{f}$ and the degradation $\mathbf{D}$ from the observation $\mathbf{g}$ is referred to as *blind deconvolution*, when $\mathbf{D}$ represents a linear and space-invariant (LSI) system. Blind deconvolution is a much harder problem than image restoration due to the interdependency of the unknown parameters.

As in image restoration, in blind deconvolution certain constraints have to be utilized for both the impulse response of the degradation system and the original image to transform the problem into a well-posed one. These constraints can be incorporated, for example, in a regularization framework, or by using Bayesian modeling techniques, as described in the previous subsection. Common constraints used for the degradation system include nonnegativity, symmetry, and smoothness, among others. For instance, a common degradation introduced in astronomical imaging is atmospheric turbulence, which can be modeled using a smoothly changing impulse response, such as a Gaussian function. On the other hand, out-of-focus blur is not smoothly varying, but has abrupt transitions. These kinds of degradations are better modeled using total-variation regularization methods [26].

Blind deconvolution methods can be classified into two main categories based on the manner the unknowns are estimated. With *a priori* blur identification methods, the degradation system is estimated separately from the original image, and then this estimate is used in any image restoration method to estimate the original image. On the other hand, joint blind deconvolution methods estimate the original image and identify the blur simultaneously. The joint estimation is typically carried out using an alternating procedure, i.e., at each iteration the unknown image is estimated using the degradation estimate in the previous iteration, and vice versa.

Assuming that the original image is known, identifying the degradation system from the observed and original images (referred to as the *system identification*) is the dual problem of image restoration, and all methods presented in this chapter, including successive approximation iterations, can be applied for identifying the unknown degradation. Based on this observation in joint identification methods, the blind deconvolution problem can be solved by composing two coupled successive approximations iterations.

As an example, blind deconvolution can be formulated by the minimization of the following functional with respect to $\mathbf{f}$ and the impulse response $\mathbf{d}$ of the degradation system:

$$M(\alpha_1, \alpha_2, \mathbf{f}, \mathbf{d}) = \|\mathbf{D}\mathbf{f} - \mathbf{g}\|^2 + \alpha_1 C_1(\mathbf{f}) + \alpha_2 C_2(\mathbf{d}), \qquad (15.56)$$

where $C_1(\mathbf{f})$ and $C_2(\mathbf{d})$ denote operators on $\mathbf{f}$ and $\mathbf{d}$, respectively, imposing constraints on the unknowns. These operators, as was the case in Section 15.4.3, are generally used to constrain the high-frequency energy of the restored image and the impulse response of the degradation system. The necessary condition for a minimum is that the gradients of $M(\alpha_1, \alpha_2, \mathbf{f}, \mathbf{d})$ with respect to $\mathbf{f}$ and $\mathbf{d}$ are equal to zero. The minimum can be found by applying two nested successive approximations iterations as follows: Once an estimate of the image $\mathbf{f}^i$ is calculated at iteration i, this estimate is used in the successive approximation iteration to calculate an estimate $\mathbf{d}^i$ by finding the root of $\Phi_{\mathbf{d}}(\mathbf{d}, \mathbf{f}^i) = \nabla_{\mathbf{d}} M(\alpha_1, \alpha_2, \mathbf{f}^i, \mathbf{d})$ according to

$$\mathbf{d}_{k+1}^i = \mathbf{d}_k^i + \beta_{\mathbf{d}} \Phi_{\mathbf{d}}(\mathbf{d}_k^i, \mathbf{f}^i). \qquad (15.57)$$

At convergence the estimate $\mathbf{d}^i$ is used to find an estimate $\mathbf{f}^{i+1}$ by finding the root of $\Phi_{\mathbf{f}}(\mathbf{f}, \mathbf{d}^i) = \nabla_{\mathbf{f}} M(\alpha_1, \alpha_2, \mathbf{f}, \mathbf{d}^i)$ according to

$$\mathbf{f}_{k+1}^{i+1} = \mathbf{f}_k^{i+1} + \beta_{\mathbf{f}} \Phi_{\mathbf{f}}(\mathbf{f}_k^{i+1}, \mathbf{d}^i). \qquad (15.58)$$

Note that although the constraints and prior knowledge about the degradation system generally differ from those about the unknown image, the tools and analysis presented earlier in this chapter can be applied to the blind deconvolution problem as well.

Overall, blind deconvolution tackles a more difficult, but also a more frequently encountered problem than image restoration. Because of its general applicability to many different areas, there has been considerable activity in developing methods for blind deconvolution, and impressive (comparable to image restoration) results can be obtained by the state-of-the-art methods (see [7] for a review of the major approaches).

15.6.4 **Additional Applications**

In this chapter we have concentrated on the application of the successive approximations iteration to the image restoration problem. However, as mentioned multiple times already a number of recovery problems can find solutions with the use of a successive approximations iteration. Three important recovery problems which have been actively pursued in the last 10–15 years due to their theoretical challenge but also their commercial significance are the removal of compression artifacts, resolution enhancement, and restoration in medical imaging.

15.6.4.1 *Removal of Compression Artifacts*

The problem of removing compression artifacts addresses the recovery of information lost due to the quantization of parameters during compression. More specifically, in the majority of existing image and video compression algorithms the image (or frame in an image sequence) is divided into square blocks which are processed independently from each other. The Discrete Cosine Transform (DCT) of such blocks (representing either the image intensity when dealing with still images or intracoding of video blocks or frames, or the displaced frame difference when dealing with intercoding of video blocks or frames) is taken and the resulting DCT coefficients are quantized. As a result of this processing, annoying blocking artifacts result, primarily at high compression ratios. A number of techniques have been developed for removing such blocking artifacts for both still images and video. For example, in [27, 28] the problem of removing the blocking artifacts is formulated as a recovery problem, according to which an estimate of the blocking artifact-free original image is estimated by utilizing the available quantized data, knowledge about the quantizer step size, and prior knowledge about the smoothness of the original image.

A deterministic formulation of the problem is followed in [27]. Two solutions are developed for the removal of blocking artifacts in still images. The first one is based on the CLS formulation and a successive approximations iteration is utilized for obtaining the solution. The second approach is based on the theory of projections onto convex sets (POCS), which has found applications in a number of recovery problems. The evidence analysis within the hierarchical Bayesian paradigm, mentioned above, is applied to the same problem in [28]. Expressions for the iterative evaluation of the unknown parameters and the reconstructed image are derived. The relationship between the CLS-iteration adaptive successive approximations solution and the hierarchical Bayesian solution discussed in the previous section is also applicable here.

15.6.4.2 *Resolution Enhancement*

Resolution enhancement (also referred to as super-resolution) is a problem which has also seen considerable activity recently (for a recent review see [29–31] and references therein). It addresses the problem of increasing the resolution of a single image utilizing multiple aliased low-resolution images of the same scene with sub-pixel shifts among them. It also addresses the problem of increasing the resolution of a video frame (and consequently the whole sequence) of a dynamic video sequence by utilizing a number of neighboring frames. In this case the shifts between any two frames are expressed by the motion field. The low-resolution images and frames can be noisy and blurred (due to the image acquisition system), or compressed, which further complicates the problem. There are a number of potential applications of this technology. It can be utilized to increase the resolution of any instrument by creating a number of images of the same scene, but also to replace an expensive high-resolution instrument by one or more low-resolution ones, or it can serve as a compression mechanism. Some of the techniques developed in the literature address, in addition to the resolution enhancement problem, the simultaneous removal of blurring and compression artifacts, i.e., they combine the objectives of multiple applications mentioned in this chapter. For illustration purposes consider the example shown in Fig. 15.10 [32]. In Fig. 15.10(a) the original high-resolution image is shown. This image is blurred with a 4×4 noncausal uniform blur function and down-sampled by a factor of 4 in each direction to generate 64 low-resolution images with global subpixel shifts which are integer multiples of $1/4$ in each direction. Noise of the same variance was added to all low-resolution images (resulting in SNR of 30 dB for this example). One of the low-resolution images (the one with zero shifts in each direction) is shown in Fig. 15.10(b). The best bilinearly interpolated image is shown in Fig. 15.10(c). A hierarchical Bayesian approach is utilized in [32] in deriving iterative algorithms for

(a)   (b)

(c)   (d)

FIGURE 15.10

Resolution enhancement example: (a) original image; (b) one of the 16 low-resolution images with SNR $= 30$ dB (all 16 images have the same SNR); (c) best bilinearly interpolated image; (d) estimated high-resolution image by the hierarchical Bayesian approach [32].

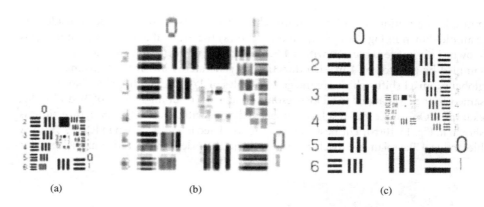

(a)   (b)   (c)

FIGURE 15.11

Resolution enhancement example: (a) one of the 16 low-resolution images (the SNR for the low-resolution images is at random either 10 dB or 20 dB or 30 dB); (b) bilinearly interpolated image; (c) estimated high-resolution image by the hierarchical Bayesian approach [32].

estimating the unknown parameters (an image model parameter similar to δ in Eq. (15.53) and the additive noise variance) and the high-resolution image by utilizing the 16 low-resolution images, assuming that the shifts and the blur are known. The resulting high-resolution image is shown in Fig. 15.10(d). Finally, the same experiment was repeated with the resolution chart image. One of the 16 low-resolution images is shown in Fig. 15.11(a). The bilinearly interpolated image is shown in Fig. 15.11(b), while the one generated by the hierarchical Bayesian approach is shown in Fig. 15.11(c). The hierarchical Bayesian approach was also used for the recovery of a high-resolution sequence from low-resolution and compressed observations [33].

15.6.4.3 Restoration in Medical Imaging

A very important emerging application area of image restoration algorithms is *medical imaging*. A number of imaging techniques are being invented in a rapid way, and medical imaging devices have many different imaging modalities. As in any other imaging system, medical image acquisition devices also introduce degradation to the images, and these degradations vary greatly among imaging applications. Currently, computer-assisted tomography (CAT or CT), positron emission tomography (PET), single photon emission computed tomography (SPECT), magnetic resonance imaging (MRI), and confocal microscopy are widely used both in medical research and diagnostics. In all these modalities, certain degradations affect the acquired images (sometimes to a hindering extent), and image restoration methods play an important role in improving their usability and extending the application of the medical imaging devices [34].

Medical imaging introduced many new challenging problems to the image restoration research both at the modeling and algorithmic level. It is usually very difficult to obtain accurate models for medical imaging devices, and much more difficult to obtain a general model for all medical imaging instruments. Moreover, common assumptions being utilized in traditional image restoration do not hold in general. For example, the noise in many medical imaging modalities is nonstationary, e.g., ultrasound speckle noise, and the degradation generally depends not only on the imaging device (as is generally the case in other imaging devices) but also on the physical location of the device and the subject being imaged.

Despite these challenges, image restoration algorithms have found great use in medical imaging. For example, in conventional fluorescence microscopy, the image is degraded by an out-of-focus blur caused by fluorescent objects not in focus. Therefore, the fluorescent objects interfere with the original object to be imaged, and they reduce the contrast. Confocal microscopes are also affected by this problem, although to a much lesser extent. To enhance the image quality, deconvolution methods are developed where first the three-dimensional point spread function (PSF) of the system is obtained through the instrument specifications and experimental measurements and second the effect of the degradation is removed from the image stack by a deconvolution algorithm. The quality of the acquired microscopic images can be significantly improved by this postprocessing step. Commercial and freeware software as well as fully assembled "deconvolution microscopes" are available and in use.

Another interesting application area is MRI with multiple surface coils. Traditionally, MR images are acquired by a single whole body coil which has a homogeneous intensity distribution across the image (bias-free) but creates images with a low signal-to-noise ratio (SNR). To increase the SNR, multiple images are acquired which take a lot of time and is inconvenient for both the patient and the medical staff. Recently, another approach, called "nuclear magnetic resonance (NMR) phased array," has been proposed and is widely adapted in current MRI technology. It is based on simultaneously acquiring multiple images with closely positioned NMR receiver coils and combining these images after the acquisition. These individual surface coil images have higher SNRs than the whole body coil images, and they shorten the acquisition time significantly. However, they are degraded by bias fields due to the locations of each surface coil since the intensity levels rapidly decrease with distance. Therefore, it is desired to combine the surface coil images to remove the bias to obtain a high SNR and bias-free image.

The bias fields in surface coil images can be seen as smoothly varying functions over the image that change the image intensity level depending on the location, that is, each surface coil image can be modeled by the product of the original image and a smoothly varying field, and can be expressed in matrix-vector notation by

$$
\begin{bmatrix} g_{i,1} \\ g_{i,2} \\ \cdot \\ \cdot \\ g_{i,m} \end{bmatrix} = \begin{bmatrix} d_{i,1} & 0 & \cdot & \cdot & \cdot & 0 \\ 0 & d_{i,2} & 0 & \cdot & \cdot & 0 \\ & & \cdot & & & \\ & & & \cdot & & \\ & & & & & d_{i,m} \end{bmatrix} \begin{bmatrix} f_1 \\ f_2 \\ \cdot \\ \cdot \\ f_m \end{bmatrix} \Rightarrow \mathbf{g}_i = \mathbf{D}_i \mathbf{f} + \mathbf{v}_i, \qquad (15.59)
$$

with $\mathbf{g}_i$ the *ith* surface coil image, $\mathbf{D}_i$ the *ith* bias field, and $f_1, f_2, \ldots, f_m$ the pixels of the original image. Assuming there are *n* coils, this imaging system can also be represented as follows:

$$
\begin{bmatrix} \mathbf{g}_1 \\ \mathbf{g}_2 \\ \cdot \\ \cdot \\ \mathbf{g}_n \end{bmatrix} = \begin{bmatrix} \mathbf{D}_1 \\ \mathbf{D}_2 \\ \cdot \\ \cdot \\ \cdot \\ \mathbf{D}_n \end{bmatrix} \mathbf{f} + \begin{bmatrix} \mathbf{v}_1 \\ \mathbf{v}_2 \\ \cdot \\ \cdot \\ \mathbf{v}_n \end{bmatrix} \Rightarrow \mathbf{g} = \mathbf{D}\mathbf{f} + \mathbf{v}, \tag{15.60}
$$

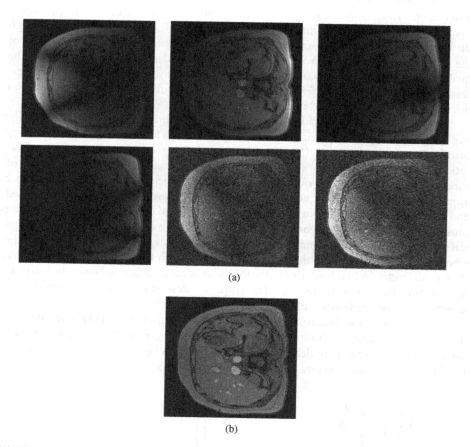

(a)

(b)

FIGURE 15.12

MR surface coil image combination example: (a) six surface coil images of a human abdomen; (b) combined image using the algorithm in [35] .

where **g** represents the observed surface coil images, **D** the bias fields, and **v** the noise in the observed images. This is the same degradation system as in (15.1), and therefore restoration algorithms similar to the ones presented so far can be applied. For instance, in [35], the spatially adaptive formulation (15.44) described in Section 15.4.4 is utilized where the weights are estimated using the local variances in the image, and the regularization parameter is calculated using (15.50) in Section 15.6.1. An example set of six images of a human abdomen acquired by a surface coil array are shown in Fig. 15.12(a). It is clear that some images are severely degraded by the bias fields, and some images contain a high level of noise. Using the algorithm in [35], these images are combined to obtain the image shown in Fig. 15.12(b). It is clear that the bias fields are removed and the noise is significantly suppressed, so that this image is considerably more useful than the individual images. Moreover, since the restoration process is performed as a postprocessing step, the acquisition duration is not affected.

15.7 DISCUSSION

In this chapter we briefly described the application of the successive approximations-based class of iterative algorithms to the problem of restoring a noisy and blurred image. We presented and analyzed in some detail the simpler forms of the algorithm, and briefly described an iteration-adaptive form of the algorithm following a deterministic approach but also a hierarchical Bayesian approach. We also provided a brief description of the blind image deconvolution problem and the use of iterative algorithms in providing solutions to this problem. Finally, we briefly described other inverse problems, i.e., the removal of blocking artifacts, the enhancement of resolution, and the removal of bias fields in NMR phased array imaging, which have been solved using the techniques described in this chapter. With this presentation we have simply touched the "tip of the iceberg." We only covered a small amount of the material on the topic. More sophisticated forms of iterative image restoration algorithms were left out, since they were deemed to be beyond the scope and level of this chapter.

It is the hope and the expectation of the authors that the presented material will form a good introduction to the topic for the engineer or the student who would like to work in this area.

REFERENCES

[1] H. C. Andrews and B. R. Hunt. *Digital Image Restoration*. Prentice-Hall, Upper Saddle River, NJ, 1977.

[2] M. R. Banham and A. K. Katsaggelos. Digital image restoration. *IEEE Signal Process. Mag.*, 14(2):24–41, 1997.

[3] G. Demoment. Image reconstruction and restoration: overview of common estimation structures and problems. *IEEE Trans. Acoust.*, 37(12):2024–2036, 1989.

[4] A. K. Katsaggelos, editor. *Digital Image Restoration*, Springer Series in Information Sciences, Vol. 23, Springer-Verlag, Heidelberg, 1991.

[5] A. N. Tikhonov and V. Y. Arsenin. *Solution of Ill-Posed Problems*. Winston, Wiley, 1977.

[6] D. Kundur and D. Hatzinakos. Blind image deconvolution. *IEEE Signal Process. Mag.*, 13(3):4364, 1996.

[7] T. E. Bishop, S. D. Babacan, B. Amizic, A. K. Katsaggelos, T. Chan, and R. Molina. Blind image deconvolution: problem formulation and existing approaches. In P. Campisi and K. Egiazarian, editors, *Blind Image Deconvolution: Theory and Applications*. CRC Press, Boca Raton, FL, 2007.

[8] J. M. Ortega and W. C. Rheinboldt. *Iterative Solution of Nonlinear Equations in Several Variables*. Academic Press, New York, 1970.

[9] R. W. Schafer, R. M. Mersereau, and M. A. Richards. Constrained iterative restoration algorithms. *Proc. IEEE*, 69:432–450, 1981.

[10] D. E. Dudgeon and R. M. Mersereau. *Multidimensional Digital Signal Processing*. Prentice-Hall, Upper Saddle River, New Jersey, 1984.

[11] J. Biemond, R. L. Lagendijk, and R. M. Mersereau. Iterative methods for image deblurring. *Proc. IEEE*, 78(5):856–883, 1990.

[12] A. K. Katsaggelos. Iterative image restoration algorithm. *Opt. Eng.*, 28(7):735–748, 1989.

[13] P. H. Van Citttert. Zum Einfluss der Spaltbreite auf die Intensitatswerteilung in Spektrallinien II. *Z. Physik*, 69:298–308, 1931.

[14] B. R. Hunt. The application of constrained least squares estimation to image restoration by digital computers. *IEEE Trans. Comput.*, C-22:805–812, 1973.

[15] N. P. Galatsanos and A.K. Katsaggelos. Methods for choosing the regularization parameter and estimating the noise variance in image restoration and their relation. *IEEE Trans. Image Process.*, 1:322–336, 1992.

[16] A. K. Katsaggelos, J. Biemond, R. M. Mersereau, and R. W. Schafer. A regularized iterative image restoration algorithm. *IEEE Trans. Signal Process.*, 39(4):914–929, 1991.

[17] G. L. Anderson and A. N. Netravali. Image restoration based on a subjective criterion. *IEEE Trans. Syst. Man Cybern.*, SMC-6:845–853, 1976.

[18] M. G. Kang and A. K. Katsaggelos. General choice of the regularization functional in regularized image restoration. *IEEE Trans. Image Process.*, 4(5):594–602, 1995.

[19] A. K. Katsaggelos and M. G. Kang. A spatially adaptive iterative algorithm for the restoration of astronomical images. *Int. J. Imaging Syst. Technol.*, 6(4):305–313, 1995.

[20] M. G. Kang and A. K. Katsaggelos. Frequency domain adaptive iterative image restoration and evaluation of the regularization parameter. *Opt. Eng.*, 33(10):3222–3232, 1994.

[21] C. E. Morris, M. A. Richards, and M. H. Hayes. Fast reconstruction of linearly distorted signals. *IEEE Trans. Acoust.*, 36:1017–1025, 1988.

[22] A. K. Katsaggelos and S. N. Efstratiadis. A class of iterative signal restoration algorithms. *IEEE Trans. Acoust.*, 38:778–786, 1990 (reprinted in *Digit. Image Process.*, R. Chellappa, editor, IEEE Computer Society Press).

[23] A. K. Katsaggelos and S. P. R. Kumar. Single and multistep iterative image restoration and VLSI implementation. *Signal Processing*, 16(1):29–40, 1989.

[24] R. Molina, A. K. Katsaggelos, and J. Mateos. Bayesian and regularization methods for hyperparameter estimation in image restoration. *IEEE Trans. Image Process.*, 8(2):231–246, 1999.

[25] G. Chantas, N. P. Galatsanos, and A. Likas. Non stationary Bayesian image restoration. In *Proc. Int. Conf. Pattern Recognit.*, Vol. 4, 689–692, August 23–26, 2004.

[26] Y. L. You and M. Kaveh. Blind image restoration by anisotropic regularization. *IEEE Trans. Image Process.*, 8(3):396–407, 1999.

[27] Y. Yang, N. P. Galatsanos, and A. K. Katsaggelos. Regularized image reconstruction from incomplete block discrete cosine transform data. *IEEE Trans. Circuits Syst. Video Technol.*, 3(6):421–432, 1993.

[28] J. Mateos, A. K. Katsaggelos, and R. Molina. A Bayesian approach for the estimation and transmission of regularization parameters for reducing blocking artifacts. *IEEE Trans. Image Process.*, 9(7):1200–1215, 2000.

[29] M. G. Kang and S. Chaudhuri, editors. Super resolution image reconstruction. *IEEE Signal Process. Mag.*, 20(3), 2003.

[30] S. Chaudhuri, editor. *Super-Resolution Imaging.* Kluwer Academic Publishers, New York, 2001.

[31] A. K. Katsaggelos, R. Molina, and J. Mateos. *Super Resolution of Images and Video.* Morgan and Claypool, San Rafael, CA, 2007.

[32] R. Molina, J. Abad, M. Vega, and A. K. Katsaggelos. Parameter estimation in Bayesian high-resolution image reconstruction with multisensors. *IEEE Trans. Image Process.*, 12(12):1642–1654, 2003.

[33] C. A. Segall, A. K. Katsaggelos, R. Molina, and J. Mateos. Bayesian resolution enhancement of compressed video. *IEEE Trans. Image Process.*, 13(7):898–911, 2004.

[34] J. Kovacevic and R. F. Murphy. Molecular and cellular bioimaging [Guest editorial]. *IEEE Signal Process. Mag.*, 23(3):19–19, 2006.

[35] S. D. Babacan, X. Yin, A. C. Larson, and A. K. Katsaggelos. Combination of MR Surface Coil Images Using Weighted Constrained Least Squares. In *IEEE Int. Conf. Image Process. 2008*, San Diego, CA, October 2008.

Lossless Image Compression

16

Lina J. Karam

Arizona State University

16.1 INTRODUCTION

The goal of lossless image compression is to represent an image signal with the smallest possible number of bits without loss of *any* information, thereby speeding up transmission and minimizing storage requirements. The number of bits representing the signal is typically expressed as an average bit rate (average number of bits per sample for still images, and average number of bits per second for video). The goal of lossy compression is to achieve the best possible fidelity given an available communication or storage bit rate capacity or to minimize the number of bits representing the image signal subject to some allowable loss of information. In this way, a much greater reduction in bit rate can be attained as compared to lossless compression, which is necessary for enabling many real-time applications involving the handling and transmission of audiovisual information. The function of *compression* is often referred to as *coding*, for short.

Coding techniques are crucial for the effective transmission or storage of data-intensive visual information. In fact, a single uncompressed color image or video frame with a medium resolution of 500×500 pixels would require 100 seconds for transmission over an Integrated Services Digital Network (ISDN) link having a capacity of 64,000 bits per second (64 Kbps). The resulting delay is intolerably large considering that a delay as small as 1 to 2 seconds is needed to conduct an interactive "slide show," and a much smaller delay (on the order of 0.1 second) is required for video transmission or playback. Although a CD-ROM device has a storage capacity of a few gigabits, its average data-read throughput is only a few Megabits per second (about 1.2 Mbps to 1.5 Mbps for the common $1\times$ read speed CLV CDs). As a result, compression is essential for the storage and real-time transmission of digital audiovisual information, where large amounts of data must be handled by devices having a limited bandwidth and storage capacity.

Lossless compression is possible because, in general, there is significant redundancy present in image signals. This redundancy is proportional to the amount of correlation among the image data samples. For example, in a natural still image, there is usually a high degree of spatial correlation among neighboring image samples. Also, for video, there is additional temporal correlation among samples in successive video frames. In color images and multispectral imagery (Chapter 8), there is correlation, known as spectral correlation, between the image samples in the different spectral components.

In lossless coding, the decoded image data should be identical both quantitatively (numerically) and qualitatively (visually) to the original encoded image. Although this requirement preserves exactly the accuracy of representation, it often severely limits the amount of compression that can be achieved to a compression factor of two or three. In order to achieve higher compression factors, perceptually lossless coding methods attempt to remove redundant as well as perceptually irrelevant information; these methods require that the encoded and decoded images be only visually, and not necessarily numerically, identical. In this case, some loss of information is allowed as long as the recovered image is perceived to be identical to the original one.

Although a higher reduction in bit rate can be achieved with lossy compression, there exist several applications that require lossless coding, such as the compression of digital medical imagery and facsimile transmissions of bitonal images. These applications triggered the development of several standards for lossless compression, including the lossless JPEG standard (Section 16.4), facsimile compression standards, and the JBIG and JBIG2 compression standards. More recently, the JPEG2000 standard was developed as a unified compression standard that integrates both lossy and lossless compression into one system for different types of images including continuous-tone, bilevel, text, and compound imagery. Furthermore, lossy coding schemes make use of lossless coding components to minimize the redundancy in the signal being compressed.

This chapter introduces the basics of lossless image coding and presents classical as well as some more recently developed lossless compression methods. This chapter is organized as follows. Section 16.2 introduces basic concepts in lossless image coding. Section 16.3 reviews concepts from information theory and presents classical lossless compression schemes including Huffman, Arithmetic, Lempel-Ziv-Welch (LZW), Elias, and Exp-Golomb codes. Standards for lossless compression are presented in Section 16.4. Section 16.5 introduces more recently developed lossless compression schemes and presents the basics of perceptually lossless image coding.

16.2 BASICS OF LOSSLESS IMAGE CODING

The block diagram of a lossless coding system is shown in Fig. 16.1. The encoder (Fig. 16.1(a)) takes as input an image and generates as output a compressed bitstream. The decoder (Fig. 16.1(b)) takes as input the compressed bitstream and recovers the original uncompressed image. In general, the encoder and decoder can each be each viewed as consisting of three main stages. In this section, only the main elements of the encoder will be discussed since the decoder performs the inverse operations of the encoder. As

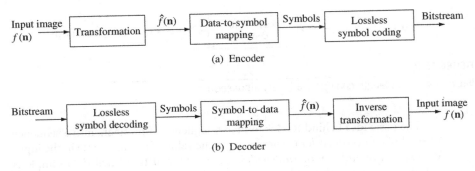

(a) Encoder

(b) Decoder

FIGURE 16.1

General lossless coding system.

shown in Fig. 16.1(a), the operations of a lossless image encoder can be grouped into three stages:

1. **Transformation:** This stage applies a reversible (one-to-one) transformation to the input image data. The purpose of this stage is to convert the input image data $f(\mathbf{n})$ into a form $\hat{f}(\mathbf{n})$ that can be compressed more efficiently. For this purpose, the selected transformation can aid in reducing the data correlation (interdependency, redundancy), alter the data statistical distribution, and/or pack a large amount of information into few data samples or subband regions. Typical transformations include differential/predictive mapping, unitary transforms such as the discrete cosine transform (DCT), subband decompositions such as wavelet transforms, and color space conversions such as conversion from the highly correlated RGB representation to the less correlated luminance-chrominance representation. A combination of the above transforms can be used at this stage. For example, an RGB color image can be transformed to its luminance-chrominance representation followed by DCT or subband decomposition followed by predictive/differential mapping. In some applications (e.g., low power), it might be desirable to operate directly on the original data without incurring the additional cost of applying a transformation; in this case, the transformation could be set to the identity mapping.

2. **Data-to-symbol mapping:** This stage converts the image data $\hat{f}(\mathbf{n})$ into entities called *symbols* that can be efficiently coded by the final stage. The conversion into symbols can be done through partitioning and/or run-length coding (RLC), for example.

   The image data can be partitioned into blocks by grouping neighboring data samples together; in this case, each data block is a symbol. Grouping several data units together allows the exploitation of any correlation that might be present between the image data, and may result in higher compression ratios at the expense of increasing the coding complexity. On the other hand, each separate data unit can be taken to be a symbol without any further grouping or partitioning.

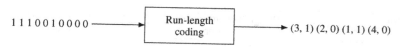

FIGURE 16.2

Illustration of run-length coding for a binary input sequence.

The basic idea behind RLC is to map a sequence of numbers into a sequence of symbol pairs (*run*, *value*), where *value* is the value of a data sample in the input data sequence and *run* or *runlength* is the number of times that data sample is contiguously repeated. In this case, each pair (*run*, *value*) is a symbol. An example illustrating RLC for a binary sequence is shown in Fig. 16.2. Different implementations might use a slightly different format. For example, if the input data sequence has long runs of zeros, some coders such as the JPEG standard (Chapter 17) use *value* to code only the value of the nonzero data samples and *run* to code the number of zeros preceding each nonzero data sample.

Appropriate mapping of the input data into symbols is very important for optimizing the coding. For example, grouping data points into small localized sets, where each set is coded separately as a symbol, allows the coding scheme to adapt to the changing local characteristics of the (transformed) image data. The appropriate data-to-symbol mapping depends on the considered application and the limitations in hardware/software complexity.

3. **Lossless symbol coding:** This stage generates a binary bitstream by assigning binary codewords to the input symbols. Lossless symbol coding is commonly referred to as noiseless coding or just lossless coding since this stage is where the actual lossless coding into the final compressed bitstream is performed. The first two stages can be regarded as preprocessing stages for mapping the data into a form that can be more efficiently coded by this lossless coding stage.

Lossless compression is usually achieved by using variable-length codewords, where the shorter codewords are assigned to the symbols that occur more frequently. This variable-length codeword assignment is known as *variable-length coding (VLC)* and also as *entropy coding*. Entropy coders, such as Huffman and arithmetic coders, attempt to minimize the average bit rate (average number of bits per symbol) needed to represent a sequence of symbols, based on the probability of symbol occurrence. Universal codes, such as Elias codes [1] and Exp-Golomb codes [2, 3], are variable-length codes that can encode positive integer values into variable-length binary codewords without knowledge of the true probability distribution of the occuring integer values. Entropy coding will be discussed in more detail in Section 16.3. An alternative way to achieve compression is to code *variable-length strings* of symbols using fixed-length binary codewords. This is the basic strategy behind dictionary (Lempel-Ziv) codes, which are also described in Section 16.3.

The generated lossless code (bitstream) should be uniquely decodable; i.e., the bitstream can be decoded without ambiguity resulting in only one unique sequence of symbols (the original one). For variable-length codes, unique decodability is achieved by imposing a prefix condition which states that no codeword can be a prefix of another codeword. Codes that satisfy the prefix condition are called *prefix codes* or *instantaneously decodable codes*, and they include Huffman, arithmetic, Elias, and Exp-Golomb codes. Binary prefix codes can be represented as a binary tree, and are also called *tree-structured codes*. For dictionary codes, unique decodability can be easily achieved since the generated code words are fixed length.

Selecting which lossless coding method to use depends on the application and usually involves a tradeoff between several factors including the implementation hardware or software, the allowable coding delay, and the required compression level. Some of the factors that need to be considered when choosing or devising a lossless compression scheme are listed below.

1. **Compression efficiency:** Compression efficiency is usually given in the form of a compression ratio C_R,

$$C_R = \frac{\text{Total size in bits of original input image}}{\text{Total size in bits of compressed bitstream}} = \frac{\text{Total size in bits of encoder input}}{\text{Total size in bits of encoder output}},$$

(16.1)

which compares the size of the original input image data with the size of the generated compressed bitstream. Compression efficiency is also commonly expressed as an average bit rate B in bits per pixel, or *bpp* for short,

$$B = \frac{\text{Total size in bits of compressed bitstream}}{\text{Total number of pixels in original input image}} = \frac{\text{Total size in bits of encoder output}}{\text{Total size in pixels of encoder input}}.$$

(16.2)

As discussed in Section 16.3, for lossless coding, the achievable compression efficiency is bounded by the *entropy* of the finite set of symbols generated as the output of Stage 2, assuming these symbols are each coded separately, on a one-by-one basis, by Stage 3.

2. **Coding delay:** The coding delay can be defined as the minimum time required to both encode and decode an input data sample. The coding delay increases with the total number of required arithmetic operations. It also usually increases with an increase in memory requirements since memory usage usually leads to communication delays. Minimizing the coding delay is especially important for real-time applications.

3. **Implementation complexity:** Implementation complexity is measured in terms of the total number of required arithmetic operations and in terms of the memory requirements. Alternatively, implementation complexity can be measured in terms of the required number of arithmetic operations per second and the memory requirements for achieving a given coding delay or real-time performance. For applications that put a limit on power consumption, the implementation complexity would also include a measure of the level of power consumption. Higher

compression efficiency can usually be achieved by increasing the implementation complexity, which would in turn lead to an increase in the coding delay. In practice, it is desirable to optimize the compression efficiency while keeping the implementation requirements as simple as possible. For some applications such as database browsing and retrieval, only a low decoding complexity is needed since the encoding is not performed as frequently as the decoding.

4. **Robustness:** For applications that require transmission of the compressed bitstream in error-prone environments, robustness of the coding method to transmission errors becomes an important consideration.

5. **Scalability:** Scalable encoders generate a layered bitstream embedding a hierarchical representation of the input image data. In this way, the input data can be recovered at different resolutions in a hierarchical manner (scalability in resolution), and the bit rate can be varied depending on the available resources using the same encoded bitstream (scalability in bit rate; the encoding does not have to be repeated to generate the different bit rates). The JPEG 2000 standard (Chapter 17) is an example of a scalable image coder that generates an embedded bitstream and that supports scalability in quality, resolution, spatial location, and image components [4, Chapter 9].

16.3 LOSSLESS SYMBOL CODING

As mentioned in Section 16.2, lossless symbol coding is commonly referred to as lossless coding or lossless compression. The popular lossless symbol coding schemes fall into one of the following main categories:

- Statistical schemes (Huffman, arithmetic): these schemes require knowledge of the source symbol probability distribution; shorter code words are assigned to the symbols with higher probability of occurrence (VLC); a statistical source model (also called probability model) gives the symbol probabilities; the statistical source model can be fixed, in which case the symbol probabilities are fixed, or adaptive, in which case the symbol probabilities are calculated adaptively; sophisticated source models can provide more accurate modeling of the source statistics and, thus, achieve higher compression at the expense of an increase in complexity.

- Dictionary-based schemes (Lempel-Ziv): these schemes do not require a priori knowledge of the source symbol probability distribution; they dynamically construct encoding and decoding tables (called dictionaries) of variable-length symbol strings as they occur in the input data; as the encoding table is constructed, fixed-length binary codewords are generated by indexing into the encoding table.

- Structured universal coding schemes (Elias codes, Exponential-Golomb codes): these schemes generate variable-length code words with a regular structure; they operate on positive integers, which necessitates first mapping the source symbols to positive integers; no a priori knowledge of the true probability distribution of the integer values is required, but these codes are constructed under the assumption of a monotonically decreasing distribution, which implies that smaller integer values are more probable than larger integer values.

The aforementioned statistical, dictionary-based, and structured universal coders attempt to minimize the average bit rate without incurring any loss in fidelity. Such lossless coding schemes are also referred to as *entropy coding* schemes. The field of information theory gives lower bounds on the achievable bit rates. This section presents the popular classical lossless symbol coding schemes, including Huffman, Arithmetic and Lempel-Ziv coding, in addition to the structured Elias and Exp-Golomb universal codes. In order to gain an insight into how the bit rate minimization is done by these different lossless coding schemes, some important basic concepts from information theory are reviewed first.

16.3.1 Basic Concepts from Information Theory

Information theory makes heavy use of probability theory since information is related to the degree of unpredictability and randomness in the generated messages. In here, the generated messages are the symbols output by Stage 2 (Section 16.2).

An information source is characterized by the set of symbols S it is capable of generating and by the probability of occurrence of these symbols. For the considered lossless image coding application, the information source is a discrete-time, discrete-amplitude source with a finite set of unique symbols; i.e., S consists of a finite number of symbols and is commonly called the *source alphabet*.

Let S consist of N symbols:

$$S = \{s_0, s_1, \ldots, s_{N-1}\}. \tag{16.3}$$

Then the information source outputs a sequence of symbols $\{x_1, x_2, x_3, \ldots, x_i, \ldots\}$ drawn from the set of symbols S, where x_1 is the first output source sample, x_2 is the second output sample, and x_i is the ith output sample from S. At any given time (given by the output sequence index), the probability that the source outputs symbol s_k is $p_k = P(s_k)$, $0 \le k \le N - 1$. Note that $\sum_{k=0}^{N-1} p_k = 1$ since it is certain that the source outputs only symbols from its alphabet S. The source is said to be *stationary* if its statistics (set of probabilities) do not change with time.

The information associated with a symbol s_k ($0 \le k \le N - 1$), also called *self-information*, is defined as

$$I_{s_k} = \log_2(1/p_k) \text{ (bits)} = -\log_2(p_k) \text{ (bits)}. \tag{16.4}$$

From (16.4), it can be seen that $I_k = 0$ if $p_k = 1$ (certain event) and $I_k \to \infty$ if $p_k = 0$ (impossible event). Also, I_k is large when p_k is small (unlikely symbols), as expected.

The information content of the source can be measured using the source *entropy* $H(S)$, which is a measure of the average amount of information per symbol. The source entropy $H(S)$, also known as *first-order entropy* or *marginal entropy*, is defined as the expected value of the self information and is given by

$$H(S) = E\{I_k\} = -\sum_{k=0}^{N-1} p_k \log_2(p_k) \ \ \text{(bits per symbol)}. \tag{16.5}$$

Note that $H(S)$ is maximal if the symbols in S are equiprobable (flat probability distribution), in which case $H(S) = \log_2(N)$ bits per symbol. A skewed probability distribution results in a smaller source entropy.

In the case of memoryless coding, each source symbol is coded separately. For a given lossless code C, let l_k denote the length (number of bits) of the code word assigned to code symbol s_k ($0 \le k \le N - 1$). Then, the resulting average bit rate B_C corresponding to code C is

$$B_C = \sum_{k=0}^{N-1} p_k l_k \ \ \text{(bits per symbol)}. \tag{16.6}$$

For any uniquely decodable lossless code C, the entropy $H(S)$ is a lower bound on the average bit rate B_C [5]:

$$B_C \ge H(S). \tag{16.7}$$

So, $H(S)$ puts a limit on the achievable average bit rate given that each symbol is coded separately in a memoryless fashion.

In addition, a uniquely decodable prefix code C can always be constructed (e.g., Huffman coding - Section 16.3.3) such that

$$H(S) \le B_C \le H(S) + 1. \tag{16.8}$$

An important result which can be used in constructing prefix codes is the Kraft inequality:

$$\sum_{k=1}^{N} 2^{-l_k} \le 1. \tag{16.9}$$

Every uniquely decodable code has codewords with lengths satisfying the Kraft inequality (16.9), and prefix codes can be constructed with any set of lengths satisfying (16.9) [6].

Higher compression can be achieved by coding a block (subsequence, vector) of M successive symbols jointly. The coding can be done as in the case of memoryless coding by regarding each block of M symbols as one compound symbol $S^{(M)}$ drawn from the alphabet:

$$S^{(M)} = \underbrace{S \times S \times \ldots \times S}_{M \text{ times}}, \tag{16.10}$$

where $\times$ in (16.10) denotes a cartesian product, and the superscript (M) denotes the size of each compound block of symbols. Therefore, $S^{(M)}$ is the set of all possible compound symbols of the form $[x_1, x_2, \ldots, x_M]$, where $x_i \in S$, $1 \le i \le M$. Since S consists of N symbols, $S^{(M)}$ will contain $L = N^M$ compound symbols:

$$S^{(M)} = \{s_0^{(M)}, s_1^{(M)}, \ldots, s_{L-1}^{(M)}\}; \quad L = N^M. \tag{16.11}$$

The previous results and definitions directly generalize by replacing S with $S^{(M)}$ and replacing the symbol probabilities $p_k = P(s_k)$ ($0 \le k \le N-1$) with the joint probabilities (compound symbol probabilities) $p_k^{(M)} = P(s_k^{(M)})$ ($0 \le k \le L-1$). So, the entropy of the set $S^{(M)}$, which is the set of all compound symbols $s_k^{(M)}$ ($0 \le k \le L-1$) is given by

$$H(S^{(M)}) = -\sum_{k=0}^{L-1} p_k^{(M)} \log_2 (p_k^{(M)}). \tag{16.12}$$

$H(S^{(M)})$ is also called the *Mth-order entropy* of S. If S corresponds to a stationary source (i.e., symbol probabilities do not change over time), $H(S^{(M)})$ is related to the source entropy $H(S)$ as follows [5]:

$$\frac{H(S^{(M)})}{M} \le H(S), \tag{16.13}$$

with equality if and only if the symbols in S are statistically independent (memoryless source). The quantity

$$\lim_{M \to \infty} \frac{H(S^{(M)})}{M} \tag{16.14}$$

is called the *entropy rate* of the source S and gives the average information per output symbol drawn from S. For a stationary source, the limit in (16.14) always exists. Also, from (16.13), the entropy rate is equal to the source entropy for the case of a memoryless source.

As before each output (compound) symbol can be coded separately. For a given lossless code C, if $l_k^{(M)}$ is the length of the codeword assigned to code symbol $s_k^{(M)}$ ($0 \le k \le L-1$), the resulting average bit rate $B_C^{(M)}$ in code bits per *compound* symbol is

$$B_C^{(M)} = \sum_{k=0}^{L-1} p_k^{(M)} l_k^{(M)} \quad (\text{bits per } compound \text{ symbol}). \tag{16.15}$$

Also, as before, a prefix code C can be constructed such that

$$H(S^{(M)}) \le B_C^{(M)} \le H(S^{(M)}) + 1, \tag{16.16}$$

where $B_C^{(M)}$ is the resulting average bit rate per *compound* symbol. The desired average bit rate B_C in bits per *source* symbol is equal to $B_C^{(M)}/M$. So, dividing the terms in (16.16) by M, we obtain

$$\frac{H(S^{(M)})}{M} \le B_C \le \frac{H(S^{(M)})}{M} + \frac{1}{M}. \tag{16.17}$$

From (16.17), it follows that, by jointly coding very large blocks of source symbols (M very large), a source code C can be found with an average bit rate B_C approaching monotonically the entropy rate of the source as M goes to infinity. For a memoryless source, (16.17) becomes

$$H(S) \leq B_C \leq H(S) + \frac{1}{M}, \tag{16.18}$$

where $B_C = B_C^{(M)}/M$.

From the discussion above, the statistics of the considered source (given by the symbol probabilities) need to be known in order to compute the lower bounds on the achievable bit rate. In addition, statistical-based entropy coding schemes, such as Huffman (Section 16.3.3) and Arithmetic (Section 16.3.4) coding, require that the source statistics be known or be estimated accurately. In practice, the source statistics can be estimated from the histogram of a set of sample source symbols. For a nonstationary source, the symbol probabilities need to be estimated adaptively since the source statistics change over time.

In the case of block coding where M successive symbols are coded jointly as a compound symbol $s^{(M)}$, the joint probabilities $P(s^{(M)})$ need to be estimated. These joint probabilities are very difficult to calculate and the computational complexity grows exponentially with the block size M, except when the source is memoryless in which case the compound source entropy $H(S^{(M)}) = MH(S)$ based on (16.13).

However, in practice, the memoryless source model is not generally appropriate when compressing digital imagery as various dependencies exist between the image data elements even after the transformation and mapping stages (Fig. 16.1). This is due to the fact that the existing practical image transforms, such as the DCT (Chapter 17) and the discrete wavelet transform (DWT) (Chapter 17), reduce the dependencies but do not eliminate them as they cannot totally decorrelate the real-world image data. Therefore, most state-of-the-art image compression coders model these dependencies by estimating the statistics of a source with memory. The statistics of the source with memory are usually estimated by computing conditional probabilities which provide context-based modeling. Context-based modeling is a key component of context-based entropy coding which has been widely adopted in image compression schemes and is at the core of the JPEG2000 image compression standard. Context-based entropy coding is discussed below in Section 16.3.2.

16.3.2 Context-Based Entropy Coding

Context-based entropy coding is a key component of the JPEG-LS standard [7] and the more recent JPEG2000 standard [4, 8] and has also been adopted as the final entropy coding stage of various other state-of-the-art image and video compression schemes [9, 10]. The contexts provide a model for estimating the probability of each possible message to be coded, and the entropy coder translates the estimated probabilities into bits.

For sources with memory, the dependencies among adjacent symbols can be captured by either estimating joint probabilities or conditional probabilities. As discussed earlier, joint probabilities are difficult to calculate due to increased computational complexity, memory requirements, and coding delay. These issues can be resolved in practice by estimating conditional probabilities as implied by the following entropy chain rule:

$$H(S_1, S_2, \ldots, S_M) = \sum_{i=1}^{M} H(S_{M-i} | S^{i-1}), \tag{16.19}$$

where $S^{i-1} = S_1, \ldots, S_{i-1}$, and where $S_i, i = 1, \ldots, M$ are random variables that can take on realizations drawn from a finite set such as the set S of (16.3). In this way, the entropy coding can be performed incrementally one symbol at a time.

One way of calculating the conditional probabilities is to estimate those based on a finite number of previous symbols in the data stream. A more common way of estimating the conditional probabilities is to assign each possible realization s^{i-1} of S^{i-1} to some conditioning class; in this latter case, the conditional probability $p(S_i | s^{i-1}) \approx p(S_i | C(s^{i-1}))$, where $C(\bullet)$ is the function that maps each possible realization s^{i-1} to a conditioning context.

Therefore, context-based entropy coding can be separated into 2 parts: a context-based modeler and a coder. At each time instance i, the modeler tries to estimate $p(S_i | C(s^{i-1}))$, the probability of the next symbol to be coded based on the observed context. Because the contexts are formed by the already coded symbols and the entropy coding is lossless, the same context is also available at the decoder, so no side information needs to be transmitted. Given the estimated $\hat{p}(S_i | C(s^{i-1}))$, an ideal entropy coder places $-\log_2(\hat{p}(S_i = s_k | C(s^{i-1})))$ bits onto its output stream if $S_i = s_k$ actually occurs.

The estimated probabilities $\hat{p}(S_i | C(s^{i-1}))$ can be precomputed through training and made available to both the encoder and decoder before the actual coding starts. Alternatively, these estimates can be adaptively computed and updated on the fly based on the past-coded symbols. Practical applications generally require adaptive, online estimation of these probabilities either because sufficient prior statistical knowledge is not available or because these statistics are time-varying. One very natural way of estimating the probabilities online is to count the number of times each symbol has occurred under a certain context. The estimates at the encoder are updated with each encoded symbol, and the estimates at the decoder are updated with each decoded symbol. This universal modeling approach requires no a priori knowledge about the data to be coded, and the coding can be completed in only one pass.

16.3.3 Huffman Coding

In [11], Huffman presented a simple technique for constructing prefix codes which results in an average bit rate satisfying (16.8) when the source symbols are coded separately, or (16.17) in the case of joint M-symbol vector coding. A tighter upper bound on the resulting average bit rate is derived in [6].

The Huffman coding algorithm is based on the following optimality conditions for a prefix code [11]: 1) If $P(s_k) > P(s_j)$ (symbol s_k more probable than symbol s_j, $k \neq j$), then $l_k \leq l_j$, where l_k and l_j are the lengths of the codewords assigned to code symbols s_k and s_j, respectively; 2) If the symbols are listed in the order of decreasing probabilities, the last two symbols in the ordered list are assigned codewords that have the same length and are alike except for their final bit.

Given a source with alphabet S consisting of N symbols s_k with probabilities $p_k = P(s_k)$ ($0 \le k \le (N-1)$), a Huffman code corresponding to source S can be constructed by iteratively constructing a binary tree as follows:

1. Arrange the symbols of S such that the probabilities p_k are in decreasing order; i.e.,

$$p_0 \ge p_1 \ge \ldots \ge p_{(N-1)} \tag{16.20}$$

and consider the ordered symbols s_k, $0 \le k \le (N-1)$ as the leaf nodes of a tree. Let T be the set of the leaf nodes corresponding to the ordered symbols of S.

2. Take the two nodes in T with the smallest probabilities and merge them into a new node whose probability is the sum of the probabilities of these two nodes. For the tree construction, make the new resulting node the "parent" of the two least probable nodes of T by connecting the new node to each of the two least probable nodes. Each connection between two nodes form a "branch" of the tree; so two new branches are generated. Assign a value of 1 to one branch and 0 to the other branch.

3. Update T by replacing the two least probable nodes in T with their "parent" node and reorder the nodes (with their subtrees) if needed. If T contains more than one node, repeat from Step 2; otherwise the last node in T is the "root" node of the tree.

4. The codeword of a symbol $s_k \in S$ ($0 \le k \le (N-1)$) can be obtained by traversing the linked path of the tree from the root node to the leaf node corresponding to s_k ($0 \le k \le (N-1)$) while reading sequentially the bit values assigned to the tree branches of the traversed path.

The Huffman code construction procedure is illustrated by the example shown in Fig. 16.3 for the source alphabet $S = \{s_0, s_1, s_2, s_3\}$ with symbol probabilities as given in Table 16.1. The resulting symbol codewords are listed in the 3rd column of Table 16.1. For this example, the source entropy is $H(S) = 1.84644$ and the resulting average bit rate is $B_H = \sum_{k=0}^{3} p_k l_k = 1.9$ (bits per symbol), where l_k is the length of the codeword assigned

TABLE 16.1 Example of Huffman code assignment.

Source symbol s_k	Probability p_k	Assigned codeword
s_0	0.1	111
s_1	0.3	10
s_2	0.4	0
s_3	0.2	110

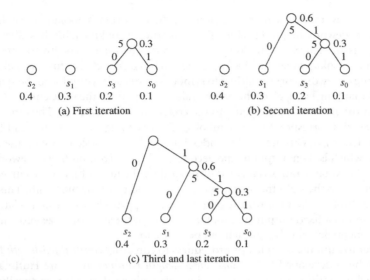

(a) First iteration

(b) Second iteration

(c) Third and last iteration

FIGURE 16.3

Example of Huffman code construction for the source alphabet of Table 16.1.

to symbol s_k of S. The symbol codewords are usually stored in a symbol-to-codeword mapping table that is made available to both the encoder and the decoder.

If the symbol probabilities can be accurately computed, the above Huffman coding procedure is optimal in the sense that it results in the minimal average bit rate among all uniquely decodable codes assuming memoryless coding. Note that, for a given source S, more than one Huffman code is possible but they are all optimal in the above sense. In fact another optimal Huffman code can be obtained by simply taking the complement of the resulting binary codewords.

As a result of memoryless coding, the resulting average bit rate is within one bit of the source entropy since integer-length codewords are assigned to each symbol separately. The described Huffman coding procedure can be directly applied to code a group of M symbols jointly by replacing S with $S^{(M)}$ of (16.10). In this case, higher compression can be achieved (Section 16.3.1), but at the expense of an increase in memory and complexity since the alphabet becomes much larger and joint probabilities need to be computed.

While encoding can be simply done by using the symbol-to-codeword mapping table, the realization of the decoding operation is more involved. One way of decoding the bitstream generated by a Huffman code is to first reconstruct the binary tree from the symbol-to-codeword mapping table. Then, as the bitstream is read one bit at a time, the tree is traversed starting at the root until a leaf node is reached. The symbol corresponding to the attained leaf node is then output by the decoder. Restarting at the root of the tree, the above tree traversal step is repeated until all the bitstream is decoded. This decoding method produces a variable symbol rate at the decoder output since the codewords vary in length.

Another way to perform the decoding is to construct a lookup table from the symbol-to-codeword mapping table. The constructed lookup table has $2^{l_{max}}$ entries, where l_{max} is the length of the longest codeword. The binary codewords are used to index into the lookup table. The lookup table can be constructed as follows. Let l_k be the length of the codeword corresponding to symbol s_k. For each symbol s_k in the symbol-to-codeword mapping table, place the pair of values (s_k, l_k) in all the table entries, for which the l_k leftmost address bits are equal to the codeword assigned to s_k. Thus there will be $2^{(l_{max}-l_k)}$ entries corresponding to symbol s_k. For decoding, l_{max} bits are read from the bitstream. These l_{max} bits are used to index into the lookup table to obtain the decoded symbol s_k, which is then output by the decoder, and the corresponding codeword length l_k. Then the next table index is formed by discarding the first l_k bits of the current index and appending to the right the next l_k bits that are read from the bitstream. This process is repeated until all the bitstream is decoded. This approach results in a relatively fast decoding and in a fixed output symbol rate. However, the memory size and complexity grows exponentially with l_{max}, which can be very large.

In order to limit the complexity, procedures to construct *constrained-length Huffman codes* have been developed [12]. *Constrained-length Huffman codes* are Huffman codes designed while limiting the maximum allowable codeword length to a specified value l_{max}. The shortened Huffman codes result in a higher average bit rate compared to the unconstrained-length Huffman code.

Since the symbols with the lowest probabilities result in the longest codewords, one way of constructing shortened Huffman codes is to group the low probability symbols into a compound symbol. The low probability symbols are taken to be the symbols in S with a probability $\leq 2^{-l_{max}}$. The probability of the compound symbol is the sum of the probabilities of the individual low-probability symbols. Then the original Huffman coding procedure is applied to an input set of symbols formed by taking the original set of symbols and replacing the low probability symbols with one compound symbol s_c. When one of the low probability symbols is generated by the source, it is encoded using the codeword corresponding to s_c followed by a second fixed-length binary code word corresponding to that particular symbol. The other "high probability" symbols are encoded as usual by using the Huffman symbol-to-codeword mapping table.

In order to avoid having to send an additional codeword for the low probability symbols, an alternative approach is to use the original unconstrained Huffman code design procedure on the original set of symbols S with the probabilities of the low probability symbols changed to be equal to $2^{-l_{max}}$. Other methods [12] involve solving a constrained optimization problem to find the optimal codeword lengths l_k $(0 \leq k \leq N-1)$ that minimize the average bit rate subject to the constraints $1 \leq l_k \leq l_{max}$ $(0 \leq k \leq N-1)$. Once the optimal codeword lengths have been found, a prefix code can be constructed using the Kraft inequality (16.9). In this case the codeword of length l_k corresponding to s_k is given by the l_k bits to the right of the binary point in the binary representation of the fraction $\sum_{1 \leq i \leq k-1} 2^{-l_i}$.

The discussion above assumes that the source statistics are described by a *fixed* (nonvarying) set of source symbol probabilities. As a result, only one fixed set of codewords need to be computed and supplied once to the encoder/decoder. This fixed model fails if the source statistics vary, since the performance of Huffman coding depends on how accurately the source statistics are modeled. For example, images can contain different data types, such as text and picture data, with different statistical characteristics. Adaptive Huffman coding changes the codeword set to match the locally estimated source statistics. As the source statistics change, the code changes, remaining optimal for the current estimate of source symbol probabilities. One simple way for adaptively estimating the symbol probabilities is to maintain a count of the number of occurrences of each symbol [6]. The Huffman code can be dynamically changed by precomputing offline different codes corresponding to different source statistics. The precomputed codes are then stored in symbol-to-codeword mapping tables that are made available to the encoder and decoder. The code is changed by dynamically choosing a symbol-to-codeword mapping table from the available tables based on the frequencies of the symbols that occurred so far. However, in addition to storage and the run-time overhead incurred for selecting a coding table, this approach requires a priori knowledge of the possible source statistics in order to predesign the codes. Another approach is to dynamically redesign the Huffman code while encoding based on the local probability estimates computed by the provided source model. This model is also available at the decoder, allowing it to dynamically alter its decoding tree or decoding table in synchrony with the encoder. Implementation details of adaptive Huffman coding algorithms can be found in [6, 13].

In the case of context-based entropy coding, the described procedures are unchanged except that now the symbol probabilities $P(s_k)$ are replaced with the symbol conditional probabilities $P(s_k|\text{Context})$ where the context is determined from previously occuring neighboring symbols, as discussed in Section 16.3.2.

16.3.4 Arithmetic Coding

As indicated in Section 16.3.3, the main drawback of Huffman coding is that it assigns an integer-length codeword to each symbol separately. As a result the bit rate cannot be less than one bit per symbol unless the symbols are coded jointly. However, joint symbol coding, which codes a block of symbols jointly as one compound symbol, results in delay and in an increased complexity in terms of source modeling, computation, and memory. Another drawback of Huffman coding is that the realization and the structure of the encoding and decoding algorithms depend on the source statistical model. It follows that any change in the source statistics would necessitate redesigning the Huffman codes and changing the encoding and decoding trees, which can render adaptive coding more difficult.

Arithmetic coding is a lossless coding method which does not suffer from the aforementioned drawbacks and which tends to achieve a higher compression ratio than Huffman coding. However, Huffman coding can generally be realized with simpler software and hardware.

In arithmetic coding, each symbol does not need to be mapped into an integral number of bits. Thus, an average fractional bit rate (in bits per symbol) can be achieved without the need for blocking the symbols into compound symbols. In addition, arithmetic coding allows the source statistical model to be separate from the structure of the encoding and decoding procedures; i.e., the source statistics can be changed without having to alter the computational steps in the encoding and decoding modules. This separation makes arithmetic coding more attractive than Huffman for adaptive coding.

The arithmetic coding technique is a practical extended version of Elias code and was initially developed by Pasco and Rissanen [14]. It was further developed by Rubin [15] to allow for incremental encoding and decoding with fixed-point computation. An overview of arithmetic coding is presented in [14] with C source code.

The basic idea behind arithmetic coding is to map the input sequence of symbols into one single codeword. Symbol blocking is not needed since the codeword can be determined and updated incrementally as each new symbol is input (symbol-by-symbol coding). At any time, the determined codeword uniquely represents all the past occurring symbols. Although the final codeword is represented using an integral number of bits, the resulting average number of bits per symbol is obtained by dividing the length of the codeword by the number of encoded symbols. For a sequence of M symbols, the resulting average bit rate satisfies (16.17) and, therefore, approaches the optimum (16.14) as the length M of the encoded sequence becomes very large.

In the actual arithmetic coding steps, the codeword is represented by a half-open subinterval $[L_c, H_c) \subset [0, 1)$. The half-open subinterval gives the set of all codewords that can be used to encode the input symbol sequence, which consists of all past input symbols. So any real number within the subinterval $[L_c, H_c)$ can be assigned as the codeword representing all the past occurring symbols. The selected real codeword is then transmitted in binary form (fractional binary representation, where 0.1 represents 1/2, 0.01 represents 1/4, 0.11 represents 3/4, and so on). When a new symbol occurs, the current subinterval $[L_c, H_c)$ is updated by finding a new subinterval $[L_c', H_c') \subset [L_c, H_c)$ to represent the new change in the encoded sequence. The codeword subinterval is chosen and updated such that its length is equal to the probability of occurrence of the corresponding encoded input sequence. It follows that less probable events (given by the input symbol sequences) are represented with shorter intervals and, therefore, require longer codewords since more precision bits are required to represent the narrower subintervals. So the arithmetic encoding procedure constructs, in a hierarchical manner, a code subinterval which uniquely represents a sequence of successive symbols.

In analogy with Huffman where the root node of the tree represents all possible occurring symbols, the interval $[0, 1)$ here represents all possible occurring sequences of symbols (all possible messages including single symbols). Also, considering the set of all possible M-symbol sequences having the same length M, the total interval $[0,1)$ can be subdivided into nonoverlapping subintervals such that each M symbol sequence is represented uniquely by one and only one subinterval whose length is equal to its probability of occurrence.

Let S be the source alphabet consisting of N symbols $s_0, \ldots, s_{(N-1)}$. Let $p_k = P(s_k)$ be the probability of symbol s_k, $0 \leq k \leq (N-1)$. Since, initially, the input sequence will consist of the first occurring symbol ($M = 1$), arithmetic coding begins by subdividing the interval $[0,1)$ into N nonoverlapping intervals, where each interval is assigned to a distinct symbol $s_k \in S$ and has a length equal to the symbol probability p_k. Let $[L_{s_k}, H_{s_k})$

TABLE 16.2   Example of code subinterval construction in arithmetic coding.

Source symbol s_k	Probability p_k	Symbol subinterval $[L_{s_k}, H_{s_k})$
s_0	0.1	$[0, 0.1)$
s_1	0.3	$[0.1, 0.4)$
s_2	0.4	$[0.4, 0.8)$
s_3	0.2	$[0.8, 1)$

denote the interval assigned to symbol s_k, where $p_k = H_{s_k} - L_{s_k}$. This assignment is illustrated in Table 16.2; the same source alphabet and source probabilities as in the example of Fig. 16.3 are used for comparison with Huffman. In practice, the subinterval limits L_{s_k} and H_{s_k} for symbol s_k can be directly computed from the available symbol probabilities and are equal to cumulative probabilities P_k as given below:

$$L_{s_k} = \sum_{i=0}^{k-1} p_k = P_{k-1}; \quad 0 \le k \le (N-1), \tag{16.21}$$

$$H_{s_k} = \sum_{i=0}^{k} p_k = P_k; \quad 0 \le k \le (N-1). \tag{16.22}$$

Let $[L_c, H_c)$ denote the code interval corresponding to the input sequence which consists of the symbols that occurred so far. Initially, $L_c = 0$ and $H_c = 1$; so the initial code interval is set to $[0, 1)$. Given an input sequence of symbols, the calculation of $[L_c, H_c)$ is performed based on the following encoding algorithm:

1. $L_c = 0$; $H_c = 1$.

2. Calculate code subinterval length,

$$\text{length} = H_c - L_c. \tag{16.23}$$

3. Get next input symbol s_k.

4. Update the code subinterval,

$$L_c = L_c + \text{length} \cdot L_{s_k},$$
$$H_c = L_c + \text{length} \cdot H_{s_k}. \tag{16.24}$$

5. Repeat from Step 2 until all the input sequence has been encoded.

As indicated before, any real number within the final interval $[L_c, H_c)$ can be used as a valid codeword for uniquely encoding the considered input sequence. The binary representation of the selected codeword is then transmitted. The above arithmetic encoding procedure is illustrated in Table 16.3 for encoding the sequence of symbols $s_1 s_0 s_2 s_3 s_3$. Another representation of the encoding process within the context of the considered

TABLE 16.3 Example of code subinterval construction in arithmetic coding.

Iteration # I	Encoded symbol s_k	Code subinterval $[L_c, H_c)$
1	s_1	$[0.1, 0.4)$
2	s_0	$[0.1, 0.13)$
3	s_2	$[0.112, 0.124)$
4	s_3	$[0.1216, 0.124)$
5	s_3	$[0.12352, 0.124)$

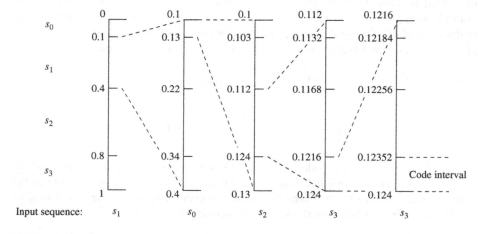

FIGURE 16.4

Arithmetic coding example.

example is shown in Fig. 16.4. Note that arithmetic coding can be viewed as remapping, at each iteration, the symbol subintervals $[L_{s_k}, H_{s_k})$ $(0 \le k \le (N-1))$ to the current code subinterval $[L_c, H_c)$. The mapping is done by rescaling the symbol subintervals to fit within $[L_c, H_c)$, while keeping them in the same relative positions. So when the next input symbol occurs, its symbol subinterval becomes the new code subinterval, and the process repeats until all input symbols are encoded.

In the arithmetic encoding procedure, the length of a code subinterval, *length* of (16.23), is always equal to the product of the probabilities of the individual symbols encoded so far, and it monotonically decreases at each iteration. As a result, the code interval shrinks at every iteration. So, longer sequences result in narrower code subintervals which would require the use of high-precision arithmetic. Also, a direct implementation of the presented arithmetic coding procedure produces an output only after all the input symbols have been encoded. Implementations that overcome these problems are presented in [14, 15]. The basic idea is to begin outputting the leading bit of the result as soon as it can be determined (incremental encoding), and then to shift out this bit (which amounts to scaling the current code subinterval by 2). In order to illustrate how incremental encoding would be possible, consider the example in Table 16.3. At the second iteration, the leading part "0.1" can be output since it is not going to be changed by the future encoding steps. A simple test to check whether a leading part can be output is to compare the leading parts of L_c and H_c; the leading digits that are the same can then be output and they remain unchanged since the next code subinterval will become smaller. For fixed-point computations, overflow and underflow errors can be avoided by restricting the source alphabet size [12].

Given the value of the codeword, arithmetic decoding can be performed as follows:

1. $L_c = 0$; $H_c = 1$.

2. Calculate the code subinterval length,

$$\text{length} = H_c - L_c.$$

3. Find symbol subinterval $[L_{s_k}, H_{s_k})$ $(0 \leq k \leq N - 1)$ such that

$$L_{s_k} \leq \frac{\text{codeword} - L_c}{\text{length}} < H_{s_k}.$$

4. Output symbol s_k.

5. Update code subinterval,

$$L_c = L_c + \text{length} \cdot L_{s_k}$$
$$H_c = L_c + \text{length} \cdot H_{s_k}.$$

6. Repeat from Step 2 until last symbol is decoded.

In order to determine when to stop the decoding (i.e., which symbol is the last symbol), a special end-of-sequence symbol is usually added to the source alphabet S and is handled like the other symbols. In the case when fixed-length blocks of symbols are encoded, the decoder can simply keep a count of the number of decoded symbols and no end-of-sequence symbol is needed. As discussed before, incremental decoding can be achieved before all the codeword bits are output [14, 15].

Context-based arithmetic coding has been widely used as the final entropy coding stage in state-of-the-art image and video compression schemes, including the JPEG-LS and the JPEG2000 standards. The same procedures and discussions hold for context-based arithmetic coding with the symbol probabilities $P(s_k)$ replaced with conditional symbol probabilities $P(s_k|\text{Context})$ where the context is determined from previously occuring neighboring symbols, as discussed in Section 16.3.2. In JPEG2000, context-based adaptive binary arithmetic coding (CABAC) is used with 17 contexts to efficiently code the binary significance, sign, and magnitude refinement information (Chapter 17). Binary arithmetic coding work with a binary (two-symbol) source alphabet, can be implemented more efficiently than nonbinary arithmetic coders, and has universal application as data symbols from any alphabet can be represented as a sequence of binary symbols [16].

16.3.5 Lempel-Ziv Coding

Huffman coding (Section 16.3.3) and arithmetic coding (Section 16.3.4) require a priori knowledge of the source symbol probabilities or of the source statistical model. In some cases, a sufficiently accurate source model is difficult to obtain, especially when several types of data (such as text, graphics, and natural pictures) are intermixed.

Universal coding schemes do not require a priori knowledge or explicit modeling of the source statistics. A popular lossless universal coding scheme is a dictionary-based coding method developed by Ziv and Lempel in 1977 [17] and known as Lempel-Ziv-77 (LZ77) coding. One year later, Ziv and Lempel presented an alternate dictionary-based method known as LZ78.

Dictionary-based coders dynamically build a coding table (called dictionary) of variable-length symbol strings as they occur in the input data. As the coding table is constructed, fixed-length binary codewords are assigned to the variable-length input symbol strings by indexing into the coding table. In Lempel-Ziv (LZ) coding, the decoder can also dynamically reconstruct the coding table and the input sequence as the code bits are received without any significant decoding delays. Although LZ codes do not explicitly make use of the source probability distribution, they asymptotically approach the source entropy rate for very long sequences [5]. Because of their adaptive nature, dictionary-based codes are ineffective for short input sequences since these codes initially result in a lot of bits being output. Short input sequences can thus result in data expansion instead of compression.

There are several variations of LZ coding. They mainly differ in how the dictionary is implemented, initialized, updated, and searched. Variants of the LZ77 algorithm have been used in many other applications and provided the basis for the development of many popular compression programs such as gzip, winzip, pkzip, and the public-domain Portable Network Graphics (PNG) image compression format.

One popular LZ coding algorithm is known as the LZW algorithm, a variant of the LZ78 algorithm developed by Welch [18]. This is the algorithm used for implementing the *compress* command in the UNIX operating system. The LZW procedure is also incorporated in the popular CompuServe GIF image format, where GIF stands for Graphics Interchange Format. However, the LZW compression procedure is patented, which decreased the popularity of compression programs and formats that make use of LZW. This was one main reason that triggered the development of the public-domain lossless PNG format.

Let S be the source alphabet consisting of N symbols s_k $(1 \leq k \leq N)$. The basic steps of the LZW algorithm can be stated as follows:

1. Initialize the first N entries of the dictionary with the individual source symbols of S, as shown below.

2. Parse the input sequence and find the longest input string of successive symbols w (including the first still unencoded symbol s in the sequence) that has a matching entry in the dictionary.

3. Encode w by outputing the index (address) of the matching entry as the codeword for w.

4. Add to the dictionary the string ws formed by concatenating w and the next input symbol s (following w).

5. Repeat from Step 2 for the remaining input symbols starting with the symbol s, until the entire input sequence is encoded.

Consider the source alphabet $S = \{s_1, s_2, s_3, s_4\}$. The encoding procedure is illustrated for the input sequence $s_1 s_2 s_1 s_2 s_3 s_2 s_1 s_2$. The constructed dictionary is shown in Table 16.4. The resulting code is given by the fixed-length binary representation of the following sequence of dictionary addresses: 1 2 5 3 6 2. The length of the generated binary codewords depends on the maximum allowed dictionary size. If the maximum dictionary size is M entries, the length of the codewords would be $\log_2 (M)$ rounded to the next smallest integer.

The decoder constructs the same dictionary (Table 16.4) as the codewords are received. The basic decoding steps can be described as follows:

1. Start with the same initial dictionary as the encoder. Also, initialize w to be the empty string.

2. Get the next "codeword" and decode it by outputing the symbol string sm stored at address "codeword" in dictionary.

3. Add to the dictionary the string ws formed by concatenating the previous decoded string w (if any) and the first symbol s of the current decoded string.

4. Set $w = m$ and repeat from Step 2 until all the codewords are decoded.

TABLE 16.4 Dictionary constructed while encoding the sequence $s_1 s_2 s_1 s_2 s_3 s_2 s_1 s_2$, which is emitted by a source with alphabet $S = \{s_1, s_2, s_3, s_4\}$.

Address	Entry	Address	Entry
1	s_1	1	s_1
2	s_2	2	s_2
3	s_3	3	s_3
4	s_4	⋮	⋮
5	$s_1 s_2$	N	s_N
6	$s_2 s_1$		
7	$s_1 s_2 s_3$		
8	$s_3 s_2$		
9	$s_2 s_1 s_2$		

Note that the constructed dictionary has a prefix property; i.e., every string w in the dictionary has its prefix string (formed by removing the last symbol of w) also in the dictionary. Since the strings added to the dictionary can become very long, the actual LZW implementation exploits the prefix property to render the dictionary construction more tractable. To add a string ws to the dictionary, the LZW implementation only stores the pair of values (c, s), where c is the address where the prefix string w is stored and s is the last symbol of the considered string ws. So the dictionary is represented as a linked list [5, 18].

16.3.6 Elias and Exponential-Golomb Codes

Similar to LZ coding, Elias codes [1] and Exponential-Golomb (Exp-Golomb) codes [2] are universal codes that do not require knowledge of the true source statistics. They belong to a class of structured codes that operate on the set of positive integers. Furthermore, these codes do not require having a finite set of values and can code arbitrary positive integers with an unknown upper bound. For these codes, each codeword can be constructed in a regular manner based on the value of the corresponding positive integer. This regular construction is formed based on the assumption that the probability distribution decreases monotonically with increasing integer values, i.e., smaller integer values are more probable than larger integer values. Signed integers can be coded by remapping them to positive integers. For example, an integer i can be mapped to the odd positive integer $2|i| - 1$ if it is negative, and to the even positive integer $2|i|$ if it is positive. Similarly, other one-to-one mapping can be formed to allow the coding of the entire integer set including zero. Noninteger source symbols can also be coded by first sorting them in the order of decreasing frequency of occurrence and then mapping the sorted set of symbols to the set of positive integers using a one-to-one (bijection) mapping, with smaller integer values being mapped to symbols with a higher frequency of occurrence. In this case, each positive integer value can be regarded as the index of the source symbol to which it is mapped, and can be referred to as the source symbol index or the codeword number or the codeword index.

Elias [1] described a set of codes including alpha (α), beta (β), gamma (γ), gamma' (γ'), delta (δ), and omega (ω) codes. For a positive integer I, the alpha code $\alpha(I)$ is a unary code that represents the value I with $(I - 1)$ 0's followed by a 1. The last 1 acts as a terminating flag which is also referred to as a comma. For example, $\alpha(1) = 1, \alpha(2) = 01$, $\alpha(3) = 001, \alpha(4) = 0001$, and so forth. The beta code of I, $\beta(I)$, is simply the natural binary representation of I with the most significant bit set to 1. For example, $\beta(1) = 1$, $\beta(2) = 10, \beta(3) = 11$, and $\beta(4) = 100$. One drawback of the beta code is that the codewords are not decodable, since it is not a prefix code and it does not contain a way to determine the length of the codewords. Thus the beta code is usually combined with other codes to form other useful codes, such as Elias gamma, gamma', delta, and omega codes, and Exp-Golomb codes. The Exp-Golomb codes have been incorporated within the H.264/AVC, also known as MPEG-4 Part 10, video coding standard to code different parameters and data values, including types of macro blocks, indices of reference frames, motion vector differences, quantization parameters, patterns for coded blocks, and others. Details about these codes are given below.

16.3.6.1 *Elias Gamma (γ) and Gamma' (γ') Codes*

The Elias γ and γ' codes are variants of each other with one code being a permutation of the other code. The γ' code is also commonly referred to as a γ code.

For a positive integer I, Elias γ' coding generates a binary codeword of the form

$$\gamma'(I) = [(L-1) \text{ zeros}][\beta(I)], \tag{16.25}$$

where $\beta(I)$ is the beta code of I which corresponds to the natural binary representation of I, and L is the length of (number of bits in) the binary codeword $\beta(I)$. L can be computed as $L = (\lfloor \log_2 (I) \rfloor + 1)$, where $\lfloor . \rfloor$ denotes rounding to the nearest smaller integer value. For example, $\gamma'(1) = 1, \gamma'(2) = 010, \gamma'(3) = 011$, and $\gamma'(4) = 00100$. In other words, an Elias γ' code can be constructed for a positive integer I using the following procedure:

1. Find the natural binary representation, $\beta(I)$, of I.

2. Determine the total number of bits, L, in $\beta(I)$.

3. The codeword $\gamma'(I)$ is formed as $(L-1)$ zeros followed by $\beta(I)$.

Alternatively, the Elias γ' code can be constructed as the unary alpha code $\alpha(L)$, where L is the number of bits in $\beta(I)$, followed by the last $(L-1)$ bits of $\beta(I)$ (i.e., $\beta(I)$ with the ommission of the most significant bit 1).

An Elias γ' code can be decoded by reading and counting the leading 0 bits until 1 is reached, which gives a count of $L-1$. Decoding then proceeds by reading the following $L-1$ bits and by appending those to 1 in order to get the $\beta(I)$ natural binary code. $\beta(I)$ is then converted into its corresponding integer value.

The Elias γ code of I, $\gamma(I)$, can be obtained as a permutation of the γ' code of I, $\gamma'(I)$, by preceding each bit of the last $L-1$ bits of the $\beta(I)$ codeword with one of the bits of the $\alpha(L)$ codeword, where L is the length of $\beta(I)$. In other words, interleave the first L bits in $\gamma'(I)$ with the last $L-1$ bits by alternating those. For example, $\gamma(1) = 1$, $\gamma(2) = 001, \gamma(3) = 011$, and $\gamma(4) = 00001$.

16.3.6.2 *Elias Delta (δ) Code*

For a positive integer I, Elias δ coding generates a binary codeword of the form:

$$\delta(I) = [(L'-1) \text{ zeros}][\beta(L)][\text{Last } (L-1) \text{ bits of } \beta(I)]$$

$$= [\gamma'(L)][\text{Last } (L-1) \text{ bits of } \beta(I)], \tag{16.26}$$

where $\beta(I)$ and $\beta(L)$ are the beta codes of I and L, respectively, L is the length of the binary codeword $\beta(I)$, and L' is the length of the binary codeword $\beta(L)$. For example,

$\delta(1) = 1, \delta(2) = 0100, \delta(3) = 0101$, and $\delta(4) = 01100$. In other words, Elias δ code can be constructed for a positive integer I using the following procedure:

1. Find the natural binary representation, $\beta(I)$, of I.

2. Determine the total number of bits, L, in $\beta(I)$.

3. Construct the γ' codeword, $\gamma'(L)$, of L, as discussed in Section 16.3.6.1.

4. The codeword $\delta(I)$ is formed as $\gamma'(L)$ followed by the last $(L-1)$ bits of $\beta(I)$ (i.e., $\beta(I)$ without the most significant bit 1).

An Elias δ code can be decoded by reading and counting the leading 0 bits until 1 is reached, which gives a count of $L'-1$. The $L'-1$ bits following the reached 1 bit are then read and appended to the 1 bit, which gives $\beta(L)$ and thus its corresponding integer value L. The next $L-1$ bits are then read and are appended to 1 in order to get $\beta(I)$. $\beta(I)$ is then converted into its corresponding integer value I.

16.3.6.3 *Elias Omega (ω) Code*

Similar to the previously discussed Elias δ code, the Elias ω code encodes the length L of the beta code, $\beta(I)$ of I, but it does this encoding in a recursive manner.

For a positive integer I, Elias ω coding generates a binary codeword of the form

$$\omega(I) = [\beta(L_N)][\beta(L_{N-1})]\dots[\beta(L_1)][\beta(L_0)][\beta(I)][0], \qquad (16.27)$$

where $\beta(I)$ is the beta code of I, $\beta(L_i)$ is the beta code of L_i, $i = 0,\dots,N$, and $(L_i + 1)$ corresponds to the length of the codeword $\beta(L_{i-1})$, for $i = 1,\dots,N$. In (16.27), $L_0 + 1$ corresponds to the length L of the codeword $\beta(I)$. The first codeword $\beta(L_N)$ can only be 10 or 11 for all positive integer values $I > 1$, and the other codewords $\beta(L_i)$, $i = 0,\dots,N-1$, have lengths greater than two. The Elias omega code is thus formed by recursively encoding the lengths of the $\beta(L_i)$ codewords. The recursion stops when the produced beta codeword has a length of two bits.

An Elias ω code, $\omega(I)$, for a positive integer I can be constructed using the following recursive procedure:

1. Set $R = I$ and set $\omega(I) = [0]$.

2. Set $C = \omega(I)$.

3. Find the natural binary representation, $\beta(R)$, of R.

4. Set $\omega(I) = [\beta(R)][C]$.

5. Determine the length (total number of bits) L_R of $\beta(R)$.

6. If L_R is greater than 2, set $R = L_R - 1$ and repeat from Step 2.

7. If L_R is equal to 2, stop.

8. If L_R is equal to 1, set $\omega(I) = [0]$ and stop.

For example, $\omega(1) = 0, \omega(2) = 100, \omega(3) = 110$, and $\omega(4) = 101000$.

An Elias ω code can be decoded by initially reading the first three bits. If the third bit is 0, then the first two bits correspond to the beta code of the value of the integer data I, $\beta(I)$. If the third bit is one, then the first two bits correspond to the beta code of a length, whose value indicates the number of bits to be read and placed following the third 1 bit in order to form a beta code. The newly formed beta code corresponds either to a coded length or to the coded data value I depending whether the next following bit is 0 or 1. So the decoding proceeds by reading the next bit following the last formed beta code. If the read bit is 1, the last formed beta code corresponds to the beta code of a length whose value indicated the number of values to read following the read 1 bit. If the read bit is 0, the last formed beta code corresponds to the beta code of I and the decoding terminates.

16.3.6.4 *Exponential-Golomb Codes*

Exponential-Golomb codes [2] are parameterized structured universal codes that encode nonnegative integers, i.e., both positive integers and zero can be encoded in contrast to the previously discussed Elias codes which do not provide a code for zero.

For a positive integer I, a kth order Exp-Golomb (Exp-Golomb) code generates a binary codeword of the form

$$EG_k(I) = [(L' - 1) \text{ zeros}][(\text{Most significant } (L - k) \text{ bits of } \beta(I)) + 1][\text{Last } k \text{ bits of } \beta(I)]$$

$$= [(L' - 1) \text{ zeros}][\beta(1 + I/2^k)][\text{Last } k \text{ bits of } \beta(I)], \tag{16.28}$$

where $\beta(I)$ is the beta code of I which corresponds to the natural binary representation of I, L is the length of the binary codeword $\beta(I)$, and L' is the length of the binary codeword $\beta(1 + I/2^k)$, which corresponds to taking the first $(L - k)$ bits of $\beta(I)$ and arithmetically adding 1. The length L can be computed as $L = (\lfloor \log_2 (I) \rfloor + 1)$, for $I > 0$, where $\lfloor . \rfloor$ denotes rounding to the nearest smaller integer. For $I = 0, L = 1$. Similarly, the length L' can be computed as $L' = (\lfloor \log_2 (1 + I/2^k) \rfloor + 1)$.

For example, for $k = 0$, $EG_0(0) = 1$, $EG_0(1) = 010$, $EG_0(2) = 011$, $EG_0(3) = 00100$, and $EG_0(4) = 00101$. For $k = 1$, $EG_1(0) = 10$, $EG_1(1) = 11$, $EG_1(2) = 0100$, $EG_1(3) = 0101$, and $EG_1(4) = 0110$.

Note that the Exp-Golomb code with order $k = 0$ of a nonnegative integer I, $EG_0(I)$, is equivalent to the Elias gamma' code of $I + 1$, $\gamma'(I + 1)$. The zeroth-order $(k = 0)$ Exp-Golomb codes are used as part of the H.264/AVC (MPEG-4 Part 10) video coding standard for coding parameters and data values related to macro blocks type, reference frame index, motion vector differences, quantization parameters, patterns for coded blocks, and other values [19].

A kth-order Exp-Golomb code can be decoded by first reading and counting the leading 0 bits until 1 is reached. Let the number of counted 0's be N. The binary codeword $\beta(I)$ is then obtained by reading the next N bits following the 1 bit, appending those read N bits to 1 in order to form a binary beta codeword, subtracting 1 from the formed binary codeword, and then reading and appending the last k bits. The obtained $\beta(I)$ codeword is converted into its corresponding integer value I.

16.4 LOSSLESS CODING STANDARDS

The need for interoperability between various systems have led to the formulation of several international standards for lossless compression algorithms targeting different applications. Examples include the standards formulated by the International Standards Organization (ISO), the International Electrotechnical Commission (IEC), and the International Telecommunication Union (ITU), which was formerly known as the International Consultative Committee for Telephone and Telegraph. A comparison of the lossless still image compression standards is presented in [20].

Lossless image compression standards include lossless JPEG (Chapter 17), JPEG-LS (Chapter 17), which supports lossless and near lossless compression, JPEG2000 (Chapter 17), which supports both lossless and scalable lossy compression, and facsimile compression standards such as the ITU-T Group 3 (T.4), Group 4 (T.6), JBIG (T.82), JBIG2 (T.88), and the Mixed Raster Content (MRC-T.44) standards [21]. While the lossless JPEG, JPEG-LS, and JPEG2000 standards are optimized for the compression of continuous-tone images, the facsimile compression standards are optimized for the compression of bilevel images except for the lastest MRC standard which is targeted for mixmode documents that can contain continuous-tone images in addition to text and line art.

The remainder of this section presents a brief overview of the JBIG, JBIG2, lossless JPEG, and JPEG2000 (with emphasis on lossless compression) standards. It is important to note that the image and video compression standards generally only specify the decoder-compatible bitstream syntax, thus leaving enough room for innovations and flexibility in the encoder and decoder design. The presented coding procedures below are popular standard implementations, but they can be modified as long as the generated bitstream syntax is compatible with the considered standard.

16.4.1 The JBIG and JBIG2 Standards

The JBIG standard (ITU-T Recommendation T.82, 1993) was developed jointly by the ITU and the ISO/IEC with the objective to provide improved lossless compression performance, for both business-type documents and binary halftone images, as compared to the existing standards. Another objective was to support progressive transmission. Grayscale images are also supported by encoding separately each bit plane. Later, the same JBIG committee drafted the JBIG2 standard (ITU-T Recommendation T.88, 2000) which provides improved lossless compression as compared to JBIG in addition to allowing lossy compression of bilevel images.

The JBIG standard consists of a context-based arithmetic encoder which takes as input the original binary image. The arithmetic encoder makes use of a context-based modeler that estimates conditional probabilities based on causal templates. A causal template consists of a set of already encoded neighboring pixels and is used as a context for the model to compute the symbol probabilities. Causality is needed to allow the decoder to recompute the same probabilities without the need to transmit side information.

JBIG supports sequential coding transmission (left to right, top to bottom) as well as progressive transmission. Progressive transmission is supported by using a layered coding scheme. In this scheme, a low resolution initial version of the image (initial layer) is first encoded. Higher resolution layers can then be encoded and transmitted in the order of increasing resolution. In this case the causal templates used by the modeler can include pixels from the previously encoded layers in addition to already encoded pixels belonging to the current layer.

Compared to the ITU Group 3 and Group 4 facsimile compression standards [12, 20], the JBIG standard results in 20% to 50% more compression for business-type documents. For halftone images, JBIG results in compression ratios that are two to five times greater than those obtained from the ITU Group 3 and Group 4 facsimile standards [12, 20].

In contrast to JBIG, JBIG2 allows the bilevel document to be partitioned into three types of regions: 1) text regions, 2) halftone regions, and 3) generic regions (such as line drawings or other components that cannot be classified as text or halftone). Both quality progressive and content progressive representations of a document are supported and are achieved by ordering the different regions in the document. In addition to the use of context-based arithmetic coding (MQ coding as in JBIG), JBIG2 allows also the use of run-length MMR (modified modified relative address designate) Huffman coding as in the Group 4 (ITU-T.6) facsimile standard, when coding the generic regions. Furthermore, JBIG2 supports both lossless and lossy compression. While the lossless compression performance of JBIG2 is slightly better than JBIG, JBIG2 can result in substantial coding improvements if lossy compression is used to code some parts of the bilevel documents.

16.4.2 **The Lossless JPEG Standard**

The JPEG standard was developed jointly by the ITU and ISO/IEC for the lossy and lossless compression of continuous-tone, color or grayscale, still images [22]. This section discusses very briefly the main components of the lossless mode of the JPEG standard (known as lossless JPEG).

The lossless JPEG coding standard can be represented in terms of the general coding structure of Fig. 16.1 as follows:

- Stage 1: Linear prediction/differential (DPCM) coding is used to form prediction residuals. The prediction residuals usually have a lower entropy than the original input image. Thus higher compression ratios can be achieved.

- Stage 2: The prediction residual is mapped into a pair of symbols (*category, magnitude*), where the symbol *category* gives the number of bits needed to encode *magnitude*.

- Stage 3: For each pair of symbols (*category, magnitude*), Huffman coding is used to code the symbol *category*. The symbol *magnitude* is then coded using a binary codeword whose length is given by the value *category*. Arithmetic coding can also be used in place of Huffman coding.

Complete details about the lossless JPEG standard and related recent developments, including JPEG-LS [23], are presented in Chapter 17.

16.4.3 The JPEG2000 Standard

JPEG2000 is the latest still image coding standard developed by the JPEG in order to support new features that are demanded by current modern applications and that are not supported by JPEG. Such features include lossy and lossless representations embedded within the same codestream, highly scalable codestreams with different progression orders (quality, resolution, spatial location, and component), region-of-interest (ROI) coding, and support for continuous-tone, bilevel, and compound image coding.

JPEG2000 is divided into 12 different parts featuring different application areas. JPEG2000 Part 1 [24] is the baseline standard and describes the minimal codestrean syntax that must be followed for compliance with the standard. All the other parts should include the features supported by this part. JPEG2000 Part 2 [25] is an extension of Part 1 and supports add-ons to improve the performance, including different wavelet filters with various subband decompositions. A brief overview of the JPEG2000 baseline (Part 1) coding procedure is presented below.

JPEG2000 [24] is a wavelet-based bit plane coding method. In JPEG2000, the original image is first divided into tiles (if needed). Each tile (subimage) is then coded independently. For color images, two optional color transforms, an irreversible color transform and a reversible color transform (RCT) are provided to decorrelate the color image components and increase the compression efficiency. The RCT should be used for lossless compression as it can be implemented using finite precision arithmetic and is perfectly invertible. Each color image component is then coded separately by dividing it first into tiles.

For each tile, the image samples are first shifted in level (if they are unsigned pixel values) such that they form a symmetric distribution of the DWT coefficients for the low-low (LL) subband. JPEG2000 (Part 1) supports two types of wavelet transforms: 1) an irreversible floating point 9/7 DWT [26], and 2) a reversible integer 5/3 DWT [27]. For lossless compression the 5/3 DWT should be used. After DC level shifting and the DWT, if lossy compression is chosen, the transformed coefficients are quantized using a deadzone scalar quantizer [4]. No quantization should be used in the case of lossless compression. The coefficients in each subband are then divided into coding blocks. The usual code block size is 64×64 or 32×32. Each coding block is then independently bit plane coded from the most significant bit plane (MSB) to the least significant bit plane using the embedded block coding with optimal truncation (EBCOT) algorithm [28].

The EBCOT algorithm consists of two coding stages known as tier-1 and tier-2 coding. In the tier-1 coding stage, each bit plane is fractionally coded using three coding passes: significant propagation, magnitude refinement, and cleanup (except the MSB, which is coded using only the cleanup pass). The significance propagation pass codes the significance of each sample based upon the significance of the neighboring eight pixels. The sign coding primitive is applied to code the sign information when a sample is coded for the first time as a nonzero bit plane coefficient. The magnitude refinement pass codes only those samples that have already become significant. The cleanup pass will code the remaining coefficients that are not coded during the first two passes. The output symbols from each pass are entropy coded using context-based arithmetic coding. At the same time, the rate increase and the distortion reduction associated with each coding pass is recorded. This information is then used by the postcompression rate-distortion (PCRD) optimization (PCRD-opt) algorithm to determine the contribution of each coding block to the different quality layers in the final bitstream. Given the compressed bitstream for each coding block and the rate allocation result, tier-2 coding is performed to form the final coded bitstream. This two-tier coding structure gives great flexibility to the final bitstream formation. By determining how to assemble the sub-bitstreams from each coding block to form the final bitstream, different progression (quality, resolution, position, component) order can be realized. More details about the JPEG2000 standard are given in Chapter 17.

16.5 OTHER DEVELOPMENTS IN LOSSLESS CODING

Several other lossless image coding systems have been proposed [7, 9, 29]. Most of these systems can be described in terms of the general structure of Fig. 16.1, and they make use of the lossless symbol coding techniques discussed in Section 16.3 or variations on those. Among the recently developed coding systems, LOCO-I [7] was adopted as part of the JPEG-LS standard (Chapter 17), since it exhibits the best compression/complexity tradeoff. Context-based, Adaptive, Lossless Image Code (CALIC) [9] achieves the best compression performance at a slightly higher complexity than LOCO-I. Perceptual-based coding schemes can achieve higher compression ratios at a much reduced complexity by removing perceptually-irrelevant information in addition to the redundant information. In this case, the decoded image is required to only be visually, and not necessarily numerically, identical to the original image. In what follows, CALIC and perceptual-based image coding are introduced.

16.5.1 CALIC

CALIC represents one of the best performing practical and general purpose lossless image coding techniques.

CALIC encodes and decodes an image in raster scan order with a single pass through the image. For the purposes of context modeling and prediction, the coding process uses a neighborhood of pixel values taken only from the previous two rows of the image. Consequently, the encoding and decoding algorithms require a buffer that holds only two rows of pixels that immediately precede the current pixel. Figure 16.5 presents a schematic description of the encoding process in CALIC. Decoding is achieved by the reverse process. As shown in Fig. 16.5, CALIC operates in two modes: binary mode and continuous-tone mode. This allows the CALIC system to distinguish between binary and continuous-tone images on a local, rather than a global, basis. This distinction between the two modes is important due to the vastly different compression methodologies employed within each mode. The former uses predictive coding, whereas the latter codes pixel values directly. CALIC selects one of the two modes depending on whether or not the local neighborhood of the current pixel has more than two distinct pixel values. The two-mode design contributes to the universality and robustness of CALIC over a wide range of images.

In the binary mode, a context-based adaptive ternary arithmetic coder is used to code three symbols, including an escape symbol. In the continuous-tone mode, the system has four major integrated components: prediction, context selection and quantization, context-based bias cancellation of prediction errors, and conditional entropy coding of prediction errors. In the prediction step, a gradient-adjusted prediction $\hat{y}$ of the current pixel y is made. The predicted value $\hat{y}$ is further adjusted via a bias cancellation procedure that involves an error feedback loop of one-step delay. The feedback value is the sample mean of prediction errors $\bar{e}$ conditioned on the current context. This results in an adaptive, context-based, nonlinear predictor $\check{y} = \hat{y} + \bar{e}$. In Fig. 16.5, these operations correspond to the blocks of "context quantization," "error modeling," and the error feedback loop.

The bias corrected prediction error $\check{y}$ is finally entropy coded based on a few estimated conditional probabilities in different conditioning states or coding contexts. A small number of coding contexts are generated by context quantization. The context quantizer partitions prediction error terms into a few classes by the expected error magnitude. The described procedures in relation to the system are identified by the blocks of "context quantization" and "conditional probabilities estimation" in Fig. 16.5. The details of this context quantization scheme in association with entropy coding are given in [9].

CALIC has also been extended to exploit interband correlations found in multiband images like color images, multispectral images, and 3D medical images. Interband CALIC

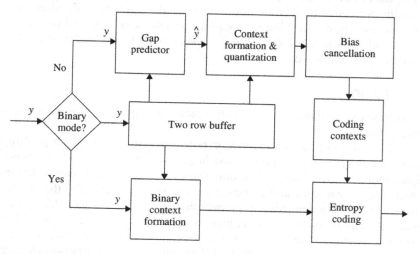

FIGURE 16.5

Schematic description of CALIC (Courtesy of Nasir Memon).

TABLE 16.5   Lossless bit rates with Intraband and Interband CALIC (Courtesy of Nasir Memon).

Image	JPEG-LS	Intraband CALIC	Interband CALIC
band	3.36	3.20	2.72
aerial	4.01	3.78	3.47
cats	2.59	2.49	1.81
water	1.79	1.74	1.51
cmpnd1	1.30	1.21	1.02
cmpnd2	1.35	1.22	0.92
chart	2.74	2.62	2.58
ridgely	3.03	2.91	2.72

can give 10% to 30% improvement over intraband CALIC, depending on the type of image. Table 16.5 shows bit rates achieved with intraband and interband CALIC on a set of multiband images. For the sake of comparison, results obtained with JPEG-LS are also included.

16.5.2 Perceptually Lossless Image Coding

The lossless coding methods presented so far require the decoded image data to be identical both quantitatively (numerically) and qualitatively (visually) to the original encoded image. This requirement usually limits the amount of compression that can be achieved to a compression factor of two or three even when sophisticated adaptive models are used as discussed in Section 16.5.1. In order to achieve higher compression factors, perceptually lossless coding methods attempt to remove redundant as well as perceptually irrelevant information.

Perceptual-based algorithms attempt to discriminate between signal components which are and are not detected by the human receiver. They exploit the spatio-temporal masking properties of the human visual system and establish thresholds of *just-noticeable distortion* (JND) based on psychophysical contrast masking phenomena. The interest is in bandlimited signals because of the fact that visual perception is mediated by a collection of individual mechanisms in the visual cortex, denoted *channels* or *filters*, that are selective in terms of frequency and orientation [30]. Mathematical models for human vision are discussed in Chapter 8.

Neurons respond to stimuli above a certain contrast. The necessary contrast to provoke a response from the neurons is defined as the *detection threshold*. The inverse of the detection threshold is the *contrast sensitivity*. Contrast sensitivity varies with frequency (including spatial frequency, temporal frequency, and orientation) and can be measured using *detection experiments* [31].

In *detection experiments*, the tested subject is presented with test images and needs only to specify whether the target stimulus is visible or not visible. They are used to derive JND or detection thresholds in the absence or presence of a masking stimulus superimposed over the target. For the image coding application, the input image is the masker and the target (to be masked) is the quantization noise (distortion). JND contrast sensitivity profiles, obtained as the inverse of the measured detection thresholds, are derived by varying the target or the masker contrast, frequency, and orientation. The common signals used in vision science for such experiments are sinusoidal gratings. For image coding, bandlimited subband components are used [31].

Several perceptual image coding schemes have been proposed [31–35]. These schemes differ in the way the perceptual thresholds are computed and used in coding the visual data. For example, not all the schemes account for contrast masking in computing the thresholds. One method called DCTune [33] fits within the framework of JPEG. Based on a model of human perception that considers frequency sensitivity and contrast masking, it designs a fixed DCT quantization matrix (3 quantization matrices in the case of color images) for each image. The fixed quantization matrix is selected to minimize an overall perceptual distortion which is computed in terms of the perceptual thresholds. In such block-based methods, a scalar value can be used for each block or macro block to uniformly scale a fixed quantization matrix in order to account for the variation in available masking (and as a means to control the bit rate) [34]. The quantization matrix and the scalar value for each block need to be transmitted, resulting in additional side information.

The perceptual image coder proposed by Safranek and Johnston [32] works in a subband decomposition setting. Each subband is quantized using a uniform quantizer with a fixed step size. The step size is determined by the JND threshold for uniform noise at the most sensitive coefficient in the subband. The model used does not include contrast masking. A scalar multiplier in the range of 2 to 2.5 is applied to uniformly scale all step sizes in order to compensate for the conservative step size selection and to achieve a good compression ratio.

Higher compression can be achieved by exploiting the varying perceptual characteristics of the input image in a locally-adaptive fashion. Locally-adaptive perceptual image coding requires computing and making use of image-dependent, locally-varying, masking thresholds to adapt the quantization to the varying characteristics of the visual data. However, the main problem in using a locally-adaptive perceptual quantization strategy is that the locally-varying masking thresholds are needed both at the encoder and at the decoder in order to be able to reconstruct the coded visual data. This, in turn, would require sending or storing a large amount of side information, which might lead to data expansion instead of compression. The aforementioned perceptual-based compression methods attempt to avoid this problem by giving up or significantly restricting the local adaptation. They either choose a fixed quantization matrix for the whole image, select one fixed step size for a whole subband, or scale all values in a fixed quantization matrix uniformly.

In [31, 35], locally-adaptive perceptual image coders are presented without the need for side information for the locally-varying perceptual thresholds. This is accomplished by using a low-order linear predictor, at both the encoder and decoder, for estimating the locally available amount of masking. The locally-adaptive perceptual image coding

(a) Original *Lena* image, 8 bpp   (b) Decoded *Lena* image at 0.361 bpp

FIGURE 16.6

Perceptually-lossless image compression [31]. The perceptual thresholds are computed for a viewing distance equal to 6 times the image height.

schemes [31, 35] achieve higher compression ratios (25% improvement on average) in comparison with the nonlocally adaptive schemes [32, 33] with no significant increase in complexity. Figure 16.6 presents coding results obtained by using the locally adaptive perceptual image coder of [31] for the *Lena* image. The original image is represented by 8 bits per pixel (bpp) and is shown in Fig. 16.6(a). The decoded perceptually-lossless image is shown in Fig 16.6(b) and requires only 0.361 bpp (compression ratio $C_R = 22$).

REFERENCES

[1] P. Elias. Universal codeword sets and representations of the integers. *IEEE Trans. Inf. Theory*, IT-21:194–203, 1975.

[2] J. Teuhola. A compression method for clustered bit-vectors. *Inf. Process. Lett.*, 7:308–311, 1978.

[3] J. Wen and J. D. Villasenor. Structured prefix codes for quantized low-shape-parameter generalized Gaussian sources. *IEEE Trans. Inf. Theory*, 45:1307–1314, 1999.

[4] D. S. Taubman and M. W. Marcellin. *JPEG2000: Image Compression Fundamentals, Standards, and Practice*. Kluwer Academic Publishers, Boston, MA, 2002.

[5] R. B. Wells. *Applied Coding and Information Theory for Engineers*. Prentice Hall, New Jersey, 1999.

[6] R. G. Gallager. Variations on a theme by Huffman. *IEEE Trans. Inf. Theory*, IT-24:668–674, 1978.

[7] M. J. Weinberger, G. Seroussi, and G. Sapiro. LOCO-I: a low complexity, context-based, lossless image compression algorithm. In *Data Compression Conference*, 140–149, March 1996.

[8] D. Taubman. Context-based, adaptive, lossless image coding. *IEEE Trans. Commun.*, 45:437–444, 1997.

[9] X. Wu and N. Memon. Context-based, adaptive, lossless image coding. *IEEE Trans. Commun.*, 45:437–444, 1997.

[10] Z. Liu and L. Karam. Mutual information-based analysis of JPEG2000 contexts. *IEEE Trans. Image Process.*, accepted for publication.

[11] D. A. Huffman. A method for the construction of minimum-redundancy codes. *Proc. IRE*, 40: 1098–1101, 1952.

[12] V. Bhaskaran and K. Konstantinides. *Image and Video Compression Standards: Algorithms and Architectures.* Kluwer Academic Publishers, Norwell, MA, 1995.

[13] W. W. Lu and M. P. Gough. A fast adaptive Huffman coding algorithm. *IEEE Trans. Commun.*, 41:535–538, 1993.

[14] I. H. Witten, R. M. Neal, and J. G. Cleary. Arithmetic coding for data compression. *Commun. ACM*, 30:520–540, 1987.

[15] F. Rubin. Arithmetic stream coding using fixed precision registers. *IEEE Trans. Inf. Theory*, IT-25:672–675, 1979.

[16] A. Said. Arithmetic coding. In K. Sayood, editor, *Lossless Compression Handbook*, Ch. 5, Academic Press, London, UK, 2003.

[17] J. Ziv and A. Lempel. A universal algorithm for sequential data compression. *IEEE Trans. Inf. Theory*, IT-23:337–343, 1977.

[18] T. A. Welch. A technique for high-performance data compression. *Computer*, 17:8–19, 1987.

[19] ITU-T Rec. H.264 (11/2007). Advanced video coding for generic audiovisual services. http://www.itu.int/rec/T-REC-H.264-200711-I/en (Last viewed: June 29, 2008).

[20] R. B. Arps and T. K. Truong. Comparison of international standards for lossless still image compression. *Proc. IEEE*, 82:889–899, 1994.

[21] K. Sayood. Facsimile compression. In K. Sayood, editor, *Lossless Compression Handbook*, Ch. 20, Academic Press, London, UK, 2003.

[22] W. Pennebaker and J. Mitchell. *JPEG Still Image Data Compression Standard.* Van Nostrand Rheinhold, New York, 1993.

[23] ISO/IEC JTC1/SC29 WG1 (JPEG/JBIG); ITU Rec. T. 87. Information technology – lossless and near-lossless compression of continuous-tone still images – final draft international standard FDIS14495-1 (JPEG-LS). Tech. Rep., ISO, 1998.

[24] ISO/IEC 15444-1. JPEG2000 image coding system – part 1: core coding system. Tech. Rep., ISO, 2000.

[25] ISO/IEC JTC1/SC20 WG1 N2000. JPEG2000 part 2 final committee draft. Tech. Rep., ISO, 2000.

[26] A. Cohen, I. Daubechies, and J. C. Feaveau. Biorthogonal bases of compactly supported wavelets. *Commun. Pure Appl. Math.*, 45:485–560, 1992.

[27] R. Calderbank, I. Daubechies, W. Sweldens, and B. L. Yeo. Wavelet transforms that map integers to integers. *Appl. Comput. Harmonics Anal.*, 5(3):332–369, 1998.

[28] D. Taubman. High performance scalable image compression with EBCOT. *IEEE Trans. Image Process.*, 9:1151–1170, 2000.

[29] A. Said and W. A. Pearlman. An image multiresolution representation for lossless and lossy compression. *IEEE Trans. Image Process.*, 5:1303–1310, 1996.

[30] L. Karam. An analysis/synthesis model for the human visual based on subspace decomposition and multirate filter bank theory. In *IEEE International Symposium on Time-Frequency and Time-Scale Analysis*, 559–562, October 1992.

[31] I. Hontsch and L. Karam. APIC: Adaptive perceptual image coding based on subband decomposition with locally adaptive perceptual weighting. In *IEEE International Conference on Image Processing*, Vol. 1, 37–40, October 1997.

[32] R. J. Safranek and J. D. Johnston. A perceptually tuned subband image coder with image dependent quantization and post-quantization. In *IEEE ICASSP*, 1945–1948, 1989.

[33] A. B. Watson. DCTune: A technique for visual optimization of DCT quantization matrices for individual images. *Society for Information Display Digest of Technical Papers XXIV*, 946–949, 1993.

[34] R. Rosenholtz and A. B. Watson. Perceptual adaptive JPEG coding. In *IEEE International Conference on Image Processing*, Vol. 1, 901–904, September 1996.

[35] I. Hontsch and L. Karam. Locally-adaptive image coding based on a perceptual target distortion. In *IEEE International Conference on Acoustics, Speech, and Signal Processing*, 2569–2572, May 1998.

JPEG and JPEG2000

17

Rashid Ansari[1], Christine Guillemot[2], Nasir Memon[3]

[1] *University of Illinois at Chicago;* [2] *TEMICS Research Group, INRIA, Rennes, France;* [3] *Polytechnic University, Brooklyn, New York*

17.1 INTRODUCTION

Joint Photographic Experts Group (JPEG) is currently a worldwide standard for compression of digital images. The standard is named after the committee that created it and continues to guide its evolution. This group consists of experts nominated by national standards bodies and by leading companies engaged in image-related work. The standardization effort is led by the International Standards Organization (ISO) and the International Telecommunications Union Telecommunication Standardization Sector (ITU-T). The JPEG committee has an official title of ISO/IEC JTC1 SC29 Working Group 1, with a web site at http://www.jpeg.org. The committee is charged with the responsibility of pooling efforts to pursue promising approaches to compression in order to produce an effective set of standards for still image compression. The lossy JPEG image compression procedure described in this chapter is part of the multipart set of ISO standards IS 10918-1,2,3 (ITU-T Recommendations T.81, T.83, T.84). A subsequent standardization effort was launched to improve compression efficiency and to support several desired features. This effort led to the JPEG2000 standard. In this chapter, the structure of the coder and decoder used in the JPEG and JPEG2000 standards and the features and options supported by these standards are described.

The JPEG standardization activity commenced in 1986, and it generated twelve proposals for consideration by the committee in March 1987. The initial effort produced consensus that the compression should be based on the discrete cosine transform (DCT). Subsequent refinement and enhancement led to the Committee Draft in 1990. Deliberations on the JPEG Draft International Standard (DIS) submitted in 1991 culminated in the International Standard (IS) being approved in 1992.

Although the JPEG and JPEG2000 standards define both lossy and lossless compression algorithms, the focus in this chapter is on the lossy compression component of the JPEG and the JPEG2000 standards. JPEG lossy compression entails an irreversible mapping of the image to a compressed bitstream, but the standard provides mechanisms for a controlled loss of information. Lossy compression produces a bitstream that is usually much smaller in size than that produced with lossless compression. Lossless image

421

compression is described in detail in Chapter 16 of this Guide [20]. The JPEG lossless standard is described in detail in The Handbook of Image and Video Processing [25].

The key features of the lossy JPEG standard are as follows:

- Both sequential and progressive modes of encoding are permitted. These modes refer to the manner in which quantized DCT coefficients are encoded. In sequential coding, the coefficients are encoded on a block-by-block basis in a single scan that proceeds from left to right and top to bottom. On the other hand, in progressive encoding only partial information about the coefficients is encoded in the first scan followed by encoding the residual information in successive scans.

- Low complexity implementations in both hardware and software are feasible.

- All types of images, regardless of source, content, resolution, color formats, etc., are permitted.

- A graceful tradeoff in bit rate and quality is offered, except at very low bit rates.

- A hierarchical mode with multiple levels of resolution is allowed.

- Bit resolution of 8 to 12 bits is permitted.

- A recommended file format, JPEG File Interchange Format (JFIF), enables the exchange of JPEG bitstreams among a variety of platforms.

A JPEG compliant decoder has to support a minimum set of requirements, the implementation of which is collectively referred to as *baseline implementation.* Additional features are supported in the extended implementation of the standard. The features supported in the baseline implementation include the ability to provide the following:

- a sequential build-up;

- custom or default Huffman tables;

- 8-bit precision per pixel for each component;

- image scans with 1-4 components;

- both interleaved and noninterleaved scans.

A JPEG extended system includes all features in a baseline implementation and supports many additional features. It allows sequential buildup as well as an optional progressive buildup. Either Huffman coding or arithmetic coding can be used in the entropy coding unit. Precision of up to 12 bits per pixel is allowed. The extended system includes an option for lossless coding.

The JPEG standard suffers from shortcomings in compression efficiency and progressive decoding. This led the JPEG committee to launch an effort in late 1996 and early 1997 to create a new image compression standard. The initiative resulted in the 15444/ITU-T Recommendation T.8000 known as the JPEG2000 standard that is based on wavelet analysis and encoding. The new standard is described in some detail in this chapter.

The rest of this chapter is organized as follows. In Section 17.2, we describe the structure of the JPEG codec and the units that it is comprised of. In Section 17.3, the role and computation of the DCT is examined. Procedures for quantizing the DCT coefficients are presented in Section 17.4. In Section 17.5, the mapping of the quantized DCT coefficients into symbols suitable for entropy coding is described. Syntactical issues and organization of data units are discussed in Section 17.6. Section 17.7 describes alternative modes of operation like the progressive and hierarchical modes. In Section 17.8, some extensions made to the standard, collectively known as JPEG Part 3, are described. Sections 17.9 and 17.10 provide a description of the new JPEG2000 standard and its coding architecture. The performance of JPEG2000 and the extensions included in part 2 of the standard are briefly described in Section 17.11. Finally, Section 17.12 lists further sources of information on the standards.

17.2 LOSSY JPEG CODEC STRUCTURE

It should be noted that in addition to defining an encoder and decoder, the JPEG standard also defines a syntax for representing the compressed data along with the associated tables and parameters. In this chapter, however, we largely ignore these syntactical issues and focus instead on the encoding and decoding procedures. We begin by examining the structure of the JPEG encoding and decoding systems. The discussion centers on the encoder structure and the building blocks that an encoder is comprised of. The decoder essentially consists of the inverse operations of the encoding process carried out in reverse.

17.2.1 Encoder Structure

The JPEG encoder and decoder are conveniently decomposed into units that are shown in Fig. 17.1. Note that the encoder shown in Fig. 17.1 is applicable in open-loop/unbuffered environments where the system is not operating under a constraint of a prescribed bit rate/budget. The units constituting the encoder are described next.

17.2.1.1 *Signal Transformation Unit: DCT*

In JPEG image compression, each component array in the input image is first partitioned into 8×8 rectangular blocks of data. A signal transformation unit computes the DCT of each 8×8 block in order to map the signal reversibly into a representation that is better suited for compression. The object of the transformation is to reconfigure the information in the signal to capture the redundancies and to present the information in a "machine-friendly" form that is convenient for disregarding the perceptually least relevant content. The DCT captures the spatial redundancy and packs the signal energy into a few DCT coefficients. The coefficient with zero frequency in both dimensions is called the direct current (DC) coefficient, and the remaining 63 coefficients are called alternating current (AC) coefficients.

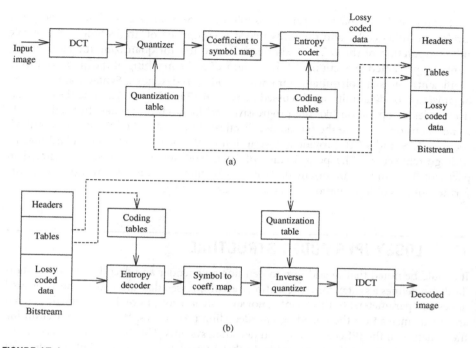

FIGURE 17.1

Constituent units of (a) JPEG encoder; (b) JPEG decoder.

17.2.1.2 *Quantizer*

If we wish to recover the original image exactly from the DCT coefficient array, then it is necessary to represent the DCT coefficients with high precision. Such a representation requires a large number of bits. In lossy compression, the DCT coefficients are mapped into a relatively small set of possible values that are represented compactly by defining and coding suitable symbols. The quantization unit performs this task of a many-to-one mapping of the DCT coefficients so that the possible outputs are limited in number. A key feature of the quantized DCT coefficients is that many of them are zero, making them suitable for efficient coding.

17.2.1.3 *Coefficient-to-Symbol Mapping Unit*

The quantized DCT coefficients are mapped to new symbols to facilitate a compact representation in the symbol coding unit that follows. The symbol definition unit can also be viewed as part of the symbol coding unit. However, it is shown here as a separate unit to emphasize the fact that the definition of symbols to be coded is an important task. An effective definition of symbols for representing AC coefficients in JPEG is the "runs" of zero coefficients followed by a nonzero terminating coefficient. For representing DC coefficients, symbols are defined by computing the difference between the DC coefficient in the current block and that in the previous block.

17.2.1.4 *Entropy Coding Unit*

This unit assigns a codeword to the symbols that appear at its input and generates the bitstream that is to be transmitted or stored. Huffman coding is usually employed for variable-length coding (VLC) of the symbols, with arithmetic coding allowed as an option.

17.2.2 Decoder Structure

In a decoder the inverse operations are performed in an order that is the reverse of that in the encoder. The coded bitstream contains coding and quantization tables which are first extracted. The coded data are then applied to the entropy decoder which determines the symbols that were encoded. The symbols are then mapped to an array of quantized and scaled values of DCT coefficients. This array is then appropriately rescaled by multiplying each entry with the corresponding entry in the quantization table to recover the approximations to the original DCT coefficients. The decoded image is then obtained by applying the inverse two-dimensional (2D) DCT to the array of the recovered approximate DCT coefficients.

In the next three sections, we consider each of the above encoder operations, DCT, quantization, and symbol mapping and coding, in more detail.

17.3 DISCRETE COSINE TRANSFORM

Lossy JPEG compression is based on the use of transform coding using the DCT [2]. In DCT coding, each component of the image is subdivided into blocks of 8×8 pixels. A 2D DCT is applied to each block of data to obtain an 8×8 array of coefficients. If $x[m, n]$ represents the image pixel values in a block, then the DCT is computed for each block of the image data as follows:

$$X[u, v] = \frac{C[u]C[v]}{4} \sum_{m=0}^{7} \sum_{n=0}^{7} x[m, n] \cos \frac{(2m + 1)u\pi}{16} \cos \frac{(2n + 1)v\pi}{16} \quad 0 \le u, v \le 7,$$

where

$$C[u] = \begin{cases} \frac{1}{\sqrt{2}} & u = 0, \\ 1 & 1 \le u \le 7. \end{cases}$$

The original image samples can be recovered from the DCT coefficients by applying the inverse discrete cosine transform (IDCT) as follows:

$$x[m, n] = \sum_{u=0}^{7} \sum_{v=0}^{7} \frac{C[u]C[v]}{4} X[u, v] \cos \frac{(2m + 1)u\pi}{16} \cos \frac{(2n + 1)v\pi}{16} \quad 0 \le m, n \le 7$$

The DCT, which belongs to the family of sinusoidal transforms, has received special attention due to its success in compression of real-world images. It is seen from the definition of the DCT that an 8×8 image block being transformed is being represented as a linear combination of real-valued basis vectors that consist of samples of a product of one-dimensional (1D) cosinusoidal functions. The 2D transform can be expressed as a product of 1D DCT transforms applied separably along the rows and columns of the image block. The coefficients $X(u, v)$ of the linear combination are referred to as the *DCT coefficients*. For real-world digital images in which the inter-pixel correlation is reasonably high and which can be characterized with first-order autoregressive models, the performance of the DCT is very close to that of the Karhunen-Loeve transform [2]. The discrete fourier transform (DFT) is not as efficient as DCT in representing an 8×8 image block. This is because when the DFT is applied to each row of the image, a periodic extension of the data, along with concomitant edge discontinuities, produces high-frequency DFT coefficients that are larger than the DCT coefficients of corresponding order. On the other hand, there is a mirror periodicity implied by the DCT which avoids the discontinuities at the edges when image blocks are repeated. As a result, the "high-frequency" or "high-order AC" coefficients are on the average smaller than the corresponding DFT coefficients.

We consider an example of the computation of the 2D DCT of an 8×8 block in the 512×512 gray-scale image, *Lena*. The specific block chosen is shown in the image in Fig. 17.2 (top) where the block is indicated with a black boundary with one corner of the 8×8 block at [209, 297]. A closeup of the block enclosing part of the hat is shown in Fig. 17.2 (bottom).

The 8-bit pixel values of the block chosen are shown in Fig. 17.3. After the DCT is applied to this block, the 8×8 DCT coefficient array obtained is shown in Fig. 17.4.

The magnitude of the DCT coefficients exhibits a pattern in their occurrences in the coefficient array. Also, their contribution to the perception of the information is not uniform across the array. The DCT coefficients corresponding to the lowest frequency basis functions are usually large in magnitude, and are also deemed to be perceptually most significant. These properties are exploited in developing methods of quantization and symbol coding. The bulk of the compression achieved in lossy transform coding occurs in the quantization step. The compression level is controlled by changing the total number of bits available to encode the blocks. The coefficients are quantized more coarsely when a large compression factor is required.

17.4 QUANTIZATION

Each DCT coefficient $X[m, n]$, $0 \le m, n \le 7$, is mapped into one of a finite number of levels determined by the compression factor desired.

17.4.1 DCT Coefficient Quantization Procedure

Quantization is done by dividing each element of the DCT coefficient array by a corresponding element in an 8×8 *quantization matrix* and rounding the result. Thus if the entry $q[m, n]$, $0 \le m, n \le 7$, in the m-th row and n-th column of the quantization matrix, is large then the corresponding DCT coefficient is coarsely quantized. The

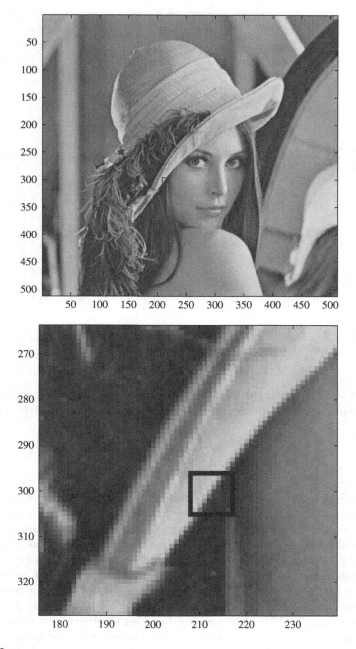

FIGURE 17.2

The original 512×512 *Lena* image (top) with an 8×8 block (bottom) identified with black boundary and with one corner at [209, 297].

187	188	189	202	209	175	66	41
191	186	193	209	193	98	40	39
188	187	202	202	144	53	35	37
189	195	206	172	58	47	43	45
197	204	194	106	50	48	42	45
208	204	151	50	41	41	41	53
209	179	68	42	35	36	40	47
200	117	53	41	34	38	39	63

FIGURE 17.3

The 8 × 8 block identified in Fig. 17.2.

915.6	451.3	25.6	−12.6	16.1	−12.3	7.9	−7.3
216.8	19.8	−228.2	−25.7	23.0	−0.1	6.4	2.0
−2.0	−77.4	−23.8	102.9	45.2	−23.7	−4.4	−5.1
30.1	2.4	19.5	28.6	−51.1	−32.5	12.3	4.5
5.1	−22.1	−2.2	−1.9	−17.4	20.8	23.2	−14.5
−0.4	−0.8	7.5	6.2	−9.6	5.7	−9.5	−19.9
5.3	−5.3	−2.4	−2.4	−3.5	−2.1	10.0	11.0
0.9	0.7	−7.7	9.3	2.7	−5.4	−6.7	2.5

FIGURE 17.4

DCT of the 8 × 8 block in Fig. 17.3.

values of $q[m,n]$ are restricted to be integers with $1 \leq q[m,n] \leq 255$, and they determine the quantization step for the corresponding coefficient. The quantized coefficient is given by

$$qX[m,n] = \left[\frac{X[m,n]}{q[m,n]} \right]_{\text{round}}.$$

A quantization table (or matrix) is required for each image component. However, a quantization table can be shared by multiple components. For example, in a luminance-plus-chrominance $Y - Cr - Cb$ representation, the two chrominance components usually share a common quantization matrix. JPEG quantization tables given in Annex K of the standard for luminance and components are shown in Fig. 17.5. These tables were obtained from a series of psychovisual experiments to determine the visibility thresholds for the DCT basis functions for a 760 × 576 image with chrominance components downsampled by 2 in the horizontal direction and at a viewing distance equal to six times the screen width. On examining the tables, we observe that the quantization table for the chrominance components has larger values in general implying that the quantization of the chrominance planes is coarser when compared with the luminance plane. This is done to exploit the human visual system's (HVS) relative insensitivity to chrominance components as compared with luminance components. The tables shown

16	11	10	16	24	40	51	61
12	12	14	19	26	58	60	55
14	13	16	24	40	57	69	56
14	17	22	29	51	87	80	62
18	22	37	56	68	109	103	77
24	35	55	64	81	104	113	92
49	64	78	87	103	121	120	101
72	92	95	98	112	100	103	99

17	18	24	47	99	99	99	99
18	21	26	66	99	99	99	99
24	26	56	99	99	99	99	99
47	66	99	99	99	99	99	99
99	99	99	99	99	99	99	99
99	99	99	99	99	99	99	99
99	99	99	99	99	99	99	99
99	99	99	99	99	99	99	99

FIGURE 17.5

Example quantization tables for luminance (left) and chrominance (right) components provided in the informative sections of the standard.

have been known to offer satisfactory performance, on the average, over a wide variety of applications and viewing conditions. Hence they have been widely accepted and over the years have become known as the "default" quantization tables.

Quantization tables can also be constructed by casting the problem as one of optimum allocation of a given budget of bits based on the coefficient statistics. The general principle is to estimate the variances of the DCT coefficients and assign more bits to coefficients with larger variances.

We now examine the quantization of the DCT coefficients given in Fig. 17.4 using the luminance quantization table in Fig. 17.5(a). Each DCT coefficient is divided by the corresponding entry in the quantization table, and the result is rounded to yield the array of quantized DCT coefficients in Fig. 17.6. We observe that a large number of quantized DCT coefficients are zero, making the array suitable for runlength coding as described in Section 17.6. The block from the *Lena* image recovered after decoding is shown in Fig. 17.7.

17.4.2 Quantization Table Design

With lossy compression, the amount of distortion introduced in the image is inversely related to the number of bits (bit rate) used to encode the image. The higher the rate, the lower the distortion. Naturally, for a given rate, we would like to incur the minimum possible distortion. Similarly, for a given distortion level, we would like to encode with the minimum rate possible. Hence lossy compression techniques are often studied in terms of their rate-distortion (RD) performance that bounds according to the highest compression achievable at a given level of distortion they introduce over different bit rates. The RD performance of JPEG is determined mainly by the quantization tables. As mentioned before, the standard does not recommend any particular table or set of tables and leaves their design completely to the user. While the image quality obtained from the use of the "default" quantization tables described earlier is very good, there is a need to provide flexibility to adjust the image quality by changing the overall bit rate. In practice, scaled versions of the "default" quantization tables are very commonly used to vary the quality and compression performance of JPEG. For example, the popular IJPEG implementation, freely available in the public domain, allows this adjustment through

57	41	2	0	0	0	0	0
18	1	−16	−1	0	0	0	0
0	−5	−1	4	1	0	0	0
2	0	0	0	−1	0	0	0
0	−1	0	0	0	0	0	0
0	0	0	0	0	0	0	0
0	0	0	0	0	0	0	0
0	0	0	0	0	0	0	0

FIGURE 17.6

8×8 discrete cosine transform block in Fig. 17.4 after quantization with the luminance quantization table shown in Fig. 17.5.

181	185	196	208	203	159	86	27
191	189	197	203	178	118	58	25
192	193	197	185	136	72	36	33
184	199	195	151	90	48	38	43
185	207	185	110	52	43	49	44
201	198	151	74	32	40	48	38
213	161	92	47	32	35	41	45
216	122	43	32	39	32	36	58

FIGURE 17.7

The block selected from the *Lena* image recovered after decoding.

the use of *quality factor Q* for scaling all elements of the quantization table. The scaling factor is computed as

$$\text{Scale factor} = \begin{cases} +\frac{5000}{Q} & \text{for} \quad 1 \leq Q < 50 \\ 200 - 2 * Q & \text{for} \quad 50 \leq Q \leq 99 \\ 1 & \text{for} \quad Q = 100 \end{cases} \qquad (17.1)$$

Although varying the rate by scaling a base quantization table according to some fixed scheme is convenient, it is clearly not optimal. Given an image and a bit rate, there exists a quantization table that provides the "optimal" distortion at the given rate. Clearly, the "optimal" table would vary with different images and different bit rates and even different definitions of distortion such as mean square error (MSE) or perceptual distortion. To get the best performance from JPEG in a given application, custom quantization tables may need to be designed. Indeed, there has been a lot of work reported in the literature addressing the issue of quantization table design for JPEG. Broadly speaking, this work can be classified into three categories. The first deals with explicitly optimizing the RD performance of JPEG based on statistical models for DCT coefficient distributions. The second attempts to optimize the visual quality of the reconstructed image at a given bit rate, given a set of display conditions and a perception model. The third addresses constraints imposed by applications, such as optimization for printers.

An example of the first approach is provided by the work of Ratnakar and Livny [30] who propose RD-OPT, an efficient algorithm for constructing quantization tables with optimal RD performance for a given image. The RD-OPT algorithm uses DCT coefficient distribution statistics from any given image in a novel way to optimize quantization tables simultaneously for the entire possible range of compression-quality tradeoffs. The algorithm is restricted to the MSE-related distortion measures as it exploits the property that the DCT is a unitary transform, that is, MSE in the pixel domain is the same as MSE in the DCT domain. The RD-OPT essentially consists of the following three stages:

1. Gather DCT statistics for the given image or set of images. Essentially this step involves counting how many times the n-th coefficient gets quantized to the value v when the quantization step size is q and what is the MSE for the n-th coefficient at this step size.

2. Use statistics collected above to calculate $R_n(q)$, the rate for the nth coefficient when the quantization step size is q and the corresponding distortion is $D_n(q)$, for each possible q. The rate $R_n(q)$ is estimated from the corresponding first-order entropy of the coefficient at the given quantization step size.

3. Compute $R(Q)$ and $D(Q)$, the rate and distortions for a quantization table Q, as

$$R(Q) = \sum_{n=0}^{63} R_n(Q[n]) \quad \text{and} \quad D(Q) = \sum_{n=0}^{63} D_n(Q[n]),$$

respectively. Use dynamic programming to optimize $R(Q)$ against $D(Q)$.

Optimizing quantization tables with respect to MSE may not be the best strategy when the end image is to be viewed by a human. A better approach is to match the quantization table to the human visual system HVS model. As mentioned before, the "default" quantization tables were arrived at in an *image independent* manner, based on the visibility of the DCT basis functions. Clearly, better performance could be achieved by an *image dependent* approach that exploits HVS properties like frequency, contrast, and texture masking and sensitivity. A number of HVS model based techniques for quantization table design have been proposed in the literature [3, 18, 41]. Such techniques perform an analysis of the given image and arrive at a set of thresholds, one for each coefficient, called the just noticeable distortion (JND) thresholds. The underlying idea being that if the distortion introduced is at or just below these thresholds, the reconstructed image will be perceptually distortion free.

Optimizing quantization tables with respect to MSE may also not be appropriate when there are constraints on the type of distortion that can be tolerated. For example, on examining Fig. 17.5, it is clear that the "high-frequency" AC quantization factors, i.e., $q[m, n]$ for larger values of m and n, are significantly greater than the DC coefficient $q[0, 0]$ and the "low-frequency" AC quantization factors. There are applications in which the information of interest in an image may reside in the high-frequency AC coefficients. For example, in compression of radiographic images [34], the critical diagnostic information is often in the high-frequency components. The size of microcalcification in mammograms is often so small that a coarse quantization of the higher AC coefficients will be unacceptable. In such cases, JPEG allows custom tables to be provided in the bitstreams.

Finally, quantization tables can also be optimized for hard copy devices like printers. JPEG was designed for compressing images that are to be displayed on devices that use cathode ray tube that offers a large range of pixel intensities. Hence, when an image is rendered through a half-tone device [40] like a printer, the image quality could be far from optimal. Vander Kam and Wong [37] give a closed-loop procedure to design a quantization table that is optimum for a given half-toning and scaling method. The basic idea behind their algorithm is to code more coarsely frequency components that are corrupted by half-toning and to code more finely components that are left untouched by half-toning. Similarly, to take into account the effects of scaling, their design procedure assigns higher bit rate to the frequency components that correspond to a large gain in the scaling filter response and lower bit rate to components that are attenuated by the scaling filter.

17.5   COEFFICIENT-TO-SYMBOL MAPPING AND CODING

The quantizer makes the coding lossy, but it provides the major contribution in compression. However, the nature of the quantized DCT coefficients and the preponderance of zeros in the array leads to further compression with the use of lossless coding. This requires that the quantized coefficients be mapped to symbols in such a way that the symbols lend themselves to effective coding. For this purpose, JPEG treats the DC coefficient and the set of AC coefficients in a different manner. Once the symbols are defined, they are represented with Huffman coding or arithmetic coding.

In defining symbols for coding, the DCT coefficients are scanned by traversing the quantized coefficient array in a zig-zag fashion shown in Fig. 17.8. The zig-zag scan processes the DCT coefficients in increasing order of spatial frequency. Recall that the quantized high-frequency coefficients are zero with high probability. Hence scanning in this order leads to a sequence that contains a large number of trailing zero values and can be efficiently coded as shown below.

The [0,0]-th element or the quantized DC coefficient is first separated from the remaining string of 63 AC coefficients, and symbols are defined next as shown in Fig. 17.9.

17.5.1   DC Coefficient Symbols

The DC coefficients in adjacent blocks are highly correlated. This fact is exploited to differentially code them. Let $qX_i[0,0]$ and $qX_{i-1}[0,0]$ denote the quantized DC coefficient in blocks i and $i-1$. The difference $\delta_i = qX_i[0,0] - qX_{i-1}[0,0]$ is computed. Assuming a precision of 8 bits/pixel for each component, it follows that the largest DC coefficient value (with $q[0,0] = 1$) is less than 2048, so that values of δ_i are in the range $[-2047, 2047]$. If Huffman coding is used, then these possible values would require a very large coding

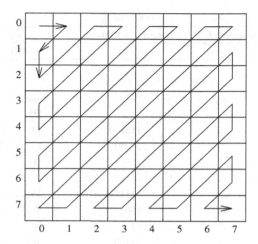

FIGURE 17.8

Zig-zag scan procedure.

table. In order to limit the size of the coding table, the values in this range are grouped into 12 size categories, which are assigned labels 0 through 11. Category k contains 2^k elements $\{\pm 2^{k-1}, \ldots, \pm (2^k - 1)\}$. The difference δ_i is mapped to a symbol described by a pair (category, amplitude). The 12 categories are Huffman coded. To distinguish values within the same category, extra k bits are used to represent a specific one of the possible 2^k "amplitudes" of symbols within category k. The amplitude of $\delta_i \{2^{k-1} \leq \delta_i \leq 2^k - 1\}$ is simply given by its binary representation. On the other hand, the amplitude of δ_i $\{-2^k - 1 \leq \delta_i \leq -2^{k-1}\}$ is given by the one's complement of the absolute value $|\delta_i|$ or simply by the binary representation of $\delta_i + 2^k - 1$.

17.5.2 Mapping AC Coefficient to Symbols

As observed before, most of the quantized AC coefficients are zero. The zig-zag scanned string of 63 coefficients contains many consecutive occurrences or "runs of zeros", making the quantized AC coefficients suitable for run-length coding (RLC). The symbols in this case are conveniently defined as [size of run of zeros, nonzero terminating value], which can then be entropy coded. However, the number of possible values of AC coefficients is large as is evident from the definition of DCT. For 8-bit pixels, the allowed range of AC coefficient values is $[-1023, 1023]$. In view of the large coding tables this entails, a procedure similar to that discussed above for DC coefficients is used. Categories are defined for suitable grouped values that can terminate a run. Thus a run/category pair together with the amplitude within a category is used to define a symbol. The category definitions and amplitude bits generation use the same procedure as in DC coefficient difference coding. Thus, a 4-bit category value is concatenated with a 4-bit run length to get an 8-bit [run/category] symbol. This symbol is then encoded using either Huffman or

(a) DC coding

Difference δ_i	[Category, Amplitude]	Code
−2	[2,−2]	01101

(b) AC coding

Terminating value	Run/ categ.	Code length	Code	Total bits	Amplitude bits
41	0/6	7	1111000	13	010110
18	0/5	5	11010	10	10010
1	1/1	4	1100	5	1
2	0/2	2	01	4	10
−16	1/5	11	11111110110	16	01111
−5	0/3	3	100	6	010
2	0/2	2	01	4	10
−1	2/1	5	11100	6	0
−1	0/1	2	00	3	0
4	3/3	12	111111110101	15	100
−1	1/1	4	1100	5	1
1	5/1	7	1111010	8	1
−1	5/1	7	1111010	8	0
EOB	EOB	4	1010	4	−
Total bits for block		112			
Rate = 112/64 = 1.75 bits per pixel					

FIGURE 17.9

(a) Coding of DC coefficient with value 57, assuming that the previous block has a DC coefficient of value 59; (b) Coding of AC coefficients.

arithmetic coding. There are two special cases that arise when coding the [run/category] symbol. First, since the run value is restricted to 15, the symbol (15/0) is used to denote fifteen zeroes followed by a zero. A number of such symbols can be cascaded to specify larger runs. Second, if after a nonzero AC coefficient, all the remaining coefficients are zero, then a special symbol (0/0) denoting an end-of-block (EOB) is encoded. Fig. 17.9 continues our example and shows the sequence of symbols generated for coding the quantized DCT block in the example shown in Fig. 17.6.

17.5.3 Entropy Coding

The symbols defined for DC and AC coefficients are entropy coded using mostly Huffman coding or, optionally and infrequently, arithmetic coding based on the probability estimates of the symbols. Huffman coding is a method of VLC in which shorter code words are assigned to the more frequently occurring symbols in order to achieve an average symbol code word length that is as close to the symbol source entropy as possible.

Huffman coding is optimal (meets the entropy bound) only when the symbol probabilities are integral powers of 1/2. The technique of *arithmetic coding* [42] provides a solution to attaining the theoretical bound of the source entropy. The baseline implementation of the JPEG standard uses Huffman coding only.

If Huffman coding is used, then Huffman tables, up to a maximum of eight in number, are specified in the bitstream. The tables constructed should not contain code words that (a) are more than 16 bits long or (b) consist of all ones. Recommended tables are listed in annex K of the standard. If these tables are applied to the output of the quantizer shown in the first two columns of Fig. 17.9, then the algorithm produces output bits shown in the following columns of the figure. The procedures for specification and generation of the Huffman tables are identical to the ones used in the lossless standard [25].

17.6 IMAGE DATA FORMAT AND COMPONENTS

The JPEG standard is intended for the compression of both grayscale and color images. In a grayscale image, there is a single "luminance" component. However, a color image is represented with multiple components, and the JPEG standard sets stipulations on the allowed number of components and data formats. The standard permits a maximum of 255 color components which are rectangular arrays of pixel values represented with 8- to 12-bit precision. For each color component, the largest dimension supported in either the horizontal or the vertical direction is $2^{16} = 65,536$.

All color component arrays do not necessarily have the same dimensions. Assume that an image contains K color components denoted by C_n, $n = 1, 2, \ldots, K$. Let the horizontal and vertical dimensions of the n-th component be equal to X_n and Y_n, respectively. Define dimensions X_{max}, Y_{max}, and X_{min}, Y_{min} as

$$X_{max} = \max_{n=1}^{K}\{X_n\}, \quad Y_{max} = \max_{n=1}^{K}\{Y_n\}$$

and

$$X_{min} = \min_{n=1}^{K}\{X_n\}, \quad Y_{min} = \min_{n=1}^{K}\{Y_n\}.$$

Each color component C_n, $n = 1, 2, \ldots, K$, is associated with relative horizontal and vertical sampling factors, denoted by H_n and V_n respectively, where

$$H_n = \frac{X_n}{X_{min}}, \quad V_n = \frac{Y_n}{Y_{min}}.$$

The standard restricts the possible values of H_n and V_n to the set of four integers $1, 2, 3, 4$. The largest values of relative sampling factors are given by $H_{max} = \max\{H_n\}$ and $V_{max} = \max\{V_n\}$.

According to the JFIF, the color information is specified by $[X_{max}, Y_{max}, H_n$ and V_n, $n = 1, 2, \ldots, K, H_{max}, V_{max}]$. The horizontal dimensions of the components are computed by the decoder as

$$X_n = \lceil X_{max} \times \frac{H_n}{H_{max}} \rceil.$$

Example 1: Consider a raw image in a luminance-plus-chrominance representation consisting of $K = 3$ components, $C_1 = Y$, $C_2 = Cr$, and $C_3 = Cb$. Let the dimensions of the luminance matrix (Y) be $X_1 = 720$ and $Y_1 = 480$, and the dimensions of the two chrominance matrices (Cr and Cb) be $X_2 = X_3 = 360$ and $Y_2 = Y_3 = 240$. In this case, $X_{max} = 720$ and $Y_{max} = 480$, and $X_{min} = 360$ and $Y_{min} = 240$. The relative sampling factors are $H_1 = V_1 = 2$ and $H_2 = V_2 = H_3 = V_3 = 1$.

When images have multiple components, the standard specifies formats for organizing the data for the purpose of storage. In storing components, the standard provides the option of using either interleaved or noninterleaved formats. Processing and storage efficiency is aided, however, by interleaving the components where the data is read in a single scan. Interleaving is performed by defining a *data unit* for lossy coding as a single block of 8×8 pixels in each color component. This definition can be used to partition the n-th color component C_n, $n = 1, 2, \ldots, K$, into rectangular blocks, each of which contains $H_n \times V_n$ data units. A *minimum coded unit* (MCU) is then defined as the smallest interleaved collection of data units obtained by successively picking $H_n \times V_n$ data units from the n-th color component. Certain restrictions are imposed on the data in order to be stored in the interleaved format:

- The number of interleaved components should not exceed four;

- An MCU should contain no more than ten data units, i.e.,

$$\sum_{n=1}^{K} H_n V_n \leq 10.$$

If the above restrictions are not met, then the data is stored in a noninterleaved format, where each component is processed in successive scans.

Example 2: Let us consider the case of storage of the Y, Cr, Cb components in Example 1. The luminance component contains 90×60 data units, and each of the two chrominance components contains 45×30 data units. Figure 17.10 shows both a noninterleaved and an interleaved arrangement of the data for $K = 3$ components, $C_1 = Y$, $C_2 = Cr$, and $C_3 = Cb$, with $H_1 = V_1 = 2$ and $H_2 = V_2 = H_3 = V_3 = 1$. The MCU in this case contains six data units, consisting of $H_1 \times V_1 = 4$ data units of the Y component and $H_2 \times V_2 = H_3 \times V_3 = 1$ each of the Cr and Cb components.

17.7 ALTERNATIVE MODES OF OPERATION

What has been described thus far in this chapter represents the JPEG *sequential DCT mode*. The sequential DCT mode is the most commonly used mode of operation of

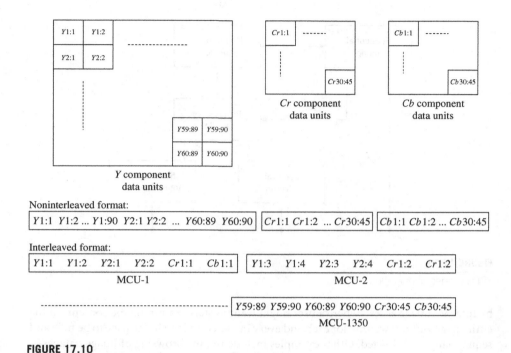

FIGURE 17.10

Organizations of the data units in the Y, Cr, Cb components into noninterleaved and interleaved formats.

JPEG and is required to be supported by any baseline implementation of the standard. However, in addition to the sequential DCT mode, JPEG also defines a *progressive DCT mode, sequential lossless mode,* and a *hierarchical mode.* In Figure 17.11 we show how the different modes can be used. For example, the hierarchical mode could be used in conjunction with any of the other modes as shown in the figure. In the lossless mode, JPEG uses an entirely different algorithm based on predictive coding [25]. In this section we restrict our attention to lossy compression and describe in greater detail the DCT-based progressive and hierarchical modes of operation.

17.7.1 Progressive Mode

In some applications it may be advantageous to transmit an image in multiple passes, such that after each pass an increasingly accurate approximation to the final image can be constructed at the receiver. In the first pass, very few bits are transmitted and the reconstructed image is equivalent to one obtained with a very low quality setting. Each of the subsequent passes contain an increasing number of bits which are used to refine the quality of the reconstructed image. The total number of bits transmitted is roughly the same as would be needed to transmit the final image by the sequential DCT mode. One example of an application which would benefit from progressive transmission is provided

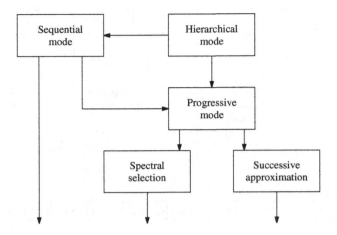

FIGURE 17.11

JPEG modes of operation.

by Internet image access, where a user might want to start examining the contents of the entire page without waiting for each and every image contained in the page to be fully and sequentially downloaded. Other examples include remote browsing of image databases, tele-medicine, and network-centric computing in general. JPEG contains a progressive mode of coding that is well suited to such applications. The disadvantage of progressive transmission, of course, is that the image has to be decoded a multiple number of times, and its use only makes sense if the decoder is faster than the communication link.

In the progressive mode, the DCT coefficients are encoded in a series of scans. JPEG defines two ways for doing this: *spectral selection* and *successive approximation*. In the spectral selection mode, DCT coefficients are assigned to different groups according to their position in the DCT block, and during each pass, the DCT coefficients belonging to a single group are transmitted. For example, consider the following grouping of the 64 DCT coefficients numbered from 0 to 63 in the zig-zag scan order,

$$\{0\}, \{1, 2, 3\}, \{4, 5, 6, 7\}, \{8, \ldots, 63\}.$$

Here, only the DC coefficient is encoded in the first scan. This is a requirement imposed by the standard. In the progressive DCT mode, DC coefficients are always sent in a separate scan. The second scan of the example codes the first three AC coefficients in zig-zag order, the third scan encodes the next four AC coefficients, and the fourth and the last scan encodes the remaining coefficients. JPEG provides the syntax for specifying the starting coefficient number and the final coefficient number being encoded in a particular scan. This limits a group of coefficients being encoded in any given scan to being successive in the zig-zag order. The first few DCT coefficients are often sufficient to give a reasonable rendition of the image. In fact, just the DC coefficient can serve to essentially identify the contents of an image, although the reconstructed image contains severe blocking artifacts. It should be noted that after all the scans are decoded, the final image quality is the same as that obtained by a sequential mode of operation. The bit rate, however, can be different as the entropy coding procedures for the progressive mode are different as described later in this section.

In successive approximation coding, the DCT coefficients are sent in successive scans with increasing level of precision. The DC coefficient, however, is sent in the first scan with full precision, just as in the case of spectral selection coding. The AC coefficients are sent bit plane by bit plane, starting from the most significant bit plane to the least significant bit plane.

The entropy coding techniques used in the progressive mode are slightly different from those used in the sequential mode. Since the DC coefficient is always sent as a separate scan, the Huffman and arithmetic coding procedures used remain the same as those in the sequential mode. However, coding of the AC coefficients is done a bit differently. In spectral selection coding (without selective refinement) and in the first stage of successive approximation coding, a new set of symbols is defined to indicate runs of EOB codes. Recall that in the sequential mode the EOB code indicates that the rest of the block contains zero coefficients. With spectral selection, each scan contains only a few AC coefficients and the probability of encountering EOB is significantly higher. Similarly, in successive approximation coding, each block consists of reduced precision coefficients, leading again to a large number of EOB symbols being encoded. Hence, to exploit this fact and achieve further reduction in bit rate, JPEG defines an additional set of fifteen symbols, EOB_n, each representing a run of 2^n EOB codes. After each EOB_i run-length code, extra i bits are appended to specify the exact run-length.

It should be noted that the two progressive modes, spectral selection and successive refinement, can be combined to give successive approximation in each spectral band being encoded. This results in quite a complex codec, which to our knowledge is rarely used.

It is possible to transcode between progressive JPEG and sequential JPEG without any loss in quality and approximately maintaining the same bit rate. Spectral selection results in bit rates slightly higher than the sequential mode, whereas successive approximation often results in lower bit rates. The differences however are small.

Despite the advantages of progressive transmission, there have not been many implementations of progressive JPEG codecs. There has been some interest in them due to the proliferation of images on the Internet.

17.7.2 **Hierarchical Mode**

The hierarchical mode defines another form of progressive transmission where the image is decomposed into a pyramidal structure of increasing resolution. The top-most layer in the pyramid represents the image at the lowest resolution, and the base of the pyramid represents the image at full resolution. There is a doubling of resolutions both in the horizontal and vertical dimensions, between successive levels in the pyramid. Hierarchical coding is useful when an image could be displayed at different resolutions in units such as handheld devices, computer monitors of varying resolutions, and high-resolution printers. In such a scenario, a multiresolution representation allows the transmission

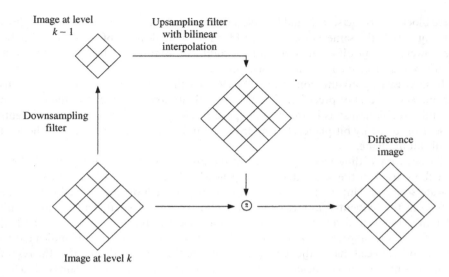

FIGURE 17.12

JPEG hierarchical mode.

of the appropriate layer to each requesting device, thereby making full use of available bandwidth.

In the JPEG hierarchical mode, each image component is encoded as a sequence of frames. The lowest resolution frame (level 1) is encoded using one of the sequential or progressive modes. The remaining levels are encoded differentially. That is, an estimate I_i' of the image, I_i, at the $i'th$ level ($i \geq 2$) is first formed by upsampling the low-resolution image I_{i-1} from the layer immediately above. Then the difference between I_i' and I_i is encoded using modifications of the DCT-based modes or the lossless mode. If lossless mode is used to code each refinement, then the final reconstruction using all layers is lossless. The upsampling filter used is a bilinear interpolating filter that is specified by the standard and cannot be specified by the user. Starting from the high-resolution image, successive low-resolution images are created essentially by downsampling by two in each direction. The exact downsampling filter to be used is not specified but the standard cautions that the downsampling filter used be consistent with the fixed upsampling filter. Note that the decoder does not need to know what downsampling filter was used in order to decode a bitstream. Figure 17.12 depicts the sequence of operations performed at each level of the hierarchy.

Since the differential frames are already signed values, they are not level-shifted prior to forward discrete cosine transform (FDCT). Also, the DC coefficient is coded directly rather than differentially. Other than these two features, the Huffman coding model in the progressive mode is the same as that used in the sequential mode. Arithmetic coding is, however, done a bit differently with conditioning states based on the use of differences with the pixel to the left as well as the one above. For details the user is referred to [28].

17.8 JPEG PART 3

JPEG has made some recent extensions to the original standard described in [11]. These extensions are collectively known as JPEG Part 3. The most important elements of JPEG part 3 are variable quantization and tiling, as described in more detail below.

17.8.1 Variable Quantization

One of the main limitations of the original JPEG standard was the fact that visible artifacts can often appear in the decompressed image at moderate to high compression ratios. This is especially true for parts of the image containing graphics, text, or some synthesized components. Artifacts are also common in smooth regions and in image blocks containing a single dominant edge. We consider compression of a 24 bits/pixel color version of the *Lena* image. In Fig. 17.13 we show the reconstructed *Lena* image with different compression ratios. At 24 to 1 compression we see few artifacts. However, as the compression ratio is increased to 96 to 1, noticeable artifacts begin to appear. Especially annoying is the "blocking artifact" in smooth regions of the image.

One approach to deal with this problem is to change the "coarseness" of quantization as a function of image characteristics in the block being compressed. The latest extension of the JPEG standard, called JPEG Part 3, allows rescaling of quantization matrix Q on a block by block basis, thereby potentially changing the manner in which quantization is performed for each block. The scaling operation is not done on the DC coefficient $Y[0,0]$ which is quantized in the same manner as in the baseline JPEG. The remaining 63 AC coefficients, $Y[u,v]$, are quantized as follows:

$$\hat{Y}[u,v] = \left[\frac{Y[u,v] \times 16}{Q[u,v] \times \text{QScale}} \right],$$

where QScale is a parameter that can take on values from 1 to 112, with a default value of 16. For the decoder to correctly recover the quantized AC coefficients, it needs to know the value of QScale used by the encoding process. The standard specifies the exact syntax by which the encoder can specify change in QScale values. If no such change is signaled, then the decoder continues using the QScale value that is in current use. The overhead incurred in signaling a change in the scale factor is approximately 15 bits depending on the Huffman table being employed.

It should be noted that the standard only specifies the syntax by means of which the encoding process can signal changes made to the QScale value. It does not specify how the encoder may determine if a change in QScale is desired and what the new value of QScale should be. Typical methods for variable quantization proposed in the literature use the fact that the HVS is less sensitive to quantization errors in highly active regions of the image. Quantization errors are frequently more perceptible in blocks that are smooth or contain a single dominant edge. Hence, prior to quantization, a few simple features for each block are computed. These features are used to classify the block as either smooth, edge, or texture, and so forth. On the basis of this classification as well as a simple activity measure computed for the block, a QScale value is computed.

FIGURE 17.13

Lena image at 24 to 1 (top) and 96 to 1 (bottom) compression ratios.

For example, Konstantinides and Tretter [21] give an algorithm for computing QScale factors for improving text quality on compound documents. They compute an activity measure M_i for each image block as a function of the DCT coefficients as follows:

$$M_i = \frac{1}{64} \left[\log_2 |Y_i[0,0] - Y_{i-1}[0,0]| + \sum_{j,k} \log_2 |Y_i[j,k]| \right].$$ (17.2)

The QScale value for the block is then computed as

$$QScale_i = \begin{cases} a \times M_i + b & \text{if } 2 > a \times M_i + b \geq 0.4 \\ 0.4 & a \times M_i + b \geq 0.4 \\ 2 & a \times M_i + b > 2. \end{cases}$$ (17.3)

The technique is only designed to detect text regions and will quantize high-activity textured regions in the image part at the same scale as text regions. Clearly, this is not optimal as high-activity textured regions can be quantized very coarsely leading to improved compression. In addition, the technique does not discriminate smooth blocks where artifacts are often the first to appear.

Algorithms for variable quantization that perform a more extensive classification have been proposed for video coding but nevertheless are also applicable to still image coding. One such technique has been proposed by Chun *et al.* [10] who classify blocks as being either smooth, edge, or texture, based on several parameters defined in the DCT domain as shown below:

E_h: horizontal energy E_v: vertical energy E_d: diagonal energy
E_a: $avg(E_h, E_v, E_d)$ E_m: $min(E_h, E_v, E_d)$ E_M: $max(E_h, E_v, E_d)$
$E_{m/M}$: ratio of E_m and E_M.

E_a represents the average high-frequency energy of the block, and is used to distinguish between low-activity blocks and high-activity blocks. Low-activity (smooth) blocks satisfy the relationship, $E_a \leq T_1$, where T_1 is a low-valued threshold. High-activity blocks are further classified into texture blocks and edge blocks. Texture blocks are detected under the assumption that they have relatively uniform energy distribution in comparison with edge blocks. Specifically, a block is deemed to be a texture block if it satisfies the conditions: $E_a > T_1, E_{min} > T_2$, and $E_{m/M} > T_3$, where T_1, T_2, and T_3 are experimentally determined constants. All blocks which fail to satisfy the smoothness and texture tests are classified as edge blocks.

17.8.2 Tiling

JPEG Part 3 defines a tiling capability whereby an image is subdivided into blocks or tiles, each coded independently. Tiling facilitates the following features:

- Display of an image region on a given screen size;
- Fast access to image subregions;

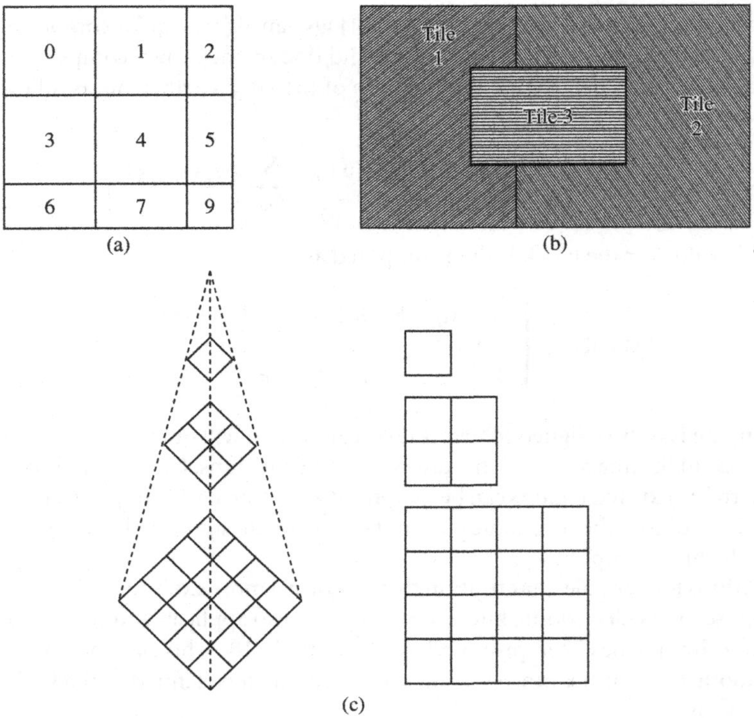

FIGURE 17.14

Different types of tilings allowed in JPEG Part 3: (a) simple; (b) composite; and (c) pyramidal.

- Region of interest refinement;
- Protection of large images from copying by giving access to only a part of it.

As shown in Fig. 17.14, the different types of tiling allowed by JPEG are as follows:

- Simple tiling: This form of tiling is essentially used for dividing a large image into multiple sub-images which are of the same size (except for edges) and are nonoverlapping. In this mode, all tiles are required to have the same sampling factors and components. Other parameters like quantization tables and Huffman tables are allowed to change from tile to tile.

- Composite tiling: This allows multiple resolutions on a single image display plane. Tiles can overlap within a plane.

- Pyramidal tiling: This is used for storing multiple resolutions of an image. Simple tiling as described above is used in each resolution. Tiles are stored in raster order, left to right, top to bottom, and low resolution to high resolution.

Another Part 3 extension is selective refinement. This feature permits a scan in a progressive mode, or a specific level of a hierarchical sequence, to cover only part of the total image area. Selective refinement could be useful, for example, in telemedicine applications where a radiologist could request refinements to specific areas of interest in the image.

17.9 THE JPEG2000 STANDARD

The JPEG standard has proved to be a tremendous success over the past decade in many digital imaging applications. However, as the needs of multimedia and imaging applications evolved in areas such as medical imaging, reconnaissance, the Internet, and mobile imaging, it became evident that the JPEG standard suffered from shortcomings in compression efficiency and progressive decoding. This led the JPEG committee to launch an effort in late 1996 and early 1997 to create a new image compression standard. The intent was to provide a method that would support a range of features in a single compressed bitstream for different types of still images such as bilevel, gray level, color, multicomponent—in particular multispectral—or other types of imagery. A call for technical contributions was issued in March 1997. Twenty-four proposals were submitted for consideration by the committee in November 1997. Their evaluation led to the selection of a wavelet-based coding architecture as the backbone for the emerging coding system. The initial solution, inspired by the wavelet trellis-coded quantization (WTCQ) algorithm [32] based on combining wavelets and trellis-coded quantization (TCQ) [6, 23], has been refined via a series of core experiments over the ensuing three years. The initiative resulted in the ISO 15444/ITU-T Recommendation T.8000 known as the JPEG2000 standard. It comprises six parts that are either complete or nearly complete at the time of writing this chapter, together with four new parts that are under development. The status of the parts is available at the official website [19].

Part 1, in the spirit of the JPEG baseline system, specifies the core compression system together with a minimal file format [13]. JPEG2000 Part 1 addresses some limitations of existing standards by supporting the following features:

- Lossless and lossy compression of continuous-tone and bilevel images with reduced distortion and superior subjective performance.

- Progressive transmission and decoding based on resolution scalability by pixel accuracy (i.e., based on quality or signal-to-noise (SNR) scalability). The bytes extracted are identical to those that would be generated if the image had been encoded targeting the desired resolution or quality, the latter being directly available without the need for decoding and re-encoding.

- Random access to spatial regions (or regions of interest) as well as to components. Each region can be accessed at a variety of resolutions and qualities.

- Robustness to bit errors (e.g., for mobile image communication).

■ Encoding capability for sequential scan, thereby avoiding the need to buffer the entire image to be encoded. This is especially useful when manipulating images of very large dimensions such as those encountered in reconnaissance (satellite and radar) images.

Some of the above features are supported to a limited extent in the JPEG standard. For instance, as described earlier, the JPEG standard has four modes of operation: sequential, progressive, hierarchical, and lossless. These modes use different techniques for encoding (e.g., the lossless compression mode relies on predictive coding, whereas the lossy compression modes rely on the DCT). One drawback is that if the JPEG lossless mode is used, then lossy decompression using the lossless encoded bitstream is not possible. One major advantage of JPEG2000 is that these four operation modes are integrated in it in a "compress once, decompress many" paradigm, with superior RD and subjective performance over a large range of RD operating points.

Part 2 specifies extensions to the core compression system and a more complete file format [14]. These extensions address additional coding features such as generalized and variable quantization offsets, TCQ, visual masking, and multiple component transformations. In addition it includes features for image editing such as cropping in the compressed domain or mirroring and flipping in a partially-compressed domain.

Parts 3, 4, and 5 provide a specification for motion JPEG 2000, conformance testing, and a description of a reference software implementation, respectively [15–17]. Four parts, numbered 8–11, are still under development at the time of writing. Part 8 deals with security aspects, Part 9 specifies an interactive protocol and an application programming interface for accessing JPEG2000 compressed images and files via a network, Part 10 deals with volumetric imaging, and Part 11 specifies the tools for wireless imaging.

The remainder of this chapter provides a brief overview of JPEG2000 Part 1 and outlines the main extensions provided in Part 2. The JPEG2000 standard embeds efficient lossy, near-lossless and lossless representations within the same stream. However, while some coding tools (e.g., color transformations, discrete wavelet transforms) can be used both for lossy and lossless coding, others can be used for lossy coding only. This led to the specification of two coding paths or options referred to as the reversible (embedding lossy and lossless representations) and irreversible (for lossy coding only) paths with common and path-specific building blocks. This chapter presents the main components of the two coding paths which can be used for lossy coding. Discussion of the components specific to JPEG2000 lossless coding can be found in [25], and a detailed description of the JPEG2000 coding tools and system can be found in [36]. Tutorials and overviews are presented in [9, 29, 33].

17.10 JPEG2000 PART 1: CODING ARCHITECTURE

The coding architecture comprises two paths, the irreversible and the reversible paths shown in Fig. 17.15. Both paths can be used for lossy coding by truncating the compressed codestream at the desired bit rate. The input image may comprise one or more (up to 16, 384) signed or unsigned components to accommodate various forms of imagery,

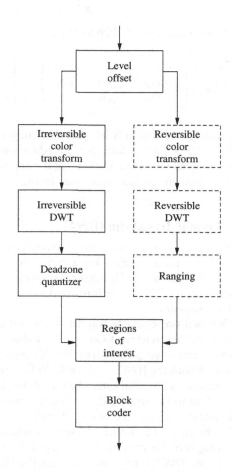

FIGURE 17.15

Main building blocks of the JPEG2000 coder. The path with boxes in dotted lines corresponds to the JPEG2000 lossless coding mode [25].

including multispectral imagery. The various components may have different bit depth, resolution, and sign specifications.

17.10.1 Preprocessing: Tiling, Level Offset, and Color Transforms

The first steps in both paths are optional and can be regarded as preprocessing steps. The image is first, optionally, partitioned into rectangular and nonoverlapping tiles of equal size. If the sample values are unsigned and represented with B bits, an offset of -2^{B-1} is added leading to a signed representation in the range $[-2^{B-1}, 2^{B-1}]$ that is symmetrically distributed about 0. The color component samples may be converted into luminance and color difference components via an irreversible color transform (ICT) or a reversible color transform (RCT) in the irreversible or reversible paths, respectively.

The ICT is identical to the conversion from RGB to YC_bC_r,

$$
\begin{bmatrix} Y \\ C_b \\ C_r \end{bmatrix} = \begin{bmatrix} 0.299 & 0.587 & 0.114 \\ -0.169 & -0.331 & 0.500 \\ 0.500 & -0.419 & -0.081 \end{bmatrix} \begin{bmatrix} R \\ G \\ B \end{bmatrix},
$$

and can be used for lossy coding only. The RCT is a reversible integer-to-integer transform that approximates the ICT. This color transform is required for lossless coding [25]. The RCT can also be used for lossy coding, thereby allowing the embedding of both a lossy and lossless representation of the image in a single codestream.

17.10.2 Discrete Wavelet Transform (DWT)

After tiling, each tile component is decomposed with a forward discrete wavelet transform (DWT) into a set of $L = 2^l$ resolution levels using a dyadic decomposition. A detailed and complete presentation of the theory and implementation of filter banks and wavelets is beyond the scope of this chapter. The reader is referred to Chapter 6 [26] and to [38] for additional insight on these issues.

The forward DWT is based on separable wavelet filters and can be irreversible or reversible. The transforms are then referred to as reversible discrete wavelet transform (RDWT) and irreversible discrete wavelet transform (IDWT). As for the color transform, lossy coding can make use of both the IDWT and the RDWT. In the case of RDWT, the codestream is truncated to reach a given bit rate. The use of the RDWT allows for both lossless and lossy compression to be embedded in a single compressed codestream. In contrast, lossless coding restricts us to the use of only RDWT.

The default RDWT is based on the spline 5/3 wavelet transform first introduced in [22]. The RDWT filtering kernel is presented elsewhere [25] in this handbook. The default irreversible transform, IDWT, is implemented with the Daubechies 9/7 wavelet kernel [4]. The coefficients of the analysis and synthesis filters are given in Table 17.1. Note however that, in JPEG2000 Part 2, other filtering kernels specified by the user can be used to decompose the image.

TABLE 17.1 Indirect discrete wavelet transform analysis and synthesis filters coefficients.

Index	Lowpass analysis filter coefficient	Highpass analysis filter coefficient	Lowpass synthesis filter coefficient	Highpass synthesis filter coefficient
0	0.602949018236360	1.115087052457000	1.115087052457000	0.602949018236360
+/−1	0.266864118442875	−0.591271763114250	0.591271763114250	−0.266864118442875
+/−2	−0.078223266528990	−0.057543526228500	−0.057543526228500	−0.078223266528990
+/−3	−0.016864118442875	0.091271763114250	−0.091271763114250	0.016864118442875
+/−4	+0.026748757410810			0.026748757410810

These filtering kernels are of odd length. Their implementation at the boundary of the image or subbands requires a symmetric signal extension. Two filtering modes are possible: convolution- and lifting-based [26].

17.10.3 Quantization and Inverse Quantization

JPEG2000 adopts a scalar quantization strategy, similar to that in the JPEG baseline system. One notable difference is in the use of a central deadzone quantizer. A detailed description of the procedure can be found in [36]. This section provides only an outline of the algorithm. In Part 1, the subband samples are quantized with a deadzone scalar quantizer with a central interval that is twice the quantization step size. The quantization of $y_i(n)$ is given by

$$\hat{y}_i(n) = \text{sign}(y_i(n))\lfloor \frac{|y_i(n)|}{\Delta_i} \rfloor, \tag{17.4}$$

where Δ_i is the quantization step size in the subband i. The parameter Δ_i is chosen so that $\Delta_i = \Delta\sqrt{\frac{1}{G_i}}$, where G_i is the squared norm of the DWT synthesis basis vectors for subband i and Δ is a parameter to be adjusted to meet given RD constraints. The step size Δ_i is represented with two bytes, and consists of a 11-bit mantissa μ_i and a 5-bit exponent ϵ_i:

$$\Delta_i = 2^{R_i - \epsilon_i} \left(1 + \frac{\mu_i}{2^{11}} \right), \tag{17.5}$$

where R_i is the number of bits corresponding to the nominal dynamic range of the coefficients in subband i. In the reversible path, the step size Δ_i is set to 1 by choosing $\mu_i = 0$ and $\epsilon_i = R_i$.

The nominal dynamic range in subband i depends on the number of bits used to represent the original tile component and on the wavelet transform used.

The choice of a deadzone that is twice the quantization step size allows for an optimal bitstream embedded structure, i.e., for SNR scalability. The decoder can, by decoding up to any truncation point, reconstruct an image identical to what would have been obtained if encoded at the corresponding target bit rate. All image resolutions and qualities are directly available from a single compressed stream (also called codestream) without the need for decoding and re-encoding the existing codestream. In Part 2, the size of the deadzone can have different values in the different subbands.

Two modes have been specified for signaling the quantization parameters: *expounded* and *derived*. In the expounded mode, the pair of values (ϵ_i, μ_i) for each subband are explicitly transmitted. In the derived mode, codestream markers quantization default and quantization coefficient supply step size parameters only for the lowest frequency subband. The quantization parameters for other subbands i are then derived according to

$$(\epsilon_i, \mu_i) = (\epsilon_0 + l_i - L, \mu_0), \tag{17.6}$$

where L is the total number of wavelet decomposition levels and l_i is the number of levels required to generate the subband i.

The inverse quantization allows for a reconstruction bias from the quantizer midpoint for nonzero indices to accommodate skewed probability distributions of wavelet coefficients. The reconstructed values are thus computed as

$$\tilde{y}_i = \begin{cases} (\hat{y}_i + \gamma)\Delta_i 2^{M_i - N_i} & \text{if } \hat{y}_i > 0, \\ (\hat{y}_i - \gamma)\Delta_i 2^{M_i - N_i} & \text{if } \hat{y}_i < 0, \\ 0 & \text{otherwise.} \end{cases} \quad (17.7)$$

Here γ is a parameter which controls the reconstruction bias; a value of $\gamma = 0.5$ results in midpoint reconstruction. The term M_i denotes the maximum number of bits for a quantizer index in subband i. N_i represents the number of bits to be decoded in the case where the embedded bitstream is truncated prior to decoding.

17.10.4 Precincts and Code-blocks

Each subband, after quantization, is divided into nonoverlapping rectangular blocks, called code-blocks, of equal size. The dimensions of the code-blocks are powers of 2 (e.g., of size 16×16 or 32×32), and the total number of coefficients in a code-block should not exceed 4096. The code-blocks formed by the quantizer indexes corresponding to the quantized wavelet coefficients constitute the input to the entropy coder. Collections of spatially consistent code-blocks taken from each subband at each resolution level are called precincts and will form a packet partition in the bitstream structure. The purpose of precincts is to enable spatially progressive bitstreams. This point is further elaborated in Section 17.10.6.

17.10.5 Entropy Coding

The JPEG2000 entropy coding technique is based on the EBCOT (Embedded Block Coding with Optimal Truncation) algorithm [35]. Each code-block B_i is encoded separately, bit plane by bit plane, starting with the most significant bit plane (MSB) with a nonzero element and progressing towards the least significant bit plane. The data in each bit plane is scanned along the stripe pattern shown in Fig. 17.16 (with a stripe height of 4 samples) and encoded in three passes. Each pass collects contextual information that first helps decide which primitives to encode. The primitives are then provided to a context-dependent arithmetic coder. The bit plane encoding procedure is well suited for creating an embedded bitstream. Note that the approach does not exploit interscale dependencies. This potential loss in compression efficiency is compensated by beneficial features such as spatial random access, geometric manipulations in the compression domain, and error resilience.

17.10.5.1 Context Formation

Let $s_i[\mathbf{k}] = s_i[k_1, k_2]$ be the subband sample belonging to the block B_i at the horizontal and vertical positions k_1 and k_2. Let $\chi_i[\mathbf{k}] \in \{-1, 1\}$ denote the sign of $s_i[\mathbf{k}]$ and $\nu_i[\mathbf{k}] = \frac{|s_i[\mathbf{k}]|}{\delta_{\beta_i}}$, the amplitude of the quantized samples represented with M_i bits, where δ_{β_i} is the

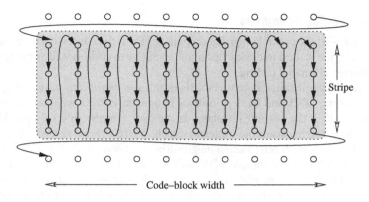

FIGURE 17.16

Stripe bit plane scanning pattern.

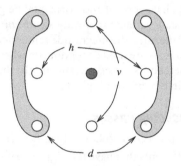

FIGURE 17.17

Neighbors involved in the context formation.

quantization step of the subband β_i containing the block B_i. Let $\nu_i^b[\mathbf{k}]$ be the bth bit of the binary representation of $\nu_i[\mathbf{k}]$.

A sample $s_i[\mathbf{k}]$ is said to be *nonsignificant* ($\sigma(s_i[\mathbf{k}]) = 0$) if the first nonzero bit $\nu_i^b[\mathbf{k}]$ of $\nu_i[\mathbf{k}]$ is yet to be encountered. The statistical dependencies between neighboring samples are captured via the formation of contexts which depend upon the *significance* state variable $\sigma(s_i[\mathbf{k}])$ associated with the eight-connect neighbors depicted in Fig. 17.17. These contexts are grouped in the following categories:

- h : number of significant horizontal neighbors, $0 \leq h \leq 2$;

- v : number of significant vertical neighbors, $0 \leq v \leq 2$;

- d : number of significant diagonal neighbors, $0 \leq d \leq 4$.

Neighbors which lie beyond the code-block boundary are considered to be nonsignificant to avoid dependence between code-blocks.

17.10.5.2 *Coding Primitives*

Different subsets of the possible *significance* patterns form the contextual information (or state variables) that is used to decide upon the primitive to code as well as the probability model to use in arithmetic coding.

If the sample significance state variable is in the *non-significant* state, a combination of the zero coding (ZC) and RLC primitives is used to encode whether the symbol is significant or not in the current bit plane. If the four samples in a column defined by the column-based stripe scanning pattern (see Fig. 17.16) have a zero significance state value ($\sigma(s_i[\mathbf{k}]) = 0$), with zero-valued neighborhoods, then the *RLC* primitive is coded. Otherwise, the value of the sample $\nu_i^b[\mathbf{k}]$ in the current bit plane b is coded with the primitive *ZC*. In other words, *RLC* coding occurs when all four locations of a column in the scan pattern are nonsignificant and each location has only nonsignificant neighbors.

Once the first nonzero bit $\nu_i^b[\mathbf{k}]$ has been encoded, the coefficient becomes *significant* and its sign $\chi_i[\mathbf{k}]$ is encoded with the sign coding (SC) primitive. The binary-valued sign bit $\chi_i[\mathbf{k}]$ is encoded conditionally to 5 different context states depending upon the sign and significance of the immediate vertical and horizontal neighbors.

If a sample significance state variable is already *significant*, i.e., ($\sigma(s_i[\mathbf{k}]) = 1$), when scanned in the current bit plane, then the magnitude refinement (MR) primitive encodes the bit value $\nu_i^b[\mathbf{k}]$. Three contexts are used depending on whether or not (a) the immediate horizontal and vertical neighbors are significant and (b) the *MR* primitive has already been applied to the sample in a previous bit plane.

17.10.5.3 *Bit Plane Encoding Passes*

Briefly, the different passes proceed as follows. In a first *significance propagation* pass ($\mathcal{P}_i^{p,1}$), the insignificant coefficients that have the highest probability of becoming significant are encoded. A nonsignificant coefficient is considered to have a high probability of becoming significant if at least one of its eight-connect neighbors is significant. For each sample $s_i[\mathbf{k}]$ that is non-significant with a significant neighbor, the primitive ZC is encoded followed by the primitive SC if $\nu_i^b[\mathbf{k}] = 1$.

Once the first nonzero bit has been encoded, the coefficient becomes *significant* and its sign is encoded. All subsequent bits are called *refinement* bits. In the second pass, referred to as the *refinement* pass ($\mathcal{P}_i^{b,2}$), the significant coefficients are refined by their bit representation in the current bit plane. Following the stripe-based scanning pattern, the primitive *MR* is encoded for each significant coefficient for which no information has been encoded yet in the current bit plane b.

In a final *normalization* or *cleanup* pass ($\mathcal{P}_3^b$), all the remaining coefficients in the bit plane (i.e., the nonsignificant samples for which no information has yet been coded) are encoded with the primitives ZC, RLC, and, if necessary, SC. The *cleanup* pass $\mathcal{P}_3^b$ corresponds to the encoding of all the bit plane b samples.

The encoding in three passes, $\mathcal{P}_1^b, \mathcal{P}_2^b, \mathcal{P}_3^b$, leads to the creation of distinct subsets in the bitstream. This structure in *partial bit planes* allows a fine granular bitstream representation providing a large number of RD truncation points. The standard allows the placement of bitstream truncation points at the end of each coding pass (this point is revisited in the sequel). The bitstream can thus be organized in such a way that the subset leading to a larger reduction in distortion is transmitted first.

17.10.5.4 *Arithmetic Coding*

Entropy coding is done by means of an arithmetic coder that encodes binary symbols (the primitives) using adaptive probability models conditioned by the corresponding contextual information. A reduced number of contexts, up to a maximum of 9, is used for each primitive. The corresponding probabilities are initialized at the beginning of each code-block and then updated using a state automaton. The reduced number of contexts allows for rapid probability adaptation. In a default operation mode, the encoding process starts at the beginning of each code-block and terminates at the end of each code-block. However, it is also possible to start and terminate the encoding process at the beginning and at the end, respectively, of a *partial* bit plane in a code-block. This allows increased error resilience of the codestream.

An arithmetic coder proceeds with recursive probability interval subdivisions. The arithmetic coding principles are described in [25]. In brief, the interval [0, 1] is partitioned into two cells representing the binary symbols of the alphabet. The size of each cell is given by the stationary probability of the corresponding symbol. The partition, and hence the bounds of the different segments, of the unit interval is given by the cumulative stationary probability of the alphabet symbols. The interval corresponding to the first symbol to be encoded is chosen. It becomes the current interval that is again partitioned into different segments. The subinterval associated with the more probable symbol (*MPS*) is ordered ahead of the subinterval corresponding to the less probable symbol (*LPS*). The symbols are thus often recognized as *MPS* and *LPS* rather than as 0 or 1. The bounds of the different segments are hence driven by the statistical model of the source. The codestream associated with the sequence of coded symbols points to the lower bound of the final subinterval. The decoding of the sequence is performed by reproducing the coder behavior in order to determine the sequence of subintervals pointed to by the codestream.

Practical implementations use fixed precision integer arithmetic with integer representations of fractional values. This potentially forces an approximation of the symbol probabilities leading to some coding suboptimality. The corresponding states of the encoder (interval values that cannot be reached by the encoder) are used to represent markers which contribute to improving the error resilience of the codestream [36]. One of the early practical implementations of arithmetic coding is known as the Q-coder [27]. The JPEG2000 standard has adopted a modified version of the Q-coder, called the MQ-coder, introduced in the JBIG2 standard [12] and available on a license and royalty-free basis. The various versions of arithmetic coders inspired from the Q-coder often differ by their stuffing procedure and the way they handle the carryover.

In order to reduce the number of symbols to encode, the standard specifies an option that allows the bypassing of some coding passes. Once the fourth bit plane has been coded, the data corresponding to the first and second passes is included as raw data without being arithmetically encoded. Only the third pass is encoded. This coding option is referred to as the *lazy* coding mode.

17.10.6 Bitstream Organization

The compressed data resulting from the different coding passes can be arranged in different configurations in order to accommodate a rich set of progression orders that are dictated by the application needs of random access and scalability. This flexible progression order is enabled by essentially four bitstream structuring components: code-block, precinct, packet, and layer.

17.10.6.1 *Packets and Layers*

The bitstream is organized as a succession of layers, each one being formed by a collection of packets. The layer gathers sets of compressed *partial* bit plane data from all the code-blocks of the different subbands and components of a tile. A packet is formed by an aggregation of compressed *partial* bit planes of a set of code-blocks that correspond to one spatial location at one resolution level and that define a precinct. The number of bit plane coding passes contained in a packet varies for different code-blocks. Each packet starts with a header that contains information about the number of coding passes required for each code-block assigned to the packet. The code-block compressed data is distributed across the different layers in the codestream. Each layer contains the additional contributions from each code-block (see Figure 17.18). The number of coding passes for a given code-block that are included in a layer is determined by RD optimization, and it defines truncation points in the codestream [35].

Notions of precincts, code-blocks, packets, and layers are well suited to allow the encoder to arrange the bitstream in an arbitrary progression manner, i.e., to accommodate the different modes of scalability that are desired. Four types of progression, namely

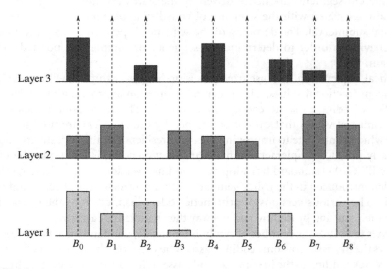

FIGURE 17.18

Code-block contributions to layers.

resolution, quality, spatial, and component, can be achieved by an appropriate ordering of the packets in the bitstream. For instance, layers and packets are key components for allowing quality scalability, i.e., packets containing less significant bits can be discarded to achieve lower bit rates and higher distortion. This flexible bitstream structuring gives application developers a high degree of freedom. For example, images can be transmitted over a network at arbitrary bit rates by using a layer-progressive order; lower resolutions, corresponding to low-frequency subbands, can be sent first for image previewing; and spatial browsing of large images is also possible through appropriate tile and/or partition selection. All these operations do not require any re-encoding but only byte-wise copy operations. Additional information on the different modes of scalability is provided in the standard.

17.10.6.2 *Truncation Points RD Optimization*

The problem now is to find the packet length for all code-blocks, i.e., define truncation point that will minimize the overall distortion. The recommended method to solve this problem, which is not part of the standard, makes use of a RD optimization procedure. Under certain assumptions about the quantization noise, the distortion is additive across code-blocks. The overall distortion can thus be written as $D = \sum_i D_i^{n_i}$. There is thus a need to search for the packet lengths n_i so that the distortion is minimized under the constraint of an overall bit rate, $R = \sum_i R_i^{n_i} \leq R^{max}$. The distortion measure $D_i^{n_i}$ is defined as the MSE weighted by the square of the L_2 norm of the wavelet basis functions used for the subband i to which the code-block B_i belongs. This optimization problem is solved using a Lagrangian formulation.

17.10.7 **Additional Features**

17.10.7.1 *Region-of-Interest Coding*

The JPEG2000 standard has a provision for defining the so-called *regions-of-interest* (ROI) in an image. The objective is to encode the ROIs with a higher quality and possibly to transmit them first in the bitstream so that they can be rendered first in a progressive decoding scenario. To allow for ROI coding, an ROI mask must first be derived. A mask is a map of the ROI in the image domain with nonzero values inside the ROI and zero values outside. The mask identifies the set of pixels (or the corresponding wavelet coefficients) that should be reconstructed with higher fidelity. ROI coding thus consists of encoding the quantized wavelet coefficients corresponding to the ROI with a higher precision. The ROI coding approach in JPEG2000 Part 1 is based on the MAXSHIFT method [8] which is an extension of the ROI scaling-based method introduced in [5]. The ROI scaling method consists of scaling up the coefficients belonging to the ROI or scaling down the coefficients corresponding to non-ROI regions in the image. The goal of the scaling operation is to place the bits of the ROI in higher bit planes than the bits associated with the non-ROI regions as shown in Fig. 17.19. Thus, the ROI will be decoded before the rest of the image, and if the bitstream is truncated, the ROI will be of higher quality. The ROI scaling method described in [5] requires the coding and transmission of the ROI shape information to the decoder. In order to minimize the decoder complexity,

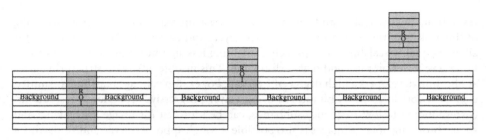

FIGURE 17.19

From left to right, no ROI coding, scaling method, and MAXSHIFT method for ROI coding.

the MAXSHIFT method adopted by JPEG2000 Part 1 shifts down all the coefficients not belonging to the ROI by a certain number s of bits chosen so that 2^s is larger than the largest non-ROI coefficients. This ensures that the minimum value contained in the ROI is higher than the maximum value of the non-ROI area. The compressed data associated with the ROI will then be placed first in the bitstream. With this approach the decoder does not need to generate the ROI mask. All the coefficients lower than the scaling value belong to the non-ROI region. Therefore the ROI shape information does not need to be encoded and transmitted. The drawback of this reduced complexity is that the ROI cannot be encoded with multiple quality differentials with respect to the non-ROI area.

17.10.7.2 *File Format*

Part 1 of the JPEG2000 standard also defines an optional file format referred to as JP2. It defines a set of data structures used to store information that may be required to render and display the image such as the colorspace (with two methods of color specification), the resolution of the image, the bit depth of the components, and the type and ordering of the components. The JP2 file format also defines two mechanisms for embedding application-specific data or metadata using either a universal unique identifier (UUID) or XML [43].

17.10.7.3 *Error Resilience*

Arithmetic coding is very sensitive to transmission noise; when some bits are altered by the channel, synchronization losses can occur at the receiver leading to error propagation that results in dramatic symbol error rates. JPEG2000 Part 1 provides several options to improve the error resilience of the codestream. First, the independent coding of the code-blocks limit error propagation across code-blocks boundaries. Certain coding options such as terminating the arithmetic coding at the end of each coding pass and reinitializing the contextual information at the beginning of the next coding pass further confine error propagation within a partial bit plane of a code-block. The optional *lazy* coding mode, that bypasses arithmetic coding for some passes, can also help to protect against error propagation. In addition, at the end of each cleanup pass, segmentation symbols are added in the codestream. These markers can be exploited for error detection.

If the segmentation symbol is not decoded properly, the data in the corresponding bit plane and of the subsequent bit planes in the code-block should be discarded. Finally, resynchronization markers, including the numbering of packets, are also inserted in front of each packet in a tile.

17.11 PERFORMANCE AND EXTENSIONS

The performance of JPEG2000 when compared with the JPEG baseline algorithm is briefly discussed in this section. The extensions included in Part 2 of the JPEG2000 standard are also listed.

17.11.1 Comparison of Performance

The efficiency of the JPEG2000 lossy coding algorithm in comparison with the JPEG baseline compression standard has been extensively studied and key results are summarized in [7, 9, 24]. The superior RD and error resilience performance, together with features such as progressive coding by resolution, scalability, and region of interest, clearly demonstrate the advantages of JPEG2000 over the baseline JPEG (with optimum Huffman codes). For coding common test images such as *Foreman* and *Lena* in the range of 0.125-1.25 bits/pixel, an improvement in the peak signal-to-noise ratio (PSNR) for JPEG2000 is consistently demonstrated at each compression ratio. For example, for the *Foreman* image, an improvement of 1.5 to 4 dB is observed as the bits per pixel are reduced from 1.2 to 0.12 [7].

17.11.2 Part 2 Extensions

Most of the technologies that have not been included in Part 1 due to their complexity or because of intellectual property rights (IPR) issues have been included in Part 2 [14]. These extensions concern the use of the following:

- different offset values for the different image components;

- different deadzone sizes for the different subbands;

- TCQ [23];

- visual masking based on the application of a nonlinearity to the wavelet coefficients [44, 45];

- arbitrary wavelet decomposition for each tile component;

- arbitrary wavelet filters;

- single sample tile overlap;

- arbitrary scaling of the ROI coefficients with the necessity to code and transmit the ROI mask to the decoder;

- nonlinear transformations of component samples and transformations to decorrelate multiple component data;

- extensions to the JP2 file format.

17.12 ADDITIONAL INFORMATION

Some sources and links for further information on the standards are provided here.

17.12.1 Useful Information and Links for the JPEG Standard

A key source of information on the JPEG compression standard is the book by Pennebaker and Mitchell [28]. This book also contains the entire text of the official committee draft international standard ISO DIS 10918-1 and ISO DIS 10918-2. The official standards document [11] contains information on JPEG Part 3.

The JPEG committee maintains an official website http://www.jpeg.org, which contains general information about the committee and its activities, announcements, and other useful links related to the different JPEG standards. The JPEG FAQ is located at http://www.faqs.org/faqs/jpeg-faq/part1/preamble.html.

Free, portable C code for JPEG compression is available from the Independent JPEG Group (IJG). Source code, documentation, and test files are included. Version 6b is available from

$$\text{ftp.uu.net:/graphics/jpeg/jpegsrc.v6b.tar.gz}$$

and in ZIP archive format at

$$\text{ftp.simtel.net:/pub/simtelnet/msdos/graphics/jpegsr6b.zip.}$$

The IJG code includes a reusable JPEG compression/decompression library, plus sample applications for compression, decompression, transcoding, and file format conversion. The package is highly portable and has been used successfully on many machines ranging from personal computers to super computers. The IJG code is free for both noncommercial and commercial use; only an acknowledgement in your documentation is required to use it in a product. A different free JPEG implementation, written by the PVRG group at Stanford, is available from http://www.havefun.stanford.edu:/pub/jpeg/JPEGv1.2.1.tar.Z. The PVRG code is designed for research and experimentation rather than production use; it is slower, harder to use, and less portable than the IJG code, but the PVRG code is easier to understand.

17.12.2 Useful Information and Links for the JPEG2000 Standard

Useful sources of information on the JPEG2000 compression standard include two books published on the topic [1, 36]. Further information on the different parts of the JPEG2000 standard can be found on the JPEG website http://www.jpeg.org/jpeg2000.html. This website provide links to sites from which various official standards and other documents can be downloaded. It also provides links to sites from which software implementations of the standard can be downloaded. Some software implementations are available at the following addresses:

- JJ2000 software that can be accessed at http://www.jpeg2000.epfl.ch. The JJ2000 software is a Java implementation of JPEG2000 Part 1.

- Kakadu software that can be accessed at http://www.ee.unsw.edu.au/taubman/kakadu. The Kakadu software is a C++ implementation of JPEG2000 Part 1. The Kakadu software is provided with the book [36].

- Jasper software that can be accessed at http://www.ece.ubc.ca/mdadams/jasper/. Jasper is a C implementation of JPEG2000 that is free for commercial use.

REFERENCES

[1] T. Acharya and P.-S. Tsai. *JPEG2000 Standard for Image Compression*. John Wiley & Sons, New Jersey, 2005.

[2] N. Ahmed, T. Natrajan, and K. R. Rao. Discrete cosine transform. *IEEE Trans. Comput.*, C-23:90–93, 1974.

[3] A. J. Ahumada and H. A. Peterson. Luminance model based DCT quantization for color image compression. *Human Vision, Visual Processing, and Digital Display III, Proc. SPIE*, 1666:365–374, 1992.

[4] A. Antonini, M. Barlaud, P. Mathieu, and I. Daubechies. Image coding using the wavelet transform. *IEEE Trans. Image Process.*, 1(2):205–220, 1992.

[5] E. Atsumi and N. Farvardin. Lossy/lossless region-of-interest image coding based on set partitioning in hierarchical trees. In *Proc. IEEE Int. Conf. Image Process.*, 1(4–7):87–91, October 1998.

[6] A. Bilgin, P. J. Sementilli, and M. W. Marcellin. Progressive image coding using trellis coded quantization. *IEEE Trans. Image Process.*, 8(11):1638–1643, 1999.

[7] D. Chai and A. Bouzerdoum. JPEG2000 image compression: an overview. *Australian and New Zealand Intelligent Information Systems Conference (ANZIIS'2001)*, Perth, Australia, 237–241, November 2001.

[8] C. Christopoulos, J. Askelof, and M. Larsson. Efficient methods for encoding regions of interest in the upcoming JPEG2000 still image coding standard. *IEEE Signal Process. Lett.*, 7(9):247–249, 2000.

[9] C. Christopoulos, A. Skodras, and T. Ebrahimi. The JPEG 2000 still image coding system: an overview. *IEEE Trans. Consum. Electron.*, 46(4):1103–1127, 2000.

[10] K. W. Chun, K. W. Lim, H. D. Cho, and J. B. Ra. An adaptive perceptual quantization algorithm for video coding. *IEEE Trans. Consum. Electron.*, 39(3):555–558, 1993.

[11] ISO/IEC JTC 1/SC 29/WG 1 N 993. Information technology—digital compression and coding of continuous-tone still images. Recommendation T.84 *ISO/IEC CD 10918-3*. 1994.

[12] ISO/IEC International standard 14492 and ITU recommendation T.88. *JBIG2 Bi-Level Image Compression Standard*. 2000.

[13] ISO/IEC International standard 15444-1 and ITU recommendation T.800. *Information Technology—JPEG2000 Image Coding System*. 2000.

[14] ISO/IEC International standard 15444-2 and ITU recommendation T.801. *Information Technology—JPEG2000 Image Coding System: Part 2, Extensions*. 2001.

[15] ISO/IEC International standard 15444-3 and ITU recommendation T.802. *Information Technology—JPEG2000 Image Coding System: Part 3, Motion JPEG2000*. 2001.

[16] ISO/IEC International standard 15444-4 and ITU recommendation T.803. *Information Technology—JPEG2000 Image Coding System: Part 4, Compliance Testing*. 2001.

[17] ISO/IEC International standard 15444-5 and ITU recommendation T.804. *Information Technology—JPEG2000 Image Coding System: Part 5, Reference Software*. 2001.

[18] N. Jayant, R. Safranek, and J. Johnston. Signal compression based on models of human perception. *Proc. IEEE*, 83:1385–1422, 1993.

[19] JPEG2000. http://www.jpeg.org/jpeg2000/.

[20] L. Karam. *Lossless Image Compression, Chapter 15, The Essential Guide to Image Processing*. Elsevier Academic Press, Burlington, MA, 2008.

[21] K. Konstantinides and D. Tretter. A method for variable quantization in JPEG for improved text quality in compound documents. In *Proc. IEEE Int. Conf. Image Process.*, Chicago, IL, October 1998.

[22] D. Le Gall and A. Tabatabai. Subband coding of digital images using symmetric short kernel filters and arithmetic coding techniques. In *Proc. Intl. Conf. on Acoust., Speech and Signal Process., ICASSP'88*, 761–764, April 1988.

[23] M. W. Marcellin and T. R. Fisher. Trellis coded quantization of memoryless and Gauss-Markov sources. *IEEE Trans. Commun.*, 38(1):82–93, 1990.

[24] M. W. Marcellin, M. J. Gormish, A. Bilgin, and M. P. Boliek. An overview of JPEG2000. In *Proc. of IEEE Data Compression Conference*, 523–541, 2000.

[25] N. Memon, C. Guillemot, and R. Ansari. *The JPEG Lossless Compression Standards. Chapter 5.6, Handbook of Image and Video Processing*. Elsevier Academic Press, Burlington, MA, 2005.

[26] P. Moulin. *Multiscale Image Decomposition and Wavelets, Chapter 6, The Essential Guide to Image Processing*. Elsevier Academic Press, Burlington, MA, 2008.

[27] W. B. Pennebaker, J. L. Mitchell, G. G. Langdon, and R. B. Arps. An overview of the basic principles of the q-coder adaptive binary arithmetic coder. *IBM J. Res. Dev.*, 32(6):717–726, 1988.

[28] W. B. Pennebaker and J. L. Mitchell. *JPEG Still Image Data Compression Standard*. Van Nostrand Reinhold, New York, 1993.

[29] M. Rabbani and R. Joshi. An overview of the JPEG2000 still image compression standard. *Elsevier J. Signal Process.*, 17:3–48, 2002.

[30] V. Ratnakar and M. Livny. RD-OPT: an efficient algorithm for optimizing DCT quantization tables. *IEEE Proc. Data Compression Conference (DCC)*, Snowbird, UT, 332–341, 1995.

[31] K. R. Rao and P. Yip. *Discrete Cosine Transform—Algorithms, Advantages, Applications*. Academic Press, San Diego, CA, 1990.

[32] P. J. Sementilli, A. Bilgin, J. H. Kasner, and M. W. Marcellin. Wavelet tcq: submission to JPEG2000. In *Proc. SPIE, Applications of Digital Processing*, 2–12, July 1998.

[33] A. Skodras, C. Christopoulos, and T. Ebrahimi. The JPEG 2000 still image compression standard. *IEEE Signal Process. Mag.*, 18(5):36–58, 2001.

[34] B. J. Sullivan, R. Ansari, M. L. Giger, and H. MacMohan. Relative effects of resolution and quantization on the quality of compressed medical images. In *Proc. IEEE Int. Conf. Image Process.*, Austin, TX, 987–991, November 1994.

[35] D. Taubman. High performance scalable image compression with ebcot. *IEEE Trans. Image Process.*, 9(7):1158–1170, 1999.

[36] D. Taubman and M.W. Marcellin. *JPEG2000: Image Compression Fundamentals: Standards and Practice.* Kluwer Academic Publishers, New York, 2002.

[37] R. VanderKam and P. Wong. Customized JPEG compression for grayscale printing. In *Proc. Data Compression Conference (DCC)*, Snowbird, UT, 156–165, 1994.

[38] M. Vetterli and J. Kovacevic. *Wavelet and Subband Coding.* Prentice-Hall, Englewood Cliffs, NJ, 1995.

[39] G. K. Wallace. The JPEG still picture compression standard. *Commun. ACM*, 34(4):31–44, 1991.

[40] P. W. Wang. *Image Quantization, Halftoning, and Printing. Chapter 8.1, Handbook of Image and Video Processing.* Elsevier Academic Press, Burlington, MA, 2005.

[41] A. B. Watson. Visually optimal DCT quantization matrices for individual images. In *Proc. IEEE Data Compression Conference (DCC)*, Snowbird, UT, 178–187, 1993.

[42] I. H. Witten, R. M. Neal, and J. G. Cleary. Arithmetic coding for data compression. *Commun. ACM*, 30(6):520–540, 1987.

[43] World Wide Web Consortium (W3C). *Extensible Markup Language (XML) 1.0*, 3rd ed., T. Bray, J. Paoli, C. M. Sperberg-McQueen, E. Maler, F. Yergeau, editors, http://www.w3.org/TR/REC-xml, 2004.

[44] W. Zeng, S. Daly, and S. Lei. Point-wise extended visual masking for JPEG2000 image compression. In *Proc. IEEE Int. Conf. Image Process.*, Vancouver, BC, Canada, vol. 1, 657–660, September 2000.

[45] W. Zeng, S. Daly, and S. Lei. Visual optimization tools in JPEG2000. In *Proc. IEEE Int. Conf. Image Process.*, Vancouver, BC, Canada, vol. 2, 37–40, September 2000.

Wavelet Image Compression 18

Zixiang Xiong[1] and Kannan Ramchandran[2]

[1] *Texas A&M University;* [2] *University of California*

18.1 WHAT ARE WAVELETS: WHY ARE THEY GOOD FOR IMAGE CODING?

During the past 15 years, wavelets have made quite a splash in the field of image compression. The FBI adopted a wavelet-based standard for fingerprint image compression. The JPEG2000 image compression standard [1], which is a much more efficient alternative to the old JPEG standard (see Chapter 17), is also based on wavelets. A natural question to ask then is why wavelets have made such an impact on image compression. This chapter will answer this question, providing both high-level intuition and illustrative details based on state-of-the-art wavelet-based coding algorithms. Visually appealing time-frequency-based analysis tools are sprinkled in generously to aid in our task.

Wavelets are tools for decomposing signals, such as images, into a hierarchy of increasing resolutions: as we consider more and more resolution layers, we get a more and more detailed look at the image. Figure 18.1 shows a three-level hierarchy wavelet decomposition of the popular test image *Lena* from coarse to fine resolutions (for a detailed treatment on wavelets and multiresolution decompositions, also see Chapter 6). Wavelets can be regarded as "mathematical microscopes" that permit one to "zoom in" and "zoom out" of images at multiple resolutions. The remarkable thing about the wavelet decomposition is that it enables this zooming feature at absolutely no cost in terms of excess redundancy: for an $M \times N$ image, there are exactly MN wavelet coefficients—exactly the same as the number of original image pixels (see Fig. 18.2).

As a basic tool for decomposing signals, wavelets can be considered as duals to the more traditional Fourier-based analysis methods that we encounter in traditional undergraduate engineering curricula. Fourier analysis associates the very intuitive engineering concept of "spectrum" or "frequency content" of the signal. Wavelet analysis, in contrast, associates the equally intuitive concept of "resolution" or "scale" of the signal. At a functional level, Fourier analysis is to wavelet analysis as spectrum analyzers are to microscopes.

As wavelets and multiresolution decompositions have been described in greater depth in Chapter 6, our focus here will be more on the image compression application. Our goal is to provide a self-contained treatment of wavelets within the scope of their role

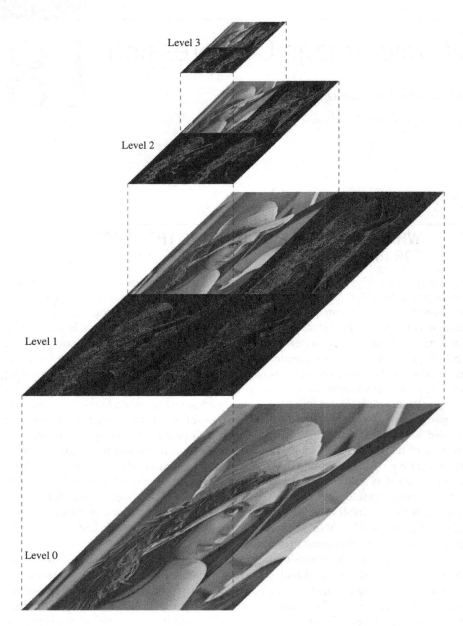

FIGURE 18.1

A three-level hierarchy wavelet decomposition of the 512×512 color *Lena* image. Level 1 (512×512) is the one-level wavelet representation of the original *Lena* at Level 0; Level 2 (256×256) shows the one-level wavelet representation of the lowpass image at Level 1; and Level 3 (128×128) gives the one-level wavelet representation of the lowpass image at Level 2.

FIGURE 18.2

A three-level wavelet representation of the *Lena* image generated from the top view of the three-level hierarchy wavelet decomposition in Fig. 18.1. It has exactly the same number of samples as in the image domain.

in image compression. More importantly, our goal is to provide a high-level explanation for why they are well suited for image compression. Indeed, wavelets have superior properties vis-a-vis the more traditional Fourier-based method in the form of the discrete cosine transform (DCT) that is deployed in the old JPEG image compression standard (see Chapter 17). We will also cover powerful generalizations of wavelets, known as wavelet packets, that have already made an impact in the standardization world: the FBI fingerprint compression standard is based on wavelet packets.

Although this chapter is about image coding,[1] which involves two-dimensional (2D) signals or images, it is much easier to understand the role of wavelets in image coding using a one-dimensional (1D) framework, as the conceptual extension to 2D is straightforward. In the interests of clarity, we will therefore consider a 1D treatment here. The story begins with what is known as the time-frequency analysis of the 1D signal. As mentioned, wavelets are a tool for changing the coordinate system in which we represent the signal: we transform the signal into another domain that is much better suited for processing, e.g., compression. What makes for a good transform or analysis tool? At the basic level, the goal is to be able to represent all the useful signal features and important phenomena in as compact a manner as possible. It is important to be able to compact the bulk of the signal energy into the fewest number of transform coefficients: this way, we can discard the bulk of the transform domain data without losing too much information. For example, if the signal is a time impulse, then the best thing is to do no transforms at

---

[1] We use the terms *image compression* and *image coding* interchangeably in this chapter.

all! Keep the signal information in its original and sparse time-domain representation, as that will maximize the temporal energy concentration or time resolution. However, what if the signal has a critical frequency component (e.g., a low-frequency background sinusoid) that lasts for a long time duration? In this case, the energy is spread out in the time domain, but it would be succinctly captured in a single frequency coefficient if one did a Fourier analysis of the signal. If we know that the signals of interest are pure sinusoids, then Fourier analysis is the way to go. But, what if we want to capture both the time impulse and the frequency impulse with good resolution? Can we get arbitrarily fine resolution in both time and frequency?

The answer is no. There exists an uncertainty theorem (much like what we learn in quantum physics), which disallows the existence of arbitrary resolution in time and frequency [2]. A good way of conceptualizing these ideas and the role of wavelet basis functions is through what is known as time-frequency "tiling" plots, as shown in Fig. 18.3, which shows where the basis functions live on the time-frequency plane: i.e., where is the bulk of the energy of the elementary basis elements localized? Consider the Fourier

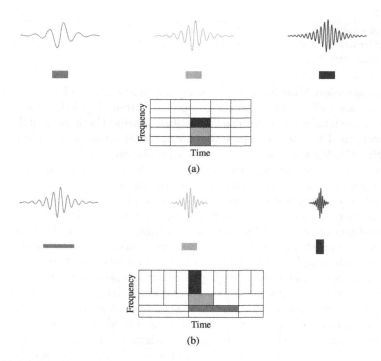

FIGURE 18.3

Tiling diagrams associated with the STFT bases and wavelet bases. (a) STFT bases and the tiling diagram associated with a STFT expansion. STFT bases of different frequencies have the same resolution (or length) in time; (b) Wavelet bases and tiling diagram associated with a wavelet expansion. The time resolution is inversely proportional to frequency for wavelet bases.

case first. As impulses in time are completely spread out in the frequency domain, all localization is lost with Fourier analysis. To alleviate this problem, one typically decomposes the signal into finite-length chunks using windows or so-called short-time Fourier transform (STFT). Then, the time-frequency tradeoffs will be determined by the window size. An STFT expansion consists of basis functions that are shifted versions of one another in both time and frequency: some elements capture low-frequency events localized in time, and others capture high-frequency events localized in time, but the resolution or window size is constant in both time and frequency (see Fig. 18.3(a)). Note that the uncertainty theorem says that the area of these tiles has to be nonzero.

Shown in Fig. 18.3(b) is the corresponding tiling diagram associated with the wavelet expansion. The key difference between this and the Fourier case, which is the critical point, is that the tiles are not all of the same size in time (or frequency). Some basis elements have short time windows; others have short frequency windows. Of course, the uncertainty theorem ensures that the area of each tile is constant and nonzero. It can be shown that the basis functions are related to one another by shifts and scales as this is the key to wavelet analysis.

Why are wavelets well suited for image compression? The answer lies in the time-frequency (or more correctly, space-frequency) characteristics of typical natural images, which turn out to be well captured by the wavelet basis functions shown in Fig. 18.3(b). Note that the STFT tiling diagram of Fig. 18.3(a) is conceptually similar to what commercial DCT-based image transform coding methods like JPEG use. Why are wavelets inherently a better choice? Looking at Fig. 18.3(b), one can note that the wavelet basis offers elements having good frequency resolution at lower frequency (the short and fat basis elements) while simultaneously offering elements that have good time resolution at higher frequencies (the tall and skinny basis elements).

This tradeoff works well for natural images and scenes that are typically composed of a mixture of important long-term low-frequency trends that have larger spatial duration (such as slowly varying backgrounds like the blue sky, and the surface of lakes) as well as important transient short duration high-frequency phenomena such as sharp edges. The wavelet representation turns out to be particularly well suited to capturing both the transient high-frequency phenomena such as image edges (using the tall and skinny tiles) and long spatial duration low-frequency phenomena such as image backgrounds (the short and fat tiles). As natural images are dominated by a mixture of these kinds of events,[2] wavelets promise to be very efficient in capturing the bulk of the image energy in a small fraction of the coefficients.

To summarize, the task of separating transient behavior from long-term trends is a very difficult task in image analysis and compression. In the case of images, the difficulty stems from the fact that statistical analysis methods often require the introduction of at least some local stationarity assumption, i.e., the image statistics do not change abruptly

---

[2]Typical images also contain textures; however, conceptually, textures can be assumed to be a dense concentration of edges, and so it is fairly accurate to model typical images as smooth regions delimited by edges.

over time. In practice, this assumption usually translates into ad hoc methods to block data samples for analysis, methods that can potentially obscure important signal features: e.g., if a block is chosen too big, a transient component might be totally neglected when computing averages. The blocking artifact in JPEG decoded images at low rates is a result of the block-based DCT approach. A fundamental contribution of wavelet theory [3] is that it provides a unified framework in which transients and trends can be simultaneously analyzed without the need to resort to blocking methods.

As a way of highlighting the benefits of having a sparse representation, such as that provided by the wavelet decomposition, consider the lowest frequency band in the top level (Level 3) of the three-level wavelet hierarchy of *Lena* in Fig. 18.1. This band is just a downsampled (by a factor of $8^2 = 64$) and smoothed version of the original image. A very simple way of achieving compression is to simply retain this lowpass version and throw away the rest of the wavelet data, instantly achieving a compression ratio of 64:1. Note that if we want a full-size approximation to the original, we would have to interpolate the lowpass band by a factor of 64—this can be done efficiently by using a three-stage synthesis filter bank (see Chapter 6). We may also desire better image fidelity, as we may be compromising high-frequency image detail, especially perceptually important high-frequency edge information. This is where wavelets are particularly attractive as they are capable of capturing most image information in the highly subsampled low-frequency band and additional localized edge information in spatial clusters of coefficients in the high-frequency bands (see Fig. 18.1). The bulk of the wavelet data is insignificant and can be discarded or quantized very coarsely.

Another attractive aspect of the coarse-to-fine nature of the wavelet representation naturally facilitates a transmission scheme that progressively refines the received image quality. That is, it would be highly beneficial to have an encoded bitstream that can be chopped off at any desired point to provide a commensurate reconstruction image quality. This is known as a progressive transmission feature or as an embedded bitstream (see Fig. 18.4). Many modern wavelet image coders have this feature, as will be covered in more detail in Section 18.5. This is ideally suited, for example, to Internet image applications. As is well known, the Internet is a heterogeneous mess in terms of the number of users and their computational capabilities and effective bandwidths. Wavelets provide a natural way to satisfy users having disparate bandwidth and computational capabilities: the low-end users can be provided a coarse quality approximation, whereas higher-end users can use their increased bandwidth to get better fidelity. This is also very useful for Web browsing applications, where having a coarse quality image with a short waiting time may be preferable to having a detailed quality with an unacceptable delay.

These are some of the high-level reasons why wavelets represent a superior alternative to traditional Fourier-based methods for compressing natural images: this is why the JPEG2000 standard [1] uses wavelets instead of the Fourier-based DCT.

In this chapter, we will review the salient aspects of the general compression problem and the transform coding paradigm in particular, and highlight the key differences between the class of early subband coders and the recent more advanced class of modern-day wavelet image coders. We pick the celebrated embedded zerotree wavelet (EZW) coder as a representative of this latter class, and we describe its operation by using a

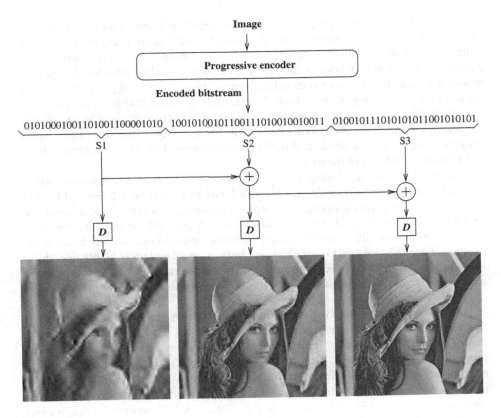

FIGURE 18.4

Multiresolution wavelet image representation naturally facilitates progressive transmission— a desirable feature for the transmission of compressed images over heterogeneous packet networks and wireless channels.

simple illustrative example. We conclude with more powerful generalizations of the basic wavelet image coding framework to wavelet packets, which are particularly well suited to handle special classes of images such as fingerprints.

18.2 THE COMPRESSION PROBLEM

Image compression falls under the general umbrella of data compression, which has been studied theoretically in the field of information theory [4], pioneered by Claude Shannon [5] in 1948. Information theory sets the fundamental bounds on compression performance theoretically attainable for certain classes of sources. This is very useful because it provides a theoretical benchmark against which one can compare the performance of more practical but suboptimal coding algorithms.

Historically, the lossless compression problem came first. Here the goal is to compress the source with no loss of information. Shannon showed that given any discrete source with a well-defined statistical characterization (i.e., a probability mass function), there is a fundamental theoretical limit to how well you can compress the source before you start to lose information. This limit is called the *entropy* of the source. In lay terms, entropy refers to the uncertainty of the source. For example, a source that takes on any of N discrete values $a_1, a_2, \ldots, a_N$ with equal probability has an entropy given by $\log_2 N$ bits per source symbol. If the symbols are not equally likely, however, then one can do better because more predictable symbols should be assigned fewer bits. The fundamental limit is the Shannon entropy of the source.

Lossless compression of images has been covered in Chapter 16. For image coding, typical lossless compression ratios are of the order of 2:1 or at most 3:1. For a 512×512 8-bit grayscale image, the uncompressed representation is 256 Kbytes. Lossless compression would reduce this to at best $\sim$80 Kbytes, which may still be excessive for many practical low-bandwidth transmission applications. Furthermore, lossless image compression is for the most part overkill, as our human visual system is highly tolerant to losses in visual information. For compression ratios in the range of 10:1 to 40:1 or more, lossless compression cannot do the job, and one needs to resort to lossy compression methods.

The formulation of the lossy data compression framework was also pioneered by Shannon in his work on rate-distortion (RD) theory [6], in which he formalized the theory of compressing certain limited classes of sources having well-defined statistical properties, e.g., independent, identically distributed (i.i.d.) sources having a Gaussian distribution subject to a fidelity criterion, i.e., subject to a tolerance on the maximum allowable loss or distortion that can be endured. Typical distortion measures used are mean square error (MSE) or peak signal-to-noise ratio (PSNR)[3] between the original and compressed versions. These fundamental compression performance bounds are called the theoretical RD bounds for the source: they dictate the minimum rate R needed to compress the source if the tolerable distortion level is D (or alternatively, what is the minimum distortion D subject to a bit rate of R). These bounds are unfortunately not constructive; i.e., Shannon did not give an actual algorithm for attaining these bounds, and furthermore, they are based on arguments that assume infinite complexity and delay, obviously impractical in real life. However, these bounds are useful in as much as they provide valuable benchmarks for assessing the performance of more practical coding algorithms. The major obstacle of course, as in the lossless case, is that these theoretical bounds are available only for a narrow class of sources, and it is difficult to make the connection to real world image sources which are difficult to model accurately with simplistic statistical models.

Shannon's theoretical RD framework has inspired the design of more practical *operational* RD frameworks, in which the goal is similar but the framework is constrained to be more practical. Within the operational constraints of the chosen coding

---

[3]The PSNR is defined as $10 \log_{10} \frac{255^2}{MSE}$ and measured in decibels (dB).

framework, the goal of operational RD theory is to minimize the rate R subject to a distortion constraint D, or vice versa. The message of Shannon's RD theory is that one can come close to the theoretical compression limit of the source if one considers *vectors* of source symbols that get infinitely large in dimension in the limit; i.e., it is a good idea not to code the source symbols one at a time, but to consider chunks of them at a time, and the bigger the chunks the better. This thinking has spawned an important field known as *vector quantization* (VQ) [7], which, as the name indicates, is concerned with the theory and practice of quantizing sources using high-dimensional VQ. There are practical difficulties arising from making these vectors too high-dimensional because of complexity constraints, so practical frameworks involve relatively small dimensional vectors that are therefore further from the theoretical bound.

Due to this difficulty, there has been a much more popular image compression framework that has taken off in practice: this is the transform coding framework [8] that forms the basis of current commercial image and video compression standards like JPEG and MPEG (see Chapters 9 and 10 in [9]). The transform coding paradigm can be construed as a practical special case of VQ that can attain the promised gains of processing source symbols in vectors through the use of efficiently implemented high dimensional source transforms.

18.3 THE TRANSFORM CODING PARADIGM

In a typical transform image coding system, the encoder consists of a linear transform operation, followed by quantization of transform coefficients, and lossless compression of the quantized coefficients using an entropy coder. After the encoded bitstream of an input image is transmitted over the channel (assumed to be perfect), the decoder undoes all the functionalities applied in the encoder and tries to reconstruct a decoded image that looks as close as possible to the original input image, based on the transmitted information. A block diagram of this transform image paradigm is shown in Fig. 18.5.

For the sake of simplicity, let us look at a 1D example of how transform coding is done (for 2D images, we treat the rows and columns separately as 1D signals). Suppose we have a two-point signal, $x_0 = 216$, $x_1 = 217$. It takes 16 bits (8 bits for each sample) to store this signal in a computer. In transform coding, we first put x_0 and x_1 in a column vector $X = \begin{bmatrix} x_0 \\ x_1 \end{bmatrix}$ and apply an orthogonal transformation T to X to get $Y = \begin{bmatrix} y_0 \\ y_1 \end{bmatrix} =$

$$TX = \begin{bmatrix} 1/\sqrt{2} & 1/\sqrt{2} \\ 1/\sqrt{2} & -1/\sqrt{2} \end{bmatrix} \begin{bmatrix} x_0 \\ x_1 \end{bmatrix} = \begin{bmatrix} (x_0 + x_1)/\sqrt{2} \\ (x_0 - x_1)/\sqrt{2} \end{bmatrix} = \begin{bmatrix} 306.177 \\ -.707 \end{bmatrix}.$$ The transform T can

be conceptualized as a counter-clockwise rotation of the signal vector X by 45° with respect to the original (x_0, x_1) coordinate system. Alternatively and more conveniently, one can think of the signal vector as being fixed and instead rotate the (x_0, x_1) coordinate system by 45° clockwise to the new (y_1, y_0) coordinate system (see Fig. 18.6). Note that the abscissa for the new coordinate system is now y_1.

Orthogonality of the transform simply means that the length of Y is the same as the length of X (which is even more obvious when one freezes the signal vector and

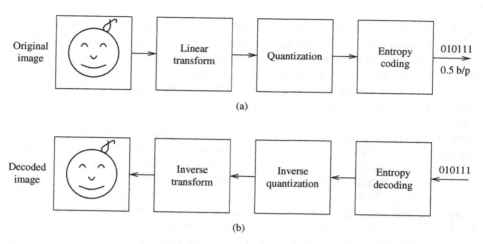

(a)

(b)

FIGURE 18.5

Block diagrams of a typical transform image coding system: (a) encoder and (b) decoder diagrams.

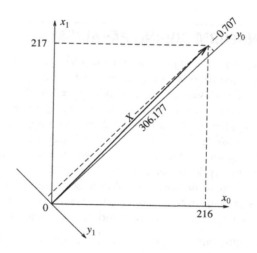

FIGURE 18.6

The transform T can be conceptualized as a counter-clockwise rotation of the signal vector X by 45° with respect to the original (x_0, x_1) coordinate system.

rotates the coordinate system as discussed above). This concept still carries over to the case of high-dimensional transforms. If we decide to use the simplest form of quantization known as uniform scalar quantization, where we round off a real number to the nearest integer multiple of a step size q (say $q = 20$), then the quantizer index vector $\hat{I}$, which captures *what* integer multiples of q are nearest to the entries of Y, is given by

$\hat{I} = \begin{bmatrix} \text{round}(y_0/q) \\ \text{round}(y_1/q) \end{bmatrix} = \begin{bmatrix} 15 \\ 0 \end{bmatrix}$. We store (or transmit) $\hat{I}$ as the compressed version of X using 4 bits, achieving a compression ratio of 4:1. To decode X from $\hat{I}$, we first multiply $\hat{I}$ by $q = 20$ to dequantize, i.e., to form the quantized approximation $\hat{Y}$ of Y with

$\hat{Y} = q \cdot \hat{I} = \begin{bmatrix} 300 \\ 0 \end{bmatrix}$, and then apply the inverse transform T^{-1} to $\hat{Y}$ (which corresponds in our example to a counter-clockwise rotation of the (y_1, y_0) coordinate system by 45°, just the reverse operation of the T operation on the original (x_0, x_1) coordinate system—see

Fig. 18.6) to get $\hat{X} = T^{-1} \begin{bmatrix} qy_0 \\ qy_1 \end{bmatrix} = \begin{bmatrix} 1/\sqrt{2} & 1/\sqrt{2} \\ 1/\sqrt{2} & -1/\sqrt{2} \end{bmatrix} \begin{bmatrix} 300 \\ 0 \end{bmatrix} = \begin{bmatrix} 212.132 \\ 212.132 \end{bmatrix}$.

We see from the above example that, although we "zero out" or throw away the transform coefficient y_1 in quantization, the decoded version $\hat{X}$ is still very close to X. This is because the transform effectively compacts most of the energy in X into the first coefficient y_0, and renders the second coefficient y_1 considerably insignificant to keep. The transform T in our example actually computes a weighted sum and difference of the two samples x_0 and x_1 in a manner that preserves the original energy. It is in fact the *simplest* wavelet transform!

The energy compaction aspect of wavelet transforms was highlighted in Section 18.1. Another goal of linear transformation is decorrelation. This can be seen from the fact that, although the values of x_0 and x_1 are very close (highly correlated) before the transform, y_0 (sum) and y_1 (difference) are very different (less correlated) after the transform. Decorrelation has a nice geometric interpretation. A cloud of input samples of length-2 is shown along the 45° line in Fig. 18.7. The coordinates (x_0, x_1) at each point of the cloud are nearly the same, reflecting the high degree of correlation among neighboring image pixels. The linear transformation T essentially amounts to a rotation of the coordinate

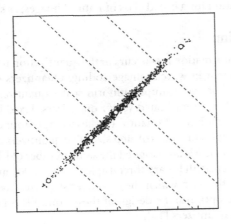

FIGURE 18.7

Linear transformation amounts to a rotation of the coordinate system, making correlated samples in the time domain less correlated in the transform domain.

system. The axes of the new coordinate system are parallel and perpendicular to the orientation of the cloud. The coordinates (y_0, y_1) are less correlated, as their magnitudes can be quite different and the sign of y_1 is random. If we assume x_0 and x_1 are samples of a stationary random sequence $X(n)$, then the correlation between y_0 and y_1 is $E\{y_0 y_1\} = E\{(x_0^2 - x_1^2)/2\} = 0$. This decorrelation property has significance in terms of how much gain one can get from transform coding than from doing signal processing (quantization and coding) directly in the original signal domain, called pulse code modulation (PCM) coding.

Transform coding has been extensively developed for coding of images and video, where the DCT is commonly used because of its computational simplicity and its good performance. But as shown in Section 18.1, the DCT is giving way to the wavelet transform because of the latter's superior energy compaction capability when applied to natural images. Before discussing state-of-the-art wavelet coders and their advanced features, we address the functional units that comprise a transform coding system, namely the transform, quantizer, and entropy coder (see Fig. 18.5).

18.3.1 Transform Structure

The basic idea behind using a linear transformation is to make the task of compressing an image in the transform domain after quantization easier than direct coding in the spatial domain. A good transform, as has been mentioned, should be able to *decorrelate* the image pixels and provide good *energy compaction* in the transform domain so that very few quantized nonzero coefficients have to be encoded. It is also desirable for the transform to be orthogonal so that the energy is conserved from the spatial domain to the transform domain, and the distortion in the spatial domain introduced by quantization of transform coefficients can be directly examined in the transform domain. What makes the wavelet transform special in all possible choices is that it offers an efficient space-frequency characterization for a broad class of natural images, as shown in Section 18.1.

18.3.2 Quantization

As the only source of information loss occurs in the quantization unit, efficient quantizer design is a key component in wavelet image coding. Quantizers come in many different shapes and forms, from very simple uniform scalar quantizers, such as the one in the example earlier, to very complicated vector quantizers. Fixed length uniform scalar quantizers are the simplest kind of quantizers: these simply round off real numbers to the nearest integer multiples of a chosen step size. The quantizers are fixed length in the sense that all quantization levels are assigned the same number of bits (e.g., an eight-level quantizer would be assigned all binary three-tuples between 000 and 111). Fixed length *nonuniform* scalar quantizers, in which the quantizer step sizes are not all the same, are more powerful: one can optimize the design of these nonuniform step sizes to get what is known as Lloyd-Max quantizers [10].

It is more efficient to do a joint design of the quantizer and the entropy coding functional unit (this will be described in the next subsection) that follows the quantizer in a lossy compression system. This joint design results in a so-called entropy-constrained quantizer that is more efficient but more complex, and results in variable length quantizers in which the different quantization choices are assigned variable codelengths. Variable length quantizers can come in either scalar, known as entropy-constrained scalar quantization (ECSQ) [11], or vector varieties, known as entropy-constrained vector quantization (ECVQ) [7]. An efficient way of implementing vector quantizers is by the use of so-called trellis coded quantization (TCQ) [12]. The performance of the quantizer (in conjunction with the entropy coder) characterizes the operational RD function of the source. The theoretical RD function characterizes the fundamental lossy compression limit theoretically attainable [13], and it is rarely known in analytical form except for a few special cases, such as the i.i.d. Gaussian source [4]:

$$D(R) = \sigma^2 2^{-2R}, \tag{18.1}$$

where the Gaussian source is assumed to have zero mean and variance σ^2 and the rate R is measured in bits per sample. Note from the formula that every extra bit reduces the expected distortion by a factor of 4 (or increases the signal to noise ratio by 6 dB). This formula agrees with our intuition that the distortion should decrease exponentially as the rate increases. In fact, this is true when quantizing sources with other probability distributions as well under high-resolution (or bit rate) conditions: the optimal RD performance of encoding a zero mean stationary source with variance σ^2 takes the form of [7]

$$D(R) = h\sigma^2 2^{-2R}, \tag{18.2}$$

where the factor h depends on the probability distribution of the source. For a Gaussian source, $h = \sqrt{3}\pi/2$ with optimal scalar quantization. Under high-resolution conditions, it can be shown that the optimal entropy-constrained scalar quantizer is a uniform one, whose average distortion is only approximately 1.53 dB worse than the theoretical bound attainable that is known as the Shannon bound [7, 11]. For low bit rate coding, most current subband coders employ a uniform quantizer with a "deadzone" in the central quantization bin. This simply means that the all-important central bin is wider than the other bins: this turns out to be more efficient than having all bins be of the same size. The performance of deadzone quantizers is nearly optimal for memoryless sources even at low rates [14]. An additional advantage of using deadzone quantization is that, when the deadzone is twice as much as the uniform step size, an embedded bitstream can be generated by successive quantization. We will elaborate more on embedded wavelet image coding in Section 18.5.

18.3.3 Entropy Coding

Once the quantization process is completed, the last encoding step is to use entropy coding to achieve the entropy rate of the quantizer. Entropy coding works like the Morse code in electric telegraph: more frequently occurring symbols are represented by short codewords, whereas symbols occurring less frequently are represented by longer codewords. On average, entropy coding does better than assigning the same codelength to all symbols. For example, a source that can take on any of the four symbols $\{A, B, C, D\}$

with equal likelihood has 2 bits of information or uncertainty, and its entropy is 2 bits per symbol (e.g., one can assign a binary code of 00 to A, 01 to B, 10 to C, and 11 to D). However if the symbols are not equally likely, e.g., if the probabilities of A, B, C, and D are $0.5, 0.25, 0.125$, and 0.125, respectively, then one can do much better on average by not assigning the same number of bits to each symbol but rather by assigning fewer bits to the more popular or predictable ones. This results in a variable length code. In fact, one can show that the optimal code would be one in which A gets 1 bit, B gets 2 bits, and C and D get 3 bits each (e.g., $A = 0, B = 10, C = 110$, and $D = 111$). This is called an entropy code. With this code, one can compress the source with an average of only 1.75 bits per symbol, a 12.5% improvement in compression over the original 2 bits per symbol associated with having fixed length codes for the symbols. The two popular entropy coding methods are Huffman coding [15] and arithmetic coding [16]. A comprehensive coverage of entropy coding is given in Chapter 16. The Shannon entropy [4] provides a lower bound in terms of the amount of compression entropy coding can best achieve. The optimal entropy code constructed in the example actually achieves the theoretical Shannon entropy of the source.

18.4 SUBBAND CODING: THE EARLY DAYS

Subband coding normally uses bases of roughly equal bandwidth. Wavelet image coding can be viewed as a special case of subband coding with logarithmically varying bandwidth bases that satisfy certain properties.[4] Early work on wavelet image coding was thus hidden under the name of subband coding [8, 17], which builds upon the traditional transform coding paradigm of energy compaction and decorrelation. The main idea of subband coding is to treat different bands differently as each band can be modeled as a statistically distinct process in quantization and coding.

To illustrate the design philosophy of early subband coders, let us again assume, for example, that we are coding a vector source $\{x_0, x_1\}$, where both x_0 and x_1 are samples of a stationary random sequence $X(n)$ with zero mean and variance σ_x^2. If we code x_0 and x_1 directly by using PCM coding, from our earlier discussion on quantization, the RD performance can be approximated as

$$D_{PCM}(R) = h\sigma_x^2 2^{-2R}. \tag{18.3}$$

In subband coding, two quantizers are designed: one for each of the two transform coefficients y_0 and y_1. The goal is to choose rates R_0 and R_1 needed for coding y_0 and y_1 so that the average distortion

$$D_{SBC}(R) = (D(R_0) + D(R_1))/2 \tag{18.4}$$

is minimized with the constraint on the average bit rate

$$(R_0 + R_1)/2 = R. \tag{18.5}$$

---

[4]Both wavelet image coding and subband coding are special cases of transform coding.

Using the high rate approximation, we write $D(R_0) = h\sigma_{y_0}^2 2^{-2R_0}$ and $D(R_1) = h\sigma_{y_1}^2 2^{-2R_1}$; then the solutions to this *bit allocation* problem are [8]

$$R_0 = R + \frac{1}{2}\log_2 \frac{\sigma_{y_0}}{\sigma_{y_1}}; \quad R_1 = R - \frac{1}{2}\log_2 \frac{\sigma_{y_0}}{\sigma_{y_1}}, \tag{18.6}$$

with the minimum average distortion being

$$D_{\text{SBC}}(R) = h\sigma_{y_0}\sigma_{y_1} 2^{-2R}. \tag{18.7}$$

Note that, at the optimal point, $D(R_0) = D(R_1) = D_{\text{SBC}}(R)$. That is, the quantizers for y_0 and y_1 give the same distortion with optimal bit allocation. Since the transform T is orthogonal, we have $\sigma_x^2 = (\sigma_{y_0}^2 + \sigma_{y_1}^2)/2$. The *coding gain* of using subband coding over PCM is

$$\frac{D_{\text{PCM}}(R)}{D_{\text{SBC}}(R)} = \frac{\sigma_x^2}{\sigma_{y_0}\sigma_{y_1}} = \frac{(\sigma_{y_0}^2 + \sigma_{y_1}^2)/2}{(\sigma_{y_0}^2\sigma_{y_1}^2)^{1/2}}, \tag{18.8}$$

the ratio of arithmetic mean to geometric mean of coefficient variances $\sigma_{y_0}^2$ and $\sigma_{y_1}^2$. What this important result states is that subband coding performs no worse than PCM coding, and that the larger the disparity between coefficient variances, the bigger the subband coding gain, because $(\sigma_{y_0}^2 + \sigma_{y_1}^2)/2 \geq (\sigma_{y_0}^2\sigma_{y_1}^2)^{1/2}$, with equality if $\sigma_{y_0}^2 = \sigma_{y_1}^2$. This result can be easily extended to the case when $M > 2$ uniform subbands (of equal size) are used instead. The coding gain in this general case is as follows:

$$\frac{D_{\text{PCM}}(R)}{D_{\text{SBC}}(R)} = \frac{\frac{1}{M}\sum_{k=0}^{M-1}\sigma_k^2}{\left(\prod_{k=0}^{M-1}\sigma_k^2\right)^{1/M}}, \tag{18.9}$$

where σ_k^2 is the sample variance of the kth band ($0 \leq k \leq M - 1$). The above assumes that all M bands are of the same size. In the case of the subband or wavelet transform, the sizes of the subbands are not the same (see Fig. 18.8), but the above formula can be generalized pretty easily to account for this. As another extension of the results given in the above example, it can be shown that the necessary condition for optimal bit allocation is that all subbands should incur the same distortion at optimality—else it is possible to steal some bits from the lower distortion bands to the higher distortion bands in a way that makes the overall performance better.

Figure 18.8 shows typical bit allocation results for different subbands under a total bit rate budget of 1 bit per pixel for wavelet image coding. Since low-frequency bands in the upper-left corner have far more energy than high-frequency bands in the lower-right corner (see Fig. 18.1), more bits have to be allocated to lowpass bands than to highpass bands. The last two frequency bands in the bottom half are not coded (set to zero) because of limited bit rate. Since subband coding treats wavelet coefficients according to their frequency bands, it is effectively a *frequency* domain transform technique.

Initial wavelet-based coding algorithms, e.g., [18], followed exactly this subband coding methodology. These algorithms were designed to exploit the energy compaction

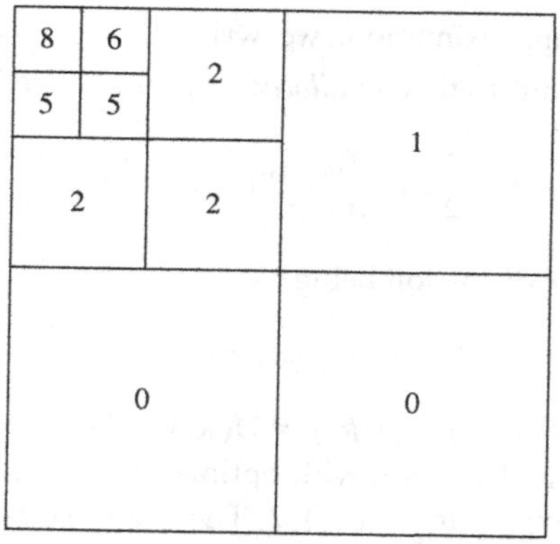

FIGURE 18.8

Typical bit allocation results for different subbands. The unit of the numbers is bits per pixel. These are designed to satisfy a total bit rate budget of 1 bit per pixel. That is, $\{[(8 + 6 + 5 + 5)/4 + 2 + 2 + 2]/4 + 1 + 0 + 0\}/4 = 1$.

properties of the wavelet transform only in the frequency domain by applying quantizers optimized for the statistics of each frequency band. Such algorithms have demonstrated small improvements in coding efficiency over standard transform-based algorithms.

18.5 NEW AND MORE EFFICIENT CLASS OF WAVELET CODERS

Because wavelet decompositions offer *space-frequency* representations of images, i.e., low-frequency coefficients have large spatial support (good for representing large image background regions), whereas high-frequency coefficients have small spatial support (good for representing spatially local phenomena such as edges), the wavelet representation calls for new quantization strategies that go beyond traditional subband coding techniques to exploit this underlying space-frequency image characterization.

Shapiro made a breakthrough in 1993 with his EZW coding algorithm [19]. Since then a new class of algorithms have been developed that achieve significantly improved performance over the EZW coder. In particular, Said and Pearlman's work on set partitioning in hierarchical trees (SPIHT) [20], which improves the EZW coder, has established zerotree techniques as the current state-of-the-art of wavelet image coding since the SPIHT algorithm proves to be very successful for both lossy and lossless compression.

18.5.1 Zerotree-Based Framework and EZW Coding

A wavelet image representation can be thought of as a tree-structured spatial set of coefficients. *A wavelet coefficient tree* is defined as the set of coefficients from different bands that represent the same spatial region in the image. Figure 18.9 shows a three-level wavelet decomposition of the *Lena* image, together with a wavelet coefficient tree

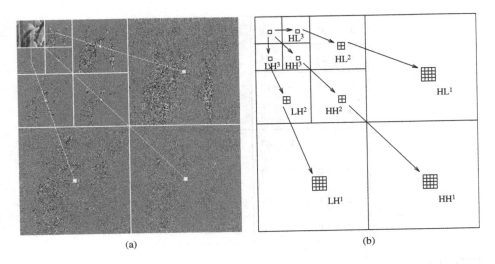

FIGURE 18.9

Wavelet decomposition offers a tree-structured image representation. (a) Three-level wavelet decomposition of the *Lena* image; (b) Spatial wavelet coefficient tree consisting of coefficients from different bands that correspond to the same spatial region of the original image (e.g., the eye of *Lena*). Arrows identify the parent-children dependencies.

structure representing the eye region of *Lena*. Arrows in Fig. 18.9(b) identify the parent-children dependencies in a tree. The lowest frequency band of the decomposition is represented by the root nodes (top) of the tree, the highest frequency bands by the leaf nodes (bottom) of the tree, and each parent node represents a lower frequency component than its children. Except for a root node, which has only three children nodes, each parent node has four children nodes, the 2×2 region of the same spatial location in the immediately higher frequency band.

Both the EZW and SPIHT algorithms [19, 20] are based on the idea of using multipass zerotree coding to transmit the largest wavelet coefficients (in magnitude) at first. We hereby use "zero coding" as a generic term for both schemes, but we focus on the popular SPIHT coder because of its superior performance. A set of tree coefficients is significant if the largest coefficient magnitude in the set is greater than or equal to a certain threshold (e.g., a power of 2); otherwise, it is insignificant. Similarly, a coefficient is significant if its magnitude is greater than or equal to the threshold; otherwise, it is insignificant. In each pass the significance of a larger set in the tree is tested at first: if the set is insignificant, a binary "zerotree" bit is used to set all coefficients in the set to zero; otherwise, the set is partitioned into subsets (or child sets) for further significance tests. After all coefficients are tested in one pass, the threshold is halved before the next pass.

The underlying assumption of the zerotree coding framework is that most images can be modeled as having decaying power spectral densities. That is, if a parent node in the wavelet coefficient tree is insignificant, it is very likely that its descendents are also

63	−34	49	10	7	13	−12	7
−31	23	14	−13	3	4	6	−1
15	14	3	−12	5	−7	3	9
−9	−7	−14	8	4	−2	3	2
−5	9	−1	47	4	6	−2	2
3	0	−3	2	3	−2	0	4
2	−3	6	−4	3	6	3	6
5	11	5	6	0	3	−4	4

FIGURE 18.10

Example of a three-level wavelet representation of an 8×8 image.

insignificant. The zerotree symbol is used very efficiently in this case to signify a spatial subtree of zeros.

We give a SPIHT coding example to highlight the order of operations in zerotree coding. Start with a simple three-level wavelet representation of an 8×8 image,[5] as shown in Fig. 18.10. The largest coefficient magnitude is 63. We can choose a threshold in the first pass between 31.5 and 63. Let $T_1 = 32$. Table 18.1 shows the first pass of the SPIHT coding process, with the following comments:

1. The coefficient value 63 is greater than the threshold 32 and positive, so a significance bit "1" is generated, followed by a positive sign bit "0." After decoding these symbols, the decoder knows the coefficient is between 32 and 64 and uses the midpoint 48 as an estimate.[6]

2. The descendant set of coefficient −34 is significant; a significance bit "1" is generated, followed by a significance test of each of its four children {49, 10, 14, −13}.

3. The descendant set of coefficient −31 is significant; a significance bit "1" is generated, followed by a significance test of each of its four children {15, 14, −9, −7}.

---

[5]This set of wavelet coefficients is the same as the one used by Shapiro in an example to showcase EZW coding [19]. Curious readers can compare these two examples to see the difference between EZW and SPIHT coding.

[6]The reconstruction value can be anywhere in the uncertainty interval (32, 64). Choosing the midpoint is the result of a simple form of minimax estimation.

TABLE 18.1 First pass of the SPIHT coding process at threshold $T_1 = 32$.

Coefficient coordinates	Coefficient value	Binary symbol	Reconstruction value	Comments
(0,0)	63	1		(1)
		0	48	
(1,0)	−34	1		
		1	−48	
(0,1)	−31	0	0	
(1,1)	23	0	0	
(1,0)	−34	1		(2)
(2,0)	49	1		
		0	48	
(3,0)	10	0	0	
(2,1)	14	0	0	
(3,1)	−13	0	0	
(0,1)	−31	1		(3)
(0,2)	15	0	0	
(1,2)	14	0	0	
(0,3)	−9	0	0	
(1,3)	−7	0	0	
(1,1)	23	0		(4)
(1,0)	−34	0		(5)
(0,1)	−31	1		(6)
(0,2)	15	0		(7)
(1,2)	14	1		(8)
(2,4)	−1	0	0	
(3,4)	47	1		
		0	48	
(2,5)	−3	0	0	
(3,5)	2	0	0	
(0,3)	−9	0		(9)
(1,3)	−7	0		

4. The descendant set of coefficient 23 is insignificant; an insignificance bit "0" is generated. This zerotree bit is the only symbol generated in the current pass for the whole descendant set of coefficient 23.

5. The grandchild set of coefficient −34 is insignificant; a binary bit "0" is generated.[7]

---

[7] In this example, we use the following convention: when a coefficient or set is significant, a binary bit "1" is generated; otherwise, a binary bit "0" is generated. In the actual SPIHT implementation [20], this convention was not always followed—when a grandchild set is significant, a binary bit "0" is generated, otherwise, a binary bit "1" is generated.

6. The grandchild set of coefficient -31 is significant; a binary bit "1" is generated.

7. The descendant set of coefficient 15 is insignificant; an insignificance bit "0" is generated. This zerotree bit is the only symbol generated in the current pass for the whole descendant set of coefficient 15.

8. The descendant set of coefficient 14 is significant; a significance bit "1" is generated, followed by a significance test of each of its four children $\{-1, 47, -3, 2\}$.

9. Coefficient -31 has four children $\{15, 14, -9, -7\}$. Descendant sets of child 15 and child 14 were tested for significance before. Now descendant sets of the remaining two children -9 and -7 are tested.

In this example, the encoder generates 29 bits in the first pass. Along the process, it identifies four significant coefficients $\{63, -34, 49, 47\}$. The decoder reconstructs each coefficient based on these bits. When a set is insignificant, the decoder knows each coefficient in the set is between -32 and 32 and uses the midpoint 0 as an estimate. The reconstruction result at the end of the first pass is shown in Fig. 18.11(a).

The threshold is halved ($T_2 = T_1/2 = 16$) before the second pass, where insignificant coefficients and sets in the first pass are tested for significance again against T_2, and significant coefficients found in the first pass are refined. The second pass thus consists of the following:

1. Significance tests of the 12 insignificant coefficients found in the first pass—those having reconstruction value 0 in Table 18.1. Coefficients -31 at (0, 1) and 23 at (1, 1) are found to be significant in this pass; a sign bit is generated for each. The

48	-48	48	0	0	0	0	0
0	0	0	0	0	0	0	0
0	0	0	0	0	0	0	0
0	0	0	0	0	0	0	0
0	0	0	48	0	0	0	0
0	0	0	0	0	0	0	0
0	0	0	0	0	0	0	0
0	0	0	0	0	0	0	0

(a)

56	-40	56	0	0	0	0	0
-24	24	0	0	0	0	0	0
0	0	0	0	0	0	0	0
0	0	0	0	0	0	0	0
0	0	0	40	0	0	0	0
0	0	0	0	0	0	0	0
0	0	0	0	0	0	0	0
0	0	0	0	0	0	0	0

(b)

FIGURE 18.11

Reconstructions after the (a) first and (b) second passes in SPIHT coding.

decoder knows the coefficient magnitude is between 16 and 32 and decode them as -24 and 24.

2. The descendant set of coefficient 23 at (1, 1) is insignificant; so are the grandchild set of coefficient 49 at (2, 0) and descendant sets of coefficients 15 at (0, 2), -9 at (0, 3), and -7 at (1, 3). A zerotree bit is generated in the current pass for each insignificant descendant set.

3. Refinement of the four significant coefficients $\{63, -34, 49, 47\}$ found in the first pass. The coefficient magnitudes are identified as being either between 32 and 48, which will be encoded with "0" and decoded as the midpoint 40, or between 48 and 64, which will be encoded with "1" and decoded as 56.

The encoder generates 23 bits (14 from step 1, 5 from step 2, and 4 from step 3) in the second pass. Along the process it identifies two more significant coefficients. Together with the four found in the first pass, the set of significant coefficients now becomes $\{63, -34, 49, 47, -31, 23\}$. The reconstruction result at the end of the second pass is shown in Fig. 18.11(b).

The above encoding process continues from one pass to another and can stop at any point. For better coding performance, arithmetic coding [16] can be used to further compress the binary bitstream out of the SPIHT encoder.

From this example, we note that when the thresholds are powers of 2, zerotree coding can be thought of as a bit-plane coding scheme. It encodes one bit-plane at a time, starting from the most significant bit. The effective quantizer in each pass is a deadzone quantizer with the deadzone being twice the uniform step size. With the sign bits and refinement bits (for coefficients that become significant in previous passes) being coded on the fly, zerotree coding generates an *embedded* bitstream, which is highly desirable for progressive transmission (see Fig. 18.4). A simple example of embedded representation is the approximation of an irrational number (say $\pi = 3.1415926535\cdots$) by a rational number. If we were only allowed two digits after the decimal point, then $\pi \approx 3.14$; if three digits after the decimal point were allowed, then $\pi \approx 3.141$; and so on. Each additional bit of the embedded bitstream is used to improve upon the previously decoded image for successive approximation, so rate control in zerotree coding is exact, and no loss is incurred if decoding stops at any point of the bitstream. The remarkable thing about zerotree coding is that it outperforms almost all other schemes (such as JPEG coding) while being embedded. This good performance can be partially attributed to the fact that zerotree coding captures across-scale interdependencies of wavelet coefficients. The zerotree symbol effectively zeros out a set of coefficients in a subtree, achieving the coding gain of VQ [7] over scalar quantization.

Figure 18.12 shows the original *Lena* and *Barbara* images and their decoded versions at 0.25 bit per pixel (32:1 compression ratio) by baseline JPEG and SPIHT [20]. These images are coded at a relatively low bit rate to emphasize coding artifacts. The *Barbara* image is known to be hard to compress because of its insignificant high-frequency content (see the periodic stripe texture on *Barbara*'s trousers and scarf, and the checkerboard texture pattern on the tablecloth). The subjective difference in reconstruction

FIGURE 18.12

Coding of the 512×512 *Lena* and *Barbara* images at 0.25 bit per pixel (compression ratio of 32:1). Top: the original *Lena* and *Barbara* images. Middle: baseline JPEG decoded images, PSNR = 31.6 dB for *Lena*, and PSNR = 25.2 dB for *Barbara*. Bottom: SPIHT decoded images, PSNR = 34.1 dB for *Lena*, and PSNR = 27.6 dB for *Barbara*.

quality between the two decoded versions of the same image is quite perceptible on a high-resolution monitor. The JPEG decoded images show highly visible blocking artifacts while the wavelet-based SPIHT decoded images have much sharper edges and preserve most of the striped texture.

18.5.2   Advanced Wavelet Coders: High-Level Characterization

We saw that the main difference between the early class of subband image coding algorithms and the zerotree-based compression framework is that the former exploits only the frequency characterization of the wavelet image representation, whereas the latter exploits both the *spatial* and *frequency* characterization. To be more precise, the early class of coders was adept at exploiting the wavelet transform's ability to concentrate the image energy disparately in the different frequency bands, with the lower frequency bands having a much higher energy density. What these coders failed to exploit was the very definite spatial characterization of the wavelet representation. In fact, this is even apparent to the naked eye if one views the wavelet decomposition of the *Lena* image in Fig. 18.1, where the spatial structure of the image is clearly exposed in the high-frequency wavelet bands, e.g., the edge structure of the hat and face and the feather texture. Failure to exploit this spatial structure limited the performance potential of the early subband coders.

In explicit terms, not only is it true that the energy density of the different wavelet subbands is highly disparate, resulting in gains by separating the data set into statistically dissimilar frequency groupings of data, but it is also true that the data in the high-frequency subbands are highly spatially structured and clustered around the spatial edges of the original image. The early class of coders exploited the conventional coding gain associated with dissimilarity in the statistics of the frequency bands, but not the potential coding gain from separating individual frequency band energy into spatially localized clusters.

It is insightful to note that unlike the coding gain based on the frequency characterization, which is statistically predictable for typical images (the low-frequency subbands have much higher energy density than the high frequency ones), there is a difficulty in going after the coding gain associated with the spatial characterization that is not statistically predictable; after all, there is no reason to expect the upper-left corner of the image to have more edges than the lower right. This calls for a drastically different way of exploiting this structure—a way of pointing to the *spatial location* of significant edge regions within each subband. At a high level, a zerotree is no more than an efficient "pointing" data structure that incorporates the spatial characterization of wavelet coefficients by identifying tree-structured collections of insignificant spatial subregions across hierarchical subbands.

Equipped with this high-level insight, it becomes clear that the zerotree approach is but only one way to skin the cat. Researchers in the wavelet image compression community have found other ways to exploit this phenomenon by using an array of creative ideas. The array of successful data structures in the research literature include (a) RD optimized zerotree-based structures, (b) morphology- or region-growing-based structures,

(c) spatial context modeling based structures, (d) statistical mixture modeling based structures, (e) classification-based structures, and so on. Due to space limitations, we omit the details of these advanced methods here.

18.6 ADAPTIVE WAVELET TRANSFORMS: WAVELET PACKETS

In noting how transform coding has become the *de facto* standard for image and video compression, it is important to realize that the traditional approach of using a transform with fixed frequency resolution (be it the logarithmic wavelet transform or the DCT) is good only in an ensemble sense for a typical *statistical* class of images. This class is well suited to the characteristics of the chosen fixed transform. This raises the natural question; is it possible to do better by being *adaptive* in the transformation so as to best match the features of the transform to the specific attributes of arbitrary individual images that may not belong to the typical ensemble?

To be specific, the wavelet transform is a good fit for typical natural images that have an exponentially decaying spectral density, with a mixture of strong stationary low-frequency components (such as the image background) and perceptually important short-duration high-frequency components (such as sharp image edges). The fit is good because of the wavelet transform's logarithmic decomposition structure, which results in its well-advertised attributes of good frequency resolution at low frequencies, and good time resolution at high frequencies (see Fig. 18.3(b)).

There are, however, important classes of images (or significant subimages) whose attributes go against those offered by the wavelet decomposition, e.g., images having strong highpass components. A good example is the periodic texture pattern in the *Barbara* image of Fig. 18.12—see the trousers and scarf textures and the tablecloth texture. Another special class of images for which the wavelet is not a good idea is the class of fingerprint images (see Fig. 18.13 for a typical example) which has periodic high-frequency ridge patterns. These images are better matched with decomposition elements that have good frequency localization at high frequencies (corresponding to the texture patterns), which the wavelet decomposition does not offer in its menu.

This motivates the search for alternative transform descriptions that are more adaptive in their representation, and that are more robust to a large class of images of unknown or mismatched space-frequency characteristics. Although the task of finding an optimal decomposition for every individual image in the world is an ill-posed problem, the situation gets more interesting if we consider a large but finite library of desirable transforms and match the best transform in the library adaptively to the individual image. In order to make this feasible, there are two requirements. First, the library must contain a good representative set of entries (e.g., it would be good to include the conventional wavelet decomposition). Second, it is essential that there exists a fast way of searching through the library to find the best transform in an image-adaptive manner.

Both these requirements are met with an elegant generalization of the wavelet transform, called the *wavelet packet* decomposition, also known sometimes as the *best basis* framework. Wavelet packets were introduced to the signal processing community by

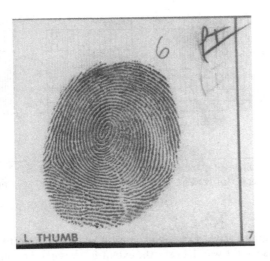

FIGURE 18.13

Fingerprint image: image coding using logarithmic wavelet transform does not perform well for fingerprint images such as this one with strong highpass ridge patterns.

Coifman and Wickerhauser [21]. They represent a huge library of orthogonal transforms having a rich time-frequency diversity that also come with an easy-to-search capability, thanks to the existence of fast algorithms that exploit the tree-structured nature of these basis expansions—the tree-structure comes from the cascading of multirate filter bank operations; see Chapter 6 and [3]. Wavelet packet bases essentially look like the wavelet bases shown in Fig. 18.3(b), but they have more oscillations.

The wavelet decomposition, which corresponds to a logarithmic tree structure, is the most famous member of the wavelet packet family. Whereas wavelets are best matched to signals having a decaying energy spectrum, wavelet packets can be matched to signals having almost arbitrary spectral profiles, such as signals having strong high-frequency or mid-frequency stationary components, making them attractive for decomposing images having significant texture patterns, as discussed earlier. There are an astronomical number of basis choices available in the typical wavelet packet library: for example, it can be shown that the library has over 10^{78} transforms for typical five-level 2D wavelet packet image decompositions. The library is thus well equipped to deal efficiently with arbitrary classes of images requiring diverse spatial-frequency resolution tradeoffs.

Using the concept of time-frequency tilings introduced in Section 18.1, it is easy to see what wavelet packet tilings look like, and how they are a generalization of wavelets. We again start with 1D signals. Tiling representations of several expansions are plotted in Fig. 18.14. Figure 18.14(a) shows a uniform STFT-like expansion, where the tiles are all of the same shape and size; Fig. 18.14(b) is the familiar wavelet expansion or the logarithmic subband decomposition; Fig. 18.14(c) shows a wavelet packet expansion where the bandwidths of the bases are neither uniformly nor logarithmically varying; and

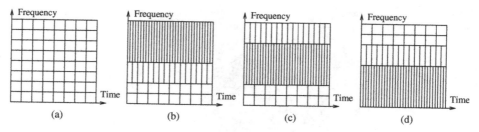

FIGURE 18.14

Tiling representations of several expansions for 1D signals. (a) STFT-like decomposition; (b) wavelet decomposition; (c) wavelet packet decomposition, and (d) "anti-wavelet" packet decomposition.

Fig. 18.14(d) highlights a wavelet packet expansion where the time-frequency attributes are exactly the reverse of the wavelet case: the expansion has good frequency resolution at higher frequencies, and good time localization at lower frequencies—we might call this the "anti-wavelet" packet. There are a plethora of other options for the time-frequency resolution tradeoff, and these all correspond to admissible wavelet packet choices.

The extra adaptivity of the wavelet packet framework is obtained at the price of added computation in searching for the best wavelet packet basis, so an efficient fast search algorithm is the key in applications involving wavelet packets. The problem of searching for the best basis from the wavelet packet library for the compression problem using an RD optimization framework and a fast tree-pruning algorithm was described in [22].

The 1D wavelet packet bases can be easily extended to 2D by writing a 2D basis function as the product of two 1D basis functions. In another words, we can treat the rows and columns of an image separately as 1D signals. The performance gains associated with wavelet packets are obviously image-dependent. For difficult images such as *Barbara* in Fig. 18.12, a wavelet packet decomposition shown in Fig. 18.15(a) gives much better coding performance than the wavelet decomposition. The wavelet packet decoded *Barbara* image at 0.1825 b/p is shown in Fig. 18.15(b), whose visual quality (or PSNR) is the same as the wavelet SPIHT decoded *Barbara* image at 0.25 b/p in Fig. 18.12. The bit rate saving achieved by using a wavelet packet basis instead of the wavelet basis in this case is 27% at the same visual quality.

An important practical application of wavelet packet expansions is the FBI wavelet scalar quantization (WSQ) standard for fingerprint image compression [23]. Because of the complexity associated with adaptive wavelet packet transforms, the FBI WSQ standard uses a fixed wavelet packet decomposition in the transform stage. The transform structure specified by the FBI WSQ standard is shown in Fig. 18.16. It was designed for 500 dots per inch fingerprint images by spectral analysis and trial and error. A total of 64 subbands are generated with a five-level wavelet packet decomposition. Trials by the FBI have shown that the WSQ standard benefited from having fine frequency partitions in the middle frequency region containing the fingerprint ridge patterns.

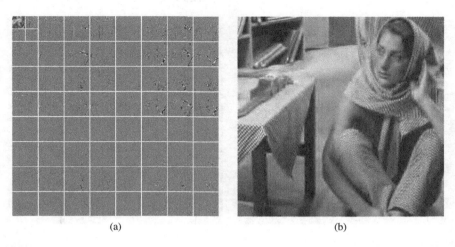

FIGURE 18.15

(a) A wavelet packet decomposition for the *Barbara* image. White lines represent frequency boundaries. Highpass bands are processed for display; (b) Wavelet packet decoded *Barbara* at 0.1825 b/p. PSNR = 27.6 dB.

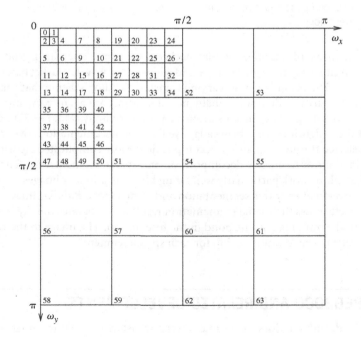

FIGURE 18.16

The wavelet packet transform structure given in the FBI WSQ specification. The number sequence shows the labeling of the different subbands.

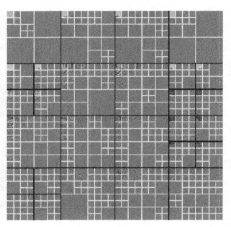

FIGURE 18.17

Space-frequency segmentation and tiling for the *Building* image. The image to the left shows that spatial segmentation separates the sky in the background from the building and the pond in the foreground. The image to the right gives the best wavelet packet decomposition of each spatial segment. Dark lines represent spatial segments; white lines represent subband boundaries of wavelet packet decompositions. Note that the upper-left corners are the lowpass bands of wavelet packet decompositions.

As an extension of adaptive wavelet packet transforms, one can introduce time-variation by segmenting the signal in time and allowing the wavelet packet bases to evolve with the signal. The result is a time-varying transform coding scheme that can adapt to signal nonstationarities. Computationally fast algorithms are again very important for finding the optimal signal expansions in such a time-varying system. For 2D images, the simplest of these algorithms performs adaptive frequency segmentations over regions of the image selected through a quadtree decomposition. More complicated algorithms provide combinations of frequency decomposition and spatial segmentation. These jointly adaptive algorithms work particularly well for highly nonstationary images. Figure 18.17 shows the space-frequency tree segmentation and tiling for the *Building* image [24]. The image to the left shows the spatial segmentation result that separates the sky in the background from the building and the pond in the foreground. The image to the right gives the best wavelet packet decomposition for each spatial segment.

18.7 JPEG2000 AND RELATED DEVELOPMENTS

JPEG2000 by default employs the dyadic wavelet transform for natural images in many standard applications. It also allows the choice of the more general wavelet packet transforms for certain types of imagery (e.g., fingerprints and radar images). Instead of using the zerotree-based SPIHT algorithm, JPEG2000 relies on embedded block coding with optimized truncation (EBCOT) [25] to provide a rich set of features such as quality scalability, resolution scalability, spatial random access, and region-of-interest coding. Besides robustness to image type changes in terms of compression performance, the main advantage of the block-based EBCOT algorithm is that it provides easier random access to local image components. On the other hand, both encoding and decoding in SPIHT require nonlocal memory access to the whole tree of wavelet coefficients, causing reduction in throughput when coding large-size images. A thorough description of the JPEG2000 standard is in [1]. Other JPEG2000 related references are Chapter 17 and [26, 27].

Although this chapter is about wavelet coding of 2D images, the wavelet coding framework and its extension to wavelet packets apply to 3D video as well. Recent research works (see [28] and references therein) on 3D scalable wavelet video coders based on the framework of motion-compensated temporal filtering (MCTF) [29] have shown competitive or better performance than the best MC-DCT-based standard video coder (e.g., H.264/AVC [30]). They have stirred considerable excitement in the video coding community and stimulated research efforts toward subband/wavelet interframe video coding, especially in the area of scalable motion coding [31] within the context of MCTF. MCTF can be conceptually viewed as the extension of wavelet-based coding in JPEG2000 from 2D images to 3D video. It nicely combines scalability features of wavelet-based coding with motion compensation, which has been proven to be very efficient and necessary in MC-DCT-based standard video coders. We refer the readers to a recent special issue [32] on the latest results and Chapter 11 in [9] for an exposition of 3D subband/wavelet video coding.

18.8 CONCLUSION

Since the introduction of wavelets as a signal processing tool in the late 1980s, a variety of wavelet-based coding algorithms have advanced the limits of compression performance well beyond that of the current commercial JPEG image coding standard. In this chapter, we have provided very simple high-level insights, based on the intuitive concept of time-frequency representations, into why wavelets are good for image coding. After introducing the salient aspects of the compression problem in general and the transform coding problem in particular, we have highlighted the key important differences between the early class of subband coders and the more advanced class of modern-day wavelet image coders. Selecting the EZW coding structure embodied in the celebrated SPIHT algorithm as a representative of this latter class, we have detailed its operation by using a simple illustrative example. We have also described the role of wavelet packets as a simple but powerful generalization of the wavelet decomposition in order to offer a more robust and adaptive transform image coding framework.

JPEG2000 is the result of the rapid progress made in wavelet image coding research in the 1990s. The triumph of wavelet transform in the evolution of the JPEG2000 standard underlines the importance of the fundamental insights provided in this chapter into why wavelets are so attractive for image compression.

REFERENCES

[1] D. Taubman and M. Marcellin. *JPEG2000: Image Compression Fundamentals, Standards, and Practice.* Kluwer, New York, 2001.

[2] G. Strang and T. Nguyen. *Wavelets and Filter Banks.* Wellesley-Cambridge Press, New York, 1996.

[3] M. Vetterli and J. Kovačević. *Wavelets and Subband Coding.* Prentice-Hall, Englewood Cliffs, NJ, 1995.

[4] T. M. Cover and J. A. Thomas. *Elements of Information Theory.* Wiley & Sons, Inc., New York, 1991.

[5] C. E. Shannon. A mathematical theory of communication. *Bell Syst. Tech. J.*, 27:379–423, 623–656, 1948.

[6] C. E. Shannon. Coding theorems for a discrete source with a fidelity criterion. *IRE Natl. Conv. Rec.*, 4:142–163, 1959.

[7] A. Gersho and R. M. Gray. *Vector Quantization and Signal Compression.* Kluwer Academic, Boston, MA, 1992.

[8] N. S. Jayant and P. Noll. *Digital Coding of Waveforms.* Prentice-Hall, Englewood Cliffs, NJ, 1984.

[9] A. Bovik, editor. *The Video Processing Companion.* Elsevier, Burlington, MA, 2008.

[10] S. P. Lloyd. Least squares quantization in PCM. *IEEE Trans. Inf. Theory*, IT-28:127–135, 1982.

[11] H. Gish and J. N. Pierce. Asymptotically efficient quantizing. *IEEE Trans. Inf. Theory*, IT-14(5): 676–683, 1968.

[12] M. W. Marcellin and T. R. Fischer. Trellis coded quantization of memoryless and Gauss-Markov sources. *IEEE Trans. Commun.*, 38(1):82–93, 1990.

[13] T. Berger. *Rate Distortion Theory.* Prentice-Hall, Englewood Cliffs, NJ, 1971.

[14] N. Farvardin and J. W. Modestino. Optimum quantizer performance for a class of non-Gaussian memoryless sources. *IEEE Trans. Inf. Theory*, 30:485–497, 1984.

[15] D. A. Huffman. A method for the construction of minimum redundancy codes. *Proc. IRE*, 40: 1098–1101, 1952.

[16] T. C. Bell, J. G. Cleary, and I. H. Witten. *Text Compression.* Prentice-Hall, Englewood Cliffs, NJ, 1990.

[17] J. W. Woods, editor. *Subband Image Coding.* Kluwer Academic, Boston, MA, 1991.

[18] M. Antonini, M. Barlaud, P. Mathieu, and I. Daubechies. Image coding using wavelet transform. *IEEE Trans. Image Process.*, 1(2):205–220, 1992.

[19] J. Shapiro. Embedded image coding using zero-trees of wavelet coefficients. *IEEE Trans. Signal Process.*, 41(12):3445–3462, 1993.

[20] A. Said and W. A. Pearlman. A new, fast, and efficient image codec based on set partitioning in hierarchical trees. *IEEE Trans. Circuits Syst. Video Technol.*, 6(3):243–250, 1996.

[21] R. R. Coifman and M. V. Wickerhauser. Entropy based algorithms for best basis selection. *IEEE Trans. Inf. Theory*, 32:712–718, 1992.

[22] K. Ramchandran and M. Vetterli. Best wavelet packet bases in a rate-distortion sense. *IEEE Trans. Image Process.*, 2(2):160–175, 1992.

[23] Criminal Justice Information Services. *WSQ Gray-Scale Fingerprint Image Compression Specification (Ver. 2.0).* Federal Bureau of Investigation, 1993.

[24] K. Ramchandran, Z. Xiong, K. Asai, and M. Vetterli. Adaptive transforms for image coding using spatially-varying wavelet packets. *IEEE Trans. Image Process.*, 5:1197–1204, 1996.

[25] D. Taubman. High performance scalable image compression with EBCOT. *IEEE Trans. Image Process.*, 9(7):1151–1170, 2000.

[26] Special Issue on JPEG2000. *Signal Process. Image Commun.*, 17(1), 2002.

[27] D. Taubman and M. Marcellin. JPEG2000: standard for interactive imaging. *Proc. IEEE*, 90(8): 1336–1357, 2002.

[28] J. Ohm, M. van der Schaar, and J. Woods. Interframe wavelet coding – motion picture representation for universal scalability. *Signal Process. Image Commun.*, 19(9):877–908, 2004.

[29] S.-T. Hsiang and J. Woods. Embedded video coding using invertible motion compensated 3D subband/wavelet filter bank. *Signal Process. Image Commun.*, 16(8):705–724, 2001.

[30] T. Wiegand, G. Sullivan, G. Bjintegaard, and A. Luthra. Overview of the H.264/AVC video coding standard. *IEEE Trans. Circuits Syst. Video Technol.*, 13:560–576, 2003.

[31] A. Secker and D. Taubman. Highly scalable video compression with scalable motion coding. *IEEE Trans. Image Process.*, 13(8):1029–1041, 2004.

[32] Special issue on subband/wavelet interframe video coding. *Signal Process. Image Commun.*, 19, 2004.

Gradient and Laplacian Edge Detection

19

Phillip A. Mlsna[1] and Jeffrey J. Rodríguez[2]

[1] *Northern Arizona University;* [2] *University of Arizona*

19.1 INTRODUCTION

One of the most fundamental image analysis operations is edge detection. Edges are often vital clues toward the analysis and interpretation of image information, both in biological vision and in computer image analysis. Some sort of edge detection capability is present in the visual systems of a wide variety of creatures, so it is obviously useful in their abilities to perceive their surroundings.

For this discussion, it is important to define what is and is not meant by the term "edge." The everyday notion of an edge is usually a physical one, caused by either the shapes of physical objects in three dimensions or by their inherent material properties. Described in geometric terms, there are two types of physical edges: (1) the set of points along which there is an abrupt change in local orientation of a physical surface and (2) the set of points describing the boundary between two or more materially distinct regions of a physical surface. Most of our perceptual senses, including vision, operate at a distance and gather information using receptors that work in, at most, two dimensions. Only the sense of touch, which requires direct contact to stimulate the skin's pressure sensors, is capable of direct perception of objects in three-dimensional (3D) space. However, some physical edges of the second type may not be perceptible by touch because material differences—for instance different colors of paint—do not always produce distinct tactile sensations. Everyone first develops a working understanding of physical edges in early childhood by touching and handling every object within reach.

The imaging process inherently performs a projection from a 3D scene to a two-dimensional (2D) representation of that scene, according to the viewpoint of the imaging device. Because of this projection process, edges in images have a somewhat different meaning than physical edges. Although the precise definition depends on the application context, an edge can generally be defined as a boundary or contour that separates adjacent image regions having relatively distinct characteristics according to some feature of interest. Most often this feature is gray level or luminance, but others, such as reflectance, color, or texture, are sometimes used. In the most common situation where luminance is of primary interest, edge pixels are those at the locations of abrupt gray level change. To eliminate single-point impulses from consideration as edge pixels, one usually requires that edges be sustained along a contour; i.e., an edge point must be part of an edge structure having some minimum extent appropriate for the scale of interest. Edge detection is the process of determining which pixels are the edge pixels. The result of the edge detection process is typically an edge map, a new image that describes each original pixel's edge classification and perhaps additional edge attributes, such as magnitude and orientation.

There is usually a strong correspondence between the physical edges of a set of objects and the edges in images containing views of those objects. Infants and young children learn this as they develop hand–eye coordination, gradually associating visual patterns with touch sensations as they feel and handle items in their vicinity. There are many situations, however, in which edges in an image do not correspond to physical edges. Illumination differences are usually responsible for this effect—for example, the boundary of a shadow cast across an otherwise uniform surface.

Conversely, physical edges do not always give rise to edges in images. This can also be caused by certain cases of lighting and surface properties. Consider what happens when one wishes to photograph a scene rich with physical edges—for example, a craggy mountain face consisting of a single type of rock. When this scene is imaged while the sun is directly behind the camera, no shadows are visible in the scene and hence shadow-dependent edges are nonexistent in the photo. The only edges in such a photo are produced by the differences in material reflectance, texture, or color. Since our rocky subject material has little variation of these types, the result is a rather dull photograph because of the lack of apparent depth caused by the missing edges. Thus images can exhibit edges having no physical counterpart, and they can also miss capturing edges that do. Although edge information can be very useful in the initial stages of such image processing and analysis tasks as segmentation, registration, and object recognition, edges are not completely reliable for these purposes.

If one defines an edge as an abrupt gray level change, then the derivative, or gradient, is a natural basis for an edge detector. Figure 19.1 illustrates the idea with a continuous, one-dimensional (1D) example of a bright central region against a dark background. The left-hand portion of the gray level function $f_c(x)$ shows a smooth transition from dark to bright as x increases. There must be a point x_0 that marks the transition from the low-amplitude region on the left to the adjacent high-amplitude region in the center. The gradient approach to detecting this edge is to locate x_0 where $|f_c'(x)|$ reaches a local maximum or, equivalently, $f_c'(x)$ reaches a local extremum, as shown in the second plot of Fig. 19.1. The second derivative, or Laplacian approach, locates x_0 where a zero-crossing of $f_c''(x)$ occurs, as in the third plot of Fig. 19.1. The right-hand side of Fig. 19.1 illustrates the case for a falling edge located at x_1.

To use the gradient or the Laplacian approaches as the basis for practical image edge detectors, one must extend the process to two dimensions, adapt to the discrete case, and somehow deal with the difficulties presented by real images. Relative to the 1D edges

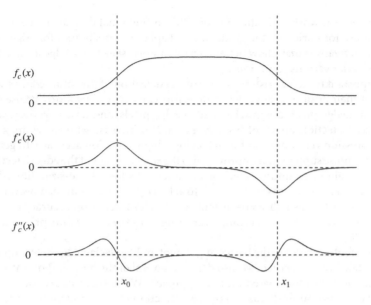

FIGURE 19.1

Edge detection in the 1D continuous case; changes in $f_c(x)$ indicate edges, and x_0 and x_1 are the edge locations found by local extrema of $f_c'(x)$ or by zero-crossings of $f_c''(x)$.

shown in Fig. 19.1, edges in 2D images have the additional quality of direction. One usually wishes to find edges regardless of direction, but a directionally sensitive edge detector can be useful at times. Also, the discrete nature of digital images requires the use of an approximation to the derivative. Finally, there are a number of problems that can confound the edge detection process in real images. These include noise, crosstalk or interference between nearby edges, and inaccuracies resulting from the use of a discrete grid. False edges, missing edges, and errors in edge location and orientation are often the result.

Because the derivative operator acts as a highpass filter, edge detectors based on it are sensitive to noise. It is easy for noise inherent in an image to corrupt the real edges by shifting their apparent locations and by adding many false edge pixels. Unless care is taken, seemingly moderate amounts of noise are capable of overwhelming the edge detection process, rendering the results virtually useless. The wide variety of edge detection algorithms developed over the past three decades exists, in large part, because of the many ways proposed for dealing with noise and its effects. Most algorithms employ noise-suppression filtering of some kind before applying the edge detector itself. Some decompose the image into a set of lowpass or bandpass versions, apply the edge detector to each, and merge the results. Still others use adaptive methods, modifying the edge detector's parameters and behavior according to the noise characteristics of the image data. Some recent work by Mathieu *et al.* [20] on fractional derivative operators shows some promise for enriching the gradient and Laplacian possibilities for edge detection. Fractional derivatives may allow better control of noise sensitivity, edge localization, and error rate under various conditions.

An important tradeoff exists between correct detection of the actual edges and precise location of their positions. Edge detection errors can occur in two forms: false positives, in which nonedge pixels are misclassified as edge pixels, and false negatives, which are the reverse. Detection errors of both types tend to increase with noise, making good noise suppression very important in achieving a high detection accuracy. In general, the potential for noise suppression improves with the spatial extent of the edge detection filter. Hence, the goal of maximum detection accuracy calls for a large-sized filter. Errors in edge localization also increase with noise. To achieve good localization, however, the filter should generally be of small spatial extent. The goals of detection accuracy and location accuracy are thus put into direct conflict, creating a kind of uncertainty principle for edge detection [28].

In this chapter, we cover the basics of gradient and Laplacian edge detection methods in some detail. Following each, we also describe several of the more important and useful edge detection algorithms based on that approach. While the primary focus is on gray level edge detectors, some discussion of edge detection in color and multispectral images is included.

19.2 GRADIENT-BASED METHODS

19.2.1 Continuous Gradient

The core of gradient edge detection is, of course, the gradient operator, ∇. In continuous form, applied to a continuous-space image, $f_c(x, y)$, the gradient is defined as

$$\nabla f_c(x, y) = \frac{\partial f_c(x, y)}{\partial x} \mathbf{i}_x + \frac{\partial f_c(x, y)}{\partial y} \mathbf{i}_y, \tag{19.1}$$

where $\mathbf{i}_x$ and $\mathbf{i}_y$ are the unit vectors in the x and y directions. Notice that the gradient is a vector, having both magnitude and direction. Its magnitude, $|\nabla f_c(x_0, y_0)|$, measures the maximum rate of change in the intensity at the location (x_0, y_0). Its direction is that of the greatest increase in intensity; i.e., it points "uphill."

To produce an edge detector, one may simply extend the 1D case described earlier. Consider the effect of finding the local extrema of $\nabla f_c(x, y)$ or the local maxima of

$$|\nabla f_c(x, y)| = \sqrt{\left(\frac{\partial f_c(x, y)}{\partial x}\right)^2 + \left(\frac{\partial f_c(x, y)}{\partial y}\right)^2}. \tag{19.2}$$

The precise meaning of "local" is very important here. If the maxima of Eq. (19.2) are found over a 2D neighborhood, the result is a set of isolated points rather than the desired edge contours. The problem stems from the fact that the gradient magnitude is seldom constant along a given edge, so finding the 2D local maxima yields only the locally strongest of the edge contour points. To fully construct edge contours, it is better to apply Eq. (19.2) to a 1D local neighborhood, namely a line segment, whose direction is chosen to cross the edge. The situation is then similar to that of Fig. 19.1, where the point of locally maximum gradient magnitude is the edge point. Now the issue becomes how to select the best direction for the line segment used for the search.

The most commonly used method of producing edge segments or contours from Eq. (19.2) consists of two stages: thresholding and thinning. In the thresholding stage, the gradient magnitude at every point is compared with a predefined threshold value, T. All points satisfying the following criterion are classified as candidate edge points:

$$|\nabla f_c(x,y)| \geq T. \tag{19.3}$$

The set of candidate edge points tends to form strips, which have positive width. Since the desire is usually for zero-width boundary segments or contours to describe the edges, a subsequent processing stage is needed to thin the strips to the final edge contours. Edge contours derived from continuous-space images should have zero width because any local maxima of $|\nabla f_c(x,y)|$, along a line segment that crosses the edge, cannot be adjacent points. For the case of discrete-space images, the nonzero pixel size imposes a minimum practical edge width.

Edge thinning can be accomplished in a number of ways, depending on the application, but thinning by nonmaximum suppression is usually the best choice. Generally speaking, we wish to suppress any point that is not, in a 1D sense, a local maximum in gradient magnitude. Since a 1D local neighborhood search typically produces a single maximum, those points that are local maxima will form edge segments only one point wide. One approach classifies an edge-strip point as an edge point if its gradient magnitude is a local maximum in at least one direction. However, this thinning method sometimes has the side effect of creating false edges near strong edge lines [17]. It is also somewhat inefficient because of the computation required to check along a number of different directions. A better, more efficient thinning approach checks only a single direction, the gradient direction, to test whether a given point is a local maximum in gradient magnitude. The points that pass this scrutiny are classified as edge points. Looking in the gradient direction essentially searches perpendicular to the edge itself, producing a scenario similar to the 1D case shown in Fig. 19.1. The method is efficient because it is not necessary to search in multiple directions. It also tends to produce edge segments having good localization accuracy. These characteristics make the gradient direction, local extremum method quite popular. The following steps summarize its implementation.

1. Using one of the techniques described in the next section, compute ∇f for all pixels.

2. Determine candidate edge pixels by thresholding all pixels' gradient magnitudes by T.

3. Thin by supressing all candidate edge pixels whose gradient magnitude is not a local maximum along its gradient direction. Those that survive nonmaximum supression are classified as edge pixels.

The order of the thinning and thresholding steps might be interchanged. If thresholding is accomplished first, the computational cost of thinning can be significantly reduced. However, it can become difficult to predict the number of edge pixels that will be produced by a given threshold value. By thinning first, there tends to be somewhat better predictability of the richness of the resulting edge map as a function of the applied threshold.

Consider the effect of performing the thresholding and thinning operations in isolation. If thresholding alone were done, the edges would show as strips or patches instead of thin segments. If thinning were done without thresholding, that is, if edge points were simply those having locally maximum gradient magnitude, many false edge points would likely result because of noise. Noise tends to create false edge points because some points in edge-free areas happen to have locally maximum gradient magnitudes. The thresholding step of Eq. (19.3) is often useful to reduce noise either prior to or following thinning. A variety of adaptive methods have been developed that adjust the threshold according to certain image characteristics, such as an estimate of local signal-to-noise ratio. Adaptive thresholding can often do a better job of noise suppression while reducing the amount of edge fragmentation.

The edge maps in Fig. 19.3, computed from the original image in Fig. 19.2, illustrate the effect of the thresholding and subsequent thinning steps.

The selection of the threshold value T is a tradeoff between the wish to fully capture the actual edges in the image and the desire to reject noise. Increasing T decreases sensitivity to noise at the cost of rejecting the weakest edges, forcing the edge segments to

FIGURE 19.2

Original cameraman image, 512 × 512 pixels.

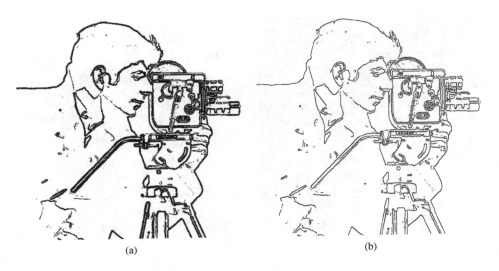

(a)   (b)

FIGURE 19.3

Gradient edge detection steps, using the Sobel operator: (a) After thresholding $|\nabla f|$; (b) after thinning (a) by finding the local maximum of $|\nabla f|$ along the gradient direction.

become more broken and fragmented. By decreasing T, one can obtain more connected and richer edge contours, but the greater noise sensitivity is likely to produce more false edges. If only thresholding is used, as in Eq. (19.3) and Fig. 19.3(a), the edge strips tend to narrow as T increases and widen as it decreases. Figure 19.4 compares edge maps obtained from several different threshold values.

Sometimes a directional edge detector is useful. One can be obtained by decomposing the gradient into horizontal and vertical components and applying them separately. Expressed in the continuous domain, the operators become:

$$\left| \frac{\partial f_c(x,y)}{\partial x} \right| \geq T \quad \text{for edges in the } y \text{ direction,}$$

$$\left| \frac{\partial f_c(x,y)}{\partial y} \right| \geq T \quad \text{for edges in the } x \text{ direction.}$$

An example of directional edge detection is illustrated in Fig. 19.5.

A directional edge detector can be constructed for any desired direction by using the directional derivative along a unit vector $\mathbf{n}$,

$$\frac{\partial f_c}{\partial \mathbf{n}} = \nabla f_c(x,y) \cdot \mathbf{n},$$

$$= \frac{\partial f_c(x,y)}{\partial x} \cos\theta + \frac{\partial f_c(x,y)}{\partial y} \sin\theta, \tag{19.4}$$

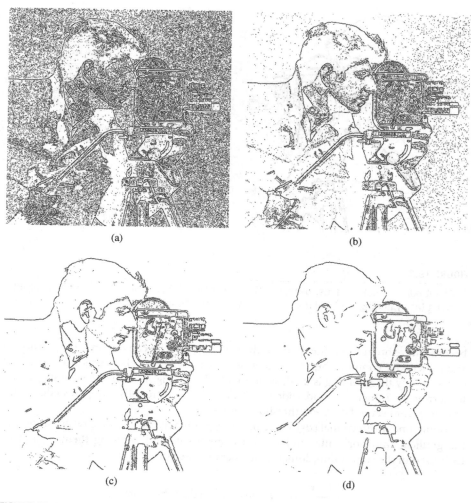

(a) (b)

(c) (d)

FIGURE 19.4

Roberts edge maps obtained by using various threshold values: (a) $T = 5$; (b) $T = 10$; (c) $T = 20$; and (d) $T = 40$. As T increases, more noise-induced edges are rejected along with the weaker real edges.

where θ is the angle of **n** relative to the positive x axis. The directional derivative is most sensitive to edges perpendicular to **n**.

The continuous-space gradient magnitude produces an isotropic or rotationally symmetric edge detector, equally sensitive to edges in any direction [17]. It is easy to show why $|\nabla f|$ is isotropic. In addition to the original X-Y coordinate system, let us introduce a new system, X'-Y', which is rotated by an angle of ϕ relative to X-Y. Let $\mathbf{n}_{x'}$ and $\mathbf{n}_{y'}$ be

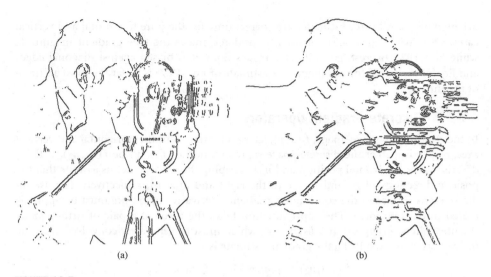

(a)   (b)

FIGURE 19.5

Directional edge detection comparison, using the Sobel operator: (a) results of horizontal difference operator; (b) results of vertical difference operator.

the unit vectors in the x' and y' directions, respectively. For the gradient magnitude to be isotropic, the same result must be produced in both coordinate systems, regardless of ϕ. Using Eq. (19.4) along with abbreviated notation, the partial derivatives with respect to the new coordinate axes are

$$f_{x'} = \nabla f \cdot \mathbf{n}_{x'} = f_x \cos \phi + f_y \sin \phi,$$
$$f_{y'} = \nabla f \cdot \mathbf{n}_{y'} = -f_x \sin \phi + f_y \cos \phi.$$

From this point, it is a simple matter of plugging into Eq. (19.2) to show the gradient magnitudes are identical in both coordinate systems, regardless of the rotation angle, ϕ.

Occasionally, one may wish to reduce the computation load of Eq. (19.2) by approximating the square root with a computationally simpler function. Three possibilities are

$$|\nabla f_c(x,y)| \approx \max\left\{ \left| \frac{\partial f_c(x,y)}{\partial x} \right|, \left| \frac{\partial f_c(x,y)}{\partial y} \right| \right\}, \tag{19.5}$$

$$\approx \left| \frac{\partial f_c(x,y)}{\partial x} \right| + \left| \frac{\partial f_c(x,y)}{\partial y} \right|, \tag{19.6}$$

$$\approx \max\left\{ \left| \frac{\partial f_c(x,y)}{\partial x} \right|, \left| \frac{\partial f_c(x,y)}{\partial y} \right| \right\} + \frac{1}{4} \min\left\{ \left| \frac{\partial f_c(x,y)}{\partial x} \right|, \left| \frac{\partial f_c(x,y)}{\partial y} \right| \right\}. \tag{19.7}$$

One should be aware that approximations of this type may alter the properties of the gradient somewhat. For instance, the approximated gradient magnitudes of Eqs. (19.5), (19.6), and (19.7) are not isotropic and produce their greatest errors for purely diagonally oriented edges. All three estimates are correct only for the pure horizontal and vertical cases. Otherwise, Eq. (19.5) consistently underestimates the true gradient magnitude while Eq. (19.6) overestimates it. This makes Eq. (19.5) biased against diagonal edges and Eq. (19.6) biased toward them. The estimate of Eq. (19.7) is by far the most accurate of the three.

19.2.2 Discrete Gradient Operators

In the continuous-space image, $f_c(x,y)$, let x and y represent the horizontal and vertical axes, respectively. Let the discrete-space representation of $f_c(x,y)$ be $f(n_1,n_2)$, with n_1 describing the horizontal position and n_2 describing the vertical. Let us assume that the positive directions of n_1 and n_2 are to the right and upward, respectively. For use on discrete-space images, the continuous gradient's derivative operators must be approximated in discrete form. The approximation takes the form of a pair of orthogonally oriented filters, $h_1(n_1,n_2)$ and $h_2(n_1,n_2)$, which must be separately convolved with the image. Based on Eq. (19.1), the gradient estimate is

$$\widehat{\nabla}f(n_1,n_2) = f_1(n_1,n_2)\,\mathbf{i}_{n_1} + f_2(n_1,n_2)\,\mathbf{i}_{n_2},$$

where

$$f_1(n_1,n_2) = f(n_1,n_2) * h_1(n_1,n_2)$$
$$f_2(n_1,n_2) = f(n_1,n_2) * h_2(n_1,n_2).$$

Two filters are necessary because the gradient requires the computation of an orthogonal pair of directional derivatives. The gradient magnitude and direction estimates can then be computed as follows:

$$\left|\widehat{\nabla}f(n_1,n_2)\right| = \sqrt{f_1^2(n_1,n_2) + f_2^2(n_1,n_2)},$$
$$\angle\widehat{\nabla}f(n_1,n_2) = \tan^{-1}\left(\frac{f_2(n_1,n_2)}{f_1(n_1,n_2)}\right). \tag{19.8}$$

Each of the filters implements a derivative and should not respond to a constant, so the sum of its coefficients must always be zero. A more general statement of this property is described later in this chapter by Eq. (19.12).

There are many possible derivative-approximation filters for use in gradient estimation. Let us start with the simplest cases. Two simple approximation schemes for the derivative in one dimension are

$$\text{First difference:}\quad f(n) - f(n-1),$$
$$\text{Central difference:}\quad \frac{1}{2}[f(n+1) - f(n-1)],$$

which result in the following 1D convolutional filters:

$$\text{First difference: } h(n) = \delta(n) - \delta(n-1) = [1\ \ -1],$$
$$\text{Central difference: } h(n) = \delta(n+1) - \delta(n-1) = \frac{1}{2}[1\ \ 0\ \ -1]. \tag{19.9}$$

It is common practice to incorporate the dimensional reversal into the filter kernel mask and then perform correlation instead of convolution. This exploits the close relationship between the discrete convolution and the correlation operators, thereby saving a bit of computation. As a practical matter, the convolution and correlation responses to a given gradient-based edge detection mask will differ only in sign. However, the magnitude of the response—not its sign—is usually the important thing. Henceforth, the filters are expressed as correlation masks and the dimensional reversal will be omitted. For instance, the convolutional filters of Eq. (19.9) are expressed in correlational form, with $h(-n)$ listed as $h(n)$ for simplicity:

$$\text{First difference:} \quad h(n) = [-1 \ 1],$$

$$\text{Central difference:} \quad h(n) = \frac{1}{2}[-1 \ 0 \ 1]. \tag{19.10}$$

The scaling factor of 1/2 for the central difference is caused by the two-pixel distance between the nonzero samples. The origin positions for both filters are usually set at (n_1, n_2), shown in boldface. The gradient magnitude threshold value can be easily adjusted to compensate for any scaling, so the scale factor will be omitted from here on.

Let us now extend Eq. (19.10) to the 2D case. These derivative approximations can be expressed as filter kernels, whose impulse responses, $h_1(n_1, n_2)$ and $h_2(n_1, n_2)$, are shown dimensionally reversed as the following correlation masks:

$$\text{First difference:} \quad h_1(n_1, n_2) = \begin{bmatrix} 0 & 0 \\ -1 & 1 \end{bmatrix}, \quad h_2(n_1, n_2) = \begin{bmatrix} 1 & 0 \\ -1 & 0 \end{bmatrix},$$

$$\text{Central difference:} \quad h_1(n_1, n_2) = \begin{bmatrix} 0 & 0 & 0 \\ -1 & 0 & 1 \\ 0 & 0 & 0 \end{bmatrix}, \quad h_2(n_1, n_2) = \begin{bmatrix} 0 & 1 & 0 \\ 0 & 0 & 0 \\ 0 & -1 & 0 \end{bmatrix}.$$

Again, the origin position is indicated by boldface. Filters $h_1(n_1, n_2)$ respond most strongly to vertical edges and do not respond to horizontal edges. The $h_2(n_1, n_2)$ counterparts respond to horizontal edges and not to vertical ones.

If used to detect edges, the pair of first difference filters above presents the problem that the zero crossings of its two $[-1 \ 1]$ derivative kernels lie at different positions. This prevents the two filters from measuring horizontal and vertical edge characteristics at the same location, causing error in the estimated gradient. The central difference, caused by the common center of its horizontal and vertical differencing kernels, avoids this position mismatch problem. This benefit comes at the costs of larger filter size and the fact that the measured gradient at a pixel (n_1, n_2) does not actually consider the value of that pixel.

Rotating the first difference kernels by an angle of $\frac{\pi}{4}$ and stretching the grid a bit produces the $h_1(n_1, n_2)$ and $h_2(n_1, n_2)$ correlation masks for the Roberts operator:

$$\begin{bmatrix} 0 & 1 \\ -1 & 0 \end{bmatrix} \quad \begin{bmatrix} 1 & 0 \\ 0 & -1 \end{bmatrix}.$$

The Roberts operator's component filters are tuned for diagonal edges rather than vertical and horizontal ones. For use in an edge detector based on the gradient magnitude, it is important only that the two filters be orthogonal. They need not be aligned with the n_1 and n_2 axes. The pair of Roberts filters has a common zero-crossing point for their differencing kernels. This common center eliminates the position mismatch error exhibited by the horizontal-vertical first difference pair, as described earlier. If the origins of the Roberts kernels are positioned on the $+1$ samples, as is sometimes found in the literature, then no common center point exists for their first differences.

The Roberts operator, like any simple first-difference gradient operator, has two undesirable characteristics. First, the zero crossing of its $[-1\ 1]$ diagonal kernel lies off grid, but the edge location must be assigned to an actual pixel location, namely the one at the filter's origin. This can create edge location bias that may lead to location errors approaching the interpixel distance. If we could use the central difference instead of the first difference, this problem would be reduced because the central difference operator inherently constrains its zero crossing to an exact pixel location.

The other difficulty caused by the first difference is its noise sensitivity. In fact, both the first- and central-difference derivative estimators are quite sensitive to noise. The noise problem can be reduced somewhat by incorporating smoothing into each filter in the direction normal to that of the difference. Consider an example based on the central difference in one direction for which we wish to smooth along the orthogonal direction with a simple three-sample average. To that end, let us define the impulse responses of two filters:

$$h_a(n_1) = \begin{bmatrix} 1 & 1 & 1 \end{bmatrix}, \quad h_b(n_2) = \begin{bmatrix} -1 & 0 & 1 \end{bmatrix}.$$

Since h_a is a function only of n_1 and h_b depends only on n_2, one can simply multiply them as an outer product to form a separable derivative filter that incorporates smoothing:

$$h_a(n_1)h_b(n_2) = h_1(n_1, n_2),$$

$$\begin{bmatrix} 1 \\ 1 \\ 1 \end{bmatrix} \begin{bmatrix} -1 & 0 & 1 \end{bmatrix} = \begin{bmatrix} -1 & 0 & 1 \\ -1 & 0 & 1 \\ -1 & 0 & 1 \end{bmatrix}.$$

Repeating this process for the orthogonal case produces the Prewitt operator mask:

$$\begin{bmatrix} -1 & 0 & 1 \\ -1 & 0 & 1 \\ -1 & 0 & 1 \end{bmatrix} \qquad \begin{bmatrix} 1 & 1 & 1 \\ 0 & 0 & 0 \\ -1 & -1 & -1 \end{bmatrix}.$$

The Prewitt edge gradient operator simultaneously accomplishes differentiation in one coordinate direction, using the central difference, and noise reduction in the orthogonal direction, by means of local averaging. Because it uses the central difference instead of the first difference, there is less edge-location bias.

In general, the smoothing characteristics can be adjusted by choosing an appropriate lowpass filter kernel in place of the Prewitt's three-sample average. One such variation is the Sobel operator, one of the most widely used gradient edge detectors:

$$\begin{bmatrix} -1 & 0 & 1 \\ -2 & 0 & 2 \\ -1 & 0 & 1 \end{bmatrix} \qquad \begin{bmatrix} 1 & 2 & 1 \\ 0 & 0 & 0 \\ -1 & -2 & -1 \end{bmatrix}.$$

Sobel's operator is often a better choice than Prewitt's because the lowpass filter produced by the [1 2 1] kernel results in a smoother frequency response compared with that of [1 1 1].

The Prewitt and Sobel operators respond differently to diagonal edges than to horizontal or vertical ones. This behavior is a consequence of the fact that their filter coefficients do not compensate for the different grid spacings in the diagonal and the horizontal directions. The Prewitt operator is less sensitive to diagonal edges than to vertical or horizontal ones. The opposite is true for the Sobel operator [23]. A variation designed for equal gradient magnitude response to diagonal, horizontal, and vertical edges is the Frei-Chen operator:

$$
\begin{bmatrix} -1 & 0 & 1 \\ -\sqrt{2} & 0 & \sqrt{2} \\ -1 & 0 & 1 \end{bmatrix}
\begin{bmatrix} 1 & \sqrt{2} & 1 \\ 0 & 0 & 0 \\ -1 & -\sqrt{2} & -1 \end{bmatrix}.
$$

However, even the Frei-Chen operator retains some directional sensitivity in gradient magnitude, so it is not truly isotropic. The residual anisotropy is caused by the fact that the difference operators used to approximate Eq. (19.1) are not rotationally symmetric. Merron and Brady [21] describe a simple method for greatly reducing the residual directional bias by using a set of four difference operators instead of two. Their operators are oriented in increments of $\pi/4$ radians, adding a pair of diagonal ones to the original horizontal and vertical pair. Averaging the gradients produced by the diagonal operators with those of the nondiagonal ones allows their complementary directional biases to reduce the overall anisotropy. However, Ziou and Wang [31] have described how an isotropic gradient applied to a discrete grid tends to introduce some anisotropy. They have also analyzed the errors of gradient magnitude and direction as a function of edge translation and orientation for several detectors. Figure 19.6 shows the results of performing edge detection on an example image by applying the discrete gradient operators discussed so far.

Haralick's facet model [10, 11] provides another way of calculating the gradient in order to perform edge detection. In the sloped facet model, a small neighborhood is parameterized by $\alpha n_2 + \beta n_1 + \gamma$, describing the plane that best fits the gray levels in that neighborhood. The plane parameters α and β can be used to compute the gradient magnitude:

$$
|\nabla f_c(n_1, n_2)| = \sqrt{\alpha^2 + \beta^2}.
$$

The facet model also provides means for computing directional derivatives, zero crossings, and a variety of other useful operations.

Improved noise suppression is possible with increased kernel size. The additional coefficients can be used to better approximate the desired continuous-space noise-suppression filter. Greater filter extent can also be used to reduce directional sensitivity by more accurately modeling an ideal isotropic filter. However, increasing the kernel size will exacerbate edge localization problems and create interference between nearby

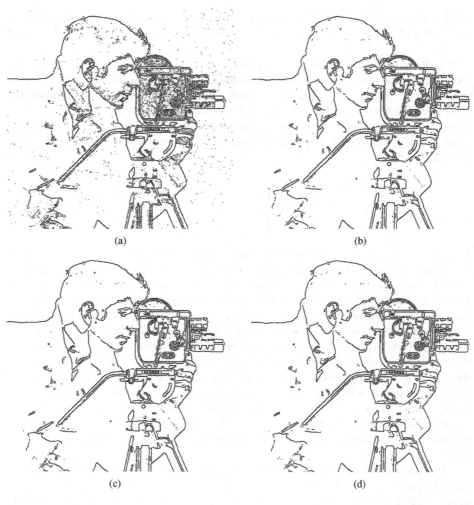

(a)

(b)

(c)

(d)

FIGURE 19.6

Comparison of edge detection using various gradient operators: (a) Roberts; (b) 3 × 3 Prewitt; (c) 3 × 3 Sobel; and (d) 3 × 3 Frei-Chen. In each case, the threshold has been set to allow a fair comparison.

edges. Noise suppression can be improved by other methods as well. Papers by Bovik *et al.* [3] and Hardie and Boncelet [12] are just two that describe the use of edge-enhancing prefilters, which simultaneously suppress noise and steepen edges prior to gradient edge detection.

19.3 LAPLACIAN-BASED METHODS

19.3.1 Continuous Laplacian

The Laplacian is defined as

$$\nabla^2 f_c(x,y) = \nabla \cdot \nabla f_c(x,y) = \frac{\partial^2 f_c(x,y)}{\partial x^2} + \frac{\partial^2 f_c(x,y)}{\partial y^2}. \tag{19.11}$$

The zero crossings of $\nabla^2 f_c(x,y)$ occur at the edge points of $f_c(x,y)$ because of the second derivative action (see Fig. 19.1). Laplacian-based edge detection has the nice property that it produces edges of zero thickness, making edge-thinning steps unnecessary. This is because the zero crossings themselves define the edge locations.

The continuous Laplacian is isotropic, favoring no particular edge orientation. Consequently, its second partial terms in Eq. (19.11) can be oriented in any direction as long as they remain perpendicular to each other. Consider an ideal, straight, noise-free edge oriented in an arbitrary direction. Let us realign the first term of Eq. (19.11) parallel to that edge and the second term perpendicular to it. The first term then generates no response at all because it acts only along the edge. The second term produces a zero crossing at the edge position along its edge-crossing profile.

An edge detector based solely on the zero crossings of the continuous Laplacian produces closed edge contours if the image, $f(x,y)$, meets certain smoothness constraints [28]. The contours are closed because edge strength is not considered, so even the slightest, most gradual intensity transition produces a zero crossing. In effect, the zero-crossing contours define the boundaries that separate regions of nearly constant intensity in the original image. The second derivative zero crossings occur at the local extrema of the first derivative (see Fig. 19.1), but many zero crossings are not local maxima of the gradient magnitude. Some local minima of the gradient magnitude give rise to phantom edges, which can be largely eliminated by appropriately thresholding the edge strength. Figure 19.7 illustrates a 1D example of a phantom edge.

Noise presents a problem for the Laplacian edge detector in several ways. First, the second-derivative action of Eq. (19.11) makes the Laplacian even more sensitive to noise than the first-derivative-based gradient. Second, noise produces many false edge contours because it introduces variation to the constant-intensity regions in the noise-free image. Third, noise alters the locations of the zero-crossing points, producing location errors along the edge contours. The problem of noise-induced false edges can be addressed by applying an additional test to the zero-crossing points. Only the zero crossings that satisfy this new criterion are considered edge points. One commonly-used technique classifies a zero crossing as an edge point if the local gray level variance exceeds a threshold amount. Another method is to select the strong edges by thresholding the gradient magnitude or the slope of the Laplacian output at the zero crossing. Both criteria serve to reject zero-crossing points that are more likely caused by noise than by a real edge in the original scene. Of course, thresholding the zero crossings in this manner tends to break up the closed contours.

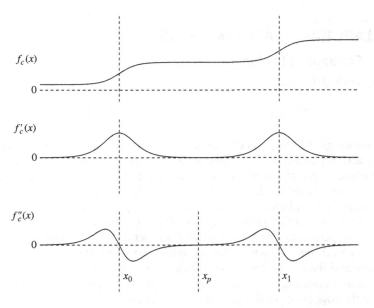

FIGURE 19.7

The zero crossing of $f_c''(x)$ at x_p creates a phantom edge.

Like any derivative filter, the continuous-space Laplacian filter, $h_c(x, y)$, has this important zero-mean property [15]:

$$\int_{-\infty}^{+\infty} \int_{-\infty}^{+\infty} h_c(x, y) \, dx \, dy = 0. \tag{19.12}$$

In other words, $h_c(x, y)$ is a surface bounding equal volumes above and below zero. Consequently, $\nabla^2 f_c(x, y)$ will also have equal volumes above and below zero. This property eliminates any response that is due to the constant or DC bias contained in $f_c(x, y)$. Without DC bias rejection, the filter's edge detection performance would be compromised. Gunn [15] has described the effect of truncation of the Laplacian and the subsequent bias introduced when Eq. (19.12) is violated.

19.3.2 Discrete Laplacian Operators

It is useful to construct a filter to serve as the Laplacian operator when applied to a discrete-space image. Recall that the gradient, which is a vector, required a pair of orthogonal filters. The Laplacian is a scalar. Therefore, a single filter, $h(n_1, n_2)$, is sufficient for realizing a Laplacian operator. The Laplacian estimate for an image, $f(n_1, n_2)$, is then

$$\widehat{\nabla}^2 f(n_1, n_2) = f(n_1, n_2) * h(n_1, n_2).$$

One of the simplest Laplacian operators can be derived as follows. First needed is an approximation to the derivative in x, so let us use a simple first difference:

$$\frac{\partial f_c(x,y)}{\partial x} \rightarrow f_x(n_1, n_2) = f(n_1 + 1, n_2) - f(n_1, n_2). \tag{19.13}$$

The second derivative in x can be built by applying the first difference to Eq. (19.13). However, we discussed earlier how the first difference produces location errors because its zero crossing lies off grid. This second application of a first difference can be shifted to counteract the error introduced by the previous one:

$$\frac{\partial^2 f_c(x,y)}{\partial x^2} \rightarrow f_{xx}(n_1, n_2) = f_x(n_1, n_2) - f_x(n_1 - 1, n_2). \tag{19.14}$$

Combining the two derivative-approximation stages from Eqs. (19.13) and (19.14) produces

$$\frac{\partial^2 f_c(x,y)}{\partial x^2} \rightarrow f_{xx}(n_1, n_2) = f(n_1 + 1, n_2) - 2f(n_1, n_2) + f(n_1 - 1, n_2) = \begin{bmatrix} 1 & -2 & 1 \end{bmatrix}. \tag{19.15}$$

Proceeding in an identical manner for y yields

$$\frac{\partial^2 f_c(x,y)}{\partial y^2} \rightarrow f_{yy}(n_1, n_2) = f(n_1, n_2 + 1) - 2f(n_1, n_2) + f(n_1, n_2 - 1) = \begin{bmatrix} 1 \\ -2 \\ 1 \end{bmatrix}. \tag{19.16}$$

Combining the x and y second partials of Eqs. (19.15) and (19.16) produces a filter, $h(n_1, n_2)$, which estimates the Laplacian:

$$\begin{aligned} \nabla^2 f_c(x,y) \rightarrow \widehat{\nabla}^2 f(n_1, n_2) &= f_{xx}(n_1, n_2) + f_{yy}(n_1, n_2) \\ &= f(n_1 + 1, n_2) + f(n_1 - 1, n_2) + f(n_1, n_2 + 1) \\ &\quad + f(n_1, n_2 - 1) - 4f(n_1, n_2) \\ &= \begin{bmatrix} 1 & -2 & 1 \end{bmatrix} + \begin{bmatrix} 1 \\ -2 \\ 1 \end{bmatrix} = \begin{bmatrix} 0 & 1 & 0 \\ 1 & -4 & 1 \\ 0 & 1 & 0 \end{bmatrix}. \end{aligned}$$

Other Laplacian estimation filters can be constructed using this method of designing a pair of appropriate 1D second derivative filters and combining them into a single 2D filter. The results depend on the choice of derivative approximator, the size of the desired filter kernel, and the characteristics of any noise-reduction filtering applied. Two other 3×3 examples are

$$\begin{bmatrix} 1 & 1 & 1 \\ 1 & -8 & 1 \\ 1 & 1 & 1 \end{bmatrix}, \quad \begin{bmatrix} -1 & 2 & -1 \\ 2 & -4 & 2 \\ -1 & 2 & -1 \end{bmatrix}.$$

In general, a discrete-space smoothed Laplacian filter can be easily constructed by sampling an appropriate continuous-space function, such as the Laplacian of Gaussian. When constructing a Laplacian filter, make sure that the kernel's coefficients sum to zero in order to satisfy the discrete form of Eq. (19.12). Truncation effects may upset this property and create bias. If so, the filter coefficients should be adjusted in a way that restores proper balance.

Locating zero crossings in the discrete-space image, $\nabla^2 f(n_1, n_2)$, is fairly straight-forward. Each pixel should be compared to its eight immediate neighbors; a four-way neighborhood comparison, while faster, may yield broken contours. If a pixel, $\mathbf{p}$, differs in sign with its neighbor, $\mathbf{q}$, an edge lies between them. The pixel, $\mathbf{p}$, is classified as a zero crossing if

$$\left| \nabla^2 f(\mathbf{p}) \right| \le \left| \nabla^2 f(\mathbf{q}) \right|. \tag{19.17}$$

19.3.3 The Laplacian of Gaussian (Marr-Hildreth Operator)

It is common for a single image to contain edges having widely different sharpnesses and scales, from blurry and gradual to crisp and abrupt. Edge scale information is often useful as an aid toward image understanding. For instance, edges at low resolution tend to indicate gross shapes while texture tends to become important at higher resolutions. An edge detected over a wide range of scale is more likely to be physically significant in the scene than an edge found only within a narrow range of scale. Furthermore, the effects of noise are usually most deleterious at the finer scales.

Marr and Hildreth [19] advocated the need for an operator that can be tuned to detect edges at a particular scale. Their method is based on filtering the image with a Gaussian kernel selected for a particular edge scale. The Gaussian smoothing operation serves to band-limit the image to a small range of frequencies, reducing the noise sensitivity problem when detecting zero crossings. The image is filtered over a variety of scales, and the Laplacian zero crossings are computed at each. This produces a set of edge maps as a function of edge scale. Each edge point can be considered to reside in a region of scale space, for which edge point location is a function of x, y, and σ. Scale space has been successfully used to refine and analyze edge maps [30].

The Gaussian has some very desirable properties that facilitate this edge detection procedure. First, the Gaussian function is smooth and localized in both the spatial and frequency domains, providing a good compromise between the need for avoiding false edges and for minimizing errors in edge position. In fact, Torre and Poggio [28] describe the Gaussian as the only real-valued function that minimizes the product of spatial- and frequency-domain spreads. The Laplacian of Gaussian essentially acts as a bandpass filter because of its differential and smoothing behavior. Second, the Gaussian is separable, which helps make computation very efficient.

Omitting the scaling factor, the Gaussian filter can be written as

$$g_c(x, y) = \exp\left(-\frac{x^2 + y^2}{2\sigma^2} \right). \tag{19.18}$$

Its frequency response, $G(\Omega_x, \Omega_y)$, is also Gaussian:

$$G(\Omega_x, \Omega_y) = 2\pi\sigma^2 \exp\left(-\frac{\sigma^2}{2} \left(\Omega_x^2 + \Omega_y^2 \right) \right).$$

The σ parameter is inversely related to the cutoff frequency.

Because the convolution and Laplacian operations are both linear and shift invariant, their computation order can be interchanged:

$$\nabla^2 \left[f_c(x,y) * g_c(x,y) \right] = \left[\nabla^2 g_c(x,y) \right] * f_c(x,y). \tag{19.19}$$

Here we take advantage of the fact that the derivative is a linear operator. Therefore, Gaussian filtering followed by differentiation is the same as filtering with the derivative of a Gaussian. The right-hand side of Eq. (19.19) usually provides for more efficient computation since $\nabla^2 g_c(x,y)$ can be prepared in advance due to its image independence. The Laplacian of Gaussian (LoG) filter, $h_c(x,y)$, therefore has the following impulse response:

$$\begin{aligned} h_c(x,y) &= \nabla^2 g_c(x,y) \\ &= \frac{x^2 + y^2 - 2\sigma^2}{\sigma^4} \exp\left(-\frac{x^2 + y^2}{2\sigma^2} \right). \end{aligned} \tag{19.20}$$

To implement the LoG in discrete form, one may construct a filter, $h(n_1, n_2)$, by sampling Eq. (19.20) after choosing a value of σ, then convolving with the image. If the filter extent is not small, it is usually more efficient to work in the frequency domain by multiplying the discrete Fourier transforms of the filter and the image, then inverse transforming the result. The fast Fourier transform is the method of choice for computing these transforms.

Although the discrete form of Eq. (19.20) is a 2D filter, Chen et al. [7] have shown that it is actually the sum of two separable filters because the Gaussian itself is a separable function. By constructing and applying the appropriate 1D filters successively to the rows and columns of the image, the computational expense of 2D convolution becomes unnecessary. Separable convolution to implement the LoG is roughly 1–2 orders of magnitude more efficient than 2D convolution. If an image is $M \times M$ in size, the number of operations at each pixel is M^2 for 2D convolution and only $2M$ if done in a separable, 1D manner.

Figure 19.8 shows an example of applying the LoG using various σ values. Figure 19.8(d) includes a gradient magnitude threshold, which suppresses noise and breaks contours. Lim [17] describes an adaptive thresholding scheme that produces better results.

Equation (19.20) has the shape of a *sombrero* or "Mexican hat." Figure 19.9 shows a perspective plot of $\nabla^2 g_c(x,y)$ and its frequency response, $F\left\{\nabla^2 g_c(x,y)\right\}$. This profile closely mimics the response of the spatial receptive field found in biological vision. Biological receptive fields have been shown to have a circularly symmetric impulse response, with a central excitory region surrounded by an inhibitory band.

When sampling the LoG to produce a discrete version, it is important to size the filter large enough to avoid significant truncation effects. A good rule of thumb is to make the filter at least three times the width of the LoG's central excitory lobe [23]. Siohan et al. [27] describe two approaches for the practical design of LoG filters. The errors in edge location produced by the LoG have been analyzed in some detail by Berzins [2].

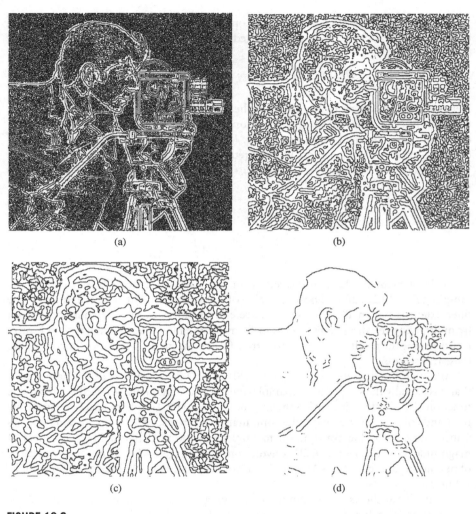

FIGURE 19.8

Zero crossings of $f * \nabla^2 g$ for several values of σ, with (d) also thresholded: (a) $\sigma = 1.0$; (b) $\sigma = 1.5$; (c) $\sigma = 2.0$; (d) $\sigma = 2.0$, and $T = 20$.

Gunn [15] has analyzed and described the relationship between the choice of LoG mask size and the resulting probability of edge detection and localization errors.

Sarkar and Boyer [25] have developed an optimal recursive filter for use as a zero-crossing edge detector. They observed that the optimal zero-crossing detector cannot generally be obtained by simply taking the derivative of the optimal gradient detector. Their filter has been optimized for step edges and to simultaneously achieve low error rate, high edge localization precision, and low rate of spurious responses. These goals

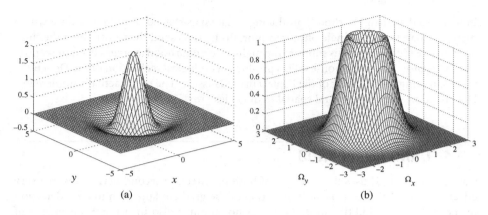

FIGURE 19.9

Plots of the LoG and its frequency response for $\sigma = 1$: (a) $-\nabla^2 g_c(x,y)$, the negative of Eq. (19.20); (b) $F\{\nabla^2 g_c(x,y)\}$, the bandpass-shaped frequency response of Eq. (19.20).

are similar to those of Canny, whose method is described later in this chapter. The Sarkar-Boyer filter is claimed to perform somewhat better than the LoG filter [25].

19.3.4 Difference of Gaussian

The Laplacian of Gaussian of Eq. (19.20) can be closely approximated by the difference of two Gaussians having properly-chosen scales. The difference of Gaussian (DoG) filter is

$$h_c(x,y) = g_{c1}(x,y) - g_{c2}(x,y),$$

where

$$\frac{\sigma_2}{\sigma_1} \approx 1.6,$$

and g_{c1}, g_{c2} are evaluated using Eq. (19.18). However, the LoG is usually preferred because it is theoretically optimal and its separability allows for efficient computation [19]. For the same accuracy of results, the DoG requires a slightly larger filter size [14].

The technique of unsharp masking, used in photography, is basically a DoG's operation done with light and negatives. Unsharp masking involves making a somewhat blurry exposure of an original negative onto a new piece of film. When the film is developed, it contains a blurred and inverted-brightness version of the original negative. Finally, a print is made from these two negatives sandwiched together, producing a sharpened image with the edges showing increased contrast.

Nature uses the DoGs as a basis for the architecture of the retina's visual receptive field. The spatial-domain impulse response of a photoreceptor cell in the mammalian retina has a roughly Gaussian shape. The photoreceptor output feeds into horizontal cells in the adjacent layer of neurons. Each horizontal cell averages the responses of the receptors in its immediate neighborhood, producing a Gaussian-shaped impulse response with a higher σ than that of a single photoreceptor. Both layers send their outputs to the third layer, where bipolar neurons subtract the high-σ neighborhood averages from the central photoreceptors' low-σ responses. This produces a biological realization of the DoG filter, approximating the behavior of the Laplacian of Gaussian. The retina actually implements DoG bandpass filters at several spatial frequencies [18].

19.4 CANNY'S METHOD

Canny's method [5] uses the concepts of both the first and second derivatives in a very effective manner. His is a classic application of the gradient approach to edge detection in the presence of additive white Gaussian noise, but it also incorporates elements of the Laplacian approach. The method has three simultaneous goals: low rate of detection errors, good edge localization, and only a single detection response per edge. Canny assumed that false-positive and false-negative detection errors are equally undesirable and so gave them equal weight. He further assumed that each edge has nearly constant cross section and orientation, but his general method includes a way to effectively deal with the cases of curved edges and corners. With these constraints, Canny determined the optimal 1D edge detector for the step edge and showed that its impulse response can be approximated fairly well by the derivative of a Gaussian.

An important action of Canny's edge detector is to prevent multiple responses per true edge. Without this criterion, the optimal step-edge detector would have an impulse response in the form of a truncated signum function. (The signum function produces $+1$ for any positive argument and -1 for any negative argument.) But this type of filter has high bandwidth, allowing noise or texture to produce several local maxima in the vicinity of the actual edge. The effect of the derivative of Gaussian is to prevent multiple responses by smoothing the truncated signum in order to permit only one response peak in the edge neighborhood. The choice of variance for the Gaussian kernel controls the filter width and the amount of smoothing. This defines the width of the neighborhood in which only a single peak is to be allowed. The variance selected should be proportional to the amount of noise present. If the variance is chosen too low, the filter can produce multiple detections for a single edge; if too high, edge localization suffers needlessly. Because the edges in a given image are likely to differ in signal-to-noise ratio, a single-filter implementation is usually not best for detecting them. Hence a thorough edge detection procedure should operate at different scales.

Canny's approach begins by smoothing the image with a Gaussian filter:

$$g_c(x,y) = \frac{1}{\sigma\sqrt{2\pi}} \exp\left(-\frac{x^2 + y^2}{2\sigma^2}\right). \tag{19.21}$$

One may sample and truncate Eq. (19.21) to produce a finite-extent filter, $g(n_1, n_2)$. At each pixel, Eq. (19.8) is used to estimate the gradient direction. From a set of prepared edge detection filter masks having various orientations, the one oriented nearest to the gradient direction for the targeted pixel is then chosen. When applied to the Gaussian-smoothed image, this filter produces an estimate of gradient magnitude at that pixel. One may instead apply simpler methods, such as the central difference operator, to estimate the partial derivatives. The initial Gaussian smoothing step makes additional smoothing along the edge, as with the Prewitt or Sobel operators, completely unnecessary. Next, the goal is to suppress nonmaxima of the gradient magnitude by testing a 3×3 neighborhood, comparing the magnitude at the center pixel with those at interpolated positions to either side along the gradient direction.

The pixels that survive to this point are candidates for the edge map. To produce an edge map from these candidate pixels, Canny applies thresholding by gradient magnitude in an adaptive manner with hysteresis. An estimate of the noise in the image determines the values of a pair of thresholds, with the upper threshold typically two or three times that of the lower. A candidate edge segment is included in the output edge map if at least one of its pixels has a gradient magnitude exceeding the upper threshold, but pixels not meeting the lower threshold are excluded. This hysteresis action helps reduce the problem of broken edge contours while improving the ability to reject noise.

A set of edge maps over a range of scales can be produced by varying the σ values used to Gaussian-filter the image. Since smoothing at different scales produces different errors in edge location, an edge segment that appears in multiple edge maps at different scales may exhibit some position shift. Canny proposed unifying the set of edge maps into a single result by a technique he called "feature synthesis," which proceeds in a fine-to-coarse manner while tracking the edge segments within their possible displacements.

The preoriented edge detection filters, mentioned previously, have some interesting properties. Each mask includes a derivative of Gaussian function to perform the nearly optimal directional derivative across the intended edge. A smooth, averaging profile appears in the mask along the intended edge direction in order to reduce noise without compromising the sharpness of the edge profile. In the smoothing direction, the filter extent is usually several times that in the derivative direction when the filter is intended for straight edges. Canny's method includes a "goodness of fit" test to determine if the selected filter is appropriate before it is applied. The test examines the gray level variance of the strip of pixels along the smoothing direction of the filter. If the variance is small, then the edge must be close to linear, and the filter is a good choice. A large variance indicates the presence of curvature or a corner, in which case a better choice of filter would have smaller extent in the smoothing direction. There were six oriented filters used in Canny's work, thus the greatest directional mismatch between the actual gradient and the nearest filter is $15°$.

As discussed previously, edges can be detected from either the maxima of the gradient magnitude or the zero crossings of the second derivative. Another way to realize the essence of Canny's method is to look for zero crossings of the second directional derivative taken along the gradient direction. Let us examine the mathematical basis for this. If $\mathbf{n}$ is a unit vector in the gradient direction, and f is the Gaussian-smoothed image, then we wish to find

$$\frac{\partial^2 f}{\partial \mathbf{n}^2} = \nabla \left(\frac{\partial f}{\partial \mathbf{n}} \right) \cdot \mathbf{n}$$

$$= \nabla (\nabla f \cdot \mathbf{n}) \cdot \mathbf{n},$$

which can be expanded to the following form:

$$\frac{\partial^2 f}{\partial \mathbf{n}^2} = \frac{f_x^2 f_{xx} + 2 f_x f_y f_{xy} + f_y^2 f_{yy}}{\sqrt{f_x^2 + f_y^2}}. \qquad (19.22)$$

In Eq. (19.22), a concise notation has been used for the partial derivatives.

Like the Laplacian approach, Canny's method looks for zero crossings of the second derivative. The Laplacian's second derivative is nondirectional; it includes a component taken parallel to the edge and another taken across it. Canny's method is evaluated only in the gradient direction, directly across the local edge. A derivative taken along an edge is counterproductive because it introduces noise without improving edge detection capability. By being selective about the direction in which its derivatives are evaluated, Canny's approach avoids this source of noise and tends to produce better results.

Figures 19.10 and 19.11 illustrate the results of applying the Canny edge detector of Eq. (19.22) after Gaussian smoothing, then looking for zero crossings. Figure 19.10 demonstrates the effect of using the same upper and lower thresholds, T_U and T_L, over a range of σ values. The behavior of hysteresis thresholding is shown in Fig. 19.11. The partial derivatives were approximated using central differences. Thresholding was performed with hysteresis, but using fixed threshold values for each image instead of Canny's noise-adaptive threshold values. Zero-crossing detection was implemented in an eight-way manner, as described by Eq. (19.17) in the earlier discussion of discrete Laplacian operators. Also, Canny's preoriented edge detection filters were not used in preparing these examples, so it was not possible to adapt the edge detection filters according to the "goodness of fit" of the local edge profile as Canny did.

Ding and Goshtasby [9] have developed refinements to Canny's method. Canny selects edge pixels by comparing gradient magnitudes of neighboring pixels along the gradient direction. Ding does this also, and classifies these as major edge pixels. He further selects minor edge pixels as those having locally maximum gradient magnitude in any direction. Next, the major-minor edge branch points are located. Branches that do not contain a major edge are eliminated. The result is a decrease in the number of missed and broken edges.

19.5 APPROACHES FOR COLOR AND MULTISPECTRAL IMAGES

Edge detection for color images presents additional challenges because of the three color components used. The most straightforward technique is to perform edge detection on the luminance component image while ignoring the chrominance information. The

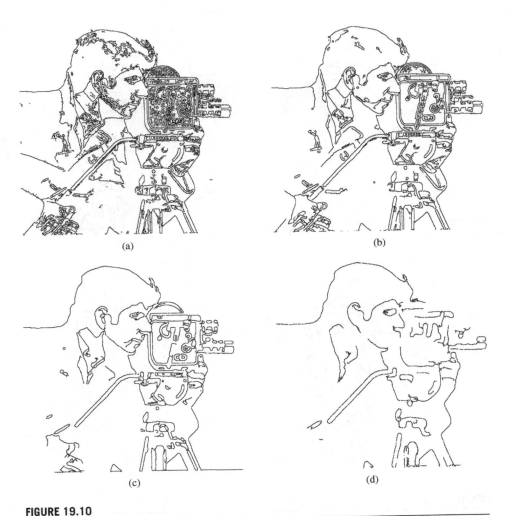

(a)

(b)

(c)

(d)

FIGURE 19.10

Canny edge detector of Eq. (19.22) applied after Gaussian smoothing over a range of σ: (a) $\sigma = 0.5$; (b) $\sigma = 1$; (c) $\sigma = 2$; and (d) $\sigma = 4$. The thresholds are fixed in each case at $T_U = 10$ and $T_L = 4$.

only computational cost beyond that for grayscale images is incurred in obtaining the luminance component image, if necessary. In many color spaces, such as *YIQ*, *HSL*, *CIELUV*, and *CIELAB*, the luminance image is simply one of the components in that representation. For others, such as *RGB*, computing the luminance image is usually easy and efficient. The main drawback to luminance-only processing is that important edges are often not confined to the luminance component. Therefore, a gray level difference in the luminance component is often not the most appropriate criterion for edge detection in color images.

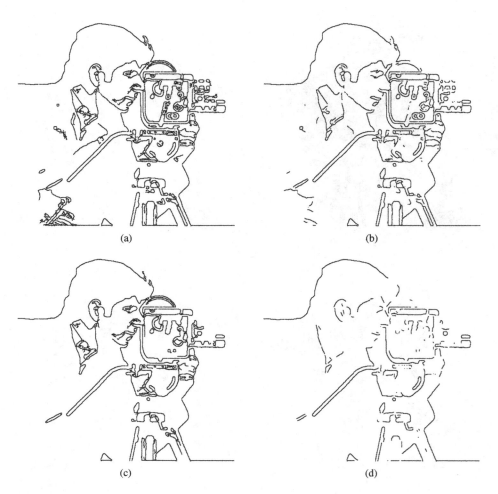

FIGURE 19.11

Canny edge detector of Eq. (19.22) applied after Gaussian smoothing with $\sigma = 2$: (a) $T_U = 10$, $T_L = 1$; (b) $T_U = T_L = 10$; (c) $T_U = 20$, $T_L = 1$; (d) $T_U = T_L = 20$. As T_L is changed, notice the effect on the results of hysteresis thresholding.

Another rather obvious approach is to apply a desired edge detection method separately to each color component and construct a cumulative edge map. One possibility for overall gradient magnitude, shown here for the RGB color space, combines the component gradient magnitudes [24]:

$$\left|\nabla f_c(x,y)\right| = \left|\nabla f_R(x,y)\right| + \left|\nabla f_G(x,y)\right| + \left|\nabla f_B(x,y)\right|.$$

The results, however, are biased according to the properties of the particular color space used. It is often important to employ a color space that is appropriate for the target application. For example, edge detection that is intended to approximate the human visual system's behavior should utilize a color space having a perceptual basis, such as *CIELUV* or perhaps *HSL*. Another complication is the fact that the components' gradient vectors may not always be similarly oriented, making the search for local maxima of $|\nabla f_c|$ along the gradient direction more difficult. If a total gradient image were to be computed by summing the color component gradient vectors, not just their magnitudes, then inconsistent orientations of the component gradients could destructively interfere and nullify some edges.

Vector approaches to color edge detection, while generally less computationally efficient, tend to have better theoretical justification. Euclidean distance in color space between the color vectors of a given pixel and its neighbors can be a good basis for an edge detector [24]. For the *RGB* case, the magnitude of the vector gradient is as follows:

$$|\nabla f_c(x,y)| = \sqrt{|\nabla f_R(x,y)|^2 + |\nabla f_G(x,y)|^2 + |\nabla f_B(x,y)|^2}.$$

Trahanias and Venetsanopoulos [29] described the use of vector order statistics as the basis for color edge detection. A later paper by Scharcanski and Venetsanopoulos [26] furthered the concept. While not strictly founded on the gradient or Laplacian, their techniques are effective and worth mention here because of their vector bases. The basic idea is to look for changes in local vector statistics, particularly vector dispersion, to indicate the presence of edges.

Multispectral images can have many components, complicating the edge detection problem even further. Cebrián *et al.* [6] describe several methods that are useful for multispectral images having any number of components. Their description uses the second directional derivative in the gradient direction as the basis for the edge detector, but other types of detectors can be used instead. The components-average method forms a grayscale image by averaging all components, which have first been Gaussian-smoothed, and then finds the edges in that image. The method generally works well because multispectral images tend to have high correlation between components. However, it is possible for edge information to diminish or vanish if the components destructively interfere.

Cumani [8] explored operators for computing the vector gradient and created an edge detection approach based on combining the component gradients. A multispectral contrast function is defined, and the image is searched for pixels having maximal directional contrast. Cumani's method does not always detect edges present in the component bands, but it better avoids the problem of destructive interference between bands.

The maximal gradient method constructs a single gradient image from the component images [6]. The overall gradient image's magnitude and direction values at a given pixel are those of the component having the greatest gradient magnitude at that pixel. Some edges can be missed by the maximal gradient technique because they may be swamped by differently oriented, stronger edges present in another band.

The method of combining component edge maps is the least efficient because an edge map must first be computed for every band. On the positive side, this method is capable of detecting any edge that is detectable in at least one component image. Combination of component edge maps into a single result is made more difficult by the edge location errors induced by Gaussian smoothing done in advance. The superimposed edges can become smeared in width because of the accumulated uncertainty in edge localization. A thinning step applied during the combination procedure can greatly reduce this edge blurring problem.

19.6 SUMMARY

Gray level edge detection is most commonly performed by convolving an image, f, with a filter that is somehow based on the idea of the derivative. Conceptually, edges can be revealed by locating either the local extrema of the first derivative of f or the zero-crossings of its second derivative. The gradient and the Laplacian are the primary derivative-based functions used to construct such edge-detection filters. The gradient, ∇, is a 2D extension of the first derivative while the Laplacian, ∇^2, acts as a 2D second derivative. A variety of edge detection algorithms and techniques have been developed that are based on the gradient or Laplacian in some way. Like any type of derivative-based filter, ones based on these two functions tend to be very sensitive to noise. Edge location errors, false edges, and broken or missing edge segments are often problems with edge detection applied to noisy images. For gradient techniques, thresholding is a common way to suppress noise and can be done adaptively for better results. Gaussian smoothing is also very helpful for noise suppression, especially when second-derivative methods such as the Laplacian are used. The Laplacian of Gaussian approach can also provide edge information over a range of scales, helping to further improve detection accuracy and noise suppression as well as providing clues that may be useful during subsequent processing.

Recent comparisons of various edge detectors have been made by Heath *et al.* [13] and Bowyer *et al.* [4]. They have concluded that the subjective quality of the results of various edge detectors applied to real images is quite dependent on the images themselves. Thus, there is no single edge detector that produces a consistently best overall result. Furthermore, they found it difficult to predict the best choice of edge detector for a given situation.

REFERENCES

[1] D. H. Ballard and C. M. Brown, *Computer Vision*. Prentice-Hall, Englewood Cliffs, NJ, 1982.

[2] V. Berzins. Accuracy of Laplacian edge detectors. *Comput. Vis. Graph. Image Process.*, 27:195–210, 1984.

[3] A. C. Bovik, T. S. Huang, and D. C. Munson, Jr. The effect of median filtering on edge estimation and detection. *IEEE Trans. Pattern Anal. Mach. Intell.*, PAMI-9:181–194, 1987.

[4] K. Bowyer, C. Kranenburg, and S. Dougherty. Edge detector evaluation using empirical ROC curves. *Comput. Vis. Image Underst.*, 84:77–103, 2001.

[5] J. Canny. A computational approach to edge detection. *IEEE Trans. Pattern Anal. Mach. Intell.*, PAMI-8:679–698, 1986.

[6] M. Cebrián, M. Perez-Luque, and G. Cisneros. Edge detection alternatives for multispectral remote sensing images. In *Proceedings of the 8th Scandinavian Conference on Image Analysis*, Vol. 2, 1047–1054. NOBIM-Norwegian Soc. Image Pross & Pattern Recognition, Tromso, Norway, 1993.

[7] J. S. Chen, A. Huertas, and G. Medioni. Very fast convolution with Laplacian-of-Gaussian masks. In *Proc. IEEE Comput. Soc. Conf. Comput. Vis. Pattern Recognit.*, 293–298. IEEE, New York, 1986.

[8] A. Cumani. Edge detection in multispectral images. *Comput. Vis. Graph. Image Process. Graph. Models Image Process.*, 53:40–51, 1991.

[9] L. Ding and A. Goshtasby. On the Canny edge detector. *Pattern Recognit.*, 34:721–725, 2001.

[10] R. M. Haralick and L. G. Shapiro. *Computer and Robot Vision*, Vol. 1. Addison-Wesley, Reading, MA, 1992.

[11] Q. Ji and R. M. Haralick. Efficient facet edge detection and quantitative performance evaluation. *Pattern Recognit.*, 35:689–700, 2002.

[12] R. C. Hardie and C. G. Boncelet. Gradient-based edge detection using nonlinear edge enhancing prefilters. *IEEE Trans. Image Process.*, 4:1572–1577, 1995.

[13] M. Heath, S. Sarkar, T. Sanocki, and K. Bowyer. Comparison of edge detectors, a methodology and initial study. *Comput. Vis. Image Underst.*, 69(1):38–54, 1998.

[14] A. Huertas and G. Medioni. Detection of intensity changes with subpixel accuracy using Laplacian-Gaussian masks. *IEEE Trans. Pattern. Anal. Mach. Intell.*, PAMI-8(5):651–664, 1986.

[15] S. R. Gunn. On the discrete representation of the Laplacian of Gaussian. *Pattern Recognit.*, 32: 1463–1472, 1999.

[16] A. K. Jain. *Fundamentals of Digital Image Processing*. Prentice-Hall, Englewood Cliffs, NJ, 1989.

[17] J. S. Lim. *Two-Dimensional Signal and Image Processing*. Prentice-Hall, Englewood Cliffs, NJ, 1990.

[18] D. Marr. *Vision*. W. H. Freeman, New York, 1982.

[19] D. Marr and E. Hildreth. Theory of edge detection. *Proc. R. Soc. Lond. B*, 270:187–217, 1980.

[20] B. Mathieu, P. Melchior, A. Oustaloup, and Ch. Ceyral. Fractional differentiation for edge detection. *Signal Processing*, 83:2421–2432, 2003.

[21] J. Merron and M. Brady. Isotropic gradient estimation. In *Proc. IEEE Comput. Soc. Conf. Comput. Vis. Pattern Recognit.*, 652–659. IEEE, New York, 1996.

[22] V. S. Nalwa and T. O. Binford. On detecting edges. *IEEE Trans. Pattern. Anal. Mach. Intell.*, PAMI-8(6):699–714, 1986.

[23] W. K. Pratt. *Digital Image Processing*, 2nd ed. Wiley, New York, 1991.

[24] S. J. Sangwine and R. E. N. Horne, editors. *The Colour Image Processing Handbook*. Chapman and Hall, London, 1998.

[25] S. Sarkar and K. L. Boyer. Optimal infinite impulse response zero crossing based edge detectors. *Comput. Vis. Graph. Image Process. Image Underst.*, 54(2):224–243, 1991.

[26] J. Scharcanski and A. N. Venetsanopoulos. Edge detection of color images using directional operators. *IEEE Trans. Circuits Syst. Video Technol.*, 7(2):397–401, 1997.

[27] P. Siohan, D. Pele, and V. Ouvrard. Two design techniques for 2-D FIR LoG filters. In M. Kunt, editor, *Proc. SPIE, Visual Communications and Image Processing*, Vol. 1360, 970–981, 1990.

[28] V. Torre and T. A. Poggio. On edge detection. *IEEE Trans. Pattern Anal. Mach. Intell.*, PAMI-8(2):147–163, 1986.

[29] P. E. Trahanias and A. N. Venetsanopoulos. Color edge detection using vector order statistics. *IEEE Trans. Image Process.*, 2(2):259–264, 1993.

[30] A. P. Witkin. Scale-space filtering. In *Proc. Int. Joint Conf. Artif. Intell.*, 1019–1022. William Kaufmann Inc., Karlsruhe, Germany, 1983.

[31] D. Ziou and S. Wang. Isotropic processing for gradient estimation. In *Proc. IEEE Comput. Soc. Conf. Comput. Vis. Pattern Recognit.*, 660–665. IEEE, New York, 1996.

Diffusion Partial Differential Equations for Edge Detection

20

Scott T. Acton

University of Virginia

20.1 INTRODUCTION AND MOTIVATION

20.1.1 Partial Differential Equations in Image and Video Processing

The collision of imaging and differential equations makes sense. Without motion or change of scene or changes within the scene, imaging is worthless. First, consider a static environment—we would not need vision in this environment, as the components of the scene are unchanging. In a dynamic environment, however, vision becomes the most valuable sense. Second, consider a constant-valued image with no internal changes or edges. Such an image is devoid of value in the information-theoretic sense.

The need for imaging is based on the presence of change. The mechanism for change in both time and space is described and governed by *differential equations*.

The partial differential equations (PDEs) of interest in this chapter enact *diffusion*. In chemistry or heat transfer, *diffusion* is a process that equilibrates concentration differences without creating or destroying mass. In image and video processing, we can consider the *mass* to be the pixel intensities or the gradient magnitudes, for example.

These important differential equations are PDEs, since they contain partial derivatives with respect to spatial coordinates and time. These equations, especially in the case of anisotropic diffusion, are nonlinear PDEs since the diffusion coefficient is typically nonlinear.

20.1.2 Edges and Anisotropic Diffusion

Sudden, sustained changes in image intensity are called *edges*. We know that the human visual system makes extensive uses of edges to perform visual tasks such as object recognition [1]. Humans can recognize complex 3D objects using only line drawings or image edge information. Similarly, the extraction of edges from digital imagery allows a valuable abstraction of information and a reduction in processing and storage costs. Most

525

definitions of image edges involve some concept of feature scale. Edges are said to exist at certain scales—edges from detail existing at fine scales and edges from the boundaries of large objects existing at large scales. Furthermore, large-scale edges exist at fine scales, leading to a notion of edge causality.

In order to locate edges of various scales within an image, it is desirable to have an image operator that computes a scaled version of a particular image or frame in a video sequence. This operator should preserve the position of such edges and facilitate the extraction of the edge map through the *scale space*. The tool of isotropic diffusion, a linear lowpass filtering process, is not able to preserve the position of important edges through the scale space. Anisotropic diffusion, however, meets this criterion and has been used effectively in conjunction with edge detection.

The main benefit of anisotropic diffusion is edge preservation through the image smoothing process. Anisotropic diffusion yields intra-region smoothing, not inter-region smoothing, by impeding diffusion at the image edges. The anisotropic diffusion process can be used to retain image features of a specified scale. Furthermore, the localized computation of anisotropic diffusion allows efficient implementation on a locally-interconnected computer architecture. Caselles *et al.* furnish additional motivation for using diffusion in image and video processing [2]. The diffusion methods use localized models where discrete filters become PDEs as the sample spacing goes to zero. The PDE framework allows various properties to be proved or disproved including stability, locality, causality, and the existence and uniqueness of solutions. Through the established tools of numerical analysis, high degrees of accuracy and stability are possible.

In this chapter, we introduce diffusion for image and video processing. We specifically concentrate on the implementation of anisotropic diffusion, providing several alternatives for the diffusion coefficient and the diffusion PDE. Energy-based variational diffusion techniques are also reviewed. Recent advances in anisotropic diffusion processes, including multiresolution techniques, multispectral techniques, and techniques for ultrasound and radar imagery, are discussed. Finally, the extraction of image edges after anisotropic diffusion is addressed, and vector diffusion processes for attracting active contours to boundaries are examined.

20.2 BACKGROUND ON DIFFUSION

20.2.1 Scale Space and Isotropic Diffusion

In order to introduce the diffusion-based processing methods and the associated processes of edge detection, let us define some notation. Let $\mathbf{I}$ represent an image with real-valued intensity $I(\mathbf{x})$ image at position $\mathbf{x}$ in the domain Ω. When defining the PDEs for diffusion, let $\mathbf{I}_t$ be the image at time t with intensities $I_t(\mathbf{x})$. Corresponding with image $\mathbf{I}$ is the edge map $\mathbf{e}$—the image of "edge pixels" $e(\mathbf{x})$ with Boolean range (0 = no edge, 1 = edge), or real-valued range $e(\mathbf{x}) \in [0, 1]$. The set of edge positions in an image is denoted by Ψ.

The concept of *scale space* is at the heart of diffusion-based image and video processing. A scale space is a collection of images that begins with the original, fine scale image and progresses toward more coarse scale representations. Using a scale space, important image processing tasks such as hierarchical searches, image coding, and image segmentation may be efficiently realized. Implicit in the creation of a scale space is the *scale generating filter*. Traditionally, linear filters have been used to scale an image. In fact, the scale space of Witkin [3] can be derived using a Gaussian filter:

$$\mathbf{I}_t = \mathbf{G}_\sigma * \mathbf{I}_0, \tag{20.1}$$

where $\mathbf{G}_\sigma$ is a Gaussian kernel with standard deviation (scale) of σ, and $\mathbf{I}_0 = \mathbf{I}$ is the initial image. If

$$\sigma = \sqrt{t}, \tag{20.2}$$

then the Gaussian filter result may be achieved through an isotropic diffusion process governed by

$$\frac{\partial \mathbf{I}_t}{\partial t} = \nabla^2 \mathbf{I}_t, \tag{20.3}$$

where $\nabla^2 \mathbf{I}_t$ is the Laplacian of $\mathbf{I}_t$ [3, 4]. To evolve one pixel of $\mathbf{I}$, we have the following PDE:

$$\frac{\partial I_t(\mathbf{x})}{\partial t} = \nabla^2 I_t(\mathbf{x}). \tag{20.4}$$

The Marr-Hildreth paradigm uses a Gaussian scale space to define multiscale edge detection. Using the Gaussian-convolved (or diffused) images, one may detect edges by applying the Laplacian operator and then finding zero-crossings [5]. This popular method of edge detection, called the Laplacian-of-a-Gaussian (LoG), is strongly motivated by the biological vision system. However, the edges detected from isotropic diffusion (Gaussian scale space) suffer from artifacts such as corner rounding and from edge localization error (deviation in detected edge position from the "true" edge position). The localization errors increase with increased scale, precluding straightforward multiscale image/video analysis. As a result, many researchers have pursued anisotropic diffusion as a viable alternative for generating images suitable for edge detection. This chapter focuses on such methods.

20.2.1.1 *Anisotropic Diffusion*

The main idea behind anisotropic diffusion is the introduction of a function that inhibits smoothing at the image edges. This function, called the diffusion coefficient $c(\mathbf{x})$, encourages intra-region smoothing over inter-region smoothing. For example, if $c(\mathbf{x})$ is constant at all locations, then smoothing progresses in an isotropic manner. If $c(\mathbf{x})$ is allowed to vary according to the local image gradient, we have anisotropic diffusion. A basic anisotropic diffusion PDE is

$$\frac{\partial I_t(\mathbf{x})}{\partial t} = \operatorname{div}\{c(\mathbf{x})\nabla I_t(\mathbf{x})\} \tag{20.5}$$

with $\mathbf{I}_0 = \mathbf{I}$ [6].

The discrete formulation proposed in [6] will be used as a general framework for implementation of anisotropic diffusion in this chapter. Here the image intensities are updated according to

$$[I(\mathbf{x})]_{t+1} = \left[I(\mathbf{x}) + (\Delta T) \sum_{d=1}^{\Gamma} c_d(\mathbf{x}) \nabla I_d(\mathbf{x}) \right]_t, \tag{20.6}$$

where Γ is the number of directions in which diffusion is computed, $\nabla I_d(\mathbf{x})$ is the directional derivative (simple difference) in direction d at location $\mathbf{x}$, and time (in iterations) is given by t. ΔT is the time step—for stability, $\Delta T \leq \frac{1}{2}$ in the 1D case, and $\Delta T \leq \frac{1}{4}$ in the 2D case using four diffusion directions. For 1D discrete-domain signals, the simple differences $\nabla I_d(\mathbf{x})$ with respect to the "western" and "eastern" neighbors, respectively (neighbors to the left and right), are defined by

$$\nabla I_1(x) = I(x - h_1) - I(x) \tag{20.7}$$

and

$$\nabla I_2(x) = I(x + h_2) - I(x). \tag{20.8}$$

The parameters h_1 and h_2 define the sample spacing used to estimate the directional derivatives. For the 2D case, the diffusion directions include the "northern" and "southern" directions (up and down), as well as the "western" and "eastern" directions (left and right). Given the motivation and basic definition of diffusion-based processing, we will now define several implementations of anisotropic diffusion that can be applied for edge extraction.

20.3 ANISOTROPIC DIFFUSION TECHNIQUES

20.3.1 The Diffusion Coefficient

The link between edge detection and anisotropic diffusion is found in the edge-preserving nature of anisotropic diffusion. The function that impedes smoothing at the edges is the diffusion coefficient. Therefore, the selection of the diffusion coefficient is the most critical step in performing diffusion-based edge detection. We will review several possible variants of the diffusion coefficient and discuss the associated positive and negative attributes.

To simplify the notation, we will denote the diffusion coefficient at location $\mathbf{x}$ by $c(\mathbf{x})$ in the continuous case. For the discrete-domain case, $c_d(\mathbf{x})$ represents the diffusion coefficient for direction d at location $\mathbf{x}$. Although the diffusion coefficients here are defined using $c(\mathbf{x})$ for the continuous case, the functions are equivalent in the discrete-domain case of $c_d(\mathbf{x})$. Typically $c(\mathbf{x})$ is a nonincreasing function of $|\nabla I(\mathbf{x})|$, the gradient magnitude at position $\mathbf{x}$. As such, we often refer to the diffusion coefficient as $c(|\nabla I(\mathbf{x})|)$. For small values of $|\nabla I(\mathbf{x})|$, $c(\mathbf{x})$ tends to unity. As $|\nabla I(\mathbf{x})|$ increases, $c(\mathbf{x})$ decreases to zero. Teboul *et al.* [7] establish three conditions for edge-preserving diffusion coefficients. These conditions are (1) $\lim_{|\nabla I(\mathbf{x})| \to 0} c(\mathbf{x}) = M$ where $0 < M < \infty$, (2) $\lim_{|\nabla I(\mathbf{x})| \to \infty} c(\mathbf{x}) = 0$, and (3) $c(\mathbf{x})$ is a strictly decreasing function of $|\nabla I(\mathbf{x})|$. Property 1 ensures isotropic smoothing in regions of similar intensity, while property 2 preserves edges. The third property is given in order to avoid numerical instability. While most of the coefficients discussed here obey the first two properties, not all formulations obey the third property.

In [6], Perona and Malik propose

$$c(\mathbf{x}) = \exp\left\{-\left[\frac{\nabla I(\mathbf{x})}{k}\right]^2\right\} \tag{20.9}$$

and

$$c(\mathbf{x}) = \frac{1}{1 + \left[\frac{\nabla I(\mathbf{x})}{k}\right]^2} \tag{20.10}$$

as diffusion coefficients. Diffusion operations using (20.9) and (20.10) have the ability to sharpen edges (backward diffusion), and are inexpensive to compute. However, these diffusion coefficients are unable to remove heavy-tailed noise and create "staircase" artifacts [8, 9]. See the example of smoothing using (20.9) on the noisy image in Fig. 20.1(a), producing the result in Fig. 20.1(b). In this case, the anisotropic diffusion operation leaves several outliers in the resultant image. A similar problem is observed in Fig. 20.2(b), using the corrupted image in Fig. 20.2(a) as input. You *et al.* have also shown that (20.9) and (20.10) lead to an ill-posed diffusion—a small perturbation in the data may cause a significant change in the final result [10].

The inability of anisotropic diffusion to denoise an image has been addressed by Catte *et al.* [11] and Alvarez *et al.* [12]. Their regularized diffusion operation uses a modification of the gradient image used to compute the diffusion coefficients. In this case, a Gaussian-convolved version of the image is employed in computing diffusion coefficients. Using the same basic form as (20.9), we have

$$c(\mathbf{x}) = \exp\left\{-\left[\frac{\nabla S(\mathbf{x})}{k}\right]^2\right\}, \tag{20.11}$$

where $\mathbf{S}$ is the convolution of $\mathbf{I}$ and a Gaussian filter with standard deviation σ:

$$\mathbf{S} = \mathbf{I} * \mathbf{G}_\sigma. \tag{20.12}$$

This method can be used to rapidly eliminate noise in the image as shown in Fig. 20.1(c). In this case, the diffusion is well posed and converges to a unique result, under certain conditions [11]. Drawbacks of this diffusion coefficient implementation include the additional computational burden of filtering at each step and the introduction of a linear filter into the edge-preserving anisotropic diffusion approach. The loss of sharpness due to the linear filter is evident in Fig. 20.2(c). Although the noise is eradicated, the edges are softened and blotching artifacts appear in the background of this example result.

Another modified gradient implementation, called *morphological anisotropic diffusion*, can be formed by substituting

$$\mathbf{S} = (\mathbf{I} \circ \mathbf{B}) \bullet \mathbf{B} \tag{20.13}$$

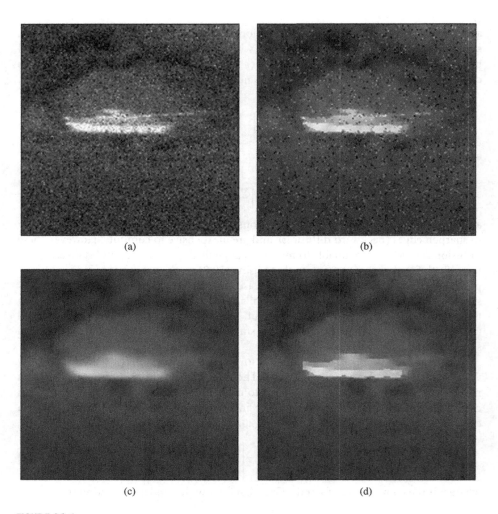

(a)

(b)

(c)

(d)

FIGURE 20.1

Three implementations of anisotropic diffusion applied to an infrared image of a tank: (a) original noisy image; (b) results obtained using anisotropic diffusion with (20.9); (c) results obtained using modified gradient anisotropic diffusion with (20.11) and (20.12); (d) results obtained using morphological anisotropic diffusion with (20.11) and (20.13).

into (20.11), where **B** is a structuring element of size $m \times m$, $\mathbf{I} \circ \mathbf{B}$ is the morphological opening of **I** by **B**, and $\mathbf{I} \bullet \mathbf{B}$ is the morphological closing of **I** by **B**. In [13], the open-close and close-open filters were used in an alternating manner between iterations, thus reducing grayscale bias of the open-close and close-open filters. As the result in Fig. 20.1(d) demonstrates, the morphological anisotropic diffusion method can be used to eliminate noise and insignificant features while preserving edges. Morphological anisotropic diffusion has the advantage of selecting feature scale (by specifying the structuring element **B**) and selecting the gradient magnitude threshold, whereas previous anisotropic diffusions, such as (20.9) and (20.10), only allowed selection of the gradient magnitude threshold.

You *et al.* introduce the following diffusion coefficient in [10]:

$$c(\mathbf{x}) = \begin{cases} 1/T + p(T + \varepsilon)^{p-1}/T, & \nabla I(\mathbf{x}) < T \\ 1/|\nabla I(\mathbf{x})| + p(|\nabla I(\mathbf{x})| + \varepsilon)^{p-1}/|\nabla I(\mathbf{x})|, & |\nabla I(\mathbf{x})| \geq T, \end{cases} \tag{20.14}$$

(a)

(b)

(c)

FIGURE 20.2

Continued

(d)   (e)

FIGURE 20.2

(a) Corrupted "cameraman" image (Laplacian noise, SNR = 13 dB) used as input for results in Figs. 20.2(b)–(e); (b) after 8 iterations of anisotropic diffusion with (20.9), $k = 25$; (c) after 8 iterations of anisotropic diffusion with (20.11) and (20.12), $k = 25$; (d) after 75 iterations of anisotropic diffusion with (20.14), $T = 6$, $e = 1$, $p = 0.5$; (e) after 15 iterations of multigrid anisotropic diffusion with (20.11) and (20.12), $k = 6$ [35].

where the parameters are constrained by $\varepsilon > 0$ and $0 < p < 1$. T is a threshold on the gradient magnitude, similar to k in (20.9). This approach has the benefits of avoiding staircase artifacts and removing impulse noise. The main drawback is computational expense. As seen in Fig. 20.2(d), anisotropic diffusion with this diffusion coefficient succeeds in removing noise and retaining important features from Fig. 20.2(a), but requires a significant number of updates.

The diffusion coefficient

$$c(\mathbf{x}) = \frac{1}{|\nabla I(\mathbf{x})|} \qquad (20.15)$$

is used in mean curvature motion formulations of diffusion [14], shock filters [15], and locally monotonic (LOMO) diffusion [16]. One may notice that this diffusion coefficient is parameter-free.

Designing a diffusion coefficient with robust statistics, Black *et al.* [17] model anisotropic diffusion as a robust estimation procedure that finds a piecewise smooth representation of an input image. A diffusion coefficient that utilizes the Tukey's biweight norm is given by

$$c(\mathbf{x}) = \frac{1}{2} \left\{ 1 - \left[\frac{\nabla I(\mathbf{x})}{\sigma} \right]^2 \right\}^2 \qquad (20.16)$$

for $|\nabla I(\mathbf{x})| \leq \sigma$ and is 0 otherwise. Here the parameter σ represents scale. Where the standard anisotropic diffusion coefficient as in (20.9) continues to smooth over edges while iterating, the robust formulation (20.16) preserves edges of a prescribed scale σ and effectively stops diffusion.

Here seven important versions of the diffusion coefficient were given that involve tradeoffs between solution quality, solution expense, and convergence behavior. Other research in the diffusion area focuses on the diffusion PDE itself. The next section reveals significant modifications to the anisotropic diffusion PDE that affect fidelity to the input image, edge quality, and convergence properties.

20.3.2 The Diffusion PDE

In addition to the basic anisotropic diffusion PDE given in Section 20.1.2, other diffusion mechanisms may be employed to adaptively filter an image for edge detection. Nordstrom [18] used an additional term to maintain fidelity to the input image, to avoid the selection of a stopping time, and to avoid termination of the diffusion at a trivial solution, such as a constant image. This PDE is given by

$$\frac{\partial I_t(\mathbf{x})}{\partial t} - \text{div}\{c(\mathbf{x})\nabla I_t(\mathbf{x})\} = I_0(\mathbf{x}) - I_t(\mathbf{x}). \tag{20.17}$$

Obviously, the right-hand side $I_0(\mathbf{x}) - I_t(\mathbf{x})$ enforces an additional constraint that penalizes deviation from the input image.

Just as Canny [19] modified the LoG edge detection technique by detecting zero-crossings of the Laplacian only in the direction of the gradient, a similar edge-sensitive approach can be taken with anisotropic diffusion. Here, the boundary-preserving diffusion is executed only in the direction orthogonal to the gradient direction, whereas the standard anisotropic diffusion schemes *impede* diffusion across the edge. If the rate of change of intensity is set proportional to the second partial derivative in the direction orthogonal to the gradient (called τ), we have

$$\frac{\partial I_t(\mathbf{x})}{\partial t} = \frac{\partial^2 I_t(\mathbf{x})}{\partial \tau^2} = |\nabla I_t(\mathbf{x})| \, \text{div}\left\{ \frac{\nabla I_t(\mathbf{x})}{|\nabla I_t(\mathbf{x})|} \right\}. \tag{20.18}$$

This anisotropic diffusion model is called *mean curvature motion*, because it induces a diffusion in which the connected components of the image level sets of the solution image move in proportion to the boundary mean curvature. Several effective edge-preserving diffusion methods have arisen from this framework including [20] and [21]. Alvarez *et al.* [12] have used the mean curvature method in tandem with the regularized diffusion coefficient of (20.11) and (20.12). The result is a processing method that preserves the causality of edges through scale space. For edge-based hierarchical searches and multiscale analyses, the edge causality property is extremely important.

The mean curvature method has also been given a graph theoretic interpretation [22, 23]. Yezzi [23] treats the image as a graph in $\Re^n$—a typical 2D grayscale image would be a surface in $\Re^3$ where the image intensity is the third parameter, and each pixel is a graph node. Hence a color image could be considered a surface in $\Re^5$. The curvature motion of the graphs can be used as a model for smoothing and edge detection. For example, let a 3D graph $\mathbf{s}$ be defined by $\mathbf{s}(\mathbf{x}) = \mathbf{s}(x, y) = [x, y, I(x, y)]$ for the 2D image $\mathbf{I}$

with $\mathbf{x} = (x, y)$. To implement mean curvature motion on this graph, the PDE is given by

$$\frac{\partial \mathbf{s}(\mathbf{x})}{\partial t} = h(\mathbf{x})\mathbf{n}(\mathbf{x}), \tag{20.19}$$

where $h(\mathbf{x})$ is the mean curvature,

$$h(x, y) =$$

$$\frac{\frac{\partial^2 I(x,y)}{\partial x^2}\left[1 + \left(\frac{\partial I(x,y)}{\partial y}\right)^2\right] - 2\left(\frac{\partial I(x,y)}{\partial x}\right)\left(\frac{\partial I(x,y)}{\partial y}\right)\left(\frac{\partial^2 I(x,y)}{\partial x \partial y}\right) + \frac{\partial^2 I(x,y)}{\partial y^2}\left[1 + \left(\frac{\partial I(x,y)}{\partial x}\right)^2\right]}{2\left[1 + \left(\frac{\partial I(x,y)}{\partial y}\right)^2 + \left(\frac{\partial I(x,y)}{\partial y}\right)^2\right]^{3/2}}, \tag{20.20}$$

and $\mathbf{n}(\mathbf{x})$ is the unit normal of the surface:

$$\mathbf{n}(x, y) = \frac{\left[-\frac{\partial I(x,y)}{\partial x}, -\frac{\partial I(x,y)}{\partial x}, 1\right]}{\sqrt{1 + \left(\frac{\partial I(x,y)}{\partial y}\right)^2 + \left(\frac{\partial I(x,y)}{\partial y}\right)^2}}. \tag{20.21}$$

For a discrete implementation, the partial derivatives of $I(x, y)$ may be approximated using simple differences. One-sided differences or central differences may be employed. For example, a one-sided difference approximation for $\partial I(x,y)/\partial x$ is $I(x + 1, y) - I(x, y)$. A central difference approximation for the same partial derivative is given by $\frac{1}{2}[I(x + 1, y) - I(x - 1, y)]$.

The standard mean curvature PDE (20.19) has the drawback of edge movement that sacrifices edge sharpness. A remedy to this undesired movement is the use of projected mean curvature vectors. Let $\mathbf{z}$ denote the unit vector in the vertical (intensity) direction on the graph $\mathbf{s}$. The projected mean curvature diffusion PDE can be formed by

$$\frac{\partial \mathbf{s}(\mathbf{x})}{\partial t} = \left\{\left[h(\mathbf{x})\mathbf{n}(\mathbf{x})\right] \cdot \mathbf{z}\right\} \mathbf{z}. \tag{20.22}$$

The PDE for updating image intensity is then

$$\frac{\partial I(\mathbf{x})}{\partial t} =$$

$$\frac{\Delta I(x,y) + k^2\left[\left(\frac{\partial I(x,y)}{\partial x}\right)^2\left(\frac{\partial^2 I(x,y)}{\partial y^2}\right) - 2\left(\frac{\partial I(x,y)}{\partial x}\right)\left(\frac{\partial I(x,y)}{\partial y}\right)\left(\frac{\partial^2 I(x,y)}{\partial x \partial y}\right) + \left(\frac{\partial I(x,y)}{\partial y}\right)^2\left(\frac{\partial^2 I(x,y)}{\partial x^2}\right)\right]}{\left\{1 + k^2\left[\nabla I(x,y)\right]^2\right\}^2}, \tag{20.23}$$

where k scales the intensity variable. When k is zero, we have isotropic diffusion, and when k becomes larger, we have a damped geometric heat equation that preserves edges but diffuses more slowly. The projected mean curvature PDE gives edge preservation through scale space.

Another anisotropic diffusion technique leads to LOMO signals [16]. Unlike previous diffusion techniques that diverge or converge to trivial signals, LOMO diffusion converges rapidly to well-defined LOMO signals of the desired degree. (A signal is LOMO of degree

d (LOMO-d) if each interval of length d is nonincreasing or nondecreasing.) The property of local monotonicity allows both slow and rapid signal transitions (ramp and step edges) while excluding outliers due to noise. The degree of local monotonicity defines the signal scale. In contrast to other diffusion methods, LOMO diffusion does not require an additional regularization step to process a noisy signal and uses no thresholds or ad hoc parameters.

On a 1D signal, the basic LOMO diffusion operation is defined by (20.6) with $\Gamma = 2$ and using the diffusion coefficient (20.15), yields

$$[I(x)]_{t+1} \leftarrow \left(I(x) + (1/2) \left\{ \text{sign}[\nabla I_1(x)] + \text{sign}[\nabla I_2(x)] \right\} \right)_t, \qquad (20.24)$$

where a time step of $\Delta T = 1/2$ is used. Equation (20.24) is modified for the case where the simple difference $\nabla I_1(x)$ or $\nabla I_2(x)$ is zero. Let $\nabla I_1(x) \leftarrow -\nabla I_2(x)$ in the case of $\nabla I_1(x) = 0$; $\nabla I_2(x) \leftarrow -\nabla I_1(x)$ when $\nabla I_2(x) = 0$. The fixed point of (20.24) is defined as ld($\mathbf{I}, h_1, h_2$), where h_1 and h_2 are the sample spacings used to compute the simple differences $\nabla I_1(x)$ and $\nabla I_2(x)$, respectively (see (20.7) and (20.8)). Let $\text{ld}_d(\mathbf{I})$ denote the LOMO diffusion sequence that gives a LOMO-d signal from the input $\mathbf{I}$. For odd values of $d = 2m + 1$,

$$\text{ld}_d(\mathbf{I}) = \text{ld}(\ldots \text{ld}(\text{ld}(\text{ld}(\mathbf{I}, m, m), m - 1, m), m - 1, m - 1) \ldots, 1, 1). \qquad (20.25)$$

In (20.25), the process commences with ld($\mathbf{I}, m, m$) and continues with spacings of decreasing widths until ld($\mathbf{I}, 1, 1$) is implemented. For even values of $d = 2m$, the sequence of operations is similar:

$$\text{ld}_d(\mathbf{I}) = \text{ld}(\ldots \text{ld}(\text{ld}(\text{ld}(\mathbf{I}, m - 1, m), m - 1, m - 1), m - 2, m - 1) \ldots, 1, 1). \qquad (20.26)$$

To extend this method to two dimensions, the same procedure may be followed using (20.6) with $\Gamma = 4$ [16]. Another possibility is diffusing orthogonal to the gradient direction at each point in the image, using the 1D LOMO diffusion. Examples of 2D LOMO diffusion and the associated edge detection results are given in Section 20.4.

20.3.3 Variational Formulation

The diffusion PDEs discussed thus far may be considered numerical methods that attempt to minimize a cost or energy functional. Energy-based approaches to diffusion have been effective for edge detection and image segmentation. Morel and Solimini [24] give an excellent overview of the variational methods. Isotropic diffusion via the heat diffusion equation leads to a minimization of the following energy:

$$E(\mathbf{I}) = \int_\Omega |\nabla I(\mathbf{x})|^2 \, d\mathbf{x}. \qquad (20.27)$$

Given an initial image $\mathbf{I}_0$, the intermediate diffusion solutions may be considered a descent on

$$E(\mathbf{I}) = \lambda^2 \int_\Omega |\nabla I(\mathbf{x})|^2 \, d\mathbf{x} + \int_\Omega \left[I(\mathbf{x}) - I_0(\mathbf{x}) \right]^2 d\mathbf{x}, \qquad (20.28)$$

where the regularization parameter λ denotes scale [24].

Likewise, anisotropic diffusion has a variational formulation. The energy associated with the Perona and Malik diffusion is

$$E(I) = \lambda^2 \int_{\Omega} C\left[|\nabla I(\mathbf{x})|^2\right] d\mathbf{x} + \int_{\Omega} [I(\mathbf{x}) - I_0(\mathbf{x})]^2 d\mathbf{x}, \tag{20.29}$$

where C is the integral of $c'(\mathbf{x})$ with respect to the independent variable $|\nabla I(\mathbf{x})|^2$. Here $c'(\mathbf{x})$, as a function of $|\nabla I(\mathbf{x})|^2$, is equivalent to the diffusion coefficient $c(\mathbf{x})$ as a function of $|\nabla I(\mathbf{x})|$, so $c'(|\nabla I(\mathbf{x})|^2) = c(|\nabla I(\mathbf{x})|)$. The Nordstrom [18] diffusion PDE (20.17) yields steepest descent on this energy functional.

Teboul *et al.* have introduced a variational method that preserves edges and is useful for edge detection. In their approach, image enhancement and edge preservation are treated as two separate processes. The energy functional is given by

$$E(I, e) = \lambda^2 \int_{\Omega} \left[e(\mathbf{x})^2 |\nabla I(\mathbf{x})|^2 + k(e(\mathbf{x}) - 1)^2\right] d\mathbf{x} + \frac{\alpha^2}{k} \int_{\Omega} \varphi(|\nabla e(\mathbf{x})|) d\mathbf{x} + \int_{\Omega} [I(\mathbf{x}) - I_0(\mathbf{x})]^2 d\mathbf{x},$$
$$\tag{20.30}$$

where the real-valued variable $e(\mathbf{x})$ is the edge strength at position $\mathbf{x}$, and $e(\mathbf{x}) \in [0, 1]$. In (20.30), the diffusion coefficient is defined by $c(|\nabla I(\mathbf{x})|) = \varphi'(|\nabla I(\mathbf{x})|)/2(|\nabla I(\mathbf{x})|)$. An additional regularization parameter α is needed, and k is essentially an edge threshold parameter.

The energy functional in (20.30) leads to a system of two coupled PDEs:

$$I_0(\mathbf{x}) - I_t(\mathbf{x}) - \lambda^2 \text{div}\{e(\mathbf{x})[\nabla I_t(\mathbf{x})]\nabla I_t(\mathbf{x})\} = 0 \tag{20.31}$$

and

$$e(\mathbf{x})\left[\frac{|\nabla I(\mathbf{x})|^2}{k} + 1\right] - 1 + \frac{\alpha^2}{k^2}\text{div}[c(|\nabla e(\mathbf{x})|)\nabla e(\mathbf{x})] = 0. \tag{20.32}$$

The coupled PDEs have the advantage of edge preservation within the adaptive smoothing process. An edge map can be directly extracted from the final state of **e**.

This edge-preserving variational method is related to the segmentation approach of Mumford and Shah [25]. The energy functional to be minimized is

$$E(I) = \lambda^2 \int_{\Omega \backslash \Psi} |\nabla I(\mathbf{x})|^2 d\mathbf{x} + \int_{\Omega \backslash \Psi} [I(\mathbf{x}) - I_0(\mathbf{x})]^2 d\mathbf{x} + \mu \lambda^2 \int_{\Psi} d\psi, \tag{20.33}$$

where $\int_{\Psi} d\psi$ is the integrated length of the edges (Hausdorff measure), $\Omega \backslash \Psi$ is the set of image locations that exclude the edge positions, and μ is an additional weight parameter. The additional edge-length term reflects the goal of computing a minimal-length edge map for a given scale λ. The Mumford-Shah functional has spurred several variational image segmentation schemes, including PDE-based solutions [24].

In edge detection, thin, contiguous edges are typically desired. With diffusion-based edge detectors, the edges may be "thick" or "broken" when a gradient magnitude threshold is applied after diffusion. The variational formulation allows the addition of additional constraints that promote edge thinning and connectivity. Black *et al.* used two additional terms, a hysteresis term for improved connectivity and a nonmaximum suppression term for thinning [17]. A similar approach was taken in [26]. The additional terms allow the effective extraction of spatially coherent outliers. This idea is also found in the design of line processes for regularization [27].

20.3.4 Multiresolution Diffusion

One drawback of diffusion-based edge detection is the computational expense. Typically, a large number (anywhere from 20 to 200) of iterative steps are needed to provide a high-quality edge map. One solution to this dilemma is the use of multiresolution schemes. Two such approaches have been investigated for edge detection: the anisotropic diffusion pyramid and multigrid anisotropic diffusion.

In the case of isotropic diffusion, the Gaussian pyramid has been used for edge detection and image segmentation [28, 29]. The basic idea is that the scale generating operator (a Gaussian filter, for example) can be used as an antialiasing filter before sampling. Then, a set of image representations of increasing scale and decreasing resolution (in terms of the number of pixels) can be generated. This image pyramid can be used for hierarchical searches and coarse-to-fine edge detection.

The anisotropic diffusion pyramids [30, 31] are born from the same fundamental motivation as their isotropic, linear counterparts. However, with a nonlinear scale-generating operator, the presampling operation is constrained morphologically, not by the traditional sampling theorem. In the nonlinear case, the scale-generating operator should remove image features not supported in the subsampled domain. Therefore, morphological methods [32, 33] for creating image pyramids have also been used in conjunction with the morphological sampling theorem [34].

The anisotropic diffusion pyramids are, in a way, ad hoc multigrid schemes. A multigrid scheme can be useful for diffusion-based edge detectors in two ways. First, like the anisotropic diffusion pyramids, the number of diffusion updates may be decreased. Second, the multigrid approach can be used to eliminate low-frequency error. The anisotropic diffusion PDEs are stiff—they rapidly reduce high-frequency error (noise, small details), but slowly reduce background variations and often create artifacts such as blotches (false regions) or staircases (false step edges). See Fig. 20.3 for an example of a staircasing artifact.

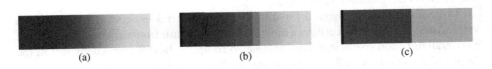

(a)   (b)   (c)

FIGURE 20.3

(a) Sigmoidal ramp edge; (b) after anisotropic diffusion with (9) ($k = 10$); (c) after multigrid anisotropic diffusion with (9) ($k = 10$) [35].

To implement a multigrid anisotropic diffusion operation [35], define **J** as an estimate of the image **I**. A system of equations is defined by $A(\mathbf{I}) = 0$ where

$$[A(\mathbf{I})](\mathbf{x}) = (\Delta T) \sum_{d=1}^{\Gamma} c_d(\mathbf{x}) \nabla I_d(\mathbf{x}), \qquad (20.34)$$

which is relaxed by the discrete anisotropic diffusion PDE (20.6). For this system of equations, the (unknown) algebraic error is $\mathbf{E} = \mathbf{I} - \mathbf{J}$ and the residual is $\mathbf{R} = -A(\mathbf{J})$ for image estimate **J**. The residual equation $A(\mathbf{E}) = \mathbf{R}$ can be *relaxed* (diffused) in the same manner as (20.34) using (20.6) to form an estimate of the error.

The first step is performing ν diffusion steps on the original input image (level $L = 0$). Then, the residual equation at the coarser grid $L + 1$ is

$$A(\mathbf{E}_{L+1}) = -A\left[(\mathbf{J}_L)_{\downarrow S}\right], \qquad (20.35)$$

where $\downarrow S$ represents downsampling by a factor of S. Now, the residual equation (20.33) can be relaxed using the discrete diffusion PDE (20.6) with an initial error estimate of $\mathbf{E}_{L+1} = 0$. The new error estimate $\mathbf{E}_{L+1}$ after relaxation can then be transferred to the finer grid to correct the initial image estimate **J** in a simple two-grid scheme. Alternatively, the process of transferring the residual to successively coarser grids can be continued until a grid is reached in which a closed form solution is possible. Then, the error estimates are propagated back to the original grid.

Additional steps may be taken to account for the nonlinearity of the anisotropic diffusion PDE, such as implementing a full approximation scheme (FAS) multigrid system, or by using a global linearization step in combination with a Newton method to solve for the error iteratively [9, 19].

The results of applying multigrid anisotropic diffusion are shown in Fig. 20.2(e). In just 15 updates, the multigrid anisotropic diffusion method was able to remove the noise from Fig. 20.2(b) while preserving the significant objects and avoiding the introduction of blotching artifacts.

20.3.5 Multispectral Anisotropic Diffusion

Color edge detection and boundary detection for multispectral imagery are important tasks in general image/video processing, remote sensing, and biomedical image processing. Applying anisotropic diffusion to each channel or spectral band separately is one possible way of processing multichannel or multispectral image data. However, this single band approach forfeits the richness of the multispectral data and provides individual edge maps that do not possess corresponding edges.

Two solutions have emerged for diffusing multispectral imagery. The first, called *vector distance dissimilarity*, utilizes a function of the gradients from each band to compute an overall diffusion coefficient. For example, to compute the diffusion coefficient in the "western" direction on an RGB color image, the following function could be applied:

$$\nabla I_1(\mathbf{x}) = \sqrt{\left[R(x - h_1, y) - R(x, y)\right]^2 + \left[G(x - h_1, y) - G(x, y)\right]^2 + \left[B(x - h_1, y) - B(x, y)\right]^2}, \qquad (20.36)$$

where $R(\mathbf{x})$ is the red band intensity at $\mathbf{x}$, $G(\mathbf{x})$ is the green band, and $B(\mathbf{x})$ is the blue band. Using the vector distance dissimilarity method, the standard diffusion coefficients such as (20.9) can be employed. This technique was used in [38] for shape-based processing and in [39] for processing remotely sensed imagery. An example of multispectral anisotropic diffusion is shown in Fig. 20.4. Using the noisy multispectral image in Fig. 20.4(a) as input, the vector distance dissimilarity method produces the smoothed result shown in Fig. 20.4(b), which has an associated image of gradient magnitude shown in Fig. 20.4(c). As can be witnessed in Fig. 20.4(c), an edge detector based on vector distance dissimilarity is sensitive to noise and does not identify the important image boundaries.

The second method uses mean curvature motion and a multispectral gradient formula to achieve anisotropic, edge-preserving diffusion. The idea behind mean curvature motion, as discussed above, is to diffuse in the direction opposite to the gradient such that the image level set objects move with a rate in proportion to their mean curvature. With a grayscale image, the gradient is always perpendicular to the level set objects of the image. In the multispectral case, this quality does not hold. A well-motivated diffusion is defined by Sapiro and Ringach [40], using DiZenzo's multispectral gradient formula [41]. In Fig. 20.4(d), results for multispectral anisotropic diffusion are shown for the mean curvature approach of [40] used in combination with the modified gradient approach of [11]. The edge map in Fig. 20.4(e) shows improved resilience to impulse noise over the vector distance dissimilarity approach.

20.3.6　Speckle Reducing Anisotropic Diffusion

The anisotropic diffusion PDE introduced in (20.5) assumes that the image is corrupted by additive noise. Speckle reducing anisotropic diffusion (SRAD) is a PDE technique for image enhancement in which signal-dependent multiplicative noise is present, as with radar and ultrasonic imaging. Whereas traditional anisotropic diffusion can be viewed as the edge-sensitive version of classical linear filters (e.g., the Gaussian filter), SRAD can be viewed as the edge-sensitive 3 version of classical speckle reducing filters that emerged from the radar community (e.g., the Lee filter). SRAD smoothes the imagery and enhances edges by inhibiting diffusion across edges and allowing isotropic diffusion within homogeneous regions. For images containing signal-dependent, spatially correlated multiplicative noise, SRAD excels over the adaptive filtering techniques designed with additive noise models in mind.

The SRAD technique employs an adaptive speckle filter that uses a local statistic for the coefficient of variation, defined as the ratio of standard deviation to mean, to measure the strength of edges in speckle imagery. A discrete form of this operator in 2D is [42]

$$q(\mathbf{x}) = \sqrt{\frac{\left|(1/2)\,|\nabla I(\mathbf{x})|^2 - (1/16)(\nabla^2 I(\mathbf{x}))^2\right|}{\left[I(\mathbf{x}) + (1/4)\nabla^2 I(\mathbf{x})\right]^2}}, \qquad (20.37)$$

where $\nabla^2 I(\mathbf{x})$ is the Laplacian of image at position $\mathbf{x}$, and $\nabla I(\mathbf{x})$ is the gradient of image at position $\mathbf{x}$.

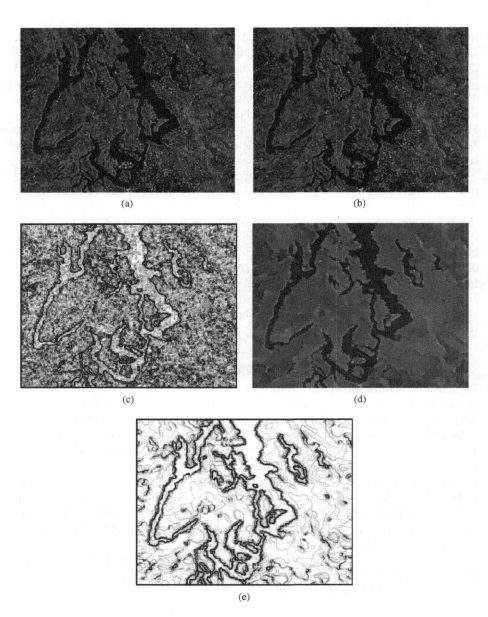

FIGURE 20.4

(a) SPOT multispectral image of the Seattle area, with additive Gaussian-distributed noise, $\sigma = 10$; (b) Vector distance dissimilarity diffusion result, using diffusion coefficient in (20.9); (c) Edges (gradient magnitude) from result in (b); (d) Mean curvature motion (20.18) result using diffusion coefficient from (20.11) and (20.12); (e) Edges (gradient magnitude) from result in (d).

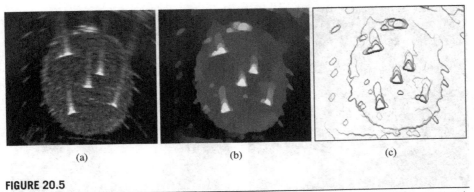

(a)   (b)   (c)

FIGURE 20.5

(a) An ultrasound image of a prostate phantom with implanted radioactive seeds; (b) corresponding SRAD-diffused image; (c) corresponding ICOV edge strength image [43].

The operator $q(\mathbf{x})$ is called the *instantaneous coefficient of variation* (ICOV). The ICOV uses the absolute value of the difference of a normalized gradient magnitude and a normalized Laplacian operator to measure the strength of edges in speckled imagery. The normalizing function $I(\mathbf{x}) + (1/4)\nabla^2 I(\mathbf{x})$ gives a smoothed version of the image at position $\mathbf{x}$. This term compensates for the edge measurement localization error. The ICOV allows for balanced and well-localized edge strength measurements in bright regions as well as in dark regions. It has been shown that the ICOV operator optimizes the edge detection in speckle imagery in terms of low false edge detection probability and high edge localization accuracy [43]. Figure 20.5 shows an example of an ultrasound image (Fig. 20.5(a)) that has been processed by SRAD (Fig. 20.5(b)) and where edges are displayed using the ICOV values (Fig. 20.5(c)).

Given an intensity image $\mathbf{I}$ having no zero-valued intensities over the image domain, the output image is evolved according to the following PDE:

$$\partial I_t(\mathbf{x})/\partial t = \mathrm{div}\big[c(q(\mathbf{x}))\,\nabla I_t(\mathbf{x})\big], \tag{20.38}$$

where ∇ is the gradient operator, div is the divergence operator, and $\|$ denotes the magnitude. The diffusion coefficient $c(q(\mathbf{x}))$ is given by

$$c\big(q(\mathbf{x})\big) = \left\{ 1 + \frac{\big[q^2(x) - \tilde{q}^2(x)\big]}{\tilde{q}^2(x)(1 + \tilde{q}^2(x))} \right\}^{-1}, \tag{20.39}$$

where $q(\mathbf{x})$ is the ICOV as determined by (20.37), and $\tilde{q}(\mathbf{x})$ is the current speckle noise level. Example input and output images from an ultrasound image of the human heart are shown in Fig. 20.6.

The diffusion coefficient $c(q(\mathbf{x}))$ is proportional to the likelihood that a point $\mathbf{x}$ is in a homogeneous speckle region at the update time. From (20.39), it is seen that the diffusion coefficient exhibits nearly zero values at edges with high contrast (i.e., $q(\mathbf{x}) \gg \tilde{q}(\mathbf{x})$), while in homogeneous speckle regions, the coefficient approaches unity. Hence, it is the

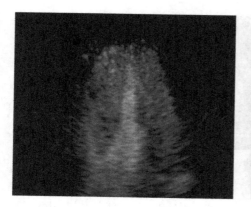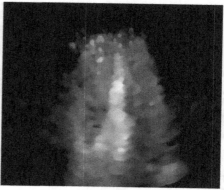

FIGURE 20.6

(Left) Speckled ultrasound image of left ventricle in a human heart prior to SRAD; (right) same image after SRAD enhancement.

diffusion coefficient that allows isotropic diffusion in homogeneous speckle regions and prohibits diffusion across the edges.

The implementation issues connected with anisotropic diffusion include specification of the diffusion coefficient and diffusion PDE, as discussed above. The anisotropic diffusion method can be expedited through multiresolution implementations. Furthermore, anisotropic diffusion can be extended to multispectral imagery and to ultrasound/radar imagery. In the following section, we discuss the specific application of anisotropic diffusion to edge detection.

20.4   APPLICATION OF ANISOTROPIC DIFFUSION TO EDGE DETECTION

20.4.1   Edge Detection by Thresholding

Once anisotropic diffusion has been applied to an image I, a procedure needs to be defined to extract the image edges $\mathbf{e}$. The most typical procedure is to simply define a gradient magnitude threshold, T, that defines the location of an edge. For example, $e(\mathbf{x}) = 1$ if $|\nabla I(x)| > T$ and $e(\mathbf{x}) = 0$ otherwise. Of course, the question becomes one of selecting a proper value for T. With typical diffusion coefficients such as (20.9) and (20.10), $T = k$ is often asserted. Another approach is to use the diffusion coefficient itself as the measure of edge strength: $e(\mathbf{x}) = 1$ if $c(x) < T$ and $e(\mathbf{x}) = 0$ otherwise.

$T = \sigma_e$ of the image, as defined in [44]: $\sigma_e = 1.4826 \, \text{med}\{|\nabla I(\mathbf{x}) - \text{med}(|\nabla I(\mathbf{x})|)|\}$ where the med operator is the median performed over the entire image domain Ω. The constant used (1.4826) is derived from the mean absolute deviation of the normal distribution with unit variance [17].

20.4.2 Edge Detection from Image Features

Aside from thresholding the gradient magnitude of a diffusion result, a feature detection approach may be used. As with Marr's classical LoG detector, the inflection points of a diffused image may be located by finding the zero-crossing in a Laplacian-convolved result. However, if the anisotropic diffusion operation produces piecewise constant images as in [10, 17], the gradient magnitude is sufficient to define thin, contiguous edges.

With LOMO diffusion, other features that appear in the diffused image may be used for edge detection. An advantage of LOMO diffusion is that no threshold is required for edge detection. LOMO diffusion segments each row and column of the image into ramp segments and constant segments. Within this framework, we can define *concave-down*, *concave-up*, and *ramp center* edge detection processes. Consider an image row or column. With a concave-down edge detection, the ascending (increasing intensity) segments mark the beginning of an object and the descending (decreasing intensity) segments terminate the object. With a concave-up edge detection, negative-going objects (in intensity) are detected. The ramp center edge detection sets the boundary points at the centers of the ramp edges, as the name implies. When no bias toward bright or dark objects is present, a ramp center edge detection can be utilized.

Figure 20.7 provides two examples of feature-based edge detection using LOMO diffusion. The images in Fig. 20.7(b) and (e) are the results of applying 2D LOMO diffusion to Fig. 20.7(a) and (d), respectively. The concave-up edge detection given in Fig. 20.7(c) reveals the boundaries of the blood cells. In Fig. 20.7(f), a ramp center edge detection is used to find the boundaries between the aluminum grains of Fig. 20.7(d).

20.4.3 Quantitative Evaluation of Edge Detection by Anisotropic Diffusion

When choosing a suitable anisotropic diffusion process for edge detection, one may evaluate the results qualitatively or use an objective measure. Three such quantitative assessment tools include the percentage of edges correctly identified as edges, the percentage of false edges, and Pratt's edge quality metric. Given ground truth edge information, usually with synthetic data, one may measure the correlation between the ideal edge map and the computed edge map. This correlation leads to a classification of "correct" edges (where the computed edge map and ideal version match) and "false" edges. Another method utilizes Pratt's edge quality measurement [45]:

$$F = \frac{\sum_{i=1}^{I_A} \frac{1}{1 + \alpha\left(\mathrm{d}(i)^2\right)}}{\max\{I_A, I_I\}}, \tag{20.40}$$

where I_A is the number of edge pixels detected in the diffused image result, I_I is the number of edge pixels existing in the original, noise free imagery, $\mathrm{d}(i)$ is the Euclidean

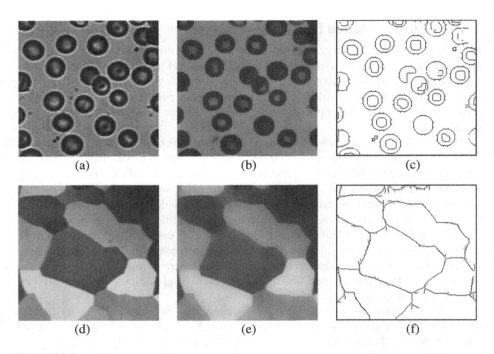

FIGURE 20.7

(a) Original "blood cells" image; (b) 2D LOMO-3 diffusion result; (c) Boundaries from concave-up edge detection of image in (b); (d) Original "Aluminum Grains" image; (e) 2D LOMO-3 diffusion result; (f) Boundaries from ramp center edge detection of image in (e).

distance between an edge location in the original image and the nearest detected edge, and α is a scaling constant (with suggested value of 1/9 [45]). A "perfect" edge detection result has value $F = 1$ in (20.40).

An example is given here where a synthetic image is corrupted by 40% salt and pepper noise (Fig. 20.8). Three versions of anisotropic diffusion are implemented on the noisy imagery using the diffusion coefficients from (20.9), from (20.11) and (20.12), and from (20.11) and (20.13). The threshold of the edge detector was defined to be equal to the gradient threshold of the diffusion coefficient, $T = k$. The results of the numerical experiment are presented in Fig. 20.9 for several solution times. It may be seen that the modified gradient coefficient [(20.11) and (20.12)] initially outperforms the other diffusion methods in the edge quality measurement, but produces the poorest identification percentage (due to the edge localization errors associated with the Gaussian filter). The morphological anisotropic diffusion method [(20.11) and (20.13)] provides significant performance improvement, providing a 70% identification of true edges and a Pratt quality measurement of 0.95.

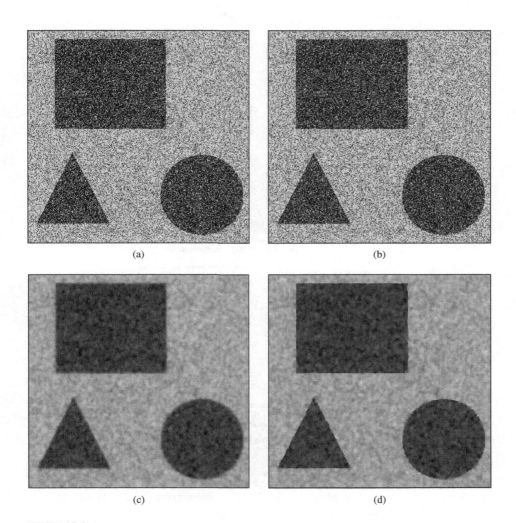

FIGURE 20.8

Three implementations of anisotropic diffusion applied to synthetic imagery: (a) original image corrupted with 40% salt and pepper noise; (b) results obtained using original anisotropic diffusion with (20.9); (c) results obtained using modified gradient anisotropic diffusion with (20.11) and (20.12); (d) results obtained using morphological anisotropic diffusion with (20.11) and (20.13) [13].

In summary, edges may be extracted from a diffused image by applying a heuristically selected threshold, by using a statistically motivated threshold, or by identifying features in the processed imagery. The success of the edge detection method can be evaluated qualitatively by visual inspection or quantitatively with edge quality metrics.

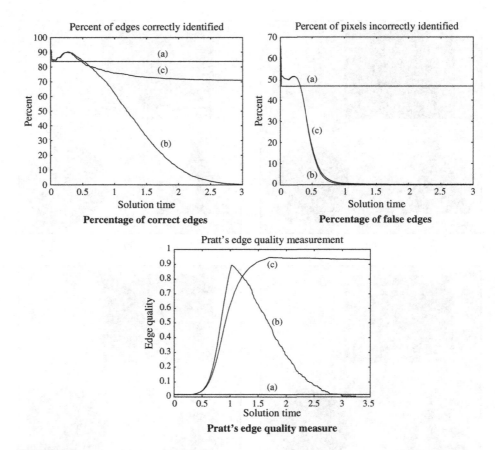

FIGURE 20.9

Edge detector performance versus diffusion time for the results shown in Fig. 20.8. In the graphs, (a) corresponds to anisotropic diffusion with (20.9); (b) corresponds to (20.11) and (20.12); (c) corresponds to (20.11) and (20.13) [13].

20.5 USING VECTOR DIFFUSION AND PARAMETRIC ACTIVE CONTOURS TO LOCATE EDGES

In this section, we discuss diffusion methods that drive a parametric active contour toward the boundary of a desired object. Instead of the diffusion of intensities or of gradient magnitude, we utilize here the diffusion of vectors that point toward strong edges.

20.5.1 Parametric Active Contours

Active contours (or snakes) may be used to detect the edges forming an object boundary, given an initial guess (i.e., an initial contour). A parametric active contour is simply a set of contour points $(X(s), Y(s))$ parameterized by $s \in [0, 1]$. Typically, parametric active contours are implemented by finding the contour that minimizes $E = E_{\text{internal}} + E_{\text{external}}$, where E_{internal} is the internal energy of the active contour that quantifies the contour smoothness [46]:

$$E_{\text{internal}} = \frac{1}{2} \int_0^1 \left\{ \alpha \left[\left(\frac{dX(s)}{ds} \right)^2 + \left(\frac{dY(s)}{ds} \right)^2 \right] + \beta \left[\left(\frac{d^2X(s)}{ds^2} \right)^2 + \left(\frac{d^2Y(s)}{ds^2} \right)^2 \right] \right\} ds.$$

$$(20.41)$$

Here, α and β are two nonnegative weighting parameters expressing, respectively, the degree of the resistance to stretching and bending of the contour. The external energy E_{external} is typically defined such that the contour seeks the edges in the image, I:

$$E_{\text{external}} = - \int_0^1 f(X(s), Y(s)) ds, \quad \text{where } f(x, y) = |\nabla I(x, y)|^2. \quad (20.42)$$

To minimize the contour energy ($= E_{\text{external}} + E_{\text{internal}}$), the calculus of variations [47, 48] is applied to obtain the following Euler equations [46]:

$$-\alpha \frac{d^2X}{ds^2} + \beta \frac{d^4X}{ds^2} - \frac{\partial f}{\partial x} = 0, \quad -\alpha \frac{d^2Y}{ds^2} + \beta \frac{d^4Y}{ds^2} - \frac{\partial f}{\partial y} = 0. \quad (20.43)$$

To solve for the active contour positions such that (20.43) is satisfied, we can use PDEs for which $(X(s), Y(s))$ are treated as a function of time as well:

$$\frac{\partial X}{\partial t} = \alpha \frac{\partial^2 X}{\partial s^2} - \beta \frac{\partial^4 X}{\partial s^2} + u, \quad \frac{\partial Y}{\partial t} = \alpha \frac{\partial^2 Y}{\partial s^2} - \beta \frac{\partial^4 Y}{\partial s^2} + v. \quad (20.44)$$

Here the external forces on the contour, $\frac{\partial f}{\partial x}$ and $\frac{\partial f}{\partial y}$, have been replaced by a force vector (u, v). It is this vector that "points" to the desired edge. Figure 20.10(b) shows the external forces generated by gradient vector flow (GVF) for the circular target boundary shown in Fig. 20.10(a).

20.5.2 Gradient Vector Flow

Diffusion can be used to capture an edge that is distant from the initial active contour. In this situation, the active contour is not driven toward the edge using (20.42), because the contour is not in "contact" with the gradient from the edge. To alleviate this problem, Xu and Prince construct a force field by diffusing the external force (u, v) away from edges into the homogeneous regions, while at the same time retaining the initial external force at the boundaries. This vector diffusion is achieved by minimizing [49]

$$E_{\text{GVF}}(u, v) = \frac{1}{2} \int \int \mu (u_x^2 + u_y^2 + v_x^2 + v_y^2) + (f_x^2 + f_y^2)((u - f_x)^2 + (v - f_y)^2) dx dy, \quad (20.45)$$

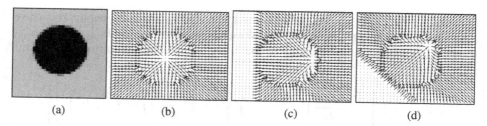

(a)	(b)	(c)	(d)

FIGURE 20.10

(a) A circle; (b) the external forces on the active contour after vector diffusion by GVF; (c) the external forces on the active contour after vector diffusion by MGVF with $(v^x, v^y) = (-1, 0)$; (d) the external forces on the active contour after vector diffusion by MGVF with $(v^x, v^y) = (-1, -1)$.

where μ is a nonnegative parameter expressing the degree of smoothness of the field (u, v) (and can be replaced by an edge-sensitive diffusion coefficient—see [50]), and f is the edge-map as defined in (20.42). The interpretation of (20.45) is straightforward—the first term keeps the vector field, (u, v), smooth, while the second term forces it to be close to the external forces near the edges (*i.e.*, where the edge-force strength is high). Variational minimization of (20.19) results in the following two Euler equations [49]:

$$\mu \nabla^2 u - (f_x^2 + f_y^2)(u - f_x) = 0, \quad \mu \nabla^2 v - (f_x^2 + f_y^2)(v - f_y) = 0. \quad (20.46)$$

Solving (20.46) for (u, v) results in a vector diffusion process called GVF. The resulting (u, v) vectors can be used as the external force in (20.44). Figures 20.10(c) and (d) show the external forces for a circular object for two directions of motion.

20.5.3 Motion Gradient Vector Flow

GVF can be modified to track a moving object boundary in a video sequence. Here we assume that the direction of object motion (v^x, v^y) is known. Instead of computing the components of external force (u and v) separately, we utilize a vector ∇w, which is the gradient of a computed quantity w (at each point in the image). To obtain w, we minimize

$$E_{\text{MGVF}}(w) = \frac{1}{2} \int\int \left[\mu H_\varepsilon(\nabla w \cdot (v^x, v^y)) |\nabla w|^2 + f(w - f)^2 \right] dx dy, \quad (20.47)$$

where H_ε is a regularized Heaviside (step) function, f is the squared image gradient magnitude as defined in (20.42), and μ is a weight on smoothness of the vector field. The first term in (20.47) encourages diffusion of motion gradient vectors in the direction of flow, and discourages diffusion in the opposite direction. The second term forces the magnitude of w to resemble that of f, the edge strength, where f is high in magnitude.

A gradient descent update for w is derived by way of variational calculus:

$$\frac{\partial w}{\partial \tau} = \mu \text{div}(H_\varepsilon(\nabla w \cdot (v^x, v^y)) \nabla w) - f(w - f). \quad (20.48)$$

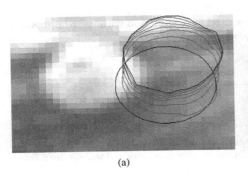 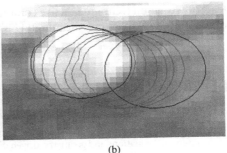

(a)   (b)

FIGURE 20.11

(a) In tracking a white blood cell, the GVF vector diffusion fails to attract the active contour; (b) successful detection is yielded by MGVF.

Thus (20.48) provides an external force that can guide an active contour to a moving object boundary. The capture range of GVF is increased using the motion gradient vector flow (MGVF) vector diffusion [51]. With MGVF, a tracking algorithm can simply use the final position of the active contour from a previous video frame as the initial contour in the subsequent frame. For an example of tracking using MGVF, see Fig. 20.11.

20.6 CONCLUSIONS

Anisotropic diffusion is an effective precursor to edge detection. The main benefit of anisotropic diffusion over isotropic diffusion and linear filtering is edge preservation. By properly specifying the diffusion PDE and the diffusion coefficient, an image can be scaled, denoised, and simplified for boundary detection. For edge detection, the most critical design step is specification of the diffusion coefficient. The variants of the diffusion coefficient involve tradeoffs between sensitivity to noise, the ability to specify scale, convergence issues, and computational cost. The diverse implementations of the anisotropic diffusion PDE result in improved fidelity to the original image, mean curvature motion, and convergence to LOMO signals. As the diffusion PDE may be considered a descent on an energy surface, the diffusion operation can be viewed in a variational framework. Recent variational solutions produce optimized edge maps and image segmentations in which certain edge-based features, such as edge length, curvature, thickness, and connectivity, can be optimized.

The computational cost of anisotropic diffusion may be reduced by using multiresolution solutions, including the anisotropic diffusion pyramid and multigrid anisotropic diffusion. Application of edge detection to multispectral imagery and to radar/ultrasound imagery is possible through techniques presented in the literature. In general, the edge detection step after anisotropic diffusion of the image is straightforward. Edges may be detected using a simple gradient magnitude threshold, using robust statistics, or using a feature extraction technique. Active contours, used in conjunction with vector diffusion, can be employed to extract meaningful object boundaries.

REFERENCES

[1] D. G. Lowe. *Perceptual Organization and Visual Recognition*. Kluwer Academic, New York, 1985.

[2] V. Caselles, J.-M. Morel, G. Sapiro, and A. Tannenbaum. Introduction to the special issue on partial differential equations and geometry-driven diffusion in image processing and analysis. *IEEE Trans. Image Process.*, 7:269–273, 1998.

[3] A. P. Witkin. Scale-space filtering. In *Proc. Int. Joint Conf. Art. Intell.*, 1019–1021, 1983.

[4] J. J. Koenderink. The structure of images. *Biol. Cybern.*, 50:363–370, 1984.

[5] D. Marr and E. Hildreth. Theory of edge detection. *Proc. R. Soc. Lond. B, Biol. Sci.*, 207:187–217, 1980.

[6] P. Perona and J. Malik. Scale-space and edge detection using anisotropic diffusion. *IEEE Trans. Pattern Anal. Mach. Intell.*, PAMI-12:629–639, 1990.

[7] S. Teboul, L. Blanc-Feraud, G. Aubert, and M. Barlaud. Variational approach for edge-preserving regularization using coupled PDE's. *IEEE Trans. Image Process.*, 7:387–397, 1998.

[8] R. T. Whitaker and S. M. Pizer. A multi-scale approach to nonuniform diffusion. *Comput. Vis. Graph. Image Process.—Image Underst.*, 57:99–110, 1993.

[9] Y.-L. You, M. Kaveh, W. Xu, and A. Tannenbaum. Analysis and design of anisotropic diffusion for image processing. In *Proc. IEEE Int. Conf. Image Process.*, Austin, Texas, November 13–16, 1994.

[10] Y.-L. You, W. Xu, A. Tannenbaum, and M. Kaveh. Behavioral analysis of anisotropic diffusion in image processing. *IEEE Trans. Image Process.*, 5:1539–1553, 1996.

[11] F. Catte, P.-L. Lions, J.-M. Morel, and T. Coll. Image selective smoothing and edge detection by nonlinear diffusion. *SIAM J. Numer. Anal.*, 29:182–193, 1992.

[12] L. Alvarez, P.-L. Lions, and J.-M. Morel. Image selective smoothing and edge detection by nonlinear diffusion II. *SIAM J. Numer. Anal.*, 29:845–866, 1992.

[13] C. A. Segall and S. T. Acton. Morphological anisotropic diffusion. In *Proc. IEEE Int. Conf. Image Process.*, Santa Barbara, CA, October 26–29, 1997.

[14] L.-I. Rudin, S. Osher, and E. Fatemi. Nonlinear total variation noise removal algorithm. *Physica D*, 60:1217–1229, 1992.

[15] S. Osher and L.-I. Rudin. Feature-oriented image enhancement using shock filters. *SIAM J. Numer. Anal.*, 27:919–940, 1990.

[16] S. T. Acton. Locally monotonic diffusion. *IEEE Trans. Signal. Process.*, 48:1379–1389, 2000.

[17] M. J. Black, G. Sapiro, D. H. Marimont, and D. Heeger. Robust anisotropic diffusion. *IEEE Trans. Image. Process.*, 7:421–432, 1998.

[18] K. N. Nordstrom. Biased anisotropic diffusion—a unified approach to edge detection. Tech. Report, Dept. of Electrical Engineering and Computer Sciences, University of California at Berkeley, Berkeley, CA, 1989.

[19] J. Canny. A computational approach to edge detection. *IEEE Trans. Pattern Anal. Mach. Intell.*, PAMI-8:679–714, 1986.

[20] A. El-Fallah and G. Ford. The evolution of mean curvature in image filtering. In *Proc. IEEE Int. Conf. Image Process.*, Austin, Texas, November 1994.

[21] S. Osher and J. Sethian. Fronts propagating with curvature dependent speed: algorithms based on the Hamilton-Jacobi formulation. *J. Comp. Phys.*, 79:12–49, 1988.

[22] N. Sochen, R. Kimmel, and R. Malladi. A general framework for low level vision. *IEEE Trans. Image Process.*, 7:310–318, 1998.

[23] A. Yezzi, Jr. Modified curvature motion for image smoothing and enhancement. *IEEE Trans. Image Process.*, 7:345–352, 1998.

[24] J.-L. Morel and S. Solimini. *Variational Methods in Image Segmentation*. Birkhauser, Boston, MA, 1995.

[25] D. Mumford and J. Shah. Boundary detection by minimizing functionals. In *IEEE Int. Conf. Comput. Vis. Pattern Recognit.*, San Francisco, 1985.

[26] S. T. Acton and A. C. Bovik. Anisotropic edge detection using mean field annealing. In *Proc. IEEE Int. Conf. Acoust., Speech and Signal Process. (ICASSP-92)*, San Francisco, March 23–26, 1992.

[27] D. Geman and G. Reynolds. Constrained restoration and the recovery of discontinuities. *IEEE Trans. Pattern Anal. Mach. Intell.*, 14:376–383, 1992.

[28] P. J. Burt, T. Hong, and A. Rosenfeld. Segmentation and estimation of region properties through cooperative hierarchical computation. *IEEE Trans. Syst. Man Cybern.*, 11(12):1981.

[29] P. J. Burt. Smart sensing within a pyramid vision machine. *Proc. IEEE*, 76(8):1006–1015, 1988.

[30] S. T. Acton. A pyramidal edge detector based on anisotropic diffusion. In *Proc. of the IEEE Int. Conf. Acoust., Speech and Signal Process. (ICASSP-96)*, Atlanta, May 7–10, 1996.

[31] S. T. Acton, A. C. Bovik, and M. M. Crawford. Anisotropic diffusion pyramids for image segmentation. In *Proc. IEEE Int. Conf. Image Process.*, Austin, Texas, November 1994.

[32] A. Morales, R. Acharya, and S. Ko. Morphological pyramids with alternating sequential filters. *IEEE Trans. Image Process.*, 4(7):965–977, 1996.

[33] C. A. Segall, S. T. Acton, and A. K. Katsaggelos. Sampling conditions for anisotropic diffusion. In *Proc. SPIE Symp. Vis. Commun. Image Process.*, San Jose, January 23–29, 1999.

[34] R. M. Haralick, X. Zhuang, C. Lin, and J. S. J. Lee. The digital morphological sampling theorem. *IEEE Trans. Acoust.*, 3720(12):2067–2090, 1989.

[35] S. T. Acton. Multigrid anisotropic diffusion. *IEEE Trans. Image. Process.*, 7:280–291, 1998.

[36] J. H. Bramble. *Multigrid Methods*. John Wiley, New York, 1993.

[37] W. Hackbush and U. Trottenberg, editors. *Multigrid Methods*. Springer-Verlag, New York, 1982.

[38] R. T. Whitaker and G. Gerig. Vector-valued diffusion. In B. ter Haar Romeny, editor, *Geometry-Driven Diffusion in Computer Vision*, 93–134. Kluwer, 1994.

[39] S. T. Acton and J. Landis. Multispectral anisotropic diffusion. *Int. J. Remote Sens.*, 18:2877–2886, 1997.

[40] G. Sapiro and D. L. Ringach. Anisotropic diffusion of multivalued images with applications to color filtering. *IEEE Trans. Image Process.*, 5:1582–1586, 1996.

[41] S. DiZenzo. A note on the gradient of a multi-image. *Comput. Vis. Graph. Image Process.*, 33:116–125, 1986.

[42] Y. Yu and S. T. Acton. Speckle reducing anisotropic diffusion. *IEEE Trans. Image Process.*, 11:1260–1270, 2002.

[43] Y. Yu and S. T. Acton. Edge detection in ultrasound imagery using the instantaneous coefficient of variation. *IEEE Trans. Image Process.*, 13(12):1640–1655, 2004.

[44] P. J. Rousseeuw and A. M. Leroy. *Robust Regression and Outlier Detection.* Wiley, New York, 1987.

[45] W. K. Pratt. *Digital Image Processing.* Wiley, New York, 495–501, 1978.

[46] M. Kass, A. Witkin, and D. Terzopoulos. Snakes: active contour models. *Int. J. Comput. Vis.*, 1(4):321–331, 1987.

[47] R. Courant and D. Hilbert. *Methods of Mathematical Physics*, Vol. 1. Interscience Publishers Inc., New York, 1953.

[48] J. L. Troutman. *Variational Calculus with Elementary Convexity.* Springer-Verlag, New York, 1983.

[49] C. Xu and J. L. Prince. Snakes, shapes, and gradient vector flow. *IEEE Trans. Image Process.*, 7: 359–369, 1998.

[50] C. Xu and J. L. Prince. Generalized gradient vector flow external force for active contours. *Signal Processing*, 71:131–139, 1998.

[51] N. Ray and S. T. Acton. Tracking rolling leukocytes with motion gradient vector flow. In *Proc. 37th Asilomar Conf. on Signals, Systems and Computers*, Pacific Grove, California, November 9–12, 2003.

Image Quality Assessment

21

Kalpana Seshadrinathan[1], Thrasyvoulos N. Pappas[2], Robert J. Safranek[3], Junqing Chen[4], Zhou Wang[5], Hamid R. Sheikh[6], and Alan C. Bovik[7]

[1] *The University of Texas at Austin;* [2] *Northwestern University;* [3] *Benevue, Inc.;* [4] *Northwestern University;* [5] *University of Waterloo;* [6] *Texas Instruments, Inc.;* [7] *The University of Texas at Austin*

21.1 INTRODUCTION

Recent advances in digital imaging technology, computational speed, storage capacity, and networking have resulted in the proliferation of digital images, both still and video. As the digital images are captured, stored, transmitted, and displayed in different devices, there is a need to maintain image quality. The end users of these images, in an overwhelmingly large number of applications, are human observers. In this chapter, we examine objective criteria for the evaluation of image quality as perceived by an average human observer. Even though we use the term image quality, we are primarily interested in image fidelity, i.e., how close an image is to a given original or reference image. This paradigm of image quality assessment (QA) is also known as full reference image QA. The development of objective metrics for evaluating image quality without a reference image is quite different and is outside the scope of this chapter.

Image QA plays a fundamental role in the design and evaluation of imaging and image processing systems. As an example, QA algorithms can be used to systematically evaluate the performance of different image compression algorithms that attempt to minimize the number of bits required to store an image, while maintaining sufficiently high image quality. Similarly, QA algorithms can be used to evaluate image acquisition and display systems. Communication networks have developed tremendously over the past decade, and images and video are frequently transported over optic fiber, packet switched networks like the Internet, wireless systems, etc. Bandwidth efficiency of applications such as video conferencing and Video on Demand can be improved using QA systems to evaluate the effects of channel errors on the transported images and video. Further, QA algorithms can be used in "perceptually optimal" design of various components of an image communication system. Finally, QA and the psychophysics of human vision are closely related disciplines. Research on image and video QA may lend deep

553

insights into the functioning of the human visual system (HVS), which would be of great scientific value.

Subjective evaluations are accepted to be the most effective and reliable, albeit quite cumbersome and expensive, way to assess image quality. A significant effort has been dedicated for the development of subjective tests for image quality [56, 57]. There has also been standards activity on subjective evaluation of image quality [58]. The study of the topic of subjective evaluation of image quality is beyond the scope of this chapter.

The goal of an objective perceptual metric for image quality is to determine the differences between two images that are visible to the HVS. Usually one of the images is the reference which is considered to be "original," "perfect," or "uncorrupted." The second image has been modified or distorted in some sense. The output of the QA algorithm is often a number that represents the probability that a human eye can detect a difference in the two images or a number that quantifies the perceptual dissimilarity between the two images. Alternatively, the output of an image quality metric could be a map of detection probabilities or perceptual dissimilarity values.

Perhaps the earliest image quality metrics were the mean squared error (MSE) and peak signal-to-noise ratio (PSNR) between the reference and distorted images. These metrics are still widely used for performance evaluation, despite their well-known limitations, due to their simplicity. Let $f(\mathbf{n})$ and $g(\mathbf{n})$ represent the value (intensity) of an image pixel at location $\mathbf{n}$. Usually the image pixels are arranged in a Cartesian grid and $\mathbf{n} = (n_1, n_2)$. The MSE between $f(\mathbf{n})$ and $g(\mathbf{n})$ is defined as

$$\mathrm{MSE}\big[f(\mathbf{n}), g(\mathbf{n})\big] = \frac{1}{N} \sum_{\mathbf{n}} \big[f(\mathbf{n}) - g(\mathbf{n})\big]^2, \tag{21.1}$$

where N is the total number of pixel locations in $f(\mathbf{n})$ or $g(\mathbf{n})$. The PSNR between these image patches is defined as

$$\mathrm{PSNR}\big[f(\mathbf{n}), g(\mathbf{n})\big] = 10 \log_{10} \frac{E^2}{\mathrm{MSE}\big[f(\mathbf{n}), g(\mathbf{n})\big]}, \tag{21.2}$$

where E is the maximum value that a pixel can take. For example, for 8-bit grayscale images, $E = 255$.

In Fig. 21.1, we show two distorted images generated from the same original image. The first distorted image (Fig. 21.1(b)) was obtained by adding a constant number to all signal samples. The second distorted image (Fig. 21.1(c)) was generated using the same method except that the signs of the constant were randomly chosen to be positive or negative. It can be easily shown that the MSE/PSNR between the original image and both of the distorted images are exactly the same. However, the visual quality of the two distorted images is drastically different. Another example is shown in Fig. 21.2, where Fig. 21.2(b) was generated by adding independent white Gaussian noise to the original texture image in Fig. 21.2(a). In Fig. 21.2(c), the signal sample values remained the same as in Fig. 21.2(a), but the spatial ordering of the samples has been changed (through a sorting procedure). Figure 21.2(d) was obtained from Fig. 21.2(b), by following the same reordering procedure used to create Fig. 21.2(c). Again, the MSE/PSNR between

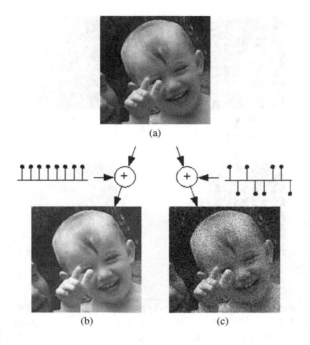

(a)

(b)   (c)

FIGURE 21.1

Failure of the Minkowski metric for image quality prediction. (a) original image; (b) distorted image by adding a positive constant; (c) distorted image by adding the same constant, but with random sign. Images (b) and (c) have the same Minkowski metric with respect to image (a), but drastically different visual quality.

Figs. 21.2(a) and 21.2(b) and Figs. 21.2(c) and 21.2(d) is exactly the same. However, Fig. 21.2(d) appears to be significantly noisier than Fig. 21.2(b).

The above examples clearly illustrate the failure of PSNR as an adequate measure of visual quality. In this chapter, we will discuss three classes of image QA algorithms that correlate with visual perception significantly better—human vision based metrics, Structural SIMilarity (SSIM) metrics, and information theoretic metrics. Each of these techniques approaches the image QA problem from a different perspective and using different first principles. As we proceed in this chapter, in addition to discussing these QA techniques, we will also attempt to shed light on the similarities, dissimilarities, and interplay between these seemingly diverse techniques.

21.2 HUMAN VISION MODELING BASED METRICS

Human vision modeling based metrics utilize mathematical models of certain stages of processing that occur in the visual systems of humans to construct a quality metric. Most HVS-based methods take an engineering approach to solving the QA problem by

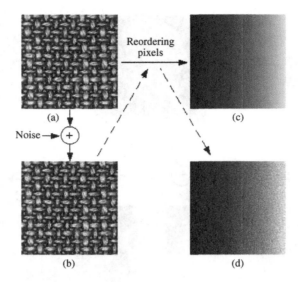

FIGURE 21.2

Failure of the Minkowski metric for image quality prediction. (a) original texture image; (b) distorted image by adding independent white Gaussian noise; (c) reordering of the pixels in image (a) (by sorting pixel intensity values); (d) reordering of the pixels in image (b), by following the same reordering used to create image (c). The Minkowski metrics between images (a) and (b) and images (c) and (d) are the same, but image (d) appears much noisier than image (b).

measuring the threshold of visibility of signals and noise in the signals. These thresholds are then utilized to normalize the error between the reference and distorted images to obtain a perceptually meaningful error metric. To measure visibility thresholds, different aspects of visual processing need to be taken into consideration such as response to average brightness, contrast, spatial frequencies, orientations, etc. Other HVS-based methods attempt to directly model the different stages of processing that occur in the HVS that results in the observed visibility thresholds. In Section 21.2.1, we will discuss the individual building blocks that comprise a HVS-based QA system. The function of these blocks is to model concepts from the psychophysics of human perception that apply to image quality metrics. In Section 21.2.2, we will discuss the details of several well-known HVS-based QA systems. Each of these QA systems is comprised of some or all of the building blocks discussed in Section 21.2.1, but uses different mathematical models for each block.

21.2.1 Building Blocks

21.2.1.1 Preprocessing

Most QA algorithms include a preprocessing stage that typically comprises of calibration and registration. The array of numbers that represents an image is often mapped to units of visual frequencies or cycles per degree of visual angle, and the calibration stage receives input parameters such as viewing distance and physical pixel spacings (screen resolution) to perform this mapping. Other calibration parameters may include fixation depth and eccentricity of the images in the observer's visual field [37, 38]. Display calibration or an accurate model of the display device is an essential part of any image quality metric [55], as the HVS can only see what the display can reproduce. Many quality metrics require that the input image values be converted to physical luminances[1] before they enter the HVS model. In some cases, when the perceptual model is obtained empirically, the effects of the display are incorporated in the model [40]. The obvious disadvantage of this approach is that when the display changes, a new set of model parameters must be obtained [43]. The study of display models is beyond the scope of this chapter.

Registration, i.e., establishing point-by-point correspondence between two images, is also necessary in most image QA systems. Often times, the performance of a QA model can be extremely sensitive to registration errors since many QA systems operate pixel by pixel (e.g., PSNR) or on local neighborhoods of pixels. Errors in registration would result in a shift in the pixel or coefficient values being compared and degrade the performance of the system.

21.2.1.2 *Frequency Analysis*

The frequency analysis stage decomposes the reference and test images into different channels (usually called subbands) with different spatial frequencies and orientations using a set of linear filters. In many QA models, this stage is intended to mimic similar processing that occurs in the HVS: neurons in the visual cortex respond selectively to stimuli with particular spatial frequencies and orientations. Other QA models that target specific image coders utilize the same decomposition as the compression system and model the thresholds of visibility for each of the channels. Some examples of such decompositions are shown in Fig. 21.3. The range of each axis is from $-u_s/2$ to $u_s/2$ cycles per degree, where u_s is the sampling frequency. Figures 21.3(a)–(c) show transforms that are polar separable and belong to the former category of decompositions (mimicking processing in the visual cortex). Figures 21.3(d)–(f) are used in QA models in the latter category and depict transforms that are often used in compression systems.

In the remainder of this chapter, we will use $f(\mathbf{n})$ to denote the value (intensity, grayscale, etc.) of an image pixel at location $\mathbf{n}$. Usually the image pixels are arranged in a Cartesian grid and $\mathbf{n} = (n_1, n_2)$. The value of the kth image subband at location $\mathbf{n}$ will be denoted by $b(\mathbf{k}, \mathbf{n})$. The subband indexing $\mathbf{k} = (k_1, k_2)$ could be in Cartesian or polar or even scalar coordinates. The same notation will be used to denote the kth coefficient of the nth discrete cosine transform (DCT) block (both Cartesian coordinate systems). This notation underscores the similarity between the two transformations,

---

[1] In video practice, the term luminance is sometimes, incorrectly, used to denote a nonlinear transformation of luminance [75, p. 24].

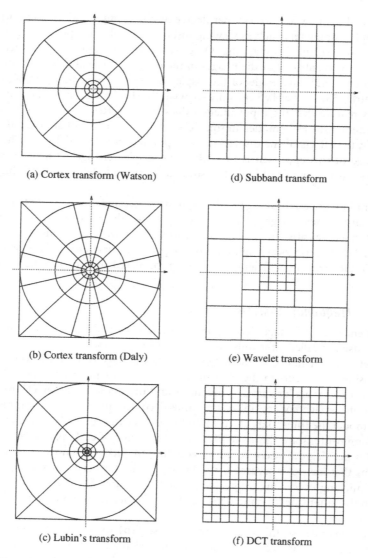

(a) Cortex transform (Watson)

(b) Cortex transform (Daly)

(c) Lubin's transform

(d) Subband transform

(e) Wavelet transform

(f) DCT transform

FIGURE 21.3

The decomposition of the frequency plane corresponding to various transforms. The range of each axis is from $-u_s/2$ to $u_s/2$ cycles per degree, where u_s is the sampling frequency.

even though we traditionally display the subband decomposition as a collection of subbands and the DCT as a collection of block transforms: a regrouping of coefficients in the blocks of the DCT results in a representation very similar to a subband decomposition.

21.2.1.3 *Contrast Sensitivity*

The HVS's contrast sensitivity function (CSF, also called the modulation transfer function) provides a characterization of its frequency response. The CSF can be thought of as a bandpass filter. There have been several different classes of experiments used to determine its characteristics which are described in detail in [59, Chapter 12].

One of these methods involves the measurement of visibility thresholds of sine-wave gratings. For a fixed frequency, a set of stimuli consisting of sine waves of varying amplitudes are constructed. These stimuli are presented to an observer, and the detection threshold for that frequency is determined. This procedure is repeated for a large number of grating frequencies. The resulting curve is called the CSF and is illustrated in Fig. 21.4. Note that these experiments used sine-wave gratings at a single orientation. To fully characterize the CSF, the experiments would need to be repeated with gratings at various orientations. This has been accomplished and the results show that the HVS is not perfectly isotropic. However, for the purposes of QA, it is close enough to isotropic that this assumption is normally used.

It should also be noted that the spatial frequencies are in units of cycles per degree of visual angle. This implies that the visibility of details at a particular frequency is a function of viewing distance. As an observer moves away from an image, a fixed size feature in the image takes up fewer degrees of visual angle. This action moves it to the right on the contrast sensitivity curve, possibly requiring it to have greater contrast to remain visible. On the other hand, moving closer to an image can allow previously imperceivable details to rise above the visibility threshold. Given these observations, it is clear that the minimum viewing distance is where distortion is maximally detectable. Therefore, quality metrics often specify a minimum viewing distance and evaluate the distortion metric at that point. Several "standard" minimum viewing distances have been established

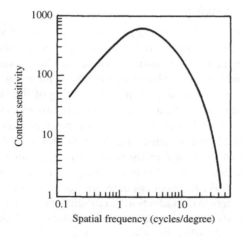

FIGURE 21.4

Spatial contrast sensitivity function (reprinted with permission from reference [63], p. 269).

for subjective quality measurement and have generally been used with objective models as well. These are six times image height for standard definition television and three times image height for high definition television.

The baseline contrast sensitivity determines the amount of energy in each subband that is required in order to detect the target in a (arbitrary or) flat mid-gray image. This is sometimes referred to as the just noticeable difference (JND). We will use $t_b(\mathbf{k})$ to denote the baseline sensitivity of the $\mathbf{k}$th band or DCT coefficient. Note that the base sensitivity is independent of the location $\mathbf{n}$.

21.2.1.4 *Luminance Masking*

It is well known that the perception of lightness is a nonlinear function of luminance. Some authors call this "light adaptation." Others prefer the term "luminance masking," which groups it together with the other types of masking we will see below [41]. It is called masking because the luminance of the original image signal masks the variations in the distorted signal.

Consider the following experiment: create a series of images consisting of a background of uniform intensity, I, each with a square of a different intensity, $I + \delta I$, inserted into its center. Show these to an observer in order of increasing δI. Ask the observer to determine the point at which she can first detect the square. Then, repeat this experiment for a large number of different values of background intensity. For a wide range of background intensities, the ratio of the threshold value δI divided by I is a constant. This equation

$$\frac{\delta I}{I} = k \tag{21.3}$$

is called *Weber's Law*. The value for k is roughly 0.33.

21.2.1.5 *Contrast Masking*

We have dealt with stimuli that are either constant or contain a single frequency in describing the luminance masking and contrast sensitivity properties of the visual system. In general, this is not characteristic of natural scenes. They have a wide range of frequency content over many different scales. Also, since the HVS is not a linear system, the CSF or frequency response does not characterize the functioning of the HVS for any arbitrary input. Study the following thought experiment: consider two images, a constant intensity field and an image of a sand beach. Take a random noise process whose variance just exceeds the amplitude and contrast sensitivity thresholds for the flat field image. Add this noise field to both images. By definition, the noise will be detectable in the flat field image. However, it will not be detectable in the beach image. The presence of the multitude of frequency components in the beach image hides or *masks* the presence of the noise field.

Contrast masking refers to the reduction in visibility of one image component caused by the presence of another image component with similar spatial location and frequency content. As we mentioned earlier, the visual cortex in the HVS can be thought of as a spatial frequency filter bank with octave spacing of subbands in radial frequency and angular bands of roughly 30 degree spacing. The presence of a signal component in one of these subbands will raise the detection threshold for other signal components in the same subband [64–66] or even neighboring subbands.

21.2.1.6 *Error Pooling*

The final step of an image quality metric is to combine the errors (at the output of the models for various psychophysical phenomena) that have been computed for each spatial frequency and orientation band and each spatial location, into a single number for each pixel of the image, or a single number for the whole image. Some metrics convert the JNDs to detection probabilities.

An example of error pooling is the following Minkowski metric:

$$E(\mathbf{n}) = \frac{1}{M} \left\{ \sum_{\mathbf{k}} \left| \frac{b(\mathbf{k},\mathbf{n}) - \hat{b}(\mathbf{k},\mathbf{n})}{t(\mathbf{k},\mathbf{n})} \right|^{Q} \right\}^{1/Q}, \tag{21.4}$$

where $b_{\mathbf{k}}(\mathbf{n})$ and $\hat{b}_{\mathbf{k}}(\mathbf{n})$ are the **n**th element of the **k**th subband of the original and coded image, respectively, $t(\mathbf{k},\mathbf{n})$ is the corresponding sensitivity threshold, and M is the total number of subbands. In this case, the errors are pooled across frequency to obtain a distortion measure for each spatial location. The value of Q varies from 2 (energy summation) to infinity (maximum error).

21.2.2 **HVS-Based Models**

In this section, we will discuss some well-known HVS modeling based QA systems. We will first discuss four general purpose QA models: the visible differences predictor (VDP), the Sarnoff JND vision model, the Teo and Heeger model, and visual signal-to-noise ratio (VSNR).

We will then discuss quality models that are designed specifically for different compression systems: the perceptual image coder (PIC) and Watson's DCT and wavelet-based metrics. While still based on the properties of the HVS, these models adopt the frequency decomposition of a given coder, which is chosen to provide high compression efficiency as well as computational efficiency. The block diagram of a generic perceptually based coder is shown in Fig. 21.5. The frequency analysis decomposes the image into several

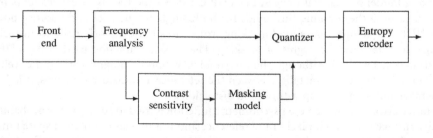

FIGURE 21.5

Perceptual coder.

components (subbands, wavelets, etc.) which are then quantized and entropy coded. The frequency analysis and entropy coding are virtually lossless; the only losses occur at the quantization step. The perceptual masking model is based on the frequency analysis and regulates the quantization parameters to minimize the visibility of the errors. The visual models can be incorporated in a compression scheme to minimize the visibility of the quantization errors, or they can be used independently to evaluate its performance. While coder-specific image quality metrics are quite effective in predicting the performance of the coder they are designed for, they may not be as effective in predicting performance across different coders [36, 83].

21.2.2.1 *Visible Differences Predictor*

The VDP is a model developed by Daly for the evaluation of high quality imaging systems [37]. It is one of the most general and elaborate image quality metrics in the literature. It accounts for variations in sensitivity due to light level, spatial frequency (CSF), and signal content (contrast masking).

To model luminance masking or amplitude nonlinearities in the HVS, Daly includes a simple point-by-point amplitude nonlinearity where the adaptation level for each image pixel is solely determined from that pixel (as opposed to using the average luminance in a neighborhood of the pixel). To account for contrast sensitivity, the VDP filters the image by the CSF before the frequency decomposition. Once this normalization is accomplished to account for the varying sensitivities of the HVS to different spatial frequencies, the thresholds derived in the contrast masking stage become the same for all frequencies.

A variation of the Cortex transform shown in Fig. 21.3(b) is used in the VDP for the frequency decomposition. Daly proposes two alternatives to convert the output of the linear filter bank to units of contrast: local contrast, which uses the value of the baseband at any given location to divide the values of all the other bands, and global contrast, which divides all subbands by the average value of the input image. The conversion to contrast is performed since to a first approximation the HVS produces a neural image of local contrast [35]. The masking stage in the VDP utilizes a "threshold elevation" approach, where a masking function is computed that measures the contrast threshold of a signal as a function of the background (masker) contrast. This function is computed for the case when the masker and signal are single, isolated frequencies. To obtain a masking model for natural images, the VDP considers the results of experiments that have measured the masking thresholds for both single frequencies and additive noise. The VDP also allows for *mutual* masking which uses both the original and distorted images to determine the degree of masking. The masking function used in the VDP is illustrated in Fig. 21.6. Although the threshold elevation paradigm works quite well in determining the discriminability between the reference and distorted images, it fails to generalize to the case of supra-threshold distortions.

In the error pooling stage, a psychometric function is used to compute the probability of discrimination at each pixel of the reference and test images to obtain a spatial map. Further details of this algorithm can be found in [37], along with an interesting discussion of different approaches used in the literature to model various stages of processing in the HVS, including their merits and drawbacks.

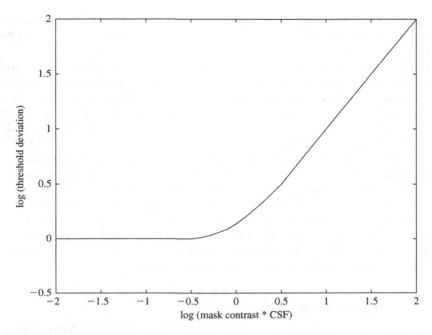

FIGURE 21.6

Contrast masking function.

21.2.2.2 *Sarnoff JND Vision Model*

The Sarnoff JND vision model received a technical Emmy award in 2000 and is one of the best known QA systems based on human vision models. This model was developed by Lubin and coworkers, and details of this algorithm can be found in [38].

Preprocessing steps in this model include calibration for distance of the observer from the images. In addition, this model also accounts for fixation depth and eccentricity of the observer's visual field. The human eye does not sample an image uniformly since the density of retinal cells drops off with eccentricity, resulting in a decreased spatial resolution as we move away from the point of fixation of the observer. To account for this effect, the Lubin model resamples the image to generate a modeled retinal image. The Laplacian pyramid of Burt and Adelson [77] is used to decompose the image into seven radial frequency bands. At this stage, the pyramid responses are converted to units of local contrast by dividing each point in each level of the Laplacian pyramid by the corresponding point obtained from the Gaussian pyramid two levels down in resolution. Each pyramid level is then convolved with eight spatially oriented filters of Freeman and Adelson [78], which constitute Hilbert transform pairs for four different orientations. The frequency decomposition so obtained is illustrated in Fig. 21.3(c). The two Hilbert transform pair outputs are squared and summed to obtain a local energy measure at each pixel location, pyramid level, and orientation. To account for the contrast sensitivity of human vision, these local energy measures are normalized by the base sensitivities for that position and pyramid level, where the base sensitivities are obtained from the CSF.

The Sarnoff model does not use the threshold elevation approach to model masking used by the VDP, instead adopting a transducer or a contrast gain control model. Gain control models a mechanism that allows a neuron in the HVS to adjust its response to the ambient contrast of the stimulus. Such a model generalizes better to the case of suprathreshold distortions since it models an underlying mechanism in the visual system, as opposed to measuring visibility thresholds. The transducer model used in [38] takes the form of a sigmoid nonlinearity. A sigmoid function starts out flat, its slope increases to a maximum, and then decreases back to zero, i.e., it changes curvature like the letter S.

Finally, a distance measure is calculated using a Minkowski error between the responses of the test and distorted images at the output of the vision model. A psychometric function is used to convert the distance measure to a probability value, and the Sarnoff JND vision model outputs a spatial map that represents the probability that an observer will be able to discriminate between the two input images (reference and distorted) based on the information in that spatial location.

21.2.2.3 *Teo and Heeger Model*

The Teo and Heeger metric uses the steerable pyramid transform [79] which decomposes the image into several spatial frequency and orientation bands [39]. A more detailed discussion of this model, with a different transform, can be found in [80]. However, unlike the other two models we saw above, it does not attempt to separate the contrast sensitivity and contrast masking effects. Instead, Teo and Heeger propose a *normalization model* that explains baseline contrast sensitivity, contrast masking, and masking that occurs when the orientations of the target and the masker are different. The normalization model has the following form:

$$R(\mathbf{k},\mathbf{n},i) = R(\rho,\theta,\mathbf{n},i) = \kappa_i \frac{[b(\rho,\theta,\mathbf{n})]^2}{\sum_\phi [b(\rho,\phi,\mathbf{n})]^2 + \sigma_i^2}, \tag{21.5}$$

where $R(\mathbf{k},\mathbf{n},i)$ is the normalized response of a sensor corresponding to the transform coefficient $b(\rho,\theta,\mathbf{n})$, $\mathbf{k} = (\rho,\theta)$ specifies the spatial frequency and orientation of the band, $\mathbf{n}$ specifies the location, and i specifies one of four different contrast discrimination bands characterized by different scaling and saturation constants, κ_i and σ_i^2, respectively. The scaling and saturation constants κ_i and σ_i^2 are chosen to fit the experimental data of Foley and Boynton. This model is also a contrast gain control model (similar to the Sarnoff JND vision model) that uses a divisive normalization model to explain masking effects. There is increasing evidence for divisive normalization mechanisms in the HVS, and this model can account for various aspects of contrast masking in human vision [18, 31–34, 80]. Finally, the quality of the image is computed at each pixel as the Minkowski error between the contrast masked responses to the two input images.

21.2.2.4 *Safranek-Johnston Perceptual Image Coder*

The Safranek-Johnston PIC image coder was one of the first image coders to incorporate an elaborate perceptual model [40]. It is calibrated for a given CRT display and viewing conditions (six times image height). The PIC coder has the basic structure shown in Fig. 21.5. It uses a separable generalized quadrature mirror filter (GQMF) bank for subband analysis/synthesis shown in Fig. 21.3(d). The baseband is coded with DPCM while all other subbands are coded with PCM. All subbands use uniform quantizers with sophisticated entropy coding. The perceptual model specifies the amount of noise that can be added to each subband of a given image so that the difference between the output image and the original is just noticeable.

The model contains the following components: the base sensitivity $t_b(\mathbf{k})$ determines the noise sensitivity in each subband given a flat mid-gray image and was obtained using subjective experiments. The results are listed in a table. The second component is a brightness adjustment denoted as $\tau_l(\mathbf{k}, \mathbf{n})$. In general this would be a two dimensional lookup table (for each subband and gray value). Safranek and Johnston made the reasonable simplification that the brightness adjustment is the same for all subbands. The final component is the texture masking adjustment. Safranek and Johnston [40] define as texture any deviation from a flat field within a subband and use the following texture masking adjustment:

$$\tau_t(\mathbf{k}, \mathbf{n}) = \max\left\{ 1, \left[\sum_{\mathbf{k}} w_{\mathrm{MTF}}(\mathbf{k}) e_t(\mathbf{k}, \mathbf{n}) \right]^{w_t} \right\}, \tag{21.6}$$

where $e_t(\mathbf{k}, \mathbf{n})$ is the "texture energy" of subband $\mathbf{k}$ at location $\mathbf{n}$, $w_{\mathrm{MTF}}(\mathbf{k})$ is a weighting factor for subband $\mathbf{k}$ determined empirically from the MTF of the HVS, and w_t is a constant equal to 0.15. The subband texture energy is given by

$$e_t(\mathbf{k}, \mathbf{n}) = \begin{cases} \text{local variance}_{\mathbf{m} \in \mathrm{N}(\mathbf{n})}(b(\mathbf{0}, \mathbf{m})), & \text{if } \mathbf{k} = \mathbf{0} \\ b(\mathbf{k}, \mathbf{n})^2, & \text{otherwise,} \end{cases} \tag{21.7}$$

where $\mathrm{N}(\mathbf{n})$ is the neighborhood of the point $\mathbf{n}$ over which the variance is calculated. In the Safranek-Johnston model, the overall sensitivity threshold is the product of three terms

$$t(\mathbf{k}, \mathbf{n}) = \tau_t(\mathbf{k}, \mathbf{n})\, \tau_l(\mathbf{k}, \mathbf{n})\, t_b(\mathbf{k}), \tag{21.8}$$

where $\tau_t(\mathbf{k}, \mathbf{n})$ is the texture masking adjustment, $\tau_l(\mathbf{k}, \mathbf{n})$ is the luminance masking adjustment, and $t_b(\mathbf{k})$ is the baseline sensitivity threshold.

A simple metric based on the PIC coder can be defined as follows:

$$E = \left\{ \frac{1}{N} \sum_{\mathbf{n}, \mathbf{k}} \left[\frac{b(\mathbf{k}, \mathbf{n}) - \hat{b}(\mathbf{k}, \mathbf{n})}{t(\mathbf{k}, \mathbf{n})} \right]^Q \right\}^{\frac{1}{Q}}, \tag{21.9}$$

where $b_{\mathbf{k}}(\mathbf{n})$ and $\hat{b}_{\mathbf{k}}(\mathbf{n})$ are the $\mathbf{n}$th element of the $\mathbf{k}$th subband of the original and coded image, respectively, $t(\mathbf{k}, \mathbf{n})$ is the corresponding perceptual threshold, and N is the

(a) Original 512 × 512 image   (b) SPIHT coder at 0.52 bits/pixel, PSNR = 33.3 dB

(c) PIC coder at 0.52 bits/pixel, PSNR = 29.4 dB   (d) JPEG coder at 0.52 bits/pixel, PSNR = 30.5 dB

FIGURE 21.7

Continued

number of pixels in the image. A typical value for Q is 2. If the error pooling is done over the subband index $\mathbf{k}$ only, as in (21.4), we obtain a spatial map of perceptually weighted errors. This map is downsampled by the number of subbands in each dimension. A full resolution map can also be obtained by doing the error pooling on the upsampled and filtered subbands.

Figures 21.7(a)–(g) demonstrate the performance of the PIC metric. Figure 21.7(a) shows an original 512 × 512 image. The grayscale resolution is 8 bits/pixel. Figure 21.7(b) shows the image coded with the SPIHT coder [84] at 0.52 bits/pixel; the PSNR is 33.3 dB. Figure 21.7(b) shows the same image coded with the PIC coder [40] at the same rate.

(e) PIC metric for SPIHT coder, perceptual PSNR = 46.8 dB   (f) PIC metric for PIC coder, perceptual PSNR = 49.5 dB

(g) PIC metric for JPEG coder, perceptual PSNR = 47.9 dB

FIGURE 21.7

The PSNR is considerably lower at 29.4 dB. This is not surprising as the SPIHT algorithm is designed to minimize the MSE and has no perceptual weighting. The PIC coder assumes a viewing distance of six image heights or 21 inches. Depending on the quality of reproduction (which is not known at the time this chapter is written), at a close viewing distance, the reader may see ringing near the edges of the PIC image. On the other hand, the SPIHT image has considerable blurring, especially on the wall near the left edge of the image. However, if the reader holds the image at the intended viewing distance (approximately at arm's length), the ringing disappears and all that remains visible is the blurring of the SPIHT image. Figures 21.7(e) and 21.7(f) show the corresponding perceptual distortion maps provided by the PIC metric. The resolution is 128×128, and the distortion increases with pixel brightness. Observe that the distortion is considerably higher for the SPIHT image. In particular, the metric picks up the blurring on the wall on the left. The perceptual PSNR (pooled over the whole image) is 46.8 dB for the SPIHT image and 49.5 dB for the PIC image, in contrast to the PSNR values. Figure 21.7(d) shows the image coded with the standard JPEG algorithm at 0.52 bits/pixel, and Fig. 21.7(g) shows the PIC metric. The PSNR is 30.5 dB and the perceptual PSNR is 47.9 dB. At the intended viewing distance, the quality of the JPEG image is higher than the SPIHT image and worse than the PIC image as the metric indicates. Note that the quantization matrix provides some perceptual weighting, which explains why the SPIHT image is superior according to PSNR and inferior according to perceptual PSNR. The above examples illustrate the power of image quality metrics.

21.2.2.5 *Watson's DCTune*

Many current compression standards are based on a DCT decomposition. Watson [6, 41] presented a model known as DCTune that computes the visibility thresholds for the DCT coefficients, and thus provides a metric for image quality. Watson's model was developed as a means to compute the perceptually optimal image-dependent quantization matrix for DCT-based image coders like JPEG. It has also been used to further optimize JPEG-compatible coders [42, 44, 81]. The JPEG compression standard is discussed in Chapter 17.

Because of the popularity of DCT-based coders and computational efficiency of the DCT, we will give a more detailed overview of DCTune and how it can be used to obtain a metric of image quality.

The original reference and degraded images are partitioned into 8×8 pixel blocks and transformed to the frequency domain using the forward DCT. The DCT decomposition is similar to the subband decomposition and is shown in Fig. 21.7(f). Perceptual thresholds are computed from the DCT coefficients of each block of data of the original image. For each coefficient $b(\mathbf{k}, \mathbf{n})$, where $\mathbf{k}$ identifies the DCT coefficient and $\mathbf{n}$ denotes the block within the reference image, a threshold $t(\mathbf{k}, \mathbf{n})$ is computed using models for contrast sensitivity, luminance masking, and contrast masking.

The baseline contrast sensitivity thresholds $t_b(\mathbf{k})$ are determined by the method of Peterson, *et al.* [85]. The quantization matrices can be obtained from the threshold matrices by multiplying by 2. These baseline thresholds are then modified to account, first for luminance masking, and then for contrast masking, in order to obtain the overall sensitivity thresholds.

Since luminance masking is a function of only the average value of a region, it depends only on the DC coefficient $b(\mathbf{0}, \mathbf{n})$ of each DCT block. The luminance-masked threshold is given by

$$t_l(\mathbf{k}, \mathbf{n}) = t_b(\mathbf{k}) \left[\frac{b(\mathbf{0}, \mathbf{n})}{\bar{b}(\mathbf{0})} \right]^{a_T}, \tag{21.10}$$

where $\bar{b}(0)$ is the DC coefficient corresponding to average luminance of the display (1024 for an 8-bit image using a JPEG compliant DCT implementation) and a_T has a suggested value of 0.649. This parameter controls the amount of luminance masking that takes place. Setting it to zero turns off luminance masking.

The Watson model of contrast masking assumes that the visibility reduction is confined to each coefficient in each block. The overall sensitivity threshold is determined as a function of a contrast masking adjustment and the luminance-masked threshold $t_l(\mathbf{k}, \mathbf{n})$:

$$t(\mathbf{k}, \mathbf{n}) = \max\left\{ t_l(\mathbf{k}, \mathbf{n}), |b(\mathbf{k}, \mathbf{n})|^{w_c(\mathbf{k})} t_l(\mathbf{k}, \mathbf{n})^{1 - w_c(\mathbf{k})} \right\}, \qquad (21.11)$$

where $w_c(\mathbf{k})$ has values between 0 and 1. The exponent may be different for each frequency, but is typically set to a constant in the neighborhood of 0.7. If $w_c(\mathbf{k})$ is 0, no contrast masking occurs and the contrast masking adjustment is equal to 1.

A distortion visibility threshold $d(\mathbf{k}, \mathbf{n})$ is computed at each location as the error at each location (the difference between the DCT coefficients in the original and distorted images) weighted by the sensitivity threshold:

$$d(\mathbf{k}, \mathbf{n}) = \frac{b(\mathbf{k}, \mathbf{n}) - \hat{b}(\mathbf{k}, \mathbf{n})}{t(\mathbf{k}, \mathbf{n})}, \qquad (21.12)$$

where $b(\mathbf{k}, \mathbf{n})$ and $\hat{b}(\mathbf{k}, \mathbf{n})$ are the reference and distorted images, respectively. Note that $d(\mathbf{k}, \mathbf{n}) < 1$ implies the distortion at that location is not visible, while $d(\mathbf{k}, \mathbf{n}) > 1$ implies the distortion is visible.

To combine the distortion visibilities into a single value denoting the quality of the image, error pooling is first done spatially. Then the pools of spatial errors are pooled across frequency. Both pooling processes utilize the same probability summation framework:

$$p(\mathbf{k}) = \left\{ \sum_{\mathbf{n}} |d(\mathbf{k}, \mathbf{n})|^{Q_s} \right\}^{\frac{1}{Q_s}} \qquad (21.13)$$

From psychophysical experiments, a value of 4 has been observed to be a good choice for Q_s.

The matrix $p(\mathbf{k})$ provides a measure of the degree of visibility of artifacts at each frequency that are then pooled across frequency using a similar procedure,

$$P = \left\{ \sum_{\mathbf{k}} p(\mathbf{k})^{Q_f} \right\}^{\frac{1}{Q_f}}. \qquad (21.14)$$

Q_f again can have many values depending on if average or worst case error is more important. Low values emphasize average error, while setting Q_f to infinity reduces the summation to a maximum operator thus emphasizing worst case error.

DCTune has been shown to be very effective in predicting the performance of block-based coders. However, it is not as effective in predicting performance across different coders. In [36, 83], it was found that the metric predictions (they used $Q_f = Q_s = 2$)

are not always consistent with subjective evaluations when comparing different coders. It was found that this metric is strongly biased toward the JPEG algorithm. This is not surprising since both the metric and the JPEG are based on the DCT.

21.2.2.6 *Visual Signal-to-Noise Ratio*

A general purpose quality metric known as the VSNR was developed by Chandler and Hemami [30]. VSNR differs from other HVS-based techniques that we discuss in this section in three main ways. Firstly, the computational models used in VSNR are derived based on psychophysical experiments conducted to quantify the visual detectability of distortions in *natural images*, as opposed to the sine wave gratings or Gabor patches used in most other models. Second, VSNR attempts to quantify the perceived contrast of *supra-threshold* distortions, and the model is not restricted to the regime of threshold of visibility (such as the Daly model). Third, VSNR attempts to capture a *mid-level* property of the HVS known as global precedence, while most other models discussed here only consider low-level processes in the visual system.

In the preprocessing stage, VSNR accounts for viewing conditions (display resolution and viewing distance) and display characteristics. The original image, $f(\mathbf{n})$, and the pixel-wise errors between the original and distorted images, $f(\mathbf{n}) - g(\mathbf{n})$, are decomposed using an M-level discrete wavelet transform using the 9/7 biorthogonal filters. VSNR defines a model to compute the average contrast signal-to-noise ratios (CSNR) at the threshold of detection for wavelet distortions in natural images for each subband of the wavelet decomposition. To determine whether the distortions are visible within each octave band of frequencies, the actual contrast of the distortions is compared with the corresponding contrast detection threshold. If the contrast of the distortions is lower than the corresponding detection threshold for all frequencies, the distorted image is declared to be of perfect quality.

In Section 21.2.1.3, we mentioned the CSF of human vision and several models discussed here attempt to model this aspect of human perception. Although the CSF is critical in determining whether the distortions are visible in the test image, the utility of the CSF in measuring the visibility of supra-threshold distortions has been debated. The perceived contrast of supra-threshold targets has been shown to depend much less on spatial frequency than what is predicted by the CSF, a property also known as contrast constancy. The VSNR assumes contrast constancy, and if the distortion is supra-threshold, the RMS contrast of the error signal is used as a measure of the perceived contrast of the distortion, denoted by d_{pc}.

Finally, the VSNR models the global precedence property of human vision—the visual system has a preference for integrating edges in a coarse to fine scale fashion. VSNR models the global precedence preserving CSNR for each octave band of spatial frequencies. This model satisfies the following property—for supra-threshold distortions, the CSNR corresponding to coarse spatial frequencies is greater than the CSNR corresponding to finer scales. Further, as the distortions become increasingly supra-threshold, coarser scales have increasingly greater CSNR than finer scales in order to preserve visual integration of edges in a coarse to fine scale fashion. For a given distortion contrast, the contrast of the distortions within each subband is compared with the corresponding global precedence preserving contrast specified by the model to compute a measure d_{gp} of the extent to which global precedence has been disrupted. The final quality metric is a linear combination of d_{pc} and d_{gp}.

21.3 STRUCTURAL APPROACHES

In this section, we will discuss structural approaches to image QA. We will discuss the SSIM philosophy in Section 21.3.1. We will show some illustrations of the performance of this metric in Section 21.3.2. Finally, we will discuss the relation between SSIM-and HVS-based metrics in Section 21.3.3.

21.3.1 The Structural Similarity Index

The most fundamental principle underlying structural approaches to image QA is that the HVS is highly adapted to extract structural information from the visual scene, and therefore a measurement of SSIM (or distortion) should provide a good approximation to perceptual image quality. Depending on how structural information and structural distortion are defined, there may be different ways to develop image QA algorithms. The SSIM index is a specific implementation from the perspective of image formation. The luminance of the surface of an object being observed is the product of the illumination and the reflectance, but the structures of the objects in the scene are independent of the illumination. Consequently, we wish to separate the influence of illumination from the remaining information that represents object structures. Intuitively, the major impact of illumination change in the image is the variation of the average local luminance and contrast, and such variation should not have a strong effect on perceived image quality.

Consider two image patches $\tilde{\mathbf{f}}$ and $\tilde{\mathbf{g}}$ obtained from the reference and test images. Mathematically, $\tilde{\mathbf{f}}$ and $\tilde{\mathbf{g}}$ denote two vectors of dimension N, where $\tilde{\mathbf{f}}$ is composed of N elements of $f(\mathbf{n})$ spanned by a window B and similarly for $\tilde{\mathbf{g}}$. To index each element of $\tilde{\mathbf{f}}$, we use the notation $\tilde{\mathbf{f}} = [\tilde{f}_1, \tilde{f}_2, \ldots, \tilde{f}_N]^T$.

First, the luminance of each signal is estimated as the mean intensity:

$$\mu_{\tilde{\mathbf{f}}} = \frac{1}{N} \sum_{i=1}^{N} \tilde{f}_i. \tag{21.15}$$

A luminance comparison function $l(\tilde{\mathbf{f}}, \tilde{\mathbf{g}})$ is then defined as a function of $\mu_{\tilde{\mathbf{f}}}$ and $\mu_{\tilde{\mathbf{g}}}$:

$$l[\tilde{\mathbf{f}}, \tilde{\mathbf{g}}] = \frac{2\mu_{\tilde{\mathbf{f}}}\mu_{\tilde{\mathbf{g}}} + C_1}{\mu_{\tilde{\mathbf{f}}}^2 + \mu_{\tilde{\mathbf{g}}}^2 + C_1}, \tag{21.16}$$

where the constant C_1 is included to avoid instability when $\mu_{\tilde{\mathbf{f}}}^2 + \mu_{\tilde{\mathbf{g}}}^2$ is very close to zero. One good choice is $C_1 = (K_1 E)^2$, where E is the dynamic range of the pixel values (255 for 8-bit grayscale images) and $K_1 << 1$ is a small constant. Similar considerations also apply to contrast comparison and structure comparison terms described below.

The contrast of each image patch is defined as an unbiased estimate of the standard deviation of the patch:

$$\sigma_{\tilde{\mathbf{f}}}^2 = \frac{1}{N-1} \sum_{i=1}^{N} (\tilde{f}_i - \mu_{\tilde{\mathbf{f}}})^2. \tag{21.17}$$

The contrast comparison $c(\tilde{\mathbf{f}}, \tilde{\mathbf{g}})$ takes a similar form as the luminance comparison function and is defined as a function of $\sigma_{\tilde{\mathbf{f}}}$ and $\sigma_{\tilde{\mathbf{g}}}$:

$$c[\tilde{\mathbf{f}}, \tilde{\mathbf{g}}] = \frac{2\sigma_{\tilde{\mathbf{f}}}\sigma_{\tilde{\mathbf{g}}} + C_2}{\sigma_{\tilde{\mathbf{f}}}^2 + \sigma_{\tilde{\mathbf{g}}}^2 + C_2}, \tag{21.18}$$

where C_2 is a nonnegative constant. $C_2 = (K_2 E)^2$, where K_2 satisfies $K_2 << 1$.

Third, the signal is normalized (divided) by its own standard deviation so that the two signals being compared have unit standard deviation. The structure comparison $s(\tilde{\mathbf{f}}, \tilde{\mathbf{g}})$ is conducted on these normalized signals. The SSIM framework uses a geometric interpretation, and the structures of the two images are associated with the direction of the two unit vectors $\tilde{\mathbf{f}} - \mu_{\tilde{\mathbf{f}}}/\sigma_{\tilde{\mathbf{f}}}$ and $\tilde{\mathbf{g}} - \mu_{\tilde{\mathbf{g}}}/\sigma_{\tilde{\mathbf{g}}}$. The angle between the two vectors provides a simple and effective measure to quantify SSIM. In particular, the correlation coefficient between $\tilde{\mathbf{f}}$ and $\tilde{\mathbf{g}}$ corresponds to the cosine of the angle between them and is used as the structure comparison function:

$$s[\tilde{\mathbf{f}}, \tilde{\mathbf{g}}] = \frac{\sigma_{\tilde{\mathbf{f}}\tilde{\mathbf{g}}} + C_3}{\sigma_{\tilde{\mathbf{f}}}\sigma_{\tilde{\mathbf{g}}} + C_3}, \tag{21.19}$$

where the sample covariance between $\tilde{\mathbf{f}}$ and $\tilde{\mathbf{g}}$ is estimated as

$$\sigma_{\tilde{\mathbf{f}}\tilde{\mathbf{g}}} = \frac{1}{N-1} \sum_{i=1}^{N} (\tilde{f}_i - \mu_{\tilde{\mathbf{f}}})(\tilde{g}_i - \mu_{\tilde{\mathbf{g}}}). \tag{21.20}$$

Finally, the SSIM index between image patches $\tilde{\mathbf{f}}$ and $\tilde{\mathbf{g}}$ is defined as

$$\text{SSIM}[\tilde{\mathbf{f}}, \tilde{\mathbf{g}}] = l[\tilde{\mathbf{f}}, \tilde{\mathbf{g}}]^\alpha \cdot c[\tilde{\mathbf{f}}, \tilde{\mathbf{g}}]^\beta \cdot s[\tilde{\mathbf{f}}, \tilde{\mathbf{g}}]^\gamma, \tag{21.21}$$

where α, β, and γ are parameters used to adjust the relative importance of the three components.

The SSIM index and the three comparison functions—luminance, contrast, and structure—satisfy the following desirable properties.

- *Symmetry*: $\text{SSIM}(\tilde{\mathbf{f}}, \tilde{\mathbf{g}}) = \text{SSIM}(\tilde{\mathbf{g}}, \tilde{\mathbf{f}})$. When quantifying the similarity between two signals, exchanging the order of the input signals should not affect the resulting measurement.

- *Boundedness*: $\text{SSIM}(\tilde{\mathbf{f}}, \tilde{\mathbf{g}}) \leq 1$. An upper bound can serve as an indication of how close the two signals are to being perfectly identical.

- *Unique maximum:* $\text{SSIM}(\tilde{\mathbf{f}}, \tilde{\mathbf{g}}) = 1$ if and only if $\tilde{\mathbf{f}} = \tilde{\mathbf{g}}$. The perfect score is achieved only when the signals being compared are identical. In other words, the similarity measure should quantify any variations that may exist between the input signals.

The structure term of the SSIM index is independent of the luminance and contrast of the local patches, which is physically sensible because the change of luminance and/or contrast has little impact on the structures of the objects in the scene. Although the SSIM index is defined by three terms, the structure term in the SSIM index is generally regarded as the most important, since variations in luminance and contrast of an image do not affect visual quality as much as structural distortions [28].

21.3.2 Image Quality Assessment Using SSIM

The SSIM index measures the SSIM between two images. If one of the images is regarded as of perfect quality, then the SSIM index can be viewed as an indication of the quality of the other image signal being compared. When applying the SSIM index approach to large-size images, it is useful to compute it locally rather than globally. The reasons are manifold. First, statistical features of images are usually spatially nonstationary. Second, image distortions, which may or may not depend on the local image statistics, may also vary across space. Third, due to the nonuniform retinal sampling feature of the HVS, at typical viewing distances, only a local area in the image can be perceived with high resolution by the human observer at one time instance. Finally, localized quality measurement can provide a spatially varying quality map of the image, which delivers more information about the quality degradation of the image. Such a quality map can be used in different ways. It can be employed to indicate the quality variations across the image. It can also be used to control image quality for space-variant image processing systems, e.g., region-of-interest image coding and foveated image processing.

In early instantiations of the SSIM index approach [28], the local statistics $\mu_{\tilde{\mathbf{f}}}$, $\sigma_{\tilde{\mathbf{f}}}$, and $\sigma_{\tilde{\mathbf{f}}\tilde{\mathbf{g}}}$ defined in Eqs. (21.15), (21.17), and (21.20) were computed within a local 8×8 square window. The window moves pixel-by-pixel from the top-left corner to the bottom-right corner of the image. At each step, the local statistics and SSIM index are calculated within the local window. One problem with this method is that the resulting SSIM index map often exhibits undesirable "blocking" artifacts as exemplified by Fig. 21.8(c). Such "artifacts" are not desirable because they are created from the choice of the quality measurement method (local square window) and not from image distortions. In [29], a circular-symmetric Gaussian weighting function $\mathbf{w} = \{w_i, i = 1, 2, \ldots N\}$ with unit sum $\left(\sum_{i=1}^{N} w_i = 1\right)$ is adopted. The estimates of $\mu_{\tilde{\mathbf{f}}}$, $\sigma_{\tilde{\mathbf{f}}}$, and $\sigma_{\tilde{\mathbf{f}}\tilde{\mathbf{g}}}$ are then modified accordingly:

$$\mu_{\tilde{\mathbf{f}}} = \sum_{i=1}^{N} w_i \tilde{f}_i, \tag{21.22}$$

$$\sigma_{\tilde{\mathbf{f}}}^2 = \sum_{i=1}^{N} w_i (\tilde{f}_i - \mu_{\tilde{\mathbf{f}}})^2, \tag{21.23}$$

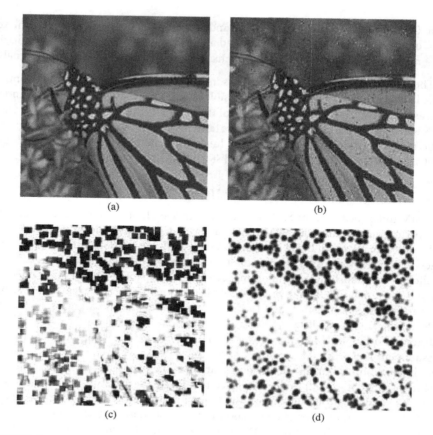

FIGURE 21.8

The effect of local window shape on SSIM index map. (a) original image; (b) impulsive noise contaminated image; (c) SSIM index map using square windowing approach; (d) SSIM index map using smoothed windowing approach. In both SSIM index maps, brighter indicates better quality.

$$\sigma_{\tilde{\mathbf{f}}\tilde{\mathbf{g}}} = \sum_{i=1}^{N} w_i(\tilde{f}_i - \mu_{\tilde{\mathbf{f}}})(\tilde{g}_i - \mu_{\tilde{\mathbf{g}}}). \tag{21.24}$$

With such a smoothed windowing approach, the quality maps exhibit a locally isotropic property as demonstrated in Fig. 21.8(d).

Figure 21.9 shows the SSIM index maps of a set of sample images with different types of distortions. The absolute error map for each distorted image is also included for comparison. The SSIM index and absolute error maps have been adjusted so that brighter always indicates better quality in terms of the given quality/distortion measure. It can be seen that the distorted images exhibit variable quality across space. For example,

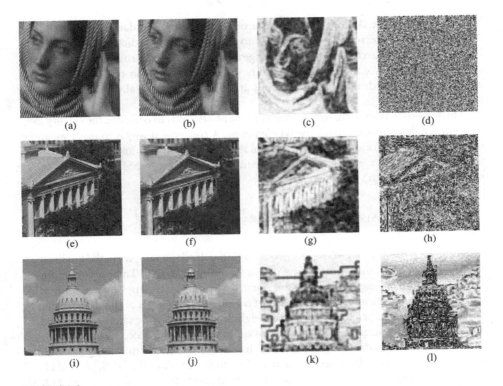

FIGURE 21.9

Sample distorted images and their quality/distortion maps (images are cropped to 160×160 for visibility); (a), (e), and (i) original images; (b) Gaussian noise contaminated image; (f) JPEG2000 compressed image; (j) JPEG compressed image; (c), (g), and (k) SSIM index maps of the distorted images, where brightness indicates the magnitude of the local SSIM index (squared for visibility); (d), (h), and (l) absolute error maps of the distorted images, where darkness indicates the absolute value of the local pixel difference. Note that in all quality/distortion maps ((c), (d), (g), (h), (k), and (l)), brighter indicates better quality in terms of the underlying quality/distortion measure.

in image 21.9(b), the noise over the face region appears to be much more significant than that in the texture regions. However, the absolute error map 21.9(d) is completely independent of the underlying image structures. By contrast, the SSIM index map 21.9(c) gives perceptually consistent prediction. In image 21.9(f), the bit allocation scheme of low bit-rate JPEG2000 compression leads to smooth representations of detailed image structures. For example, the texture information of the roof of the building and the trees is lost. This is very well indicated by the SSIM index map 21.9(g), but cannot be predicted from the absolute error map 21.9(h). Some different types of distortions are caused by low bit-rate JPEG compression. In image 21.9(j), the major distortions we observe are the blocking effect in the sky and the artifacts around the outer boundaries of the building. Again, the absolute error map 21.9(l) fails to provide useful prediction, and the SSIM index map 21.9(k) successfully predicts image quality variations across space. From these sample images, we see that an image quality measure as simple as the SSIM index can adapt to various kinds of image distortions and provide perceptually consistent quality predictions.

The final step of an image quality measurement system is to combine the quality map into one single quality score for the whole image. A convenient way is to use a weighted summation. Let $f(\mathbf{n})$ and $g(\mathbf{n})$ be the two images being compared, and $\mathrm{SSIM}[f(\mathbf{n}), g(\mathbf{n})]$ the local SSIM index evaluated at location $\mathbf{n}$. Then, the mean SSIM (MSSIM) index between $f(\mathbf{n})$ and $g(\mathbf{n})$ is defined as:

$$\mathrm{MSSIM}[f(\mathbf{n}), g(\mathbf{n})] = \frac{\sum_{\mathbf{n}} W(\mathbf{n})\, \mathrm{SSIM}[f(\mathbf{n}), g(\mathbf{n})]}{\sum_{\mathbf{n}} W(\mathbf{n})}, \tag{21.25}$$

where $W(\mathbf{n})$ is the weight given to the pixel location $\mathbf{n}$. If all the samples in the quality map are equally weighted, this results in the measure employed in [29]. There are two cases in which nonuniform weighting is desirable. First, depending on the application, some prior knowledge about the importance of different regions in the image is available, and such an importance map can be converted into a weighting function. For example, object-based region-of-interest image processing systems often segment the objects in the scene and give different objects different importance. In a foveated image processing system, the weighting function can be determined according to foveation in the HVS, i.e., the visual resolution decreases gradually as a function of the distance from the fixation point [2, 3]. Note that the weighting function here is determined only by the spatial location and is independent of the local image content. In the second case, the image content also plays a role. It has been observed that different image textures attract human fixations with varying degrees, and therefore different weights can be assigned. In [1], local variance-weighted, local quality/distortion-weighted, and information-content-weighted pooling functions were used. It was observed that these weighting functions may improve the performance of image QA algorithms. For example, local variance-weighting is useful to balance the extreme case where severe high-variance distortions concentrate at some small areas in the image.

21.3.3 Relation to HVS-Based Models

In this section, we will relate the structure term in the SSIM index to MSE and human vision based metrics [13, 17].

The MSE between image patches $\tilde{\mathbf{f}}$ and $\tilde{\mathbf{g}}$ is

$$\mathrm{MSE}[\tilde{\mathbf{f}}, \tilde{\mathbf{g}}] = \frac{1}{N} \sum_{i=1}^{N} (\tilde{f}_i - \tilde{g}_i)^2.$$

We define normalized random variables,

$$\tilde{\mathbf{f}} = \frac{\tilde{\mathbf{f}} - \mu_{\tilde{\mathbf{f}}}}{\sigma_{\tilde{\mathbf{f}}}}, \tag{21.26}$$

$$\tilde{\mathbf{g}} = \frac{\tilde{\mathbf{g}} - \mu_{\tilde{\mathbf{g}}}}{\sigma_{\tilde{\mathbf{g}}}}. \tag{21.27}$$

Observe that

$$\mathrm{MSE}\left(\frac{\tilde{\mathbf{f}} - \mu_{\tilde{\mathbf{f}}}}{\sigma_{\tilde{\mathbf{f}}}}, \frac{\tilde{\mathbf{g}} - \mu_{\tilde{\mathbf{g}}}}{\sigma_{\tilde{\mathbf{g}}}}\right) = 2\left(1 - \frac{\sigma_{\tilde{\mathbf{f}}\tilde{\mathbf{g}}}}{\sigma_{\tilde{\mathbf{f}}}\sigma_{\mathbf{g}}}\right). \tag{21.28}$$

Thus, the structure term in the SSIM index essentially computes an MSE between image patches, after normalizing them for their mean and standard deviations. This is not surprising in view of the fact that the structure term of the SSIM index is defined to be independent of the mean and standard deviation of the image intensity values. This relationship between the structure term of SSIM and MSE illustrates that simple modifications to the MSE computation can help overcome some of its drawbacks such as failure to detect brightness and contrast changes (illustrated in Fig. 21.1(b)).

We will now discuss the relation of SSIM to HVS-based metrics. We first look at the structure term of the SSIM index, which is arguably the most important term in the SSIM index [13, 17]. It is evident that the definition of the normalized variables in (21.26) and (21.27) is very similar to divisive normalization models of contrast gain control in HVS-based metrics. In fact, a SSIM contrast masking model can be defined by:

$$R[\tilde{f}_i] = \frac{\tilde{f}_i - \mu_{\tilde{\mathbf{f}}}}{\sqrt{\frac{1}{N}\sum_{i=1}^{N}\left[\tilde{f}_i - \mu_{\tilde{\mathbf{f}}}\right]^2}}. \tag{21.29}$$

We discussed in Section 21.2.1.6 that most HVS-based QA systems compute a Minkowski error between the outputs of the contrast gain control model (as well as models of other aspects of the HVS incorporated in QA) to the reference and test image patches as an index of quality, often with a Minkowski exponent of 2 [38, 39, 41]. Similarly, observe that the structure term of SSIM in (21.28) is a monotonic function of the square of the Minkowski error between the outputs of the SSIM contrast gain control model in (21.29) with exponent 2.

It is important to note that the SSIM indices perform the gain control normalization in the *image pixel* domain. In (21.29), the contrast gain control model divisively inhibits each pixel by pooling the responses of a local spatial neighborhood of pixels. However, contrast masking in the HVS is a phenomenon that is better modeled in a frequency-orientation decomposed domain. For example, the masking effect is maximum when the orientation of the masker and the target are parallel and decreases when their orientations are perpendicular [27]. The contrast masking models that we saw in Section 21.2.2 can capture these effects since they normalize each filter output by pooling the responses at the same location of *other filters* tuned to different orientations. However, the SSIM metric will not be able to account for such effects. Improved versions of the SSIM index that use a frequency decomposition have been proposed [19, 20], and our discussion of the relation between the SSIM index and the contrast masking models helps us understand the reasons for the improved performance of these metrics. Interestingly, the square of the response of the SSIM contrast gain control model defined by (21.29) is equal to the response of the Teo and Heeger gain control model defined by (21.5) with $\kappa_i = N$ and $\sigma_i^2 = 0$ for all i. Thus, if the vectors $\tilde{\mathbf{f}}$ and $\tilde{\mathbf{g}}$ at each spatial location $\mathbf{n}$ are defined using the responses of a filter bank at that location (as is done in the Complex Wavelet SSIM model [20]), the SSIM contrast masking model is just the square root of the Teo and Heeger contrast masking model.

Also, (21.16) is connected with Weber's law which was discussed in Section 21.2.1.4 [29]. According to Weber's law, the HVS is sensitive to the relative rather than the absolute luminance change. Letting R represent the ratio of the luminance of the distorted signal relative to the reference signal, then we can write $\mu_{\tilde{\mathbf{g}}} = R\mu_{\tilde{\mathbf{f}}}$. Substituting this into (21.16) gives

$$l[\tilde{\mathbf{f}}, \tilde{\mathbf{g}}] = \frac{2R}{1 + R^2 + \frac{C_1}{\mu_{\tilde{\mathbf{f}}}^2}}. \tag{21.30}$$

If we assume C_1 is small enough (relative to $\mu_{\tilde{\mathbf{f}}}^2$) to be ignored, then $l[\tilde{\mathbf{f}}, \tilde{\mathbf{g}}]$ is a function only of R. In this sense, it is qualitatively consistent with Weber's law.

This discussion shows that although the SSIM index was derived from very different first principles, at least part of the reasons for its success can be attributed to similarities it shares with models of the HVS.

21.4   INFORMATION THEORETIC APPROACHES

In this section, we discuss information theoretic approaches to image QA. We will discuss the information theoretic metrics in Section 21.4.1. We will show some illustrations of the performance of this metric in Section 21.4.2. Finally, we will discuss the relation between SSIM and information theoretic metrics in Section 21.4.3.

21.4.1   Information Theoretic Metrics

In the information-theoretic approach to QA, the QA problem is viewed as an information fidelity problem rather than a signal fidelity problem. An image source communicates to a receiver through a channel that limits the amount of information that could flow through it, thereby introducing distortions. The output of the image source is the reference image, the output of the channel is the test image, and the visual quality of the test image is computed as the amount of information shared between the test and the reference signals, i.e., the mutual information between them. Thus, information fidelity methods exploit the relationship between statistical image information and visual quality.

Statistical models for signal sources and transmission channels are at the core of information theoretic analysis techniques. A fundamental component of information fidelity based QA methods is a model for image sources. Images and videos whose quality needs to be assessed are usually optical images of the 3D visual environment or *natural scenes*. Natural scenes form a very tiny subspace in the space of all possible image signals, and researchers have developed sophisticated models that capture key statistical features of natural images.

In this chapter, we present two full-reference QA methods based on the information-fidelity paradigm. Both methods share a common mathematical framework. The first method, the information fidelity criterion (IFC) [26], uses a distortion channel model as depicted in Fig. 21.10. The IFC quantifies the information shared between the test image and the distorted image. The other method we present in this chapter is the visual information fidelity (VIF) measure [25], which uses an additional HVS channel model and utilizes two aspects of image information for quantifying perceptual quality: the information shared between the test and the reference images and the information content of the reference image itself. This is depicted pictorially in Fig. 21.11.

Images and videos of the visual environment captured using high-quality capture devices operating in the visual spectrum are broadly classified as natural scenes. This differentiates them from text, computer-generated graphics scenes, cartoons and animations, paintings and drawings, random noise, or images and videos captured from

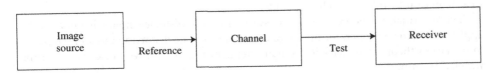

FIGURE 21.10

The information-fidelity problem: a channel distorts images and limits the amount of information that could flow from the source to the receiver. Quality should relate to the amount of information about the reference image that could be extracted from the test image.

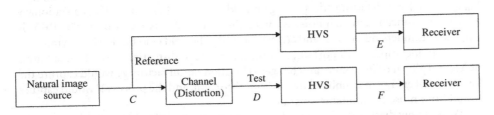

FIGURE 21.11

An information-theoretic setup for quantifying visual quality using a distortion channel model as well as an HVS model. The HVS also acts as a channel that limits the flow of information from the source to the receiver. Image quality could also be quantified using a relative comparison of the information in the upper path of the figure and the information in the lower path.

nonvisual stimuli such as radar and sonar, X-rays, and ultrasounds. The model for natural images that is used in the information theoretic metrics is the Gaussian scale mixture (GSM) model in the wavelet domain.

A GSM is a random field (RF) that can be expressed as a product of two independent RFs [14]. That is, a GSM $\mathcal{C} = \{\vec{C}_{\mathbf{n}} : \mathbf{n} \in \mathcal{N}\}$, where $\mathcal{N}$ denotes the set of spatial indices for the RF, can be expressed as:

$$\mathcal{C} = \mathcal{S} \cdot \mathcal{U} = \{S_{\mathbf{n}} \cdot \vec{U}_{\mathbf{n}} : \mathbf{n} \in \mathcal{N}\}, \qquad (21.31)$$

where $\mathcal{S} = \{S_{\mathbf{n}} : \mathbf{n} \in \mathcal{N}\}$ is an RF of positive scalars also known as the mixing density and $\mathcal{U} = \{\vec{U}_{\mathbf{n}} : \mathbf{n} \in \mathcal{N}\}$ is a Gaussian vector RF with mean zero and covariance matrix $\mathbf{C}_U$. $\vec{C}_{\mathbf{n}}$ and $\vec{U}_{\mathbf{n}}$ are M dimensional vectors, and we assume that for the RF $\mathcal{U}$, $\vec{U}_{\mathbf{n}}$ is independent of $\vec{U}_{\mathbf{m}}$, $\forall \mathbf{n} \neq \mathbf{m}$. We model each subband of a scale-space-orientation wavelet decomposition (such as the steerable pyramid [15]) of an image as a GSM. We partition the subband coefficients into nonoverlapping blocks of M coefficients each, and model block $\mathbf{n}$ as the vector $\vec{C}_{\mathbf{n}}$. Thus image blocks are assumed to be uncorrelated with each other, and any linear correlations between wavelet coefficients are modeled only through the covariance matrix $\mathbf{C}_U$.

One could easily make the following observations regarding the above model: $\mathcal{C}$ is normally distributed given $\mathcal{S}$ (with mean zero, and covariance of $\vec{C}_{\mathbf{n}}$ being $S_{\mathbf{n}}^2 \mathbf{C}_U$), that given $S_{\mathbf{n}}$, $C_{\mathbf{n}}$ are independent of $S_{\mathbf{m}}$ for all $\mathbf{n} \neq \mathbf{m}$, and that given $\mathcal{S}$, $\vec{C}_{\mathbf{n}}$ are conditionally independent of $\vec{C}_{\mathbf{m}}$, $\forall \mathbf{n} \neq \mathbf{m}$ [14]. These properties of the GSM model make analytical treatment of information fidelity possible.

The information theoretic metrics assume that the distorted image is obtained by applying a distortion operator on the reference image. The distortion model used in the information theoretic metrics is a signal attenuation and additive noise model in the wavelet domain:

$$\mathcal{D} = \mathcal{G}\mathcal{C} + \mathcal{V} = \{g_{\mathbf{n}} \vec{C}_{\mathbf{n}} + \vec{V}_{\mathbf{n}} : \mathbf{n} \in \mathcal{N}\}, \qquad (21.32)$$

where $\mathcal{C}$ denotes the RF from a subband in the reference signal, $\mathcal{D} = \{\vec{D}_{\mathbf{n}} : \mathbf{n} \in \mathcal{N}\}$ denotes the RF from the corresponding subband from the test (distorted) signal, $\mathcal{G} = \{g_{\mathbf{n}} : \mathbf{n} \in \mathcal{N}\}$ is a deterministic scalar gain field, and $\mathcal{V} = \{\vec{V}_{\mathbf{n}} : \mathbf{n} \in \mathcal{N}\}$ is a stationary additive zero-mean Gaussian noise RF with covariance matrix $\mathbf{C}_V = \sigma_V^2 \mathbf{I}$. The RF $\mathcal{V}$ is white and is independent of $\mathcal{S}$ and $\mathcal{U}$. We constrain the field $\mathcal{G}$ to be slowly varying.

This model captures important, and complementary, distortion types: blur, additive noise, and global or local contrast changes. The attenuation factors $g_{\mathbf{n}}$ would capture the loss of signal energy in a subband due to blur distortion, and the process $\mathcal{V}$ would capture the additive noise components separately.

We will now discuss the IFC and the VIF criteria in the following sections.

21.4.1.1 *The Information Fidelity Criterion*

The IFC quantifies the information shared between a test image and the reference image. The reference image is assumed to pass through a channel yielding the test image, and the mutual information between the reference and the test images is used for predicting visual quality.

Let $\vec{C}^N = \{\vec{C}_1, \vec{C}_2, \ldots, \vec{C}_N\}$ denote N elements from $\mathcal{C}$. Let S^N and $\vec{D}^N$ be correspondingly defined. The IFC uses the mutual information between the reference and test images conditioned on a fixed mixing multiplier in the GSM model, i.e., $I(\vec{C}^N; \vec{E}^N | \vec{S}^N = s^N)$, as an indicator of visual quality. With the stated assumptions on $\mathcal{C}$ and the distortion model, it can easily be shown that [26]

$$I(\vec{C}^N; \vec{D}^N | s^N) = \frac{1}{2} \sum_{n=1}^{N} \sum_{k=1}^{M} \log_2 \left(1 + \frac{g_{\mathbf{n}}^2 s_{\mathbf{n}}^2 \lambda_k}{\sigma_V^2} \right), \tag{21.33}$$

where λ_k are the eigenvalues of $\mathbf{C}_U$.

Note that in the above treatment it is assumed that the model parameters s^N, $\mathcal{G}$, and σ_V^2 are known. Details of practical estimation of these parameters are given in Section 21.4.1.3. In the development of the IFC, we have so far only dealt with one subband. One could easily incorporate multiple subbands by assuming that each subband is completely independent of others in terms of the RFs as well as the distortion model parameters. Thus the IFC is given by:

$$\text{IFC} = \sum_{j \in \text{subbands}} I(\vec{C}^{N,j}; \vec{D}^{N,j} | s^{N,j}), \tag{21.34}$$

where the summation is carried over the subbands of interest, and $\vec{C}^{N,j}$ represent N_j elements of the RF $\mathcal{C}_j$ that describes the coefficients from subband j, and so on.

21.4.1.2 *The Visual Information Fidelity Criterion*

In addition to the distortion channel, VIF assumes that both the reference and distorted images pass through the HVS, which acts as a "distortion channel" that imposes limits on how much information could flow through it. The purpose of the HVS model in the information fidelity setup is to quantify the uncertainty that the HVS adds to the signal that flows through it. As a matter of analytical and computational simplicity, we lump all sources of HVS uncertainty into one additive noise component that serves as a *distortion baseline* in comparison to which the distortion added by the distortion channel could be evaluated. We call this lumped HVS distortion *visual noise* and model it as a stationary, zero mean, additive white Gaussian noise model in the wavelet domain. Thus, we model the HVS noise in the wavelet domain as stationary RFs $\mathcal{H} = \{\vec{H}_{\mathbf{n}} : \mathbf{n} \in \mathcal{N}\}$ and $\mathcal{H}' = \{\vec{H}'_{\mathbf{n}} : \mathbf{n} \in \mathcal{N}\}$, where $\vec{H}_i$ and $\vec{H}'_i$ are zero-mean uncorrelated multivariate Gaussian with the same dimensionality as $\vec{C}_{\mathbf{n}}$:

$$\mathcal{E} = \mathcal{C} + \mathcal{H} \quad \text{(reference image)}, \tag{21.35}$$

$$\mathcal{F} = \mathcal{D} + \mathcal{H}' \quad \text{(test image)}, \tag{21.36}$$

where $\mathcal{E}$ and $\mathcal{F}$ denote the visual signal at the output of the HVS model from the reference and test images in one subband, respectively (Fig. 21.11). The RFs $\mathcal{H}$ and $\mathcal{H}'$ are assumed to be independent of $\mathcal{U}$, $\mathcal{S}$, and $\mathcal{V}$. We model the covariance of $\mathcal{H}$ and $\mathcal{H}'$ as

$$\mathbf{C}_H = \mathbf{C}_{H'} = \sigma_H^2 \mathbf{I}, \tag{21.37}$$

where σ_H^2 is an HVS model parameter (variance of the visual noise).

It can be shown [25] that

$$I(\vec{C}^N; \vec{E}^N | s^N) = \frac{1}{2} \sum_{\mathbf{n}=1}^{N} \sum_{k=1}^{M} \log_2 \left(1 + \frac{s_{\mathbf{n}}^2 \lambda_k}{\sigma_H^2} \right), \tag{21.38}$$

$$I(\vec{C}^N; \vec{F}^N | s^N) = \frac{1}{2} \sum_{\mathbf{n}=1}^{N} \sum_{k=1}^{M} \log_2 \left(1 + \frac{g_{\mathbf{n}}^2 s_{\mathbf{n}}^2 \lambda_k}{\sigma_V^2 + \sigma_H^2} \right), \tag{21.39}$$

where λ_k are the eigenvalues of $\mathbf{C}_U$.

$I(\vec{C}^N; \vec{E}^N | s^N)$ and $I(\vec{C}^N; \vec{F}^N | s^N)$ represent the information that could ideally be extracted by the brain from a particular subband of the reference and test images, respectively. A simple *ratio* of the two information measures relates quite well with visual quality [25]. It is easy to motivate the suitability of this relationship between image information and visual quality. When a human observer sees a distorted image, she has an idea of the amount of information that she expects to receive in the image (modeled through the known $\mathcal{S}$ field), and it is natural to expect the fraction of the expected information that is actually received from the distorted image to relate well with visual quality.

As with the IFC, the VIF could easily be extended to incorporate multiple subbands by assuming that each subband is completely independent of others in terms of the RFs as well as the distortion model parameters. Thus, the VIF is given by

$$\text{VIF} = \frac{\sum_{j \in \text{subbands}} I(\vec{C}^{N,j}; \vec{F}^{N,j} | s^{N,j})}{\sum_{j \in \text{subbands}} I(\vec{C}^{N,j}; \vec{E}^{N,j} | s^{N,j})}, \tag{21.40}$$

where we sum over the subbands of interest, and $\vec{C}^{N,j}$ represent N elements of the RF $\mathcal{C}_j$ that describes the coefficients from subband j, and so on.

The VIF given in (21.40) is computed for a collection of wavelet coefficients that could represent either an entire subband of an image or a spatially localized set of subband coefficients. In the former case, the VIF is a single number that quantifies the information fidelity for the entire image, whereas in the latter case, a sliding-window approach could be used to compute a *quality map* that could visually illustrate how the visual quality of the test image varies over space.

21.4.1.3 *Implementation Details*

The source model parameters that need to be estimated from the data consist of the field $\mathcal{S}$. For the vector GSM model, the maximum-likelihood estimate of s_n^2 can be found as follows [21]:

$$\widehat{s_n^2} = \frac{\vec{C}_n^T \mathbf{C}_U^{-1} \vec{C}_n}{M}.$$
(21.41)

Estimation of the covariance matrix $\mathbf{C}_U$ is also straightforward from the reference image wavelet coefficients [21]:

$$\widehat{\mathbf{C}}_U = \frac{1}{N} \sum_{n=1}^{N} \vec{C}_n \vec{C}_n^T.$$
(21.42)

In (21.41) and (21.42), $\frac{1}{N} \sum_{n=1}^{N} s_n^2$ is assumed to be unity without loss of generality [21].

The parameters of the distortion channel are estimated locally. A spatially localized block-window centered at coefficient $\mathbf{n}$ could be used to estimate g_n and σ_V^2 at $\mathbf{n}$. The value of the field $\mathcal{G}$ over the block centered at coefficient $\mathbf{n}$, which we denote as g_n, and the variance of the RF $\mathcal{V}$, which we denote as $\sigma_{V,n}^2$, are fairly easy to estimate (by linear regression) since both the input (the reference signal) and the output (the test signal) of the system (21.32) are available:

$$\widehat{g_n} = \widehat{\mathrm{Cov}}(C,D)\widehat{\mathrm{Cov}}(C,C)^{-1},$$
(21.43)

$$\widehat{\sigma}_{V,n}^2 = \widehat{\mathrm{Cov}}(D,D) - \widehat{g_n}\widehat{\mathrm{Cov}}(C,D),$$
(21.44)

where the covariances are approximated by sample estimates using sample points from the corresponding blocks centered at coefficient $\mathbf{n}$ in the reference and the test signals.

For VIF, the HVS model is parameterized by only one parameter: the variance of visual noise σ_H^2. It is easy to hand-optimize the value of the parameter σ_H^2 by running the algorithm over a range of values and observing its performance.

21.4.2 Image Quality Assessment Using Information Theoretic Metrics

Firstly, note that the IFC is bounded below by zero (since mutual information is a nonnegative quantity) and bounded above by ∞, which occurs when the reference and test images are identical. One advantage of the IFC is that like the MSE, it does not depend upon model parameters such as those associated with display device physics, data from visual psychology experiments, viewing configuration information, or stabilizing constants.

Note that VIF is basically IFC normalized by the reference image information. The VIF has a number of interesting features. Firstly, note that VIF is bounded below by zero, which indicates that all information about the reference image has been lost in the distortion channel. Secondly, if the test image is an exact copy of the reference image, then VIF is *exactly* unity (this property is satisfied by the SSIM index also). For many distortion types,

VIF would lie in the interval [0, 1]. Thirdly, a linear contrast enhancement of the reference image that does not add noise would result in a VIF value *larger* than unity, signifying that the contrast-enhanced image has a *superior* visual quality than the reference image! It is common observation that contrast enhancement of images increases their perceptual quality unless quantization, clipping, or display nonlinearities add additional distortion. This improvement in visual quality is captured by the VIF.

We now illustrate the performance of VIF by an example. Figure 21.12 shows a reference image and three of its distorted versions that come from three different types of

(a) Reference image   (b) Contrast enhancement

(c) Blurred   (d) JPEG compressed

FIGURE 21.12

The VIF has an interesting feature: it can capture the effects of linear contrast enhancements on images and quantify the improvement in visual quality. A VIF value greater than unity indicates this improvement, while a VIF value less than unity signifies a loss of visual quality. (a) Reference Lena image (VIF = 1.0); (b) contrast stretched Lena image (VIF = 1.17); (c) Gaussian blur (VIF = 0.05); (d) JPEG compressed (VIF = 0.05).

distortion, all of which have been adjusted to have about the same MSE with the reference image. The distortion types illustrated in Fig. 21.12 are contrast stretch, Gaussian blur, and JPEG compression. In comparison with the reference image, the contrast-enhanced image has a better visual quality despite the fact that the "distortion" (in terms of a perceivable difference with the reference image) is clearly visible. A VIF value larger than unity indicates that the perceptual difference in fact constitutes improvement in visual quality. In contrast, both the blurred image and the JPEG compressed image have clearly visible distortions and poorer visual quality, which is captured by a low VIF measure.

Figure 21.13 illustrates spatial quality maps generated by VIF. Figure 21.13(a) shows a reference image and Fig. 21.13(b) the corresponding JPEG2000 compressed image in which the distortions are clearly visible. Figure 21.13(c) shows the reference image information map. The information map shows the spread of statistical information in the reference image. The statistical information content of the image is low in flat image regions, whereas in textured regions and regions containing strong edges, it is high. The quality map in Fig. 21.13(d) shows the proportion of the image information that has been lost to JPEG2000 compression. Note that due to the nonlinear normalization in the denominator of VIF, the scalar VIF value for a reference/test pair is *not* the mean of the corresponding VIF-map.

21.4.3 Relation to HVS-Based Metrics and Structural Similarity

We will first discuss the relation between IFC and SSIM index [13, 17]. First of all, the GSM model used in the information theoretic metrics results in the subband coefficients being Gaussian distributed, when conditioned on a fixed mixing multiplier in the GSM model. The linear distortion channel model results in the reference and test images being jointly Gaussian. The definition of the correlation coefficient in the SSIM index in (21.19) is obtained from regression analysis and implicitly assumes that the reference and test image vectors are jointly Gaussian [22]. In fact, (21.19) coincides with the maximum likelihood estimate of the correlation coefficient only under the assumption that the reference and distorted image patches are jointly Gaussian distributed [22]. These observations hint at the possibility that the IFC index may be closely related to SSIM. A well-known result in information theory states that when two variables are jointly Gaussian, the mutual information between them is a function of just the correlation coefficient [23, 24]. Thus, recent results show that a scalar version of the IFC metric is a monotonic function of the square of the structure term of the SSIM index when the SSIM index is applied on subband filtered coefficients [13, 17]. The reasons for the monotonic relationship between the SSIM index and the IFC index are the explicit assumption of a Gaussian distribution on the reference and test image coefficients in the IFC index (conditioned on a fixed mixing multiplier) and the implicit assumption of a Gaussian distribution in the SSIM index (due to the use of regression analysis). These results indicate that the IFC index is equivalent to multiscale SSIM indices since they satisfy a monotonic relationship.

Further, the concept of the correlation coefficient in SSIM was generalized to vector valued variables using canonical correlation analysis to establish a monotonic relation between the squares of the canonical correlation coefficients and the vector IFC index

(a) Reference image   (b) JPEG2000 compressed

(c) Reference image info. map   (d) VIF map

FIGURE 21.13

Spatial maps showing how VIF captures spatial information loss.

[13, 17]. It was also established that the VIF index includes a structure comparison term and a contrast comparison term (similar to the SSIM index), as opposed to just the structure term in IFC. One of the properties of the VIF index observed in Section 21.4.2 was the fact that it can predict improvement in quality due to contrast enhancement. The presence of the contrast comparison term in VIF explains this effect [13, 17].

We showed the relation between SSIM- and HVS-based metrics in Section 21.3.3. From our discussion here, the relation between IFC-, VIF-, and HVS-based metrics is also immediately apparent. Similarities between the scalar IFC index and the HVS-based metrics were also observed in [26]. It was shown that the IFC is functionally similar to HVS-based FR QA algorithms [26]. The reader is referred to [13, 17] for a more thorough treatment of this subject.

Having discussed the similarities between the SSIM and the information theoretic frameworks, we will now discuss the differences between them. The SSIM metrics use a measure of *linear* dependence between the reference and test image pixels, namely the Pearson product moment correlation coefficient. However, the information theoretic metrics use the mutual information, which is a more general measure of correlation that can capture nonlinear dependencies between variables. The reason for the monotonic relation between the square of the structure term of the SSIM index applied in the subband filtered domain and the IFC index is due to the assumption that the reference and test image coefficients are jointly Gaussian. This indicates that the structure term of SSIM and IFC is equivalent under the statistical source model used in [26], and more sophisticated statistical models are required in the IFC framework to distinguish it from the SSIM index.

Although the information theoretic metrics use a more general and flexible notion of correlation than the SSIM philosophy, the form of the relationship between the reference and test images might affect visual quality. As an example, if one test image is a deterministic linear function of the reference image, while another test image is a deterministic parabolic function of the reference image, the mutual information between the reference and the test image is identical in both cases. However, it is unlikely that the visual quality of both images is identical. We believe that further investigation of suitable models for the distortion channel and the relation between such channel models and visual quality are required to answer this question.

21.5 PERFORMANCE OF IMAGE QUALITY METRICS

In this section, we present results on the validation of some of the image quality metrics presented in this chapter and present comparisons with PSNR. All results use the LIVE image QA database [8] developed by Bovik and coworkers and further details can be found in [7]. The validation is done using subjective quality scores obtained from a group of human observers, and the performance of the QA algorithms is evaluated by comparing the quality predictions of the algorithms against subjective scores.

In the LIVE database, 20–28 human subjects were asked to assign each image with a score indicating their assessment of the quality of that image, defined as the extent to which the artifacts were visible and annoying. Twenty-nine high-resolution 24-bits/pixel RGB color images (typically 768×512) were distorted using five distortion types: JPEG2000, JPEG, white noise in the RGB components, Gaussian blur, and transmission errors in the JPEG2000 bit stream using a fast-fading Rayleigh channel model. A database was derived from the 29 images to yield a total of 779 distorted images, which, together with the undistorted images, were then evaluated by human subjects. The raw scores were processed to yield difference mean opinion scores for validation and testing.

TABLE 21.1 Performance of different QA methods

	Performance	
Model	LCC	SROCC
PSNR	0.8709	0.8755
Sarnoff JND	0.9266	0.9291
Multiscale SSIM	0.9393	0.9527
IFC	0.9441	0.9459
VIF	0.9533	0.9584
VSNR	0.9233	0.9278

Usually, the predicted quality scores from a QA method are fitted to the subjective quality scores using a monotonic nonlinear function to account for any nonlinearities in the objective model. Numerical methods are used to do this fitting. For the results presented here, a five-parameter nonlinearity (a logistic function with additive linear term) was used, and the mapping function used is given by

$$\text{Quality}(x) = \beta_1 \text{logistic}(\beta_2, (x - \beta_3)) + \beta_4 x + \beta_5, \quad (21.45)$$

$$\text{logistic}(\tau, x) = \frac{1}{2} - \frac{1}{1 + \exp(\tau x)}. \quad (21.46)$$

Table 21.1 quantifies the performance of the various methods in terms of well-known validation quantities: the linear correlation coefficient (LCC) between objective model prediction and subjective quality and the Spearman rank order correlation coefficient (SROCC) between them. Clearly, several of these quality metrics correlate very well with visual perception. The performance of IFC and multiscale SSIM indices is comparable, which is not surprising in view of the discussion in Section 21.4.3. Interestingly, the SSIM index correlates very well with visual perception despite its simplicity and ease of computation.

21.6 CONCLUSION

Hopefully, the reader has captured an understanding of the basic principles and difficulties underlying the problem of image QA. Even when there is a reference image available, as we have assumed in this chapter, the problem remains difficult owing to the subtleties and remaining mysteries of human visual perception. Hopefully, the reader has also found that recent progress has been significant, and that image QA algorithms exist that correlate quite highly with human judgments. Ultimately, it is hoped that confidence in these algorithms will become high enough that image quality algorithms can be used as surrogates for human subjectivity.

Naturally, significant problems remain. The use of partial image information instead of a reference image—so-called reduced reference image QA—presents interesting opportunities where good performance can be achieved in realistic applications where only partial data about the reference image may be available. More difficult yet is the situation where no reference image information is available. This problem, called no-reference or blind image QA, is very difficult to approach unless there is at least some information regarding the types of distortions that might be encountered [5].

An interesting direction for future work is the further use of image QA algorithms as objective functions for image optimization problems. For example, the SSIM index has been used to optimize several important image processing problems, including image restoration, image quantization, and image denoising [9–12]. Another interesting line of inquiry is the use of image quality algorithms—or variations of them—for other purposes than image quality assessment—such as speech quality assessment [4].

Lastly, we have not covered methods for assessing the quality of digital videos. There are many sources of distortion that may occur owing to time-dependent processing of videos, and interesting aspects of spatio-temporal visual perception come into play when developing algorithms for video QA. Such algorithms are by necessity more involved in their construction and complex in their execution. The reader is encouraged to read Chapter 14 of the companion volume, *The Essential Guide to Video Processing*, for a thorough discussion of this topic.

REFERENCES

[1] Z. Wang and X. Shang. Spatial pooling strategies for perceptual image quality assessment. In *IEEE International Conference on Image Processing*. IEEE, Dept. of EE, Texas Univ. at Arlington, TX, USA, January 1996.

[2] Z. Wang and A. C. Bovik. Embedded foveation image coding. *IEEE Trans. Image Process.*, 10(10):1397–1410, 2001.

[3] Z. Wang, L. Lu, and A. C. Bovik. Foveation scalable video coding with automatic fixation selection. *IEEE Trans. Image Process.*, 12(2):243–254, 2003.

[4] Z. Wang and A. C. Bovik. Mean squared error: love it or leave it? A new look at signal fidelity measures. *IEEE Signal Process. Mag.*, to appear, January 2009.

[5] Z. Wang and A. C. Bovik. *Modern Image Quality Assessment*. Morgan and Claypool Publishing Co., San Rafael, CA, 2006.

[6] A. B. Watson. DCTune: a technique for visual optimization of dct quantization matrices for individual images. *Soc. Inf. Disp. Dig. Tech. Pap.*, 24:946–949, 1993.

[7] H. R. Sheikh, M. F. Sabir, and A. Bovik. A statistical evaluation of recent full reference image quality assessment algorithms. *IEEE Trans. Image Process.*, 15(11):3440–3451, 2006.

[8] LIVE image quality assessment database. 2003. http://live.ece.utexas.edu/research/quality/subjective.htm

[9] S. S. Channappayya, A. C. Bovik, and R. W. Heath, Jr. A linear estimator optimized for the structural similarity index and its application to image denoising. In *IEEE Intl. Conf. Image Process.*, Atlanta, GA, January 2006.

[10] S. S. Channappayya, A. C. Bovik, C. Caramanis, and R. W. Heath, Jr. Design of linear equalizers optimized for the structural similarity index. *IEEE Trans. Image Process.*, to appear, 2008.

[11] S. S. Channappayya, A. C. Bovik, and R. W. Heath, Jr. Rate bounds on SSIM index of quantized images. *IEEE Trans. Image Process.*, to appear, 2008.

[12] S. S. Channappayya, A. C. Bovik, R. W. Heath, Jr., and C. Caramanis. Rate bounds on the SSIM index of quantized image DCT coefficients. In *Data Compression Conf.*, Snowbird, Utah, March 2008.

[13] K. Seshadrinathan and A. C. Bovik. Unifying analysis of full reference image quality assessment. To appear in *IEEE Intl. Conf. on Image Process.*, 2008.

[14] M. J. Wainwright, E. P. Simoncelli, and A. S. Willsky. Random cascades on wavelet trees and their use in analyzing and modeling natural images. *Appl. Comput. Harmonic Anal.*, 11(1):89–123, 2001.

[15] E. P. Simoncelli and W. T. Freeman. The steerable pyramid: a flexible architecture for multi-scale derivative computation. In *Proc. Intl. Conf. on Image Process.*, Vol. 3, January 1995.

[16] Z. Wang and E. P. Simoncelli. Stimulus synthesis for efficient evaluation and refinement of perceptual image quality metrics. *Proc. SPIE*, 5292(1):99–108, 2004.

[17] K. Seshadrinathan and A. C. Bovik. Unified treatment of full reference image quality assessment algorithms. Submitted to the *IEEE Trans. on Image Process.*

[18] D. J. Heeger. Normalization of cell responses in cat striate cortex. *Vis. Neurosci.*, 9(2):181–197, 1992.

[19] Z. Wang, E. P. Simoncelli, and A. C. Bovik. Multiscale structural similarity for image quality assessment. In *Thirty-Seventh Asilomar Conf. on Signals, Systems and Computers*, Pacific Grove, CA, 2003.

[20] Z. Wang and E. P. Simoncelli. Translation insensitive image similarity in complex wavelet domain. In *IEEE Intl. Conf. Acoustics, Speech, and Signal Process.*, Philadelphia, PA, 2005.

[21] M. J. Wainwright and E. P. Simoncelli. Scale mixtures of gaussians and the statistics of natural images. In S. A. Solla, T. Leen, and S.-R. Muller, editors, *Advance Neural Information Processing Systems*, 12:855–861, MIT Press, Cambridge, MA, 1999.

[22] T. W. Anderson. *An Introduction to Multivariate Statistical Analysis*. John Wiley and Sons, New York, 1984.

[23] I. M. Gelfand and A. M. Yaglom. Calculation of the amount of information about a random function contained in another such function. *Amer. Math. Soc. Transl.*, 12(2):199–246, 1959.

[24] S. Kullback. *Information Theory and Statistics*. Dover Publications, Mineola, NY, 1968.

[25] H. R. Sheikh and A. C. Bovik. Image information and visual quality. *IEEE Trans. Image Process.*, 15(2):430–444, 2006.

[26] H. R. Sheikh, A. C. Bovik, and G. de Veciana. An information fidelity criterion for image quality assessment using natural scene statistics. *IEEE Trans. Image Process.*, 14(12):2117–2128, 2005.

[27] J. Ross and H. D. Speed. Contrast adaptation and contrast masking in human vision. *Proc. Biol. Sci.*, 246(1315):61–70, 1991.

[28] Z. Wang and A. C. Bovik. A universal image quality index. *IEEE Signal Process. Lett.*, 9(3):81–84, 2002.

[29] Z. Wang, A. C. Bovik, H. R. Sheikh, and E. P. Simoncelli. Image quality assessment: from error visibility to structural similarity. *IEEE Trans. Image Process.*, 13(4):600–612, 2004.

[30] D. M. Chandler and S. S. Hemami. VSNR: a wavelet-based visual signal-to-noise ratio for natural images. *IEEE Trans. Image Process.*, 16(9):2284–2298, 2007.

[31] A. Watson and J. Solomon. Model of visual contrast gain control and pattern masking. *J. Opt. Soc. Am. A Opt. Image Sci. Vis.*, 14(9):2379–2391, 1997.

[32] O. Schwartz and E. P. Simoncelli. Natural signal statistics and sensory gain control. *Nat. Neurosci.*, 4(8):819–825, 2001.

[33] J. Foley. Human luminance pattern-vision mechanisms: masking experiments require a new model. *J. Opt. Soc. Am. A Opt. Image Sci. Vis.*, 11(6):1710–1719, 1994.

[34] D. G. Albrecht and W. S. Geisler. Motion selectivity and the contrast-response function of simple cells in the visual cortex. *Vis. Neurosci.*, 7(6):531–546, 1991.

[35] R. Shapley and C. Enroth-Cugell. Visual adaptation and retinal gain controls. *Prog. Retin. Res.*, 3:263–346, 1984.

[36] T. N. Pappas, T. A. Michel, and R. O. Hinds. Supra-threshold perceptual image coding. In *Proc. Int. Conf. Image Processing (ICIP-96)*, Vol. I, 237–240, Lausanne, Switzerland, September 1996.

[37] S. Daly. The visible differences predictor: an algorithm for the assessment of image fidelity. In A. B. Watson, editor, *Digital Images and Human Vision*, 179–206. The MIT Press, Cambridge, MA, 1993.

[38] J. Lubin. The use of psychophysical data and models in the analysis of display system performance. In A. B. Watson, editor, *Digital Images and Human Vision*, 163–178. The MIT Press, Cambridge, MA, 1993.

[39] P. C. Teo and D. J. Heeger. Perceptual image distortion. In *Proc. Int. Conf. Image Processing (ICIP-94)*, Vol. II, 982–986, Austin, TX, November 1994.

[40] R. J. Safranek and J. D. Johnston. A perceptually tuned sub-band image coder with image dependent quantization and post-quantization data compression. In *Proc. ICASSP-89*, Vol. 3, Glasgow, Scotland, 1945–1948, May 1989.

[41] A. B. Watson. DCT quantization matrices visually optimized for individual images. In J. P. Allebach and B. E. Rogowitz, editors, *Human Vision, Visual Processing, and Digital Display IV*, Proc. SPIE, 1913, 202–216, San Jose, CA, 1993.

[42] R. J. Safranek. A JPEG compliant encoder utilizing perceptually based quantization. In B. E. Rogowitz and J. P. Allebach, editors, *Human Vision, Visual Processing, and Digital Display V*, Proc. SPIE, 2179, 117–126, San Jose, CA, 1994.

[43] D. L. Neuhoff and T. N. Pappas. Perceptual coding of images for halftone display. *IEEE Trans. Image Process.*, 3:341–354, 1994.

[44] R. Rosenholtz and A. B. Watson. Perceptual adaptive JPEG coding. In *Proc. Int. Conf. Image Processing (ICIP-96)*, Vol. I, 901–904, Lausanne, Switzerland, September 1996.

[45] I. Höntsch and L. J. Karam. Apic: adaptive perceptual image coding based on subband decomposition with locally adaptive perceptual weighting. In *Proc. Int. Conf. Image Processing (ICIP-97)*, Vol. I, 37–40, Santa Barbara, CA, October 1997.

[46] I. Höntsch, L. J. Karam, and R. J. Safranek. A perceptually tuned embedded zerotree image coder. In *Proc. Int. Conf. Image Processing (ICIP-97)*, Vol. I, 41–44, Santa Barbara, CA, October 1997.

[47] I. Höntsch and L. J. Karam. Locally adaptive perceptual image coding. *IEEE Trans. Image Process.*, 9:1472–1483, 2000.

[48] I. Höntsch and L. J. Karam. Adaptive image coding with perceptual distortion control. *IEEE Trans. Image Process.*, 9:1472–1483, 2000.

[49] A. B. Watson, G. Y. Yang, J. A. Solomon, and J. Villasenor. Visibility of wavelet quantization noise. *IEEE Trans. Image Process.*, 6:1164–1175, 1997.

[50] P. G. J. Barten. The SQRI method: a new method for the evaluation of visible resolution on a display. In *Proc. Society for Information Display*, Vol. 28, 253–262, 1987.

[51] J. Sullivan, L. Ray, and R. Miller. Design of minimum visual modulation halftone patterns. *IEEE Trans. Syst., Man, Cybern.*, 21:33–38, 1991.

[52] M. Analoui and J. P. Allebach. Model based halftoning using direct binary search. In B. E. Rogowitz, editor, *Human Vision, Visual Processing, and Digital Display III*, Proc. SPIE, 1666, 96–108, San Jose, CA, 1992.

[53] J. B. Mulligan and A. J. Ahumada, Jr. Principled halftoning based on models of human vision. In B. E. Rogowitz, editor, *Human Vision, Visual Processing, and Digital Display III*, Proc. SPIE, 1666, 109–121, San Jose, CA, 1992.

[54] T. N. Pappas and D. L. Neuhoff. Least-squares model-based halftoning. In B. E. Rogowitz, editor, *Human Vision, Visual Processing, and Digital Display III*, Proc. SPIE, 1666, 165–176, San Jose, CA, 1992.

[55] T. N. Pappas and D. L. Neuhoff. Least-squares model-based halftoning. *IEEE Trans. Image Process.*, 8:1102–1116, 1999.

[56] R. Hamberg and H. de Ridder. Continuous assessment of time-varying image quality. In B. E. Rogowitz and T. N. Pappas, editors, *Human Vision and Electronic Imaging II*, Proc. SPIE, 3016, 248–259, San Jose, CA, 1997.

[57] H. de Ridder. Psychophysical evaluation of image quality: from judgement to impression. In B. E. Rogowitz and T. N. Pappas, editors, *Human Vision and Electronic Imaging III*, Proc. SPIE, 3299, 252–263, San Jose, CA, 1998.

[58] ITU/R Recommendation BT.500-7, 10/1995. http://www.itu.ch.

[59] T. N. Cornsweet. *Visual Perception*. Academic Press, New York, 1970.

[60] C. F. Hall and E. L. Hall. A nonlinear model for the spatial characteristics of the human visual system. *IEEE Trans. Syst., Man. Cybern.*, SMC-7:162–170, 1977.

[61] T. J. Stockham. Image processing in the context of a visual model. *Proc. IEEE*, 60:828–842, 1972.

[62] J. L. Mannos and D. J. Sakrison. The effects of a visual fidelity criterion on the encoding of images. *IEEE Trans. Inform. Theory*, IT-20:525–536, 1974.

[63] J. J. McCann, S. P. McKee, and T. H. Taylor. Quantitative studies in the retinex theory. *Vision Res.*, 16:445–458, 1976.

[64] J. G. Robson and N. Graham. Probability summation and regional variation in contrast sensitivity across the visual field. *Vision Res.*, 21:419–418, 1981.

[65] G. E. Legge and J. M. Foley. Contrast masking in human vision. *J Opt Soc Am*, 70(12):1458–1471, 1980.

[66] G. E. Legge. A power law for contrast discrimination. *Vision Res.*, 21:457–467, 1981.

[67] B. G. Breitmeyer. *Visual Masking: An Integrative Approach*. Oxford University Press, New York, 1984.

[68] A. J. Seyler and Z. L. Budrikas. Detail perception after scene change in television image presentations. *IEEE Trans. Inform. Theory*, IT-11(1):31–43, 1965.

[69] Y. Ninomiya, T. Fujio, and F. Namimoto. Perception of impairment by bit reduction on cut-changes in television pictures. (in Japanese). *Electr. Commun. Assoc. Essay Periodical*, J62-B(6):527–534, 1979.

[70] W. J. Tam, L. Stelmach, L. Wang, D. Lauzon, and P. Gray. Visual masking at video scene cuts. In B. E. Rogowitz and J. P. Allebach, editors, *Proceedings of the SPIE Conference on Human Vision, Visual Processing and Digital Display VI*, Proc. SPIE, 2411, 111–119, San Jose, CA, 1995.

[71] D. H. Kelly. Visual response to time-dependent stimuli. *J. Opt. Soc. Am.*, 51:422–429, 1961.

[72] D. H. Kelly. Flicker fusion and harmonic analysis. *J. Opt. Soc. Am.*, 51:917–918, 1961.

[73] D. H. Kelly. Flickering patterns and lateral inhibition. *J. Opt. Soc. Am.*, 59:1361–1370, 1961.

[74] D. A. Silverstein and J. E. Farrell. The relationship between image fidelity and image quality. In *Proc. Int. Conf. Image Processing (ICIP-96)*, Vol. II, 881–884, Lausanne, Switzerland, September 1996.

[75] C. A. Poynton. *A Technical Introduction to Digital Video*. Wiley, New York, 1996.

[76] A. B. Watson. The cortex transform: rapid computation of simulated neural images. *Comput. Vision, Graphics, and Image Process.*, 39:311–327, 1987.

[77] P. J. Burt and E. H. Adelson. The Laplacian pyramid as a compact image code. *IEEE Trans. Commun.*, 31:532–540, 1983.

[78] W. T. Freeman and E. H. Adelson. The design and use of steerable filters. *IEEE Trans. Pattern Anal. Mach. Intell.*, 13:891–906, 1991.

[79] E. P. Simoncelli, W. T. Freeman, E. H. Adelson, and D. J. Heeger. Shiftable multiscale transforms. *IEEE Trans. Inform. Theory*, 38:587–607, 1992.

[80] P. C. Teo and D. J. Heeger. Perceptual image distortion. In B. E. Rogowitz and J. P. Allebach, editors, *Human Vision, Visual Processing, and Digital Display V*, Proc. SPIE, 2179, 127–141, San Jose, CA, 1994.

[81] R. J. Safranek. A comparison of the coding efficiency of perceptual models. In *Human Vision, Visual Processing, and Digital Display VI*, Proc. SPIE, 2411, 83–91, San Jose, CA, February 1995.

[82] C. J. van den Branden Lambrecht and O. Verscheure. Perceptual quality measure using a spatio-temporal model of the human visual system. In V. Bhaskaran, F. Sijstermans, and S. Panchanathan, editors, *Digital Video Compression: Algorithms and Technologies*, Proc. SPIE, 2668, 450–461, San Jose, CA, January/February 1996.

[83] J. Chen and T. N. Pappas. Perceptual coders and perceptual metrics. In B. E. Rogowitz and T. N. Pappas, editors, *Human Vision and Electronic Imaging VI*, Proc. SPIE, 4299, 150–162, San Jose, CA, January 2001.

[84] A. Said and W. A. Pearlman. A new fast and efficient image codec based on set partitioning in hierarchical trees. *IEEE Trans. Circuits Syst. Video Technol.*, 6:243–250, 1996.

[85] H. A. Peterson, A. J. Ahumada, Jr., and A. B. Watson. An improved detection model for DCT coefficient quantization. In J. P. Allebach and B. E. Rogowitz, editors, *Human Vision, Visual Processing, and Digital Display IV*, Proc. SPIE, 1913, 191–201, San Jose, CA, 1993.

[86] B. E. Usevitch. A tutorial on modern lossy wavelet image compression: foundations of JPEG 2000. *IEEE Signal Process. Mag.*, 18:22–35, 2001.

[87] A. Skodras, C. Christopoulos, and T. Ebrahimi. The JPEG 2000 still image compression standard. *IEEE Signal Process. Mag.*, 18:36–58, 2001.

[88] D. S. Taubman and M. W. Marcellin. JPEG2000: standard for interactive imaging. *Proc. IEEE*, 90:1336–1357, 2002.

[89] J. M. Shapiro. Embedded image coding using zerotrees of wavelet coefficients. *IEEE Trans. Signal Process.*, SP-41:3445–3462, 1993.

[90] D. Taubman. High performance scalable image compression with ebcot. *IEEE Trans. Image Process.*, 9:1158–1170, 2000.

[91] A. Cohen, I. Daubechies, and J. C. Feauveau. Biorthogonal bases of compactly supported wavelets. *Commun. Pure Appl. Math.*, 45:485–560, 1992.

[92] A. J. Ahumada, Jr. and H. A. Peterson. A visual detection model for DCT coefficient quantization. In *AIAA Computing in Aerospace 9: A Collection of Technical Papers*, 314–317, San Diego, CA, October 1993.

[93] M. P. Eckert and A. P. Bradley. Perceptual quality metrics applied to still image compression. *Signal Process.*, 70:177–200, 1998.

[94] S. Daly. Subroutine for the generation of a two dimensional human visual contrast sensitivity function. Technical Report 233203Y, Eastman Kodak, Rochester, NY, 1987.

[95] D. A. Silverstein and S. A. Klein. A DCT image fidelity metric for application to a text-based scheme for image display. In J. P. Allebach and B. E. Rogowitz, editors, *Human Vision, Visual Processing, and Digital Display IV*, Proc. SPIE, 1913, 229–239, San Jose, CA, 1993.

[96] S. J. P. Westen, R. L. Lagendijk, and J. Biemond. Perceptual image quality based on a multiple channel HVS model. In *Proc. ICASSP-95*, Vol. 4, 2351–2354, Detroit, MI, May 1995.

[97] M. J. Horowitz and D. L. Neuhoff. Image coding by perceptual pruning with a cortical snapshot indistinguishability criterion. In B. E. Rogowitz and T. N. Pappas, editors, *Human Vision and Electronic Imaging III*, Proc. SPIE, 3299, 330–339, San Jose, CA, 1998.

[98] N. Bekkat and A. Saadane. Coded image quality assessment based on a new contrast masking model. *J. Electron. Imaging*, 13:341–348, 2004.

[99] S. Winkler and S. Süsstrunk. Visibility of noise in natural images. In *Human Vision and Electronic Imaging IX*, Proc. SPIE, 5292, San Jose, CA, January 2004.

[100] C. Fenimore, B. Field, and C. V. Degrift. Test patterns and quality metrics for digital video compression. In B. E. Rogowitz and T. N. Pappas, editors, *Human Vision and Electronic Imaging II*, Proc. SPIE, 3016, 269–276, San Jose, CA, February 1997.

[101] J. M. Libert and C. Fenimore. Visibility thresholds for compression-induced image blocking: measurement and models. In B. E. Rogowitz and T. N. Pappas, editors, *Human Vision and Electronic Imaging IV*, Proc. SPIE, 3644, 197–206, San Jose, CA, January 1999.

[102] E. M. Yeh, A. C. Kokaram, and N. G. Kingsbury. A perceptual distortion measure for edge-like artifacts in image sequences. In B. E. Rogowitz and T. N. Pappas, editors, *Human Vision and Electronic Imaging III*, Proc. SPIE, 3299, 160–172, San Jose, CA, January 1998.

[103] P. J. Hahn and V. J. Mathews. An analytical model of the perceptual threshold function for multichannel image compression. In *Proc. Int. Conf. Image Processing (ICIP-98)*, Vol. III, 404–408, Chicago, IL, October 1998.

[104] M. G. Ramos and S. S. Hemami. Suprathreshold wavelet coefficient quantization in complex stimuli: psychophysical evaluation and analysis. *J. Opt. Soc. Am. A*, 18:2385–2397, 2001.

[105] D. M. Chandler and S. S. Hemami. Additivity models for suprathreshold distortion in quantized wavelet-coded images. In B. E. Rogowitz and T. N. Pappas, editors, *Human Vision and Electronic Imaging VII*, Proc. SPIE, 4662, 105–118, San Jose, CA, January 2002.

[106] D. M. Chandler and S. S. Hemami. Effects of natural images on the detectability of simple and compound wavelet subband quantization distortions. *J. Opt. Soc. Am. A*, 20(7):1164–1180, 2003.

[107] D. M. Chandler and S. S. Hemami. Suprathreshold image compression based on contrast allocation and global precedence. In B. E. Rogowitz and T. N. Pappas, editors, *Human Vision and Electronic Imaging VIII*, Proc. SPIE, 5007, Santa Clara, CA, January 2003.

[108] D. M. Chandler and S. S. Hemami. Contrast-based quantization and rate control for wavelet-coded images. In *Proc. Int. Conf. Image Processing (ICIP-02)*, Rochester, NY, September 2002.

[109] T. N. Pappas, J. P. Allebach, and D. L. Neuhoff. Model-based digital halftoning. *IEEE Signal Process. Mag.*, 20:14–27, 2003.

[110] W. Qian and B. Kimia. On the perceptual notion of scale for halftone representations: nonlinear diffusion. In B. E. Rogowitz and T. N. Pappas, editors, *Human Vision and Electronic Imaging*, Proc. SPIE, 3299, 473–481, San Jose, CA, January 1998.

[111] P. Lindh and C. J. van den Branden Lambrecht. Efficient spatio-temporal decomposition for perceptual processing of video sequences. In *Proc. Int. Conf. Image Processing (ICIP-96)*, Vol. III, 331–334, Lausanne, Switzerland, September 1996.

[112] S. Winkler. Quality metric design: a closer look. In B. E. Rogowitz and T. N. Pappas, editors, *Human Vision and Electronic Imaging V*, Proc. SPIE, 3959, San Jose, CA, January 2000.

[113] S. Winkler. Visual fidelity and perceived quality: towards comprehensive metrics. In B. E. Rogowitz and T. N. Pappas, editors, *Human Vision and Electronic Imaging VI*, Proc. SPIE, 4299, San Jose, CA, January 2001.

[114] A. M. Rohaly, J. Lu, N. R. Franzen, and M. K. Ravel. Comparison of temporal pooling methods for estimating the quality of complex video sequences. In B. E. Rogowitz and T. N. Pappas, editors, *Human Vision and Electronic Imaging IV*, Proc. SPIE, 3644, 218–225, San Jose, CA, January 1999.

[115] D. Pearson. Viewer response to time-varying video quality. In B. E. Rogowitz and T. N. Pappas, editors, *Human Vision and Electronic Imaging III*, Proc. SPIE, 3299, 16–25, San Jose, CA, January 1998.

[116] A. B. Watson. Toward a perceptual video quality metric. In B. E. Rogowitz and T. N. Pappas, editors, *Human Vision and Electronic Imaging III*, Proc. SPIE, 3299, 139–147, San Jose, CA, January 1998.

[117] A. B. Watson, J. Hu, and J. F. McGowan, III. Digital video quality metric based on human vision. *J. Electron. Imaging*, 10:20–29, 2001.

[118] A. B. Watson, J. Hu, J. F. McGowan, and J. B. Mulligan. Design and performance of a digital video quality metric. In B. E. Rogowitz and T. N. Pappas, editors, *Human Vision and Electronic Imaging IV*, Proc. SPIE, 3644, 168–174, San Jose, CA, January 1999.

[119] R. O. Hinds and T. N. Pappas. Effect of concealment techniques on perceived video quality. In B. E. Rogowitz and T. N. Pappas, editors, *Human Vision and Electronic Imaging IV*, Proc. SPIE, 3644, 207–217, San Jose, CA, January 1999.

[120] K. Brunnström and B. N. Schenkman. Quality of video affected by packet loss distortion, compared to the predictions of a spatio-temporal model. In B. E. Rogowitz and T. N. Pappas, editors, *Human Vision and Electronic Imaging VII*, Proc. SPIE, 4662, San Jose, CA, January 2002.

Image Watermarking: Techniques and Applications

22

Anastasios Tefas, Nikos Nikolaidis, and Ioannis Pitas

Aristotle University of Thessaloniki

22.1 INTRODUCTION

Digital watermarking is a relatively new research area that has attracted the interest of numerous researchers both in academia and industry and has become one of the hottest research topics in the multimedia signal processing community. Although the term watermarking has slightly different meanings in the literature, one definition that seems to prevail is the following [1]: Watermarking is the practice of imperceptibly altering a piece of data in order to embed information about the data. The above definition reveals two important characteristics of watermarking. First, information embedding should not cause perceptible changes to the host medium (sometimes called cover medium or cover data). Second, the message should be related to the host medium. In this sense, the watermarking techniques form a subset of information hiding techniques, which also include cases where the hidden information is not related to the host medium (e.g., in covert communications). However, certain authors use the term watermarking with a meaning equivalent to that of information hiding in the general sense.

A watermarking system should consist of two distinct modules: a module that embeds the information in the host data and a module that detects if a given piece of data hosts a watermark and subsequently retrieves the conveyed information. Depending on the type, the amount, and the properties of the embedded information (e.g., robustness to host signal alterations), as well as the type of host data, watermarking can serve a multitude of applications as will be described in Section 22.2.

The first handful of papers on digital watermarking appeared in the late 1980s-early 1990s but very soon the area witnessed a tremendous growth and an explosion in the number of published papers, mainly due to the fact that people believed, at that stage, that watermarking could be a significant weapon in the battle against continuously increasing digital media piracy. During the early days, researchers focused mainly on a limited range of host data, that is, digital image, video, and audio data. Later on, watermarking techniques that are applicable to other media types appeared in the corresponding literature. Such media types include but are not limited to voxel-based 3D images, 3D models represented as polygonal meshes or parametric surfaces (e.g., NURBS surfaces), vector graphics, GIS data (e.g., isosurface contours), animation parameters, object-based video representations (e.g., MPEG 4 video objects), symbolic description of audio (e.g., MIDI files), text (either in ASCII format or as a binary image), software source code, binary executables, java byte code, and numeric data sets (stock market data, scientific data). This chapter will focus on still image watermarking. However, most of the principles and techniques that will be presented are readily applicable to other media types.

Although, in its first steps, watermarking was dominated by heuristic approaches without significant theoretical background and justification, soon researchers recognized that solid theoretical foundations had to be set and worked toward this direction by adopting and utilizing successful techniques, principles, and theoretical results from several scientific areas like communications (detection theory, error correction codes, spread spectrum communications), information theory (channel capacity), signal processing (signal transforms, compression techniques), and cryptography. Today, although the optimism of the early years is over, watermarking is still a very active research area, despite the failure of the currently available watermarking technology to serve the needs of the industry (as made clear by Secure Digital Music Initiative case [2]). Researchers are now very well aware that devising effective watermarking schemes, especially for the so-called security oriented applications (e.g., copyright protection, copy control, etc.), is an extremely difficult task. However, the introduction of new application scenarios and business models along with the small but steady steps toward solid theoretical foundations of this discipline and the combination of watermarking with other technologies like cryptography and perceptual hashing [3–5] are expected to keep the interest in this new area alive [6, 7]. For a thorough review of existing schemes and a detailed discussion on the main requirements of a watermarking scheme, the interested reader may consult books [1, 8–11] and several review papers and journal special issues [12–17] that have been published on this topic. The *IEEE Transactions on Information Forensics and Security* is another excellent source of information regarding the latest developments in the field.

This chapter is organized as follows. The main application domains of watermarking are reviewed in Section 22.2. Properties and classification schemes of watermarking techniques are presented in Section 22.3, whereas Section 22.4 presents the basic functional modules of a watermarking scheme. Finally Sections 22.5 and 22.6 delve in more detail into principles and techniques devised for two major application areas, namely copyright protection and authentication.

22.2 APPLICATIONS OF WATERMARKING TECHNIQUES

Watermarking can be the enabling technology for a number of important applications [18–20]. Obviously, each application imposes different requirements on the watermarking system. As a consequence, watermarking algorithms targeting different applications might be very different in nature. Furthermore, devising an efficient watermarking scheme might be much more difficult for certain applications. In the remainder of this section, we will briefly review some of the main application domains of watermarking.

- *Owner identification and proof of ownership.* This class of applications was the first to be considered in the watermarking literature. In this case, the embedded data can carry information about the legal owner or distributor or any rights holder of a digital item and be used for notifying/warning a user that the item is copyrighted, for tracking illegal copies of the item, or for possibly proving the ownership of the item in the case of a legal dispute.

- *Broadcast monitoring.* In this case, the embedded information is utilized for various functions that are related to digital media (audio, video) broadcasting. The embedded data can be used to verify whether the actual broadcasting of commercials took place as scheduled, i.e., whether proper airtime allocation occurred, for devising an automated royalty collection scheme for copyrighted material (songs, movies) that is aired by broadcasting operators, or to collect information about the number of people who watched/listened to a certain broadcast (audience metering). Broadcast monitoring is usually performed by automated monitoring stations and is one of the watermarking applications that has found its way toward successful commercialization.

- *Transaction tracking.* In this application, each copy of a digital item that is distributed as part of a transaction bears a different watermark. The aim of this watermark is not only to carry information about the legal owner/distributor of the digital item but also to mark the specific transaction copy. As a consequence, the embedded information can be used for the identification of entities that illegally distributed the digital item or did not adopt efficient security measures for preventing the item from being copied or distributed and for deterring such actions. Identification of movie theaters where illegal recording of a movie with a handheld camera took place is a scenario that belongs to this category of applications. The watermarks used in such cases are often termed fingerprints and the corresponding application fingerprinting. However, the same term is sometimes used for the class of techniques that try to extract a unique descriptor (fingerprint) for each digital item, which is invariant to content manipulation [3–5, 21]. Obviously these techniques (which are sometimes called perceptual or robust hashing or replica detection techniques) are totally different from watermark-based fingerprinting, since they do not embed any data on the digital item, i.e., they are passive techniques.

- *Usage control.* In contrast to the applications mentioned above, where watermarking is used to deter intellectual rights infringement or to help in identifying such infringements, in usage control applications, the watermarking plays an active protection role by controlling the terms of use of the digital content. The embedded information can be used in conjunction with appropriate compliant devices to prohibit unauthorized recording of a digital item (copy control) or playback of unauthorized copies (playback control). The DVD copy and playback control using watermarking complemented by content scrambling is an example of this application [20, 22].

- *Authentication and tamper-proofing.* In this case, the role of the watermark is to verify the authenticity and integrity of a digital item for the benefit of either the owner/distributor or the user. Example applications include the authentication of surveillance videos in case their integrity is disputed [23], the authentication of critical documents (e.g., passports), and the authentication of news photos distributed by a news agency. In this context, the watermarking techniques can either signal an authentication violation even when the digital item is slightly altered or tolerate certain manipulations (e.g., valid mainstream lossy content compression) and declare an item as nonauthentic only when "significant" alterations have occurred (e.g., content editing). Certain watermarking methods used for authentication can provide tampered region localization, e.g., can detect the image regions that have been modified/edited.

- *Persistent item identification.* According to this concept, watermarking is used for associating an identifier with a digital item in a way that resists certain content alterations. This identifier can be used, in conjunction with appropriate databases, to convey various information about the digital item. Depending on the related information, persistent identification can be the vehicle for some of the applications presented above, e.g., owner identification, or usage control. Furthermore, the attached information can be used both for carrying copyright information and for enhancing the host data functionalities, e.g., by providing access to free services and products, thus, implicitly, discouraging the user from removing the watermark or illegally distributing the item and thus losing the added value provided by the watermark. Persistent association is dealt with in the MPEG-21 standard.

- *Enhancement of legacy systems.* Data embedded through watermarking can be used for the enhancement of information or functionalities carried/provided by legacy systems while ensuring backwards compatibility. For example, using techniques capable of generating watermarks that are robust to analog to digital and digital to analog conversion, one can embed in a digital image URLs that are related to the depicted objects. When such an image is printed (e.g., in a magazine) and then scanned by a reader, the embedded URL can be used for connecting her automatically to the corresponding webpage [24]. Digital data embedding in conventional analog PAL/SECAM signals is another application in this category. In a more "futuristic" scenario, one can envision that information capable of enabling stereoscopic viewing to stereo-enabled receivers could be embedded through watermarking in conventional digital TV broadcasts. Using such an approach, conventional TV receivers would continue to receive the conventional signal with—hopefully—nonperceptible degradations.

22.3 CLASSIFICATION OF WATERMARKING ALGORITHMS

Various types of watermarking techniques each with their own distinct properties and characteristics can be found in the watermarking literature. In the following, we will review the basic categories of watermarking schemes and provide descriptions for the properties that distinguish each class from the rest.

A first classification of watermarking schemes can be organized on the basis of their resistance to host medium modifications. Such modifications can either be the result of common signal processing operations (e.g., lossy compression) or be specifically devised and applied in order to render the watermark undetectable or affect the credibility and reliability of a watermarking system in other ways. Such modifications are usually referred to as *attacks*. Attacks for intellectual property rights (IPR) protection watermarking systems will be discussed in Section 22.5.2. The degree of resistance of a watermarking method to host medium modifications is usually called robustness. Depending on the level of robustness offered, one can distinguish between the following categories of watermarking techniques:

- *Robust.* In this class, the watermarks are designed so as to resist host signal manipulations and are usually employed in IPR protection applications. Obviously, no watermarking scheme can resist all types of modifications, regardless of their severity. As a consequence, robustness refers to a subset of all possible manipulations and up to a certain degree of host signal degradation.

- *Fragile.* In this case, the watermarks are designed to be vulnerable to all modifications, i.e., they become undetectable by even the slightest host data modification. Fragile watermarks are more easy to devise than robust ones and are usually applied in authentication scenarios.

- *Semifragile.* This class of watermarks provides selective robustness to a certain set of manipulations which are considered as legitimate and allowable, while being vulnerable (fragile) to others. Such watermarks can also be used in authentication cases instead of fragile ones. In practice, all robust watermarks are essentially semifragile, but in the former case, the selective robustness is not a requirement imposed by the system designer but rather something that cannot be avoided.

In order to achieve a sufficient level of security, watermark embedding and detection are usually controlled by a (usually secret) key K (see Section 22.4). In a way analogous to cryptographic systems, the watermarking schemes can be distinguished in two classes on the basis of whether the same key is used during embedding and detection:

- *Symmetric or privatekey.* In such schemes, both watermark embedding and detection are performed using the same key K.

- *Asymmetric or publickey.* In contrast to the previous class, these watermarks can be detected with a key that is different than the one that was used in the embedding stage [25, 26]. Actually, a pair of keys is used in this case: a private key to generate the watermark for embedding, and a public one for detection. For each private key, many public keys may be produced. Despite their advantages over their symmetric counterparts, asymmetric schemes are much more difficult to devise.

In terms of the information taken into account during embedding, the watermarking methods can be broadly classified in two categories:

- *Blind embedding schemes.* Schemes belonging to this category consider the host data as noise or interference. Therefore, these techniques essentially treat watermarking like the classical communications problem of signal transmission over a noisy channel, the only difference being that, in the case of watermarking, restrictions on the amount of distortions imposed on the channel (i.e., the host medium) by the signal (the watermark) should be taken into consideration. In most cases, these methods rely implicitly or explicitly on a certain degree of knowledge of the host signal statistics, thus leading to the subclass of "known host statistics" methods. Essentially all methods developed in the first years of watermarking research belong to this category, most of them revolving around the spread spectrum principle where the watermark signal consists of a pseudorandom sequence embedded, usually in an additive way, in the host signal.

- *Informed coding/embedding schemes.* These schemes emerged after the work of Cox *et al.* [27] and exploit the fact that during embedding, not only the statistics of the host data but also the actual host data themselves are known. Knowledge of the host data can be utilized to improve watermark detection performance through interference cancellation. These methods are also known as known host state methods and treat watermarking as a problem of communication with side information at the transmitter. Many of these schemes make use of the quantization index modulation (QIM) principle [28] for message coding where embedding is achieved by quantizing the host signal or certain derived features using appropriately selected quantizers. Quantizer selection is controlled by the signal to be embedded and aims at minimizing host signal interference. Perceptual masking, i.e., utilization of the host signal along with principles of human perception in order to modify the watermark in a way that renders it imperceptible, is another form of informed embedding. Both informed coding/embedding and perceptual masking will be reviewed later on in this chapter.

With respect to the information conveyed by the watermark, watermarking systems can be classified to one of the following two classes:

- *Zero-bit systems:* Watermarking systems of this type can only check whether the data under consideration host a watermark generated by a specific key K, i.e., verify whether the data are watermarked or not. Certain authors use the term single-bit when referring to systems of this category, implying that the existence or absence of a specific watermark in the data essentially conveys one bit of information. The term watermark detection is used in this chapter to denote the procedure used to declare the presence of a watermark when one is indeed present in the host data and come up with a "no watermark present" answer when applied to data hosting no watermark or hosting a different watermark than the one under investigation.

- *Multiple-bit systems*: These systems are capable of encoding a multiple-bit message in the host data. For systems of this type, we make the distinction between watermark detection and message decoding. The data under investigation are first tested to verify whether they host a watermark or not. This procedure is identical to the detection procedure described above for zero-bit watermarks. As soon as the detection algorithm declares that the data are indeed watermarked, the embedded message is decoded. Thus, for multiple-bit systems, watermark detection and message decoding should be considered as two distinct steps that are performed in cascade, the message decoding step taking place only if a watermark has been found to reside in the data.

When it comes to watermark detection, watermarking methods can be categorized into two main classes:

- Techniques that require that the original signal is available during the detection phase. These schemes are referred to as private, nonblind, or nonoblivious schemes (see, for example, [29, 30]). Nonblind schemes can be considered as the extremum of a more general category, that of informed detection schemes (e.g., [31]), which include methods that require that some sort of information related to the host signal (e.g., its original dimensions or a feature vector derived from the host signal) is available at the detector.

- Techniques that do not require the original signal (or other information about it) for watermark detection. These techniques are called oblivious or blind. Due to their wider scope of application, blind techniques received much more attention among researchers. Obviously, the lack of knowledge on the original host signal makes blind detection a much more difficult task than nonblind detection. Correlation-based detection, where the decision on the watermark presence is obtained by evaluating the correlation between the watermark and the signal under investigation, is an approach that belongs in this category. Correlation detection schemes implicitly assume that the host signal is Gaussian. Due to their simplicity, they were very popular in the early days of watermarking (see, for example, [32–34]). Later on, a number of researchers tried to devise optimal detectors for a number of situations, where the Gaussianity assumption does not hold [35–41]. Both correlation and optimal detectors will be reviewed later on in this chapter.

With respect to the output of the watermark detection procedure, systems are categorized as follows:

- *Hard decision detectors* generate a binary output (watermark detected, watermark not detected).

- *Soft decision detectors* provide along with the binary output a real number which is essentially the value of the test statistic used for detection (e.g., the value of the correlation between the signal under investigation and the watermark) and is related to detection reliability. In this case, the binary decision is obtained by internally thresholding this number using an appropriately selected threshold.

22.4 WATERMARK EMBEDDING, DETECTION, AND DECODING

Having described the main categories of watermarking algorithms along with their characteristic properties, we can now proceed in providing more formal definitions of the watermark embedding, detection, and decoding procedures.

Watermark embedding can be performed in the spatial domain [32, 42–44] by modulating the intensity of preselected samples or by modifying the magnitude of selected coefficients in an appropriate transform domain, e.g., the discrete cosine transform (DCT) [29, 45–47], discrete fourier transform (DFT) [34, 48], or wavelet transform [33, 49, 50] domain. Watermark embedding can be considered as a function that involves the host medium f_o, the embedding key K (usually an integer), a set of parameters U that control the embedding procedure, and, in the case of multiple-bit schemes, the message m that is to be embedded in the data. The message can be a character string, a number, or even multimedia data (audio, images). However at this stage it suffices to consider the message as a sequence of bits. The set of parameters U can contain, among other things, the so-called watermark embedding factor, i.e., a parameter that controls the amount of degradation that will be inflicted to the host signal by the watermark. The output of the watermark embedding function consists of the watermarked data f_w. Thus, for multiple-bit schemes, the watermark embedding function is of the following form:

$$f_w = E(f_o, K, m, U),\tag{22.1}$$

whereas for zero-bit schemes m is not an input parameter of the function.

In certain cases, it is much more intuitive to view watermark embedding as a two-step procedure, i.e., a watermark generation step that results in the watermark signal w, followed by a watermark embedding step that aims at actually embedding w in the host data. For an informed embedding multiple-bit watermarking scheme, these two functions are of the following form:

$$w = E_1(f_o, K, m, U),\tag{22.2}$$

$$f_w = E_2(f_o, w, U).\tag{22.3}$$

Watermark detection, in the way that is defined in this chapter, can be considered as a function that receives as input the data f' under investigation, a key K' (which, depending on whether the system is a symmetric or an asymmetric one, can be the same as the embedding key or a different, public key) and, in case of nonblind schemes, the original data f_o. The output of this function is a binary digit d (0: watermark has been detected, 1: watermark has not been detected), complemented, in the case of soft decision detectors, by a value r (usually in the range $[0,1]$) that corresponds to the detection reliability. Therefore, the detection function takes the following form in the case of blind and nonblind schemes, respectively:

$$\{d,r\} = D(f',K'), \tag{22.4}$$

$$\{d,r\} = D(f',f_0,K'), \tag{22.5}$$

where, as mentioned before, the reliability value r is available only in the case of soft decision detectors.

In the case of multiple-bit watermarking systems, whenever $D()$ declares that the data are watermarked, message decoding takes place. In the case of blind schemes, this operation can be expressed as follows:

$$m = Dec(d,f',K'). \tag{22.6}$$

The detection output d has been included among the arguments of function Dec to denote the fact that Dec is called only if $d = 1$.

A schematic representation of the above procedures can be seen in Fig. 22.1.

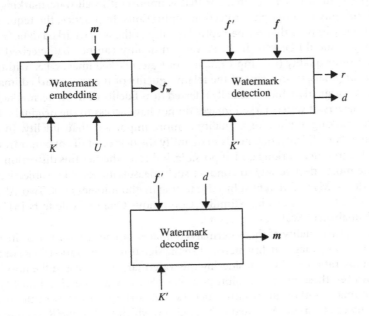

FIGURE 22.1

Modules of a watermarking system. Dashed lines indicate optional inputs or outputs.

22.5 COPYRIGHT PROTECTION WATERMARKING

Copyright protection or more generally digital rights protection and management is a major application domain of watermarking. It is actually an umbrella of applications that encompasses owner identification, proof of ownership, transaction tracking, copy and playback control, automated collection of royalties, etc. Watermarking techniques aiming at copyright protection have to face some very difficult challenges, since they have to cope with attempts to infringe digital rights through illegal copying, distribution, or playback of multimedia data. Copyright protection applications belong to the class of security-related applications, which are generally considered as the toughest watermarking applications. In this section, we will present the basic requirements of a copyright protection watermarking scheme, namely imperceptibility, cryptographical security, and robustness; briefly describe methods, metrics, and benchmarking platforms for measuring its performance; review the main categories of attacks against such a scheme; and then proceed with providing information about the most important approaches that have been proposed in the literature.

22.5.1 Requirements and Metrics

22.5.1.1 Imperceptibility—Visual Quality

By definition, host data alterations imposed by watermarks should be imperceptible. Thus, imperceptibility is a requirement that is important in all watermarking applications and not only in copyright protection applications. In practice, the requirement of imperceptibility implies that the perceptual quality of the watermarked data, in our case digital images, should be kept high. Perceptual quality can be characterized either in terms of absolute quality (or simply quality) of watermarked images, i.e., without reference to the originals, or in terms of the relative quality of the watermarked images with respect to the originals, which is usually referred to as fidelity of the watermarked images. Normally, viewers of watermarked images do not have access to the originals. Thus, for those watermarking applications, quality is more important than fidelity. In order to measure quality or fidelity, one needs to quantify the degree of distortion introduced to an image due to watermarking and, if possible, indicate whether this distortion is visible or not. The most effective way to conduct such measurements is by subjective evaluation procedures. Many different subjective testing methodologies exist: Two-Alternative, Forced-Choice tests [1], Double Stimulus Continuous Quality Scale tests [51], Double Stimulus Impairment Scale tests [51], etc.

The perceptual quality of the watermarked images can be also measured in a quantitative way by using image quality metrics like the signal-to-noise ratio (SNR) or the peak signal-to-noise ratio (PSNR), considering the watermark as noise and the host image as signal. However, these metrics exhibit poor correlation with the visual quality as perceived by humans. Other quantitative metrics that correlate better with the perceptual image quality can be used. Weighted PSNR [52, 53] which equals PSNR weighted at each image pixel by the local noise visibility function (local signal activity) could be such a metric. A wealth of image quality metrics are described in [54–56]. Moreover, a quality metric for images distorted by geometric attacks has been proposed in [57]. However, no globally agreeable and effective visual quality metric currently exists.

Obviously, the notion of "high quality" is application-dependent. For example, a certain amount of distortion on a movie might be perfectly acceptable for a television broadcast but unacceptable if the movie is to be displayed in cinema. It should also be noted that certain applications require that the alterations imposed on the image are not only perceptually insignificant but also very small in a numerical sense, i.e., they require that watermarking preserves the "numerical" quality of the data. Watermarking of medical images that are to be used as input in diagnosis procedures whose performance critically depends on the pixel intensities is such an example.

22.5.1.2 *Robustness*

As already mentioned in Section 22.3, robustness can be defined as the degree of resistance of a watermarking method to modifications of the host signal due to either common signal processing operations or operations devised specifically in order to render the watermark undetectable. Watermarking systems aiming at copyright protection should ideally exhibit high resistance to all attacks that might occur in the host data in a specific application. This means that the detection performance of the system, i.e., its ability to declare correctly the presence/absence of a watermark in an image, and in the case of multiple-bit systems, its decoding performance, i.e., its ability to retrieve successfully the hidden message bits, should not degrade significantly when data are altered due to intentional or unintentional attacks. Naturally, the set of manipulations that the watermark should be able to withstand and the severity of degradations that should be handled successfully depend on the target application. For example, a watermarking method designed to protect a database of high-quality/resolution images that are to be used in desktop publishing need not be able to withstand high compression as such a manipulation would make the images practically unusable and, thus, is not very likely to occur. In order to measure the robustness of a watermarking method, one should be able to measure the detection performance of the algorithm, usually in relation to the severity of the degradation imposed by a certain attack. Furthermore, in the case of multiple-bit algorithms, the decoding performance should be quantified [10].

Watermark Detection Performance Watermark detection can be considered as a hypothesis testing problem, the two hypotheses (events) being

- H_0: the image under test hosts the watermark under investigation.

- H_1: the image under test does not host the watermark under investigation.

Hypothesis H_1 can be further divided into two subhypotheses:

- H_{1a}: the image under test is not watermarked.

- H_{1b}: the image under test hosts a watermark different than the one under investigation.

Thus, the detection performance can be characterized by the false alarm (or false positive) error and its corresponding probability P_{fa}, i.e., the probability to detect a watermark in an image that is not watermarked or is watermarked with a different watermark than the one under investigation, and the false rejection (or false negative) error, described by the false rejection probability P_{fr}, i.e., the probability of not detecting a watermark in an image that is indeed watermarked with the watermark under investigation. Depending on the application, these two types of errors might have different importance. P_{fa} can be evaluated using detection trials with erroneous watermarks (hypothesis H_{1b}) or detection trials on nonwatermarked images (hypothesis H_{1a}). The former might sometimes be preferable, since it usually corresponds to the worst case scenario. Furthermore, false alarm probability evaluated on images watermarked by a different key than the one used for detection provides an indication on whether the keys in the algorithm "keyspace" are able to generate distinct "nonoverlapping" watermarks and, thus, lead to estimates of the "effective keyspace." One can distinguish between three types of false alarms and false rejections [1]: those evaluated on a single image using multiple keys, those evaluated on multiple images using a single key, and those evaluated on multiple images using multiple keys.

In the case of soft decision detectors (see Section 22.3), one can derive the empirical probability distribution functions (histograms) of the detection test statistic for both hypotheses H_0 and H_{1b} (or H_{1a}). By utilizing these empirical distributions, the probabilities of false alarm $P_{fa}(T_k)$ and false rejection $P_{fr}(T_k)$ as a function of the detection threshold T can be extracted. Using $P_{fa}(T_k)$ and $P_{fr}(T_k)$, we can plot the Receiver Operating Characteristic (ROC) curve, i.e., the plot of P_{fa} versus P_{fr} (Fig. 22.2). The ROC curve provides an overall view of the watermark detection performance in various operating conditions. Using the ROC curve, one can select the threshold value that gives a (P_{fa}, P_{fr}) pair satisfying the application requirements. Furthermore, the ROC curve can be used for the evaluation of other performance metrics, like the P_{fa} for a fixed, user-defined P_{fr}, the P_{fr} for a fixed, user-defined P_{fa}, and the equal error rate (EER), i.e., the point on the ROC where $P_{fa} = P_{fr}$ (Fig. 22.2).

Message Decoding Performance　The decoding performance of a watermarking method that supports message encoding can be characterized by the bit error rate (BER), i.e., the probability of erroneously decoding one message bit. Since message decoding is assumed to take place only in the case of successful detection, there is a close relation between the decoding and detection performance. As a consequence, a BER value should only be referenced along with the corresponding detection error probabilities, i.e., the probabilities of false alarm and false rejection.

Another metric that is related to the decoding performance of a watermarking algorithm is its payload, which can be defined as the maximum number of bits that can be encoded in a fixed amount of host data and decoded with a prespecified BER or alternatively as the amount of data required to host a fixed number of bits so that they can be decoded with a prespecified BER. Thus, the payload is expressed in information bits that can be embedded per host image pixel. The performance of the watermark decoder

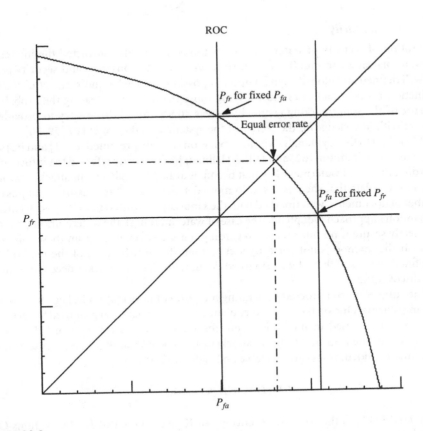

FIGURE 22.2

Receiver Operating Characteristic curve.

can be also measured by the probability of error P_e, defined as the probability of getting a wrong estimate of the hidden message m:

$$P_e \triangleq Pr\{\hat{m} \neq m\}. \tag{22.7}$$

Error correction codes can be used for correcting the watermark decoding errors. The probability of getting a correct estimate of the message by using an error correction code (e.g., BCH) that corrects R bit errors in M bits when the BER is q is given by [58]

$$P_d = \sum_{i=R}^{M} \frac{M!}{i!(M-i)!} q^i (1-q)^{M-i}. \tag{22.8}$$

22.5.1.3 *Security*

The notion of security of watermarking methods has recently attracted the interest of the watermarking community. The distinction between robustness and security is still not well defined and globally agreed upon. A possible, somewhat indirect, definition and distinction is that attacks to robustness are those that aim at increasing the probability of error of the watermarking channel whereas attacks to robustness try to provide an attacker with knowledge on the secrets of the system, e.g., the secret key [59, 60].

According to the cryptanalysis point of view on security presented in [61] and inspired by the works of Shannon [62] and Diffie-Hellman [63], security refers to the information regarding the secret watermark key that becomes available (leaks) to an attacker through watermarked data that she possesses. In more detail, the attacker is assumed to posses a number of documents, watermarked with the same key and different messages. According to Shannon's approach (adapted to the case of watermarking), the watermarking method is perfectly secure if no information regarding the secret key leaks from these "observations." If the method is not perfectly secure, then the security level of the method can be defined as the number of watermarked documents that an attacker needs in order to fully discover the key.

The authors in [61] proceed in defining measures of information leakage for a watermarking scheme. One such measure is equivocation which has been proposed by Shannon [62] and can be used in methods where the secret key is a binary word. Equivocation measures the uncertainty of an attacker on the key value $\mathbf{K}$ when N observations (watermarked documents) are available and is defined as:

$$H(\mathbf{K} \mid \mathbf{O}^N) = H(\mathbf{K}) - I(\mathbf{K}; \mathbf{O}^N), \tag{22.9}$$

where $H(\mathbf{K} \mid \mathbf{O}^N)$ is the conditional entropy of $\mathbf{K}$ given the set of N observations $\mathbf{O}^N$, $H(\mathbf{K})$ is the entropy of $\mathbf{K}$ (uncertainty on the value of the key when no observations are available), and $I(\mathbf{K}; \mathbf{O}^N)$ is the mutual information between the key and the N observations, which is the measure of information leakage due to the available observations. An equivocation equal to zero corresponds to exact knowledge of the key. The minimum number of observations that are required to achieve zero equivocation can be thought of as a measure of the security level of the algorithm. Another measure of information leakage is based on Fisher's information matrix that measures the information provided by a number of observations (in our case, watermarked data) about an unknown parameter (the watermark key). More details on this measure as well as information on how this framework can be used to calculate the security level of some standard watermarking methods can be found in [61].

It is important to note that, in compliance with one of the basic principles of cryptography, namely Kerckhoff's principle, the security of a copyright protection watermarking system should be based on the secrecy of keys that are used to embed/detect the watermark rather than on the secrecy of the algorithms. This means that designers of a watermarking system should assume that the embedding and detection algorithm (and perhaps their software implementations) will be available to users of this system and the fact that these users cannot detect or remove the watermark should be based solely on their lack of knowledge of the correct keys.

A thorough review on the topic of security can be found in [64].

22.5.2 Attacks Against Copyright Protection Watermarking Systems

As mentioned in the previous sections, a copyright protection watermarking system should exhibit a significant degree of robustness to attacks. The most obvious effect of an attack in a watermarking system is to render the watermark undetectable. Such attacks can be classified into two categories [53]: removal attacks and desynchronization (or geometrical) attacks. As implied by their name, removal attacks result in the removal of the watermark from the host image or in a significant decrease of its energy relative to the energy of the host signal. In most cases, removal attacks affect the amplitude of the watermarked signal, i.e., in the case of images, the pixel intensity or color. Removal attacks include linear or nonlinear filtering (e.g., arithmetic mean, median, Gaussian, Wiener filtering), sharpening, contrast enhancement (e.g., through histogram equalization), gamma correction, color quantization or color subsampling (e.g., due to format conversion), lossy compression (JPEG, JPEG2000, etc.), and other common image processing operations. Additive or multiplicative noise (Gaussian, uniform, salt and pepper noise), insertion of multiple watermarks on a single image, or image printing and rescanning (essentially a D/A-A/D conversion) are some additional examples of removal attacks. Finally, intentional removal attacks, i.e., attacks that have been devised with the intention to remove the watermark include, among others, the *averaging attack* where N instances of the same image, each hosting a different watermark, are averaged in order to obtain a watermark-free image, and the *collusion attack* where N images hosting the same watermark are averaged to obtain a (noisy) version of the watermark signal. This watermark estimate can be subsequently subtracted from each of the images to obtain watermark-free images.

Contrary to removal attacks, desynchronization attacks do not remove the watermark but cause a loss of synchronization (usually loss of the image coordinates) between the watermark signal embedded in the host signal and the watermark signal (see Section 22.5.4 for an example illustrating such a case). In other words, the watermark signal is still embedded in the host signal (with its energy almost intact) but cannot be detected. Desynchronization attacks usually involve global geometric distortions (i.e., distortions that are applied on the entire image using the same set of parameters) like translation, rotation, mirroring, scaling and shearing (i.e., general affine transformations), cropping, line or column removal, projective distortions (e.g., through a perspective transformation), etc. Local geometric distortions, i.e., distortions that affect subsets of an image, thus allowing an attacker to apply different operations with different parameters on each subset, can also be very effective in inducing loss of synchronization. The family of random bending attacks [65] which were first used in the Stirmark benchmark [66, 67] belong to this category. This family includes the bilinear transformation, which changes the shape of a regular rectangular sampling grid into a generic quadrilateral, the random jitter attack, which changes the positions of the sampling points by a small random amount, and the global bending attack, which displaces the locations of the sampling points by amounts that are sinusoidal functions of the points coordinates. The mosaic attack [67] that involves cutting an image into nonoverlapping pieces can also be considered a desynchronization attack. The small image tiles can be easily assembled and displayed so as to be perceptually identical to the original image using appropriate commands on the display software (e.g., the web browser). However, a detector applied on each image tile separately will fail to detect the watermark due to cropping. Template removal attacks is another category of desynchronization attacks that are only applicable to systems using a synchronization template (see Section 22.5.4.4) to regain synchronization in case of geometric distortions. Such attacks first estimate and remove the synchronization template from an image and then apply a geometric distortion to render the watermark undetectable. A review of geometric attacks and the approaches that have been proposed in order to cope with them is provided in [65].

Apart from the two attack categories described above, which are the most studied in the watermarking literature, other attacks can be devised that do not aim at making the watermark undetectable but try to harm a watermarking system or render the watermarking concept unreliable by other means [1]. Such attacks include unauthorized embedding attacks and unauthorized detection or decoding attacks. The copy attack [68] is an attack that illustrates the concept of unauthorized embedding. Using this attack, an attacker who is in possession of a method that can estimate the watermark that is embedded in an image or a set of images (e.g., through the collusion attack mentioned above) can subsequently embed this watermark in other watermark-free images. Thus, a claim from a copyright owner that images bearing her watermark are her property can be confronted by the attacker, who can show that this watermark exists in images that do not belong to her, i.e., in the fake watermarked images that the attacker has created. The single watermarked image counterfeit original (SWICO) and TWICO attacks [69] also belong to this category. In short, the SWICO attack involves the creation of a fake original image f by subtracting a watermark w from an image f_w watermarked by another person. The attacker can then claim that she has both the original image $f = f_w - w$ and an image $f_w = f + w$ watermarked with her own watermark, thus causing an ownership dispute.

Unauthorized detection attacks include attacks that aim at providing the attacker with information on whether an image is watermarked and perhaps reveal the encoded message (if any). Unauthorized detection is not a threat for all copyright protection applications. An example of an unauthorized detection attack is a brute force, exhaustive search approach where an attacker in possession of the detection algorithm checks successively all keys in the key space in order to findout whether an image is watermarked.

In order to measure the effect of a certain attack on the detection or decoding performance of an algorithm, plots of an appropriate performance metric (e.g., BER or probability of false alarm) versus the attack severity can be constructed. For attacks whose impact on the host image varies monotonically with respect to a certain parameter, it might be sufficient for the user to know only the most severe attack that the algorithm can withstand [10]. For a chosen performance metric, the "breakdown point" of the algorithm for this attack can be evaluated by increasing the attack severity (e.g., decreasing the JPEG quality factor) in appropriately selected steps until the detector output does not satisfy the chosen performance criterion. The strongest attack, for which the algorithm performance is above the selected threshold, is the algorithm breakdown point for this attack.

With respect to attacks that target the security of a watermarking system (see Section 22.5.1.3), the authors in [61] (based on the Diffie-Hellman approach [63]) define the following categories, on the basis of the information available to the attacker:

- Watermark-only attacks, where the attacker has access to a number of watermarked documents.

- Known message attacks, where the attacker has access to a number of watermarked documents and the messages that are hidden in them.

- Known original attacks, where the attacker has access to a number of watermarked documents as well as to the original, not watermarked documents.

The authors proceed in using the security framework that they developed to devise attacks against the security of spread spectrum watermarking algorithms.

22.5.3 Benchmarking of Copyright Protection Image Watermarking Algorithms

A benchmarking tool for image watermarking methods should be able to pinpoint the advantages and disadvantages of such methods and enable the user to perform efficient comparison of methods [10, 70]. Unfortunately, benchmarking of image watermarking algorithms is not an easy task since it requires the cross-examination of a set of dependent performance indicators like algorithmic complexity, decoding/detection performance, and perceptual quality of watermarked images. As a consequence, one cannot derive a single figure of merit but should deal with a set of performance indicators. An efficient benchmarking system should be able to quantify and present in an intuitive way the relations among the various performance indicators, e.g., the relation between watermark detection performance and perceptual quality. A small number of attempts to create benchmarking systems has taken place over the last few years, but this field is still in need of more efficient methodologies and actual implementations. Three benchmarking systems are presented below. OpenWatermark [71] and Watermark Evaluation Testbed [72] are two additional benchmarking systems.

22.5.3.1 *Stirmark*

Stirmark [66, 73] is the first benchmarking tool that was developed. The source code of the benchmark (version 4.0) is publicly available, and thus users can program their own attacks in addition to those provided by the benchmark (sharpening, JPEG compression, noise addition, filtering, scaling, cropping, shearing, rotation, column and line removal, flipping, and "Stirmark" attack). The user should provide, apart from the embedding and detection algorithms, appropriate command files (evaluation profiles) that define the tests or the attacks that will be performed. One can perform tests for measuring how the embedding strength influences the PSNR of the watermarked image, tests for the evaluation of the time required to perform embedding, and tests for measuring the influence of attacks on the detection and decoding performance. In this last category of tests, Stirmark performs for each attack parameter within a certain range embedding and detection with a random key and message and measures the detection certainty or the BER.

22.5.3.2 *Checkmark*

Checkmark [74] is essentially a successor of the previous Stirmark version (namely, Stirmark version 3.1). In addition to the attacks implemented in Stirmark, Checkmark provides a number of new attacks that include wavelet compression, projective transformations, modelling of video distortions, image warping, copy attack, template removal attack, denoising, nonlinear line removal, collage attack, down/up sampling, dithering, and thresholding. The developers of Checkmark provide the MATLAB source code of the application and thus one can add new attacks to the existing ones. The benchmark provides a number of "application templates" which are essentially lists of attacks related to a specific application. In addition, Checkmark incorporates two new objective quality metrics: the weighted PSNR and the so-called Watson metric. Despite the major improvements that have been introduced, the basic principles of Checkmark are very similar to those of Stirmark 3.1. In both cases, the user should provide the benchmark with a set of watermarked images and a detection routine along with a user-defined detection rule. The attacks that are included in the application template that has been selected by the user are applied in every watermarked image, and the detection routine is used to provide the detection result. It should be noted that Checkmark was last updated in 2001.

22.5.3.3 *Optimark*

Optimark [75] is a benchmarking platform that provides a graphical user interface and incorporates the same attacks as Stirmark 3.1. These attacks can be performed either one at a time or as a cascade. The user should supply embedding and detection/decoding routines in the form of executable files. Optimark supports hard and soft decision detectors. The user selects the set of test images, the set of keys and messages that will be used in the trials, and the attacks that will be performed on the watermarked images. Furthermore, she provides the set of PSNR values for the watermarked images, along with the embedding factors that the embedding software should use in order to achieve these PSNR values. Optimark performs in an automated way multiple trials using the selected images, embedding strengths, attacks, keys, and messages. Detection using both correct and erroneous keys (which are necessary for the evaluation of the probability of false alarms) is performed. Message decoding performance is evaluated separately from watermark detection. The "raw" results are processed by the benchmark in order to provide the user with a number of performance metrics and plots, depending on the type of algorithm being tested. For example, when testing a multiple-bit algorithm that employs a soft decision detector, the user can obtain the following metrics: ROC, EER, probability of false alarm for a user-defined probability of false rejection, probability of false rejection for a user-defined probability of false alarm, plots of BER and percentage of perfectly decoded messages versus the detection threshold (for a specific message length), and plot of payload versus the detection threshold (for a specific BER). The software evaluates various complexity metrics like average embedding, detection, and decoding time and provides an option to evaluate the algorithm breakdown point for a given attack. Finally, it can summarize the results in various ways, e.g., provide average results for a set of images and a specific attack or average results over a number of different attacks for a specific image.

A thorough treatment of the subject of performance evaluation of watermarking algorithms can be found in [1, 10].

22.5.4 Spread Spectrum Watermarking

Spread spectrum watermarking draws its name from spread spectrum communication techniques [76] that are used to achieve secure signal transmission in the presence of noise and/or interception attacks that generate an appropriate jamming signal to interfere with the transmission. In such a situation, one can spread the energy of a symbol to be transmitted either in the time domain by multiplying it by a pseudorandom sequence, or in the frequency domain by spreading its energy over a large part of the signal spectrum.

22.5.4.1 *Blind Additive Embedding with Correlation Detection*

In this section, a simple zero-bit spread spectrum watermarking system that consists of a blind additive embedder and a blind correlation detector will be presented. Despite its simplicity, this methodology has been utilized extensively, in many variations, in the early days of watermarking [77, 78]. Means of improving or creating variants of the basic algorithm will also be presented in this section.

The embedding procedure of this system employs the addition of a white, zero-mean pseudorandom signal w (generated by using a secret key K in conjunction with the appropriate generation function) on the host signal f_o:

$$f_w = f_o + pw, \tag{22.10}$$

where f_w is the watermarked signal and $p > 0$ is a constant that controls the watermark embedding energy (watermark embedding factor). Obviously, p is closely related to the watermark perceptibility. On a per-sample basis, the above equation can be stated as follows:

$$f_w(n) = f_o(n) + pw(n), \quad n = 0,\ldots,N-1, \tag{22.11}$$

where N denotes the signal length. In the following, we will assume that Eqs. (22.10) and (22.11) refer to the spatial domain. In case of image watermarking, the watermark modifies the intensity or color of the image pixels, and f_o, w, and f_w are 2D signals.

As has already been mentioned, the watermark detection aims at verifying whether a given watermark w_d is embedded in the test signal f_t. During detection, f_t can be represented in the following form:

$$f_t = f_o + pw_e. \tag{22.12}$$

This equation can summarize all three possible detection hypotheses, namely:

- the watermark w_d is indeed embedded in the signal (event H_0), which corresponds to $p \neq 0$ and $w_e = w_d$

- the watermark w_d is not embedded in the signal (event H_1), which can imply either that no watermark is present (event H_{1a}), or that the signal bears a different watermark than the one under investigation (event H_{1b}). In the equation above, event H_{1a} corresponds to $p = 0$, whereas event H_{1b} corresponds to $w_e \neq w_d$.

In order to decide which event holds, i.e., which is the valid hypothesis, the correlation between the signal under investigation and the watermark is evaluated:

$$c = \frac{1}{N} \sum_{n=0}^{N-1} f_t[n]w_d[n] = \frac{1}{N} \sum_{n=0}^{N-1} \left(f_o[n]w_d[n] + pw_e[n]w_d[n] \right). \tag{22.13}$$

Such a detection scheme is usually called a *correlation detector* (also known as a matched filter). By assuming statistical independence between the host signal f_o and both watermarks w_e and w_d, an expression for the mean of the correlation c can be derived in a straightforward manner [79]:

$$\mu_c = E[c] = E\left[\frac{1}{N} \sum_{n=0}^{N-1} \left(f_o[n]w_d[n] + pw_e[n]w_d[n] \right) \right]$$

$$= \frac{1}{N} \sum_{n=0}^{N-1} E[f_o[n]]E[w_d[n]] + \frac{1}{N}p \sum_{n=0}^{N-1} E[w_e[n]w_d[n]]. \tag{22.14}$$

Since the watermark has been chosen to be a zero-mean random signal, the first term of the expression will be zero and, therefore, μ_c will depend only on the second term. When the signal bears no watermark, i.e., when $p = 0$, the second term is also zero and the mean value of the correlation is zero. Furthermore, when the signal bears a different watermark than the one under investigation ($w_e \neq w_d$), the second term will obtain a small value, close to zero, as two watermarks generated using two different keys are expected to be almost orthogonal to each other. When the signal hosts the watermark under investigation, i.e., when $p \neq 0$ and $w_e = w_d$, the mean value of c can be easily shown to be equal to $p\sigma_w^2$, where σ_w^2 is the variance of the watermark signal. Thus, the conditional probability distributions $p_{c|H_0}$, $p_{c|H_1}$ of the correlation value c under the two hypotheses H_0 and H_1 will be centered around $p\sigma_w^2$ and 0, respectively (Fig. 22.3). Furthermore, for the case under study, these distributions will be approximately Gaussian. For suitable values of p, σ_w^2 and by assuming that the variances $\sigma_{c|H_0}^2$, $\sigma_{c|H_1}^2$ of c under the two hypotheses are reasonably small, a decision on the valid hypothesis can be obtained by comparing c against a suitably selected threshold $T > 0$ that lies between 0 and $p\sigma_w^2$.

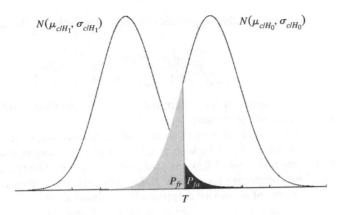

FIGURE 22.3

Conditional pdfs of the correlation value c under hypotheses H_0, H_1.

More specifically, a decision to accept hypothesis H_0 or H_1 is taken when $c > T$ and $c < T$, respectively.

For a given threshold, the probabilities of false alarm $P_{fa}(T)$ and false rejection $P_{fr}(T)$ which characterize the performance of this system can be evaluated as follows:

$$P_{fa}(T) = Prob\{c > T|H_1\} = \int_{T}^{\infty} p_{c|H_1}(t)dt, \tag{22.15}$$

$$P_{fr}(T) = Prob\{c < T|H_0\} = \int_{-\infty}^{T} p_{c|H_0}(t)dt. \tag{22.16}$$

Obviously, these two probabilities depend on $\mu_{c|H_0}$, $\mu_{c|H_1}$, $\sigma_{c|H_0}^2$, and $\sigma_{c|H_1}^2$. By observing Fig. 22.3, one can conclude that the system performance improves (i.e., the probabilities of false alarm and false rejection for a certain threshold decrease) as the two distributions come further apart, i.e., as the difference $\mu_{c|H_0} - \mu_{c|H_1}$ increases. Furthermore, the performance improves as the variances of the two distributions $\sigma_{c|H_0}^2$, $\sigma_{c|H_1}^2$ decrease.

Provided that the additive embedding model (22.10) has been used and under the assumptions that no attacks have been applied on the signal and that the host signal f_o is Gaussian, the detection theory states that the correlation detector described above is optimal with respect to the Neyman-Pearson criterion, i.e., it minimizes the probability of false rejection P_{fr} subject to a fixed probability of false alarm P_{fa}.

A variant of the above algorithm that employs nonblind detection can be easily devised by subtracting the original signal f_o from the signal under investigation before evaluating the correlation c. It can be proven that such a substraction drastically improves the performance of the algorithm by reducing the variance of the correlation distribution. Instead of the correlation (22.13), one can also use the normalized correlation, i.e., the correlation normalized by the magnitudes of the watermark and the watermarked signal:

$$c = \frac{\sum_{n=0}^{N-1} f_t[n]w_d[n]}{\sqrt{\sum_{n=0}^{N-1} f_t^2[n]\sum_{n=0}^{N-1} w_d^2[n]}}.$$ (22.17)

Normalized correlation can grant the system robustness to operations such as increase or decrease of the overall image intensity.

The zero-bit system presented above can be easily extended to a system capable of embedding one bit of information. In such a system, symbol 1 is embedded by using a positive value of p whereas symbol 0 is embedded by using $-p$. Watermark detection can be performed by comparing $|c|$ against T, i.e., a watermark presence is declared when $|c| > T$. In the case of a positive detection, the embedded bit can be decoded by comparing c against T and $-T$, i.e., 0 is decoded if $c < -T$ and 1 if $c > T$.

Another popular approach for embedding the watermark in the host signal is multiplicative embedding:

$$f_w(n) = f_o(n) + pf_o(n)w(n).$$ (22.18)

Using such an embedding law, the embedded watermark $pf_o(k)w(k)$ becomes image-dependent, thus providing an additional degree of robustness, e.g., against the collusion attack. Furthermore, by modifying the magnitude of a watermark sample proportionally to the magnitude of the corresponding signal sample (be it pixel intensity or magnitude of a transform coefficient), i.e., by imposing larger modifications to large amplitude signal samples, a form of elementary perceptual masking can be achieved.

The spectral characteristics and the spatial structure of the watermark play a very important role to robustness against several attacks. These characteristics can be controlled in the watermark generation procedure and affect the more general characteristics of the watermarking system, like robustness and perceptual invisibility. In the following sections, we will see the basic categories of watermarks as they are derived by the various existing watermark generation techniques.

22.5.4.2 *Chaotic Watermarks*

Chaotic watermarks have been introduced as a promising alternative to pseudorandom signals [79–84]. An overview of chaotic watermarking techniques can be found in [85,86]. Sequences generated by chaotic maps constitute an efficient alternative to pseudorandom watermark sequences. A chaotic discrete-time signal $x[n]$ can be generated by a chaotic system with a single state variable by applying the recursion:

$$x[n] = T(x[n-1]) = T^n(x[0]) = \underbrace{T(T(\ldots(T(x[0]))\ldots))}_{n\ \text{times}},$$ (22.19)

where $T(\cdot)$ is a nonlinear transformation that maps scalars to scalars and $x[0]$ is the system initial condition. The notation $T^n(x[0])$ is used to denote the nth application of the map. It is obvious that a chaotic sequence x is fully described by the map $T(\cdot)$ and the initial condition $x[0]$. By imposing certain constraints on the map or the initial condition, chaotic sequences of infinite period can be obtained.

A performance analysis of watermarking systems that use sequences generated by piecewise-linear Markov maps and correlation detection is presented in [79]. One property of these sequences is that their spectral characteristics are controlled by the parameters of the map. That is, watermark sequences having uniform distribution and controllable spectral characteristics can be generated using piecewise-linear Markov maps. An example of a piecewise-linear Markov map is the *skew tent map* given by:

$$\mathcal{T} : [0,1] \to [0,1],$$

$$\text{where } \mathcal{T}(x) = \begin{cases} \frac{1}{\alpha} x & , \quad 0 \le x \le \alpha \\ \frac{1}{\alpha-1} x + \frac{1}{1-\alpha} & , \quad \alpha < x \le 1 \end{cases} \quad, \alpha \in (0,1). \qquad (22.20)$$

The autocorrelation function (ACF) of skew tent sequences depends only on the parameter α of the skew tent map. Thus, by controlling the parameter α, we can generate sequences having any desirable exponential ACF. The power spectral density of the skew tent map can be easily derived [79]:

$$S_t(\omega) = \frac{1 - (2\alpha - 1)^2}{12(1 + (2\alpha - 1)^2 - 2(2\alpha - 1)\cos\omega)}. \qquad (22.21)$$

By varying the parameter α, either highpass ($\alpha < 0.5$) or lowpass ($\alpha > 0.5$) sequences can be produced. For $\alpha = 0.5$, the symmetric tent map is obtained. Sequences generated by the symmetric tent map possess a white spectrum, since the ACF becomes the Dirac delta function. The control over the spectral properties is very useful in watermarking applications, since the spectral characteristics of the watermark sequence are directly related to watermark robustness against attacks, such as filtering and compression.

The statistical analysis of chaotic watermarking systems that use a correlation detector was undertaken leading to a number of important observations on the watermarking system detection performance [79]. Highpass chaotic watermarks prove to perform better than white ones, whereas lowpass watermarks have the worst performance when no distortion is inflicted on the watermarked signal. The controllable spectral/correlation properties of Markov chaotic watermarks prove to be very important for the overall system performance. Moreover, Markov maps that have appropriate second- and third-order correlation statistics, like the skew tent map, perform better than sequences with the same spectral properties generated by either Bernoulli or pseudorandom number generators [79].

The simple watermarking systems presented above using either pseudorandom or chaotic generators and either additive or multiplicative embedding would not be robust to geometric transformations, e.g., a slight image rotation or cropping, as such attacks would cause a "loss of synchronization" (see Section 22.5.2) between the watermark signal embedded in the host image and the watermark signal used for the correlation evaluation. This happens because the success of the correlation detection method relies on our ability to correlate the watermarked signal f_t with the watermark w_d in a way that ensures that the n-th sample $w_d(n)$ of the watermark signal will be multiplied in Eq. (22.13) with the watermarked signal sample $f_t(n)$ that hosts the same sample of the watermark. In the case of geometric distortions, this "synchronization" will be lost and chances are that the correlation c will be below T, i.e., a false rejection will occur.

A brute force approach could involve the evaluation of the correlation between the watermarked signal and all transformed versions of the watermark. For example, if the image has been subject to rotation by an unknown angle, one can evaluate its correlation with all rotated versions of the watermark and decide that the image is watermarked if the correlation of one of these versions with the signal is above the threshold T. Obviously, this approach has extremely large computational complexity, especially when the image has been subject to a cascade of transforms (e.g., rotation and scaling). Multiple remedies to this problem have been proposed that will be presented in detail in the following sections.

22.5.4.3 *Transformed Watermarks*

The idea of transformed watermarks is to construct watermarks transformed in a specific domain whose detection performance is invariant to the geometric distortions of the watermarked image. For example, it is well known that the amplitude of the Fourier transform is translation invariant:

$$f(x_1 + a, x_2 + b) \leftrightarrow F(k_1, k_2)e^{-i(ak_1 + bk_2)}. \tag{22.22}$$

Therefore, if the watermark is embedded in the amplitude of the Fourier transform, it will be insensitive to a spatial shift of the image. The transform space of Mellin-Fourier is one such invariant space. It has been proposed for watermark embedding because, when the watermark is applied to the amplitude of the Fourier transform, it is invariant to translation, rotation, and scale of the watermarked image [30]. In order to become invariant to translation, the image is transformed in the Fourier domain and the amplitude of the Fourier is transformed using the log-polar mapping (LPM) defined as follows:

$$x = e^\rho \cos\theta, \tag{22.23}$$

$$y = e^\rho \sin\theta, \tag{22.24}$$

with $\rho \in \mathcal{R}$ and $\theta \in [0, 2\pi]$. Any rotation in the spatial domain will cause rotation of the Fourier amplitude and translation in the polar coordinate system. Similarly, a scaling of the spatial domain will result in a translation in the logarithmic coordinate system. That is, both rotation and scaling in the spatial domain are mapped to translation in the LPM domain. Invariance to these translations is achieved by taking again the amplitude of the Fourier of the LPM. Taking the Fourier of a LPM is equivalent to computing the Mellin-Fourier transform. Combining the DFT and the Mellin-Fourier transform results in rotation, scale, and translation transformation invariance.

The major drawback of the method above is that the various transforms decrease the embedded watermark power. That is, the interpolation applied during the various transforms constitutes an attack to the watermark, thus making it usually undetectable even without any further distortion of the watermarked image. Indeed, in the watermark embedding procedure, the watermark undergoes two inverse DFTs and one inverse LPM along with the corresponding interpolations needed. In the detection procedure, two DFTs and one LPM are needed as well. Thus, the watermark should be very strong in order to resist all these transforms. Of course, stronger embedding means the possiblity of visually perceptible watermarks, i.e., quality reduction for the host image. To overcome all these problems, iterative embedding has been proposed in [87]. The watermark is embedded iteratively until it can be reliably detected after the transforms needed in the detection procedure. Even in that case, reliable watermark detection demands very strong embedding and the results are not very promising.

A transform-based blind watermarking approach with improved robustness against geometric distortions has been proposed in [88]. The method is based on geometric image normalization that takes place before both watermark embedding and extraction. The image is normalized in order to meet a set of predefined moment criteria. The normalization procedure makes the method invariant to affine transform attacks. However, the proposed scheme is not robust against cropping or line-column removal.

Another reason the transform domains have been proposed for watermark embedding is that they also provide robustness against other intentional or unintentional attacks, such as filtering and compression. In such a case, the watermark affects the value of certain transform coefficients and the watermarked signal is obtained by applying the inverse transform on the watermarked coefficients. Transform domain watermarking allows system designers to exploit the transform properties for the benefit of the system. For this purpose, embedding, e.g., in the DFT, DWT, and DCT domains, has been proposed. For example, one can embed the watermark signal in the low-to-middle frequency coefficients of the DCT transform applied on small image blocks. By doing so, one can ensure that the watermark will remain essentially intact by lowpass operations, e.g., JPEG lossy compression or lowpass filtering, since these operations suppress mainly the higher frequencies. Moreover, the distortions imposed on the signal due to watermarking can be held at a reasonably low level as the lower frequencies, whose alterations are known to cause visible distortions, will be kept intact. Embedding in the DWT transform domain has been proposed for increased robustness against JPEG2000 compression. The 2D Radon Wigner transform has been used for watermark embedding in order to obtain robustness against geometrical attacks in [89].

Recently, a transform domain watermarking for color images has been proposed [90]. The method considers color information in the $L\ a^*\ b^*$ domain and treats colors as quaternions. Quaternions have one real and three orthogonal imaginary components and can be expressed in the form:

$$Q = a + ib + jc + kd, \tag{22.25}$$

where a, b, c, and d are real numbers and i, j, and k are imaginary operators. Therefore, quaternions can be used to represent data with up to four components and thus are sufficient to represent the three-component color information in a single, vectorial format. Color images represented in quaternion format are transformed to the "frequency" domain by using the discrete quaternion fourier transform (QFT) [91]. The watermark, which is also represented in quaternion form, is additively embedded in the transform domain. By imposing certain conditions, the modifications on the color of the image can be restricted to yellow-blue component which ensures invisibility due to the low sensitivity of the human visual system (HVS) to these colors. Detection is performed in a nonblind way. From the above short description, it is obvious that in this case the main reason to resort to a transform domain (QFT domain) is to ensure watermark invisibility.

22.5.4.4 *Template Watermarks*

Another class of watermarks that have been proposed to cope with the problem of geometrical transformations is the *template watermarks* class. A template is a structured pattern that is embedded in the image and usually conveys geometrical information. Basically, it is an additional signal that is used as a tool for recovering possible geometrical transformations of the watermarked image. The template is usually a set of peaks in the DFT domain [92–94]. The peaks are embedded in specific locations so as to define a certain structure that can be easily recovered in the detection procedure. Templates that can be used for watermarking applications are shown in Fig. 22.4.

As an example, we shall describe in more detail the template proposed in [92]. The template peaks are distributed uniformly along two lines in the DFT domain at certain angles. The angles and radii are chosen pseudorandomly by using a secret key. The strength of the template can be determined adaptively. Inserting points at a strength equal to the local average value of DFT points plus two standard deviations yields a good compromise between visibility and robustness during decoding. Peaks in the high frequencies are constructed to be less strong since, in these regions, the average spectra power is usually lower than that of the low frequencies. This type of template is applicable for all images. If someone uses more than two peak lines to construct the template, the cost of the detection algorithm is increased. However, at least two peak lines are required in order to resolve ambiguities arising from the symmetry of the magnitude of the DFT. In particular, after a rotation in the spatial domain, an ambiguity will exist as to whether the rotation was clockwise or counter-clockwise. Depending on features of the specific watermark technology, there are different strategies for template generation.

The watermark detection process consists of two phases. First the affine transformation (if any) undergone by the watermarked image is determined, then the transformation is inverted, and the watermark is detected. For detecting the template, some approaches transform the template matching problem into a point-matching problem. After this problem has been solved, the best candidates for the template points are identified. If an

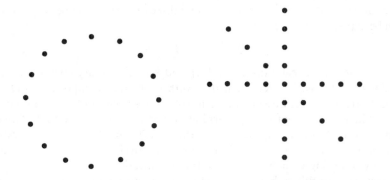

FIGURE 22.4

Templates proposed for watermarking applications.

affine transformation has been applied, the identified template points will differ from the original ones. This change is exploited to estimate the applied affine transformation. The corresponding inverse affine transformation is then applied for a better synchronization of the watermark.

The major drawback of template watermarking is its vulnerability against the template removal attack [95]. The main goal of this attack is to destroy, without any key knowledge, the synchronization pattern of the watermark in order to fool the detection process after an affine transformation of the image. An attacker does not need to know how the specific template in a domain is constructed, since the template applied will always generate some peaks in the target domain used for the template.

The attack can be easily applied by a attacker. In the first phase of the attack, the watermarked image f_w is filtered using a Wiener or median filter and an estimate of the original image $\hat{f}_o$ is derived. Then the watermark estimate $\hat{w}$ is obtained by subtracting the image $\hat{f}_o$ from the watermarked image. Using the estimate of the watermark, the peaks of the template are extracted in the appropriate transform domain (e.g., DFT). The amplitude of the extracted peaks is modified by replacing the specific amplitude of the watermarked image with the average amplitude value of the neighbors within a certain window. In general, the attacker can apply the same procedure as the template detector in order to extract the template and then she can remove it from the watermarked image. Once the template is removed, the watermark is vulnerable to any geometric attack.

22.5.4.5 *Special Structure Watermarks*

To solve the problems of template-based watermarking, a different approach has been proposed that involves watermarks whose spatial structure can provide either an invariance to certain transforms or a significant reduction in the size of the parameter search space (e.g., the space of possible rotation angles) that has to be searched during detection in order to reestablish synchronization. In this case, self-reference watermarks are mostly used in practice. The self-reference watermarks do not use any additional template to cope with the geometrical transforms. Instead, the watermark itself is constructed so as to attain a special spatial structure with desired statistical properties. The most often used watermarks within this approach have self-similar features, i.e., repetition of the same watermark in many spatial directions depending on the final goal and the targeting attack. Spatial self-similarity of the watermarks reduces the search space parameters in case of an affine transformation of the watermarked image [82, 96].

An example of self-similar watermarks is the so-called *circularly-symmetric* watermark. It is defined by [34, 96]:

$$\mathcal{W}(r,\theta) = \begin{cases} 0, & r < r_{\min} \quad \text{or} \quad r > r_{\max} \\ \pm b, & r_{\min} < r < r_{\max} \end{cases}, \qquad (22.26)$$

where (n_1, n_2) represent the spatial coordinates and $r = \sqrt{n_1^2 + n_2^2}$, $\theta = \arctan\left(\frac{n_2}{n_1}\right)$. $r_{\min}$ and $r_{\max}$ are the minimum and maximum radii that define the watermark's circular or ring-like support region. b is an integer representing the embedding level or watermark strength. In order for $\mathcal{W}(r,\theta)$ to attain sufficient lowpass characteristics and, thus, be more robust to compression or lowpass filtering, its cyclic or ring-like support domain (22.26) is additionally divided to a number of s sectors having an extend of $\frac{s}{360}$ degrees. All watermark samples inside a sector for a constant radius are set equal to b or $-b$ according to a pseudorandom number generator initialized with a random key.

A circularly-symmetric watermark has the advantage of robustness against rotation with angles less than $\frac{s}{360}$ degrees. In this case detection is possible without rotating the watermark. When rotation angles are bigger, detection is performed faster since watermark rotation is only needed for multiples of $\frac{s}{360}$ degrees. Spatial self-similarity with respect to the cartesian grid is accomplished by repeating the basic circularly-symmetric watermark at different positions in the image. Additionally, the shifted versions of the basic watermark can also be scaled versions in order to cope with scaling attacks [97] or rotated versions in order to cope with rotation attacks [42].

Another type of self-similar watermarks has been proposed in [98]. The basic watermark is replicated in the image in order to create four repetitions of the same watermark. This enables nine peaks in the ACF that are used in order to recover geometrical transformations. The descending character of the ACF peaks shaped by a triangular envelope reduces the robustness of this approach to the geometrical attacks accompanied by a lossy compression. The need for computing two DFTs of double image size to estimate the ACF also creates some problems for fast embedding/detection in the case of large images.

The known fact that periodic signals have a power spectrum containing peaks can be used to obtain a regular grid of reference points that can easily be employed for recovering from general affine transformation attacks. The existence of many peaks in the magnitude spectrum of the periodically repeated watermark increases the probability of detecting geometrical transforms even after lossy compression [94]. This fact indicates the enhanced robustness of these watermarks. Furthermore, it is more difficult to remove the peaks in the magnitude spectrum based on a local interpolation in comparison with a template scheme. Such an attack would create considerable visible distortions in the attacked image. A practical algorithm based on the magnitude spectrum of the periodical watermarks is described in [94] for spatial, wavelet, or any transform domain. First, the magnitude spectrum is computed from the estimated watermark. Due to the periodicity of the embedded information, the estimated watermark spectrum possesses a discrete structure. Assuming that the watermark is white noise within a block, the power spectrum of the watermark will be uniformly distributed. Therefore, the magnitude spectrum shows aligned and regularly spaced peaks. If an affine distortion was applied to the host image, the peaks layout will be rescaled, rotated, and/or sheared, but alignments will be preserved. Therefore, it is easy to estimate any affine geometrical distortion from these peaks by fitting alignments and estimating periods of the peaks.

Of course, as in the case of using a watermark template, the attacker may try to estimate the watermark, i.e., to find the peaks on the magnitude spectrum and then remove them by interpolation. Another possible attack is to perform an affine transformation and, afterwards, to embed a periodical signal that will create another regular grid of peaks that may deceive the detector. Finally, a critical issue in this approach is to find the minimum number of watermark samples needed in order to have reliable detection.

22.5.4.6 *Feature-Based Watermarks*

Another approach for image watermarking robust against geometric distortions is based on image feature detection. The methods that belong to this category try to find robust feature points (also referred to as *salient* features) in a given image that can be detected even after geometric attacks. These feature points are used to form local regions for embedding. At the detection end, the feature points are (hopefully) robustly detected and used for resynchronization.

One of the first methods that exploited this idea was based on facial features [99]. In this method, the image to be watermarked was supposed to contain a face in frontal view and the selected salient features were the eyes and mouth of the face. The watermark was embedded in the facial region. In the detection phase, the facial features were localized and used to compensate for the geometric attacks. Subsequently, watermark detection was performed. Although the idea was interesting, its applicability was limited to facial images.

The use of the Gaussian kernel as a preprocessing filter to extract and stabilize image feature points for watermarking has been proposed in [100]. The Gaussian kernel is a circular and symmetric filter, leading to rotation-invariant filtering. The method selects as feature points the locations having maximum filter response. Afterwards, the set of feature points is used to form a mesh by means of Delaunay triangulation. This mesh is then used to guide the watermarking process. In order to resist watermark-estimation attacks, an image hashing is extracted and combined with the watermarks in order to generate hash-based, content-dependent watermarks. The major weakness of the proposed method is its high complexity due to mesh warping, which makes the method unsuitable for real-time applications.

Another feature-based image watermarking scheme that is robust against desynchronization attacks has been proposed in [101]. Robust feature points, which can survive various signal processing and affine transformations, are extracted by using the Harris-Laplace detector. These feature points define neighborhood regions for watermark embedding. The digital watermark is repeatedly embedded on the scale-space representation of the image (i.e., the extracted feature regions) by modulating the magnitudes of DFT coefficients. In watermark detection, the salient features are recovered and the geometric distortions are compensated. The digital watermark is recovered using a maximum membership criterion applied to the watermarked regions. Simulation results have shown that the proposed scheme is robust against common signal processing and desynchronization attacks.

22.5.4.7 *Watermarking Systems Involving Optimal Detectors*

As mentioned in the beginning of this section, the correlation detector is optimal only in the case of additive watermarks and host signals following a Gaussian distribution. In the case of watermarks embedded in the signal in a multiplicative way or when the host signal follows a different distribution, detectors that are optimal in a certain sense can be constructed using the statistical detection and estimation theory. According to this theory, a decision over the two hypotheses H_0, H_1 can be obtained by evaluating the likelihood ratio $L(f_t)$:

$$L(f_t) = \frac{p_{f_t}(f_t|H_0)}{p_{f_t}(f_t|H_1)}. \tag{22.27}$$

In the previous formula, $p_{f_t}(f_t|H_0), p_{f_t}(f_t|H_1)$ are the probability density functions of the random vector f_t, i.e., the watermarked signal, conditioned on the hypotheses H_0, H_1, respectively. A decision is obtained by comparing $L(f_t)$ against a properly selected threshold T. More specifically, a decision to accept hypotheses H_0 or H_1 is taken when $L(f_t) > T$ and $L(f_t) < T$, respectively. The appropriate value of T depends on the performance criterion that we wish to optimize. If an optimal detector with respect to the Neyman-Pearson criterion (i.e., a detector that minimizes the probability of false rejection P_{fr} subject to a fixed probability of false alarm P_{fa}) is to be designed, the threshold value that achieves this minimization is the one evaluated by solving the following equation for T, assuming a fixed, user-provided probability of false alarm $P_{fa} = e$:

$$P_{fa} = e = Prob\{L(f_t) > T|H_1\} = \int_T^{+\infty} p_L(L|H_1)dL. \tag{22.28}$$

In the previous equation, $p_L(L|H_1)$ is the pdf of the likelihood ratio L conditioned on the hypothesis H_1.

Various optimal detection schemes, for different situations, have been proposed in the literature. Obviously, in order to obtain an analytic expression for the likelihood ratio (22.27) and evaluate the threshold T using (22.28), one has to obtain analytic expressions for the probability density functions that appear in these expressions. To proceed with such derivations, certain assumptions about the statistics of the involved quantities should be adopted. Optimal detectors for watermarks that have been embedded with a multiplicative rule on the magnitude of the DFT coefficients are derived in [39]. The watermark samples $w(n)$ are assumed to be bivalued, $\{-1, 1\}$, each value having equal probability 0.5. The assumptions adopted in this chapter are that the host signal is an ergodic, wide-sense stationary process that follows a first-order seperable ACF and that the DFT coefficients are independent Gaussian random variables, each with a different mean and variance. The authors verify the Gaussianity assumption through a Kolmogorov-Smirnov test. Based on these assumptions, the authors proceed in deriving an analytic expression for the probability distribution of the magnitude of the DFT coefficients. Using this expression, expressions for $p_{f_t}(f_t|H_0), p_{f_t}(f_t|H_1)$ are derived and exploited to obtain the following analytic formula for $L(f_t)$:

$$L(f_t) = \prod_{k=1}^{N-1} \frac{\frac{2}{(1+w(k)p)^2} I_0\left(0, \frac{\sigma_I^2(k)-\sigma_R^2}{4\sigma_R^2\sigma_I^2(k)}\frac{f_t(k)^2}{(1+w(k)p)^2}\right)}{\frac{1}{(1+p)^2}\exp\left(-\frac{\sigma_R^2(k)+\sigma_I^2(k)}{4\sigma_R^2(k)\sigma_I^2(k)}\frac{2p(w(k)-1)f_t(k)^2}{(1+w(k)p)^2(1+p)^2}\right)I_0\left(0, \frac{\sigma_I^2(k)-\sigma_R^2(k)}{4\sigma_R^2(k)\sigma_I^2(k)}\frac{f_t(k)^2}{(1+p)^2}\right)+}$$
$$\frac{1}{(1-p)^2}\exp\left(-\frac{\sigma_R^2(k)+\sigma_I^2(k)}{4\sigma_R^2(k)\sigma_I^2(k)}\frac{2p(w(k)+1)f_t(k)^2}{(1+w(k)p)^2(1-p)^2}\right)I_0\left(0, \frac{\sigma_I^2(k)-\sigma_R^2(k)}{4\sigma_R^2(k)\sigma_I^2(k)}\frac{f_t(k)^2}{(1-p)^2}\right).$$

$$\tag{22.29}$$

In the previous expression, p is the embedding factor, $I_0()$ is the modified Bessel function, and $\sigma_I^2(k)$, $\sigma_R^2(k)$ are the variances of the imaginary and the real part of the kth DFT coefficient for which analytical expressions have been also derived in [39]. It should be also noted that f_t is the host signal which, in this case, consists of the magnitudes of the DFT coefficients of the host signal. In order to evaluate the threshold in (22.28) for a certain $P_{fa} = e$, the pdf $p_L(L|H_1)$ of the likelihood ratio L conditioned on the hypotheses H_1 needs to be derived. The authors assume that L attains a Gaussian distribution and proceed in experimental evaluation of its mean $\widehat{\mu}$ and variance $\widehat{\sigma}$. Using these quantities, the threshold T can be found to be

$$T = \widehat{\mu} - \widehat{\sigma}\sqrt{2}\ erf^{-1}(2\,P_{fa} - 0.5). \tag{22.30}$$

Similar approaches have been proposed by others in order to derive optimal detectors under other optimality criteria or embedding domains with different distributions [35–41, 47].

22.5.5 Watermarking with Side Information
22.5.5.1 *Informed Embedding Watermarking*
A watermarking system that uses the information of the host signal (also called cover signal) in the coding/embedding procedure is called an *informed coding/embedding watermarking* system and is described in Fig. 22.5. An informed embedding scheme exploits the information conveyed in the host signal f for generating a robust watermark w in the spatial domain. The watermark generation can be seen as an optimization problem with respect to a detection statistic or to a robustness measure as it will be described in the following.

The simplest way to construct an informed embedding watermarking system is the so-called *precancellation*, which has been proposed in communication systems [102]. In such a system, the interference of the host signal is completely removed by setting the watermark to be embedded as $w_a = w - f_o$. Thus, after embedding in the host signal, the watermarked signal to be transmitted is equal to $f_w = f_o + w_a = w$. That is, the host signal is completely replaced by the watermark. It is obvious that this watermarking

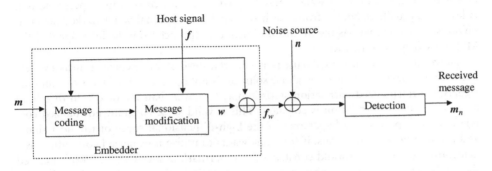

FIGURE 22.5

Watermarking with informed coding and informed embedding.

system is optimal with respect to host signal cancellation, but it is unacceptable for real applications, since the fidelity constraint is violated. That is, the host signal is not preserved at all. In most cases where fidelity constraints are imposed in the watermark generation and embedding procedure, the problem of generating an informed embedding watermarking system is dealt with as an optimization problem. The problem can be defined in two alternative ways:

- Construct a watermark to be embedded in the given host signal that maximizes watermark robustness, under the constraint that the perceptual distortion will be inside a prescribed limit.

- Construct a watermark to be embedded in the given host signal that minimizes the perceptual distortion, under the constraint that the robustness of the resulting watermark will be inside a prescribed limit.

More complicated optimization problems can be set where both perceptual distortion and robustness are optimized inside prescribed limits. It is obvious that, in order to solve the above optimization problems, someone has to define measures of robustness and perceptual distortion. This task is not trivial and many researchers proposed such measures under rather simplistic assumptions. For example, a detection statistic can be assumed to measure the robustness of a watermarking system. This assumption is not valid in most cases, since the robustness should be mostly determined for specific attacks.

If someone considers a watermarking system that uses linear correlation as a detection statistic, then it can be assumed that increased robustness means higher correlation value between the watermarked signal and the watermark. Thus, imposing a constraint on robustness can be considered as generating watermarks that have constant correlation value with the watermarked signal (i.e., $f_w w^T = c$). This can be interpreted as moving all the signals to be watermarked to a hyperplane in the high-dimensional space of the host signal that is perpendicular to the watermark vector. This hyperplane corresponds to constant correlation with the watermark signal. In such a watermarking scheme, the more distant the hyperplane is from the detection threshold, the more robust the watermarking scheme. Having constant linear correlation between the watermark and the watermarked signal corresponds to constant robustness and, thus, the constrained optimization problem is reduced to finding a watermarked signal, i.e., a vector on this hyperplane that is less perceptually distorted from the host signal. The optimal watermarked signal is obviously the one that has the minimum Euclidean distance from the host signal, if the MSE is used as a fidelity measure.

However, if someone wants to use normalized instead of linear correlation as a watermark detection statistic, then the above solution is not optimal. Indeed, for certain host signals, the normalized correlation statistic may be lower than the detection threshold even if the linear correlation is constant. This is valid since constant normalized correlation corresponds to a hypercone in the high-dimensional space of the host signal and not to a hyperplane. Thus, if someone wants to utilize normalized correlation as a detection statistic, they should construct the watermark in a way that all watermarked signals lie on a hypercone. The corresponding optimization problem (i.e., minimize perceptual distortion having constant robustness) is solved by finding the vector on the hypercone that has minimum Euclidean distance from the host signal and thus minimizes the MSE.

In the above analysis, the robustness measure that has been used was the detection statistic. However, this assumption is rather simplistic, and in most cases, optimizing the detection statistic prior to an attack does not guarantee optimal behavior against this attack. Thus, someone may design the informed embedding watermarking algorithm having as robustness measure an estimate of the robustness against a specific attack. One simple attack that can be used for this purpose is the additive white Gaussian noise (AWGN), which has been widely used in the literature to model the attacks incurred in a transmission channel. In this case, the optimization problem can be described as the minimization of the perceptual distortion after watermark embedding having constant robustness against AWGN. Assuming that the detection statistic is the normalized correlation, it can be easily proven that the optimal solution lies on a hyperboloid inside the hypercone of the normalized correlation [27]. Finding the optimal watermarked vector corresponds to finding the vector that lies on the hyperboloid that has minimum distance from the host signal. The analytical solution of this problem involves the solution of a quartic equation and, thus, for simplicity reasons the solution can be found by exhaustive search.

From the above analysis, it is obvious that the watermarking scheme designer who wants to use informed embedding has to decide on the robustness and perceptual distortion measures that will be used and on the appropriate detection statistic. Unfortunately, no robustness measure can be defined for the majority of common attacks. Furthermore, if someone can define a robustness measure against one attack, it is most probable that this measure will not be valid for other attacks. Another problem is that effective perceptual distortion measures are a research subject themselves since, in the case of images, measuring the distortion involves the modelling of the HVS. For the above reasons, the informed embedding watermarking systems cannot outperform the practical, often heuristically designed, watermarking systems under common watermark attacks such as image rotation, scaling, and cropping, despite their solid mathematical foundation.

A simple watermarking technique to utilize the interference of the host signal to watermark detection, for improved performance, has been introduced in [103]. The difference from the previous schemes is that it does not reject the host interference. Instead, in the watermark embedding phase, the correlation output prior to embedding is exploited in order to increase the embedding power and consequently the detector response. Another advantage of the proposed method is the perceptual analysis step that improves the perceptual quality of the watermarked content. The proposed technique is effectively employed in both additive and multiplicative spread spectrum schemes.

22.5.5.2 *Informed Coding Watermarking*

Informed embedding algorithms yield better performance than their blind competitors. However, the previously described algorithms do not fully exploit the information of the host signal in that they do not use this information during the message coding procedure.

That is, when multiple-bit watermarks are considered, the message coding procedure can use the information of the host signal in order to select an appropriate code vector for the specific host signal that minimizes the perceptual distortion of the watermarked signal and maximizes the robustness in terms of appropriate measures. These algorithms that take advantage of the side information during the message encoding procedure are called *informed coding* watermarking algorithms. It is possible to combine informed embedding and informed coding in order to achieve significantly better performance.

Informed coding has been described based on a theoretical result obtained by Costa [104]. This result implies that the capacity of a watermarking system might not depend on the distribution of the host signal but rather on the distortions the watermark must survive (i.e., the desired level of robustness). Costa introduced the idea of *"writing on dirty paper"* which can be described as follows: *Having a piece of paper covered with independent dirty spots of normally distributed intensity, someone has to write a message using a limited amount of ink. The dirty paper, with the message on it, is then transmitted to someone else, and acquires more normally distributed dirt on the way. If the receiver cannot distinguish between the ink and the dirt, how much information can be reliably sent?* Costa presented this problem as an analogy to the communication channel shown in Fig. 22.6. The channel has two independent white Gaussian noise sources. The first one represents the host signal and the second one represents the attacks that the watermarked signal may face. Before the transmitter chooses a signal to send, it is informed that the host signal is f_o. The transmitter should send a signal that is limited by a power constraint that corresponds to the limited ink on the dirty paper example or to the perceptual distortions constraint in a real watermarking system. Costa showed that, for the above communication scheme, the first noise source has no effect on the channel capacity. That is, if a watermarking scheme behaves enough like this dirty paper channel, its maximum message payload should not depend on the host media.

Unfortunately, a real watermarking scheme has substantial differences from the dirty paper communication scheme:

- The two noise sources, especially the host signal, are rarely Gaussian.

- The distortions implied by the channel (i.e., the second AWGN source) are dependent on the watermarked signal.

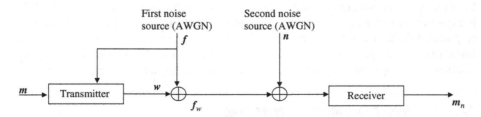

FIGURE 22.6

Communication channel model proposed by Costa.

- The fidelity constraint (i.e., the constraint on perceptual distortions) cannot be represented by the constrained power, since this is analogous to the MSE, which is a poor perceptual distortion estimate.

For the above reasons, it is not straightforward to prove that the capacity of watermarking is independent of the distribution of the host signal. Other researchers studied channels that better model a real watermarking scheme [105]. They have shown that, even if the second noise source is not AWGN but a hostile adversary that intentionally tries to remove the watermark, the dirty paper theory result still holds. They have also found that, even when the unwatermarked signal is not Gaussian, the result is approximately true.

The basic idea of applying Costa's theory to watermarking is using several alternative code words for each message instead of a single one. The code word that will be used for transmitting the message is the one that minimizes the interference of the host signal. Let us consider a simple channel in which only two different host signals, f_1 and f_2, can be transmitted. That is, the first AWGN source can generate only these two signals, while the second AWGN source can generate any noise signal. Suppose that the transmitter should send one of the M possible messages m. If the host signal to be transmitted is f_1, then the optimal solution for encoding the message m is to select $w = m - f_1$ as the signal to be transmitted, since it is known to the transmitter that the interference will be f_1. The constraint of the limited power of the watermark w defines a hypersphere in the multidimensional space of the host signal that is centered in f_1. All the watermarked versions of f_1 containing the different encoded messages should lie inside the hypersphere and should be as far as possible from each other. If the host signal to be transmitted is the signal f_2, then the optimal solution is to construct the watermark as $w = m - f_2$. Again all the encoded messages should lie inside a hypersphere centered in f_2. It is obvious that, if the two host signals are sufficiently different from each other, the two hyperspheres do not overlap and two different sets of code vectors can be used for encoding the messages depending on the host signal. The receiver also uses two sets of code vectors and selects the code vector that is closer, in terms of Euclidean distance, to the received signal. This communication scheme is reliable even though the receiver does not have any side information about the host signal.

The measures that indicate the robustness of the described scheme are firstly the radius of the hyperspheres, which allows for many different messages to be encoded (i.e., be included and sufficiently scattered inside the hypersphere), and secondly, the distance between the host signals which indicates the amount of noise that can de dealt with without causing shifting to a set of code vectors different from the one used for encoding.

In a more realistic situation, the hyperspheres of the host signals will overlap and, more generally, the number of different host signals will be unlimited, and thus generating a set of code vectors for each host signal will be infeasible. In order to solve these problems and design a scheme that can deal with the constraints of a real watermarking system, the QIM method has been proposed [28]. QIM refers to embedding information by first modulating an index or a sequence of indices with the embedded information and then quantizing the host signal with the associated quantizer or sequence of quantizers.

A simple example with two quantizers that are used to encode two different messages is shown in Fig. 22.7. The size of the quantization cells, one of which is shown in this figure, determines the distortion that can be tolerated. In QIM, each message is encoded using a different quantizer that covers the entire multidimensional space of the host signal in a structured manner. During the encoding procedure, the host signal is quantized using the quantizer that corresponds to the message to be embedded. The distortion induced by the quantization step is associated with the fidelity constraint, whereas the minimum distance between the nodes of two different quantizers is associated with the maximum perturbation that can be tolerated by the watermarking scheme.

A practical implementation of QIM is *Dither Modulation*, where the so-called dither quantizers are used during message encoding. The property of these quantizers that renders them appropriate for QIM is that the quantization cells and reconstruction points of any given quantizer are shifted versions of the quantization cells and the reconstruction points of any other quantizer. This property is exploited in the message encoding procedure, where each possible message maps uniquely onto a different dither vector (i.e., quantizer) and the host signal is quantized using the resulting quantizer. The purpose of dither modulation, which is commonly used to deal with quantization noise, is threefold [106]. First, a pseudorandom dither signal can reduce quantization artifacts to produce a perceptually superior quantized signal. Second, dither ensures that the quantization noise is independent of the host signal. Third, the pseudorandom dither signal can be

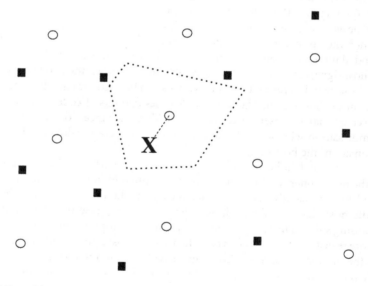

FIGURE 22.7

Quantization index modulation for information embedding. Two different quantizers are depicted. The input signal is quantized with the 0-quantizer that corresponds to the first of the two possible messages.

considered to act as a key which is only known to the embedder and detector, thereby improving the security of the system.

Many practical realizations of QIM have been proposed to deal with the problem of the increased computational complexity of the original approach [107]. In most cases, regular lattices are used as realization of the quantizers. The drawback of these lattice codes is that they are not robust against valumetric scaling (i.e., multiplication of the host signal by a gain factor) such as contrast changes in images. A small change on the contrast of a watermarked image may result in errors in message decoding. If the gain is constant throughout the feature sequence, the so-called fixed gain attack (FGA) results. If the gain is not constant (e.g., γ-correction), the attack is more difficult to deal with. Weakness against FGA is a serious drawback, e.g., with respect to classical spread spectrum methods, since in many cases, multiplication by a constant factor does not introduce any annoying artifacts, yet it results in a dramatic performance degradation of the QIM-based methods.

Various approaches that try to remedy the valumetric scaling attack vulnerability of QIM have been proposed. To this end, the embedding of an auxiliary pilot signal, to be used by the decoder in order to recover from amplitude scaling, has been proposed in [108]. However, this approach poses several problems, since the embedding of this signal, which is deterministically known to both transmitting and receiving ends, reduces the available payload and introduces an additional source of weakness against malicious attacks because attackers can decide to attack either the watermark or the pilot signal.

Another class of methods is based on the idea of estimating the valumetric scaling factor prior to detection. The estimated factor can be used either for reversing the attack or for adapting the watermark detection parameters to agree with the valumetric characteristics of the attacked image. In [106], the use of perceptual models has been proposed in order to improve fidelity and provide resistance to valumetric scaling. The Watson's perceptual model is used in order to provide adaptive quantization steps, instead of the constant step used in standard QIM, based on the perceptual characteristics of the image to be watermarked. Moreover, the use of the perceptual model allows for compensation for the valumetric scaling attack by estimating adaptively the quantization step in the watermark detection phase, taking into account the undergone valumetric scaling.

A third way of dealing with valumetric scaling is by embedding the watermark in a domain that is invariant to valumetric scaling. One such method has been presented in [109], where a quantization-based data-hiding method, called Rational Dither Modulation, was proposed. The proposed method modifies the conventional dither modulation in such a way that the result becomes invariant to gain attacks. The method utilizes a gain-invariant adaptive quantization step-size at both the embedder and decoder. Invariance against FGA is obtained by applying conventional dither modulation to the ratio of the current host sample and the previously generated watermarked sample. It is obvious that the ratio of two consecutive pixel values is invariant to the FGA. The Rational Dither Modulation method has been compared with improved spread-spectrum methods and achieved much higher information embedding rates for the same bit error probability.

Finally, adoption of spherical codewords together with correlation decoding is another way to deal with the valumetric scaling attack. A dirty paper coding technique of this category has been presented in [110]. Robustness against the gain attack is obtained by adopting a set of (orthogonal) equienergetic codewords and a correlation-based decoder. The performance of the dirty coding algorithm is further improved by replacing orthogonal with quasi-orthogonal codes, namely, Gold sequences, and by concatenating them with an outer turbo code. The use of spherical codes to cope with the FGA relies on the observation that using a minimum distance decoder on a set of codewords lying on the surface of a sphere ensures that multiplication by a constant factor does not move a point from one decoding region to the other. The problem with spherical codes is that watermark embedding and recovery get very complicated, thus losing the simplicity of lattice-based watermarking. An attempt, also in [110], to develop a simple watermarking scheme relying on the properties of orthogonal spherical codes has given good results. However, such results have been obtained at the expense of watermark payload.

22.5.5.3 *Perceptual Masking*

As we have previously seen, one of the watermark characteristics is imperceptibility which can be defined using two different terms: *fidelity* and *quality*. Fidelity is a measure of the similarity between the original host signal and the watermarked signal and, thus, should be as high as possible. Image quality is a measure of the viewer's satisfaction when viewing an image. A high-quality image looks good (i.e., it has no obvious processing artifacts). Thus, fidelity is a two argument measure, whereas quality is a single argument one.

From the above definitions, it is obvious that image quality may be reduced after watermark embedding, only if the quality of the original host signal is high. However, there are many cases where the quality of the original host signal is medium or low (for example, surveillance images). In these cases, the objective is to maintain the image quality. Preserving the quality after watermark embedding is achieved when the fidelity is high, since in this case, the watermarked signal is indistinguishable from the original host signal. High fidelity of the watermarked image assures quality preservation, whereas high quality of the watermarked image does not assure high fidelity.

In the design of watermarking schemes using informed embedding or coding, we have seen that the MSE is the simplest measure of the perceptual distortion implied by the watermark. Constructing watermarks resulting in constant MSE between the watermarked and the original signal was considered as a way of keeping constant fidelity. However, in a real situation, two different watermarked signals that have the same MSE when compared to their originals are not perceived as having the same fidelity, because MSE does not correlate well with subjective fidelity. Instead of the MSE, other more complex models have been proposed for measuring visual fidelity that is closer to the way HVS works.

The properties of the HVS that are taken into account when constructing a watermarking scheme are the following:

- *Sensitivity*, which is related to the eye response to direct stimuli. For example, a common sensitivity measure is the minimum brightness required for an eye to perceive certain spatial or temporal image frequencies.

- *Masking,* which is a measure of the observer's response to one stimulus (watermark), when a second "masking" stimulus (host image) is also present. Thus, by using visual masking, we can hide more information in a textured image region than in a uniform region since texture provides a better mask.

- *Pooling,* which is a single estimate of the overall change in the visual appearance of an image that is caused by combining the perception of separate distortions.

The important issue is the adaptation of the watermark to the properties of the HVS, i.e., content-adaptive watermarking. Assuming we are given a masking function constructed according to the HVS properties, we wish to embed the watermark into the host image by keeping it under the threshold of visual imperceptibility. Perceptually adaptive watermarking was mainly inspired by the achievements in perceptual-based image/video compression, especially in the early stages of watermarking technology development. Therefore, perceptual masking was mainly addressed in the transform domain.

One of the first approaches to content-adaptive digital watermarking was proposed in [111]. This approach was based on the just noticeable difference, which was computed in the DCT domain using a mask developed for lossy JPEG compression by Watson [112]. Another sophisticated approach takes into account the luminance and contrast sensitivity of the HVS, and differentiates between the visibility of the edge and texture regions. The resulting mask was used in the spatial image.

The next generation of content-adaptive digital watermarking methods utilizes the idea that perceptual masking should be performed directly in the transform domain (mostly wavelet domain) to be matched with the JPEG2000 image compression standard. Perceptual masking performed in the transform domain has a number of advantages. First, the watermark embedding process is accomplished in the same domain as image/video compression, which renders watermarking-on-the-fly possible. Second, it allows JPEG2000 compliance that ensures additional robustness of the watermarking algorithms to lossy JPEG2000 compression. It should be noted that the mask and watermark energy allocation should be simultaneously adapted to the other types of lossy compression algorithms such as DCT-based JPEG.

The factors that should be taken into account for the mask design to match HVS properties are the following:

- Background luminance sensitivity;

- Contrast sensitivity depending on image subband or resolution;

- Orientation sensitivity (anisotropy);

- Edge and pattern (texture) masking.

The background luminance sensitivity is described according to Weber's law: the eye is less sensitive to noise in bright regions. For the subband sensitivity, experiments have proven that the eye is differently stimulated depending on the frequency, orientation, and luminance of the image content. Finally, the higher the texture energy of an image region is, the lower the visual sensitivity is with respect to content changes in this region. The combination of high-frequency edges and low-frequency luminance variance is used to calculate the texture masking factor. The edge proximity factor refers to the fact that the higher the luminance difference of an edge is, the less the visual sensitivity is at the vicinity of the edge. However, as we get away from the edge, the visual sensitivity is reestablished.

To establish a bridge between the rate-distortion theory and content-adaptive watermarking, a stochastic texture perceptual mask was prepared based on a noise visibility function (NVF) developed earlier only for the spatial image domain [113]. This result was also extended to the wavelet domain aiming at including the multi-resolution paradigm in the stochastic framework to take into account a modulation transfer function (MTF) of the HVS and to match the proposed watermarking algorithm with the image compression standard JPEG2000. Practically, this means that the assigned watermark strength depends on the image subband. Such a modification leads to a nonwhite watermark spectrum matched with the MTF, which was not the case for the spatial domain-based version of the NVF. The second reason to use wavelet domain embedding is the desire to incorporate the HVS anisotropy to different spatial directions in the perceptual mask. The spatial domain version of the NVF uses isotropic image decomposition based on the extraction of a local mean from the original image or its highpass filter output. In the wavelet domain, the image coefficients in three basic spatial directions, i.e., vertical, horizontal, and diagonal, are received as a result of the decomposition that better reflects the HVS anisotropy properties. As a result, the watermark strength varies with orientations in the proposed mask.

Recent efforts to incorporate perceptual masking principles in image watermarking algorithms are described in [106, 114, 115].

22.6 IMAGE CONTENT INTEGRITY AND AUTHENTICATION WATERMARKING

In this section, we focus on how watermarks can be used to assist in maintaining and verifying image content integrity. Image editing is in many cases legitimate and desirable. However, for many applications, modifications introduced by image editing may be malicious or may affect image interpretation, thus resulting in image tampering. For example, tampering with images acquired for medical applications may result in misdiagnosis. Changes in images that are used as evidence in a criminal trial can result in a wrongful conviction. Image content authentication mainly focuses on the development of *fragile* or *semifragile* watermarks.

In a real world application scenario, the image owner embeds a watermark so that either she or the user is able to check if the image has been manipulated by someone. Sometimes, the owner can subsequently become the user as well. The information conveyed by the watermark is related to the authenticity and/or integrity of the product. In certain cases, it is interesting to detect which image regions have been altered. In general, the attacker intends to alter the content of the product without distorting the watermark.

This goal is the primary distinction between watermarking for image authentication and for copyright protection. Alternatively, the attacker may try to copy the watermark from a watermarked image to another image. This is the *forgery attack.* The watermark should generally be destroyed by any authentication attack on the image. In some cases, it is desirable that the watermark withstands certain attacks that do not degrade image quality (e.g., high-quality lossy compression). A usual application occurs when photographs or movies are sent by news agencies (owners) to newspapers or TV channels (users). The news broadcaster wants to be assured that the images are authentic before broadcasting. Thus, the authenticity check is a procedure that concerns primarily the user of the image. If someone wants to use an image authentication algorithm in front of a court, then the user is the judge who needs to be assured of the image authenticity.

The purpose of an image authentication algorithm is to provide the user with all the information needed about image authenticity. Image authentication algorithms usually produce an authenticity measure in the range $[0,1]$. If is required that the image data should not change at all, any image alteration is prohibited and the threshold for accepting an image as authentic should be very high (or close to 1). Thus, the use of an authenticity measure is enough to assure authenticity. This type of authentication is called *complete* or *exact* or *hard* authentication. However, in certain cases of image content authentication (e.g., in the case of stored surveillance images), legitimate distortions such as high-quality image compression are allowed, i.e., an image is considered authentic even after high-quality lossy compression. In such cases, the image authenticity measure is lower than 1 even for authentic images, where the image content remains unaltered. Thus, the use of the authenticity measure alone is not adequate for deciding about the image authenticity. In this case, the image authentication algorithm should have the ability of tamper proofing, i.e., to detect the tampered image regions that change the image content. This type of authentication is called *content* or *selective* or *soft* authentication. We shall use only the terms complete and content authentication for the rest of this section.

It is obvious that the requirements of an image authentication scheme are different, depending on whether we are considering complete or content authentication. In complete authentication, the watermark should be destroyed even if a single bit has been changed in the watermarked image. Localization of the tampered regions is a desirable feature in complete authentication. In content authentication schemes, the basic requirement is that the watermark should be destroyed whenever an illegitimate distortion is applied to the watermarked image and should not be destroyed when the distortions are legitimate. It is obvious that the critical point is the characterization of a distortion as legitimate or illegitimate. This issue cannot be dealt with universally but rather is specific for each particular application. For example, uniform geometric distortions applied on an image may be legitimate for a press agency but illegitimate for a medical application. The distortions that are usually considered legitimate for many applications are those that do not change in any manner the content of the image. Such distortions are high-quality lossy compression, mild noise filtering, geometric transformations, etc. In any case, the content authentication should have the ability of tampering localization. That is, the image regions that have been subject to any illegitimate distortion should be detected.

In the extreme case of content authentication, only the semantic meaning of an image should be preserved for an image to be considered as authentic. The semantic content of the image can be represented in the form of a feature vector. In this case, content authentication means that the feature vector of the image should always be extracted unaltered. In this case, the image is considered authentic even if heavy distortions have significantly reduced the image quality. Another characteristic of some content authentication algorithms is the property of self restoration. Self restoration is ability of the algorithm to restore the image to its original content even if the content has been distorted by illegitimate manipulations. To this end, *self-embedding* has been proposed in [116]. In this approach, a low-quality version of the image is embedded in the original image using least significant bit (LSB) modulation. The low-quality information of each image block is embedded in the LSBs of a different image block so as to assure that either the original image block or the embedded one will be available at the restoration phase.

A general watermarking framework used for image authentication was presented in [12]. This framework is comprised of three steps: the watermark generation, embedding, and detection procedures. Let $\mathcal{R}$ and $\mathcal{Z}$ denote the set of real and integer numbers, respectively, and $\mathcal{U} = \{0, 1, 2\}$. Given an image $f(\mathbf{x}) : \mathcal{D} \subseteq \mathcal{Z}^2 \rightarrow \mathcal{Z}$ and a watermark key $k \in \mathcal{Z}$, a ternary watermark $w(\mathbf{x}) : \mathcal{G} \subseteq \mathcal{Z}^2 \rightarrow \mathcal{U}$ can be produced by the watermark generation procedure, where $\mathbf{x}$ denotes the coordinate vector of a pixel. In watermark generation, either chaotic techniques or pseudorandom number generators can be used for producing the watermark $w(\mathbf{x})$. The watermark embedding procedure of the watermark $w(\mathbf{x})$ in the image $f(\mathbf{x})$ is given by

$$f_w(\mathbf{x}) = f(\mathbf{x}) \otimes w(\mathbf{x}), \tag{22.31}$$

where $f_w(\mathbf{x})$ is the watermarked image and $\otimes$ is a generalized superposition operator which includes appropriate data truncation and quantization, if needed. The watermark embedding procedure can be applied either in the spatial domain or in the other domains (e.g., in the DFT, DWT, DCT domains). Watermark detection produces a binary decision on image authenticity and the watermark detection credibility measured by the watermark detection ratio. Additionally, watermark detection produces an image denoting the unaltered (authentic) regions of the watermarked image.

In the following, a brief description of the most important schemes proposed for image authentication is presented. The first class of authentication methods is based on using a separate header (*digital signature*) which must be known to the receiver that will perform the authenticity check. These schemes, although not based on watermarking, will be briefly reviewed since they appeared first in the literature. The development of a "trustworthy digital camera" was proposed in [117]. A digital image is captured by a camera and then is passed through a hash function. The output of the hash function is encrypted by the photographer's private key, and a separate authentication signal is created. In order to ensure image authentication, the encrypted signal is decrypted by the photographer's public key and the hashed version of the original image is compared with that of the received image. A compression tolerant method for image authentication is proposed in [118]. The proposed scheme is based on the extraction of feature points that are almost unaffected by lossy compression. The major drawbacks of this method are the need of a separate header for storing the digital signature and the low accuracy in the detection of the tampered regions.

An authentication method that gives a distortion measurement instead of a binary decision on image authenticity is proposed in [119]. It does not require a separate signature file or header for image authentication, but it cannot detect the image regions that are authentic, if selective modifications to fine image details have been made. A method for image authentication that is based on changing the LSB of the image pixels is proposed in [120]. The method can detect alterations that are made in several image regions. However, it is not robust to high-quality lossy compression.

Another class of authentication methods uses a transform domain such as DFT, DWT, DCT, etc. for watermark embedding. These methods are usually more robust against attacks such as filtering and compression [121]. A method in which a look-up table is needed for image authentication is proposed in [122]. The watermark is embedded in the DCT domain and attains robustness to lossy compression. This method is based on [123], where the LUT is used for watermark embedding in the spatial domain. However, it is not robust against lossy compression and, furthermore, is very sensitive to the forgery attack presented in [124]. A method for wavelet transform domain authentication watermarking is proposed in [125]. The detection of the tampered image regions is not addressed, and the compression of the watermarked image is not supported. Instead, the authentication algorithm is integrated within the SPIHT codec. This means that the authentication algorithm protects the already compressed image. An image watermarking method for tamper proofing and authentication is proposed in [126]. The watermark is embedded in the discrete wavelet transform domain by quantizing the corresponding wavelet coefficients. The method is robust against compression and succeeds in detecting alterations of the host image, whether they are produced by changing fine image details or by high-quality image compression. The major drawback of the proposed method is that, when combined attacks are made to the watermarked image, the authenticity check is based on the low DWT levels, i.e., levels 4 and 5. These levels contain a small number of coefficients, and a decision about image authenticity based on the watermark detection at these levels is unreliable, since each coefficient represents a region of 16×16 or 32×32 image pixels. Robustness against combined attacks is not supported.

An attack that has been developed for image authentication algorithms is presented in [124] and succeeds in copying the watermark from a watermarked image to another image that is not watermarked. The attack has been successfully applied to the methods presented in [120, 127, 128]. Knowledge of the logo to be embedded is assumed in order to forge the watermark. The main disadvantage that renders these methods vulnerable to the attack is that they are blockwise independent. That is, the watermark in a certain block depends neither on the entire image nor on other watermarked blocks. Another fact that helps the attacker to estimate the inserted watermark statistically is that critical watermark parameters, such as the positions of the watermarked pixels and the embedded logo are known to the attacker.

Another technique for image authentication has been proposed in [129, 130]. It is based on a modification of an established watermarking technique for copyright protection [12]. The image authentication algorithm generates a watermark according to the owner's private key. Subsequently, the watermark is imperceptibly embedded in the image. In the authentication detection procedure, the watermark is extracted from the image and a measure of tampering is produced for the entire image. The algorithm detects the regions of the image that are altered/unaltered and, thus, are considered nonauthentic/authentic, respectively. The alterations that are produced by a relatively mild compression and do not change significantly the quality of the image are also detected. An example of an image authentication procedure using the image "Opera of Lyon" (*http://www.petitcolas.net/fabien/watermarking/image_database/index.html*), which has been used as a reference image for watermark benchmarking, is depicted in Fig. 22.8.

The method in [12] has been extended to support tampering detection using a hierarchical structure in the detection phase that ensures accurate tamper localization [131].

A novel framework for lossless (invertible) authentication watermarking, which enables zero-distortion reconstruction of the original image upon verification, has been proposed in [132]. The framework allows authentication of the watermarked images before recovery of the original image. This reduces computational requirements in situations where either the verification step fails or the zero-distortion reconstruction is not needed. The framework also enables public-key authentication without granting access to the original and allows for efficient tamper localization. Effectiveness of the framework is demonstrated by implementing it using hierarchical image authentication along with lossless generalized-least significant bit data embedding.

A blind image watermarking method based on a multistage vector quantizer structure, which can be used simultaneously for both image authentication and copyright protection, has been proposed in [133]. In this method, the semifragile watermark and the robust watermark are embedded in different vector quantization stages using different techniques. Simulation results demonstrated the effectiveness of the proposed algorithm in terms of robustness and fragility. Another semifragile watermarking method that is

(a) (b) (c)

FIGURE 22.8

(a) Original watermarked image; (b) tampered watermarked image; (c) tampered regions.

robust against lossy compression has been proposed in [134]. The proposed method uses random bias and nonuniform quantization to improve the performance of the methods proposed in [121].

Differentiating between malicious and incidental manipulations in content authentication remains an open issue. Exploitation of robust watermarks with self-restoration capabilities for image authentication is another research topic. The authentication of certain regions instead of the whole image when only some regions are tampered with has also attracted the attention of the watermarking community.

ACKNOWLEDGMENT

The authoring of this chapter has been supported in part by the European Commission through the IST Programme under Contract IST-2002-507932 ECRYPT.

REFERENCES

[1] I. Cox, M. Miller, J. Bloom, J. Fridrich, and T. Kalker. *Digital Watermarking and Steganography, 2nd ed.* Morgan Kaufmann Publishers, Burlington, MA, 2007.

[2] S. Craver and J. Stern. Lessons Learned from SDMI. In *IEEE Workshop on Multimedia Signal Processing, MMSP 01*, pp. 213–218, Cannes, France, October 2001.

[3] R. Venkatesan, S. M. Koon, M. H. Jakubowski, and P. Moulin. Robust image hashing. In *IEEE International Conference on Image Processing*, Vancouver, Canada, October 2000.

[4] V. Monga and B. Evans. Robust perceptual image hashing using feature points. In *IEEE International Conference on Image Processing, ICIP 04*, Singapore, October 2004.

[5] N. Nikolaidis and I. Pitas. Digital rights management of images and videos using robust replica detection techniques. In L. Drossos, S. Sioutas, D. Tsolis, and T. Papatheodorou, editors, *Digital Rights Management for E- Commerce Systems*, Idea Group Publishing, Hershey, PA, 2008.

[6] M. Barni. What is the future for watermarking? (part i). *IEEE Signal Process. Mag.*, 20(5):55–59, 2003.

[7] M. Barni. What is the future for watermarking? (part ii). *IEEE Signal Process. Mag.*, 20(6):53–59, 2003.

[8] M. Barni and F. Bartolini. *Watermarking Systems Engineering: Enabling Digital Assets Security and Other Applications.* Marcel Dekker, New York, 2004.

[9] S. Katzenbeisser and F. Petitcolas. *Information Hiding Techniques for Steganography and Digital Watermarking.* Artech House, Norwood, MA, 2000.

[10] N. Nikolaidis and I. Pitas. Benchmarking of watermarking algorithms. In J.-S. Pan, H.-C. Huang, and L. Jain, editors, *Intelligent Watermarking Techniques*, 315–347. World Scientific Publishing, Hackensack, NJ, 2004.

[11] B. Furht and D. Kirovski, editors. *Multimedia Watermarking Techniques and Applications.* Auerbach, Boca Raton, FL, 2006.

[12] G. Voyatzis and I. Pitas. The use of watermarks in the protection of digital multimedia products. *Proc. IEEE*, 87(7):1197–1207, July 1999.

[13] N. Nikolaidis and I. Pitas. Digital image watermarking: an overview. In *Int. Conf. on Multimedia Computing and Systems (ICMCS'99)*, Vol. I, 1–6, Florence, Italy, June 7–11, 1999.

[14] Special issue on identification & protection of multimedia information. *Proc. IEEE*, 87(7):1999.

[15] Special issue on signal processing for data hiding in digital media and secure content delivery. *IEEE Trans. Signal Process.*, 51(4):2003.

[16] F. A. P. Petitcolas, R. J. Anderson, and M. G. Kuhn. Information hiding - a survey. *Proc. IEEE*, 87(7):1062–1078, July 1999.

[17] Special issue on authentication copyright protection and information hiding. *IEEE Trans. Circuits Syst Video Technol.*, 13(8):2003.

[18] I. Cox, M. Miller, and J. Bloom. Watermarking applications and their properties. In *International Conference on Information Technology: Coding and Computing 2000*, 6–10. Las Vegas, March 2000.

[19] A. Nikolaidis, S. Tsekeridou, A. Tefas, and V. Solachidis. A survey on watermarking application scenarios and related attacks. In *2001 IEEE International Conference on Image Processing (ICIP 2001)*, Thessaloniki, Greece, October 7–10, 2001.

[20] M. Barni and F. Bartolini. Data hiding for fighting piracy. *IEEE Signal Process. Mag.*, 21(2):28–39, 2004.

[21] S. Nikolopoulos, S. Zafeiriou, P. Sidiropoulos, N. Nikolaidis, and I. Pitas. Image replica detection using R-trees and linear discriminant analysis. In *IEEE International Conference on Multimedia and Expo*, 1797–1800. Toronto, Canada, 2006.

[22] J. A. Bloom, I. J. Cox, T. Kalker, M. I. Miller, and C. B. S. Traw. Copy protection for dvd video. *Proc. IEEE*, 87(7):1267–1276, 1999.

[23] F. Bartolini, A. Tefas, M. Barni, and I. Pitas. Image authentication techniques for surveillance applications. *Proc. IEEE*, 89(10):1403–1418, 2001.

[24] A. M. Alattar. Smart images using digimarcs watermarking technology. In *IS&T/SPIEs 12th International Symposium on Electronic Imaging*, Vol. 3971, San Jose, CA, January 25, 2000.

[25] T. Furon, I. Venturini, and P. Duhamel. A unified approach of asymmetric watermarking schemes. In *SPIE Electronic Imaging, Security and Watermarking of Multimedia Contents*, Vol. 4314, 269–279. San Jose, CA, January 2001.

[26] T. Furon and P. Duhamel. An asymmetric watermarking method. *IEEE Trans. Signal Process.*, 51(4):981–995, 2003.

[27] I. J. Cox, M. I. Miller, and A. L. McKellips. Watermarking as communications with side information. *Proc. IEEE*, 87(7):1127–1141, 1999.

[28] B. Chen and G. Wornell. Quantization index modulation: a class of provably good methods for digital watermarking and information embedding. *IEEE Trans. Inf. Theory*, 47(4):1423–1443, 2001.

[29] I. J. Cox, J. Kilian, F. T. Leighton, and T. Shamoon. Secure spread spectrum watermarking for multimedia. *IEEE Trans. Image Process.*, 6(12):1673–1687, 1997.

[30] J. K. Ruanaidh and T. Pun. Rotation, scale and translation invariant spread spectrum digital image watermarking. *Elsevier Signal Process., Sp. Issue on Copyright Protection and Access Control*, 66(3):303–317, 1998.

[31] J. Cannons and P. Moulin. Design and statistical analysis of a hash-aided image watermarking system. *IEEE Trans. Image Process.*, 13(10):1393–1408, 2004.

[32] A. Nikolaidis and I. Pitas. Region-based image watermarking. *IEEE Trans. Image Process.*, 10(11):1726–1740, 2001.

[33] M. Barni, F. Bartolini, and A. Piva. Improved wavelet-based watermarking through pixel-wise masking. *IEEE Trans. Image Process.*, 10(5):783–791, 2001.

[34] V. Solachidis and I. Pitas. Circularly symmetric watermark embedding in 2D DFT domain. *IEEE Trans. Image Process.*, 10(11):1741–1753, 2001.

[35] J. R. Hernandez, M. Amado, and F. Perez-Gonzalez. DCT-domain watermarking techniques for still images: detector performance analysis and a new structure. *IEEE Trans. Image Process.*, 9(1):55–68, 2000.

[36] M. Barni, F. Bartolini, A. DeRosa, and A. Piva. A new decoder for the optimum recovery of nonadditive watermarks. *IEEE Trans. Image Process.*, 10(5):755–766, 2001.

[37] M. Barni, F. Bartolini, A. DeRosa, and A. Piva. Optimum decoding and detection of multiplicative watermarks. *IEEE Trans. Signal Process.*, 51(4):1118–1123, 2003.

[38] Q. Cheng and T. S. Huang. Robust optimum detection of transform domain multiplicative watermarks. *IEEE Trans. Signal Process.*, 51(4):906–924, 2003.

[39] V. Solachidis and I. Pitas. Optimal detector for multiplicative watermarks embedded in the DFT domain of non-white signals. *J. Appl. Signal Process.*, 2004(1):2522–2532, 2004.

[40] M. Yoshida, T. Fujita, and T. Fujiwara. A new optimum detection scheme for additive watermarks embedded in spatial domain. In *International Conference on Intelligent Information Hiding and Multimedia Signal Processing*, 2006.

[41] J. Wang, G. Liu, Y. Dai, J. Sun, Z. Wang, and S. Lian. Locally optimum detection for Barni's multiplicative watermarking in DWT domain. *Signal Process.*, 88(1):117–130, 2008.

[42] A. Tefas and I. Pitas. Robust spatial image watermarking using progressive detection. In *Proc. of 2001 IEEE Int. Conf. on Acoustics, Speech and Signal Processing (ICASSP 2001)*, 1973–1976, Salt Lake City, Utah, USA, 7–11 May, 2001.

[43] D. P. Mukherjee, S. Maitra, and S. T. Acton. Spatial domain digital watermarking of multimedia objects for buyer authentication. *IEEE Trans. Multimed.*, 6(1):1–15, 2004.

[44] H. S. Kim and H.-K. Lee. Invariant image watermark using Zernike moments. *IEEE Trans. Circuits Syst. Video Technol.*, 13(8):766–775, 2003.

[45] M. Barni, F. Bartolini, V. Cappelini, and A. Piva. A DCT-domain system for robust image watermarking. *Elsevier Signal Process.*, 66(3):357–372, 1998.

[46] W. C. Chu. DCT-based image watermarking using subsampling. *IEEE Trans. Multimed.*, 5(1): 34–38, 2003.

[47] A. Briassouli, P. Tsakalides, and A. Stouraitis. Hidden messages in heavy-tails: DCT-domain watermark detection using alpha-stable models. *IEEE Trans. Multimed.*, 7(4):700–715, 2005.

[48] C.-Y. Lin, M. Wu, J. Bloom, I. Cox, M. Miller, and Y. Lui. Rotation, scale, and translation resilient watermarking for images. *IEEE Trans. Image Process.*, 10(5):767–782, 2001.

[49] Y. Wang, J. Doherty, and R. Van Dyck. A wavelet-based watermarking algorithm for ownership verification of digital images. *IEEE Trans. Image Process.*, 11(2):77–88, 2002.

[50] P. Bao and M. Xiaohu. Image adaptive watermarking using wavelet domain singular value decomposition. *IEEE Trans. Circuits Syst. Video Technol.*, 15(1):96–102, 2005.

[51] ITU. Methodology for the subjective assessment of the quality of television pictures. *ITU-R Recommendation BT.500-11*, 2002.

[52] A. Netravali and B. Haskell. *Digital Pictures, Representation and Compression*. Plenum Press, New York, 1988.

[53] S. Voloshynovskiy, S. Pereira, V. Iquise, and T. Pun. Attack modelling: towards a second generation benchmark. *Signal Process.*, 81(6):1177–1214, 2001.

[54] T. Pappas, R. Safranek, and J. Chen. Perceptual criteria for image quality evaluation. In A. Bovik, editor, *Handbook of Image and Video Processing*, 2nd ed, 939–960. Academic Press, Burlington, MA, 2005.

[55] Z. Wang, A. Bovik, and E. Simoncelli. Structural approaches to image quality assessment. In A. Bovik, editor, *Handbook of Image and Video Processing*, 961–974. Academic Press, Burlington, MA, 2005.

[56] H. Sheikh and A. Bovik. Information theoretic approaches to image quality assessment. In A. Bovik, editor, *Handbook of Image and Video Processing*, 975–992. Academic Press, Burlington, MA, 2005.

[57] I. Setyawan, D. Delannay, B. Macq, and R. L. Lagendijk. Perceptual quality evaluation of geometrically distorted images using relevant geometric transformation modeling. In *SPIE/IST 15th Electronic Imaging*, Santa Clara, 2003.

[58] A. Tefas and I. Pitas. Multi-bit image watermarking robust to geometric distortions. In *CD-ROM Proc. of IEEE Int. Conf. on Image Processing (ICIP'2000)*, 710–713. Vancouver, Canada, 10–13 September, 2000.

[59] P. Comesana, L. Perez-Freire, and F. Perez-Gonzalez. An information-theoretic framework for assessing security in practical water-marking and data hiding scenarios. In *6th International Workshop on Image Analysis for Multimedia Interactive Services*, Montreux, Switzerland, 2005.

[60] P. Comesana, L. Perez-Freire, and F. Perez-Gonzalez. Fundamentals of data hiding security and their application to spread-spectrum analysis. In *7th Information Hiding Workshop*, 146–160, Barcelona, Spain, 2005.

[61] F. Cayre, C. Fontaine, and T. Furon. Watermarking security: theory and practice. *IEEE Trans. Signal Process.*, 53(10):3976–3987, 2005.

[62] C. E. Shannon. Communication theory of secrecy systems. *Bell Syst. Tech. J.*, 28:656–715, 1949.

[63] W. Diffie and M. Hellman. New directions in cryptography. *IEEE Trans. Inf. Theory*, 22(6):644–654, 1976.

[64] L. Perez-Freire, P. Comesana, J. R. Troncoso-Pastoriza, and F. Perez-Gonzalez. Watermarking security: a survey. *Trans. Data Hiding Multimed. Secur. I*, 4300:41–72, 2006.

[65] V. Licks and R. Jordan. Geometric attacks on image watermarking systems. *IEEE Multimed.*, 12(3):68–78, 2005.

[66] F. A. P. Petitcolas. Watermarking schemes evaluation. *IEEE Signal Process. Mag.*, 17(5):58–64, 2000.

[67] F. Petitcolas, R. Anderson, and M. Kuhn. Attacks on copyright marking systems. In *Second Int. Workshop on Information Hiding*, Vol. LNCS 1525, 219–239, Portland, Oregon, April 1998.

[68] M. Kutter, S. Voloshynovskiy, and A. Herrigel. The watermark copy attack. In *Electronic Imaging 2000, Security and Watermarking of Multimedia Content II*, Vol. 3971, 2000.

[69] S. Craver, N. Memon, B.-L. Yeo, and M. Yeung. Resolving rightful ownerships with invisible watermarking techniques: limitations, attacks and implications. *IEEE J. Sel. Areas Commun.*, 16(4):573–586, 1998.

[70] B. Macq, J. Dittmann, and E. Delp. Benchmarking of image watermarking algorithms for digital rights management. *Proc. IEEE*, 92(6):971–984, 2004.

[71] B. Michiels and B. Macq. Benchmarking image watermarking algorithms with OpenWatermark. In *European Signal Process. Conf.*, Florence, Italy, September 2006.

[72] O. Guitart, H. Kim, and E. Delp. The watermark evaluation testbed (WET): new functionalities. In *SPIE Int. Conf. on Security and Watermarking of Multimedia Contents*, January 2006.

[73] F. Petitcolas, M. Steinebach, F. Raynal, J. Dittmann, C. Fontaine, and N. Fates. A public automated web-based evaluation service for watermarking schemes: Stirmark benchmark. In *SPIE Electronic Imaging 2001, Security and Watermarking of Multimedia Contents*, Vol. 4314, San Jose, CA, USA, January 2001.

[74] S. Pereira, S. Voloshynovskiy, M. Madueno, S. Marchand-Maillet, and T. Pun. Second generation benchmarking and application oriented evaluation. In *Information Hiding Workshop III*, Pittsburgh, PA, USA, April 2001.

[75] V. Solachidis, A. Tefas, N. Nikolaidis, S. Tsekeridou, A. Nikolaidis, and I. Pitas. A benchmarking protocol for watermarking methods. In *Proc. of ICIP'01*, 1023–1026, Thessaloniki, Greece, October 7–10, 2001.

[76] M. Simon, J. Omura, R. Scholtz, and B. Levitt. *Spread Spectrum Communications Handbook*. McGraw-Hill, Columbus, OH, 1994.

[77] I. Pitas and T. H. Kaskalis. Applying signatures on digital images. In *Proc. of 1995 IEEE Workshop on Nonlinear Signal and Image Process. (NSIP'95)*, 460–463, N. Marmaras, Greece, June 20–22, 1995.

[78] N. Nikolaidis and I. Pitas. Copyright protection of images using robust digital signatures. In *Proc. of ICASSP'96*, Vol. 4, 2168–2171, Atlanta, USA, 1996.

[79] A. Tefas, A. Nikolaidis, N. Nikolaidis, V. Solachidis, S. Tsekeridou, and I. Pitas. Performance analysis of correlation-based water-marking schemes employing markov chaotic sequences. *IEEE Trans. Signal Process.*, 51(7):1979–1994, 2003.

[80] G. Voyatzis and I. Pitas. Chaotic watermarks for embedding in the spatial digital image domain. In *Proc. of ICIP'98*, Vol. II, 432–436, Chicago, USA, October 4–7, 1998.

[81] A. Nikolaidis and I. Pitas. Comparison of different chaotic maps with application to image watermarking. In *Proc. of ISCAS'00*, Vol. V, 509–512, Geneva, Switzerland, May 28–31, 2000.

[82] S. Tsekeridou, N. Nikolaidis, N. Sidiropoulos, and I. Pitas. Copyright protection of still images using self-similar chaotic watermarks. In *Proc. of ICIP'00*, 411–414, Vancouver, Canada, September 10–13, 2000, to appear.

[83] A. Tefas, A. Nikolaidis, N. Nikolaidis, V. Solachidis, S. Tsekeridou, and I. Pitas. Statistical analysis of Markov chaotic sequences for watermarking applications. In *2001 IEEE Int. Symposium on Circuits and Systems (ISCAS 2001)*, 57–60, Sydney, Australia, May 6–9, 2001.

[84] A. Mooney, J. Keating, and I. Pitas. A comparative study of chaotic and white noise signals in digital watermarking. *Chaos, Solitons and Fractals*, 35:913–921, 2008.

[85] N. Nikolaidis, S. Tsekeridou, A. Nikolaidis, A. Tefas, V. Solachidis, and I. Pitas. Applications of chaotic signal processing techniques to multimedia watermarking. In *Proc. of the IEEE Workshop on Nonlinear Dynamics in Electronic Systems*, 1–7, Catania, Italy, May 18–20, 2000.

[86] N. Nikoladis, A. Tefas, and I. Pitas. Chaotic sequences for digital watermarking. In S. Marshall, and G. Sicuranza, editors, *Advances in Nonlinear Signal and Image Processing*, 205–238. Hindawi Publishing, New York, 2006.

[87] C. Lin, M. Wu, J. A. Bloom, I. J. Cox, M. L. Miller, and Y. M. Lui. Rotation, scale, and translation resilient public watermarking for images. In *Proc. of SPIE:Electronic Imaging 2000*, Vol. 3971, 90–98, January 2000.

[88] P. Dong, J. G. Brankov, N. P. Galatsanos, Y. Yang, and F. Davoine. Digital watermarking robust to geometric distortions. *IEEE Trans. Image Process.*, 14(12):2140–2150, 2005.

[89] I. Djurovic, S. Stankovic, and I. Pitas. Watermarking in the space/spatial domain using two-dimensional Radon Wigner distribution. *IEEE Trans. Image Process.*, 10(4):650–658, 2001.

[90] T. K. Tsui, X.-P. Zhang, and D. Androutsos. Color image watermarking using multidimensional Fourier transforms. *IEEE Trans. Inf. Forensics Secur.*, 3(1):16–28, 2008.

[91] S. Sangwine and T. Ell. The discrete Fourier transform of a colour image. In *Second IMA Conf. on Image Process.*, 430–441, Leicester, UK, September 1998.

[92] S. Pereira and T. Pun. Robust template matching for affine resistant image watermarks. *IEEE Trans. on Image Process.*, 9(6):1123–1129, 2000.

[93] S. Pereira, J. Ruanaidh, F. Deguillaume, G. Csurka, and T. Pun. Template based recovery of Fourier-based watermarks using log-polar and log-log maps. In *Proc. of ICMCS'99*, Florence, Italy, Vol. I, 870–874, June 7–11, 1999.

[94] S. Voloshynovskiy, F. Deguillaume, and T. Pun. Multibit digital watermarking robust against local non-linear geometrical transformations. In *ICIP 2001, Int. Conf. Image Process.*, October 2001.

[95] A. Herrigel, S. Voloshynovskiy, and Y. Rytsar. Template removal attack. In *Proc. of SPIE: Electronic Imaging 2001*, January 2001.

[96] V. Solachidis and I. Pitas. Circularly symmetric watermark embedding in 2-D DFT domain. In *Proc. of ICASPP'99*, 3469–3472, Phoenix, Arizona, USA, March 15–19, 1999.

[97] S. Tsekeridou and I. Pitas. Embedding self-similar watermarks in the wavelet domain. In *Proc. of ICASSP'00*, 1967–1970, Istanbul, Turkey, June 5–9, 2000.

[98] M. Kutter. Watermarking resisting to translation, rotation and scaling. *Proc. SPIE*, 3528:423–431, 1998.

[99] A. Nikolaidis and I. Pitas. Robust watermarking of facial images based on salient geometric pattern matching. *IEEE Trans. Multimed.*, 2(3):172–184, 2000.

[100] C. S. Lu, S. W. Sun, C. Y. Hsu, and P. C. Chang. Media hash-dependent image watermarking resilient against both geometric attacks and estimation attacks based on false positive-oriented detection. *IEEE Trans. Multimed.*, 8(4):668–685, 2006.

[101] X. Wang, J. Wu, and P. Niu. A new digital image watermarking algorithm resilient to desynchronization attacks. *IEEE Trans. Inf. Forensics Secur.*, 2(4):655–663, 2007.

[102] H. C. Papadopoulos and C. E. W. Sundberg. Simultaneous broadcasting of analogus fm and digital audio signals by means of precanceling techniques. In *Proc. of the IEEE Int. Conf. on Communications*, 728–732, 1998.

[103] J. Zhong and S. Huang. Double-sided watermark embedding and detection. *IEEE Trans. Inf. Forensics Secur.*, 2(3):297–310, 2007.

[104] M. Costa. Writing on dirty paper. *IEEE Trans. Inf. Theory*, 29:439–441, 1983.

[105] P. Moulin and J. A. O'Sullivan. Information-theoretic analysis of information hiding. *IEEE Trans. Inf. Theory*, 49(3):563–593, 2003.

[106] Q. Li and I. Cox. Using perceptual models to improve fidelity and provide resistance to valu-metric scaling for quantization index modulation watermarking. *IEEE Trans. Inf. Forensics Secur.*, 2(2):127–139, 2007.

[107] B. Chen and G. Wornell. Quantization index modulation methods for digital watermarking and information embedding in multimedia. *J. VLSI Signal Process.*, 27:7–33, 2001.

[108] J. J. Eggers, R. Buml, and B. Girod. Estimation of amplitude modifications before scs watermark detection. In *Proc. SPIE Security Watermarking Multimedia Contents*, Vol. 4675, 387–398, San Jose, CA, January 2002.

[109] F. Prez-Gonzlez, C. Mosquera, M. Barni, and A. Abrardo. Rational dither modulation: a high-rate data-hiding method invariant to gain attacks. *IEEE Trans. Signal Process.*, 53(10):3960–3975, 2005.

[110] A. Abrardo and M. Barni. Informed watermarking by means of orthogonal and quasi-orthogonal dirty paper coding. *IEEE Trans. on Signal Process.*, 53(2):824–833, 2005.

[111] C. I. Podilchuk and W. Zeng. Image-adaptive watermarking using visual models. *IEEE J. Sel. Areas Commun.*, 16(4):525–539, 1998.

[112] A. B. Watson. DCT quantization matrices optimized for individual images. *Human Vision, Visual Process. Digital Disp.*, SPIE-1903:202–216, 1993.

[113] N. Baumgartner, S. Voloshynovskiy, A. Herrigel, and T. Pun. A stochastic approach to content adaptive digital image watermarking. In *Proc. of Int. Workshop on Information Hiding*, 211–236, Dresden, Germany, September 1999.

[114] I. Karybali and K. Berberidis. Efficient spatial image watermarking via new perceptual masking and blind detection schemes. *IEEE Trans. Inf. Forensics Secur.*, 1(2):256–274, 2006.

[115] P. Comesaa and F. Perez-Gonzalez. On a watermarking scheme in the logarithmic domain and its perceptual advantages. In *IEEE Int. Conf. on Image Process.*, Vol. 2, 145–148, September 2007.

[116] J. Fridrich and A. Goljan. Images with self-correcting capabilities. In *Proc. IEEE Conf. on Image Process.*, 25–28, October 1999.

[117] G. L. Friedman. The trustworthy digital camera: restoring credibility to the photographic image. *IEEE Trans. Consum. Electron.*, 39(4):905–910, 1993.

[118] S. Bhattacharjee and M. Kutter. Compression tolerant image authentication. In *Proc. of ICIP'98*, Chicago, USA, Vol. I, 425–429, October 4–7, 1998.

[119] B. Zhu, M. D. Swanson, and A. H. Tewfik. Transparent robust authentication and distortion measurement technique for images. In *Proc. of DSP'96*, 45–48, Loen, Norway, September 1996.

[120] P. W. Wong. A public key watermark for image verification and authentication. In *Proc. of ICIP'98*, Chicago, USA, Vol. I, 425–429, October 4–7, 1998.

[121] C.-Y. Lin and S.-F. Chang. Semi-fragile watermarking for authenticating jpeg visual content. In *SPIE EI'00*, 3971, 2000.

[122] M. Wu and B. Liu. Watermarking for image authentication. In *Proc. of ICIP'98*, Vol. I, 425–429, Chicago, USA, October 4–7, 1998.

[123] M. M. Yeung and F. Mintzer. An invisible watermarking technique for image verification. In *Proc. of ICIP'97*, Vol. II, 680–683, Atlanta, USA, October 1997.

[124] M. Holliman and N. Memon. Counterfeiting attacks on oblivious blockwise independent invisible watermarking schemes. *IEEE Trans. Image Process.*, 9(3):432–441, 2000.

[125] L. Xie and G. R. Arce. Joint wavelet compression and authentication watermarking. In *Proc. of ICIP'98*, Vol. II, 427–431, Chicago, Illinois, USA, October 4–7, 1998.

[126] D. Kundur and D. Hatzinakos. Digital watermarking for telltale tamper proofing and authentication. *Proc. IEEE*, 87(7):1167–1180, 1999.

[127] F. Mintzer, G. W. Braudaway, and M. M. Yeung. Effective and ineffective digital watermarks. In *Proc. of ICIP'97*, Vol. III, 9–12, Atlanta, USA, October 1997.

[128] P. W. Wong. A watermark for image integrity and ownership verification. In *Proc. of IS&T PINC Conf.*, Portland, 1997.

[129] A. Tefas and I. Pitas. Image authentication based on chaotic mixing. In *CD-ROM Proc. of IEEE Int. Symposium on Circuits and Systems (ISCAS'2000)*, 216–219, Geneva, Switzerland, May 28–31, 2000.

[130] A. Tefas and I. Pitas. Image authentication and tamper proofing using mathematical morphology. In *Proc. of European Signal Processing Conf. (EUSIPCO'2000)*, Tampere, Finland, September 5–8, 2000.

[131] P. L. Lina, C. K. Hsiehb, and P. W. Huangb. A hierarchical digital watermarking method for image tamper detection and recovery. *Pattern Recognit.*, 38(12):2519–2529, 2005.

[132] M. U. Celik, G. Sharma, and A. M. Tekalp. Lossless watermarking for image authentication: a new framework and an implementation. *IEEE Trans. on Image Process.*, 15(4):1042–1049, 2006.

[133] Z. M. Lu, D. G. Xu, and S. H. Sun. Multipurpose image watermarking algorithm based on multistage vector quantization. *IEEE Trans. on Image Process.*, 14(6):822–911, 2005.

[134] K. Maeno, Q. Sun, S. F. Chang, and M. Suto. New semi-fragile image authentication watermarking techniques using random bias and nonuniform quantization. *IEEE Trans. Multimed.*, 8(1): 32–45, 2006.

Fingerprint Recognition

23

Anil Jain[1] and Sharath Pankanti[2]

[1] *Michigan State University;* [2] *IBM T. J. Watson Research Center, New York*

23.1 INTRODUCTION

The problem of resolving the identity of a person can be categorized into two fundamentally distinct types of problems with different inherent complexities [1]: (i) verification and (ii) recognition. Verification (authentication) refers to the problem of confirming or denying a person's claimed identity (Am I who I claim I am?). Recognition (Who am I?) refers to the problem of establishing a subject's identity.[1] A reliable personal identification is critical in many daily transactions. For example, access control to physical facilities and computer privileges are becoming increasingly important to prevent their abuse. There is an increasing interest in inexpensive and reliable personal identification in many emerging civilian, commercial, and financial applications.

Typically, a person could be identified based on (i) a person's possession ("something that you possess"), e.g., permit physical access to a building to all persons whose identity could be authenticated by possession of a key; (ii) a person's knowledge of a piece of information ("something that you know"), e.g., permit login access to a system to a person who knows the user id and a password associated with it. Another approach to identification is based on identifying physical characteristics of the person. The characteristics could be either a person's anatomical traits, e.g., fingerprints and hand geometry, or his behavioral characteristics, e.g., voice and signature. This method of identification of a person based on his anatomical/behavioral characteristics is called *biometrics*. Since these physical characteristics cannot be forgotten (like passwords) and cannot be easily shared or misplaced (like keys), they are generally considered to be a more reliable approach to solving the personal identification problem.

23.2 EMERGING APPLICATIONS

Accurate identification of a person could deter crime and fraud, streamline business processes, and save critical resources. Here are a few mind boggling numbers: about one

---

[1] Often, recognition is also referred to as identification.

billion dollars in welfare benefits in the United States are annually claimed by "double dipping" welfare recipients with fraudulent multiple identities [44]. MasterCard estimates credit card fraud at $450 million per annum which includes charges made on lost and stolen credit cards: unobtrusive personal identification of the legitimate ownership of a credit card at the point of sale would greatly reduce credit card fraud. About 1 billion dollars worth of cellular telephone calls are made by cellular bandwidth thieves—many of which are made using stolen PINs and/or cellular phones. Again an identification of the legitimate ownership of a cellular phone would prevent loss of bandwidth. A reliable method of authenticating the legitimate owner of an ATM card would greatly reduce ATM-related fraud worth approximately $3 billion annually [9]. A method of identifying the rightful check payee would also save billions of dollars that are misappropriated through fraudulent encashment of checks each year. A method of authentication of each system login would eliminate illegal break-ins into traditionally secure (even federal government) computers. The United States Immigration and Naturalization Service has stated that each day it could detect/deter about 3,000 illegal immigrants crossing the Mexican border without delaying legitimate persons entering the United States if it had a quick way of establishing personal identification.

High-speed computer networks offer interesting opportunities for electronic commerce and electronic purse applications. Accurate authentication of identities over networks is expected to become one of the most important applications of biometric-based authentication.

Miniaturization and mass-scale production of relatively inexpensive biometric sensors (e.g., solid state fingerprint sensors) has facilitated the use of biometric-based authentication in asset protection (laptops, PDAs, and cellular phones).

23.3 FINGERPRINT AS A BIOMETRIC

A smoothly flowing pattern formed by alternating crests (ridges) and troughs (valleys) on the palmar aspect of a hand is called a palmprint. Formation of a palmprint depends on the initial conditions of the embryonic mesoderm from which they develop. The pattern on the pulp of each terminal phalanx (finger) is considered as an individual pattern and is commonly referred to as a *fingerprint* (see Fig. 23.1). A fingerprint is believed to be unique to each person (and each finger). Even the fingerprints of identical twins are different.

Fingerprints are one of the most mature biometric technologies and are considered legitimate evidence in courts of law all over the world. Fingerprints are, therefore, routinely used in forensic divisions worldwide for criminal investigations. More recently, an increasing number of civilian and commercial applications are either using or actively considering the use of fingerprint-based identification because of a better understanding of fingerprints as well as demonstrated matching performance better than any other existing biometric technology.

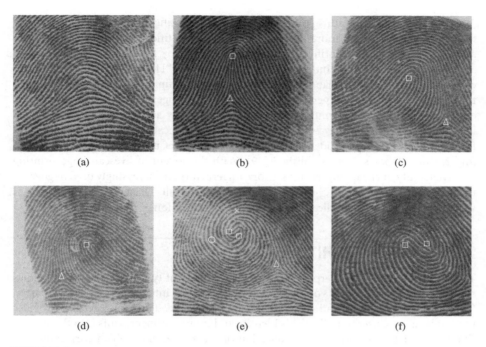

FIGURE 23.1

Fingerprints and a fingerprint classification schema involving six categories: (a) arch; (b) tented arch; (c) right loop; (d) left loop; (e) whorl; and (f) twin loop. Critical points in a fingerprint, called core and delta, are marked as squares and triangles, respectively. Note that an arch type fingerprint does not have a delta or a core. One of the two deltas in (e) and both the deltas in (f) are not imaged. A sample minutiae ridge ending (○) and ridge bifurcation (×) is illustrated in (e). Each image is 512 × 512 with 256 gray levels and is scanned at 512 dpi resolution. All feature points were manually extracted by one of the authors.

23.4 HISTORY OF FINGERPRINTS

Humans have used fingerprints for personal identification for a very long time [29]. Modern fingerprint matching techniques were initiated in the late 16th century [10]. Henry Fauld, in 1880, first scientifically suggested the individuality and uniqueness of fingerprints. At the same time, Herschel asserted that he had practiced fingerprint identification for about 20 years [29]. This discovery established the foundation of modern fingerprint identification. In the late 19th century, Sir Francis Galton conducted an extensive study of fingerprints [29]. He introduced the minutiae features for fingerprint classification in 1888. The discovery of the uniqueness of fingerprints caused an immediate decline in the prevalent use of anthropometric methods of identification and led to the adoption of fingerprints as a more efficient method of identification [36]. An important advance in fingerprint identification was made in 1899 by Edward Henry, who (actually his two assistants from India) established the famous "Henry system" of fingerprint classification [10, 29]: an elaborate method of indexing fingerprints very much tuned to facilitating the human experts performing (manual) fingerprint identification. In the early 20th century, fingerprint identification was formally accepted as a valid personal identification method by law enforcement agencies and became a standard procedure in forensics [29]. Fingerprint identification agencies were set up worldwide and criminal fingerprint databases were established [29]. With the advent of livescan fingerprinting and availability of cheap fingerprint sensors, fingerprints are increasingly used in government (US-VISIT program [40]) and commercial (Walt Disney World fingerprint verification system [8]) applications for person identification.

23.5 SYSTEM ARCHITECTURE

The architecture of a fingerprint-based automatic identity authentication system is shown in Fig. 23.2. It consists of four components: (i) user interface, (ii) system database, (iii) enrollment module, and (iv) authentication module. The user interface provides mechanisms for a user to indicate his identity and input his fingerprints into the system. The system database consists of a collection of records, each of which corresponds to an authorized person that has access to the system. In general, the records contain

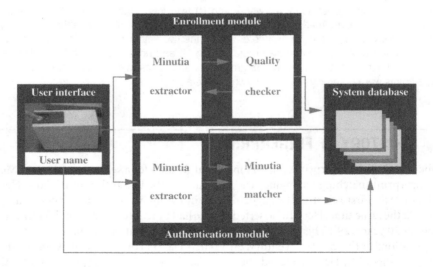

FIGURE 23.2

Architecture of an automatic identity authentication system [22]. © IEEE.

the following fields which are used for authentication purposes: (i) user name of the person, (ii) minutiae templates of the person's fingerprint, and (iii) other information (e.g., specific user privileges).

The task of the enrollment module is to enroll persons and their fingerprints into the system database. When the fingerprint images and the user name of a person to be enrolled are fed to the enrollment module, a minutiae extraction algorithm is first applied to the fingerprint images and the minutiae patterns are extracted. A quality checking algorithm is used to ensure that the records in the system database only consist of fingerprints of good quality, in which a significant number (default value is 25) of genuine minutiae are detected. If a fingerprint image is of poor quality, it is enhanced to improve the clarity of ridge/valley structures and mask out all the regions that cannot be reliably recovered. The enhanced fingerprint image is fed to the minutiae extractor again.

The task of the authentication module is to authenticate the identity of the person who intends to access the system. The person to be authenticated indicates his identity and places his finger on the fingerprint scanner; a digital image of the fingerprint is captured; minutiae pattern is extracted from the captured fingerprint image and fed to a matching algorithm which matches it against the person's minutiae templates stored in the system database to establish the identity.

23.6 FINGERPRINT SENSING

There are two primary methods of capturing a fingerprint image: inked (offline) and live scan (inkless) (see Fig. 23.3). An inked fingerprint image is typically acquired in the following way: a trained professional[2] obtains an impression of an inked finger on a paper, and the impression is then scanned using a flat bed document scanner. The live scan fingerprint is a collective term for a fingerprint image directly obtained from the finger without the intermediate step of getting an impression on a paper. Acquisition of inked fingerprints is cumbersome; in the context of an identity authentication system, it is both infeasible and socially unacceptable. The most popular technology to obtain a live-scan fingerprint image is based on the optical frustrated total internal reflection (FTIR) concept [28]. When a finger is placed on one side of a glass platen (prism), ridges of the finger are in contact with the platen, while the valleys of the finger are not in contact with the platen (see Fig. 23.4). The rest of the imaging system essentially consists of an assembly of an LED light source and a CCD placed on the other side of the glass platen. The light source illuminates the glass at a certain angle, and the camera is placed such that it can capture the light reflected from the glass. The light that incidents on the platen at the glass surface touched by the ridges is randomly scattered while the

---

[2]Possibly, for reasons of expediency, MasterCard sends fingerprint kits to their credit card customers. The kits are used by the customers themselves to create an inked fingerprint impression to be used for enrollment.

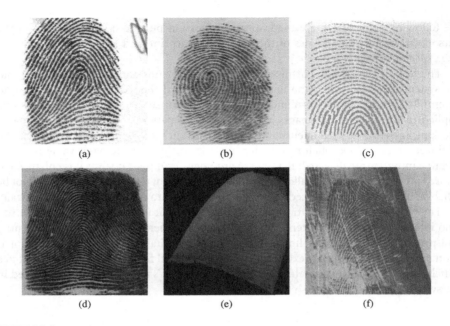

(a) (b) (c)

(d) (e) (f)

FIGURE 23.3

Fingerprint sensing: (a) An inked fingerprint image could be captured from the inked impression of a finger; (b) a live-scan fingerprint is directly imaged from a live finger based on optical total internal reflection principle: the light scatters where the finger (e.g., ridges) touches the glass prism and the light reflects where the finger (e.g., valleys) does not touches the glass prism; (c) fingerprints captured using solid state sensors show a smaller area of the finger than a typical fingerprint dab captured using optical scanners; (d) rolled fingerprints are images depicting the nail-to-nail area of a finger; (e) a 3D fingerprint is reconstructed from touchless fingerprint sensors (adopted from [33]); (f) a latent fingerprint refers to a partial print typically lifted from the scene of a crime.

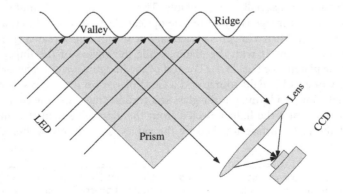

FIGURE 23.4

FTIR-based fingerprint sensing [30].

FIGURE 23.5

Fingerprint sensors can be embedded in many consumer products.

light that incidents at the glass surface corresponding to valleys suffers total internal reflection. Consequently, portions of the image formed on the imaging plane of the CCD corresponding to ridges are dark and those corresponding to valleys are bright. In recent years, capacitance-based solid state live-scan fingerprint sensors are gaining popularity since they are very small in size and can be easily embedded into laptop computers, mobile phones, computer peripherals, and the like (see Fig. 23.5). A capacitance-based fingerprint sensor essentially consists of an array of electrodes. The fingerprint skin acts as the other electrode, thereby, forming a miniature capacitor. The capacitance due to the ridges is higher than those formed by valleys. This differential capacitance is the basis of operation of a capacitance-based solid state sensor [45]. More recently, multispectral sensors [38] and touchless sensors [33] have been invented.

23.7 FINGERPRINT FEATURES

Fingerprint features are generally categorized into three levels. Level 1 features are the macro details of the fingerprint such as ridge flow, pattern type, and singular points (e.g., core and delta). Level 2 features refer to minutiae such as ridge bifurcations and endings. Level 3 features include all dimensional attributes of the ridge such as ridge path deviation, width, shape, pores, edge contour, incipient ridges, breaks, creases, scars, and other permanent details (see Fig. 23.6). While Level 1 features are mainly used for fingerprint classification, Level 2 and Level 3 features can be used to establish the individuality of fingerprints. Minutiae-based representations are the most commonly used

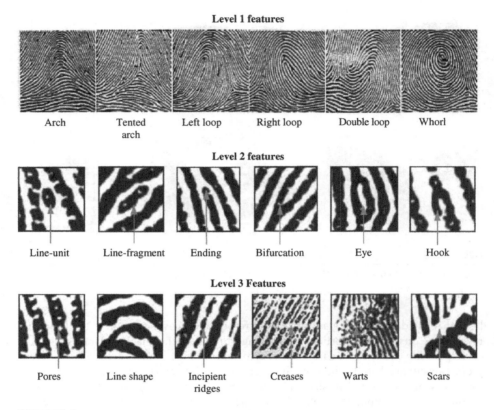

Level 1 features

Arch   Tented arch   Left loop   Right loop   Double loop   Whorl

Level 2 features

Line-unit   Line-fragment   Ending   Bifurcation   Eye   Hook

Level 3 Features

Pores   Line shape   Incipient ridges   Creases   Warts   Scars

FIGURE 23.6

Fingerprint features at three levels [25]. © IEEE.

representation, primarily due to the following reasons: (i) minutiae capture much of the individual information, (ii) minutiae-based representations are storage efficient, and (iii) minutiae detection is relatively robust to various sources of fingerprint degradation. Typically, minutiae-based representations rely on locations of the minutiae and the directions of ridges at the minutiae location. In recent years, with the advances in fingerprint sensing technology, many sensors are now equipped with dual resolution (500 ppi/1000 ppi) scanning capability. Figure 23.7 shows the images captured at 500 ppi and 1000 ppi by a CrossMatch L SCAN 1000P optical scanner for the same portion of a fingerprint. Level 3 features are receiving more and more attention [6, 25] due to their importance in matching latent fingerprints which generally contain much fewer minutiae than rolled or plain fingerprints.

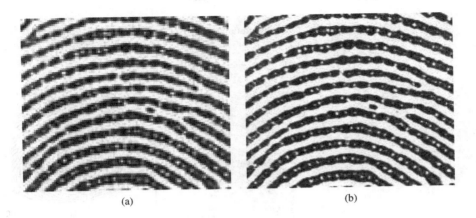

(a)   (b)

FIGURE 23.7

Local fingerprint images captured at (a) 500 ppi; and (b) 1000 ppi. Level 3 features such as pores are more clearly visible in a higher resolution image.

23.8 FEATURE EXTRACTION

A feature extractor finds the ridge endings and ridge bifurcations from the input fingerprint images. If ridges can be perfectly located in an input fingerprint image, then minutiae extraction is a relatively simple task of extracting singular points in a thinned ridge map. However, in practice, it is not always possible to obtain a perfect ridge map. The performance of currently available minutiae extraction algorithms depends heavily on the quality of the input fingerprint images. Due to a number of factors (aberrant formations of epidermal ridges of fingerprints, postnatal marks, occupational marks, problems with acquisition devices, etc.), fingerprint images may not always have well-defined ridge structures.

A reliable minutiae extraction algorithm is critical to the performance of an automatic identity authentication system using fingerprints. The overall flowchart of a typical algorithm [22, 35] is depicted in Fig. 23.8. It mainly consists of three components: (i) Orientation field estimation, (ii) ridge extraction, and (iii) minutiae extraction and postprocessing.

1. **Orientation Estimation:** The orientation field of a fingerprint image represents the directionality of ridges in the fingerprint image. It plays a very important role in fingerprint image analysis. A number of methods have been proposed to estimate the orientation field of fingerprint images [28]. A fingerprint image is typically divided into a number of nonoverlapping blocks (e.g., 32×32 pixels), and an orientation representative of the ridges in the block is assigned to the block based on an analysis of grayscale gradients in the block. The block orientation could

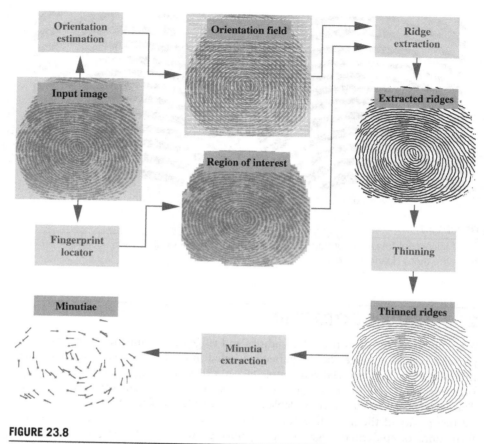

FIGURE 23.8

Flowchart of the minutiae extraction algorithm [22]. © IEEE.

be determined from the pixel gradient orientations based on, say, averaging [28], voting [31], or optimization [35]. We have summarized the orientation estimation algorithm in Fig. 23.9.

2. **Segmentation:** It is important to localize the portions of the fingerprint image depicting the finger (foreground). The simplest approach segments the foreground by global or adaptive thresholding. A novel and reliable approach to segmentation by Ratha *et al.* [35] exploits the fact that there is significant difference in the magnitudes of variance in the gray levels along and across the flow of a fingerprint ridge. Typically, block size for variance computation spans 1-2 inter-ridge distances.

3. **Ridge Detection:** Common approaches to ridge detection use either simple or adaptive thresholding. These approaches may not work for noisy and low-contrast

(a) Divide the input fingerprint image into blocks of size $W \times W$.

(b) Compute the gradients G_x and G_y at each pixel in each block [4].

(c) Estimate the local orientation at each pixel (i, j) using the following equations [35]:

$$V_x(i, j) = \sum_{u=i-\frac{W}{2}}^{i+\frac{W}{2}} \sum_{v=j-\frac{W}{2}}^{j+\frac{W}{2}} 2G_x(u, v) G_y(u, v), \qquad (23.1)$$

$$V_y(i, j) = \sum_{u=i-\frac{W}{2}}^{i+\frac{W}{2}} \sum_{v=j-\frac{W}{2}}^{j+\frac{W}{2}} (G_x^2(u, v) - G_y^2(u, v)), \qquad (23.2)$$

$$\theta(i, j) = \frac{1}{2} \tan^{-1} \left(\frac{V_x(i, j)}{V_y(i, j)} \right), \qquad (23.3)$$

where W is the size of the local window; G_x and G_y are the gradient magnitudes in x and y directions, respectively.

(d) Compute the consistency level of the orientation field in the local neighborhood of a block (i, j) with the following formula:

$$C(i, j) = \frac{1}{N} \sqrt{\sum_{(i', j') \in D} |\theta(i', j') - \theta(i, j)|^2}, \qquad (23.4)$$

$$|\theta' - \theta| = \begin{cases} d & \text{if } (d = (\theta' - \theta + 360) \bmod 360) < 180, \\ d - 180 & \text{otherwise,} \end{cases} \qquad (23.5)$$

where D represents the local neighborhood around the block (i, j) (in our system, the size of D is 5×5); N is the number of blocks within D; $\theta(i', j')$ and $\theta(i, j)$ are local ridge orientations at blocks (i', j') and (i, j), respectively.

(e) If the consistency level Eq. (23.5) is above a certain threshold T_c, then the local orientations around this region are re-estimated at a lower resolution level until $C(i, j)$ is below a certain level.

FIGURE 23.9

Hierarchical orientation field estimation algorithm [22]. © IEEE.

portions of the image. An important property of the ridges in a fingerprint image is that the gray level values on ridges attain their local maxima along a direction normal to the local ridge orientation [22, 35]. Pixels can be identified to be ridge pixels based on this property. The extracted ridges may be thinned/cleaned using standard thinning [32] and connected component algorithms [34].

4. **Minutiae Detection:** Once the thinned ridge map is available, the ridge pixels with three ridge pixel neighbors are identified as ridge bifurcations and those with one ridge pixel neighbor identified as ridge endings. However, all the minutiae thus detected are not genuine due to image processing artifacts and the noise in the fingerprint image.

5. **Postprocessing:** In this stage, typically, genuine minutiae are gleaned from the extracted minutiae using a number of heuristics. For instance, too many minutiae in a small neighborhood may indicate noise and they could be discarded. Very close ridge endings oriented antiparallel to each other may indicate spurious minutia generated by a break in the ridge due to either poor contrast or a cut in the finger. Two very closely located bifurcations sharing a common short ridge often suggest extraneous minutia generated by bridging of adjacent ridges as a result of dirt or image processing artifacts.

23.9 FINGERPRINT ENHANCEMENT

The performance of a fingerprint image matching algorithm relies critically on the quality of the input fingerprint images. In practice, a significant percentage of acquired fingerprint images (approximately 10% according to our experience) is of poor quality. The ridge structures in poor-quality fingerprint images are not always well defined and hence they cannot be correctly detected. This leads to the following problems: (i) a significant number of spurious minutiae may be created, (ii) a large percentage of genuine minutiae may be ignored, and (iii) large errors in minutiae localization (position and orientation) may be introduced. In order to ensure that the performance of the minutiae extraction algorithm will be robust with respect to the quality of fingerprint images, an enhancement algorithm which can improve the clarity of the ridge structures is necessary.

Typically, fingerprint enhancement approaches [7, 14, 18, 26] employ frequency domain techniques [14, 15, 26] and are computationally demanding. In a small local neighborhood, the ridges and furrows approximately form a two-dimensional sinusoidal wave along the direction orthogonal to local ridge orientation. Thus, the ridges and furrows in a small local neighborhood have well-defined local frequency and local orientation properties. The common approaches employ bandpass filters which model the frequency domain characteristics of a good-quality fingerprint image. The poor-quality fingerprint image is processed using the filter to block the extraneous *noise* and pass the fingerprint *signal*. Some methods may estimate the orientation and/or frequency of ridges in each block in the fingerprint image and adaptively tune the filter characteristics to match the ridge characteristics.

One typical variation of this theme segments the image into nonoverlapping square blocks of widths larger than the average inter-ridge distance. Using a bank of directional bandpass filters, each filter is matched to a predetermined model of generic fingerprint ridges flowing in a certain direction; the filter generating a strong response indicates the dominant direction of the ridge flow in the finger in the given block. The resulting orientation information is more accurate, leading to more reliable features. A single block direction can never truly represent the directions of the ridges in the block and may consequently introduce filter artifacts.

For instance, one common directional filter used for fingerprint enhancement is a Gabor filter [21]. Gabor filters have both frequency-selective and orientation-selective properties and have optimal joint resolution in both spatial and frequency domains. The even-symmetric Gabor filter has the general form [21]:

$$h(x,y) = \exp\left\{-\frac{1}{2}\left[\frac{x^2}{\delta_x^2} + \frac{y^2}{\delta_y^2}\right]\right\}\cos(2\pi u_0 x), \qquad (23.6)$$

where u_0 is the frequency of a sinusoidal plane wave along the x-axis, and δ_x and δ_y are the space constants of the Gaussian envelope along the x and y axes, respectively. Gabor filters with arbitrary orientation can be obtained via a rotation of the $x - y$ coordinate system. The modulation transfer function (MTF) of the Gabor filter can be represented as

$$H(u,v) = 2\pi\delta_x\delta_y\left(\exp\left\{-\frac{1}{2}\left[\frac{(u-u_0)^2}{\delta_u^2} + \frac{v^2}{\delta_v^2}\right]\right\} + \exp\left\{-\frac{1}{2}\left[\frac{(u-u_0)^2}{\delta_u^2} + \frac{v^2}{\delta_v^2}\right]\right\}\right), \quad (23.7)$$

where $\delta_u = 1/2\pi\delta_x$ and $\delta_v = 1/2\pi\delta_y$. Figure 23.10 shows an even-symmetric Gabor filter and its MTF. Typically, in a 500 dpi, 512 × 512 fingerprint image, a Gabor filter with $u_0 = 60$ cycles per image width (height), the radial bandwidth of 2.5 octaves, and orientation θ models the fingerprint ridges flowing in the direction $\theta + \pi/2$.

We summarize a novel approach to fingerprint enhancement proposed by Hong [16] (see Fig. 23.11). It decomposes the given fingerprint image into several component images using a bank of directional Gabor bandpass filters and extracts ridges from each of the filtered bandpass images using a typical feature extraction algorithm [22]. By integrating information from the sets of ridges extracted from filtered images, the enhancement

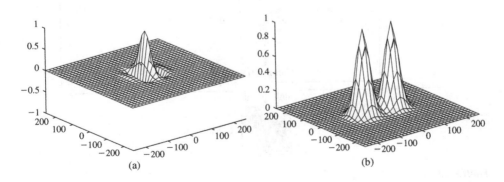

(a)

(b)

FIGURE 23.10

An even-symmetric Gabor filter: (a) Gabor filter tuned to 60 cycles/width and 0° orientation; (b) corresponding modulation transfer function.

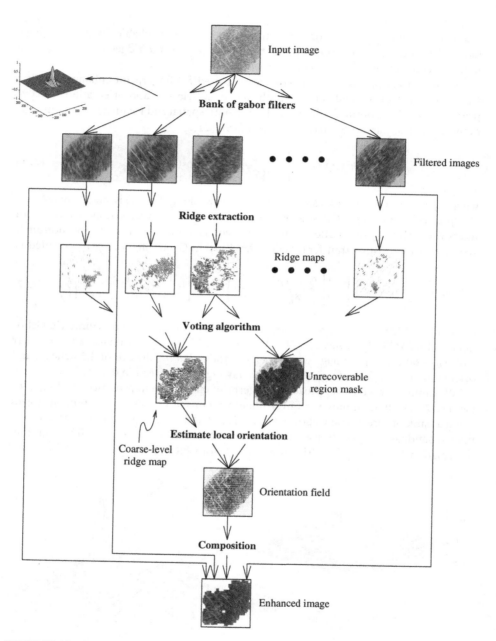

FIGURE 23.11

Fingerprint enhancement algorithm [16].

algorithm infers the region of a fingerprint where there is sufficient information to be considered for enhancement (recoverable region) and estimates a coarse-level ridge map for the recoverable region. The information integration is based on the observation that genuine ridges in a region evoke a strong response in the feature images extracted from the filters oriented in the direction parallel to the ridge direction in that region and at most a weak response in feature images extracted from the filters oriented in the direction orthogonal to the ridge direction in that region. The coarse ridge map thus generated consists of the ridges extracted from each filtered image which are mutually consistent, and portions of the image where the ridge information is consistent across the filtered images constitutes a *recoverable* region. The orientation field estimated from the coarse ridge map (see Section 23.8) is more reliable than the orientation estimation from the input fingerprint image.

After the orientation field is obtained, the fingerprint image can then be adaptively enhanced by using the local orientation information. Let $f_i(x,y)$ (i = 0, 1, 2, 3, 4, 5, 6, 7) denote the gray level value at pixel (x,y) of the filtered image corresponding to the orientation θ_i, $\theta_i = i * 22.5°$. The gray level value at pixel (x,y) of the enhanced image can be interpolated according to the following formula:

$$f_{enh}(x,y) = a(x,y)f_{p(x,y)}(x,y) + (1 - a(x,y))f_{q(x,y)}(x,y), \qquad (23.8)$$

where $p(x,y) = \lfloor \frac{\theta(x,y)}{22.5} \rfloor$, $q(x,y) = \lceil \frac{\theta(x,y)}{22.5} \rceil$ mod 8, $a(x,y) = \frac{\theta(x,y) - p(x,y)}{22.5}$, and $\theta(x,y)$ represents the value of the local orientation field at pixel (x,y). The major reason that we interpolate the enhanced image directly from the limited number of filtered images is that the filtered images are already available and the above interpolation is computationally efficient.

An example illustrating the results of a minutiae extraction algorithm on a noisy input image and its enhanced counterpart is shown in Fig. 23.12. The improvement in performance due to image enhancement was evaluated using the fingerprint matcher described in Section 23.11. Figure 23.13 shows improvement in accuracy of the matcher

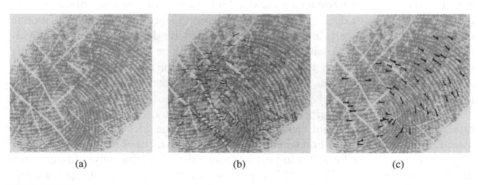

| (a) | (b) | (c) |

FIGURE 23.12

Fingerprint enhancement results: (a) a poor-quality fingerprint; (b) minutia extracted without image enhancement; (c) minutiae extracted after image enhancement [16].

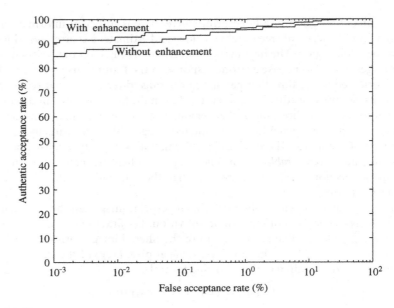

FIGURE 23.13

Performance of fingerprint enhancement algorithm.

with and without image enhancement on the MSU database consisting of 700 fingerprint images of 70 individuals (10 fingerprints per finger per individual).

23.10 FINGERPRINT CLASSIFICATION

Fingerprints have been traditionally classified into categories based on information in the global patterns of ridges. In large-scale fingerprint identification systems, elaborate methods of manual fingerprint classification systems were developed to index individuals into bins based on classification of their fingerprints; these methods of binning eliminate the need to match an input fingerprint(s) to the entire fingerprint database in identification applications and significantly reduce the computing requirements [3, 13, 23, 27, 42].

Efforts in automatic fingerprint classification have been exclusively directed at replicating the manual fingerprint classification system. Figure 23.1 shows one prevalent manual fingerprint classification scheme that has been the focus of many automatic fingerprint classification efforts. It is important to note that the distribution of fingers into the six classes (shown in Fig. 23.1) is highly skewed. Three fingerprint types, namely left loop, right loop, and whorl, account for over 93% of the fingerprints. A fingerprint classification system should be invariant to rotation, translation, and elastic distortion of the frictional skin. In addition, often a significant part of the finger may not be imaged (e.g., dabs frequently miss deltas), and the classification methods requiring information from the entire fingerprint may be too restrictive for many applications.

A number of approaches to fingerprint classification have been developed. Some of the earliest approaches did not make use of the rich information in the ridge structures and exclusively depended on the orientation field information. Although fingerprint landmarks provide very effective fingerprint class clues, methods relying on the fingerprint landmarks alone may not be very successful due to lack of availability of such information in many fingerprint images and due to the difficulty in extracting the landmark information from the noisy fingerprint images. As a result, successful approaches need to (i) supplement the orientation field information with ridge information; (ii) use fingerprint landmark information when available but devise alternative schemes when such information cannot be extracted from the input fingerprint images; and (iii) use reliable structural/syntactic pattern recognition methods in addition to statistical methods.

We summarize a method of classification [17] which takes into consideration the above-mentioned design criteria that has been tested on a large database of realistic fingerprints to classify fingers into five major categories: right loop, left loop, arch, tented arch, and whorl.[3]

The orientation field determined from the input image may not be very accurate, and the extracted ridges may contain many artifacts and, therefore, cannot be directly used for fingerprint classification. A ridge verification stage assesses the reliability of the extracted ridges based upon the length of each connected ridge segment and its alignment with other adjacent ridges. Parallel adjacent subsegments typically indicate a good-quality fingerprint region; the ridge/orientation estimates in these regions are used to refine the estimates in the orientation field/ridge map.

1. Singular points: The Poincare index [28] on the orientation field is used to determine the number of delta (N_D) and core (N_C) points in the fingerprint. A digital closed curve, Ψ, about 25 pixels long, around each pixel is used to compute the Poincare index as defined below:

$$\text{Poincare}(i,j) = \frac{1}{2\pi} \sum_{k=0}^{N_\Psi} \Delta(k),$$

where

$$\Delta(k) = \begin{cases} \delta(k), & \text{if } |\delta(k)| < \pi/2, \\ \pi + \delta(k), & \text{if } \delta(k) \le -\pi/2, \\ \pi - \delta(k), & \text{otherwise}, \end{cases}$$

$$\delta(k) = \mathcal{O}'(\Psi_x(i'), \Psi_y(i')) - \mathcal{O}'(\Psi_x(i), \Psi_y(i)),$$

$$i' = (i+1) \bmod N_\Psi,$$

---

[3]Other types of prints, e.g., twin-loop, are not considered here but, in principle, could be lumped into "other" or "reject" category.

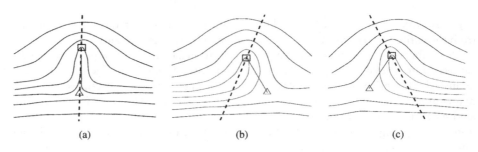

FIGURE 23.14

The axis of symmetry for (a) tented arch; (b) left loop; and (c) right loop.

$\mathcal{O}$ is the orientation field, and $\Psi_x(i)$ and $\Psi_y(i)$ denote the coordinates of the ith point on the arc length parameterized closed curve Ψ.

2. Symmetry: The feature extraction stage also estimates an axis locally symmetric to the ridge structures at the core (see Fig. 23.14) and computes (i) α, the angle between the symmetry axis and the line segment joining core and delta, (ii) β, the average angle difference between the ridge orientation and the orientation of the line segment joining the core and delta, and (iii) γ, the number of ridges crossing the line segment joining core and delta. The relative position, R, of the delta with respect to the symmetry axis is determined as follows: $R = 1$ if the delta is on the right side of the symmetry axis; $R = 0$, otherwise.

3. Ridge structure: The classifier not only uses the orientation information but also utilizes the structural information in the extracted ridges. This feature summarizes the overall nature of the ridge flow in the fingerprint. In particular, it classifies each ridge of the fingerprint into three categories (as illustrated in Fig. 23.15):

   ■ Nonrecurring ridges: ridges which do not curve very much.

   ■ Type-1 recurring ridges: ridges which curve approximately π.

   ■ Type-2 fully recurring ridges: ridges which curve by more than π.

The classification algorithm summarized here essentially devises a sequence of tests for determining the class of a fingerprint and conducts simpler tests earlier. For instance, two core points are typically detected for a whorl which is an easier condition to verify than detecting the number of Type-2 recurring ridges. The algorithm detects (i) whorl based upon detection of either two core points or a sufficient number of Type-2 recurring ridges; (ii) arch based upon the inability to detect either delta or core points; (iii) left (right) loop based on the characteristic tilt of the symmetric axis, detection of a core point, and detection of either a delta point or a sufficient number of Type-1 recurring curves; and (iv) tented arch based on a relatively upright symmetric axis, detection of a core point, and detection of either a delta point or a sufficient number of Type-1 recurring

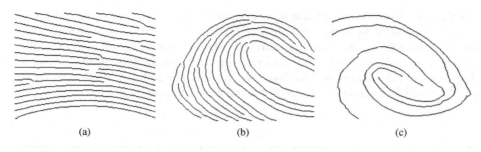

FIGURE 23.15

Three types of ridges. (a) Nonrecurring ridges; (b) Type-1 recurring ridges; and (c) Type-2 fully recurring ridges.

curves. Another highlight of the algorithm is that it does not detect the salient character-istics of any category from features detected in a fingerprint; it recomputes the features with a different preprocessing method. For instance, in the current implementation, the differential preprocessing consists of a different method/scale of smoothing.

Table 23.1 shows the results of the fingerprint classification algorithm on the NIST-4 database which contains 4,000 images (image size is 512×480) taken from 2,000 different fingers, 2 images per finger. Five fingerprint classes are defined: (i) Arch, (ii) Tented arch, (iii) Left loop, (iv) Right loop, and (v) Whorl. Fingerprints in this database are uniformly distributed among these five classes (800 per class). The five-class error rate in classifying these 4,000 fingerprints is 12.5%. The confusion matrix is given in Table 23.1; numbers shown in bold font are correct classifications. Since a number of fingerprints in the NIST-4 database are labeled (by human experts) as belonging to possibly two different classes, each row of the confusion matrix in Table 23.1 does not sum up to 800. For the five-class problem, most of the classification errors are due to misclassifying a tented arch as an arch. By combining these two arch categories into a single class, the error rate drops from 12.5% to 7.7%. Besides the tented arch-arch errors, the other errors

TABLE 23.1   Five-class classification results on the NIST-4 database; A-Arch, T-Tented Arch, L-Left loop, R-Right loop, W-Whorl.

True class	Assigned class				
	A	T	L	R	W
A	**885**	13	10	11	0
T	179	**384**	54	14	5
L	31	27	**755**	3	20
R	30	47	3	**717**	16
W	6	1	15	15	**759**

mainly come from misclassifications between arch/tented arch and loops and due to poor image quality. Recently, Cappelli *et al.* [5] proposed a two-stage algorithm where the Multispace Karhunen-Loeve transformation-based classifier is first used to transfer the five-class problem into ten two-class problems and then the Subspace-based pattern discrimination classifier is used for final classification. The accuracy in [5] for five-class problems and four-class problems is 95.2% and 96.3%, respectively.

23.11 FINGERPRINT MATCHING

Given two (input and template) sets of features originating from two fingerprints, the objective of the feature matching system is to determine whether or not the prints represent the same finger. Fingerprint matching has been approached from several different strategies [30], like image-based [2], ridge pattern-based, and point (minutiae) pattern-based fingerprint representations. There also exist graph-based schemes [19, 20, 39] for fingerprint matching. Image-based matching may not tolerate large amounts of nonlinear distortion in the fingerprint ridge structures. Matchers critically relying on extraction of ridges or their connectivity information may display drastic performance degradation with a deterioration in the quality of the input fingerprints. It is generally agreed that the point pattern matching (minutiae matching) approach facilitates the design of a robust, simple, and fast verification algorithm while maintaining a small template size.

The matching phase typically defines the similarity (distance) metric between two fingerprint representations and determines whether a given pair of representations is captured from the same finger (mated pair) based on whether this quantified (dis)similarity is greater (less) than a certain (predetermined) threshold. The similarity metric is based on the concept of correspondence in minutiae-based matching. A minutiae in the input fingerprint and a minutiae in the template fingerprint are said to be corresponding if they represent the identical minutiae scanned from the same finger.

In order to match two fingerprint representations, the minutiae-based matchers first transform (*register*) the input and template fingerprint features into a common frame of reference. The registration essentially involves alignment based on rotation/translation and may optionally include scaling. The parameters of alignment are typically estimated either from (i) singular points in the fingerprints, e.g., core and delta locations; (ii) pose clustering based on minutia distribution [35]; or (iii) any other landmark features. For example, Jain *et al.* [22] use a rotation/translation estimation method based on properties of ridge segments associated with ridge ending minutiae.[4] Feng [11] estimates the transformation by using minutia descriptor which captures texture information and neighboring minutiae information around a minutia.

---

[4]The input and template minutiae used for the alignment will be referred to as reference minutiae below.

FIGURE 23.16

Two different fingerprint impressions of the same finger. In order to know the correspondence between the minutiae of these two fingerprint images, all the minutiae must be precisely localized and the deformation must be recovered [22]. © IEEE.

There are two major challenges involved in determining the correspondence between two aligned fingerprint representations (see Fig. 23.16): (i) dirt/leftover smudges on the sensing device and the presence of scratches/cuts on the finger either introduce spurious minutiae or obliterate the genuine minutiae; (ii) variations in the area of the finger being imaged and its pressure on the sensing device affect the number of genuine minutiae captured and introduce displacements of the minutiae from their "true" locations due to elastic distortion of the fingerprint skin. Consequently, a fingerprint matcher should not only assume that the input fingerprint is a transformed template fingerprint by a similarity transformation (rotation, translation, and scale), but it should also tolerate both spurious minutiae as well as missing genuine minutiae and accommodate perturbations of minutiae from their true locations. Figure 23.17 illustrates a typical situation of aligned ridge structures of mated pairs. Note that the best alignment in one part (top left) of the image may result in a large amount of displacements between the corresponding minutiae in other regions (bottom right). In addition, observe that the distortion is nonlinear: given the amount of distortions at two arbitrary locations on the finger, it is not possible to predict the distortions at all the intervening points on the line joining the two points.

The adaptive elastic string matching algorithm [22] summarized in this chapter uses three attributes of the aligned minutiae for matching: its distance from the reference minutiae (*radius*), angle subtended to the reference minutiae (*radial angle*), and local direction of the associated ridge (*minutiae direction*). The algorithm initiates the matching by first representing the aligned input (template) minutiae as an input

FIGURE 23.17

Aligned ridge structures of mated pairs. Note that the best alignment in one part (top left) of the image results in large displacements between the corresponding minutiae in the other regions (bottom right) [22]. © IEEE.

(template) minutiae string. The string representation is obtained by imposing a linear ordering based on radial angles and radii. The resulting input and template minutiae strings are matched using an inexact string matching algorithm to establish the correspondence.

The inexact string matching algorithm essentially transforms (*edits*) the input string to template string, and the number of edit operations is considered as a metric of the (dis)similarity between the strings. While permitted edit operators model the impression variations in a representation of a finger (deletion of the genuine minutiae, insertion of spurious minutiae, and perturbation of the minutiae), the penalty associated with each edit operator models the likelihood of that edit. The sum of penalties of all the edits (*edit distance*) defines the similarity between the input and template minutiae strings. Among several possible sets of edits that permit the transformation of the input minutiae string into the reference minutiae string, the string matching algorithm chooses the transform associated with the minimum cost based on dynamic programming.

The algorithm tentatively considers a candidate (aligned) input and a candidate template minutiae in the input and template minutiae string to be a mismatch if their attributes are not within a tolerance window (see Fig. 23.18) and penalizes them for

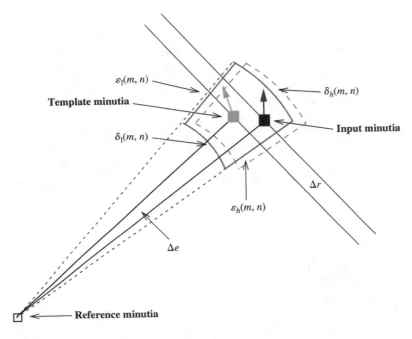

$\varepsilon_1(m, n)$

$\delta_h(m, n)$

Template minutia

Input minutia

$\delta_1(m, n)$

Δr

$\varepsilon_h(m, n)$

Δe

Reference minutia

FIGURE 23.18

Bounding box and its adjustment [22]. ©IEEE.

deletion/insertion edits. If the attributes are within the tolerance window, the amount of penalty associated with the tentative match is proportional to the disparity in the values of the attributes in the minutiae. The algorithm accommodates for the elastic distortion by adaptively adjusting the parameters of the tolerance window based on the most recent successful tentative match. The tentative matches (and correspondences) are accepted if the edit distance for those correspondences is smaller than any other correspondences.

Figure 23.19 shows the results of applying the matching algorithm to an input and a template minutiae set pair. The outcome of the matching process is defined by a matching score. Matching score is determined from the number of mated minutia from the correspondences associated with the minimum cost of matching input and template minutiae strings. The raw matching score is normalized by the total number of minutia in the input and template fingerprint representations and is used for deciding whether input and template fingerprints are mates. The higher the normalized score, the larger the likelihood that the test and template fingerprints are the scans of the same finger.

The results of performance evaluation of the fingerprint matching algorithm are illustrated in Fig. 23.20 for 1,698 fingerprint images in the NIST 9 database [41] and in Fig. 23.13 for 490 images of 70 individuals in the MSU database. Some sample points

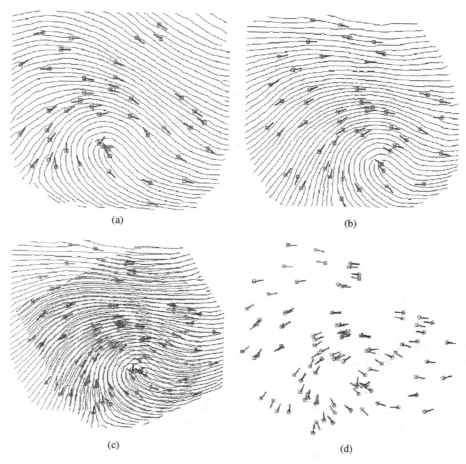

(a)

(b)

(c)

(d)

FIGURE 23.19

Results of applying the matching algorithm to an input minutiae set and a template; (a) input minutiae set; (b) template minutiae set; (c) alignment result based on the minutiae marked with green circles; (d) matching result where template minutiae and their correspondences are connected by green lines [22]. © IEEE.

on the receiver operating characteristics curve are tabulated in Table 23.2. The accuracy of fingerprint matching alogirthms heavily depends on the testing samples. For instance, the best matcher in FpVTE2003 [43] achieved 99.9% true accept rate (TAR) at 1% false accept rate (FAR), while the best matcher in FVC2006 [12] achieved only 91.8% TAR at 1% FAR on the first database in the test (DB1). Commercial fingerprint matchers are very efficient. For instance, it takes about 32 ms for the best matcher in FVC2006 to extract features and perform matching on a PC with an Intel Pentium IV 3.20 GHz.

TABLE 23.2 False acceptance and false reject rates on two data sets with different threshold values [22]. © IEEE.

Threshold value	False acceptance rate (MSU)	False reject rate (MSU)	False acceptance rate (NIST 9)	False reject rate (NIST 9)
7	0.07%	7.1%	0.073%	12.4%
8	0.02%	9.4%	0.023%	14.6%
9	0.01%	12.5%	0.012%	16.9%
10	0	14.3%	0.003%	19.5%

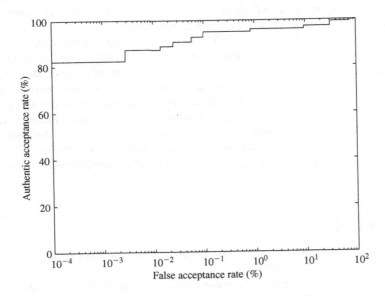

FIGURE 23.20

Receiver operating characteristic curve for NIST 9 (CD No. 1) [22]. © IEEE.

23.12 SUMMARY AND FUTURE PROSPECTS

With recent advances in fingerprint sensing technology and improvements in the accuracy and matching speed of the fingerprint matching algorithms, automatic personal identification based on a fingerprint is becoming an attractive alternative/complement to the traditional methods of identification. We have provided an overview of fingerprint-based identification and summarized algorithms for fingerprint feature extraction, enhancement, matching, and classification. We have also presented a performance evaluation of these algorithms.

The critical factor for the widespread use of fingerprints is in meeting the performance (e.g., matching speed and accuracy) standards demanded by emerging civilian identification applications. Unlike an identification based on passwords or tokens, the accuracy of the fingerprint-based identification is not perfect. There is a growing demand for faster and more accurate fingerprint matching algorithms which can (particularly) handle poor-quality images. Some of the emerging applications (e.g., fingerprint-based smartcards) will also benefit from a compact representation of a fingerprint and more efficient algorithms. The design of highly reliable, accurate, and foolproof biometric-based identification systems may warrant effective integration of discriminatory information contained in several different biometrics and/or technologies. The issues involved in integrating fingerprint-based identification with other biometric or nonbiometric technologies constitute an important research topic [24, 37].

As biometric technology matures, there will be an increasing interaction among the (biometric) market, (biometric) technology, and the (identification) applications. The emerging interaction is expected to be influenced by the added value of the technology, the sensitivities of the population, and the credibility of the service provider. It is too early to predict where, how, and which biometric technology will evolve and be mated with which applications. But it is certain that biometrics-based identification will have a profound influence on the way we conduct our daily business. It is also certain that, as the most mature and well-understood biometric, fingerprints will remain an integral part of the preferred biometrics-based identification solutions in the years to come.

REFERENCES

[1] A. K. Jain, R. Bolle, and S. Pankanti, editors. *Biometrics: Personal Identification in Networked Society.* Springer-Verlag, New York, 2005.

[2] R. Bahuguna. Fingerprint verification using hologram matched filterings. In *Proc. Biometric Consortium Eighth Meeting,* San Jose, CA, June 1996.

[3] G. T. Candela, P. J. Grother, C. I. Watson, R. A. Wilkinson, and C. L. Wilson. PCASYS: a pattern-level classification automation system for fingerprints. *NIST Tech. Report NISTIR 5647,* August 1995.

[4] J. Canny. A computational approach to edge detection. *IEEE Trans. PAMI,* 8(6):679–698, 1986.

[5] R. Cappelli, D. Maio, D. Maltoni, and L. Nanni. A two-stage fingerprint classification system. In *Proc. 2003 ACM SIGMM Workshop on Biometrics Methods and Applications,* 95–99, 2003.

[6] CDEFFS: the ANIS/NIST committee to define an extended fingerprint feature set. http://fingerprint.nist.gov/standard/cdeffs/index.html.

[7] L. Coetzee and E. C. Botha. Fingerprint recognition in low quality images. *Pattern Recognit.,* 26(10):1441–1460, 1993.

[8] Walt DisneyWorld fingerprints visitors. http://www.boingboing.net/2006/09/01/walt-disneyworld-fi.html.

[9] L. Lange and G. Leopold. Digital identification: it's now at our fingertips. *Electronic Engineering Times*, No. 946, March 24, 1997.

[10] Federal Bureau of Investigation. *The Science of Fingerprints: Classification and Uses*. U. S. Government Printing Office, Washington, DC, 1984.

[11] J. Feng. Combining minutiae descriptors for fingerprint matching. *Pattern Recognit.*, 41(1):342–352, 2008.

[12] FVC2006: the fourth international fingerprint verification competition. http://bias.csr.unibo.it/fvc2006/.

[13] R. Germain, A Califano, and S. Colville. Fingerprint matching using transformation parameter clustering. *IEEE Comput Sci. Eng.*, 4(4):42–49, 1997.

[14] L. O'Gorman and J. V. Nickerson. An approach to fingerprint filter design. *Pattern Recognit.*, 22(1):29–38, 1989.

[15] L. Hong, A. K. Jain, S. Pankanti, and R. Bolle. Fingerprint enhancement. In *Proc. IEEE Workshop on Applications of Computer Vision*, Sarasota, FL, 202–207, 1996.

[16] L. Hong. Automatic personal identification using fingerprints. *PhD Thesis*, Michigan State University, East Lansing, MI, 1998.

[17] L. Hong and A. K. Jain. Classification of fingerprint images. *MSU Technical Report*, MSU Technical Report MSUCPS:TR98–18, June 1998.

[18] D. C. D. Hung. Enhancement and feature purification of fingerprint images. *Pattern Recognit.*, 26(11):1661–1671, 1993.

[19] A. K. Hrechak and J. A. McHugh. Automated fingerprint recognition using structural matching. *Pattern Recognit.*, 23(8):893–904, 1990.

[20] D. K. Isenor and S. G. Zaky. Fingerprint identification using graph matching. *Pattern Recognit.*, 19(2):113–122, 1986.

[21] A. K. Jain and F. Farrokhnia. Unsupervised texture segmentation using Gabor filters. *Pattern Recognit.*, 24(12):1167–1186, 1991.

[22] A. K. Jain, L. Hong, S. Pankanti, and R. Bolle. On-line identity-authentication system using fingerprints. *Proc. IEEE (Special Issue on Automated Biometrics)*, 85:1365–1388, 1997.

[23] A. K. Jain, S. Prabhakar, and L. Hong. A multichannel approach to fingerprint classification. In *Proc. Indian Conf. Comput. Vis., Graphics, and Image Process. (ICVGIP'98)*, New Delhi, India, December 21–23, 1998.

[24] A. K. Jain, S. C. Dass, and K. Nandakumar. Soft biometric traits for personal recognition systems. In *Proc. Int. Conf. Biometric Authentication (ICBA)*, Hong Kong, LNCS 3072, 731–738, July 2004.

[25] A. K. Jain, Y. Chen, and M. Demirkus. Pores and ridges: high resolution fingerprint matching using level 3 features. *IEEE Trans.* PAMI, 29(1):15–27, 2007.

[26] T. Kamei and M. Mizoguchi. Image filter design for fingerprint enhancement. In *Proc. ISCV' 95*, Coral Gables, FL, 109–114, 1995.

[27] K. Karu and A. K. Jain. Fingerprint classification. *Pattern Recognit.*, 29(3):389–404, 1996.

[28] M. Kawagoe and A. Tojo. Fingerprint pattern classification. *Pattern Recognit.*, 17(3):295–303, 1984.

[29] H. C. Lee and R. E. Gaensslen. *Advances in Fingerprint Technology*. CRC Press, Boca Raton, FL, 2001.

[30] D. Maltoni, D. Maio, A. K. Jain, and S. Prabhakar. *Handbook of Fingerprint Recognition*. Springer Verlag, New York, 2003.

[31] B. M. Mehtre and B. Chatterjee. Segmentation of fingerprint images—a composite method. *Pattern Recognit.*, 22(4):381–385, 1989.

[32] N. J. Naccache and R. Shinghal. An investigation into the skeletonization approach of Hilditch. *Pattern Recognit.*, 17(3):279–284, 1984.

[33] G. Parziale, E. Diaz-Santana, and R. Hauke. The surround Imager*TM*: a multi-camera touchless device to acquire 3d rolled-equivalent fingerprints. In *Proc. IAPR Int. Conf. Biometrics*, Hong Kong, January 2006.

[34] T. Pavlidis. *Algorithms for Graphics and Image Processing*. Computer Science Press, New York, 1982.

[35] N. Ratha, K. Karu, S. Chen, and A. K. Jain. A real-time matching system for large fingerprint database. *IEEE Trans. PAMI*, 18(8):799–813, 1996.

[36] H. T. F. Rhodes. *Alphonse Bertillon: Father of Scientific Detection*. Abelard-Schuman, New York, 1956.

[37] A. Ross, K. Nandakumar, and A. K. Jain. *Handbook of Multibiometrics*. Springer Verlag, New York, 2006.

[38] R. K. Rowe, U. Uludag, M. Demirkus, S. Parthasaradhi, and A. K. Jain. A multispectral whole-hand biometric authentication system. In *Proc. Biometric Symp., Biometric Consortium Conf.*, Baltimore, September 2007.

[39] M. K. Sparrow and P. J. Sparrow. A topological approach to the matching of single fingerprints: development of algorithms for use of rolled impressions. *Tech.* Report, National Bureau of Standards, Gaithersburg, MD, May 1985.

[40] US-VISIT. http://www.dhs.gov.

[41] C. I. Watson. *NIST Special Database 9, Mated Fingerprint Card Pairs*. National Institute of Standards and Technology, Gaithersburg, MD, 1993.

[42] C. L. Wilson, G. T. Candela, and C. I. Watson. Neural-network fingerprint classification. *J. Artif. Neural Netw.*, 1(2):203–228, 1994.

[43] C. Wilson *et al.* Fingerprint vendor technology evaluation 2003: summary of results and analysis report. *NIST Tech. Report NISTIR* 7123, 2004.

[44] J. D. Woodward. Biometrics: privacy's foe or privacy's friend? *Proc. IEEE (Special Issue on Automated Biometrics)*, 85:1480–1492, 1997.

[45] N. D. Young, G. Harkin, R. M. Bunn, D. J. McCulloch, R. W. Wilks, and A. G. Knapp. Novel fingerprint scanning arrays using polysilicon TFT's on glass and polymer substrates. *IEEE Electron Device Lett.*, 18(1):19–20, 1997.

Unconstrained Face Recognition from a Single Image

24

Shaohua Kevin Zhou[1], Rama Chellappa[2], and Narayanan Ramanathan[2]

[1] *Siemens Corporate Research, Princeton;* [2] *University of Maryland*

24.1 INTRODUCTION

In most situations, identifying humans using faces is an effortless task for humans. Is this true for computers? This very question defines the field of automatic face recognition [1–3], one of the most active research areas in computer vision, pattern recognition, and image understanding. Over the past decade, the problem of face recognition has attracted substantial attention from various disciplines and has witnessed a skyrocketing growth of the literature. Below, we mainly emphasize some key perspectives of the face recognition problem.

24.1.1 Biometric Perspective

Face is a biometric. As a consequence, face recognition finds wide applications in authentication, security, and so on. One recent application is the US-VISIT system by the Department of Homeland Security (DHS), collecting foreign passengers' fingerprints and face images.

Biometric signatures of a person characterize their physiological or behavioral characteristics. Physiological biometrics are innate or naturally occuring, while behavioral biometrics arise from mannerisms or traits that are learned or acquired. Table 24.1 lists commonly used biometrics. Biometric technologies provide the foundation for an extensive array of highly secure identification and personal verification solutions. Compared with conventional identification and verification methods based on personal identification numbers (PINs) or passwords, biometric technologies offer many advantages. First, biometrics are individualized traits while passwords may be used or stolen by someone other than the authorized user. Also, biometrics are very convenient since there is nothing

TABLE 24.1   A list of physiological and behavioral biometrics.

Type	Examples
Physiological biometrics	DNA, face, fingerprint, hand geometry, iris, pulse, retinal, and body odor
Behavioral biometrics	Face, gait, handwriting, signature, and voice

to carry or remember. In addition, biometric technologies are becoming more accurate and less expensive.

Among all biometrics listed in Table 24.1, face is a very unique one because it is the only biometric belonging to both the physiological and behavioral categories. While the physiological part of the face has been widely exploited for face recognition, the behavioral part has not yet been fully investigated. In addition, as reported in [4, 5], face enjoys many advantages over other biometrics because it is a natural, nonintrusive, and easy-to-use biometric. For example [4], among the six biometrics of face, finger, hand, voice, eye, and signature, face biometric ranks the first in the compatibility evaluation of a machine readable travel document (MRTD) system in terms of six criteria: enrollment, renewal, machine-assisted identity verification requirements, redundancy, public perception, and storage requirements and performance. Probably the most important feature of acquiring the face biometric signature is that no cooperation is required during data acquisition.

Besides applications related to identification and verification such as access control, law enforcement, ID and licensing, surveillance, etc., face recognition is also useful in human-computer interaction, virtual reality, database retrieval, multimedia, computer entertainment, etc. See [2, 3] for recent summaries on face recognition applications.

24.1.2 Experimental Perspective

Face recognition mainly involves the following three tasks [6]:

- Verification: The recognition system determines if the query face image and the claimed identity match.

- Identification: The recognition system determines the identity of the query face image.

- Watch list: The recognition system first determines if the identity of the query face image is in the watch list and, if yes, then identifies the individual.

Figure 24.1 illustrates the above three tasks and corresponding metrics used for evaluation. Among these tasks, the watch list task is the most difficult one.

This chapter focuses only on the identification task. We follow the face recognition test protocol FERET [7] widely used in the face recognition literature. FERET stands for "facial recognition technology." FERET assumes the availability of the following three sets, namely a training set, a gallery set, and a probe set. The training set is provided for the recognition algorithm to learn the features that are capable of characterizing the whole

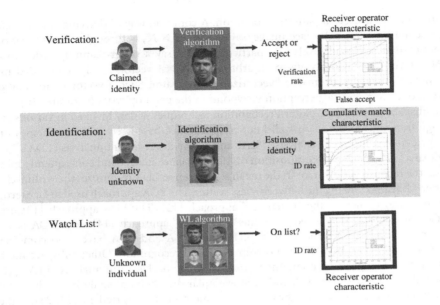

FIGURE 24.1

Three face recognition tasks: verification, identification, and watch list (courtesy of P. J. Phillips).

human face space. The gallery and probe sets are used in the testing stage. The gallery set contains images with known identities and the probe set with unknown identities. The algorithm associates descriptive features with images in the gallery and probe sets and determines the identities of the probe images by comparing their associated features with features associated with gallery images.

24.1.3 Theoretical Perspective

Face recognition is by nature an interdisciplinary research area, involving researchers from pattern recognition, computer vision and graphics, image processing/understanding, statistical computing, and machine learning. In addition, automatic face recognition algorithms/systems are often guided by psychophysics and neural studies on how humans perceive faces. A good summary of research on face perception is presented in [8]. We now focus on the theoretical implication of pattern recognition for the task of face recognition.

We present a hierarchical study of face pattern. There are three levels forming the hierarchy: pattern, visual pattern, and face pattern, each associated with a corresponding theory of recognition. Accordingly, face recognition approaches can be grouped into three categories.

Pattern and pattern recognition: Because face is first a pattern, any pattern recognition theory [9] can be directly applied to a face recognition problem. In general, a vector representation is used in pattern recognition. A common way of deriving such a vector representation from a 2D face image, say of size $M \times N$, is through a "vectorization" operator that stacks all pixels in a particular order, say a raster-scanning order, to an $MN \times 1$ vector. Obviously, given an arbitrary $MN \times 1$ vector, it can be decoded into an $M \times N$ image by an inverse-"vectorization" operator. Such a vector representation corresponds to a holistic perception viewpoint in the psychophysics literature [10].

Subspace methods are pattern recognition techniques widely invoked in various face recognition approaches. Two well-known appearance-based recognition schemes utilize principal component analysis (PCA) and linear discriminant analysis (LDA). PCA performs [11] an eigen-decomposition of the covariance matrix and consequently minimizes the reconstruction error in the mean square sense. LDA minimizes the within-class scatter while maximizing the between-class scatter. The PCA approach used in face recognition is also known as the "Eigenface" approach [12]. The LDA approach [13] used in face recognition is referred to as the "Fisherface" approach [14] since LDA is also known as Fisher discriminant analysis. Further, PCA and LDA have been combined (LDA after PCA) as in [14, 15] to obtain improved recognition. Other subspace methods such as independent component analysis [16], local feature analysis (LFA) [17], probabilistic subspace [18, 19], and multiexemplar discriminant analysis [20] have also been used. A comparison of these subspace methods is reported in [19, 21]. Other than subspace methods, classical pattern recognition tools such as neural networks [22], learning methods [23], and evolutionary pursuit/genetic algorithms [24] have also been applied.

One concern in a regular pattern recognition problem is the "curse of dimensionality" since usually M and N themselves are quite large numbers. In face recognition, because of limitations in image acquisition, practical face recognition systems store only a small number of samples per subject. This further aggravates the curse of dimensionality problem.

Visual pattern and visual recognition: In the middle of the hierarchy sits the visual pattern. Face is a visual pattern in the sense that it is a 2D appearance of a 3D object captured by an imaging system. Certainly, visual appearance is affected by the configuration of the imaging system. An illustration of the imaging system is presented in Fig. 24.2.

There are two distinct characteristics of the imaging system: photometric and geometric.

- The photometric characteristics are related to the lighting source distribution in the scene. Figure 24.3 shows the face images of a person captured under varying illumination conditions. Numerous models have been proposed to describe the illumination phenomenon, i.e., how the light travels when it hits the object. In addition to its relationship with the light distribution such as light direction and intensity, an illumination model is in general also relevant to surface material properties of the illuminated object.

- The geometric characteristic is about the camera properties and the relative positioning of the camera and the object. Camera properties include camera intrinsic

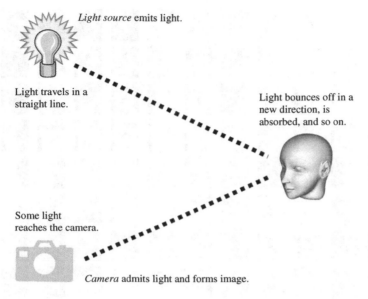

Light source emits light.

Light travels in a
straight line.

Light bounces off in a
new direction, is
absorbed, and so on.

Some light
reaches the camera.

Camera admits light and forms image.

FIGURE 24.2

An illustration of the imaging system.

parameters and camera imaging models. The imaging models widely studied in the computer vision literature are orthographic, scale orthographic, and perspective models. Due to the projective nature of the perspective model, the orthographic or scale-orthograhic models are used in the face recognition community. The relative positioning of the camera and the object results in pose variation, a key factor in determining how the 2D appearances are produced. Figure 24.3 shows the face images of a person captured at varying poses and illuminating conditions.

Studying photometric and geometric characteristics is one of the key problems in the object recognition literature, and consequently visual recognition under illumination and pose variations is the main challenge for object recognition. A full review of the visual recognition literature is beyond the scope of the chapter. However, face recognition addressing the photometric and geometric challenges is still an open question.

Approaches to face recognition under illumination variation are usually treated as extensions of research efforts on illumination models. For example, if a simplified Lambertian reflectance model ignoring the shadow pixels [26, 27] is used, rank-3 subspace can be constructed to cover appearances arbitrarily illuminated by a distant point source. A low-dimensional subspace [28] can be found in the Lambertian model that includes attached shadows. Face recognition is conducted by checking if a query face image lies in the object-specific illumination subspace. To generalize from the object-specific illumination subspace to class-specific illumination subspace, bilinear models are used in [29–31]. Most face recognition approaches addressing pose variations use view-based

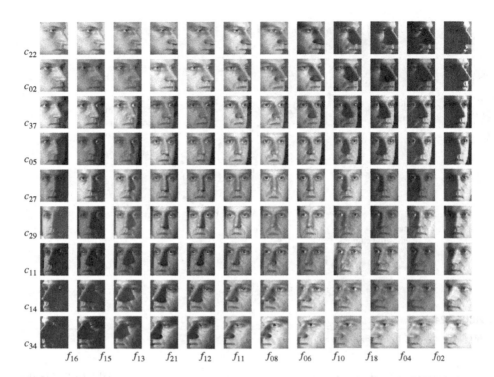

c_{22} c_{02} c_{37} c_{05} c_{27} c_{29} c_{11} c_{14} c_{34}

f_{16} f_{15} f_{13} f_{21} f_{12} f_{11} f_{08} f_{06} f_{10} f_{18} f_{04} f_{02}

FIGURE 24.3

Examples of the face images of one PIE [25] object with different illumination and poses. The images are of size 48×40.

appearance representations [32–34]. Face recognition under variations in illumination and poses is more difficult compared with recognition when only one variation is present. Proposed approaches in the literature include [35–39], among which the 3D morphable model [35] yields the best recognition performance. The feature-based approach [40] is reported to be partially robust to illumination and pose variations. Sections 24.2 and 24.3 present detailed reviews of the related literature.

Another important extension of visual pattern recognition is in exploiting video. The ubiquitousness of video sequences calls upon novel recognition algorithms based on videos. Because a video sequence is a collection of still images, face recognition from still images certainly applies to video sequences. However, an important property of a video sequence is its temporal dimension or dynamics. Recent psychophysical and neural studies [41] demonstrate the role of movement in face recognition: Famous faces are easier to recognize when presented in moving sequences than in still photographs, even under a range of different types of degradations. Computational approaches utilizing such temporal information include [42–46]. Clearly, due to the free movement of human faces and uncontrolled environments, issues like illumination and pose variations still exist. Besides these issues, localizing faces or face segmentation in a cluttered environment in video sequences is very challenging too.

In surveillance scenarios, further challenges include poor video quality and lower resolution. For example, the face region can be as small as 15×15. Most feature-based approaches [35, 40] need face images of size as large as 128×128. The attractiveness of the video sequence is that the video provides multiple observations with temporal continuity.

Face pattern and face recognition: At the top of the hierarchy lies the face pattern. The face pattern specializes the visual pattern by specializing the object to be a human face. Therefore, face-specific properties or characteristics should be taken into account when performing face recognition.

- *Expression and deformation.* Humans exhibit emotions. The natural way to express the emotions is through facial expressions, yielding patterns under nonrigid deformations. The nonrigidity introduces very high degrees of freedom and perplexes the recognition task. Figure 24.4(a) shows the face images of a person exhibiting different expressions. While face expression analysis has attracted a lot of attention [49, 50], recognition under facial expression variation has not been fully explored.

- *Aging.* Face appearances vary significantly with age and such variations are specific to an individual. Theoretical modeling of aging [48] is very difficult due to the individualized variation. Figure 24.4(b) shows the face images of a person at different ages.

- *Face surface.* One speciality of the face surface is its bilateral symmetry. The symmetry constraint has been widely exploited in [31, 51, 52]. In addition, surface integrability is an inherent property of any surface, which has also been used in [27, 31, 53].

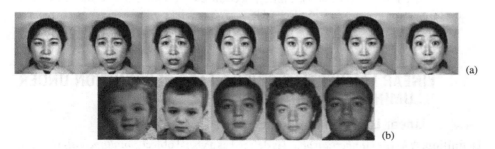

(a)

(b)

FIGURE 24.4

(a) Appearances of a person with different facial expressions (from [47]). (b) Appearances of a person at different ages (from [48]).

■ *Self-similarity.* There is a strong visual similarity among face images of different individuals. Geometric positioning of facial features such as eyes, noses, and mouths is similar across individuals. Early face recognition approaches in the 1970s [54, 55] used the distances between feature points to describe the face and achieved some success. Also, the properties of face surface materials are similar within the same race. As a consequence of visual similarity, the "shapes" of the face appearance manifolds belonging to different subjects are similar. This is the foundation of approaches [19, 20] that attempt to capture the "shape" characteristics using the so-called intraperson space.

■ *Makeup, cosmetic, etc.* These factors are very individualized and unpredictable. Other than the effect of glasses which has been studied in [14], effects induced by other factors are not widely understood in the recognition literature. However, modeling these factors can be useful for face animation in the computer graphics literature.

24.1.4 Unconstrained Face Recognition

A wide array of face recognition approaches has been proposed in the literature. Early face recognizers [11–14, 16–19] yielded unsatisfactory results especially when confronted with unconstrained scenarios such as varying illumination, varying poses, expression, and aging. In addition, the recognizers have been further hampered by the registration requirement as the images that the recognizers process contain transformed appearances of the object. Recent advances in face recognition have focused on face recognition under illumination and pose variations [28, 30–33, 35, 39, 56]. Face recognition under variations in expression [57] and aging [58–61] have been investigated too.

In this chapter, we attempt to present some representative face recognition works that deal with illumination, pose, and aging variations. Section 24.2 describes the linear Lambertian object approach [31, 62] for face recognition under illumination variation and Section 24.3 the illuminating light field approach [39, 63] for face recognition under illumination and pose variations. Finally, Section 24.4 elaborates face modeling and verification across age progression [59–61].

24.2 LINEAR LAMBERTIAN OBJECT: FACE RECOGNITION UNDER ILLUMINATION VARIATION

24.2.1 Linear Lambertian Objects

Definition: A *linear Lambertian object* is defined as a visual object *simultaneously* obeying the following two properties:

■ It is linearly spanned by basis objects.

■ It follows the Lambertian reflectance model with a varying albedo field.

A good example of a linear Lambertian object is the human face.[1] Linearity [12, 14, 56] characterizes the appearances of an image ensemble for a class of objects. It assumes that an image h is expressed as a linear combination of basis images h_i, i.e.,

$$h = \sum_{i=1}^{m} f_i h_i, \tag{24.1}$$

where f_i's are blending coefficients. In other words, the basis images span the image ensemble. Typically, the basis images are learned using images not necessarily illuminated under the same lighting condition. This forces the learned basis images to inadequately cover variations in both identity and illumination.

The Lambertian reflectance model [26, 28, 64] with varying albedo field is widely used in the literature to depict the appearance of certain reasonably matte objects such as faces. It assumes that a pixel h is represented as

$$h = \max(p\, \mathsf{n}_{3\times1}^{\mathsf{T}} \mathsf{s}_{3\times1}, 0), \tag{24.2}$$

where $[.]^{\mathsf{T}}$ denotes the transpose, p is the albedo at the pixel, n is the unit surface normal vector at the pixel, and s (a 3×1 unit vector multiplied by its intensity) specifies the direction and magnitude of a light source. When the pixel is not directly facing the light source, an attached shadow occurs with zero intensity. Another kind of shadow is the cast shadow that is generated when the light source is blocked due to object geometry.

An image h is a collection of d pixels $\{h_i, i = 1, \ldots, d\}$.[2] By stacking all the pixels into a column vector, we have

$$\mathsf{h}_{d\times1} = [h_1, h_2, \ldots, h_d]^{\mathsf{T}} = \max([(p_1\mathsf{n}_1^{\mathsf{T}})\mathsf{s}, \ldots, (p_d\mathsf{n}_d^{\mathsf{T}})\mathsf{s}]^{\mathsf{T}}, 0) = \max(\mathsf{T}_{d\times3}\, \mathsf{s}_{3\times1}, 0), \tag{24.3}$$

where the $\mathsf{T} = [p_1\mathsf{n}_1, p_2\mathsf{n}_2, \ldots, p_d\mathsf{n}_d]^{\mathsf{T}}$ matrix encodes the product of albedos and surface normal vectors for all d pixels. This Lambertian model is specific to the object, and consequently, we call the T matrix an *object-specific albedo-shape* matrix.

The process of combining the above two properties is equivalent to imposing the restriction of the same light source on the basis images, with each basis image expressed as $h_i(\mathsf{s}) = \max(\mathsf{T}_i\mathsf{s}, 0)$. Therefore, Eq. (24.1) becomes

$$h = \sum_{i=1}^{m} f_i \max(\mathsf{T}_i\mathsf{s}, 0). \tag{24.4}$$

This is the *generative* model for the linear Lambertian object. It is evident that the key difference between employing the linear Lambertian property and conventional subspace

---

[1] One may argue that specular properties of skin and eyes and the reflectance properties of hair violate the Lambertian assumption. However, the hair is excluded by preprocessing and the pixels in specular regions are excluded in our algorithm using the indicator function. Except under extreme lighting conditions, the number of pixels in specular regions is not significantly large.

[2] The index i corresponds to a spatial position $\mathsf{x} = (x, y)$. We will interchange both notations. For instance, we might also use $\mathsf{x} = 1, \ldots, d$.

analysis is in the basis image h$_i$: for a linear Lambertian object, it is now a function of the light source s through the matrix T$_i$. We denote all the matrices T$_i$ compactly by W = [T$_1$, T$_2$, . . . , T$_m$]. Since the W matrix encodes all albedos and surface normals for a class of objects, we call it a *class-specific albedo-shape* matrix.

The linear Lambertian object concept provides a unique opportunity to study the appearances of an image ensemble for a class of objects under illumination variations and opens the door to many applications. In [62], two important applications, namely *generalized photometric stereo* and *illumination-invariant face recognition*, are considered. Generalized photometric stereo studies the modeling part or the inverse problem; in particular, it attempts to recover the class-specific albedo-shape matrix. An illumination-invariant face recognition experiment validates the assumption of a linear Lambertian object.

Here we focus on the face recognition part. Section 24.2.2 reviews the literature on face recognition under illumination variations, and Section 24.2.3 discusses the effect of attached shadow. We then address face recognition in the presence of a single light source in Section 24.2.4 and in the presence of multiple light sources in Section 24.2.5.

24.2.2 Literature Review and Proposed Approach

Face recognition under illumination variation is a very challenging problem. The key is to successfully separate the illumination source from the observed appearance. Once separated, what remains is illuminant-invariant and is appropriate for recognition. In addition to variations due to changes in illumination, various issues embedded in the recognition setting make recognition even more difficult. Different recognition settings can be formed in terms of the identity and illumination overlaps among the training, gallery, and probe sets. The most difficult setting, which is the focus here, is obviously the one in which there is no overlap at all among the three sets in terms of both identity and illumination, except the identity overlap between the gallery and the probe sets. In this setting, generalizations from known illuminations to unknown illuminations and from known identities to unknown identities are particularly desired.

State-of-the-art research efforts can be grouped into three streams: subspace methods, reflectance-model methods, and 3D-model-based methods. (i) The first approach is very popular for the recognition problem. After removing the first three eigenvectors, PCA was reported to be more robust to illumination variation than the ordinary PCA or the "Eigenface" approach [12]. LDA [14, 37] has also been modified to handle illumination variations. In general, subspace learning methods are able to capture the generic face space and thus recognize objects given images not present in the training set. The disadvantage is that subspace learning is actually tuned to the lighting conditions of the training set; therefore if the illumination conditions are not similar among the training, gallery, and probe sets, recognition performance may deteriorate. (ii) The second approach [26, 28, 30, 36, 52, 65] employs a Lambertian reflectance model with a varying albedo field, mostly ignoring both attached and cast shadows. The main disadvantage of this approach is the lack of generalization from known objects to unknown objects, with the exception of [30, 65]. In [30], Shashua and Raviv used an ideal-class assumption. All objects belonging to the ideal class are assumed to have the same shape. The work of Zhang and Samaras [65] utilized the regularity of the harmonic image exemplars to perform face recognition under varying lighting conditions. (iii) The third approach employs 3D models. The "Eigenhead" approach [66] assumes that the 3D geometry (or 3D depth information) of any face lies in a linear space spanned by the 3D geometry of the training ensemble and uses a constant albedo field. The morphable model approach [35] is based on a synthesis-and-analysis strategy, assuming both geometry and texture are linearly spanned by those of the training ensemble. It is able to handle both illumination and pose variations. The weakness of 3D model-based approaches is that they require complicated fitting algorithms and images of relatively big size.

Compared with the above, the proposed recognition scheme possesses the following properties: (i) it is able to recognize new objects not present in the training set; (ii) it is able to handle new lighting conditions not present in the training set; and (iii) no explicit 3D model or prior knowledge about illumination conditions is needed. In other words, we combine the advantages of subspace learning and reflectance model-based methods.

24.2.3 The Importance of the Attached Shadow

In general, objects like faces do not have all the surface points facing the illumination source, which leads to the formation of attached shadows. The cast and attached shadows are often ignored to keep the subspace of the observed images a three [26] or with the addition of an ambient component [27], a four dimensional subspace. Inherent nonlinearity in Lambert's law can account for the formation of attached shadows. Therefore, these generative approaches either ignore this nonlinearity completely or try to somehow ignore the shadow pixels. Here we present some illustrations to highlight the role attached shadows can play.

Consider the problem of illumination estimation given only one image of a 3D surface and the following three options:

$$\text{Completely linear:} \quad \varepsilon(\mathsf{s}) = \| \, \mathsf{h} - pn^\mathsf{T}\mathsf{s} \, \|^2; \tag{24.5}$$

$$\text{Shadow pixels ignored:} \quad \varepsilon(\mathsf{s}) = \| \, \tau \circ (\mathsf{h} - pn^\mathsf{T}\mathsf{s}) \, \|^2; \tag{24.6}$$

$$\text{Nonlinear rule:} \quad \varepsilon(\mathsf{s}) = \| \, \mathsf{h} - \max(pn^\mathsf{T}\mathsf{s}, 0) \, \|^2 . \tag{24.7}$$

The error curves for the linear method and the method that ignores shadows keep getting worse as the number of shadow pixels increases. Clearly, the linear method penalizes the correct illumination at the shadow pixels by having nonzero error values for those pixels. On the other hand, when shadows are ignored, the illuminations that produce wrong values for the shadow pixels do not get penalized there. Figure 24.5 shows the error surfaces for the three methods for a given face image of known shape and albedo. For each hypothesized illumination direction s, we compute the cost function $\varepsilon(\mathsf{s})$ defined in (24.5), (24.6), and (24.7); therefore, we have an error surface for each method. The lower the error is for a hypothesized illumination direction s, the darker the surface looks at the corresponding point on the sphere. The global minimum is far from the true value using the first approach but is correct up to a discretization error for the second and third

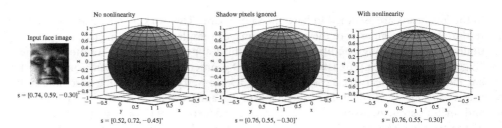

FIGURE 24.5

The error surfaces for the estimation of the light source direction given a face image of known shape and albedo. The three plots correspond to the three approaches described in the text. The lower the error is for a particular illumination direction, the darker the error sphere looks at the point corresponding to that direction. The true and estimated values of the illumination direction are listed along with their plots.

approaches. In fact, the second and third methods will always produce the same global minimum (assuming τ is correct), but the global minimum will always be less ambiguous in the third case because several wrongly hypothesized illumination directions do not get penalized enough in the second approach due to the exclusion of the shadow pixels (Fig. 24.5).

Therefore, given the class-specific albedo-shape matrix $W = [T_1, T_2, \ldots, T_m]$, the recovery of the identity vector f and illumination s can be posed as the following optimization problem:

$$[\text{Problem-I}] \min_{f,s} \varepsilon(f,s) = \| h - \sum_{i=1}^{m} f_i \max(T_i s, 0) \|^2 + (1^T f - 1)^2. \qquad (24.8)$$

Please note that s is not a unit vector as it also contains the intensity of the illumination source.

The minimization of (24.8) is performed using an iterative approach, fixing f for optimizing ε w.r.t. s and fixing s for optimization w.r.t. f. In each iteration, f can be estimated by solving a linear least-squares (LS) problem but a nonlinear LS solution is required to estimate s. The nonlinear optimization is performed using the *lsqnonlin* function in MATLAB which is based on the interior-reflective Newton method. For most faces, the function value did not change much after 4-5 iterations. Therefore, the iterative optimization was always stopped after five iterations. The whole process took about 5-7 seconds per image on a normal desktop.

24.2.4 Recognition in the Presence of a Single Light Source

The linear Lambertian object renders an illumination-invariant signature that is the blending linear coefficient f_i in (24.4), since the single light source is fully described by the s vector. Therefore, even based on a single image, we can achieve robust face recognition under illumination variation.

Recognition setting: As mentioned earlier, we study an extreme recognition setting with the following features: there is no identity overlap between training, gallery, and probe sets; only one image per object is stored in the gallery set; the lighting conditions for the training, gallery, and probe sets are completely unknown. The class-specific albedo-shape matrix W is derived from the training set. For simplicity, we assume that W is readily constructed from 3D face models.

Our recognition strategy is as follows:

- With W given, learn the identity signature f for both gallery and probe sets, assuming no knowledge of illumination directions.

- Perform recognition using the nearest correlation coefficient (or vector cosine). Suppose that a gallery image g has its signature f_g and a probe image p has its signature f_p, their vector cosine is defined as

$$cos(p,g) = (f_p, f_g) / \sqrt{(f_p, f_p)(f_g, f_g)},$$

where (x, y) is an inner-product such as $(x, y) = x^T \Sigma y$ with Σ learned or given. We simply set Σ as an identity matrix.

We use the pose, illumination, and expression (PIE) database [25] in our experiment.[3] Figure 24.6(a) shows the distribution of all 21 flashes used in PIE and their estimated positions using our algorithm. Since the flashes are almost symmetrically distributed about the head position, we only use 12 of them distributed on the right half of the unit sphere in Figure 24.6(a). More specifically, the flashes used are f_{08}, f_{09}, f_{11}-f_{17}, and f_{20}-f_{22}. In total, we used $68 \times 12 = 816$ images in a fixed view as there are 68 subjects in the PIE database. Figure 24.6(b) displays one PIE object under the selected 12 illuminants. We use only the frontal view, but we do not assume that the directions and intensities of the illuminants are known.

Affine registration is performed by aligning the eyes and mouth to desired positions. No flow computation is carried on for further pixelwise alignment as opposed to [35]. After the preprocessing step, the cropped out face image is of size 48 by 40 (i.e., $d = 1920$). Also, we only process gray images by taking the average of the red, green, and blue channels of their color versions. All 68 images under one illumination are used to form a gallery set and under another illumination to form a probe set. The training set is taken from sources other than the PIE dataset. Thus, we have $12 \times 11 = 132$ tests, each test giving rise to a recognition score.

Experiments and results: Method-I models the attached shadow using nonlinearity (i.e., the Problem-I). We form the W matrix using Vetter's 3D face database [35] that

---

[3]We use the "illum" part of the PIE database that is close to the Lambertian model as in [37], while the "light" part that includes an ambient light is used in [35].

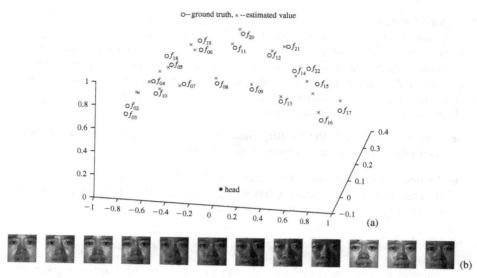

FIGURE 24.6

(a) Flash distribution in the PIE database. For illustrative purposes, we move their positions on a unit sphere as only the illuminant directions matter. "o" means the ground truth and "x" the estimated values. (b) One PIE object under the selected 12 illuminants (from left to right, f_{08}, f_{09}, $f_{11} - f_{17}$, and $f_{20} - f_{22}$).

contains 100 3D face scans. Table 24.2 shows the recognition results of Method-I. The average rate is 96%, which is significantly better than that of PCA (51%) and LDA (64%).[4]

One of the key assumptions of the linear Lambertian object is the linearity in the identity. In the above two methods, the linearity is assumed for the holistic object; their difference lies in handling shadows in the Lambertian model. In Method-II, we generalized the holistic linearity to piecewise linearity. We simply divided the face image into 16 regions with equal size. For each region, we used Method-I for recovering the blending coefficient; however, the blending coefficients for different regions can be different. Table 24.2 reports the recognition score for all possible recognition settings; on the average, we achieved almost perfect performance, 99.1% in recognition accuracy!

In the above methods, we also achieved an estimate of illuminant direction. Figure 24.5 shows the estimated illuminant directions. It is quite accurate for estimation of directions of flashes near a frontal pose. But when the flashes are significantly off-frontal, accuracy goes down slightly.

---

[4]To train the PCA and LDA basis images, we used all $68 \times 12 = 816$ images in the gallery set, while in Method-I and Method-II, we used the Vetter's dataset as the training set, which is different from the PIE.

TABLE 24.2 Recognition rate obtained by Method-I and Method-II. 'F' means 100 and "f_{nn}" means flash no. nn. In each cell, the left number is from Method-I and the right number from Method-II.

Gallery	f_{08}	f_{09}	f_{11}	f_{12}	f_{13}	f_{14}	f_{15}	f_{16}	f_{17}	f_{20}	f_{21}	f_{22}	Average
Probe
f_{08}	-/-	F/F	F/F	F/F	96/F	97/F	81/97	72/96	50/81	F/F	97/F	84/99	90/98
f_{09}	F/F	-/-	F/F	F/F	F/F	99/F	97/F	96/99	75/96	F/F	F/F	97/F	97/F
f_{11}	F/F	F/F	-/-	F/F	F/F	97/F	94/F	78/91	63/F	F/F	99/F	94/F	94/99
f_{12}	F/99	F/99	F/F	-/-	F/F	F/F	F/F	99/F	90/97	F/F	F/F	F/F	99/F
f_{13}	97/99	F/99	F/F	F/F	-/-	F/F	F/F	F/F	96/F	F/F	F/F	F/F	99/F
f_{14}	94/99	F/99	F/F	F/F	F/F	-/-	F/F	F/F	99/99	F/F	F/F	F/F	99/F
f_{15}	88/99	97/99	97/F	F/F	F/F	F/F	-/-	F/F	F/F	97/F	F/F	F/F	98/F
f_{16}	74/97	90/99	81/F	93/F	F/F	F/F	F/F	-/-	F/99	76/F	97/F	F/F	93/F
f_{17}	59/87	74/94	63/85	87/F	99/F	99/F	F/F	F/F	-	71/88	94/F	F/F	87/96
f_{20}	99/99	F/F	F/F	F/F	F/F	99/F	96/F	82/F	71/93	-/-	F/F	97/F	95/99
f_{21}	97/99	F/99	F/F	F/F	F/F	F/F	F/F	99/F	96/F	F/F	-	F/F	99/F
f_{22}	93/99	F/99	99/F	F/F	F/F	F/F	F/F	F/F	99/99	99/F	F/F	-	99/F
Average	92/98	97/99	95/99	98/F	F/F	99/F	97/F	94/F	87/96	95/99	99/F	98/F	96/99.1

24.2.5 Recognition in the Presence of Multiple Light Sources

Most of the cited face recognition approaches (except [28, 36, 65, 67]), including those proposed in the above section, assume that the face is illuminated by a single dominant light source which does not hold in most real conditions. This probably is one of the major reasons that these approaches have not been applied for the face recognition problem using a large database of realistic images. In this section, we modify our formulation so that it can handle faces illuminated by arbitrarily placed multiple light sources. It turns out that the nonlinearity in Lambert's law is very important to this task. We extend our earlier analysis to directly incorporate the attached shadows rather than excluding them from computation.

We use a result that an image of an arbitrarily illuminated face can be approximated by low dimensional linear subspace [28] that can be generated by a linear combination of the images of the same face in the same pose, illuminated by nine different light sources placed at preselected positions [67]. Lee et al [67] show that this approximation is quite good for a wide range of illumination conditions. Hence, a face image can be written as

$$h = \sum_{j=1}^{9} \alpha_j \max(T\hat{s}_j, 0), \tag{24.9}$$

where $\{\hat{s}_1, \hat{s}_2, \ldots, \hat{s}_9\}$ are the prespecified illumination directions. As proposed in [67], we use the following directions for $\{\hat{s}_1, \hat{s}_2, \ldots, \hat{s}_9\}$:

$$\phi = \{0, 49, -68, 73, 77, -84, -84, 82, -50\}°;$$

$$\theta = \{0, 17, 0, -18, 37, 47, -47, -56, -84\}°. \tag{24.10}$$

Under this formulation, Eq. (24.8) changes to

$$[\text{Problem-III}] \min_{f,\alpha} \varepsilon(f,\alpha) = \| h - \sum_{i=1}^{m} f_i \sum_{j=1}^{9} \alpha_j \max(T_i \hat{s}_j, 0) \|^2 + (1^T f - 1)^2, \tag{24.11}$$

where $f = [f_1, f_2, \ldots, f_m]^T$ and $\alpha_{9\times1} = [\alpha_1, \alpha_2, \ldots, \alpha_9]^T$. Thus one can potentially recover the illumination-free identity vector f without using any prior knowledge of the number of light sources or any need to check different hypotheses about them.

In (24.11), the objective function is minimized with respect to f and α. This gives us the illumination-free identity vector f which is used for recognition. The optimization is done in an iterative fashion by fixing one parameter and estimating the other and *vice versa*. We will identify this method as Method-III.

By defining a $d \times m$ matrix W_f as

$$W_f = \left[\sum_{j=1}^{9} \alpha_j \max(T_1 \hat{s}_j, 0), \sum_{j=1}^{9} \alpha_j \max(T_2 \hat{s}_j, 0), \ldots, \sum_{j=1}^{9} \alpha_j \max(T_m \hat{s}_j, 0) \right],$$

it is easy to show that

$$f = \begin{bmatrix} W_f \\ 1^T \end{bmatrix}^\dagger \begin{bmatrix} h \\ 1 \end{bmatrix}, \tag{24.12}$$

where $h_{d\times1}$ is the vectorized input face image, $[.]^\dagger$ is the Moore-Penrose inverse, and $1_{m\times1}$ is the m-dimensional vector of ones, included to handle scale ambiguity between f and α.

Looking carefully at the objective function (e.g., Eq. (24.11)), one can easily observe that α too can be estimated by solving a linear LS problem (as $\{\hat{s}_1, \hat{s}_2, \ldots \hat{s}_9\}$ is known). This avoids the need for any nonlinear optimization here. Recall that nonlinear LS was required to estimate s in the approach proposed for the single light source case. The expression for α can be written as:

$$\alpha = W_\alpha^\dagger h, \tag{24.13}$$

where

$$W_\alpha = \left[\sum_{i=1}^{m} f_i \max(T_i \hat{s}_1, 0), \sum_{i=1}^{m} f_i \max(T_i \hat{s}_2, 0), \ldots, \sum_{i=1}^{m} f_i \max(T_i \hat{s}_9, 0) \right]_{d\times9}. \tag{24.14}$$

For most of the face images, the iterative optimization converged within 5-6 iterations. As there is no nonlinear optimization involved, it took just 2-3 seconds to recover f and α from a given face image using a desktop. More experimental results can be found in [62].

24.3 ILLUMINATING LIGHT FIELD: FACE RECOGNITION UNDER ILLUMINATION AND POSE VARIATIONS

24.3.1 Literature Review

In addition to the two factors of illumination and identity involved in face recognition under illumination variation, face recognition under illumination and pose variations needs to address the third factor of face pose. The issue of pose essentially amounts to a correspondence problem. If dense correspondences across poses are available and if a Lambertian reflectance model is further assumed, a rank-1 constraint is implied because theoretically a 3D model can be recovered and used to render novel poses. However, recovering a 3D model from 2D images is a difficult task. There are two types of approaches: model-based and image-based. Model-based approaches [35, 68–70] require explicit knowledge of prior 3D models, while image-based approaches [71–74] do not use prior 3D models. In general, model-based approaches [35, 68–70] register the 2D face image to 3D models that are given beforehand. In [68, 69], a generative face model is deformed through bundle adjustment to fit 2D images. In [70], a generative face model is used to regularize the 3D model recovered using the structure from motion (SfM) algorithm. In [35], 3D morphable models are constructed based on many prior 3D models. There are mainly three types of image-based approaches: SfM [71], visual hull [72, 73], and light field rendering [74, 75] methods. The SfM approach [71] using sparse correspondence does not reliably recover the 3D model amenable for practical use. The visual hull methods [72, 73] assume that the shape of the object is convex, which is not always satisfied by the human face, and further require accurate calibration information. The light field rendering methods [74, 75] relax the requirement of calibration by a fine quantization of the pose space and recover a novel view by sampling the captured data that form the so-called light field. The approach herein is image-based, so no prior 3D models are used. It handles a given set of views through an analysis analogous to the light field concept. However, no novel poses are rendered.

The literature on face recognition under illumination variation has been reviewed in the previous section. Face recognition under pose variation takes into account the two factors of identity and pose. As mentioned earlier, pose variation essentially amounts to a correspondence problem. If dense correspondences across poses are available and a Lambertian reflectance is assumed, then a rank-1 constraint is implied. Unfortunately, finding correspondences is a very difficult task and, therefore, there exists no subspace based on an appearance representation when confronted with pose variation. Approaches to face recognition under pose variation [33, 34, 36] avoid the correspondence problem by sampling the continuous pose space into a set of poses, *viz* storing multiple images at different poses for each person, at least in the training set. In [34], view-based "Eigenfaces" are learned from the training set and used for recognition. In [36], a denser sampling is used to cover the pose space. However, [36] uses object-specific images and hence appearances belonging to a novel object (i.e., not in the training set) cannot be handled. In [33], the concept of light field [74] is used to characterize the continuous pose space. "Eigen" light fields are learned from the training set. However, the implementation of [33] still discretizes the pose space and recognition can be based on probe images at poses in the discretized set. One should note that the light field is not related to illumination variation. Face recognition approaches to handling both illumination and pose variations include [35, 37, 76, 77]. The approach [35] uses morphable 3D models to characterize human faces. Both geometry and texture are linearly spanned by those of the training ensemble consisting of 3D prior models. It is able to handle both illumination and pose variations. Its only weakness is a complicated fitting algorithm. Recently, a fitting algorithm more efficient than [35] has been proposed in [78]. In [37], the Fisher light field is proposed to handle both illumination and pose variations, where the light field is used to cover the pose variation and the Fisher discriminant analysis to cover the illumination variation. Since discriminant analysis is a statistical analysis tool which minimizes the within-class scatter while maximizing the between-class clatter and has no relationship with any physical illumination model, it is questionable that discriminant analysis is able to generalize to new lighting conditions. Instead, this generalization may be inferior because discriminant analysis tends to overly tune to the lighting conditions in the training set. The "Tensorface" approach [76] uses a multilinear analysis to handle various factors such as identity, illumination, pose, and expression. The factors of identity and illumination are suitable for linear analysis, as evidenced by the "Eigenface" approach (assuming a fixed illumination and a fixed pose) and the subspace induced by the Lambertian model, respectively. However, the factor of expression is arguably amenable for linear analysis and the factor of pose is not amenable for linear analysis. In [77], preliminary results are reported by first warping the albedo and surface normal fields at the desired pose and then carrying on recognition as usual.

24.3.2   Illumination- and Pose-Invariant Identity Signature

The light field measures the radiance in free space (free of occluders) as a 4D function of position and direction. An image is a 2D slice of the 4D light field. If the space is only 2D, the light field is then a 2D function. This is illustrated in Fig. 24.7 (also see another illustration in [33]), where a camera conceptually moves along a circle, within which a square object with four differently colored sides resides. The 2D light field L is a function of θ and ϕ as properly defined in Fig. 24.7. The image of the 2D object is just a vertical line. If the camera is allowed to leave the circle, then a curve is traced out in the light field to form the image, i.e., the light field is accordingly sampled.

Constructing a light field is a practically difficult task. However, in the context of view-based face recognition, where only some specific poses are of interest with each pose sampling a subset of the light field, we can only focus on the portion of the light field that is equivalent to the union of these subsets. Suppose that the K poses of interest are $\{v_1, \ldots, v_K\}$ and the images at these poses are $\{h(v_1), h(v_2), \ldots, h(v_K)\}$. The portion of

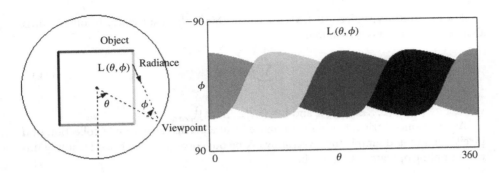

FIGURE 24.7

This figure illustrates the 2D light field of a 2D object (a square with four differently colored sides), which is placed within a circle. The angles θ and ϕ are used to relate the viewpoint with the radiance from the object. The right image shows the actual light field for the square object. See another illustration in [33].

the light field of focus is nothing but $\mathsf{L} \doteq [h(v_1)^\mathsf{T}, h(v_2)^\mathsf{T}, \ldots, h(v_K)^\mathsf{T}]^\mathsf{T}$, which is a "long" $Kd \times 1$ vector obtained by stacking all the images at all these poses. The introduction of such a "long" vector eases our computation: (i) If we are interested in a particular view v, we just simply take out those rows corresponding to this view. (ii) This allows us to perform the PCA as discussed below on the ensemble consisting of a collection of such "long" vectors, yielding a pose-invariant identity encoding.

Before presenting the details of the PCA implementation, we make the following remarks. While characterizing the appearances of one object at given views using the concept of light field is legitimate, generalizing this to many objects is questionable since the light fields belonging to different objects are not in correspondence, i.e., they are not shape-free in the terminology of [56, 79]. The mismatch in correspondence arises from differences among different objects in head sizes, head locations in world coordinate systems, positions of facial features (e.g., eye, lip, nose tip), etc. Typically, correspondences between different objects are established using face normalization, or registration is performed. Unfortunately, the normalization step that cannot be modeled by a simple camera movement ruins the static scene requirement in the light field theory. On the other hand, as argued in [56, 79], since the shape-free appearance is amenable for linear analysis, we can pursue PCA on the shape-free vector L, similar to the "Eigen" light field approach [33].

Starting from $\{\mathsf{L}_n; \ n = 1, \ldots, N\}$ of the training samples, the PCA finds eigenvectors $\{e_i; \ i = 1, \ldots, m\}$ which span a rank-m subspace. Using the fact [11, 12] that if $\mathsf{Y}^\mathsf{T}\mathsf{Y}$ has an eigenpair (λ, v), then $\mathsf{Y}\mathsf{Y}^\mathsf{T}$ has a corresponding eigenpair $(\lambda, \mathsf{Y}\mathsf{v})$, we know that e_i is just a linear combination of the L_ns, i.e., there exist a_{in}s such that

$$e_i = \sum_n a_{in}\mathsf{L}_n. \tag{24.15}$$

For an arbitrary subject, its L vector lies in this rank-m subspace. In other words, there exist coefficients f_i such that

$$L = \sum_{i=1}^{m} f_i e_i = Ef,$$

(24.16)

where $E \doteq [e_1, e_2, \ldots, e_m]_{Kd \times m}$ and $f = [f_1, f_2, \ldots, f_m]_{m \times 1}^T$.

As mentioned earlier, to obtain an image h^v at a particular pose v (a collection of d pixels), one should sample the L vector by taking out corresponding rows. Denoting this row sampling operator as $\mathcal{R}_v$, we have

$$h^v = \mathcal{R}_v[L] = \mathcal{R}_v[Ef] = \mathcal{R}_v[E]f = E^v f,$$

(24.17)

where $E^v \doteq [\mathcal{R}_v[e_1], \mathcal{R}_v[e_2], \ldots, \mathcal{R}_v[e_m]]_{d \times m}$. Eq. (24.17) has an important implication: f is a pose-invariant identity signature such that when $K = 1$, the above analysis simply reduces to the regular "Eigenface" approach [12] at a fixed pose. The assumption of a set of fixed poses can be "relaxed" by registering an image at an arbitrary pose not in the set to the nearest pose in the set using flow computation techniques or rendering a novel poses using given poses.

The light field vector L is constructed under fixed illumination s and now denoted by L^s. It also follows the Lambertian reflectance model:

$$L^s = \max(Us, 0),$$

(24.18)

where U is a matrix encoding the products of the albedos and the surface normals of all pixels present in L^s and does not depend on s. Following the assumption of the linear Lambertian object, we have

$$L^s = \sum_{i=1}^{m} f_i L_i^s = \sum_{i=1}^{m} f_i \max(U_i s, 0),$$

(24.19)

where L_i^s is the basis light field vector with U_i being its corresponding albedo and surface normal matrix.

An image h^{vs} under pose v and illumination s is

$$h^{vs} = \mathcal{R}_v[L^s] = \sum_{i=1}^{m} f_i \max(\mathcal{R}_v[U_i]s, 0).$$

(24.20)

Equation (24.20) has an important implication: the coefficient vector f provides an illumination- and pose-invariant identity signature because the pose is absorbed in U_i and the illumination is absorbed in s.

The remaining questions are how to learn the basis matrix W from a given training ensemble and how to compute the blending coefficient vector f as well as s for an arbitrary image h^{vs}. This is similar to what is presented in Section 24.2. The detailed solutions are available in [63].

24.3.3 **Implementations and Experiments**

We use the "illum" subset of the PIE database [25] in our experiments. This subset has 68 subjects under 21 illuminations and 13 poses. Out of the 21 illuminations, we select 12 of them, $f_{16}, f_{15}, f_{13}, f_{21}, f_{12}, f_{11}, f_{08}, f_{06}, f_{10}, f_{18}, f_{04}$, and f_{02}, as in [37], which typically span the set of variations. Out of the 13 poses, we select 9 of them, $c_{22}, c_{02}, c_{37}, c_{05}, c_{27}, c_{29}, c_{11}, c_{14}$, and c_{34}, which cover from the left profile to the frontal to the right profile. In total, we have $68 \times 12 \times 9 = 7,344$ images. Figure 24.3 displays one PIE object under illumination and pose variations.

Registration is performed by aligning the eyes and mouth to desired positions. No flow computation is carried on for further alignment. After the preprocessing step, the face image is of size 48 by 40, i.e., $d = 1920$. Also, we only use grayscale images by taking the average of the red, green, and blue channels of their color versions. We believe that our recognition rates can be boosted by using color images and finer registrations.

We randomly divide the 68 subjects into two parts. The first 34 subjects are used in the training set (i.e., $m = 34$) and the remaining 34 subjects are used in the gallery and probe sets. It is guaranteed that there is no identity overlap between the training set and the gallery and probe sets. To form the L vector, we use images at all available poses. Since the illumination model has generalization capability, we can select a minimum of three lighting sources in the training set. In our experiments, the training set includes only nine selected lighting sources (i.e., $r = 9$) to cover the second-order harmonic components [28]. We do not exhaust all the 12 lighting sources during training; hence the generalization capability is required in the testing stage. Notice that this is not possible in the Fisher light field approach [37] that exhausts all illumination.

The images belonging to the remaining 34 subjects are used in the gallery and probe sets. We use all 34 images under one illumination s_p and one pose v_p to form a gallery set and under the other illumination s_g and the other pose v_g to form a probe set. There are three cases of interest: *same pose but different illumination, different pose but same illumination*, and *different pose and different illumination*. We mainly concentrate on the third case with $s_p \neq s_g$ and $v_p \neq v_g$. Thus, we have $(9 \times 12)^2 - (9 \times 12) = 11,556$ tests, with each test giving rise to a recognition score. To make the recognition more difficult, we assume that the lighting conditions for the training, gallery, and probe sets are completely unknown when recovering the identity signatures. However, the pose information is known.

Our strategy is to

1. Learn U_i from the training set using the bilinear learning algorithm [29, 31].

2. With U_i given, learn the identity signature f (as well as s) for all the gallery and probe elements. Learning f and s from one single image takes about 1-2 seconds in a MATLAB implementation.

3. Perform recognition using the nearest correlation coefficient.

Table 24.3 shows the recognition results for all probe sets with a fixed gallery set (c_{27}, f_{11}), whose gallery images are at a frontal pose and under frontal illumination. Using this table we make the following observations. The case of *same pose but different*

TABLE 24.3 In each cell, the left number is the recognition rate for all the probe sets with a fixed gallery set (c_{27}, f_{11}) and the right number is the average recognition rates for all the gallery sets. Say the gallery is set at $(v_g = c_{27}, s_g = f_{12})$, the average rate is taken over all the probe sets (v_p, s_p) where $v_p \neq v_g$ and $s_p \neq s_g$. For example, the average rate for (c_{27}, f_{11}) is the average of the rates in Table 24.3 excluding the row c_{27} and the column f_{11}.

	f_{16}	f_{15}	f_{13}	f_{21}	f_{12}	f_{11}	f_{08}	f_{06}	f_{10}	f_{18}	f_{04}	f_{02}	avg
c_{22}	56/44	41/44	62/46	68/45	71/46	71/49	53/46	65/49	41/44	44/32	38/30	21/14	52/41
c_{02}	71/55	76/58	76/59	91/62	88/63	94/62	94/60	94/60	85/54	71/48	50/40	32/22	77/54
c_{37}	79/56	82/59	82/61	94/64	94/65	97/62	94/60	94/58	76/51	65/47	65/45	50/34	81/55
c_{05}	68/56	85/63	97/66	100/67	100/68	97/65	97/59	97/58	91/54	82/51	71/45	44/36	86/57
c_{27}	94/62	100/66	100/69	100/70	100/70	-/70	100/65	100/69	100/68	97/67	94/65	76/54	97/66
c_{29}	74/46	82/53	91/53	100/61	100/60	100/63	97/59	97/62	94/66	91/68	88/62	65/60	90/60
c_{11}	50/41	53/43	68/50	79/53	85/55	97/61	97/57	88/58	79/56	82/61	71/58	62/51	76/54
c_{14}	15/19	24/24	44/39	71/49	76/53	82/53	74/58	82/61	82/60	74/61	79/57	56/48	63/49
c_{34}	18/16	18/21	47/38	50/44	56/46	65/51	62/48	56/51	44/46	44/45	41/45	38/42	45/41
avg	58/44	62/48	74/53	84/57	86/59	88/60	85/57	86/59	77/56	72/53	66/50	49/40	74/53

illumination has an average rate 97% (i.e., the average of all 11 cells on the row c_{27}), the case of *different pose but same illumination* has an average rate 88% (i.e., the average of all 8 cells on the column f_{11}), and the case of *different pose and different illumination* has an average rate 70% (i.e., the average of all 88 cells excluding the row c_{27} and the column f_{11}). This shows that illumination variation is easier to handle than pose variation, and variations in both pose and illumination are the most difficult to deal with.

We now focus on the case of *different pose and different illumination*. For each gallery set, we average the recognition scores of all the probe sets with both pose and illumination different from the gallery set. Table 24.3 also shows the average recognition rates for all the gallery sets. As an interesting comparison, the "grand" average is 53% (the last cell in Table 24.3) while that of the Fisher light field approach [37] is 36%. In general, when the poses and lighting sources of the gallery and probe sets become far apart, the recognition rates decrease. The best gallery sets for recognition are those in frontal poses and under frontal illumination, and the worst gallery sets are those in profile views and off-frontal illumination. As shown in Fig. 24.3, the worst gallery sets consist of face images almost invisible (see for example the images (c_{22}, f_{02}) and (c_{34}, f_{16})) on which recognition can be hardly performed.

Figure 24.8 presents the curves of the average recognition rates (i.e., the last columns and last rows of Table 24.3) across poses and illumination. Clearly the effect of illumination variation is not as strong as due to pose variation in the sense that the curves of average recognition rates across illumination are flatter than those across poses. Figure 24.8 also shows the curves of the average recognition rates obtained based on the top 3 and top 5 matches. Using more matches increases the recognition rates significantly, which demonstrates the efficiency of our recognition scheme. For comparison, Fig. 24.8 also plots the average rates obtained using the baseline PCA or the "Eigenface" approach [12]. These rates are well below ours. The "grand" average for the PCA approach is below 10% if the top 1 match is used.

24.4 FACE MODELING AND VERIFICATION ACROSS AGE PROGRESSION

Apart from factors such as illumination, pose variations, and facial expressions that result in appearance changes, facial aging effects induce notable variations in one's appearance. Facial aging effects are often perceived in the form of gradual changes in facial shape and facial texture. Further, factors such as age group, gender, ethnicity, and weight loss/gain are said to influence the manifestation of facial aging effects. Developing a face recognition system that is well equipped in handling aging effects involves building computational models for facial shape and textural variations with age. In this section, we discuss computational models that characterize facial aging effects during different ages and briefly illustrate their application in performing face recognition across age progression.

Though face recognition systems have been in research for decades, it is only recently that there has been an enhanced interest in developing face recognition systems that are

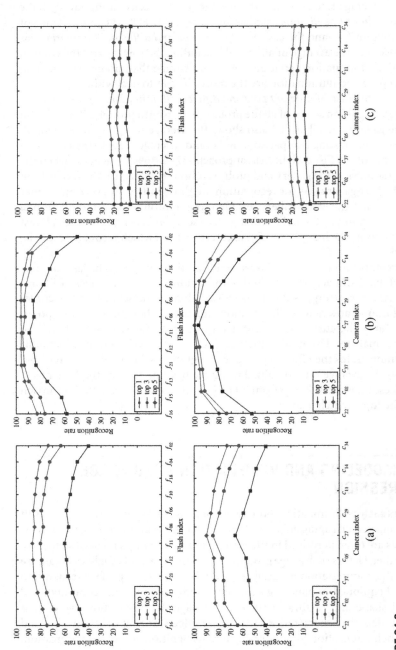

FIGURE 24.8

The average recognition rates across illumination (the top row) and across poses (the bottom row) for three cases. Case (a) shows the average recognition rate (averaging over all illumination/poses and all gallery sets) obtained by the proposed algorithm using the top n matches. Case (b) shows the average recognition rate (averaging over all illumination/poses for the gallery set (c_{27}, f_{11}) only) obtained by the proposed algorithm using the top n matches. Case(c) shows the average recognition rate (averaging over all illumination/poses and all gallery sets) obtained by the "Eigenface" algorithm using the top n matches.

robust to aging effects. Researchers from psychophysics laid the foundations for studies related to facial aging effects. D'arcy Thompson studied morphogenesis by means of geometric transformation functions applied on biological forms. Pittenger and Shaw [80] and Todd *et al* [81] identified certain forms of force configurations that when applied on 2D face profiles induce facial aging effects. Figure 24.9 illustrates the effect of applying the "revised" cardioidal strain transformation model on profile faces. The aforementioned transformation model is said to reflect the remodeling of fluid filled spherical objects with applied pressure. O'Toole *et al* [82] studied the effects of facial wrinkles in increasing the perceived age of faces. Ramanathan and Chellappa [59] developed a Bayesian age-difference classifier with the objective of developing systems that could perform face verification across age progression. The results from many such studies highlight the importance of developing computational models that characterize both growth-related shape variations and textural variations, such as wrinkles and other skin artifacts, in developing a facial aging model.

In this section, we shall present computational models that characterize shape variations that faces undergo during different stages of growth. Facial shape variations due to aging can be observed by means of facial feature drifts and progressive variations in the shape of facial contours, across ages. While facial shape variations during formative years are primarily due to craniofacial growth, during adulthood, facial shape variations are predominantly driven by the changing physical properties of facial muscles. Hence, we propose shape variation models for each of the age groups that best account for the factors that induce such variations.

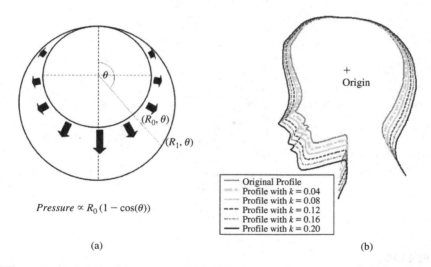

(a)   (b)

FIGURE 24.9

(a) Remodeling of a fluid filled spherical object; (b) facial growth simulated on the profile of a child's face using the "revised" cardioidal strain transformations.

24.4.1 Shape Transformation Model for Young Individuals [60]

Drawing inspiration from the "revised" cardioidal strain transformation model proposed in psychophysics [81], we propose a craniofacial growth model that assumes the underlying form:

$$P \propto R_0(1 - \cos(\theta_0)),$$

$$R_1 = R_0 + k(R_0 - R_0\cos(\theta_0)), \tag{24.21}$$

$$\theta_1 = \theta_0.$$

The model described above characterizes facial feature drifts as caused by internal pressures that are resultants of craniofacial growth. In the above Eq. (24.21), P denotes the pressure at the particular point on the object surface acting radially outward. (R_0, θ_0) and (R_1, θ_1) denote the angular coordinates of a point on the surface of the object before and after the transformation. k denotes a growth-related constant. Face anthropometric studies [83] come in handy in providing dense facial measurements extracted across different facial features across ages. Age-based facial measurements extracted across different facial features play a crucial role in developing the proposed growth model. Figure 24.10

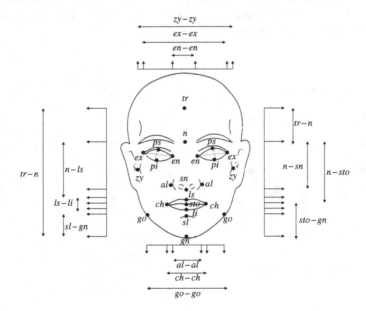

FIGURE 24.10

Face anthropometry: of the 57 facial landmarks defined in [83], we choose 24 landmarks illustrated above for our study. We further illustrate some of the key facial measurements that were used to develop the growth model.

illustrates the 24 facial landmarks and some of the important facial measurements that were used in our study.

Given the pressure model (24.21) and the age-based facial measurements, developing the craniofacial growth model amounts to identifying the growth parameters associated with different facial features. Let the facial growth parameters of the "revised" cardioidal strain transformation model that correspond to facial landmarks designated by $[n, sn, ls, sto, li, sl, gn, en, ex, ps, pi, zy, al, ch, go]$ be $[k_1, k_2, \ldots k_{15}]$. The facial growth parameters for different age transformations can be computed using anthropometric constraints on facial proportions. The computation of facial growth parameters is formulated as a nonlinear optimization problem. We identified 52 facial proportions that can be reliably estimated using the photogrammetry of frontal face images. Anthropometric constraints based on proportion indices translate into linear and nonlinear constraints on selected facial growth parameters. While constraints based on proportion indices such as the *intercanthal index* and *nasal index* result in linear constraints on the growth parameters, constraints based on proportion indices such as the *eye fissure index* and *orbital width index* result in nonlinear constraints on the growth parameters.

Let the constraints derived using proportion indices be denoted as $r_1(\mathbf{k}) = \beta_1, r_2(\mathbf{k}) = \beta_2, \ldots, r_N(\mathbf{k}) = \beta_N$. The objective function $f(\mathbf{k})$ that needs to be minimized w.r.t $\mathbf{k}$ is defined as

$$f(\mathbf{k}) = \frac{1}{2} \sum_{i=1}^{N} (r_i(\mathbf{k}) - \beta_i)^2. \tag{24.22}$$

The following equations illustrate the constraints that were derived using different facial proportion indices.

$$r_1 : \left[\frac{n-gn}{zy-zy} = c_1 \right] \equiv \alpha_1^{(1)} k_1 + \alpha_2^{(1)} k_7 + \alpha_3^{(1)} k_{12} = \beta_1$$

$$r_2 : \left[\frac{al-al}{ch-ch} = c_2 \right] \equiv \alpha_1^{(2)} k_{13} + \alpha_2^{(2)} k_{14} = \beta_2$$

$$r_3 : \left[\frac{li-sl}{sto-sl} = c_3 \right] \equiv \alpha_1^{(3)} k_4 + \alpha_2^{(3)} k_5 + \alpha_3^{(3)} k_6 = \beta_3$$

$$r_4 : \left[\frac{sto-gn}{gn-zy} = c_4 \right] \equiv \alpha_1^{(4)} k_5 + \alpha_2^{(4)} k_7 + \alpha_3^{(4)} k_{12} + \alpha_4^{(4)} k_4^2 + \alpha_5^{(4)} k_7^2$$
$$+ \alpha_6^{(4)} k_{12}^2 + \alpha_7^{(4)} k_4 k_7 + \alpha_8^{(4)} k_7 k_{12} = \beta_4$$

(α_j^i and β_i are constants. c_i is an age-based proportion index obtained from [83].)

We use the Levenberg-Marquardt nonlinear optimization algorithm [84] to compute the growth parameters that minimize the objective function in an iterative fashion. Next, using the growth parameters computed over selected facial landmarks, we compute the growth parameters over the entire face region. This is formulated as a scattered data interpolation problem [85]. Figure 24.11 shows some of the age transformation results obtained using the proposed model.

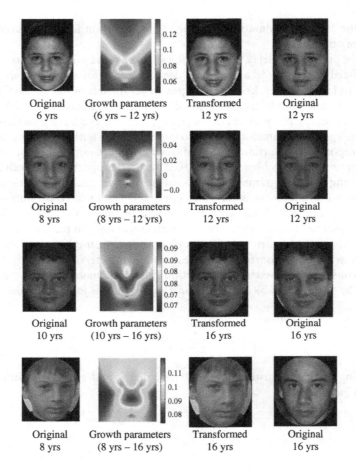

FIGURE 24.11

Age transformation results on different individuals. (The original images shown above were taken from the FG-Net database [86].)

24.4.2 Shape Transformation Model for Adults [61]

We propose a facial shape variation model that represents facial feature deformations observed during adulthood as that driven by the changing physical properties of the underlying facial muscles. The model is based on the assumption that the degrees of freedom associated with facial feature deformations are directly related to the physical properties and geometric orientations of the underlying facial muscles.

$$x_{t_1}^{(i)} = x_{t_0}^{(i)} + k^{(i)} \, [P_{t_0}^{(i)}]_x, \tag{24.23}$$

$$y_{t_1}^{(i)} = y_{t_0}^{(i)} + k^{(i)} \, [P_{t_0}^{(i)}]_y,$$

where $(x_{t_0}^{(i)}, y_{t_0}^{(i)})$ and $(x_{t_1}^{(i)}, y_{t_1}^{(i)})$ correspond to the cartesian coordinates of the ith facial feature at ages t_0 and t_1, $k^{(i)}$ corresponds to a facial growth parameter, and $[P_{t_0}^{(i)}]_x$, $[P_{t_0}^{(i)}]_y$ corresponds to the orthogonal components of the pressure applied on the ith facial feature at age t_0.

We propose a physically based parametric muscle model for human faces that implicitly accounts for the physical properties, geometric orientations, and functionalities of each of the individual facial muscles. Drawing inspiration from Waters' muscle model [87], we identify three types of facial muscles, namely linear muscles, sheet muscles, and sphincter muscles, based on their functionalities. Further, we propose transformation models for each muscle type.

The following factors are to be taken into consideration while developing the pressure models. (i) *Muscle functionality and gravitational forces*: The proposed pressure models reflect the muscle functionalities such as the "stretch" operation and the "contraction" operation. The direction of applied pressure reflects the effects of gravitational forces. (ii) *Points of origin and insertion for each muscle*: The degrees of freedom associated with muscle deformations are minimum at their points of origin (fixed end) and maximum at their points of insertion (free end). Hence, the deformations induced over a facial feature directly depend on the distance of the facial feature from the point of origin of the underlying muscle. The transformation models proposed on each muscle type are illustrated below.

1. Linear muscle (α, ϕ)

   Linear muscles correspond to the "stretch operation." These muscles are described by their attributes namely the muscle length (α) and the muscle orientation w.r.t to the facial axis (ϕ). The farther a feature is from the muscle's point of origin, the greater the chances that the feature undergoes deformation. Hence, the pressure is modeled such that $P^{(i)} \propto \alpha^{(i)}$. $(\alpha^i$ is the distance of the ith feature from the point of origin.) The corresponding shape transformation model is described below:

   $$x_{t_1}^{(i)} = x_{t_0}^{(i)} + k\,[\alpha^{(i)} \sin \phi],$$

   $$y_{t_1}^{(i)} = y_{t_0}^{(i)} + k\,[\alpha^{(i)} \cos \phi].$$

2. Sheet muscle $(\alpha, \phi, \theta, \omega)$

   Sheet muscles correspond to the "stretch operation" as well. They are described by four of their attributes (muscle length, angles subtended, etc.). The pressure applied on a fiducial feature is modeled as $P^{(i)} \propto \alpha^{(i)} \sec \theta^{(i)}$, the distance of the ith feature from the point(s) of origin of the underlying muscles. The shape transformation model is described below:

   $$x_{t_1}^{(i)} = x_{t_0}^{(i)} + k\,[\alpha^{(i)} \sec \theta^{(i)} \sin(\phi + \theta^{(i)})],$$

   $$y_{t_1}^{(i)} = y_{t_0}^{(i)} + k\,[\alpha^{(i)} \sec \theta^{(i)} \cos(\phi + \theta^{(i)})].$$

3. Sphincter muscle (α, β)

The sphincter muscle corresponds to the "contraction/expansion" operation and is described by two attributes. The pressure modeled as a function of the distance from the point of origin, $P^{(i)} \propto r^{(i)}(\phi^{(i)}) \cos \phi^{(i)}$, is directed radially inward/outward:

$$x_{t_1}^{(i)} = x_{t_0}^{(i)} + k\,[r^{(i)}(\phi^{(i)}) \cos^2 \phi^{(i)}],$$

$$y_{t_1}^{(i)} = y_{t_0}^{(i)} + k\,[r^{(i)}(\phi^{(i)}) \cos \phi^{(i)} \sin \phi^{(i)}].$$

Figure 24.12 illustrates the muscle-based pressure distributions described above.

From a database that comprises 1200 pairs of age separated face images (predominantly Caucasian), we selected 50 pairs of face images each undergoing the following age transformations (in years): $20s \rightarrow 30s$, $30s \rightarrow 40s$, $40s \rightarrow 50s$, $50s \rightarrow 60s$, and $60s \rightarrow 70s$. We selected 48 facial features from each image pair and extracted 44 projective measurements (21 horizontal measurements and 23 vertical measurements) across the facial features. We analyze the intrapair shape transformations from the perspective of weight-loss, weight-gain, and weight-retention and select the appropriate training sets for each case. Again, following an approach similar to that described in the previous section, we compute the muscle parameters by studying the transformation of ratios of facial distances across age transformations.

24.4.3 Texture Transformation Model

From a modeling perspective, facial wrinkles and other forms of textural variations observed in aging faces can be characterized on the image domain by means of image gradients. Let $(I_{t_1}^{(i)}, I_{t_2}^{(i)})$, $1 \leq i \leq N$ correspond to pairs of age-separated face images of N individuals undergoing similar age transformations ($t_1 \rightarrow t_2$). In order to study the facial wrinkle variations across age transformation, we identify four facial regions which tend to have a high propensity toward developing wrinkles, namely the forehead region (W_1), the eye-burrow region (W_2), the nasal region (W_3), and the lower chin region (W_4). W_n, $1 \leq n \leq 4$ corresponds to the facial mask that helps isolate the desired facial region. Let $\nabla I_{t_1}^{(i)}$ and $\nabla I_{t_2}^{(i)}$ correspond to the image gradients of the ith image at t_1 and t_2 years, $1 \leq i \leq N$. Given a test image J_{t_1} at t_1 years, the image gradient of which is ∇J_{t_1}, we induce textural variations by incorporating the region-based gradient differences that were learned from the set of training images discussed above:

$$\nabla J_{t_2} = \nabla J_{t_1} + \frac{1}{N} \sum_{i=1}^{N} \sum_{n=1}^{4} W_n \cdot \left(\nabla I_{t_2}^{(i)} - \nabla I_{t_1}^{(i)} \right). \tag{24.24}$$

The transformed image J_{t_2} is obtained by solving the Poisson equation corresponding to image reconstructions from gradient fields [88]. Figure 24.13 provides an overview of the proposed facial aging model.

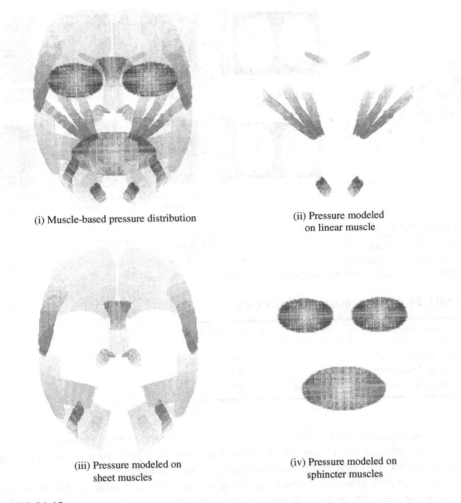

(i) Muscle-based pressure distribution

(ii) Pressure modeled
on linear muscle

(iii) Pressure modeled on
sheet muscles

(iv) Pressure modeled on
sphincter muscles

FIGURE 24.12

Muscle-based pressure illustration.

The proposed facial aging models were used to perform face recognition across age transformations on two databases. The first database was comprised of age-separated face images of individuals under 18 years of age and the second comprised of age-separated face images of adults. On a database that comprises 260 age-separated image pairs of adults, we perform face recognition across age progression. The image pairs were compiled from both the Passport database [59] and the FG-NET database [86]. We adopt PCA to perform recognition across ages under the following three settings: no transformation in shape and texture, performing shape transformation, and performing

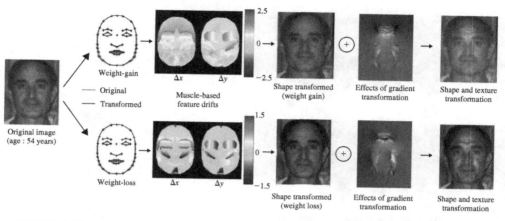

FIGURE 24.13

An overview of the proposed facial aging model: facial shape variations induced for the cases of weight-gain and weight-loss are illustrated. Further, the effects of gradient transformations in inducing textural variations are illustrated as well.

TABLE 24.4 Face recognition across ages.

Experimental Setting	Rank 1 (%)
No transformations	38
Shape transformations	41
Shape and texture transformations	51

shape and textural transformation. Table 24.4 reports the rank 1 recognition score under the three settings. The experimental results highlight the significance of transforming shape and texture when performing face recognition across ages.

A similar performance improvement was observed on the face database that comprises individuals under 18 years of age. For a more detailed account on the experimental results, we refer the readers to our earlier works [60, 61].

24.5 CONCLUSIONS

This chapter presented a hierarchical framework for face pattern and face recognition theory. Current face recognition approaches are classified according to their placements in this framework. We then presented the linear Lambertian object model for face recognition under illumination variation and the illuminating light field algorithm for face recognition under both illumination and pose variations. Finally, we discussed methods for face recognition and modeling across age progression.

REFERENCES

[1] R. Chellappa, C. L. Wilson, and S. Sirohey. Human and machine recognition of faces: a survey. *Proc. IEEE*, 83:705–740, 1995.

[2] S. Z. Li and A. K. Jain, editors. *Handbook of Face Recognition*. Springer-Verlag, 2004.

[3] W. Zhao, R. Chellappa, A. Rosenfeld, and P. J. Phillips. Face recognition: a literature survey. *ACM Comput. Surv.*, 12, 2003.

[4] R. Hietmeyer. Biometric identification promises fast and secure processings of airline passengers. *I. C. A. O. J.*, 55(9):10–11, 2000.

[5] P. J. Phillips, R. M. McCabe, and R. Chellappa. Biometric image processing and recognition. *Proc. European Signal Process. Conf.*, Rhodes, Greece, 1998.

[6] P. J. Phillips, P. Grother, R. J. Micheals, D. M. Blackburn, E. Tabbssi, and M. Bone. Face recognition vendor test 2002: evaluation report. *NISTIR 6965*, http://www.frvt.org, 2003.

[7] P. J. Phillips, H. Moon, S. Rizvi, and P. J. Rauss. The FERET evaluation methodology for face-recognition algorithms. *IEEE Trans. Pattern Anal. Mach. Intell.*, 22:1090–1104, 2000.

[8] A. J. O'Toole. Psychological and neural perspectives on human faces recognition. In S. Z. Li and A. K. Jain, editors, *Handbook of Face Recognition*, Springer, 2004.

[9] R. O. Duda, P. E. Hart, and D. G. Stork. *Pattern Classification*. Wiley-Interscience, 2001.

[10] V. Bruce. *Recognizing Faces*. Lawrence Erlbaum Associates, London, UK, 1988.

[11] M. Kirby and L. Sirovich. Application of Karhunen-Loéve procedure of the characterization of human faces. *IEEE Trans. Pattern Anal. Mach. Intell.*, 12:103–108, 1990.

[12] M. Turk and A. Pentland. Eigenfaces for recognition. *J. Cogn. Neurosci.*, 3:72–86, 1991.

[13] K. Etemad and R. Chellappa. Discriminant analysis for recognition of human face images. *J. Opt. Soc. Am. A*, 14(8):1724–1733, 1997.

[14] P. N. Belhumeur, J. P. Hespanha, and D. J. Kriegman. Eigenfaces vs. Fisherfaces: recognition using class specific linear projection. *IEEE Trans. Pattern Anal. Mach. Intell.*, 19:711–720, 1997.

[15] W. Zhao, R. Chellappa, and A. Krishnaswamy. Discriminant analysis of principal components for face recognition. In *Proc. Int. Conf. Automatic Face and Gesture Recognit.*, 361–341, Nara, Japan, 1998.

[16] M. S. Barlett, H. M. Ladesand, and T. J. Sejnowski. Independent component representations for face recognition. *Proc. Soc. Photo Opt. Instrum. Eng.*, 3299:528–539, 1998.

[17] P. Penev and J. Atick. Local feature analysis: a general statistical theory for object representation. *Netw.: Comput. Neural Syst.*, 7:477–500, 1996.

[18] B. Moghaddam and A. Pentland. Probabilistic visual learning for object representation. *IEEE Trans. Pattern Anal. Mach. Intell.*, PAMI-19(7):696–710, 1997.

[19] B. Moghaddam. Principal manifolds and probabilistic subspaces for visual recognition. *IEEE Trans. Pattern Anal. Mach. Intell.*, 24:780–788, 2002.

[20] S. Zhou and R. Chellappa. Multiple-exemplar discriminant analysis for face recognition. In *Proc. Int. Conf. Pattern Recognit.*, Cambridge, UK, August 2004.

[21] J. Li, S. Zhou, and C. Shekhar. A comparison of subspace analysis for face recognition. In *IEEE Int. Conf. Acoustics, Speech, and Signal Process. (ICASSP)*, Hongkong, China, April 2003.

[22] S. H. Lin, S. Y. Kung, and J. J. Lin. Face recognition/detection by probabilistic decision based neural network. *IEEE Trans. Neural Netw.*, 9:114–132, 1997.

[23] P. J. Phillips. Support vector machines applied to face recognition. *Adv. Neural Inf. Process. Syst.*, 11:803–809, 1998.

[24] C. Liu and H. Wechsler. Evolutionary pursuit and its applications to face recognition. *IEEE Trans. Pattern Anal. Mach. Intell.*, 22:570–582, 2000.

[25] T. Sim, S. Baker, and M. Bast. The CMU pose, illuminatin, and expression (PIE) database. In *Proc. Automatic Face and Gesture Recognit.*, 53–58, Washington, DC, 2002.

[26] A. Shashua. On photometric issues in 3d visual recognition from a single 2d image. *Int. J. Comput. Vis.*, 21:99–122, 1997.

[27] A. L. Yuille, D. Snow, R. Epstein, and P. N. Belhumeur. Determining generative models of objects under varying illumination: shape and albedo from multiple images using svd and integrability. *Int. J. Comput. Vis.*, 35:203–222, 1999.

[28] R. Basri and D. Jacobs. Lambertian reflectance and linear subspaces. *IEEE Trans. Pattern Anal. Mach. Intell.*, 25:218–233, 2003.

[29] W. T. Freeman and J. B. Tenenbaum. Learning bilinear models for two-factor problems in vision. In *Proc. IEEE Comput. Soc. Conf. Comput. Vis. Pattern Recognit.*, Puerto Rico, 1997.

[30] A. Shashua and T. R. Raviv. The quotient image: class based re-rendering and recognition with varying illuminations. *IEEE Trans. Pattern Anal. Mach. Intell.*, 23:129–139, 2001.

[31] S. Zhou, R. Chellappa, and D. Jacobs. Characterization of human faces under illumination variations using rank, integrability, and symmetry constraints. In *European Conf. on Comput. Vis.*, Prague, Czech, May 2004.

[32] T. Cootes, K. Walker, and C. Taylor. View-based active appearance models. In *Proc. of Int. Conf. on Automatic Face and Gesture Recognition*, Grenoble, France, 2000.

[33] R. Gross, I. Matthews, and S. Baker. Eigen light-fields and face recognition across pose. In *Proc. Int. Conf. Automatic Face and Gesture Recognit.*, Washington, DC, 2002.

[34] A. Pentland, B. Moghaddam, and T. Starner. View-based and modular eigenspaces for face recognition. In *Proc. of IEEE Computer Society Conf. on Computer Vision and Pattern Recognition*, Seattle, WA, 1994.

[35] V. Blanz and T. Vetter. Face recognition based on fitting a 3D morphable model. *IEEE Trans. Pattern Anal. Mach. Intell.*, 25:1063–1074, 2003.

[36] A. S. Georghiades, P. N. Belhumeur, and D. J. Kriegman. From few to many: illumination cone models for face recognition under variable lighting and pose. *IEEE Trans. Pattern Anal. Mach. Intell.*, 23:643–660, 2001.

[37] R. Gross, I. Matthews, and S. Baker. Fisher light-fields for face recognition across pose and illumination. In *Proc. of the German Symposium on Pattern Recognit.*, 2002.

[38] M. Vasilescu and D. Terzopoulos. Multilinear image analysis for facial recognition. In *Proc. of Int. Conf. on Pattern Recognit.*, Quebec City, Canada, 2002.

[39] S. Zhou and R. Chellappa. Illuminating light field: image-based face recognition across illuminations and poses. In *Proc. of Int. Conf. on Automatic Face and Gesture Recognit.*, Seoul, Korea, May 2004.

[40] M. Lades, J. C. Vorbruggen, J. Buhmann, J. Lange, C. V. D. Malsburg, R. P. Wurtz, and W. Konen. Distortion invariant object recognition in the dynamic link architecture. *IEEE Trans. Comput.*, 42(3):300–311, 1993.

[41] B. Knight and A. Johnston. The role of movement in face recognition. *Vis. Cogn.*, 4:265–274, 1997.

[42] B. Li and R. Chellappa. A generic approach to simultaneous tracking and verification in video. *IEEE Trans. Image Process.*, 11(5):530–554, 2002.

[43] V. Krueger and S. Zhou. Exemplar-based face recognition from video. In *European Conference on Computer Vision*, Copenhagen, Denmark, 2002.

[44] S. Zhou and R. Chellappa. Probabilistic human recognition from video. In *European Conf. on Comput. Vis.*, Vol. 3, 681–697, Copenhagen, Denmark, May 2002.

[45] S. Zhou, V. Krueger, and R. Chellappa. Probabilistic recognition of human faces from video. *Comput. Vis. Image Underst.*, 91:214–245, 2003.

[46] S. Zhou, R. Chellappa, and B. Moghaddam. Visual tracking and recognition using appearance-adaptive models in particle filters. *IEEE Trans. Image Process.*, 11:1434–1456, 2004.

[47] M. J. Lyons, J. Biudynek, and S. Akamatsu. Automatic classification of single facial images. *IEEE Trans. Pattern Anal. Mach. Intell.*, 21(12):1357–1362, 1999.

[48] A. Lanitis, C. J. Taylor, and T. F. Cootes. Toward automatic simulation of aging affects on face images. *IEEE Trans. Pattern Anal. Mach. Intell.*, 24:442–455, 2002.

[49] M. J. Black and Y. Yacoob. Recognizing facial expressions in image sequences using local parameterized models of image motion. *Int. J. Comput. Vis.*, 25:23–48, 1997.

[50] Y. Tian, T. Kanade, and J. Cohn. Recognizing action units of facial expression analysis. *IEEE Trans. Pattern Anal. Mach. Intell.*, 23:1–19, 2001.

[51] I. Shimshoni, Y. Moses, and M. Lindenbaum. Shape reconstruction of 3D bilaterally symmetric surfaces. *Int. J. Comput. Vis.*, 39:97–100, 2000.

[52] W. Zhao and R. Chellappa. Symmetric shape from shading using self-ratio image. *Int. J. Comput. Vis.*, 45:55–752, 2001.

[53] R. T. Frankot and R. Chellappa. A method for enforcing integrability in shape from shaging problem. *IEEE Trans. Pattern Anal. Mach. Intell.*, 10:439–451, 1988.

[54] T. Kanade. *Computer Recognition of Human Faces.* Birhauser, Basel, Switzerland, and Stuggart, Germany, 1973.

[55] M. D. Kelly. Visual identification of people by computer. *Tech. Rep. AI-130*, Stanford AI project, Stanform, CA, 1970.

[56] T. Vetter and T. Poggio. Linear object classes and image synthesis from a single example image. *IEEE Trans. Pattern Anal. Mach. Intell.*, 11:733–742, 1997.

[57] M. Ramachandran, S. Zhou, D. Jhalani, and R. Chellappa. A method for converting smiling face to a neutral face with applications to face recognition. In *IEEE Int. Conf. Acoustics, Speech, and Signal Process. (ICASSP)*, Philadelphia, USA, March 2005.

[58] H. Ling, S. Soatto, N. Ramanathan, and D. Jacobs. A study of face recognition as people age. In *Proc. IEEE Int. Conf. on Comput. Vis. (ICCV)*, Rio De Janeiro, October 2007.

[59] N. Ramanathan and R. Chellappa. Face verification across age progression. *IEEE Trans. Image. Process.*, 15(11):3349–3362, 2006.

[60] N. Ramanathan and R. Chellappa. Modeling age progression in young faces. In *Proc. IEEE Comput. Vis. Pattern Recognit. (CVPR)*, Vol. 1, 387–394, New York, June 2006.

[61] N. Ramanathan and R. Chellappa. Modeling shape and textural variations in aging adult faces. In *IEEE Conf. Automatic Face and Gesture Recognition*, 2008.

[62] S. Zhou, G. Aggarwal, R. Chellappa, and D. Jacobs. Appearance characterization of linear Lambertian objects, generalized photometric stereo and illumination-invariant face recognition. *IEEE Trans. Pattern Anal. Mach. Intell.*, 29:230–245, 2007.

[63] S. Zhou and R. Chellappa. Image-based face recognition under illumination and pose variations. *J. Opt. Soc. Am. A*, 22:217–229, 2005.

[64] H. Hayakawa. Photometric stereo under a light source with arbitrary motion. *J. Opt. Soc. Am. A*, 11(11):3079–3089, 1994.

[65] L. Zhang and D. Samaras. Face recognition under variable lighting using harmonic image exemplars. In *IEEE Conf. Comput. Vis. Pattern Recognit.*, 1925, 2003.

[66] J. Atick, P. Griffin, and A. Redlich. Statistical approach to shape from shading: reconstruction of 3-dimensional face surfaces from single 2-dimentional images. *Neural Comput.*, 8:1321–1340, 1996.

[67] K. C. Lee, J. Ho, and D. Kriegman. Nine points of light: acquiring subspaces for face recognition under variable lighting. In *IEEE Conf. Comput. Vis. Pattern Recognit.*, 519–526, December 2001.

[68] P. Fua. Regularized bundle adjustment to model heads from image sequences without calibrated data. *Int. J. Comput. Vis.*, 38:153–157, 2000.

[69] Y. Shan, Z. Liu, and Z. Zhang. Model-based bundle adjustment with application to face modeling. In *Proc. Int. Conf. Comput. Vis.*, 645651, 2001.

[70] A. R. Chowdhury and R. Chellappa. Face reconstruction from video using uncertainty analysis and a generic model. *Comput. Vis. Image Underst.*, 91:188–213, 2003.

[71] G. Qian and R. Chellappa. Structure from motion using sequential Monte Carlo methods. In *Proc. IEEE Int. Conf. Comput. Vis.*, 2:614–621, Vancouver, Canada, 2001.

[72] A. Laurentini. The visual hull concept for silhouette-based image understanding. *IEEE Trans. Pattern Anal. Mach. Intell.*, 16(2):150–162, 1994.

[73] W. Matusik, C. Buehler, R. Raskar, S. Gortler, and L. McMillan. Image-based visual hulls. In *Proc. SIGGRAPH*, 369–374, New Orleans, LA, USA, 2000.

[74] M. Levoy and P. Hanrahan. Light field rendering. In *Proc. SIGGRAPH*, 31–42, New Orleans, LA, USA, 1996.

[75] S. J. Gortler, R. Grzeszczuk, R. Szeliski, and M. Cohen. The lumigraph. In *Proc. SIGGRAPH*, 43–54, New Orleans, LA, USA, 1996.

[76] M. A. O. Vasilescu and D. Terzopoulos. Multilinear analysis of image ensembles: Tensorfaces, In *European Conf. Comput. Vis.*, 2350:447–460, Copenhagen, Denmark, May 2002.

[77] S. Zhou and R. Chellappa. Rank constrained recognition under unknown illumination. In *IEEE Int. Workshop on Analysis and Modeling of Faces and Gestures*, 2003.

[78] S. Romdhani and T. Vetter. Efficient, robust and accurate fitting of a 3D morphable model. In *Proc. IEEE Int. Conf. Comput. Vis.*, 59–66, Nice, France, 2003.

[79] A. Lanitis, C. J. Taylor, and T. F. Cootes. Automatic interpretation and coding of face images using flexible models. *IEEE Trans. Pattern Anal. Mach. Intell.*, 19(7):743–756, 1997.

[80] J. B. Pittenger and R. E. Shaw. Aging faces as viscal-elastic events: Implications for a theory of nonrigid shape perception. *J. Exp. Psychol. Hum. Percept. Perform.*, 1(4):374–382, 1975.

[81] J. T. Todd, L. S. Mark, R. E. Shaw, and J. B. Pittenger. The perception of human growth. *Sci. Am.*, 242(2):132–144, 1980.

[82] A. J. O'Toole, T. Vetter, H. Volz, and E. M. Salter. Three-dimensional caricatures of human heads: distinctiveness and the perception of facial age. *Perception*, 26:719–732, 1997.

[83] L. G. Farkas. *Anthropometry of the Head and Face*. Raven Press, New York, 1994.

[84] D. M. Bates and D. G. Watts. *Nonlinear Regression and its Applications*. Wiley, New York, 1988.

[85] Fred L. Bookstein. Principal warps: thin-plate splines and the decomposition of deformations. *IEEE Trans. Pattern Anal. Mach. Intell.*, 11(6):567–585, 1989.

[86] A. Lanitis. FG-Net aging database.

[87] K. Waters. A muscle model for animating three-dimensional facial expression. *Comput. Graph. (ACM)*, 21:17–24, 1987.

[88] P. Perez, M. Gangnet, and A. Blake. Poisson image editing. *Comput. Graph. (ACM)*, 313–318, 2003.

How Iris Recognition Works 25

John Daugman

University of Cambridge

Algorithms developed by the author for recognizing persons by their iris patterns [1–3] are the basis of all current public deployments of iris-based automated biometric identification. To date, many millions of persons across many countries have enrolled with this system, most commonly for expedited border crossing in lieu of passport presentation but also in government security watch-list screening programs. The recognition principle is the failure of a test of statistical independence on iris phase structure, as encoded by multiscale quadrature Gabor wavelets. The combinatorial complexity of this phase information across different persons spans about 249 degrees of freedom and generates a discrimination entropy of about 3.2 bits/mm^2 over the iris, enabling real-time decisions about personal identity with extremely high confidence. These high confidence levels are important because they allow very large databases to be searched exhaustively (one-to-many "identification mode"), without making false matches, despite so many chances. Biometrics that lack this property can only survive one-to-one ("verification") or few comparisons. This chapter explains the iris recognition algorithms and presents results of 9.1 million comparisons among eye images from trials in Britain, the USA, Japan, and Korea.

25.1 INTRODUCTION

Reliable automatic recognition of persons has long been an attractive goal. As in all pattern recognition problems, the key issue is the relation between interclass and intraclass variability: objects can be reliably classified only if the variation among different instances of a given class is less than the variation between different classes. For example in face recognition, difficulties arise from the fact that the face is a changeable social organ displaying a variety of expressions, as well as being an active 3D object whose image varies with viewing angle, pose, illumination, accoutrements, and age [4, 5]. It has been shown that even for "mug shot" (pose-invariant) images taken at least one year apart, even the best algorithms have unacceptably large error rates [6–8]. Against this intraclass (same face) variation, interclass variation is limited because different faces possess the same basic set of features, in the same canonical geometry.

715

Following the fundamental principle that interclass variation should be larger than intraclass variation, iris patterns offer a powerful alternative approach to reliable visual recognition of persons when imaging can be done at distances of about a meter or less, and especially when there is a need to search very large databases without incurring any false matches despite a huge number of possibilities. Although small (11 mm diameter) and sometimes problematic to image, the iris has the great mathematical advantage that its pattern variability among different persons is enormous. In addition, as an internal (yet externally visible) organ of the eye, the iris is well protected from the environment and stable over time. As a planar object its image is relatively insensitive to angle of illumination, and changes in viewing angle cause only affine transformations; even the nonaffine pattern distortion caused by pupillary dilation is readily reversible in the image coding stage. Finally, the ease of localizing eyes in faces, and the distinctive annular shape of the iris, facilitates reliable and precise isolation of this feature and the creation of a size-invariant representation.

The iris begins to form in the third month of gestation [9] and the structures creating its pattern are largely complete by the eighth month, although pigment accretion can continue into the first postnatal years. Its complex pattern can contain many distinctive features such as arching ligaments, furrows, ridges, crypts, rings, corona, freckles, and a zigzag collarette, some of which may be seen in Fig. 25.1. Iris color is determined mainly by the density of melanin pigment [10] in its anterior layer and stroma, with blue irises resulting from an absence of pigment: long wavelength light penetrates while shorter wavelengths are scattered by the stroma. The striated trabecular meshwork of elastic pectinate ligament creates the predominant texture under visible light, whereas in the near infrared (NIR) wavelengths used for unobtrusive imaging, deeper and somewhat more slowly modulated stromal features dominate the iris pattern. In NIR wavelengths, even darkly pigmented irises reveal rich and complex features.

25.2 LOCALIZING THE IRIS AND ITS BOUNDARIES

To capture the rich details of iris patterns, an imaging system should resolve a minimum of 70 pixels in iris radius. In most deployments of these algorithms to date, the resolved iris radius has typically been 80 to 130 pixels. Monochrome CCD cameras (480 × 640) have been used because NIR illumination in the 700 nm–900 nm band was required for imaging to be unintrusive to humans. Some imaging platforms deployed a wide-angle camera for coarse localization of eyes in faces to steer the optics of a narrow-angle pan/tilt camera that acquired higher resolution images of eyes. There exist many alternative methods for finding and tracking facial features such as the eyes, and this well-researched topic will not be discussed further here. Most images in the present database were acquired without active pan/tilt camera optics, instead exploiting visual feedback via a mirror or video image to enable cooperating subjects to position their own eyes within the field of view of a single narrow-angle camera.

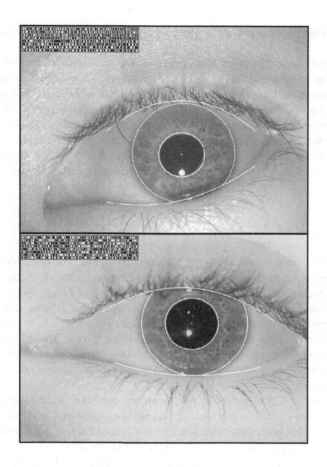

FIGURE 25.1

Examples of human iris patterns, imaged monochromatically at a distance of about 35 cm. The outline overlays show the results of the iris and pupil localization and eyelid detection steps. One cannot assume that the iris boundaries are concentric nor even circular. The bit streams pictured are the result of demodulation with complex-valued 2D Gabor wavelets to encode the phase sequence of each iris pattern.

Image focus assessment is performed in real-time (faster than video frame rate) by measuring spectral power in middle and upper frequency bands of the 2D Fourier spectrum of each image frame and seeking to maximize this quantity either by moving an active lens or by providing audio feedback to subjects to adjust their range appropriately. The video rate execution speed of focus assessment (i.e., within a few milliseconds on an ARM device) is achieved by using a bandpass 2D filter kernel requiring only summation and differencing of pixels, and no multiplications, within the 2D convolution necessary to estimate power in the selected 2D spectral bands. Details are provided in Section 25.9 (Appendix).

Images satisfying a minimum focus criterion are then analyzed to find the iris, with initially circular approximations of its boundaries using a coarse-to-fine strategy terminating in estimates of the center coordinates and radius of both the iris and the pupil. Although the results of the iris search greatly constrain the pupil search, concentricity of these boundaries cannot be assumed. Very often the pupil center is nasal, and inferior, to the iris center. Its radius can range from 0.1 to 0.8 of the iris radius. Thus, all three parameters defining the pupil when approximated as a circle must be estimated separately from those of the iris. A very effective integrodifferential operator for determining these parameters is:

$$\max_{(r,x_0,y_0)} \left| G_\sigma(r) * \frac{\partial}{\partial r} \oint_{r,x_0,y_0} \frac{I(x,y)}{2\pi r} ds \right|, \tag{25.1}$$

where $I(x,y)$ is an image such as Fig. 25.1 containing an eye. The operator searches over the image domain (x,y) for the maximum in the blurred partial derivative with respect to increasing radius r of the normalized contour integral of $I(x,y)$ along a circular arc ds of radius r and center coordinates (x_0,y_0). The symbol $*$ denotes convolution, and $G_\sigma(r)$ is a smoothing function such as a Gaussian of scale σ. The complete operator behaves as a circular edge detector, blurred at a scale set by σ, searching iteratively for the maximal contour integral derivative at successively finer scales of analysis through the three parameter space of center coordinates and radius (x_0,y_0,r) defining a path of contour integration.

The operator in (25.1) serves to find both the pupillary boundary and the outer (limbus) boundary of the iris in a mutually reinforcing manner. Once the coarse-to-fine iterative searches for both these boundaries have reached single pixel precision, then a similar approach to detecting curvilinear edges is used to localize both the upper and lower eyelid boundaries. The path of contour integration in (25.1) is changed from circular to arcuate, with spline parameters fitted by statistical estimation methods to model each eyelid boundary. Images with less than 50% of the iris visible between the fitted eyelid splines are deemed inadequate, e.g., in blink. The result of all these localization operations is the isolation of iris tissue from other image regions, as illustrated in Fig. 25.1 by the graphical overlay on the eye.

Because the inner and outer boundaries of the iris are often not actually circles, performance in iris recognition is significantly improved by relaxing both of those assumptions, replacing them with more disciplined methods for faithfully detecting and modeling those boundaries whatever their shapes, and defining a more flexible and generalized coordinate system on their basis. Because the iris outer boundary is often partly occluded by eyelids, and the iris inner boundary may be partly occluded by reflections from illumination, and sometimes both boundaries may also be partly occluded by reflections from eyeglasses, it is necessary to fit flexible contours that can tolerate interruptions and continue their trajectory across them on a principled basis, driven somehow by the data that exists elsewhere. A further constraint is that both the inner and outer boundary models must form closed curves. A final goal is that we would like to impose a constraint on smoothness, based on the credibility of any evidence for nonsmooth curvature.

An excellent way to achieve all these goals is to describe the iris inner and outer boundaries in terms of "active contours" based on discrete Fourier series expansions of the contour data. By employing Fourier components whose frequencies are integer multiples of $1/(2\pi)$, closure, orthogonality, and completeness are ensured. Selecting the number of frequency components allows control over the degree of smoothness that is imposed and over the fidelity of the approximation. In essence, truncating the discrete Fourier series after a certain number of terms amounts to lowpass filtering the boundary curvature data in the active contour model. The estimation procedure is to compute a Fourier expansion of N regularly-spaced angular samples of radial gradient edge data $\{r_\theta\}$ for $\theta = 0$ to $\theta = N - 1$. A set of M discrete Fourier coefficients $\{C_k\}$, for $k = 0$ to $k = M - 1$, are computed from the data sequence $\{r_\theta\}$ as follows:

$$C_k = \sum_{\theta=0}^{N-1} r_\theta e^{-2\pi i k\theta/N}.$$

Note that the zeroth-order coefficient or "DC term" C_0 extracts information about the average curvature of the (pupil or outer iris) boundary, in other words, about its radius when it is approximated just as a simple circle.

From these M discrete Fourier coefficients, an approximation to the corresponding iris boundary (now without interruptions, and at a resolution determined by M) is obtained as the new sequence $\{R_\theta\}$ for $\theta = 0$ to $\theta = N - 1$:

$$R_\theta = \frac{1}{N} \sum_{k=0}^{M-1} C_k e^{2\pi i k\theta/N}.$$

As is generally true of active contour methods, there is a tradeoff between how precisely one wants the model to fit all the data (improved by increasing M), versus how much one wishes to impose constraints such as keeping the model simple and of low-dimensional curvature (achieved by reducing M, for example $M = 1$ enforces a circular model). Thus the number M of activated Fourier coefficients is a specification for the number of degrees-of-freedom in the shape model. A good choice of M for capturing the true pupil boundary with appropriate fidelity is $M = 17$, whereas a good choice for the iris outer boundary where the data is often much weaker is $M = 5$. It is also useful to impose monotonically decreasing weights on the computed Fourier coefficients $\{C_k\}$ as a further control on the resolution of the approximation $\{R_\theta\} \approx \{r_\theta\}$, which amounts to lowpass filtering the curvature map in its Fourier representation. Altogether these manipulations, particularly the two different choices for M, implement the computer vision principle that strong data (the pupil boundary) may be modeled with only weak contraints, whereas weak data (the outer boundary) should be modeled with strong constraints.

The active contour models for the inner and outer iris boundaries support an isometric mapping of the iris tissue between them, regardless of the actual shapes of the contours, and with robustness to gaps such as are caused by eyelid interruptions. This dimensionless, pseudopolar mapping and its invariances for recognition will be discussed in Section 25.5.

25.3 IRIS FEATURE ENCODING BY 2D GABOR WAVELET DEMODULATION

Each isolated iris pattern is then demodulated to extract its phase information using quadrature 2D Gabor wavelets [11–13]. This encoding process is illustrated in Fig. 25.2. It amounts to a patchwise phase quantization of the iris pattern, by identifying in which quadrant of the complex plane each resultant phasor lies when a given area of the iris is projected onto complex-valued 2D Gabor wavelets:

$$h_{\{Re,Im\}} = sgn_{\{Re,Im\}} \int_{\rho} \int_{\phi} I(\rho,\phi) e^{-i\omega(\theta_0 - \phi)} \cdot e^{-(r_0 - \rho)^2/\alpha^2} e^{-(\theta_0 - \phi)^2/\beta^2} \rho d\rho d\phi, \quad (25.2)$$

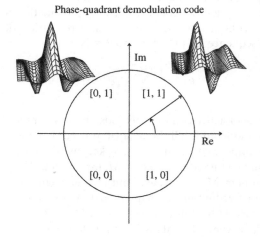

Phase-quadrant demodulation code

FIGURE 25.2

The phase demodulation process used to encode iris patterns. Local regions of an iris are projected (25.2) onto quadrature 2D Gabor wavelets, generating complex-valued coefficients whose real and imaginary parts specify the coordinates of a phasor in the complex plane. The angle of each phasor is quantized to one of the four quadrants, setting 2 bits of phase information. This process is repeated all across the iris with many wavelet sizes, frequencies, and orientations to extract 2,048 bits.

where $h_{\{Re,Im\}}$ can be regarded as a complex-valued bit whose real and imaginary parts are either 1 or 0 (sgn) depending on the sign of the 2D integral; $I(\rho,\phi)$ is the raw iris image in a dimensionless polar coordinate system that is size- and translation-invariant, and which also corrects for pupil dilation as explained in a later section; α and β are the multiscale 2D wavelet size parameters, spanning an 8-fold range from 0.15 mm to 1.2 mm on the iris; ω is wavelet frequency, spanning 3 octaves in inverse proportion to β; and (r_0, θ_0) represent the polar coordinates of each region of iris for which the phasor coordinates $h_{\{Re,Im\}}$ are computed. Such phase quadrant coding sequences are illustrated for two irises by the bit streams shown graphically in Fig. 25.1. A desirable feature of the phase code definition given in Fig. 25.2 is that it is a cyclic or gray code: in rotating between any adjacent phase quadrants, only a single bit changes, unlike a binary code in which 2 bits may change, making some errors arbitrarily more costly than others. Altogether 2048 such phase bits (256 bytes) are computed for each iris, but in a major improvement over the author's earlier [1] algorithms, now an equal number of masking bits are also computed to signify whether any iris region is obscured by eyelids, contains any eyelash occlusions, specular reflections, boundary artifacts of hard contact lenses, or contains a poor signal-to-noise ratio and thus should be ignored in the demodulation code as artifact.

The 2D Gabor wavelets were chosen for the extraction of iris information because of the nice optimality properties of these wavelets. Following the Heisenberg Uncertainty Principle as it applies generally to mathematical functions, filters that are well localized in frequency are poorly localized in space (or time), and *vice versa*. The 2D Gabor wavelets have the maximal joint resolution in the two domains simultaneously [11, 12], which means that both "what" and "where" information about iris features is extracted with optimal simultaneous resolution. A further nice property of 2D Gabor wavelets is that because they are complex-valued, they allow the definition and assignment of phase variables to any point in the image.

Only phase information is used for recognizing irises because amplitude information is not very discriminating, and it depends upon extraneous factors such as imaging contrast, illumination, and camera gain. The phase bit settings which code the sequence of projection quadrants as shown in Fig. 25.2 capture the information of wavelet zero-crossings, as is clear from the sign operator in (25.2). The extraction of phase has the further advantage that phase angles remain defined regardless of how poor the image contrast may be, as illustrated by the extremely out-of-focus image in Fig. 25.3. Its phase bit stream has statistical properties such as run lengths similar to those of the codes for the properly focused eye images in Fig. 25.1. (Fig. 25.3 also illustrates the robustness of the iris- and pupil-finding operators and the eyelid detection operators, despite poor focus.) The benefit which arises from the fact that phase bits are set also for a poorly focused image as shown here, even if based only on random CCD thermal noise, is that different poorly focused irises never become confused with each other when their phase codes are compared. By contrast, images of different faces look increasingly alike when poorly resolved, and can be confused with each other by appearance-based face recognition algorithms.

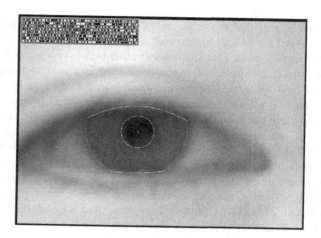

FIGURE 25.3

Illustration that even for poorly focused eye images, the bits of a demodulation phase sequence are still set, primarily by random CCD noise. This prevents poorly focused eye images from being falsely matched, as they may be in amplitude-based representations.

25.4 THE TEST OF STATISTICAL INDEPENDENCE: COMBINATORICS OF PHASE SEQUENCES

The key to iris recognition is the failure of a test of statistical independence, which involves so many degrees-of-freedom that this test is virtually guaranteed to be passed whenever the phase codes for two different eyes are compared, but to be uniquely failed when any eye's phase code is compared with another version of itself.

The test of statistical independence is implemented by the simple Boolean Exclusive-OR operator (XOR) applied to the 2048 bit phase vectors that encode any two iris patterns, masked (AND'ed) by both of their corresponding mask bit vectors to prevent noniris artifacts from influencing iris comparisons. The XOR operator $\otimes$ detects disagreement between any corresponding pair of bits, while the AND operator $\cap$ ensures that the compared bits are both deemed to have been uncorrupted by eyelashes, eyelids, specular reflections, or other noise. The norms ($\| \ \ \|$) of the resultant bit vector and of the AND'ed mask vectors are then measured in order to compute a fractional Hamming Distance (HD) as the measure of the dissimilarity between any two irises, whose two phase code bit vectors are denoted {codeA, codeB} and whose mask bit vectors are denoted {maskA, maskB}:

$$HD = \frac{\|(codeA \otimes codeB) \cap maskA \cap maskB\|}{\|maskA \cap maskB\|}. \tag{25.3}$$

The denominator tallies the total number of phase bits that mattered in iris comparisons after artifacts such as eyelashes and specular reflections were discounted, so the resulting

HD is a fractional measure of dissimilarity; 0 would represent a perfect match. The Boolean operators $\otimes$ and $\cap$ are applied in vector form to binary strings of length up to the word length of the CPU, as a single machine instruction. Thus for example on an ordinary 32-bit machine, any two integers between 0 and 4 billion can be XOR'ed in a single machine instruction to generate a third such integer, each of whose bits in a binary expansion is the XOR of the corresponding pair of bits of the original two integers. This implementation of (25.3) in parallel 32-bit chunks enables extremely rapid comparisons of iris codes when searching through a large database to find a match. On a 300 MHz CPU, such exhaustive searches are performed at a rate of about 100,000 irises per second; on a 3 GHz server, about a million iris comparisons can be performed per second.

Because any given bit in the phase code for an iris is equally likely to be 1 or 0, and different irises are uncorrelated, the expected proportion of agreeing bits between the codes for two different irises is HD = 0.500. The histogram in Fig. 25.4 shows the distribution of HDs obtained from 9.1 million comparisons between different pairings

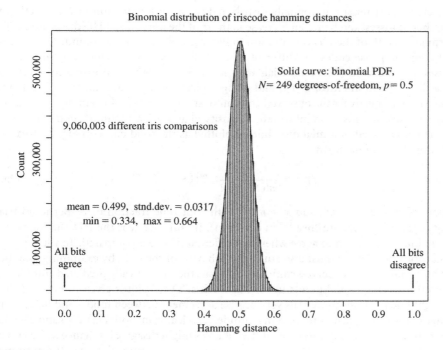

Binomial distribution of iriscode hamming distances

Solid curve: binomial PDF,
$N = 249$ degrees-of-freedom, $p = 0.5$

9,060,003 different iris comparisons

mean = 0.499, stnd.dev. = 0.0317
min = 0.334, max = 0.664

All bits
agree

All bits
disagree

Hamming distance

FIGURE 25.4

Distribution of Hamming distances from all 9.1 million possible comparisons between different pairs of irises in the database. The histogram forms a perfect binomial distribution with $p = 0.5$ and $N = 249$ degrees-of-freedom, as shown by the solid curve (25.4). The data implies that it is extremely improbable for two different irises to disagree in less than about a third of their phase information.

of iris images acquired by licensees of these algorithms in the UK, the USA, Japan, and Korea. There were 4258 different iris images, including 10 each of one subset of 70 eyes. Excluding those duplicates of (700×9) same-eye comparisons, not double-counting pairs, and not comparing any image with itself, the total number of unique pairings between different eye images whose HDs could be computed was $((4258 \times 4257 - 700 \times 9)/2) = 9,060,003$. Their observed mean HD was $p = 0.499$ with standard deviation $\sigma = 0.0317$; their full distribution in Fig. 25.4 corresponds to a binomial having $N = p(1 - p)/\sigma^2 = 249$ degrees-of-freedom, as shown by the solid curve. The extremely close fit of the theoretical binomial to the observed distribution is a consequence of the fact that each comparison between two phase code bits from two different irises is essentially a Bernoulli trial, albeit with correlations between successive "coin tosses."

In the phase code for any given iris, only small subsets of bits are mutually independent due to the internal correlations, especially radial, within an iris. (If all $N = 2048$ phase bits were independent, then the distribution in Fig. 25.4 would be very much sharper, with an expected standard deviation of only $\sqrt{p(1 - p)/N} = 0.011$, and so the HD interval between 0.49 and 0.51 would contain most of the distribution.) Bernoulli trials that are correlated [14] remain binomially distributed but with a reduction in N, the effective number of tosses, and hence an increase in the σ of the normalized HD distribution. The form and width of the HD distribution in Fig. 25.4 tell us that the amount of difference between the phase codes for different irises is distributed equivalently to runs of 249 tosses of a fair coin (Bernoulli trials with $p = 0.5, N = 249$). Expressing this variation as a discrimination entropy [15] and using typical iris and pupil diameters of 11 mm and 5 mm, respectively, the observed amount of statistical variability among different iris patterns corresponds to an information density of about 3.2 bits/mm^2 on the iris.

The theoretical binomial distribution plotted as the solid curve in Fig. 25.4 has the fractional functional form

$$f(x) = \frac{N!}{m!(N - m)!} \, p^m (1 - p)^{(N-m)}, \tag{25.4}$$

where $N = 249$, $p = 0.5$, and $x = m/N$ is the outcome fraction of N Bernoulli trials (e.g., coin tosses that are "heads" in each run). In our case, x is the HD, the fraction of phase bits that happen to agree when two different irises are compared. To validate such a statistical model, we must also study the behavior of the tails, by examining quantile-quantile plots of the observed cumulatives versus the theoretically predicted cumulatives from 0 up to sequential points in the tail. Such a "Q-Q" plot is given in Fig. 25.5. The straight line relationship reveals very precise agreement between model and data, over a range of more than three orders of magnitude. It is clear from both Figs. 25.4 and 25.5 that it is extremely improbable that two different irises might disagree by chance in fewer than at least a third of their bits. (Of the 9.1 million iris comparisons plotted in the histogram of Fig. 25.4, the smallest HD observed was 0.334.) Computing the cumulative of $f(x)$ from 0 to 0.333 indicates that the probability of such an event is about 1 in 16 million. The cumulative from 0 to just 0.300 is 1 in 10 billion. Thus, even the observation of a relatively poor degree of match between the phase codes for two different iris images (say,

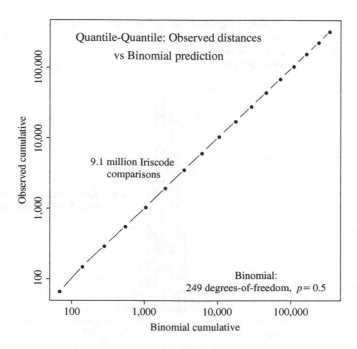

FIGURE 25.5

Quantile-quantile plot of the observed cumulatives under the left tail of the histogram in Fig. 25.4 versus the predicted binomial cumulatives. The close agreement over several orders of magnitude strongly confirms the binomial model for phase bit comparisons between different irises.

70% agreement or HD = 0.300) would still provide extraordinarily compelling evidence of identity, because the test of statistical independence is still failed so convincingly.

Genetically identical eyes were compared in the same manner, in order to discover the degree to which their textural patterns were correlated and hence genetically determined. A convenient source of genetically identical irises is the right and left pair from any given person; such pairs have the same genetic relationship as the four irises of monozygotic twins or indeed the prospective $2N$ irises of N clones. Although eye color is of course strongly determined genetically, as is overall iris appearance, the detailed patterns of genetically identical irises appear to be as uncorrelated as they are among unrelated eyes. Using the same methods as described above, 648 right/left iris pairs from 324 persons were compared pairwise. Their mean HD was 0.497 with standard deviation 0.031, and their distribution (Fig. 25.6) was statistically indistinguishable from the distribution for unrelated eyes (Fig. 25.4). A set of six pairwise comparisons among the eyes of actual monozygotic twins also yielded a result (mean $HD = 0.507$) expected for unrelated eyes. It appears that the phenotypic random patterns visible in the human iris are almost entirely epigenetic [16].

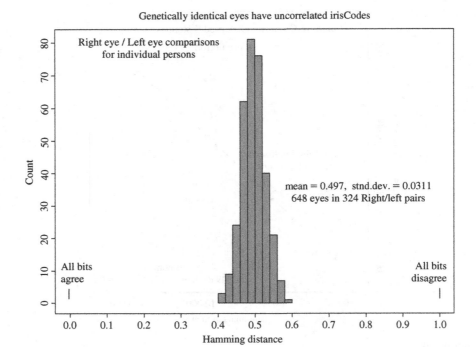

FIGURE 25.6

Distribution of Hamming distances between genetically identical irises in 648 paired eyes from 324 persons. The data are statistically indistinguishable from that shown in Fig. 25.4 comparing unrelated irises. Unlike eye color, the phase structure of iris patterns therefore appears to be epigenetic, arising from random events and circumstances in the morphogenesis of this tissue.

25.5 RECOGNIZING IRISES REGARDLESS OF SIZE, POSITION, AND ORIENTATION

Robust representations for pattern recognition must be invariant to changes in the size, position, and orientation of the patterns. In the case of iris recognition, this means we must create a representation that is invariant to the optical size of the iris in the image (which depends upon the distance to the eye and the camera optical magnification factor); the size of the pupil within the iris (which introduces a nonaffine pattern deformation); the location of the iris within the image; and the iris orientation, which depends upon head tilt, torsional eye rotation within its socket (cyclovergence), and camera angles, compounded with imaging through pan/tilt eye-finding mirrors that introduce additional image rotation factors as a function of eye position, camera position, and mirror angles. Fortunately, invariance to all these factors can readily be achieved.

For on-axis but possibly rotated iris images, it is natural to use a projected pseudopolar coordinate system. The coordinate grid is not necessarily polar nor symmetric, since circular boundaries are not assumed and in most eyes the pupil is not central in the iris; it is not unusual for its nasal displacement to be as much as 15%. This coordinate system can be described as doubly-dimensionless: the angular variable is inherently dimensionless, but in this case the radial variable is also dimensionless, because it ranges from the pupillary boundary to the limbus always as a unit interval $[0, 1]$. The dilation and constriction of the elastic meshwork of the iris when the pupil changes size is intrinsically modeled by this coordinate system as the stretching of a homogeneous rubber sheet, having the topology of an annulus anchored along its outer perimeter, with tension controlled by an (off-centered) interior ring of variable radius.

The homogeneous rubber sheet model assigns to each point on the iris, regardless of its size and pupillary dilation, a pair of real coordinates (r,θ) where r is on the unit interval $[0, 1]$ and θ is angle $[0, 2\pi]$. This normalization or remapping of the iris image $I(x,y)$ from raw cartesian coordinates (x,y) to the dimensionless pseudopolar coordinate system (r,θ) can be represented as

$$I(x(r,\theta), y(r,\theta)) \rightarrow I(r,\theta), \tag{25.5}$$

where $x(r,\theta)$ and $y(r,\theta)$ are defined as linear combinations of both the set of pupillary boundary points $(x_p(\theta), y_p(\theta))$ and the set of limbus boundary points along the outer perimeter of the iris $(x_s(\theta), y_s(\theta))$ bordering the sclera, as determined by the active contour models that were initialized by the maxima of the operator (25.1):

$$x(r,\theta) = (1 - r)x_p(\theta) + rx_s(\theta), \tag{25.6}$$

$$y(r,\theta) = (1 - r)y_p(\theta) + ry_s(\theta). \tag{25.7}$$

Since the radial coordinate ranges from the iris inner boundary to its outer boundary as a unit interval, it inherently corrects for the elastic pattern deformation in the iris when the pupil changes in size.

The localization of the iris and the coordinate system described above achieve invariance to the 2D position and size of the iris and to the dilation of the pupil within the iris. However, it would not be invariant to the orientation of the iris within the image plane. The most efficient way to achieve iris recognition with orientation invariance is not to rotate the image itself using the Euler matrix but rather to compute the iris phase code in a single canonical orientation and then to compare this very compact representation at many discrete orientations by cyclic scrolling of its angular variable. The statistical consequences of seeking the best match after numerous relative rotations of two iris codes are straightforward. Let $f_0(x)$ be the raw density distribution obtained for the HDs between different irises after comparing them only in a single relative orientation; for example, $f_0(x)$ might be the binomial defined in (25.4). Then $F_0(x)$, the cumulative of $f_0(x)$ from 0 to x, becomes the probability of getting a false match in such a test when using HD acceptance criterion x:

$$F_0(x) = \int_0^x f_0(x)\,\mathrm{d}x \tag{25.8}$$

or, equivalently,

$$f_0(x) = \frac{d}{dx} F_0(x).$$ (25.9)

Clearly, then, the probability of not making a false match when using criterion x is $1 - F_0(x)$ after a single test, and it is $[1 - F_0(x)]^n$ after carrying out n such tests independently at n different relative orientations. It follows that the probability of a false match after a "best of n" test of agreement, when using HD criterion x, regardless of the actual form of the raw unrotated distribution $f_0(x)$, is

$$F_n(x) = 1 - [1 - F_0(x)]^n$$ (25.10)

and the expected density $f_n(x)$ associated with this cumulative is

$$f_n(x) = \frac{d}{dx} F_n(x)$$
$$= n f_0(x) [1 - F_0(x)]^{n-1}.$$ (25.11)

Each of the 9.1 million pairings of different iris images whose HD distribution was shown in Fig. 25.4 was submitted to further comparisons in each of seven relative orientations. This generated 63 million HD outcomes, but in each group of seven associated with any one pair of irises, only the best match (smallest HD) was retained. The histogram of these new 9.1 million best HDs is shown in Fig. 25.7. Since only the smallest value in each group of seven samples was retained, the new distribution is skewed and biased to a lower mean value (HD = 0.458), as expected from the theory of extreme value sampling. The solid curve in Fig. 25.7 is a plot of (25.11), incorporating (25.4) and (25.8) as its terms, and it shows an excellent fit between theory (binomial extreme value sampling) and data. The fact that the minimum HD observed in all these millions of rotated comparisons was about 0.33 illustrates the extreme improbability that the phase sequences for two different irises might disagree in fewer than a third of their bits. This suggests that in order to identify people by their iris patterns with high confidence, we need to demand only a very forgiving degree of match (say, HD $\leq$ 0.32).

25.6 UNIQUENESS OF FAILING THE TEST OF STATISTICAL INDEPENDENCE

The statistical data and theory presented above show that we can perform iris recognition successfully just by a test of statistical independence. Any two different irises are statistically "guaranteed" to pass this test of independence; and any two images that fail this test must be images of the same iris. Thus, it is the unique failure of the test of independence, which is the basis for iris recognition.

It is informative to calculate the significance of any observed HD matching score, in terms of the likelihood that it could have arisen by chance from two different irises. These probabilities give a confidence level associated with any recognition decision. Figure 25.8

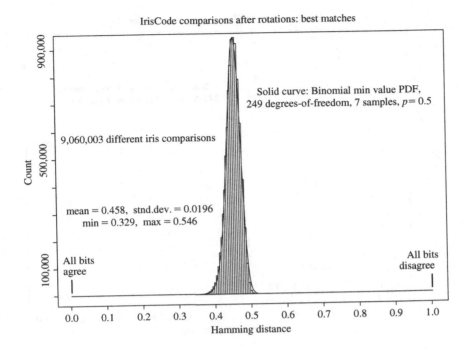

FIGURE 25.7

Distribution of Hamming distances for the same set of 9.1 million comparisons shown in Fig. 25.4, but allowing for seven relative rotations and preserving only the best match found for each pair. This "best of n" test skews the distribution to the left and reduces its mean from about 0.5 to 0.458. The solid curve is the theoretical prediction for such "extreme-value" sampling, as described by (25.4) and (25.8)–(11).

shows the false match probabilities marked off in cumulatives along the tail of the distribution presented in Fig. 25.7 (same theoretical curve (25.11) as plotted in Fig. 25.7 and with the justification presented in Figs. 25.4 and 25.5.) Table 25.1 enumerates false match probabilities, the cumulatives of (25.11), as a more fine-grained function of HD decision criterion between 0.26 and 0.35.

Calculation of the large factorial terms in (25.4) was done with Stirling's approximation which errs by less than 1% for $n \geq 9$:

$$n! \approx \exp(n \ln(n) - n + \frac{1}{2} \ln(2\pi n)). \tag{25.12}$$

The practical importance of the astronomical odds against a false match when the match quality is better than about HD ≤ 0.32, as shown in Fig. 25.8 and Table 25.1, is that such high confidence levels allow very large databases to be searched exhaustively without succumbing to any of the many opportunities for suffering a false match. The requirements of operating in one-to-many "identification" mode are vastly more demanding

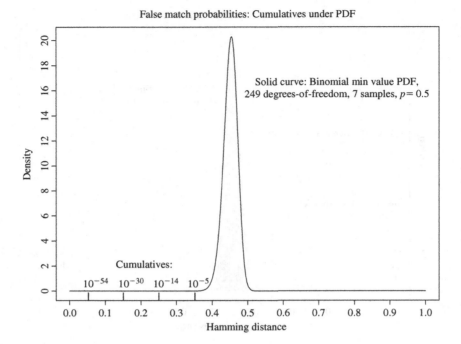

False match probabilities: Cumulatives under PDF

Solid curve: Binomial min value PDF,
249 degrees-of-freedom, 7 samples, $p = 0.5$

Cumulatives:

10^{-54} 10^{-30} 10^{-14} 10^{-5}

Density

Hamming distance

FIGURE 25.8

Calculated cumulatives under the left tail of the distribution seen in Fig. 25.7, up to sequential points, using the functional analysis described by (25.4) and (25.8)–(11). The extremely rapid attenuation of these cumulatives reflects the binomial combinatorics that dominate (25.4). This accounts for the very high confidence levels against a false match, when executing this test of statistical independence.

than operating merely in one-to-one "verification" mode (in which an identity must first be explicitly asserted, which is then verified in a yes/no decision by comparison against just the single nominated template).

If P_1 is the false match probability for single one-to-one verification trials, then clearly P_N, the probability of making at least one false match when searching a database of N unrelated patterns, is:

$$P_N = 1 - (1 - P_1)^N \tag{25.13}$$

because $(1 - P_1)$ is the probability of not making a false match in single comparisons; this must happen N independent times; and so $(1 - P_1)^N$ is the probability that such a false match never occurs.

It is interesting to consider how a seemingly impressive biometric one-to-one "verifier" would perform in exhaustive search mode once databases become larger than about 100, in view of (25.13). For example, a face recognition algorithm that truly achieved 99.9% correct rejection when tested on nonidentical faces, hence making only 0.1% false

TABLE 25.1 Cumulatives under (25.11) giving single false match probabilities for various HD criteria.

HD Criterion	Odds of False Match
0.26	1 in 10^{13}
0.27	1 in 10^{12}
0.28	1 in 10^{11}
0.29	1 in 13 billion
0.30	1 in 1.5 billion
0.31	1 in 185 million
0.32	1 in 26 million
0.33	1 in 4 million
0.34	1 in 690,000
0.35	1 in 133,000

matches, would seem to be performing at a very impressive level because it must confuse no more than 10% of all identical twin pairs (since about 1% of all persons in the general population have an identical twin). But even with its $P_1 = 0.001$, how good would it be for searching large databases?

Using (25.13) we see that when the search database size has reached merely $N = 200$ unrelated faces, the probability of at least one false match among them is already 18%. When the search database is just $N = 2000$ unrelated faces, the probability of at least one false match has reached 86%. Clearly, identification is vastly more demanding than one-to-one verification, and even for moderate database sizes, merely "good" verifiers are of no use as identifiers. Observing the approximation that $P_N \approx NP_1$ for small $P_1 << \frac{1}{N} << 1$, when searching a database of size N, an identifier needs to be roughly N times better than a verifier to achieve comparable odds against making false matches.

The algorithms for iris recognition exploit the extremely rapid attenuation of the HD distribution tail created by binomial combinatorics to accommodate very large database searches without suffering false matches. The HD threshold is adaptive to maintain $P_N < 10^{-6}$ regardless of how large the search database size N is. As Table 25.1 illustrates, this means that if the search database contains 1 million different iris patterns, it is only necessary for the HD match criterion to adjust downwards from 0.33 to 0.27 in order to maintain still a net false match probability of 10^{-6} for the entire database.

25.7 DECISION ENVIRONMENT FOR IRIS RECOGNITION

The overall "decidability" of the task of recognizing persons by their iris patterns is revealed by comparing the HD distributions for same versus different irises. The left distribution in Fig. 25.9 shows the HDs computed between 7,070 different pairs of same-eye images at different times, under different conditions, and usually with different cameras; and the right distribution gives the same 9.1 million comparisons among

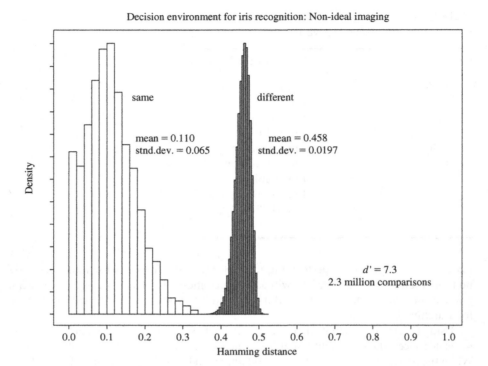

Decision environment for iris recognition: Non-ideal imaging

same

different

mean = 0.110
stnd.dev. = 0.065

mean = 0.458
stnd.dev. = 0.0197

$d' = 7.3$
2.3 million comparisons

Density

Hamming distance

FIGURE 25.9

The decision environment for iris recognition under relatively unfavorable conditions, using images acquired at different distances and by different optical platforms.

different eyes shown earlier. To the degree that one can confidently decide whether an observed sample belongs to the left or the right distribution in Fig. 25.9, iris recognition can be successfully performed. Such a dual distribution representation of the decision problem may be called the "decision environment," because it reveals the extent to which the two cases (same versus different) are separable and thus how reliably decisions can be made, since the overlap between the two distributions determines the error rates.

Whereas Fig. 25.9 shows the decision environment under less favorable conditions (images acquired by different camera platforms), Fig. 25.10 shows the decision environment under ideal (almost artificial) conditions. Subjects' eyes were imaged in a laboratory setting using always the same camera with fixed zoom factor and at fixed distance and with fixed illumination. Not surprisingly, more than half of such image comparisons achieved an HD of 0.00, and the average HD was a mere 0.019. It is clear from comparing Figs. 25.9 and 25.10 that the "authentics" distribution for iris recognition (the similarity between different images of the same eye, as shown in the left-side distributions) depends very strongly upon the image acquisition conditions. However, the measured similarity for "imposters" (the right-side distribution) is almost completely independent

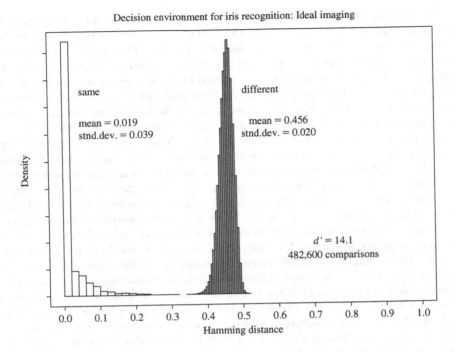

Decision environment for iris recognition: Ideal imaging

same

mean = 0.019
stnd.dev. = 0.039

different

mean = 0.456
stnd.dev. = 0.020

$d' = 14.1$
482,600 comparisons

FIGURE 25.10

The Decision Environment for iris recognition under very favorable conditions, always using the same camera, distance, and lighting.

of imaging factors. Instead, it just reflects the combinatorics of Bernoulli trials, as bits from independent binary sources (the phase codes for different irises) are compared.

For two-choice decision tasks (e.g., same versus different), such as biometric decision making, the "decidability" index d' is one measure of how well separated the two distributions are, since recognition errors would be caused by their overlap. If their two means are μ_1 and μ_2 and their two standard deviations are σ_1 and σ_2, then d' is defined as

$$d' = \frac{|\mu_1 - \mu_2|}{\sqrt{(\sigma_1^2 + \sigma_2^2)/2}}. \tag{25.14}$$

This measure of decidability is independent of how liberal or conservative the acceptance threshold used is. Rather, by measuring separation, it reflects the degree to which any improvement in (say) the false match error rate must be paid for by a worsening of the failure-to-match error rate. The performance of any biometric technology can be calibrated by its d' score, among other metrics. The measured decidability for iris recognition is $d' = 7.3$ for the nonideal (crossed platform) conditions presented in Fig. 25.9, and it is $d' = 14.1$ for the ideal imaging conditions presented in Fig. 25.10.

Based on the left-side distributions in Figs. 25.9 and 25.10, one could calculate a table of probabilities of failure to match, as a function of HD match criterion, just as we did earlier in Table 25.1 for false match probabilities based on the right-side distribution. However, such estimates may not be stable because the "authentics" distributions depend strongly on the quality of imaging (e.g., motion blur, focus, and noise) and would be different for different optical platforms. As illustrated earlier by the badly defocused image of Fig. 25.3, phase bits are still set randomly with binomial statistics in poor imaging, and so the right distribution is the stable asymptotic form both in the case of well-imaged irises (Fig. 25.10) and poorly imaged irises (Fig. 25.9). Imaging quality determines how much the same-iris distribution evolves and migrates leftward, away from the asymptotic different-iris distribution on the right. In any case, we note that for the 7,070 same-iris comparisons shown in Fig. 25.9, their highest HD was 0.327 which is below the smallest HD of 0.329 for the 9.1 million comparisons between different irises. Thus a decision criterion slightly below 0.33 for the empirical datasets shown can perfectly separate the dual distributions. At this criterion, using the cumulatives of (25.11) as tabulated in Table 25.1, the theoretical false match probability is 1 in 4 million.

Notwithstanding this diversity among iris patterns and their apparent singularity because of so many dimensions of random variation, their utility as a basis for automatic personal identification would depend upon their relative stability over time. There is a popular belief that the iris changes systematically with one's health or personality, and even that its detailed features reveal the states of individual organs ("iridology"); but such claims have been discredited (e.g., [17, 18]) as medical fraud. In any case, the recognition principle described here is intrinsically tolerant of a large proportion of the iris information being corrupted, say up to about a third, without significantly impairing the inference of personal identity by the simple test of statistical independence.

25.8 SPEED PERFORMANCE SUMMARY

On a low-cost 300 MHz reduced instruction set (RISC) processor, the execution times for the critical steps in iris recognition are as shown in Table 25.2, using optimized integer code.

The search engine can perform about 100,000 full comparisons between different irises per second on each such 300 MHz CPU, or 1 million in about a second on a 3 GHz server, because of the efficient implementation of the matching process in terms of elementary Boolean operators $\otimes$ and $\cap$ acting in parallel on the computed phase bit sequences. If a database contained many millions of enrolled persons, then the inherent parallelism of the search process should be exploited for the sake of speed by dividing up the full database into smaller chunks to be searched in parallel. The confidence levels shown in Table 25.1 indicate how the decision threshold should be adapted for each of these parallel search engines, in order to ensure that no false matches were made despite several large-scale searches being conducted independently. The mathematics of the iris recognition algorithms, particularly the binomial-class distributions (25.4) (25.11) that they generate when comparing different irises, make it clear that databases the size of

TABLE 25.2 Execution speeds of various stages in the iris recognition process on a 300 MHz RISC processor.

Operation	Time
Assess image focus	15 msec
Scrub specular reflections	56 msec
Localize eye and iris	90 msec
Fit pupillary boundary	12 msec
Detect and fit both eyelids	93 msec
Remove lashes and contact lens edges	78 msec
Demodulation and IrisCode creation	102 msec
XOR comparison of two IrisCodes	10 μs

an entire country's population could be searched in parallel to make confident and rapid identification decisions using parallel banks of inexpensive CPUs, if such iris code databases existed.

25.9 APPENDIX: 2D FOCUS ASSESSMENT AT THE VIDEO FRAME RATE

The acquisition of iris images in good focus is made difficult by the optical magnification requirements, the restrictions on illumination, and the target motion, distance, and size. All these factors act to limit the possible depth of field of the optics, because they create a requirement for a lower F number to accommodate both the shorter integration time (to reduce motion blur) and the light dilution associated with long focal length. The iris is a 1 cm target within a roughly 3 cm wide field that one would like to acquire at a range of about 30 cm to 50 cm and with a resolution of about 5 line pairs per mm. In a fixed-focus optical system, the acquisition of iris images almost always begins in poor focus. It is therefore desirable to compute focus scores for image frames very rapidly, either to control a moving lens element or to provide audible feedback to the subject for range adjustment, or to select which of several frames in a video sequence is in best focus.

Optical defocus can be fully described as a phenomenon of the 2D Fourier domain. An image represented as a 2D function of the real plane, $I(x, y)$, has a 2D Fourier transform $F(\mu, \nu)$ defined as

$$F(\mu, \nu) = \frac{1}{(2\pi)^2} \int \int I(x, y) \exp(-i(\mu x + \nu y)) dx dy. \qquad (25.15)$$

In the image domain, defocus is normally represented as convolution of a perfectly focused image by the 2D point-spread function of the defocused optics. This point-spread function is often modeled as a Gaussian whose space constant is proportional to the degree of defocus. Thus for perfectly focused optics, this optical point-spread function shrinks almost to a delta function, and convolution with a delta function has no effect on the image. Progressively defocused optics equates to convolving with ever wider point-spread functions.

If the convolving optical point-spread function causing defocus is an isotropic Gaussian whose width represents the degree of defocus, it is clear that defocus is equivalent to multiplying the 2D Fourier transform of a perfectly focused image with the 2D Fourier transform of the "defocusing" (convolving) Gaussian. This latter quantity is itself just another 2D Gaussian within the Fourier domain, and its spread constant there (σ) is the reciprocal of that of the image-domain convolving Gaussian that represented the optical point-spread function. Thus the 2D Fourier transform $D_\sigma(\mu, \nu)$ of an image defocused by degree $1/\sigma$ can be related to $F(\mu, \nu)$, the 2D Fourier transform of the corresponding perfectly focused image, by a simple model such as

$$D_\sigma(\mu, \nu) = \exp\left(-\frac{\mu^2 + \nu^2}{\sigma^2}\right) F(\mu, \nu). \tag{25.16}$$

This expression reveals that the effect of defocus is to attenuate primarily the highest frequencies in the image and that lower frequency components are affected correspondingly less, since the exponential term approaches unity as the frequencies (μ, ν) become small. (For simplicity, this analysis has assumed isotropic optics and isotropic blur, and the optical point-spread function has been described as a Gaussian just for illustration. But the analysis can readily be generalized to non-Gaussian and to anisotropic optical point-spread functions.)

This spectral analysis of defocus suggests that an effective way to estimate the quality of focus of a broadband image is simply to measure its total power in the 2D Fourier domain at higher spatial frequencies, since these are the most attenuated by defocus. One may also perform a kind of "contrast normalization" to make such a spectrally-based focus measure independent of image content, by comparing the ratio of power in higher frequency bands to that in slightly lower frequency bands. Such spectrally-based measurements are facilitated by exploiting Parseval's Theorem for conserved total power in the two domains:

$$\int \int |I(x, y)|^2 dx dy = \int \int |F(\mu, \nu)|^2 d\mu d\nu. \tag{25.17}$$

Thus, highpass filtering an image, or bandpass filtering it within a ring of high spatial frequency (requiring only a 2D convolution in the image domain), and integrating the power contained in it, is equivalent to computing the actual 2D Fourier transform of the image (a more costly operation) and performing the corresponding explicit measurement in the selected frequency band. Since the computational complexity of a fast Fourier transform on $n \times n$ data is $\mathcal{O}(n^2 \log_2 n)$, some 3 million floating-point operations are avoided which would otherwise be needed to compute the spectral measurements explicitly. Instead, only about 6,000 integer multiplications per image are needed by this algorithm, and no floating-point operations. Computation of focus scores is based only on simple algebraic combinations of pixel values within local closed neighborhoods, repeated across the image.

Pixels are combined according to the following (8 × 8) convolution kernel:

−1	−1	−1	−1	−1	−1	−1	−1
−1	−1	−1	−1	−1	−1	−1	−1
−1	−1	+3	+3	+3	+3	−1	−1
−1	−1	+3	+3	+3	+3	−1	−1
−1	−1	+3	+3	+3	+3	−1	−1
−1	−1	+3	+3	+3	+3	−1	−1
−1	−1	−1	−1	−1	−1	−1	−1
−1	−1	−1	−1	−1	−1	−1	−1

The simple weights mean that the sum of the central (4 × 4) pixels can just be tripled, and then the outer 48 pixels subtracted from this quantity; the result is squared and accumulated as per (25.17); and then the kernel moves to the next position in the image, selecting every 4th row and 4th column. This highly efficient discrete convolution has a simple 2D Fourier analysis.

The above kernel is equivalent to the superposition of two centered square box functions, one of size (8 × 8) and amplitude −1, and the other one of size (4 × 4) and amplitude +4. (For the central region in which they overlap, the two therefore sum to +3.) The 2D Fourier transform of each of these square functions is a 2D "sinc" function, whose size parameters differ by a factor of two in each of the dimensions and whose amplitudes are equal but opposite, since the two component boxes have equal but opposite volumes. Thus the overall kernel has a 2D Fourier transform $K(\mu, \nu)$ which is the difference of two, differently-sized, 2D sinc functions:

$$K(\mu, \nu) = \frac{\sin(\mu)\sin(\nu)}{\pi^2 \mu \, \nu} - \frac{\sin(2\mu)\sin(2\nu)}{4\pi^2 \mu \, \nu}. \tag{25.18}$$

The square of this function of μ and ν in the 2D Fourier domain is plotted in Fig. 25.11, revealing $K^2(\mu, \nu)$, the convolution kernel's 2D power spectrum.

Clearly, low spatial frequencies (near the center of the power spectral plot in Fig. 25.11) are ignored, reflecting the fact that the pixel weights in the convolution kernel all sum to zero, while a bandpass ring of upper frequencies is selected by this filter. The total power in that band is the spectral measurement of focus. Finally, this summed 2D spectral power is passed through a compressive nonlinearity of the form: $f(x) = 100 \cdot x^2 / (x^2 + c^2)$ (where parameter c is the half-power corresponding to a focus score of 50%), in order to generate a normalized focus score in the range of 0 to 100 for any image. The complete execution time of this 2D focus assessment algorithm, implemented in C using pointer arithmetic and operating on a (480 × 640) image, is 15 msec on a 300 MHz RISC processor.

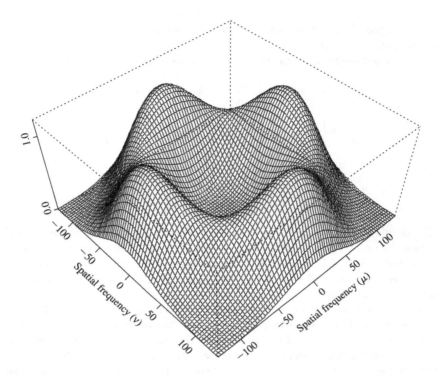

FIGURE 25.11

The 2D Fourier power spectrum of the convolution kernel used for rapid focus assessment.

REFERENCES

[1] J. Daugman. High confidence visual recognition of persons by a test of statistical independence. *IEEE Trans. Pattern Anal. Mach. Intell.*, 15(11):1148–1161, 1993.

[2] J. Daugman. *Biometric Personal Identification System Based on Iris Analysis.* U.S. Patent No. 5,291,560, US Government Printing Office, Washington, DC, 1994.

[3] J. Daugman. Statistical richness of visual phase information: update on recognizing persons by their iris patterns. *Int. J. Comput. Vis.*, 45(1):25–38, 2001.

[4] Y. Adini, Y. Moses, and S. Ullman. Face recognition: the problem of compensating for changes in illumination direction. *IEEE Trans. Pattern Anal. Mach. Intell.*, 19(7):721–732, 1997.

[5] P. N. Belhumeur, J. P. Hespanha, and D. J. Kriegman. Eigenfaces vs. Fisherfaces: recognition using class-specific linear projection. *IEEE Trans. Pattern Anal. Mach. Intell.*, 19(7):711–720, 1997.

[6] A. Pentland and T. Choudhury. Face recognition for smart environments. *Computer*, 33(2):50–55, 2000.

[7] P. J. Phillips, A. Martin, C. L. Wilson, and M. Przybocki. An introduction to evaluating biometric systems. *Computer*, 33(2):56–63, 2000.

[8] P. J. Phillips, H. Moon, S. A. Rizvi, and P. J. Rauss. The FERET evaluation methodology for face-recognition algorithms. *IEEE Trans. Pattern Anal. Mach. Intell.*, 22(10):1090–1104, 2000.

[9] P. Kronfeld. Gross anatomy and embryology of the eye. In: H. Davson, editor, *The Eye*. Academic Press, London, 1962.

[10] M. R. Chedekel. Photophysics and photochemistry of melanin. In: *Melanin: Its Role in Human Photoprotection*, 11–23. Valdenmar, Overland Park, KS, 1995.

[11] J. Daugman. Uncertainty relation for resolution in space, spatial frequency, and orientation optimized by two-dimensional visual cortical filters. *J. Opt. Soc. Am. A*, 2(7):1160–1169, 1985.

[12] J. Daugman. Complete discrete 2D Gabor transforms by neural networks for image analysis and compression. *IEEE Trans. Acoust.*, 36(7):1169–1179, 1988.

[13] J. Daugman and C. Downing. Demodulation, predictive coding, and spatial vision. *J. Opt. Soc. Am. A*, 12(4):641–660, 1995.

[14] R. Viveros, K. Balasubramanian, and N. Balakrishnan. Binomial and negative binomial analogues under correlated Bernoulli trials. *Am Stat*, 48(3):243–247, 1984.

[15] T. Cover and J. Thomas. *Elements of Information Theory*. Wiley, New York, 1991.

[16] J. Daugman and C. Downing. Epigenetic randomness, complexity, and singularity of human iris patterns. *Proc. R. Soc. Lond., B, Biol. Sci.*, 268:1737–1740, 2001.

[17] L. Berggren. Iridology: a critical review. *Acta Ophthalmol.*, 63(1):1–8, 1985.

[18] A. Simon, D. M. Worthen, and J. A. Mitas. An evaluation of iridology. *J. Am. Med. Assoc.*, 242:1385–1387, 1979.

Computed Tomography

R.M. Leahy[1], R. Clackdoyle[2], and Frédéric Noo[3]

[1] *University of Southern California;* [2] *CNRS et Université Jean Monnet;*
[3] *University of Utah*

26.1 INTRODUCTION

The term *tomography* refers to the general class of devices and procedures for producing two dimensional (2D) cross-sectional images of a three dimensional (3D) object. Tomographic systems make it possible to image the internal structure of objects in a noninvasive and nondestructive manner. By far the best known application is the computer assisted tomography (CAT or simply CT) scanner for X-ray imaging of the human body. Other medical imaging devices, including positron emission tomography (PET), single photon emission computed tomography (SPECT) and magnetic resonance imaging (MRI) systems, also make use of tomographic principles. Outside of the biomedical realm, tomography is used in diverse applications such as microscopy, nondestructive testing, radar imaging, geophysical imaging, and radio astronomy.

We will restrict our attention here to image reconstruction methods for X-ray CT, PET, and SPECT. In all three modalities, the data can be modeled as a collection of line integrals of the unknown image. Many of the methods described here can also be applied to other tomographic problems.

We describe 2D image reconstruction from parallel and fan-beam projections and 3D reconstruction from sets of 2D projections. Algorithms derived from the analytic relationships between functions and their line integrals, the so-called "direct methods," are described in Sections 26.3–26.5. In Section 26.6 we describe the class of "iterative methods" that are based on a finite dimensional discretization of the problem. We will include key results and algorithms for a range of imaging geometries, including systems currently in development. References to the appropriate sources for a complete development are also included. Our objective is to convey the wide range of methods available for reconstruction from projections and to highlight some recent developments in what remains a highly active area of research.

26.2 BACKGROUND

26.2.1 X-ray Computed Tomography

In conventional X-ray radiography, a stationary source and a planar detector are used to produce a 2D projection image of the patient. The image has intensity proportional to the amount by which the X-rays are attenuated as they pass through the body, i.e., the 3D spatial distribution of X-ray attenuation coefficients is projected into a 2D image. The resulting image provides important diagnostic information due to differences in the attenuation coefficients of bone, muscle, fat, and other tissues in the 40–120 keV range used in clinical radiography [1].

X-rays passing through an object experience exponential attenuation proportional to the linear attenuation coefficient of the object. The intensity of a collimated beam of monoenergetic X-radiation exiting a uniform block of material with linear attenuation coefficient μ and depth d is given by $I = I_0 e^{-\mu d}$, where I_0 is the intensity of the incident beam. For objects with spatially variant attenuation $\mu(z)$ along the path length z, this relationship generalizes to

$$I = I_0 e^{-\int \mu(z)\mathrm{d}z},\tag{26.1}$$

where $\int \mu(z)\mathrm{d}z$ is a *line integral* through $\mu(z)$.

Let $\mu(x,y,z)$ represent the 3D distribution of attenuation coefficients within the human body. Consider a simplified model of a radiography system that produces a broad parallel beam of X-rays passing through the patient in the z direction. An ideal 2D detector array or film in the (x,y)-plane would produce an image with intensity proportional to the negative logarithm of the attenuated X-ray beam, i.e., $-\log(I/I_0)$. The following projection image would then be formed at the ideal detector:

$$r(x,y) = \int \mu(x,y,z)\mathrm{d}z.\tag{26.2}$$

The utility of conventional radiography is limited due to the projection of 3D anatomy into a 2D image, causing certain structures to be obscured. For example, lung tumors, which have a higher density than the surrounding normal tissue, may be obscured by a more dense rib that projects into the same area in the radiograph. CT systems overcome this problem by reconstructing 2D cross sections of the 3D attenuation coefficient distribution.

The concept of the line integral is common to both the radiographic projection (26.2) and CT. Consider the first clinical X-ray CT system for which the inventor, G. Hounsfield, received the 1979 Nobel prize in medicine (the prize was shared with the mathematician A. Cormack) [2]. A collimated X-ray source and detector are translated on either side of the patient so that a single plane is illuminated, as illustrated in Fig. 26.1(a). After applying a logarithmic transformation, the detected X-ray measurements are a set of line integrals representing a 1D parallel projection of the 2D X-ray attenuation coefficient distribution in the illuminated plane. By rotating the source and detector around the patient other 1D projections can be measured in the same plane. The image can then

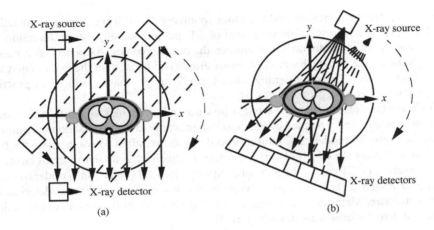

FIGURE 26.1

(a) Schematic representation of a first-generation CT scanner that uses translation and rotation of the source and a single detector to collect a complete set of 1D parallel projections; (b) The current generation of CT scanners use a fan X-ray beam and an array of detectors, which requires rotation only.

be reconstructed from these *parallel-beam* projections using the methods described in Section 26.3.1.

One major limitation of the first generation of CT systems was that the translation and rotation of the detectors was slow and a single scan would take several minutes. X-ray projection data can be collected far more quickly using the *fan-beam* X-ray source geometry employed in the current generation of CT scanners as illustrated in Fig. 26.1(b). Since an array of detectors is used, the system can simultaneously collect data for all projection paths that pass through the current location of the X-ray source. In this case, the X-ray source need not be translated, and a complete set of data is obtained through a single rotation of the source around the patient. Using this configuration, modern scanners can scan a single plane in less than one second. Methods for reconstruction from fan-beam data are described in Section 26.3.2.

Recently developed spiral CT systems allow continuous acquisition of data as the patient bed is moved through the scanner [3]. The detector traces out a helical orbit with respect to the patient allowing rapid collection of projections over a 3D volume. These data require special reconstruction algorithms as described in Section 26.4.2. In an effort to simultaneously collect fully 3D CT data, a number of systems have been developed that use a *cone-beam* of X-rays and a 2D rather than 1D array of detectors [3]. Cone-beam systems are now widespread in clinical CT and they also play an important role in industrial applications. Methods for cone-beam reconstruction are described in Sections 26.5.2 and 26.5.3.

The above descriptions can only be considered approximate because a number of factors complicate the X-ray CT problem. For example, the X-ray beam typically contains a broad spectrum of energies and therefore an energy dependence should be included in (26.1) [1]. The theoretical development of CT methods usually assumes a monoenergetic source. For broadband X-ray sources, the beam becomes "hardened" as it passes through the object, i.e., the lower energies are attenuated faster than the higher energies. This effect causes a beam hardening artifact in CT images that is reduced in practice using a data calibration procedure [4, 5].

In X-ray CT data the high photon flux produces relatively high signal-to-noise ratio. However, the data are corrupted by the detection of scattered X-rays that do not conform to the line integral model. Calibration procedures are required to compensate for this effect as well as for the effects of variable detector sensitivity. A final important factor in the acquisition of CT data is the issue of sampling. Each 1D projection is undersampled by approximately a factor of two in terms of the attainable resolution as determined by detector size. Methods to compensate for this problem in fan-beam systems using fractional detector offsets are described in [3].

26.2.2 Nuclear Imaging Using PET and SPECT

PET and SPECT are methods for producing images of the spatial distribution of biochemical tracers or probes that have been tagged with radioactive isotopes [1]. By tagging different molecules with positron or gamma-ray emitters, PET and SPECT can be used to reconstruct images of the spatial distribution of a wide range of biochemical probes. Applications include the use of tracers to measure glucose metabolism, angiogenesis, and cell proliferation for the detection and staging of cancer, imaging of cardiac function, imaging of gene expression, and studies of neurochemistry using a range of neuro-receptors and transmitters [6–8]. In recent years, smaller animal versions of clinical scanners have been developed for research applications in drug development, studies of animal models of human disease, and genomic and proteomic studies in live animals [9].

SPECT systems detect emissions using a "gamma camera." This camera is a combination of a sodium iodide scintillation crystal and an array of photomultiplier tubes (PMTs). The PMTs measure the location on the camera surface at which each gamma ray photon is absorbed by the scintillator [1]. A mechanical collimator, consisting of a sheet of dense metal in which a large number of parallel holes have been drilled, is attached to the front of the camera as illustrated in Fig. 26.2(a). The collimated camera is only sensitive to gamma rays traveling in a direction parallel to the holes in the collimator. The total number of gamma rays detected at a given pixel in the camera will be approximately proportional to the total activity (or line integral) along the line that passes through the patient and is parallel to the holes in the collimator. Thus when viewing a patient from a fixed camera position, we collect a 2D projection image of the 3D distribution of the tracer. By collecting data as the camera is rotated to multiple positions around the patient, we obtain *parallel-beam* projections for a contiguous set of parallel 2D slices through the patient, Fig. 26.2(b). The distribution can be reconstructed slice-by-slice using the same parallel-beam reconstruction methods as are used for X-ray CT.

Other collection geometries can be realized by modifying the collimator design [6]. For imaging an organ, such as the brain or heart, that is smaller than the surface area

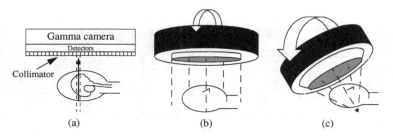

FIGURE 26.2

Schematic representation of a SPECT system: (a) cross-sectional view of a system with parallel hole collimator: gamma rays normally incident to the camera surface are detected, others are stopped by the collimator so that the camera records parallel projections of the source distribution; (b) Rotation of the camera around the patient produces a complete set of parallel projections; (c) Different collimators can be used to collect converging or diverging fan and cone-beam projections; shown is a converging cone-beam collimator.

of the camera, improved sensitivity can be realized by using converging *fan-beam* or *cone-beam* collimators as illustrated in Fig. 26.2(c). Similarly, diverging collimators can be used for imaging larger objects. Images are reconstructed from these fan-beam and cone-beam data using the methods in Sections 26.3.2 and 26.5.2, respectively. Cone-beam methods have not been widely used in clinical SPECT because of the practical difficulties in acquiring complete data (see Section 26.5.2), however the new generation of small animal SPECT systems does make use of cone-beam methods for data aquired using pinhole collimators. While the vast majority of SPECT systems use rotating planar gamma cameras, other systems have been constructed using a cylindrical scintillation detector that surrounds the patient. A rotating cylindrical collimator defines the projection geometry. Although the physical design of these cylindrical systems is quite different from that of the rotating camera, in most cases the reconstruction problem can still be reduced to one of the three basic forms: parallel, fan or cone-beam.

PET is based on the physical property that a positron produced by a radioactive nucleus travels a very short distance and then annihilates with an electron to form a pair of high-energy (511 keV) photons [7]. The pair of photons travel in opposite directions along a straight line path. Detection of the positions at which the photon pair intersects a ring of detectors allows us to approximately define a line that contains the positron emitter, as illustrated in Fig. 26.3(a). The total number of photon pairs measured by a detector pair will be proportional to the total number of positron emissions along the line joining the detectors, i.e., the number of detected events between a detector pair is an approximate line integral of the tracer density.

A PET scanner requires one or more rings of photon detectors coupled to a timing circuit that detects coincident photon pairs by checking that both photons arrive at the detectors within a few nanoseconds of each other. PET detectors are usually constructed using a combination of scintillation crystals and PMTs. A unique aspect of PET is that the ring of detectors surrounding the subject allows simultaneous acquisition of a

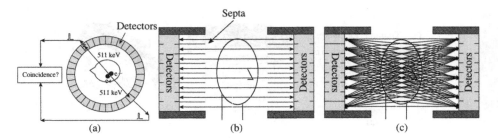

FIGURE 26.3

(a) Schematic showing how coincidence detection of photon pair produced by electron-positron annihilation determines the line along which the positron was annihilated; (b) In 2D systems, septa between adjacent rings of detectors prevent coincidence detection between rings; (c) Removal of the septa produces a fully 3D PET system in which cross-plane coincidences are collected and used to reconstruct the source distribution.

complete data set; no rotation of the detector system is required. A schematic view of two PET scanners is shown in Fig. 26.3. In the 2D scanner, multiple rings of detectors surround the patient with dense material, or "septa," separating each ring. These septa stop photons traveling between rings so that coincidence events are collected only between pairs of detectors in a single ring. We refer to this configuration as a 2D scanner since the data are separable and the image can be reconstructed as a series of 2D sections. In contrast, the 3D scanners have no septa so that coincidence photons can be detected between planes. In this case the reconstruction problem is not separable and must be treated directly in 3D.

PET data can be viewed as sets of approximate line integrals. In 2D mode, the data are sets of *parallel-beam* projections and the image can be reconstructed using methods equivalent to those in parallel-beam X-ray CT. In the 3D case, the data are still line integrals, but new algorithms are required to deal with the between-plane coincidences that represent *incomplete* projections through the patient. These methods are described in Sections 26.4 and 26.5.

As with X-ray CT, the line integral model is only approximate. Finite and spatially variant detector resolution is not accounted for in the line integral model and has a major impact on image quality [10]. The number of photons detected in PET and SPECT is relatively small so that photon-limited noise is also a factor limiting image quality. The data are further corrupted by additional noise due to scattered photons. Also, in both PET and SPECT, the probability of detecting an emission is reduced by the relatively high probability of Compton scatter of photons before they reach the detector. These attenuation effects can be quantified by performing a separate "transmission" scan in which the scattering properties of the body are measured. This information must then be incorporated into the reconstruction algorithm [7, 10]. While all these effects can, to some degree, be compensated for within the framework of analytic reconstruction from line integrals, they are more readily and accurately dealt with using the finite dimensional statistical formulations described in Section 26.6.

26.2.3 Mathematical Preliminaries

Since we deal with both 2D and 3D reconstruction problems here, we will use the following unified definition of the line integrals of an image $f(\underline{x})$:

$$g(\underline{a},\theta) = \int_{-\infty}^{\infty} f(\underline{a} + t\underline{\theta})\mathrm{d}t \quad ||\underline{\theta}|| = 1. \tag{26.3}$$

Here g is the integral of f over the line passing through $\underline{a}$ and oriented in the direction $\underline{\theta}$.

For parallel projections we consider only a fixed $\underline{\theta}$ (the projection direction). To avoid redundant parameterization of line integrals, we only consider those $\underline{a}$ perpendicular to $\underline{\theta}$ (i.e., $\underline{a} \cdot \underline{\theta} = 0$). We say a parallel projection $g(\cdot, \underline{\theta})$ is truncated if some nonzero line integrals are not measured. Generally, truncation is due to the use of a finite detector system which may be too small to gather a complete projection of the object at some orientation $\underline{\theta}$.

For fan-beam and cone-beam systems, we consider $\underline{a}$ to be fixed for a single projection; $\underline{a}$ is the fan-vertex or cone-vertex, which in practice would be the position of the X-ray source or the focal point of a converging collimator. Again, truncation of a projection $g(\underline{a}, \cdot)$ refers to line integrals which are not available, typically due to the limited extent of the detector.

26.2.4 Examples

We conclude this introductory section with examples of CT, PET, and SPECT images collected from the current generation of scanners (see Fig. 26.4). These images clearly

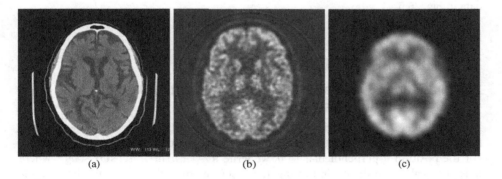

(a)   (b)   (c)

FIGURE 26.4

Examples of brain scans using: (a) X-ray CT, a nonlinear grayscale is used to enhance contrast between soft tissue regions within the brain; (b) PET, this image shows an image of glucose metabolism obtained using an analog of glucose labelled with the positron emitting isotope, fluorine-18; (c) SPECT, this is a brain perfusion scan using a technitium-99m ligand (image Courtesy of J.E. Bowsher, Duke University Medical Center).

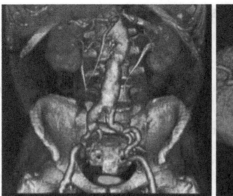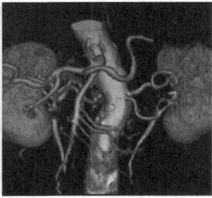

FIGURE 26.5

Volume rendering from a sequence of X-ray CT images showing the abdominal cavity and kidneys (CT images courtesy of G.E. Medical Systems).

reveal the differences between the high resolution, low-noise images produced by X-ray CT scanners and the lower resolution and noisy images produced by the nuclear imaging instruments. These differences are primarily due to the photon flux in X-ray CT which is many orders of magnitude higher than the individually detected photons in nuclear medicine imaging. In diagnostic imaging, these modalities are highly complementary since X-ray CT reveals information about the patients *anatomy* while PET and SPECT images contain *functional* information. For further insight into the ability of X-ray CT to produce high-resolution anatomical images, we show a set of 3D renderings from CT data in Fig. 26.5.

26.3 2D IMAGE RECONSTRUCTION

26.3.1 Fourier Space and Filtered Backprojection Methods for Parallel-Beam Projections

For 2D parallel-beam projections the general notation of (26.1) can be refined as illustrated in Fig. 26.6. We parameterize the direction of the rays using ϕ, so $\underline{\theta} = (\cos\phi, \sin\phi)$. For the position $\underline{a}$ perpendicular to $\underline{\theta}$, we write $\underline{a} = (-u\sin\phi, u\cos\phi) = u\underline{\theta}^{\perp}$ where u is the scalar coordinate indicating the distance from the origin to the integration line, or equivalently, the projection element index for the ϕ-projection. Since $\underline{\theta}$ depends only on ϕ and $\underline{a}$ then depends on u, we simplify the notation by writing $g(u,\phi) = g(\underline{a},\underline{\theta}) = \int f(\underline{a} + t\underline{\theta})\mathrm{d}t$. For the parallel-beam case, the function g is the Radon transform of the image f [4].

Practical inversion methods can be developed using the relationship between the Radon and Fourier transforms. The projection slice theorem is the basic result that is

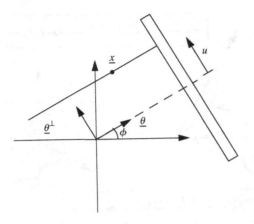

FIGURE 26.6

The coordinate system used to describe parallel-beam projection data.

used in developing these methods [4]. This theorem states that the 1D Fourier transform of the parallel projection at angle ϕ is equal to the 2D image Fourier transform evaluated along the line through the origin in the direction $\phi + \pi/2$, i.e.;

$$G(U,\phi) = \int_{-\infty}^{\infty} g(u,\phi)e^{-juU}\,du = F(\underline{X})\big|_{\underline{X}=U\underline{\theta}^{\perp}} = F(-U\sin\phi, U\cos\phi), \qquad (26.4)$$

where $F(\underline{X}) = F(X,Y)$ is the 2D image Fourier transform,

$$F(\underline{X}) = F(X,Y) = \int_{-\infty}^{\infty}\int_{-\infty}^{\infty} f(x,y)e^{-jxX}e^{-jyY}\,dxdy = \int\int_{R^2} f(\underline{x})e^{-j(\underline{X}\cdot\underline{x})}\,d\underline{x}. \qquad (26.5)$$

This result, illustrated in Fig. 26.7, can be employed in a number of ways. The discrete Fourier transform (DFT, see Chapter 5) of the samples of each 1D projection can be used to compute approximate values of the image Fourier transform. If the angular projection spacing is $\Delta\phi$, then the DFTs of all projections will produce samples of the 2D image Fourier transform on a polar sampling grid. The samples' loci lie at the intersections of radial lines, spaced by $\Delta\phi$, with circles of radii equal to integer multiples of the DFT frequency sampling interval. Once these samples are computed, the image can be reconstructed by first interpolating these values onto a regular cartesian grid and then applying an inverse 2D DFT. Design of these Fourier reconstruction methods involves a tradeoff between computational complexity and accuracy of the interpolating function [4].

A more elegant solution can be found by reworking (26.4) into a spatial domain representation. It is then straightforward to show that the image can be recovered using

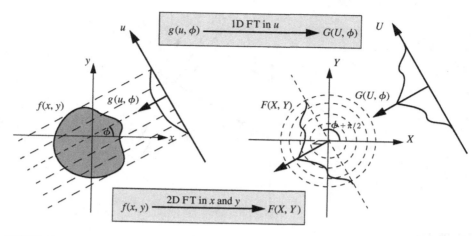

FIGURE 26.7

Illustration of the projection slice theorem. The 2D image at left is projected at angle ϕ to produce the 1D projection $g(u,\phi)$. The 1D Fourier transform $G(U,\phi)$ of this projection is equal to the 2D image Fourier transform, $F(X,Y)$ along the radial line at angle $\phi + \pi/2$.

the following equations [11]:

$$f(\underline{x}) = \frac{1}{2} \int_0^{2\pi} \tilde{g}(u,\phi)\big|_{u=u_{\underline{x},\phi}} \, d\phi, \qquad (26.6)$$

where

$$\tilde{g}(u,\phi) = \frac{1}{4\pi^2} \int_{-\infty}^{\infty} G(U,\phi)|U|e^{juU} \, dU, \qquad (26.7)$$

and $u_{x,\phi} = \underline{x} \cdot \underline{\theta}^{\perp}$ is the u-value of the parallel projection at angle ϕ of the point $\underline{x}$, see Fig. 26.6.

These two equations form the basis of the widely used filtered backprojection algorithm. Equation (26.7) is a linear shift-invariant filtering of the projection data with a filter with frequency response $H(U) = |U|$. The gain of this filter increases monotonically with frequency and the reconstruction is therefore unstable. However by assuming that the data $g(u,\phi)$, and hence the corresponding image, are bandlimited to a maximum frequency $U = U_{max}$, we need only consider the finite bandwidth filter with impulse response:

$$h(u) = \int_{-U_{max}}^{U_{max}} |U|e^{juU} \, dU. \qquad (26.8)$$

The filtered projections $\tilde{g}(u,\phi)$ are found by convolving $g(u,\phi)$ with $h(u)$ scaled by $1/4\pi^2$. To reduce effects of noise in the data, the response of this filter can be tapered off at higher frequencies [4, 11].

The integrand $\tilde{g}(u_{x,\phi},\phi)$ in (26.6) can be viewed as an image with constant values along lines in the ϕ direction that is formed by "backprojecting" the filtered projection at angle ϕ. Summing (or in the limit, integrating) these backprojected images for all ϕ produces the reconstructed image. Although this summation involves $\phi \in [0, 2\pi]$, in practice only 180 degrees of projection measurements are collected because opposing parallel-beam projections contain indentical information. In (26.6), the integration limits can be replaced with $\phi \in [0, \pi]$ and the factor of $1/2$ can be removed. This filtered backprojection method, or the modification described below for the fan-beam geometry, was the basis for image reconstruction in almost all commercially available CT systems, at least until the end of the twentieth century.

26.3.2 Fan-Beam Filtered Backprojection

X-ray CT data can be collected more rapidly using an array of detectors and a fan-beam X-ray source so that all elements in the array are simultaneously exposed to the X-rays. This arrangement gives rise to a natural fan-beam data collection geometry as illustrated in Fig. 26.1(b). The source and detector array are rotated around the patient and a set of fan-beam projections, $g(\underline{a},\theta)$, are collected, where $\underline{a}$ represents the position of the source and $\underline{\theta}$ specifies the individual line integrals in the projections. For a radius of rotation A, we parameterize the motion of the source as $\underline{a} = (A\cos\phi, A\sin\phi)$. For the case of a circular arc of detectors whose center is the fan-source and which rotates with the source, a particular detector element is conveniently specified using the relative angle β as shown in Fig. 26.8(a). The fan-beam projection notation is then simplified to $g(\phi,\beta) = g(\underline{a},\underline{\theta}) = \int f(\underline{a} + t\underline{\theta})dt$, where $\underline{\theta} = (-\cos(\phi - \beta), -\sin(\phi - \beta))$.

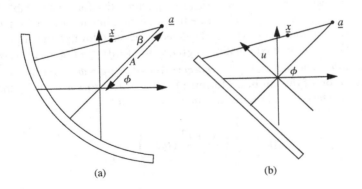

(a)   (b)

FIGURE 26.8

Illustration of the coordinate system for fan-beam tomography using (a) circular arc and (b) linear detector array arrangements.

The projection data could be resorted into equivalent parallel projections and the above reconstruction methods applied. Fortuitously this resorting is unnecessary. It can be shown [12] that reconstruction of the image can be performed using a fan-beam version of the filtered backprojection method. Development of this inverse method involves substitution of the fan-beam data in the parallel-beam formulae, (26.6) and (26.7), and applying a change of variables with the appropriate Jacobian. After some manipulation, the equations can be reduced to the form

$$f(\underline{x}) = \frac{1}{2} \int_0^{2\pi} \frac{1}{r^2} \tilde{g}(\phi,\beta)\big|_{\beta_{\underline{x},\phi=\theta}} d\phi, \tag{26.9}$$

where $r = ||\underline{x} - \underline{a}||$ is the distance from the point $\underline{x}$ to the fan-beam source,

$$\tilde{g}(\phi,\beta) = \frac{1}{4\pi^2} \int_{-\gamma}^{\gamma} (A\cos\beta' g(\varphi,\beta')) h(\sin(\beta - \beta'))d\beta', \tag{26.10}$$

where γ is the maximum value of β required to ensure that data are not truncated. In (26.9), $\beta_{x},\phi = \cos^{-1}((A^2 - (\underline{x} \cdot \underline{a}))/(rA))$ indicates the value in the ϕ-projection for the line passing through the point $\underline{x}$.

As in the parallel-beam case, this reconstruction method involves a two step procedure: filtering, in this case with a preweighting factor $A\cos\beta$, and backprojection. The backprojection for fan-beam data is performed along the paths converging at the location of the X-ray source and includes an inverse square-distance weighting factor. The filter $h(u)$ was given in (26.8) and, as before, can include a smoothing window tailored to the expected noise in the measured data.

In some fan-beam tomography applications the detector bank might be linear rather than curved. In principle, the same formula could be used by interpolating to obtain values sampled evenly in β. However, there is an alternative formula suitable for linear detectors. In this case, we use u to indicate the projection line for a scaled version of the flat detector corresponding to a virtual flat detector passing through the origin as shown in Fig. 26.8(b). The simplified notation is $g(\phi, u) = g(\underline{a}, \underline{\theta}) = \int f(\underline{a} + t\underline{\theta})dt$, where $\underline{a} = (A\cos\phi, A\sin\phi)$ as before, and $\underline{\theta} = (u\sin\phi - A\cos\phi, -u\cos\phi, -A\sin\phi)/\sqrt{u^2 + A^2}$.

The derivation of the fan-beam formula for linear detectors is virtually the same as for the curved detectors and results in equations of the form

$$f(\underline{x}) = \frac{1}{2} \int_0^{2\pi} \left(\frac{A^2 + u^2}{r^2} \tilde{g}(\underline{a}, u) \right)\bigg|_{u=u_{\underline{x},\phi}} d\phi, \tag{26.11}$$

$$\tilde{g}(\underline{a}, u) = \frac{1}{4\pi^2} \int_{-\infty}^{\infty} \left(\frac{A}{\sqrt{A^2 + u'^2}} g(\underline{a}, u') \right) h(u - u')du', \tag{26.12}$$

where, as before, $r = \|\underline{x} - \underline{a}\|$ is the distance between $\underline{x}$ and the source point $\underline{a}$, and $u_{\underline{x},\phi} = A\tan\beta$ specifies the line passing through $\underline{x}$ in the ϕ projection. The limits of integration $[-\infty,\infty]$ in the filtering step (26.12) are replaced in practice with the finite range of u corresponding to nonzero values of the projection data $g(\phi,u)$.

The existence of a filtered backprojection algorithm for these two fan-beam geometries is quite fortuitous, and does not occur for all detector sampling schemes. In fact these are two of only four sampling arrangements that admit this convenient reconstruction form [13]. In general, the filtering step must be replaced with a more general linear operation on the weighted projection values, and results in a more computationally intensive algorithm.

For the fan-beam geometry, opposing projections do not contain the same information, although all line integrals are measured twice over the range of 2π measurements. The redundancy is interwoven in the projections. An angular range of $\pi + 2\gamma$ can be used with careful adjustments to (26.11) and (26.12) to obtain a fast "short scan" reconstruction [4]. These short scan modes are used in clinical CT systems including spiral CT systems as discussed in Section 26.4.2.

26.3.3 Region of Interest Reconstruction

Important new developments in fan-beam image reconstruction occurred at the turn of the century. New image reconstruction formulas were established that admitted partial reconstruction of the object if less than a short scan of fan-beam measurements had been collected. Observe from equations (26.11) and (26.12) that reconstruction at the point $\underline{x}$ uses all the data, namely all lines passing through the image. From these formulas, it is not possible to obtain even a partial image if some of the data are missing. The new image reconstruction formulas however, are able to recover part of the image, often referred to as a region of interest, from part of the data. For fan-beam scanning with a contiguous circular motion of the source, any region of interest inside the convex hull of the source trajectory can be recovered using the formulas. For noncontiguous trajectories and for noncircular motions, the description of which regions of interest are recoverable are more complicated [14].

These and other developments in fan-beam reconstruction have ignited new research into region of interest reconstruction. The general region of interest problem can be stated as follows. Given a subset of the measured data, what is the largest region of interest that can be reliably reconstructed? This problem occurs naturally in the case of truncated projections where the object is too large to be viewed by the detector and some line integral measurements are missing in the data. As a simple guide to establishing an appropriate region of interest, any part of the object that is not subjected to measurements over a full 180 degrees cannot be accurately reconstructed. The converse however does not always hold.

The existence of inversion formulas that can achieve image reconstruction from subsets of the data immediately implies that for the complete data problem, there must be many reconstruction formulas, not just those described in Sections 26.3.1 and 26.3.2.

One class of such inversion formulas [15] involves a concept called the virtual fan-beam method, and is based on the "less than short scan" fan-beam developments [14]. Another class of inversion formulas is based on the concept of differentiated backprojection (DBP) with a subsequent inversion of the truncated Hilbert transform (ITHT). This DBP+ITHT method has useful generalizations to region of interest reconstruction in 3D cone-beam imaging. The 2D version of DBP+ITHT for parallel beam geometries is described here; more details, including the fan-beam version can be found in [16].

The DBP step consists of performing a backprojection of the derivative of the parallel projection data $g(u, \phi)$ to obtain an intermediate image $b(\underline{x}) = b(x, y)$:

$$b(x, y) = \int_{-\pi/2}^{\pi/2} \frac{\partial}{\partial u} g(u, \phi) \Big|_{u=u_{\underline{x}, \phi}} d\phi. \qquad (26.13)$$

The ITHT step performs a one-dimensional filtering in the vertical (y) direction of the DBP image $b(x, y)$. The reconstruction can be obtained along the line $x = x_0$ according to the following formula,

$$f(x_0, y) = \frac{1}{2\pi \sqrt{(y - L)(U - y)}} \left(\int_{L}^{U} \sqrt{(s - L)(U - s)} \frac{b(x_0, s)}{\pi(y - s)} ds - C \right). \qquad (26.14)$$

The lower and upper bounds of integration, L and U, are selected freely subject to the constraints that (i) $L < y < U$, (ii) the DBP image $b(x_0, s)$ is available from Eq. (26.13) for all $s \in (L, U)$, and (iii) $f(x_0, y) = 0$ for all $y \notin (L, U)$. The constant C must be determined from the data. One simple method is to use a priori information such as a known value of $f(x_0, y^*)$ with $L < y^* < U$, and applying Eq. (26.14). Typically, the point (x_0, y^*) is known to be outside the object and therefore $f(x_0, y^*) = 0$, from which one would obtain

$$C = \int_{L}^{U} \sqrt{(s - L)(U - s)} \frac{b(x_0, s)}{\pi(y^* - s)} ds. \qquad (26.15)$$

As an illustration of the DBP+ITHT method, consider an elliptical object that extends past the circular field of view of the scanner. Only those lines crossing the dashed circle in Fig. 26.9 are measured, so the DBP image $b(\underline{x})$ is only available inside this circle. Full angular coverage (180 degrees) occurs for those points inside the dashed circle. However reconstruction can only be performed using the DBP+ITHT method inside the shaded region of the figure, due to condition (iii) on the selection of L and U. For all vertical lines in the shaded region, reconstruction is achieved by selecting L, U, and y^* following the generic example in Fig. 26.9.

General region of interest reconstruction in two dimensions is still an open research area. Reconstruction formulas only exist for certain special cases, and the existence of reliable region of interest reconstruction from missing data subsets is not yet resolved in general.

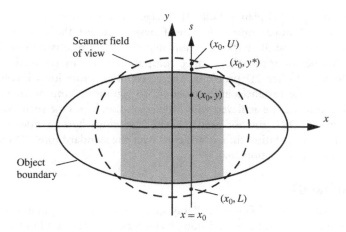

FIGURE 26.9

Illustration of the DBP + ITHT method applied to the elliptical object with truncated measurements. Only lines passing through the dotted circle are measured. Any region of interest inside the shaded part of the diagram can be reconstructed using DBP + ITHT, by filtering along vertical lines such as $x = x_0$. All points of the form $f(x_0,y)$ can be found by choosing L and U outside the ellipse but inside the circle, and y^* inside (L, U) but outside the ellipse.

26.4 EXTENDING 2D METHODS INTO 3D

26.4.1 Extracting 2D Data from 3D

A full 3D image can be built up by repeatedly performing 2D image reconstruction on a set of parallel contiguous slices. In X-ray CT, SPECT, and PET, this has been a standard method for volume tomographic reconstruction. Mathematically we use $f_z(\underline{x}) = f_z(x, y)$ to represent the z-slice of $f(x, y, z)$, and $g_z(u, \underline{\theta})$ to represent the line integrals in this z-slice. Reconstruction for each z is performed sequentially using techniques described in Section 26.3.

More sophisticated methods of building 3D tomographic images have been developed for a number of applications. For example, in spiral X-ray CT, the patient is moved continuously through the scanner so no fixed discrete set of tomographic slices is defined. In this case, there is flexibility in choosing the slice spacing and the absolute slice location, however there is no slice position for which a complete set of projection data is measured. We describe image reconstruction for spiral CT in Section 26.4.2.

In a more general framework, we call an image reconstruction problem *fully 3D* if the data cannot be separated into a set of parallel contiguous and independent 2D slices. An example is 3D PET which allows measurement of oblique coincidence events and therefore must handle line integrals that cross multiple transverse planes as shown in Fig. 26.3(c). Other examples of fully 3D problems include cone-beam SPECT and cone-beam X-ray CT where the diverging geometry of the rays precludes any sorting arrangement into parallel planes. Fully 3D image reconstruction is described in more detail in Section 26.5, but a common feature of these methods is the heavy computational load associated with the 3D backprojection step. Since 2D reconstruction is generally very fast, a number of approaches reduce computation cost by converting a fully 3D problem into a multislice 2D problem. These *rebinning* procedures involve approximations that in some instances are very good, so significant improvements in image reconstruction time can be achieved with little resolution loss. One such example is the Fourier rebinning (FORE) method used in 3D PET imaging where an order of magnitude improvement in computation time is achieved over the standard fully 3D methods; the method is described in Section 26.4.3.

26.4.2 Spiral CT

In spiral CT, a conventional fan-beam X-ray source and detector system rotates around the patient while the bed is translated along its long axis, as illustrated in Fig. 26.10. This supplementary motion, although it complicates the image reconstruction algorithms and results in slightly blurred images, provides the capability to scan large regions of the patient in a single breath hold.

The helical motion is characterized by the pitch P which is the amount of translation in the axial or z-direction for a full rotation of the source and detector assembly. Therefore $\phi = 2\pi z/P$ and we can write $g(z,\beta) = g(\underline{a},\underline{\theta}) = \int f(\underline{a},t\underline{\theta})dt$ which is similar to the fan-beam geometry of Section 26.3.2, with $\underline{a} = (A\cos\phi, A\sin\phi, z)$ and $\underline{\theta} = (-\cos(\phi - \beta), -\sin(\phi - \beta), 0)$. Note that ϕ now ranges from 0 to $2\pi n$ as z ranges from 0 to nP, where n is the number of turns of the helix. The usual method of reconstruction involves estimating a full set of fan-beam projections $g_z(\phi,\beta)$ for each transverse plane, using the available projections at other points on the helix. If the reconstruction on transverse

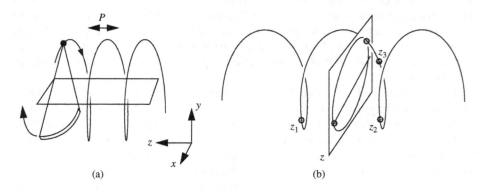

(a)   (b)

FIGURE 26.10

Illustration of spiral or helical CT geometry. (a) Relative to a stationary bed, the source and detector circle the patient in a helical fashion with pitch P; (b) To reconstruct cross section $f_z(x,y)$, missing projections are interpolated from neighboring points on the helix at which data were collected.

plane z is required, the standard fan-beam CT algorithm is used,

$$f_z(\underline{x}) = \frac{1}{2} \int\limits_0^{2\pi} \frac{1}{r^2} \, \tilde{g}_z(\phi,\beta)\big|_{\beta=\beta_{z,\underline{x},\phi}} \, d\phi \qquad (26.16)$$

$$\tilde{g}_z(\phi,\beta) = \frac{1}{4\pi^2} \int\limits_{-\gamma}^{\gamma} (A\cos\beta' g_z(\phi,\beta')) h(\sin(\beta-\beta')) \, d\beta', \qquad (26.17)$$

where $\beta_{z,\underline{x},\phi}$ indicates the relative projection angle β found by projecting in the z plane at angle ϕ through the point (x,y,z), and r is the distance (in the z plane) between the point (x,y,z) and the virtual source at angular position ϕ. In the simplest case, the z-plane projections $g_z(\phi,\beta)$ are estimated by a weighted sum of the measured projections at the same angular position above and below z on the helix:

$$g_z(\varphi,\beta) \approx \frac{w_1 g(z_1,\beta) + w_2 g(z_2,\beta)}{w_1 + w_2} \qquad (26.18)$$

for some suitable weights w_1 and w_2, and where, as illustrated in Fig. 26.10, z_1 and z_2 lie within one pitch P of the reconstruction plane z. Note that in (26.18), z_1/P differs from ϕ by some multiple of 2π, and similarly for z_2.

Various schemes for choosing the weights w_1 and w_2 exist [17]. Each weighting scheme establishes a tradeoff between increased image noise from unbalanced contributions, and inherent axial blurring artifacts from the geometric approximation of the estimation process. When the image noise is particularly low, a short-scan version of the fan-beam reconstruction algorithm might be used. This version reduces the range of contributing projections to $\pi + \gamma$ from 2π and correspondingly reduces the maximum distance required to estimate a projection $g_z(\phi, \beta)$. Even more elaborate estimation schemes exist, such as approximating $g_z(\phi, \beta)$ on a line-by-line basis. Figure 26.10(b) illustrates how the line integral $g_z(\phi, \beta)$ could be estimated from a value in the z_3 projection.

The choice of pitch P represents a compromise between maximizing the axial coverage of the patient, and avoiding unacceptable artifacts from the geometric estimation. Generally the pitch is chosen between one and two times the thickness of the detector in the axial direction [17].

In recent years, "multislice" CT scanners have been constructed with multiple rows of detectors. For four detector rows or less, the typical approach has been to ignore the divergence of the rays in the axial direction and treat the system as several interwoven helices so that the methods described above can be applied. The pitch is usually tunable, and does not normally exceed the axial extent of the detectors. However, newer systems with a greater number of detector rows involve a nonnegligible axial divergence of the X-rays and need to be treated using cone-beam techniques to correctly handle the divergence issue. Section 26.5.2 describes cone-beam tomography in general and Section 26.5.3 discusses the situation for spiral (or helical) CT.

26.4.3 Rebinning Methods in 3D PET

PET data is generally sorted into parallel-beam projections, $g(u, \phi)$ as described in Section 26.3.1. For a multiring 2D PET scanner, the data are usually processed slice-by-slice. Using z to denote the axis of the scanner, the data are reconstructed using (26.6) and (26.7) applied to each z-slice as follows:

$$f_z(\underline{x}) = \frac{1}{2} \int_0^{2\pi} \tilde{g}_z(u, \phi)\Big|_{u=u_{z,\underline{x},\varphi}} d\varphi, \qquad (26.19)$$

$$\tilde{g}_z(u, \varphi) = \frac{1}{4\pi^2} \int_{-\infty}^{\infty} g_z(u', \varphi)h(u - u')du, \qquad (26.20)$$

with $u_{z,\underline{x},\phi} = (x,y,z) \cdot (-\sin\phi, \cos\phi, 0)$. The data $g_z(u, \phi)$ are found from sampled values of u and ϕ determined by the ring geometry: the radius R and the number of crystals (typically several hundred). In (26.19) and (26.20), z is usually chosen to match the center of each detector ring. In practice, 2D scanners do allow detection of coincidences between adjacent rings. By using the single-slice rebinning (SSRB) technique described below, slices midway between adjacent detector rings can also be reconstructed from 2D scanner data.

Current commercial 3D PET scanners usually consist of a few tens of detector rings and have the capability to detect oblique photon pairs that strike detectors in different rings. These fully 3D data require more advanced reconstruction techniques. The fully 3D version of (26.19) and (26.20) is given in Section 26.5. In this section, we describe two popular rebinning methods, where the data are first processed to form independent 2D projections $g_z(u, \phi)$ from which (26.19) and (26.20) are then used for image reconstruction.

Let Δ denote the spacing between rings. Let $g_{l,m}(u, \phi)$ denote the resulting line integral, with endpoints on rings l and m, whose 2D projection variables are (u, ϕ) when the line is projected onto the $x - y$ plane, as shown in Fig. 26.11. In the SSRB method

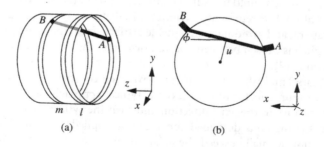

(a)   (b)

FIGURE 26.11

Oblique line integrals (along the path AB in (a)) between different rings of detectors can be rebinned into equivalent in plane data either directly using SSRB, or indirectly using FORE; (b) The relationship between the projected line integral path and the parameters (u, ϕ).

[18], all line integral data are reassigned to the slice midway between the rings where the detection occurred. Thus

$$g_z(u,\phi) \approx \sum_{(l,m)\in\Omega_z} g_{l,m}(u,\phi), \quad \text{where} \quad \Omega_z = \left\{ (l,m) : \frac{(l+m)\Delta}{2} = z \right\}, \tag{26.21}$$

and reconstruction proceeds according to (26.19) and (26.20).

A more sophisticated method, known as FORE [19], effectively performs the rebinning operation in the 2D frequency domain. The rebinned data $g_z(u,\phi)$ are found using the following transformations:

$$\tilde{g}_{l,m}(u,\phi) = g_{l,m}(u,\phi) \frac{2\sqrt{R^2 - u^2}}{\sqrt{4R^2 - 4u^2 + (l-m)^2\Delta^2}}, \tag{26.22}$$

$$\tilde{G}_{l,m}(U,k_\phi) = \int\limits_{-\infty}^{\infty} \int\limits_{0}^{2\pi} \tilde{g}_{l,m}(u,\phi) e^{-j(Uu+k_\phi\phi)} \, d\phi \, du, \tag{26.23}$$

$$G_z(U,k_\phi) \approx \sum_{(l,m)\in\Omega_z} \tilde{G}_{l,m}(U,k_\phi), \tag{26.24}$$

$$\text{where} \quad \Omega_z = \left\{ l,m : \frac{(l+m)\Delta}{2} + \frac{(l-m)\Delta}{2} \frac{k_\phi}{U\sqrt{R^2 - U^2}} = z \right\} \tag{26.25}$$

and

$$g_z(u,\phi) = \sum_{k_\phi} \int\limits_{-\infty}^{\infty} G_z(U,k_\phi) e^{j(Uu+k_\phi\phi)} \, dU. \tag{26.26}$$

Both SSRB and FORE are approximate techniques, and the geometrical misplacement of the data can cause artifacts in the reconstructed images. However, FORE is far more accurate than SSRB yet almost as fast computationally (compared to the subsequent reconstruction time using (26.19) and (26.20)). In [19] a mathematically exact rebinning formula is presented, and it is shown that SSRB and FORE represent zeroth- and first-order versions of this formula. However algorithms using the exact version are less practical than SSRB or FORE.

26.5 3D IMAGE RECONSTRUCTION

26.5.1 Fully 3D Reconstruction with Missing Data

In 3D image reconstruction, parallel-beam projection data can be specified using $\underline{\theta}$, the direction of the line integrals, and two scalars (u,v) that indicate offsets in directions

$\underline{\theta}^1$ and $\underline{\theta}^2$ perpendicular to $\underline{\theta}$. Therefore $g(u, v, \underline{\theta}) = g(\underline{a}, \underline{\theta}) = \int f(\underline{a} + t\underline{\theta}) dt$, where $\underline{a} = u\underline{\theta}^1 + u\underline{\theta}^2$, and $\{\underline{\theta}, \underline{\theta}^1, \underline{\theta}^2\}$ is an orthonormal system. Note that all vectors in this section are 3D.

Image reconstruction can be performed using a 3D version of the filtered backprojection formulas (26.6) and (26.7) given in Section 26.3.1:

$$f(\underline{x}) = \frac{1}{2} \int_{\Omega} \tilde{g}(u, v, \underline{\theta}) \Big|_{\substack{u = u_{\underline{x}, \underline{\theta}} \\ v = v_{\underline{x}, \underline{\theta}}}} d\underline{\theta}, \qquad (26.27)$$

$$\tilde{g}(u, v, \underline{\theta}) = \frac{1}{8\pi^3} \int_{-\infty}^{\infty} \int_{-\infty}^{\infty} G(U, V, \underline{\theta}) H_\Omega(U, V, \underline{\theta}) e^{j(uU + uV)} dU dV, \qquad (26.28)$$

where, in the surface integral of (26.23), $d\underline{\theta}$ can be written as $\sin\theta d\theta d\phi$ for $\underline{\theta}$ having polar angle θ and azimuthal angle ϕ; and where $(u_{\underline{x},\theta}, v_{\underline{x},\theta}) = (\underline{x} \cdot \underline{\theta}^1, \underline{x} \cdot \underline{\theta}^2)$ represents the (u, v) coordinates of the line with orientation $\underline{\theta}$ passing through $\underline{x}$. The subset Ω of the unit sphere represents the measured directions $\{\underline{\theta}\}$ and must satisfy Orlov's condition for data completeness in order for (26.27) and (26.28) to be valid. Orlov's condition requires that every great circle on the unit sphere intersects the region Ω. The tomographic reconstruction filter $H_\Omega(U, V, \underline{\theta})$ depends on the measured dataset [20]. In the special case that Ω is the whole sphere S^2, then $H_\Omega(U, V, \underline{\theta}) = \sqrt{U^2 + V^2}$.

In 3D PET imaging, data can be sorted according to the parameterization $g(u, v, \underline{\theta})$. The set of measured projections can be described by $\Omega_\Psi = \{\underline{\theta} = (\theta_x, \theta_y, \theta_z) : |\theta_z| \leq \sin\Psi\}$, where Ψ represents the most oblique line integral possible, $\Psi = \text{atan}(L/(2R))$, for a scanner radius of R and axial extent L. Provided none of the projections are truncated, reconstruction can be performed according to (26.23) and (26.24) using the Colsher filter $H_{\Omega_\Psi}(U, V, \underline{\theta}) = H^\Psi(U, V, \underline{\theta})$ given by

$$H^\Psi(U, V, \underline{\theta}) = \begin{cases} \sqrt{U^2 + V^2} & \text{if } \sqrt{U^2 + \theta_z^2 V^2} < (\sin\Psi)\sqrt{U^2 + V^2} \\ \dfrac{\pi}{2} \dfrac{\sqrt{U^2 + V^2}}{\sin^{-1}\left((\sin\Psi)\sqrt{U^2 + V^2} / \sqrt{U^2 + \theta_z^2 V^2}\right)} & \text{otherwise.} \end{cases}$$

$$(26.29)$$

In practice the object occupies most of the axial extent of the scanner so nearly all projections are truncated. However, there is always a subset $\Omega_{\Psi'}$ of the projections which are not truncated, and from these projections a reconstruction can be performed to obtain the image $f^{\Psi'}(\underline{x})$ using (26.28) and (26.29) with $H^{\Psi'}(U, V, \underline{\theta})$. In the absence of noise, this reconstruction would be sufficient, but to include partially measured projections a technique known as the "reprojection method" is used. All truncated projections are completed by estimating the missing line integrals based on the initial reconstruction $f^{\Psi'}(\underline{x})$. Then, in a second step, reconstruction from the entire data set is performed using $H^\Psi(U, V, \underline{\theta})$ to obtain the final image $f^\Psi(\underline{x})$ [21].

26.5.2 Cone-Beam Tomography

For cone-beam projections, it is convenient to use the general notation $g(\underline{a}, \underline{\theta})$, where $\underline{a}$ and $\underline{\theta}$ are 3D vectors. For an X-ray system, the position of the source would be represented by $\underline{a}$, and the direction of individual ray-sums obtained from that source would be indicated by $\underline{\theta}$. Due to difficulties obtaining sufficient tomographic data (see below), the source and/or detector may follow elaborate trajectories in space relative to the object; therefore a general description of their orientations is required. For applications involving a planar detector, we replace $\underline{\theta}$ with (u, v), coordinates on an imaginary detector centered at the origin and lying in the plane perpendicular to $\underline{a}$. The source point $\underline{a}$ is assumed never to lie on the scanner axis $\underline{e}_3$. The u-axis lies in the detector plane in the direction $\underline{e}_3 \times \underline{a}$. The v-direction is perpendicular to u and points in the same direction as $\underline{e}_3$ as shown in Fig. 26.12(b).

In the simplest applications, the detector and source rotate in a circle about the scanner axis $\underline{e}_3$. If the radius of rotation is A, the source trajectory is parameterized by $\phi_g \in [0, 2\pi]$ as $\underline{a} = (A\cos\phi, A\sin\phi, 0)$. In this case, the v-axis in the detector stays aligned with $\underline{e}_3$ and the u-axis points in the tangent direction to the motion of the source. Physical detector measurements can easily be scaled to this virtual detector system, just as for the fan-beam example of Section 26.3.2. Thus $g(\phi, u, v) = g(\underline{a}, \underline{\theta}) = \int f(\underline{a} + t\,\underline{\theta})dt$, where $\underline{\theta} = (-u\sin\phi - A\cos\phi, u\cos\phi - A\sin\phi, v)$.

The algorithm of Feldkamp *et al.* [22] is based on the fan-beam formula for flat detectors (see Section 26.3.2) and collapses to this formula in the central plane $z = 0$ where only fan-beam measurements are taken:

$$f_{\text{FDK}}(\underline{x}) = \frac{1}{2} \int_0^{2\pi} \left(\frac{A^2 + u^2 + v^2}{r^2} \tilde{g}(\phi, u, v) \right)\Bigg|_{\substack{u = u_{\underline{x},\phi} \\ v = v_{\underline{x},\phi}}} d\phi \qquad (26.30)$$

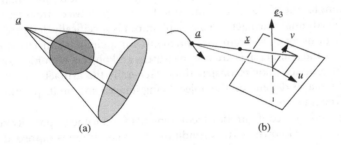

(a)   (b)

FIGURE 26.12

(a) 2D planar projections of a 3D object are collected in a cone-beam system as line integrals through the object from the cone vertex a to the detector; (b) The coordinate system for the cone-beam geometry—the cone vertices $\underline{a}$ can follow an arbitrary trajectory provided Tuy's condition is satisfied.

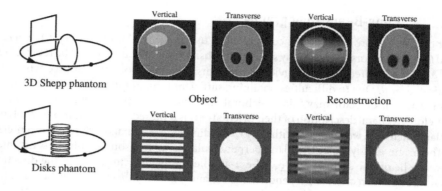

3D Shepp phantom

Object

Reconstruction

Disks phantom

FIGURE 26.13

Example of cone-beam reconstructions from a circular orbit—the obvious artifacts are a result of the incompleteness in the data. Other trajectories, such as a helix or a circle plus line, give complete data and artifact free reconstructions.

$$\tilde{g}(\phi, u, v) = \frac{1}{4\pi^2} \int_{-\infty}^{\infty} \left(\frac{A}{\sqrt{A^2 + u'^2 + v'^2}} g(\phi, u', v') \right) h(u - u') \mathrm{d}u'. \qquad (26.31)$$

Similarly to (26.11), $r = \|\underline{x} - \underline{a}\|$ is the distance between $\underline{x}$ and the source position $\underline{a}$, and $(u_{\underline{x},\phi}, v_{\underline{x},\phi})$ are the coordinates on the detector of the cone-beam projection of $\underline{x}$, see Fig. 26.12.

Figure 26.13 shows two images of reconstructions from mathematically simulated data. Using a magnified grayscale to reveal the 1% contrast structures, the top images show both a high quality reconstruction in the horizontal transverse slice at the level of the circular trajectory, and apparent decreased intensity on planes above and below this level. These artifacts are characteristic of the Feldkamp algorithm. The bottom images, showing reconstructions for the "disks" phantom, exhibit cross-talk between transverse planes and some other less dramatic artifacts. The disk phantom is specifically designed to illustrate the difficulty in using cone-beam measurements for a circular trajectory. Frequencies along and near the scanner axis are not measured, and objects with high amplitudes in this direction produce poor reconstructions. Generally the artifacts manifested in the reconstructed images depend on the object being imaged and on its position relative to the plane of the trajectory.

For the cone-beam configuration, requirements for a tomographically complete set of measurements are known as Tuy's condition. Tuy's condition is expressed in terms of a geometric relationship among the trajectory of the cone-beam vertex point (the source point) and the size and position of the object being scanned. Tuy's condition requires that every plane that cuts through the object must also contain some point of the vertex trajectory. Furthermore, it is assumed that the detector is large enough to measure the entire object at all positions of the trajectory, i.e., the projections should not be truncated. For the examples given in Fig. 26.13, the artifacts arose because the circular trajectory did not satisfy Tuy's condition (even though the projections were not truncated). In this sense, the measurements were incomplete and artifacts were inevitable.

Analytic reconstruction methods for cone-beam configurations satisfying Tuy's completeness condition are generally based on a transform pair that plays a similar role to the Fourier transform in the projection slice theorem for classical parallel-beam tomography. A mathematical result due to Grangeat [23] links the information in a single cone-beam projection to a subset of the transform domain, just as the Fourier slice theorem links a parallel projection to a certain subset of the Fourier domain. This relationship is defined through the "B transform," the derivative of the 3D Radon transform:

$$Bf(s, \underline{\gamma}) = p(s, \underline{\gamma}) = \frac{1}{2\pi} \int\limits_{R^3} f(\underline{x}) \delta'(\underline{x} \cdot \underline{\gamma} - s) d\underline{x}, \tag{26.32}$$

$$B^{-1} p(\underline{x}) = f(\underline{x}) = \frac{1}{4\pi} \int\limits_{S^2} \int\limits_{R} p(s, \underline{\gamma}) \delta'(s - \underline{x} \cdot \underline{\gamma}) ds d\underline{\gamma}, \tag{26.33}$$

where s is a scalar and $\|\underline{\gamma}\| = 1$. The symbol δ' represents the derivative of the Dirac delta function whose action is defined by $\int_{-\infty}^{\infty} f(s) \delta'(s_0 - s) ds = f'(s_0)$.

Grangeat's formula can be written as

$$\frac{1}{4\pi} \int\limits_{S^2} g(\underline{a}, \underline{\theta}) \delta'(\underline{\theta} \cdot \underline{\gamma}) d\underline{\theta} = p(s, \underline{\gamma})|_{s=\underline{a} \cdot \underline{\gamma}}. \tag{26.34}$$

An analysis of (26.34) shows that if Tuy's condition is satisfied, then all values are available in the B domain representation of $f(\underline{x})$, namely $p(s, \underline{\gamma})$ [23].

Equations (26.33) and (26.34) form the basis for a reconstruction algorithm. All values in the B domain can be found from cone-beam projections, and $f(\underline{x})$ can be recovered from the inverse transform B^{-1}. Care must be taken to ensure that the B domain is sampled uniformly in s and $\underline{\gamma}$, and that if two different cone-beam projections provide the same value of $p(s, \underline{\gamma})$, the contributions must be normalized. The method follows the concept of direct Fourier reconstruction described in Section 26.3.1.

A filtered backprojection type of formulation for cone-beam reconstruction is also possible, see, for example, [27]. If the trajectory is a piecewise smooth path, parameterized mathematically by $\phi \in \Phi \subset R$, a reconstruction formula similar to filtered backprojection can be derived from Eqs. (26.33) and (26.34):

$$f(\underline{x}) = \frac{1}{2} \int\limits_{\Phi} \frac{\tilde{g}(\underline{a}(\phi), \underline{\theta})}{r^2} \bigg|_{\underline{\theta}=\underline{\theta}_{\underline{x}, \phi}} d\phi, \tag{26.35}$$

$$\tilde{g}(\underline{a}(\phi), \underline{\theta}) = \frac{-1}{8\pi^2} \int\limits_{S^2} \left(\int\limits_{S^2} g(\underline{a}(\phi), \underline{\theta}') \delta'(\underline{\theta}' \cdot \underline{r}) d\underline{\theta}' \right) \delta'(\underline{\theta}' \cdot \underline{\gamma}) M(\underline{\gamma}, \phi) |\underline{a}'(\phi) \cdot \underline{\gamma}| d\underline{\gamma}. \tag{26.36}$$

Here $r = \|\underline{x} - \underline{a}(\phi)\|$, $\underline{\theta}_{\underline{x}, \phi} = (\underline{x} - \underline{a}(\phi))/\|\underline{x} - \underline{a}(\phi)\|$ is the line passing through $\underline{x}$ for the $\underline{a}(\phi)$ projection, and the function M must be chosen to normalize multiple contributions in the B domain [27]. The normalization condition is $1 = \sum_{k=1}^{n(\gamma,s)} M(\underline{\gamma}, \phi_k)$, where $n(\underline{\gamma}, s)$ is the number of vertices lying in the plane with unit normal $\underline{\gamma}$ and displacement s, and $\phi_1, \phi_2 \ldots \phi_{n(\gamma,s)}$ indicate the vertex locations where the path $\underline{a}(\Phi)$ intersects the plane. By Tuy's condition, $n(\underline{\gamma}, s) > 0$ for $|s| < R$.

These equations must be tailored to the specific application. When the variables are changed to reflect the planar detector arrangement specified at the beginning of this subsection, the above equations resemble the Feldkamp algorithm with a much more complicated "filtering" step. To simplify notation, we write A for the varying distance $\|\underline{a}(\phi)\|$ of the vertex from the origin.

$$f(\underline{x}) = \frac{1}{2} \int_{\Phi} \left(\frac{A^2 + u^2 + v^2}{r^2} \tilde{g}(\underline{a}(\phi), u, v) \right) \Bigg|_{\substack{u = u_{\underline{x}, \phi} \\ v = v_{\underline{x}, \phi}}} d\phi, \tag{26.37}$$

$$\tilde{g}(\underline{a}(\phi), u, v) = \frac{-1}{4\pi^2} \int_0^\pi \left(\frac{d}{dt} T(\phi, \underline{\gamma}) \frac{\sqrt{A^2 + t^2}}{A^2} \left(\frac{d}{dt} \int_R \frac{A}{\sqrt{A^2 + u'^2 + v'^2}} g(\underline{a}, u', v') dl \right) \right) \Bigg|_{\substack{t = u \cos \mu \\ + v \sin \mu}} d\mu, \tag{26.38}$$

where, in the innermost integration, $(u', v') = (t \cos \mu - l \sin \mu, t \sin \mu + l \cos \mu)$; the function $T(\phi, \underline{\gamma}) = |\underline{a}'(\phi) \cdot \underline{\gamma}| M(\underline{\gamma}, \phi)$ contains all the dependency on the particular trajectory. Note that in (26.38), $\underline{\gamma} = (A \cos \mu \underline{e}_u + A \sin \mu \underline{e}_v + t \underline{e}_w)/(\sqrt{A^2 + t^2})$ where the detector coordinate axes are $\underline{e}_u$ and $\underline{e}_v$, and $\underline{e}_w = -\underline{a}(\phi)/A$.

Although these equations are only valid when the cone-beam configuration satisfies Tuy's condition, the algorithm of Eqs. (26.37) and (26.38) collapses to the Feldkamp algorithm when a circular trajectory is specified. This general algorithm has been refined and tailored for specific applications involving truncated projections. Practical methods have been published for the case of source trajectories containing a circle.

26.5.3 Helical Multi-Slice CT Imaging

The case of a helical trajectory is particularly important for the lastest generation of CT scanners with many detector rows. The framework of (26.35) and (26.36) does not apply directly to helical CT scanning because the projections in any practical CT scanner will always be truncated in the axial direction. Furthermore, different segments of the helical trajectory are used to reconstruct different parts of the patient. The methods used to treat these problems appeal to specific properties of the helical trajectory, such as the PI-line, which is the unique line passing through a specified point in the image and connecting two trajectory points within a single turn of the helix. Using these properties, recent theoretical developments in helical image reconstruction are being described in a framework similar to (26.35) and (26.36), with careful choices for the function M. The goal of helical cone-beam image reconstruction is to find a fast, accurate filtered back-projection algorithm that can handle the axial truncation with the minimum detector usage yet with the flexibility to incorporate redundant information when a small pitch is used for improved signal-to-noise ratios in the images.

In the late 1990s, advances in detector technology allowed complete transitioning of X-ray CT from a 2D tool toward a fully 3D imaging technique. This transition had been initiated in 1990 with the spiral data acquisition concept, which allowed faster axial coverage than the old step-and-shoot mode. However, the gain was modest until 1998, when the single detector row could finally be replaced by four detector rows that quadrupled the axial coverage without compromising temporal resolution. Soon after, systems with 16, 32, 64, and even 320 detector rows (Toshiba, 2008) found their way onto the medical market so that imaging the full thorax in less than 5 seconds is nowadays easily achieved by some systems.

In a multirow scanner, the detector rows are stacked one above the other in the direction of the rotation axis and data acquisition proceeds with translation of the patient bed while the source detector assembly rotates, thus defining a helical cone-beam data acquisition geometry. The collected data is a function of three parameters and is denoted here as $g(\lambda, \alpha, w)$, with w giving the z position of the detector row relative to the z position of the X-ray source, while λ and α are the usual parameters of fan-beam data acquisition with equiangular rays. (Note that we call the z-axis the translation axis of the patient bed.) Formally,

$$g(\lambda, \alpha, w) = \int_0^\infty f\left(a(\lambda) + \frac{t}{\sqrt{w^2 + D^2}}(D\sin\alpha\, e_u(\lambda) + D\cos\alpha\, e_v(\lambda) + we_w(\lambda))\right) dt, \quad (26.39)$$

where $a(\lambda) = (R\cos\lambda, R\sin\lambda, h\lambda)$ is the source position with $2\pi h$ the distance covered by turn, where D is the curvature radius of the detector rows, and where f is the linear attenuation coefficient to be reconstructed. Also,

$$e_u(\lambda) = (-\sin\lambda, \cos\lambda, 0),$$
$$e_v(\lambda) = (-\cos\lambda, -\sin\lambda, 0), \quad\quad\quad (26.40)$$
$$e_w(\lambda) = (0, 0, 1),$$

are a set of unit orthogonal vectors that are convenient for a mathematical description of the data.

Many reconstruction techniques have been developed for accurate reconstruction of f from g, some approximate, others theoretically-exact. We present here the theoretically-exact method of Katsevich [24], which represented a breakthrough and has led to the development of many reconstruction algorithms since its development in 2002. This method efficiently yields f at any given location P as a backprojection of filtered data. The backprojection is performed only over a finite interval in λ that varies with P and is denoted as $I(P)$ here. The endpoints, $\lambda_i(P)$ and $\lambda_o(P)$, of this interval define two source positions that lie on a line through P and are separated by less than a helix turn (i.e., $|\lambda_i(P) - \lambda_o(P)| < 2\pi$). Danielsson showed in 1997 that this particular backprojection interval is uniquely defined for each P within the field-of-view of the scanner, while Tam showed at the same time how masking the data so as to constrain the backprojection to $I(P)$ can improve the reconstruction quality. Nowadays, the backprojection interval $I(P)$ is commonly called the Tam-Danielsson window [25].

Katsevich's method takes the form

$$f(P) = \frac{1}{2\pi} \int\limits_{\lambda_i(P)}^{\lambda_o(P)} \frac{g_F(\lambda,\alpha^*,w^*)}{v^*} d\lambda, \tag{26.41}$$

where α^* and w^* are the detector coordinates of the line through P, and v^* is the distance between P and the source after projection in a plane orthogonal to the z axis. When denoting the coordinates of P as x, y, z, these quantities are

$$v^* = R - x\cos\lambda - y\sin\lambda,$$

$$\alpha^* = \operatorname{atan}\left(\frac{1}{v^*}(-x\sin\lambda + y\cos\lambda)\right), \tag{26.42}$$

$$w^* = \frac{D\cos\alpha^*}{v^*}(z - h\lambda).$$

The filtered data, g_F, is obtained by convolving a differentiated version of the data, g', along a family of curves on the detector surface. These curves can be parameterized by an angle ψ following the curve equation

$$\hat{w}(\alpha;\psi) = \frac{Dh}{R}\left(\psi\cos\alpha + \frac{\psi}{\tan\psi}\sin\alpha\right). \tag{26.43}$$

The required range for ψ is $\psi \in [-\pi/2 - \alpha_m, \pi/2 + \alpha_m]$ where α_m is the maximum fan angle. The computation of g_F from g' proceeds in four steps:

- Step 1: Apply a length-correction weight to obtain $g_2(\lambda,\alpha,w) = \frac{Dg'(\lambda,\alpha,w)}{\sqrt{w^2+D^2}}$.

- Step 2: Interpolate the data so as to create data onto the ψ curves, according to $g_3(\lambda,\alpha,\psi) = g_2(\lambda,\alpha,w(\hat{\alpha};\psi))$.

- Step 3: Apply an angular 1D Hilbert transform in α to get

$$g_4(\lambda,\alpha,\psi) = \int\limits_{-\pi/2}^{\pi/2} h_H(\sin(\alpha - \gamma))g_3(\lambda,\gamma,\psi)d\gamma, \tag{26.44}$$

where h_H is the 1D inverse Fourier transform of $-i\operatorname{sgn} v$.

- Step 4: Rebin the convolved data back into the original detector geometry and postweight the result with $\cos\alpha$, to get

$$g_F(\lambda,\alpha,w) = \cos\alpha\, g_4(\lambda,\alpha,\Psi(\alpha,w)), \tag{26.45}$$

where $\psi(\alpha, w)$ is the solution of smallest absolute value of the equation

$$w = \frac{Dh}{R}\left(\psi\cos\alpha + \frac{\psi}{\tan\psi}\sin\alpha\right) \tag{26.46}$$

in ψ.

The differentiated data from which g_F is calculated is obtained by differentiating the original data while maintaining constant the ray direction, and may be expressed in the form

$$g'(\lambda, \alpha, w) = \left(\frac{\partial}{\partial \lambda} + \frac{\partial}{\partial \alpha} \right) g(\lambda, \alpha, w). \tag{26.47}$$

Details on how to achieve a numerically robust implementation of the various steps involved in the Katsevich method can be found in [25] and [26].

26.6 ITERATIVE RECONSTRUCTION METHODS

26.6.1 Finite Dimensional Formulations and Algebraic Reconstruction Technique

As noted above, the line integral model on which all of the preceding methods are based is only approximate. Furthermore, there is no explicit modeling of noise in these approaches; noise in the data is typically reduced by tapering off the response of the projection filters before backprojection. In X-ray CT, the beam is highly collimated, the detectors are high resolution, and the number of photons per measurement is very large; consequently the line integral approximation is adequate to produce low noise images at submillimeter resolution in humans. However, this may not be the case in industrial and other nonmedical applications, and these systems may benefit from more accurate modeling of the data and noise. In the case of PET and SPECT, the often low intrinsic resolution of detectors, depth dependent and geometric resolution losses, and the typically low photon count, can lead to rather poor resolution at acceptable noise levels when using direct reconstruction methods. An alternative to the direct approach is to use a finite dimensional model in which the detection system and the noise statistics can be modeled more accurately. Research in this area has lead to the development of a large class of reconstruction methods that often outperform the direct methods that often outperform the direct methods, as illustrated in Fig. 26.15.

We will assume that the image is adequately represented using a finite set of basis functions. While there has been some interest in alternative basis elements, almost all researchers currently use a cubic voxel basis function. Each voxel is an indicator function on a cubic region centered at one of the image sampling points in a regular 2D or 3D lattice. The image value at each voxel is proportional to the quantity being imaged integrated over the volume spanned by the voxel. To allow a unified treatment of 2D and 3D problems, a single index will be used to represent the lexicographically ordered elements of the image $f = \{f_1, f_2 \ldots f_N\}$. Similarly, the elements of the measured projections will be represented in lexicographically ordered form as $y = \{y_1, y_2 \ldots y_M\}$.

In X-ray CT, we can model the attenuation of a finite width X-ray beam as the integral of the linear attenuation coefficient over the path (or strip) through which the

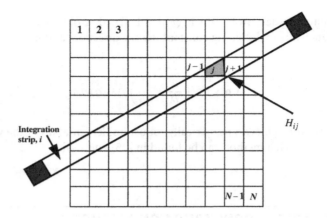

FIGURE 26.14

Example of a PET scan of metabolic activity using FDG, an F-18 tagged analog of glucose. This tracer is used in detection of malignant tumors. The left image shows a reconstruction of a slice through the chest of a patient with breast cancer; the tumor is visible in the bright region in the upper left region of the chest. This image was reconstructed using the Bayesian method described in [33] and illustrates the improvement in image quality that can be realized using a statistically based approach, in comparison to a direct reconstruction method which was used to reconstruct the right image from the same data.

beam passes. Thus the measurements can be written as

$$y_i = \int\!\!\int_{x,y \in \text{strip}\{i\}} f(x,y)\,dx dy = \sum_{j=1}^{N} H_{ij}f_j, \tag{26.48}$$

where f_j is the attenuation coefficient at the jth voxel. The elements $H(i,j)$ of the projection matrix H are equal to the area of intersection of the ith strip with the indicator function on the jth voxel. Equation (26.48) represents a huge set of simultaneous linear equations, $y = Hf$, that can be solved to compute the CT image f (see Fig. 26.14). In principle the system can be solved using standard methods. However, the size of these systems coupled with the special structure of H motivated research into more efficient specialized numerical procedures. These methods exploit the key property that H is very sparse, i.e., most elements in the matrix are zero, since the path along which each integration is performed intersects only a small fraction of the image pixels.

One algorithm that makes good use of the sparseness property is the algebraic reconstruction technique (ART) [28]. This method finds the solution to the set of equations in an iterative fashion through successive orthogonal projection of the current image estimate onto hyperplanes defined by each row of H. If this procedure converges, the solution will be a point where all of the hyperplanes intersect, i.e., a solution to (26.48). Let f^n represent the vector of image pixel values at the nth iteration, and let h_i^T represent

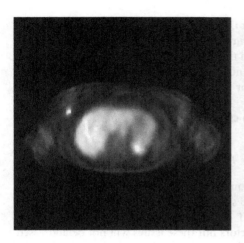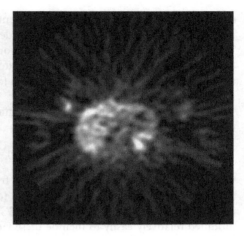

FIGURE 26.15

Illustration of the pixel-based finite dimensional formulation used in iterative X-ray CT reconstruction. The matrix element H_{ij} gives the contribution of the jth voxel to the ith measurement and is proportional to the areas of intersection of the voxel with the strip that joins the source and detector.

the ith row of H. The ART method has the following form:

$$f^{n+1} = f^n + \left(\frac{y_i - h_i^T f^n}{h_i^T h_i} \right) h_i \qquad i = (n \bmod N) + 1. \tag{26.49}$$

ART can also be viewed in terms of the backprojection operator used in filtered backprojection: each iteration of (26.49) is equivalent to adding to the current image estimate f^n the weighted backprojection of the error between the ith measured projection sample and the projection corresponding to f^n. ART will converge to a solution of (26.48) provided the system of equations is consistent. In the inconsistent case, the iterations will not converge to a single solution, and the properties of the image at a particular stopping point will be dependent on the sequence in which the data are ordered. Many variations of the ART can be found in the literature. These variations exhibit differences in convergence behavior, sensitivity to noise, and optimality properties [29].

26.6.2 **Statistical Formulations**

The ART method does not directly consider the presence of noise in the data. While acceptable in high SNR X-ray CT data, the low photon counting statistics found in PET and SPECT should be explicitly considered. The finite dimensional formulation in Section 26.6.1 can be extended to model both the physics of PET and SPECT detection and the statistical fluctuations due to noise.

Rather than simply assume a strip integral model as in (26.48), we can instead use the matrix relating image and data to more exactly model the probability, P_{ij}, of detecting an emission from voxel site j at detector element i[30, 33]. To differentiate this probabilistic model from the strip integral one, we will denote the detection probability matrix by P. The elements of this matrix are dependent on the specific data acquisition geometry and other factors such as detector efficiency, attenuation effects within the subject, and the underlying physics of gamma ray emission for SPECT or positron-electron annihilation for PET.

In PET and SPECT, the mean of the data can be estimated for a particular image as the linear transformation

$$E(y) = Pf, \tag{26.50}$$

where f represents the mean emission rates from each image voxel. In practice, these data are corrupted by additive noise terms due to scatter and either "random coincidences" in PET [7] or background radiation in SPECT [6]. The methods described below can be modified relatively easily to include these factors but these issues will not be addressed further here. See [31] and [32] for a more in depth review of these and other factors that affect statistical reconstruction methods in PET and SPECT.

In both PET and SPECT, external radiation sources are used to perform transmission measurements. These are used to correct for attenuation in the emission studies. Just as in X-ray CT, it is possible to reconstruct an image of attenuation coefficients from these transmission measurements. In this case, the unknown image f is the set of attenuation coefficients for each voxel and the transition probability matrix P, in the simplest case, has elements P_{ij} equal to the fractional intersection of the ith projection ray with the jth voxel. If $E(y)$ represents the mean value of the transmission measurement, and assuming that the source intensity is a constant α, then we can model the mean of the transmission data as

$$E(y_i) = \alpha \exp\left\{ -\sum_j P_{ij} f_j \right\}. \tag{26.51}$$

For both emission and transmission measurements, the data can be modeled as collections of independent Poisson random variables, mean Ey, with joint probability:

$$p(y/f) = \prod_{i=1}^{M} \frac{E(y_i)^{y_i} e^{-E(y_i)}}{y_i!}. \tag{26.52}$$

The physical model for the detection system is included in the likelihood function in the mapping from the image f to the mean of the detected events $E(y)$ using (26.50) and (26.51) for the transmission and emission case, respectively. Using this basic model, we can develop estimators based on maximum likelihood (ML) or Bayesian image estimation principles.

26.6.3 Maximum Likelihood Methods

The ML estimator is the image that maximizes the likelihood (26.52) over the set of feasible images, $f \geq 0$. The EM (expectation maximization) algorithm can be applied to the emission CT problem resulting in an iterative algorithm which has the elegant closed form update equation [34]:

$$f_j^{n+1} = \frac{f_j^n}{\sum_i P_{ij}} \sum_i \frac{P_{ij} y_i}{\sum_l P_{il} f_l^n}. \tag{26.53}$$

This algorithm has a number of interesting properties including the fact that the solution is naturally constrained by the iteration to be nonnegative. Unfortunately the method tends to exhibit very slow convergence and is often unstable at higher iterations. The variance problem is inherent in the ill-conditioned Fisher information matrix. This effect can be reduced using ad hoc stopping rules where the iterations are terminated before convergence. An alternative approach to reducing variance is through penalized ML or Bayesian methods as described in Section 26.6.4.

A number of modifications of the EM algorithm have been proposed to speedup convergence. The most widely used of these is the ordered subsets EM (OSEM) algorithm in which each iteration uses only a subset of the data [35]. Let $\{S_k\}$, $k = 1, \dots Q$, be a disjoint partition of the set $\{1, 2, \dots M\}$ representing the indices of the data. Let n denote the iteration number, defined as the number of complete cycles through the Q subsets, and define $f_j^{(n,0)} = f_j^{(n-1,Q)}$. Then one complete iteration of OSEM is given by

$$f_j^{(n,k)} = \frac{f_i^{(n,k-1)}}{\sum_{i \in S_k} P_{ij}} \sum_{i \in S_k} \frac{P_{ij} y_i}{\sum_l P_{il} f_l^{(n,k-1)}} \quad \text{for } j = 1, \dots N; \; k = 1, \dots Q. \tag{26.54}$$

In the early iterations, OSEM produces remarkable improvements in convergence rates compared to EM, although subsequent iterations over the entire data is required for ultimate convergence. The OSEM algorithm has now been widely adopted for reconstruction of clinical PET scans since it achieves significant improvements in image quality for photon-limited data when compared to analytic methods, and does so in clinically acceptable reconstruction times. In the case of 3D PET, computation cost is further reduced by using the Fourier rebinning method of Section 26.4.3 to reduce the data to a set of 2D sinograms (one per axial slice) and then applying OSEM to each 2D slice in turn [36].

The corresponding ML problem for transmission data does not have a closed form EM algorithm. However, both emission and transmission ML problems can be solved effectively using standard gradient ascent aproaches such as the conjugate gradient method. In fact it is easily shown that the emission EM algorithm can also be written as a steepest descent algorithm with a diagonal preconditioner equal to the current image estimate [37].

In recent years, a great deal of progress has been made in developing fast algorithms for emission and transmission tomography that borrow ideas from both the EM and OSEM approaches to generate convergent algorithms. The EM algorithm can be viewed as one of a general class of "optimization transfer" methods [38]. These methods replace the original objective function $\Phi(f; y)$ at each iteration with a surrogate function $\Psi(f, f^n; y)$

that satisfies the following properties:

$$\Phi(f^n;y) = \Psi(f^n,f^n;y), \qquad (26.55)$$

$$\Phi(f;y) \geq \Psi(f,f^n;y). \qquad (26.56)$$

These two conditions in combination can be used to ensure that if f^{n+1} is chosen as the maximizer of $\Psi(f,f^n;y)$, then $\Phi(f^n;y)$ is a non-decreasing sequence in f^n. The surrogates are chosen so that they are more easily optimized than the original objective at each iteration. Careful choice of these objective functions can lead to closed form update equations, even for the tramission reconstruction problem, and to faster convergence than that exhibited by the original EM algorithm.

The second approach to development of more rapidly converging algorithms is to use subsets of the data at each iteration. As implemented in the original OSEM algorithm, this does not lead to convergence. However, it is possible to use this concept within a globally convergent framework. Methods of this type are referred to as "incremental gradient" methods and are based on rewriting the objective function as a sum over subobjective functions [39]:

$$\Phi(f;y) = \sum_{k=1}^{Q} \Phi_k(f;y_k), \qquad (26.57)$$

where $\Phi_k(f;y_k)$ represents the portion of the objective function that is dependent only on the subset of the data: $\{y_i : (i \in S_k)\}$. Iteratively updating the image estimate with respect to each subobjective in turn and using an appropriate relaxation scheme can lead to rapid initial convergence and guarantee global convergence. Since the surrogate and incremental gradient approaches are general, they are applicable to both emission and transmission reconstruction, and also to both the ML problem described above and the Bayesian or penalized ML formulations described below.

26.6.4 Bayesian Reconstruction Methods

As noted above, direct ML estimates of PET images exhibit high variance due to ill-conditioning. Some form of regularization is required to produce acceptable images. Often regularization is accomplished simply by starting with a smooth initial estimate and terminating a ML search before convergence. Here we consider explicit regularization procedures in which a prior distribution is introduced through a Bayesian reformulation of the problem. Some authors prefer to present these regularization procedures as penalized ML methods but the differences are largely semantic.

By introduction of random field models for the unknown image, Bayesian methods can address the ill-posedness inherent in PET image estimation. In an attempt to capture the locally structured properties of images, researchers in emission tomography, and many other image processing applications, have adopted Gibbs distributions as a suitable class of prior [40]. The Markovian properties of these distributions make them both theoretically attractive as a formalism for describing empirical local image properties, as well as computationally appealing, since the local nature of their associated energy functions results in computationally efficient update strategies. The majority of work using Gibbs distributions in tomographic applications involves relatively simple pairwise interaction models where the Gibbs energy function is formed as a sum of potentials, each defined on neighboring pairs of pixels. These potential functions can be chosen to reflect the piecewise smooth property of many images. The existence of sharp intensity changes, corresponding to the edges of objects in the image, can also be modeled using more complex MRF (Markov random field) models. The Bayesian formulation also offers the potential for combining data from multiple modalities. For example, high resolution anatomical X-ray CT or MR images can be used to improve the quality of reconstructions from low resolution PET or SPECT data [41].

Let $p(f)$ denote the Gibbs prior that captures the expected statistical characteristics of the image. The posterior probability for the image conditioned on the data is then given by Bayes theorem:

$$p(f|y) = \frac{p(y|f)p(f)}{p(y)}. \tag{26.58}$$

Bayesian estimators in tomography are usually of the maximum a posteriori (MAP) type. The MAP solution is given by maximizing the posterior probability $p(f|y)$ with respect to f. For each dataset, the denominator of the right-hand side of (26.58) is a constant so that the MAP solution can be found by maximizing the log of the numerator, i.e.,

$$\max_{f} \quad \mathrm{Ln}\, p(y|f) + \mathrm{Ln}\, p(f). \tag{26.59}$$

A large number of algorithms have been developed for computing the MAP solution. The EM algorithm (26.53) can be extended to include a prior term (see for example [41]) and hence maximize (26.59). This algorithm suffers from the same slow convergence problems as (26.53). Alternatively, (26.59) can be maximized using standard nonlinear optimization algorithms such as the preconditioned conjugate gradient method [33] or coordinate-wise optimization [42]. The specific algorithmic form is found by applying these standard methods to (26.59) after substituting both the log of the likelihood function (26.52) in place of Ln $p(y|f)$ and the log of the Gibbs density in place of Ln $p(y|f)$. Alternatively, optimization transfer and incremental gradient methods can be modified to include prior distribution and have been used effectively to design rapidly converging algorithms. For compound Gibbs priors that involve line processes, mean field annealing techniques can be combined with any of the above methods [41],[32].

26.7 SUMMARY

We have summarized direct and iterative approaches to 2D and 3D tomographic reconstruction for X-ray CT, PET, and SPECT. With the exception of the rebinning algorithms, which can be used in place of fully 3D reconstruction methods, the choice of direct reconstruction algorithm is determined primarily by the data collection geometry. On the other hand, the iterative approaches (ART, ML, and MAP) can be applied to any collection geometry in PET and SPECT. Furthermore, after appropriate modifications to account for differences in the mapping from image to data, these methods are also applicable to transmission PET and SPECT data. X-ray CT data are not Poisson so a different likelihood model is required if ML or MAP methods are to be used.

Image processing for CT remains an active area of research. In large part, development is driven by construction of new imaging systems which are continuing to improve the resolution of these technologies. Carefully tailored reconstruction algorithms will help to realize the full potential of these new systems. In the realm of X-ray CT, new spiral and cone-beam systems are extending the capabilities of CT systems to allow fast volumetric imaging for medical and other applications. In PET and SPECT, recent developments are also aimed at achieving high-resolution volumetric imaging through combinations of new detector and collimator designs with fast, accurate reconstruction algorithms. In addition to advances resulting from new instrumentation developments, current areas of intense research activity include theoretical analysis of algorithm performance, combining accurate modeling with fast implementations of iterative methods, direct methods that account for factors not included in the line integral model, and development of methods for fast dynamic volumetric (4D) imaging.

REFERENCES

[1] H. H. Barrett and W. Swindell. *Radiological Imaging*, Vols. I and II. Academic Press, New York, 1981.

[2] S. Webb. *From the Watching of Shadows: The Origins of Radiological Tomography*. Institute of Physics, 1990.

[3] H. P. Hiriyannaiah. X-ray computed tomography. *IEEE Signal Process. Mag.*, 14(2):42–59, 1997.

[4] A. C. Kak and M. Slaney. *Principles of Computerized Tomographic Imaging*. IEEE Press, New York, 1988.

[5] S. Webb, editor. *The Physics of Medical Imaging*. Institute of Physics, London, 1988.

[6] G. Gullberg, G. Zeng, F. Datz, R. Christian, C. Tung, and H. Morgan. Review of convergent beam tomography in single photon emission computed tomography. *Phys. Med. Biol.*, 37:507–534, 1992.

[7] S. R. Cherry and M. E. Phelps. Imaging brain function with positron emission tomography. In A. W. Toga and J. C. Mazziotta, editors, *Brain Mapping: the Methods*. Academic Press, New York, 1996.

[8] D. MacLaren, T. Toyokuni, S. Cherry, J. Barrio, M. E. Phelps, H. R. Herschman, and S. S. Gambhir. PET imaging of transgene expression. *Biol. Psychiatry*, 48(5):337–348, 2000.

[9] S. R. Cherry, Y. Shao, R. W. Silverman, K. Meadors, S. Siegel, A. Chatziioannou, J. W. Young, W. F. Jones, J. C. Moyers, D. Newport, A. Boutefnouchet, T. H. Farquhar, M. Andreaco, M. J. Paulus, D. M. Binkley, R. Nutt, and M. E. Phelps. MicroPET: a high resolution PET scanner for imaging small animals. *IEEE Trans. Nucl. Sci.*, 44:1161–1166, 1997.

[10] B. Bendriem and D. W. Townsend, editors. *The Theory and Practice of 3D PET*. Kluwer, 1998.

[11] L. A. Shepp and B. F. Logan. The Fourier reconstruction of a head section. *IEEE Trans. Nucl. Sci.*, NS-21:21–33, 1974.

[12] B. K. Horn. Fan-beam reconstruction methods. *Proc. IEEE*, 67:1616–1623, 1979.

[13] G. Besson. CT fan-beam parameterizations leading to shift-invariant filtering. *Inverse Probl.*, 12:815–833, 1996.

[14] F. Noo, M. Defrise, R. Clackdoyle, and H. Kudo. Image reconstruction from fan-beam projections on less than a short scan. *Phys. Med. Biol.*, 47:2525–2546. 2002.

[15] R. Clackdoyle and F. Noo. A large class of inversion formulae for the Radon transform of functions of compact support. *Inverse Probl.*, 20:1281–1291, 2004.

[16] F. Noo, R. Clackdoyle, J. D. Pack. A two-step Hilbert transform method for 2D image reconstruction. *Phys. Med. Biol.*, 49:3903–3923, 2004.

[17] C. Crawford and K. King. Computed tomography scanning with simultaneous patient translation. *Med. Phys.*, 17:967–982, 1990.

[18] M. Daube-Witherspoon and G. Muehllehner. An iterative image space reconstruction algorithm suitable for volume ECT. *IEEE Trans. Med. Imaging*, 5:61–66, 1986.

[19] M. Defrise, P. Kinahan, D. Townsend, C. Michel, M. Sibomana, and D. Newport. Exact and approximate rebinning algorithms for 3D PET data. *IEEE Trans. Med. Imaging*, 16:145–158, 1997.

[20] J. Colsher. Fully three dimensional positron emission tomography. *Phys. Med. Biol.*, 25:103–115, 1980.

[21] P. Kinahan and L. Rogers. Analytic 3D image reconstruction using all detected events. *IEEE Trans. Nucl. Sci.*, NS-36:964–968, 1996.

[22] L. Feldkamp, L. Davis, and J. Kress. Practical cone-beam algorithm. *J. Opt. Soc. Am. A*, 1:612–619, 1984.

[23] P. Grangeat. Mathematical framework for cone-beam three-dimensional image reconstruction via the first derivative of the Radon transform. In G. Herman, A. Louis, and F. Natterer, editors, *Mathematical Methods in Tomography (Lecture Notes in Mathematics)*, Vol. 1497, 66–97. Springer, Berlin, 1991.

[24] A. Katsevich. Theoretically exact filtered backprojection-type inversion algorithm for spiral CT. *SIAM J. Appl. Math.*, 62:2012–2026, 2002.

[25] F. Noo, J. D. Pack, and D. Heuscher. Exact helical reconstruction using native cone-beam geometries. *Phys. Med. Biol.*, 48:3787–3818, 2003.

[26] F. Noo, S. Hoppe, F. Dennerlein, G. Lauritsch, and J. Hornegger. A new scheme for view-dependent data differentiation in fan-beam and cone-beam computed tomography. *Phys. Med. Biol.*, 52: 5393–5414, 2007.

[27] M. Defrise and R. Clack. A cone-beam reconstruction algorithm using shift-variant filtering and cone-beam backprojection. *IEEE Trans. Med. Imaging*, 13:186–195, 1994.

[28] G. T. Herman. *Image Reconstruction From Projections: The Fundamentals of Computerized Tomography*. Academic Press, New York, 1980.

[29] Y. Censor. Finite series-expansion reconstruction methods. *Proc. IEEE*, 71:409–419, 1983.

[30] M. Smith, C. Floyd, R. Jaszczak, and E. Coleman. Three-dimensional photon detection kernels and their application to SPECT reconstruction. *Phys. Med. Biol.*, 37:605–622, 1992.

[31] J. M. Ollinger and J. A. Fessler. Positron emission tomography. *IEEE Signal Process. Mag.*, 14(1): 43–55, 1997.

[32] R. Leahy and J. Qi. Statistical approaches in quantitative PET, to appear. *Stat. Comput.*, 10:147–165, 2000.

[33] J. Qi, R. Leahy, S. Cherry, A. Chatziioannou, and T. Farquhar. High resolution 3D bayesian image reconstruction using the microPET small animal scanner. *Phys. Med. Biol.*, 43(4):1001–1013, 1998.

[34] A. Shepp and Y. Vardi. Maximum likelihood reconstruction for emission tomography. *IEEE Trans. Med. Imaging*, 2:113–122, 1982.

[35] H. Hudson and R. Larkin. Accelerated image reconstruction using ordered subsets of projection. *IEEE Trans. Med. Imaging*, 13:601–609, 1994.

[36] C. Comtat, P. E. Kinahan, M. Defrise, C. Michel, and D. W. Townsend. Fast reconstruction of 3D PET data with accurate statistical modeling. *IEEE Trans. Nucl. Sci.*, 45:1083–1089, 1998.

[37] L. Kaufman. Maximum likelihood, least squares, and penalized least squares for PET. *IEEE Trans. Med. Imaging*, 12:200–214, 1993.

[38] K. Lange, D. R. Hunter, and I. Yang. Optimization transfer using surrogate objective functions (with discussion). *J. Comput. Graph. Stat.*, 9:1–59, 2000.

[39] S. Ahn and J. A. Fessler. Globally convergent image reconstruction for emission tomography using relaxed ordered subsets algorithms. *IEEE Trans. Med. Imaging*, 22(5):613–626, 2003.

[40] S. Geman and D. McClure. Statistical methods for tomographic image reconstruction. *Proc. 46th Session of the International Statistical Institute, Bulletin of the ISI*, 52:4–20, 1987.

[41] G. Gindi, M. Lee, A. Rangarajan, and I. G. Zubal. Bayesian reconstruction of functional images using anatomical information as priors. *IEEE Trans. Med. Imaging*, 12:670–680, 1993.

[42] J. Fessler. Hybrid polynomial objective functions for tomographic image reconstruction from transmission scans. *IEEE Trans. Image Process.*, 4:1439–1450, 1995.

Computer-Assisted Microscopy

27

Fatima A. Merchant[1] and Kenneth R. Castleman[2]

[1] *University of Houston;* [2] *The University of Texas at Austin*

27.1 INTRODUCTION

Recent advances in the field of microscopy have been driven not only by technological innovations in engineering, optics, computer science, and precision manufacturing but also by fundamental discoveries in chemical and biological sciences. The last two decades have seen various changes to the microscope, resulting in the current modern microscope, which is a powerful tool in biological research and development. Although most of the changes pertain to advances in optics, today's microscopes are also fitted with ergonomic features, and they include automation of several manual functions. Microscopes today offer automated focusing, selection of objectives and filters, as well as light control and a wide range of other features. Moreover, the increasing complexity of biological experiments performed with optical microscopy, and the complexity of data that can be obtained, has resulted in even more sophisticated optical microscopy systems [1–5].

Traditionally, microscopy-based research has relied on visual interpretation and qualitative descriptions of the observed samples by experts. The most important benefit of this approach is that the robustness of the human visual system, coupled with advanced cognitive abilities, allows trained personnel to perceive variations in illumination and contrast, recognize significant objects or events based on appearance, and differentiate among a large variety of normal and abnormal phenotypes. The limiting factors are that manual analysis can become tedious and time consuming, and it is not scalable for large studies.

Although a completely automated microscope is not required in most applications, there are a growing number of applications for which it is, such as medium-large scale screening of specimens with different protocols, deconvolution, fluorescence resonance energy transfer (FRET) imaging, multi-spectral imaging, and ion ratio imaging [6]. Furthermore, computer technology is changing the ways we access equipment, view samples, and record, manage, and disseminate images. Digital imaging has created the need for archiving, managing, manipulating, and quantifying images. The coupling of computers to microscope systems has resulted in the development of optical imaging systems that can perform complicated experiments, and provide more data, convenient storage of and access to the data, and, perhaps most importantly, substantial aid in the analysis of large volumes of data.

Quantitative methods for computer-assisted morphometry and cytometry have become well established during the past decade [7–9]. In complex experiments involving a myriad of peripheral devices, computer technology is useful for controlling filter wheels, shutters, automated stages, and cameras. Computer-assisted microscopy allows the extraction of quantitative, reproducible, diagnostically relevant information while reducing subjective influences of human operators. It increases the quality and relevance of data and enables comparison and verification of the results. Standardized and automated systems are needed to generate consistent results among different experiments and laboratories, while enhancing productivity and realizing cost savings. The advantages of computerized microscopy include the ability to (1) control and synchronize the functions of multiple peripheral devices, (2) extract information using quantitative methods, (3) analyze data to determine statistical significance, (4) reduce cost by relieving humans of tedious duties, and (5) support significantly expanded testing volumes without concomitant requirements for additional personnel and floor space.

Computerized microscopy is useful in (1) color and monochrome brightfield applications with thick specimens that require multiple optical sections (i.e., z-planes), (2) applications in fluorescence microscopy that require the acquisition of more than one wavelength, (3) applications that require large specimens to be scanned, (i.e., when adjacent areas need be scanned and "tiled" together to create a much larger image), and (4) time-lapse or motion analysis applications. Although computerized microscopy systems expedite data collection by automating the process, the real power of the system is the extraction from the images of useful, quantitative information that previously required human interpretation.

Generic digital image processing algorithms can often be applied successfully, but they sometimes fail to exploit fully the information contained in microscope images. This is because the optics that produced a particular image are poorly understood. Computerized microscopy systems need to tailor image processing algorithms to particular forms of microscopy such as brightfield, darkfield, phase contrast, interference contrast, fluorescence, and confocal systems [10]. Further, the theory of microscopy is complicated, and agreement with data is less than perfect. The challenge is to synthesize theoretical models and empirical evidence in order to tackle particular image analysis problems [11].

Automated image analysis system design begins by specifying requirements for the system and then selecting image processing functions that meet those requirements. It is difficult to describe a generic set of algorithms for object identification and classification, and actual systems are often limited to one or a few specific applications. Universal systems, capable of automatically learning and finding desired user information, do not exist. In this chapter, we describe an automated imaging system for use in clinical cytogenetics. Image analysis algorithms specific to particular applications are discussed separately.

We present an overview of what constitutes a microscope imaging system, with special emphasis on automation of image acquisition and analysis, and imaging techniques used in the context of biological applications—specifically clinical cytogenetics. Instrumentation is discussed, including motorized stages and filters, cameras and digitizer boards, software algorithms (including image enhancement, autofocusing, object detection and relocation), and features for operator review and data analysis for computerized microscopy.

27.2 COMPUTER-ASSISTED MICROSCOPY SYSTEMS

The modern microscopy imaging system is composed of five basic components: the microscope, a lighting source, a specimen stage control, an image acquisition device, and a postacquisition image analysis system. The microscope can be practically any high-end optical device, such as those offered by Leica, Meiji, Nikon, Olympus, and Zeiss, which meets the initial resolution and functionality requirements of the experiments for which the instrument will be used. The components of a computer-assisted microscopy system are depicted in Fig. 27.1. The core of any computer-assisted microscopy system is the microscope, which creates an image of the specimen. An image sensor, such as a CCD camera or a photomultiplier tube (PMT), converts the magnified image to an analog voltage signal. The signal is then digitized and stored in a computer for later retrieval or analysis. Several peripheral devices, such as motorized X, Y stages, Z-axis motors, and filter wheels, are fitted to the microscope to perform different kinds of experiments. A host computer interfaces the separate hardware components to the microscope. Communication with the peripheral devices is through hardware boards that reside in the system's host computer and use the computer's internal bus. The PCI bus is well suited for imaging applications due to its high speed of operation and support from many manufacturers. Other types of connections include serial lines (RS-232, RS-485), USB, and the GPIB bus.

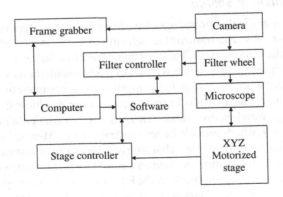

FIGURE 27.1

Components of a computer-assisted microscopy system.

Automation is implemented via software tailored to control the experiment, for example, by automatically selecting the wavelength and intensity of excitation light, positioning the specimen in the field of view (FOV), autofocusing, and capturing single- or multi-channel images. The microscope and the associated optics and detectors determine the quality of the image data obtained, and care must be taken to match the components to obtain the best possible results. Overall, the system contains computer software and hardware that can automatically scan microscope slides, find the optimal focus position, capture images, manipulate those images via preprocessing to remove noise, segment the images to identify objects therein, extract relevant information about each object (e.g., size, presence or absence of a diagnostic marker), and finally classify them according to the experimental specifications.

27.2.1 Hardware

The basic components of a computerized microscopy system can be broken down into three fundamental areas. These are (1) automation of the microscope's X-, Y- and Z-axes, (2) control of fluorescence excitation or brightfield illumination, and (3) control of the image sensing device. In most instances, these components can be added easily to any commercially available microscope. Motorized stages and filter wheels are readily available and can be custom fitted to a variety of different microscopes. Similarly, focus drives are available to control the Z-axis position of the stage (or the microscope head in the case of fixed stage microscopes). Automated X, Y stages allow movement of the stage in the X- and Y-axes, shutters control the exposure of both the fluorescent and incandescent light, monochromators or filter wheels are used to select a specific excitation wavelength, and image sensors, such as CCD cameras or PMTs, capture images. Given the number of hardware options available, the selection of components depends on the efficiency and precision requirements of the particular application. Cortese presents a listing of commercial vendors, for a variety of microscope components [12].

27.2.1.1 Illumination Source

Most commercially available microscopes have built-in light sources for brightfield and fluorescence microscopy. For transmitted light microscopy, an incandescent tungsten-halogen bulb supplies illumination. These lamps operate in a "halogen regenerative cycle" wherein the evaporated metal is returned back to the filament, thereby keeping the glass bulb clean and maintaining a constant light output and color temperature throughout the life of the lamp. These lamps are low in cost and have extended lifespan, but are not very bright at the shorter wavelengths. Fluorescence microscopy uses either mercury (Hg) or xenon (Xe) arc lamps, which provide better spectral quality. Mercury lamps operate on vaporized metal plasma and provide a characteristic spectrum of distinct lines at specific wavelengths, whereas xenon lamps provide an output that is constant, with more spectral lines but slightly reduced brightness. Both the Hg and Xe arc lamps require careful alignment for uniform sample illumination. Most recently, EFOS Inc. (Missisauga, Ontario) has developed a novel fluorescence illuminator, the X-Cite^{TM}, which delivers a uniform FOV and maximum energy spectrum from a 50 W metal-halide lamp, outperforming

100 W mercury-arc lamps with a 2,000-hour lifespan. Alternatively, Photon Technology International Inc. (Lawrenceville, NJ) has developed the NovaLight™, an illuminator that delivers up to seven times the output of conventional arc lamps. These new arc lamp systems offer longer lifespan, do not require lamp alignment, and provide stable, ozone-free output.

In fluorescence illumination, the excitation light and emission light are controlled by filters inserted in the light path between the illuminator and the specimen, and again between the specimen and the image sensor. A wide variety of filter cubes is available from major manufacturers, who produce filter sets suitable for most of the common fluorophores in use today. Other illumination sources include arc-discharge lamps and light emitting diodes (LEDs). Nonlaser LEDs provide a cheaper and more compact option because of their small size. They can provide light at various wavelengths with low power consumption and low heat generation. Lasers can also be used as light sources, but they are expensive and are commonly employed only for specialized techniques such as confocal fluorescence microscopy, optical trapping, lifetime imaging studies, photobleaching recovery studies, and total internal reflection fluorescence imaging. The most common types of lasers used in microscopy are continuous wave lasers (for confocal microscopy) and pulsed lasers (for multiphoton microscopy) [13].

Mechanical or electronic optical shutters can be used to control the amount of time that the specimen is exposed to illumination. They typically allow various opening and closing patterns and can allow 100% transmission or complete extinction of the light. The minimum opening speed of most commercially available shutters is on the order of 10 milliseconds. In multimode light microscopy applications, they are often used to switch between transmitted and fluorescence imaging, with shutters in both the transmitted light path and the fluorescence path. Shutters come with computer interface controllers that allow both manual and remote opening and closing. Controlling the exposure time can reduce photobleaching in fluorescence imaging applications.

27.2.1.2 *Filter Control*

Filter control on microscopes can be automated in two ways: (1) a motorized filter turret (available on most motorized microscopes) that moves an entire filter cube assembly (including the excitation and emission filters and dichroic mirror) or (2) filters mounted on wheels positioned in either the excitation or emission path, or both.

Filter wheels are used on microscopes, cameras, and video systems to position a selected filter in the imaging path quickly and accurately. Typically, a filter or combination of filters is used to attenuate the light intensity or to prevent unwanted spectral wavelengths from contaminating the recorded image. They are used in machine vision, parts inspection, and research applications, especially those involving fluorescent microscopy, spectrophotometry, photometry, color photography, and optical and infrared imaging.

Fluorescence illumination and observation is the most rapidly expanding microscopy technique employed today in the medical and biological sciences. Fluorescence probes (fluorochromes) have unique individual excitation and emission spectra. Typically, there is overlap between the higher wavelength end of the excitation spectrum and the lower wavelength end of the emission spectrum. A combination of filters (made up of an excitation filter, a dichromatic beamsplitter, and a barrier filter) is used to separate the excitation and emission light and improve image contrast. This is achieved by the proper selection of filters to block or pass specific wavelength bands of the spectrum. Chroma Technology Corp. (Brattleboro, VT) and Omega Optical Inc. (Brattleboro) provide a number of filter sets designed for specific combinations of fluorescence probes, for a variety of single- and multiple-labeling fluorescence microscopy applications.

In automated microscopy, a convenient approach to providing the required wavelengths of excitation light is to use a filter wheel coupled with a multispectral lamp (e.g., mercury or xenon). Filter wheels can be fitted to either the excitation or emission (or both) ports of a microscope. Mounting flanges allow filter wheels to be effectively inserted into the optical path of most microscopes without affecting camera or eyepiece focus. In epifluorescence microscopy, excitation filter wheels control the illumination spectrum before it enters the objective, whereas emission filter wheels are mounted on the camera port and select a unique emission wavelength.

Another common application of a filter wheel is to introduce neutral density (ND) filters into the optical path. ND filters attenuate the light passing through the microscope optics. This can allow the lamp to operate at a higher brightness level (to provide enhanced color balance), and they afford aperture adjustment for depth-of-field effects without saturating the camera. Typically, ND filters are graduated in degrees of light absorption.

Filter wheels are generally available with mountings for between three and twelve filters. The filters in the wheel may be of different colors, thickness, or materials. For multispectral applications requiring a large number of wavelengths, two or more filter wheels may be used to allow for more filters or to control separately the upper and lower cutoff wavelengths of a passband. For example, a dual filter wheel system with two six-filter wheels could provide ten individual filters (one open position in each wheel), with the potential for as many as thirty-six combinations. The most common use of dual filter wheels is to combine a bandpass filter with a ND filter to control both the wavelength and intensity of the fluorescence excitation.

For automated microscopy applications, motorized filter wheels with controller systems are interfaced to a computer, permitting direct communications via RS-232, GPIB, or USB serial ports. Typically, such systems also offer optional filter wheel control via a keypad. Automation and programming of unique filter sequences are conducted through the RS-232 interface. A simple command language (provided by the manufacturer) is used to retrieve filter status and to select filters. Often there is also an integrated shutter that can be opened and closed by commands sent to the serial port. The typically adjacent filter switching speed of commercially available shutters is on the order of 30–100 milliseconds. Filter wheels are provided by vendors such as Prior Scientific, Inc. (Rockland, MA), Ludl Electronic Products, Ltd. (Hawthorne, NY), Applied Scientific Instruments (Eugene, OR), and Marzhauser Wetzlar (Germany).

27.2.1.3 *X, Y Stage Positioning and Z-axis motors*

X, Y motorized stages and Z-axis (focus drive) motors form an integral part of any computer-assisted microscopy system. They allow a specimen to be positioned accurately and automatically relative to the microscope objective. They also allow optical sectioning of thick specimens and automatic focus. The stages can be customized to hold specimens on microscope slides, petri dishes, or cell plates. X, Y linear translation stages are typically driven by a stepper motor and a controller that interfaces to the host computer. Typically, the X, Y resolution of such systems ranges from $0.1-5\,\mu m$.

Translation stages are important because they allow automated scanning of specimen slides, thus increasing the speed of data collection. An important feature of computer-controlled motorized stages is the ability to move the slide to a previously located position. This is useful in experiments where serial observations of a specimen require removal, restaining, and subsequent return of specimens to the microscope stage. Specimen positioning provides the ability to return to specific coordinates and thus to view the same region of the slide that was observed prior to removal of the slide.

Z-axis motors allow controlled movement of the specimen in the vertical (Z or axial) direction, thus providing the capability for automatic focusing, and collection of optical sections of specimens (i.e., sets of images taken at different focal planes). The minimum Z-axis step size of typical focus motors can range from $0.01-0.05\,\mu m$.

Burleigh Instruments Inc. (Fishers, NY) offers an extensive line of micropositioning devices for the life sciences. Prior Scientific Inc. (Rockland) has the OptiScan^{TM} and ProScan^{TM} stage products that fit most upright and inverted microscopes and include a controller capable of managing the stage. Other vendors include Ludl Electronic Products, Ltd. (Hawthorne), Applied Scientific Instruments (Eugene), Sutter Instrument Company (Novato, CA), and Marzhauser Wetzlar (Germany).

27.2.1.4 *Image Sensors*

With the advent of high-resolution solid state cameras, such as the slow scan, cooled charge coupled device (CCD) camera, photographic film cameras are now being replaced by digital imaging systems. CCDs are the most frequently used image detectors as they have many advantages, including larger dynamic range, good quantum efficiency, low noise, linear response, and negligible geometrical distortion. CCD cameras are extensively used because of their low-light image capturing ability. A CCD camera is a light-sensitive, silicon, solid-state device. This imaging technology is based on collecting photons on the surface of a silicon grid containing many collection wells or pixels (for example, a 512×512-pixel grid). The photons within a well are subsequently converted to an electrical charge via the photoelectric effect, which is then passed to an image digitizer for conversion into an electrical signal that represents the intensity of the pixel. The intensities for all pixels over the grid are organized as a digital image and projected as a single video frame.

The variety of light detection methods and the number of imaging devices currently available [12] to the researcher make the selection process complex and often confusing. Many types of image sensors exist, and the choice is often dictated by the requirements of a given experiment. CCD camera selection typically involves the choice of appropriate parameters for several features. Two operational modes, interlaced scan and progressive scan, are available. Interlaced cameras scan an image in two steps, usually odd and even horizontal lines, and then reconstruct it in a buffer, which decreases monitor flickering.

Progressive scan cameras transfer an entire image without interlacing. Another design issue in CCD cameras is the number of chips. Single-chip cameras generate color using specialized sensor areas that detect R (red), G (green), and B (blue) components, whereas multiple-chip designs (3CCD) use a beam-splitter and R-, G-, and B-chips to detect each color component. The resolution is higher for the 3CCD cameras when compared with single-chip cameras. The sensor size of the camera determines its sensitivity and resolution. CCD cameras also have a frame rate of over 30 frames per second, and a digital output rate (usually in cycles/second or Hz) dependent on the analog-digital converter. CCD cameras can be "cooled" to sub-zero temperatures, thus minimizing noise and thermal variations, which reduces the dark current (i.e., the charge accumulated within the CCD, in the absence of light) and allows longer integration times for image capture. The digital dynamic range or bit depth is another important feature when choosing a CCD camera. For example, an 8-bit image can store 256 shades of gray and a 12-bit image can store 4,096. A 12-bit image can reveal greater detail than an 8-bit one; and quantification is also more precise in 12-bit images. The level of overall noise (in decibels, or dB) for a CCD camera is usually expressed as a signal/noise (S/N) ratio: S/N (dB) = $20 * \log$ (S/N). Thus, a S/N ratio of 100/1 is equivalent to 40 dB. Due to the low-light conditions inherent in fluorescence imaging, camera sensitivity is more critical here than in transmitted light microscopy. To increase sensitivity for low-light observation, image intensifiers may be used to amplify light before it reaches the camera faceplate. CCDs that are not intensified may increase sensitivity using "binning" (or "super-pixeling"). This technique increases sensitivity and frame rate by grouping pixel intensities.

Specialty cameras are particularly well suited for scientific applications in computerized microscopy systems since they offer unique features like large chip area, resolution up to 4k × 4k pixels, digital output, cooling to reduce noise and dark current, flexible timing, and full computer control. Suppliers of CCD cameras include Cooke Corporation (Auburn Hills, MI), Diagnostic Instruments (Sterling Heights, MI), Dage-MTI Inc. (Michigan City, IN), Optronics Inc. (Muskogee, OK), Cohu Inc. (San Diego, CA), and Roper Scientific Inc. (Duluth, GA). Cortese [12] provides a comprehensive listing of camera vendors.

Frame grabbers are image processing computer boards that capture and store image data. In a computerized microscopy system, digitizer boards (frame grabbers) are typically used in conjunction with CCD cameras for digitizing the microscope image for further analysis and permanent storage. The type of imaging board must be matched to the camera in terms of type of signal format, resolution, and precision. Acquisition formats include RS-170, CCIR, RS-330, RS-422, NTSC, Y/C, PAL, and RGB. For frame grabbers that can handle camera outputs in digital format, the input pixel acquisition depth is important to consider. The pixel depth refers to the number of bits used to the store the gray level at each pixel. Increasing pixel depth increases the amount of detail that can be reproduced in the scanned image. Frame grabbers are available in 8-, 12-, 16-, and 24-bit pixel depth formats. The 8-, 12-, and 16-bit frame grabbers allow digitization of monochrome images, whereas 24-bit frame grabbers are designed to work with RGB cameras and can be used to acquire three monochrome video signals synchronously. Frame grabbers are provided by Scion Corporation (Frederick, MD), MuTech Corporation (Billerica, MA), and National Instruments (Austin, TX).

27.2.2 Image Capture

Capturing a digital image of a specimen should be done in such a way that a high-quality analog image of that specimen can be recovered by interpolation. The sampling theorem tells us that a band-limited analytic function can be recovered, without error, from its sample points if (1) interpolation is done with a suitable interpolation function and (2) the function contains no energy at frequencies above one-half the sampling frequency. That is, the function is band-limited at f_c, and the sample spacing is $\delta x \leq 1/(2f_c)$.

Since the imaging optics are fundamentally band-limited, the latter requirement is rather easily met. The former, however, is frankly impossible. Any interpolating function that satisfies the sampling theorem, such as $\sin(\alpha x)/\alpha x$, is of infinite extent. Thus, for practical reasons, one cannot hope to recover the function exactly. The amount of reconstruction error, however, can be reduced by oversampling. Using a sample spacing that is one-half to one-third of that required by the sampling theorem will reduce interpolation error significantly but at the cost of larger image files.

Each component in the image capture chain (optics, image sensor, A-to-D converter, etc.) has its own transfer function. Each of these tends to fall off with increasing frequency. Since the overall transfer function is the product of all these, it will generally be worse (i.e., fall off faster) than any of the individual transfer functions. Thus no one component single-handedly limits resolution, and all must be considered in the system design.

The factors that degrade an image in the digitizing process are (1) blurring, (2) noise, (3) aliasing (4) shading, (5) photometric nonlinearity, and (6) geometric distortion. If the level of each of these can be kept low enough, then the digital images obtained from the microscope will be useful for the task at hand. Each type of degradation should be considered in the system design.

Blurring is minimized by keeping the overall system transfer function relatively flat over the frequency range of interest. The noise level can be controlled by choosing quality components and by cooling the image sensor, if necessary. The RMS noise level should be no more than about one gray level (e.g., $1/256 = 0.39\%$ for 8-bit data). Aliasing is avoided if the sample spacing satisfies the sampling theorem, as stated above. Shading can be minimized by proper alignment of the microscope components and further corrected with background subtraction. Photometric nonlinearity, if problematic, can be corrected by a suitably designed grayscale transformation. Geometric distortion, if problematic, can be corrected by a suitably designed geometric transformation.

Each application has its own specific requirements in each of these six areas. While none of the six sources of degradation can be eliminated completely, each can be controlled by proper system design, instrument setup, and postprocessing where necessary.

27.2.3 Imaging Software

In computerized microscopy systems, the complexity and number of variables that need to be configured, monitored, and adjusted prior to and during data collection are enormous, but they can easily be managed by software created specifically for this purpose. For example, to create a 3D image of a thick specimen, images of the sample are taken as the specimen stage is systematically and carefully moved along the Z-axis. For multiple fluorescent probe analyses on the same sample, the change in excitation wavelength and collection of emission data must be precisely controlled. These and other experimental operations are controlled and managed by the software. Software plays a major role in computer-assisted microscopy.

The software of an automated microscopy system has three functions: (1) control of peripheral devices (stage, filter, camera, etc.), (2) image processing and analysis, and (3) the user-interface. Peripheral devices are controlled by integrated computer interfaces (controllers) or proprietary interface cards. This component includes the core functions of the automated imaging system. It involves the control of the motorized stage for slide scanning, the focus motor for autofocusing and optical sectioning, the filter wheel control for filter selection, shutter control, camera integration time, and image capture and storage.

The second software component involves image preprocessing and analysis. Typical algorithms integrated in this component include image editing functions, geometrical manipulations, morphological filtering, smoothing, background subtraction, color separation, object segmentation, and classification. This software component is also responsible for collecting and storing additional information about the experiment, such as excitation wavelength, number of frames averaged, time, date, and other information pertinent to the experiment.

The user interface component is critical for user control and communicating with the other components of the software package, as well as providing feedback and reporting results. Commonly used graphical user interface (GUI) elements include (1) buttons for simple actions such as grabbing an image, (2) input fields for numerical values and textual information, (3) menus for choices, (4) windows for displaying acquired images, and (5) windows presenting extracted information about the current experiment.

27.3 SOFTWARE FOR HARDWARE CONTROL

Devices such as cameras, filter wheels, and microscope stages can be controlled by commands sent through an RS-232 serial connection. Typically, the device will respond with confirmations or status information. Every device uses a particular communication protocol and has its own set of commands. Software designed to integrate such equipment needs to send the appropriate commands and acts on any information that comes back from the device. Typical operations include transmitting and receiving serial data, correctly terminating commands, recognizing incoming commands, flushing the serial buffers, and waiting while a device is busy. Stage controllers, for example, move samples under a microscope, in one or more axes. The commands sent to an axis motor typically include operations such as the following: Home (move the stage to the home position in this axis), Move (move to an absolute position), and Relative move (move to a position relative to the current position). These commands are defined by the manufacturer of the device and can be easily integrated into the algorithms that interact with the peripheral devices. Some of the key algorithms used in computerized microscopy are described in the following sections.

27.3.1 Automated Slide Scanning

Three important parameters must be considered when designing software for stage control: (1) speed of scanning is important when a large area of the slide is to be viewed to capture numerous cells for automated analysis, (2) range of scanning is useful to determine the number of slides that need to be analyzed (single slide versus multiple slide holding stages), and (3) precision of movement is critical for automated tasks such as autofocusing and relocation of cells. Typically a stepping accuracy of a few micrometers is needed for precise relocation and focusing.

Scanning a specimen slide involves repeating a series of steps until the entire slide, or a user-defined area thereof, has been scanned. In a typical scanning algorithm, each FOV is focused automatically, after which an image is acquired. When image capture is complete, the stage is moved to the next FOV. Generally, an algorithm to implement automated slide scanning moves the slide in a raster pattern. For example, it moves vertically down the user-selected area and then retraces back to the top, and moves a predetermined fixed distance across and then starts another scan vertically downward. This process is continued until the entire user-defined area has been scanned. The step sizes in the X- and Y-directions are adjusted (depending on the pixel size for the objective in use) so that there is no overlap between the adjacent fields.

Scanning speed is an important feature in a computerized microscopy system. Factors that influence the scanning rate include stage movement, filter movement, the readout rate of the CCD chip, and the time required for focusing. The scanning rate is computed as the total area scanned per unit time. It also depends upon the sampling density of the image, varying inversely with the square of the sampling density and the integration time [10].

The cutoff frequency ϖ_c for an imaging system is given by

$$\varpi_c = \frac{4\pi \cdot NA}{\lambda}, \tag{27.1}$$

where NA is the numerical aperture, and λ is the wavelength.

The Nyquist sampling theorem states that an image can be reconstructed from its samples without error if the sampling frequency is at least $2 \cdot \varpi_c$. Sampling density is defined as the number of pixels per unit length. The choice of sampling density can be based on the Nyquist sampling theorem or on the required measurement precision. The decision should be based on the requirements of the system. For autofocusing, depth-of-focus, or image restoration applications, the Nyquist theorem should be used. If measurements derived from microscope images are required, then the sampling frequency should be derived from the measurement specifications [11].

27.3.2 Autofocusing

Automatic focusing is essential for automated microscopy when fully automated image acquisition in unattended operation is required. It also facilitates objective and consistent image measurements for quantitative analysis.

Two approaches are used for autofocusing. One, known as "active" focusing, is based on surface sensing, while the other, known as "passive" focusing, is based on image analysis. Surface sensing methods require a single surface from which light or sound is reflected and calibration of the in-focus location. These methods are generally impractical for light microscopy since there is usually more than one reflective surface (e.g., slide and coverslip), and specimens vary in thickness. The image analysis approach depends only on focus measurements from the image itself. The in-focus position is found by locating the maximum of a focus function computed on a series of images acquired at different Z-axis positions. The focus function is a measure of image sharpness, as a function of Z-axis position. A sequence of images taken along the Z-axis has to be acquired to find the focus position. The value of the focus function is computed for each image captured at a different Z-position. The maximum value of the focus function identifies the optimal focal plane. These methods are not sensitive to multiple reflective surfaces and therefore are widely used in light microscopy.

27.3.2.1 *Focus Functions*

Several groups have developed and tested focus functions for automated microscopy. The autofocus functions can be broadly categorized into methods based on (1) edge strength, (2) image contrast, (3) autocorrelation, and (4) the Fourier spectrum of an image [14]. Johnson and Goforth at JPL [15] and Dew *et al.* [16] used hardware that integrates the highpass filtered video signal in a real-time autofocusing system. Harms and Aus [17] evaluated a Laplacian-based autofocus parameter that mimics the receptive fields in the human eye. Groen *et al.* advanced eight criteria for evaluating autofocus parameters and compared eleven parameters [18]. Their results varied somewhat with different types of specimens, but two "gradient squared" functions, along with gray level variance, proved best in their testing on brightfield images. Boddeke *et al.* argued that the most effective autofocus parameters emphasize the mid-frequency range, since the optical transfer function (OTF) does not change with defocus at either the high- or low-frequency end [19]. They also argued for "binning" to reduce noise and for fitting a parabola to the focus parameter values to interpolate the exact point of best focus. Firestone *et al.* [20] tested autofocus functions and found those that sum gradients to be effective. Vollath presented new autofocus functions based on autocorrelation [21] and considered the effects of noise [22]. Brenner *et al.* [23], Erteza [24, 25], and Muller and Buffington [26] also advanced and tested autofocus functions based on the sum of squared gradients. Price and Gough [27] advanced a new prefiltered 2D Laplacian parameter and evaluated 11 functions in fluorescence and phase-contrast microscopy. They found that, in fluorescence microscopy, (1) the gradient-based parameters gave sharper peaks, and thus had narrower operating ranges than the statistical parameters, (2) the gradient and statistical techniques did not agree exactly on the point of best focus, and (3) autocorrelation-based parameters performed similarly to gradient-based ones. Geusebroek *et al.* proposed an autofocusing algorithm based on the signal power after convolving the image with a first-order derivative of a Gaussian filter [28]. This technique was found to be generally applicable to several light microscopy modalities including fluorescence, brightfield, and phase-contrast microscopy on a wide variety of preparation and specimen types. An exhaustive search for finding the maximum focus position was used to circumvent problems caused by noisy focus function curves. Santos *et al.* systematically evaluated 13 autofocus functions for analytical fluorescent image cytometry studies of counterstained nuclei [29]. Using a proposed figure-of-merit criterion that weights five different features of a focus function, the test results suggested that focus functions based on correlation measures had the best performance for the type of images they studied. Selecting an appropriate focus function is critical for autofocusing, and it depends largely on the requirements of the system, specifically the mode of microscopy and the type of specimens being examined.

The step size for obtaining a sequence of images along the Z-axis can be computed using the focal depth of a microscope objective, which is approximately

$$DOF = \frac{\lambda}{2 \cdot NA^2}, \qquad (27.2)$$

where λ is the wavelength, and NA is the numerical aperture of the objective. One can take a series of images spaced by approximately that distance along the optical axis of the microscope to effect the optical equivalent of serial sectioning.

27.3.2.2 *Autofocus Speed*

An important factor in most autofocusing is the time-consuming nature of the procedure. To reduce autofocusing time when scanning a slide, the number of FOVs that are focused should be reduced. This may be achieved by extrapolation or interpolation strategies to estimate the focus position of most fields based on the measured focus positions of a few fields. Interpolation can be used to estimate the focus position of a particular field from previously measured positions from at least three adjacent fields. This estimation is thus based on local information, and a focusing error in one field will only influence the estimated focus position of neighboring FOVs. In the interpolation approach, the focus positions of a number of FOVs across the entire slide have to be measured initially. A linear surface ($z = ax + by + e$) is then fitted through these measured values. During scanning, the fitted linear surface is used to calculate the focus position of each field. A disadvantage of this approach arises when the area to be scanned is large. For increased scanning times, the linear fit may not reliable because the position of the focal plane may change over time due to temperature-induced sample variations.

Another approach to improve the speed of autofocusing algorithms is to use a two-phase focusing approach. Initially, a coarse focus phase may be implemented using larger Z-step sizes. The first two samples are used to determine the direction of the focus. The algorithm then steps in the determined direction until the last focus value is lower than the previous one. The in-focus Z-position is then located between adjacent Z-positions with the highest and second highest focus value. A second, fine focus phase then repeats the same procedure with smaller Z-step sizes. The starting point is the optimal focus position determined by the coarse focus phase. The fine focus phase then samples the focus function along N positions equidistant about the coarse phase in-focus Z-position. The final estimation of the in-focus Z-position is calculated through a quadratic fit of the N points. The outcome of a robust autofocus algorithm should be a sharp peak in the focus function at the in-focus position as shown in Fig. 27.2. The selected autofocus algorithm for computerized microscopy should be generally applicable on a large variety of microscopic modes and a large variety of preparation techniques and specimen types.

27.3.3 Image Capture

Most frame grabber boards come with source code examples designed to illustrate important features of the board. The features covered include initialization, triggering, and real-time control of I/O lines. Customized software for image capture can be developed around the drivers and libraries provided by the manufacturers of the camera and digitizer boards.

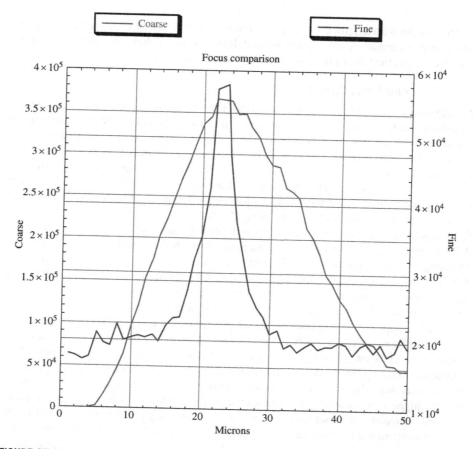

FIGURE 27.2

Two-phase approach to autofocusing, using a coarse focus and a separate fine focus function.

27.4 IMAGE PROCESSING AND ANALYSIS SOFTWARE

In any computerized microscopy system, quantitative data is obtained, and certain tasks are automated, using image processing and analysis techniques. Most automated microscopy applications are based on the measurement of features that cannot be obtained reliably by visual inspection. While visualization of images allows a qualitative analysis of the objects of interest, quantitative measures are usually essential to define and understand the biological processes. Digital image processing theory provides a number of tools to enhance images and to measure size, shape, and intensity features accurately. In addition to autofocusing, scanning, and image capture algorithms, computerized microscopy requires specific image processing algorithms to obtain the desired results. We discuss some of those here.

27.4.1 Correction of Instrumentation-based Errors

Most modern microscopy equipment is constructed to minimize phase distortion and optical aberrations. Nevertheless, no optical system is completely free of distortion, and, in practice, aberrations are always present to some extent. The choice and configuration of an automated microscopy system has a profound effect on the quality of the acquired image. For instance, changing filters in multicolor fluorescence microscopy may cause shifts in the spatial location of the imaged objects due to changes in the optical path. This effect can be corrected by choosing multiband filter cubes (including the excitation, emission, and dichroic components), which can simultaneously transmit and/or reflect multiple wavelengths of light, thereby eliminating aberrations introduced by the physical switching of filters. Alternatively, the images acquired at different wavelengths can be digitally corrected by estimating the amount of pixel shift occurring at each wavelength. Calibration images can be obtained using colored microspheres that are excited using various filter combinations, and the pixel shift occurring at different wavelengths for an imaged microsphere can be computed from the sequentially recorded images obtained by switching filters. The amount of pixel shift thus calculated can be used to correct images subsequently acquired.

Alignment of the light source for uniform sample illumination is also critical. Even with appropriately adjusted lamps, variation in the light intensity across an image is typical of most microscope images. Image processing algorithms that correct background intensity variations are described in the following sections.

27.4.2 Background Shading Correction

Quantitative image analysis typically involves measuring the brightness of regions in the image as a means of identification, that is it is often assumed that the same type of feature will be similar in intensity wherever it appears in the image. Differences in intensity values among different object features can then be used to differentiate between several features for counting, measurement, or identification. In most imaging processes, illumination sources can create intensity variations across the FOV. Even with proper calibration, one can only approximate uniform illumination of the scene being imaged.

This is particularly true for the microscope system. Both the halogen (transmitted light) and mercury (fluorescence light) lamps have to be adjusted for uniform illumination of the FOV prior to use. Moreover, microscope optics and/or cameras may also show vignetting, in which the corners of the image are darker than the center because the light is partially absorbed. The process of eliminating these defects by application of image processing to facilitate object segmentation or to obtain accurate quantitative measurements is known as background correction or background flattening.

27.4.2.1 *Background Subtraction*

For microscopy applications, there are two approaches that are popular for background flattening [30]. In the first approach, a "background" image is acquired in which a uniform reference surface or specimen is inserted in place of actual samples to be viewed, and an image of the FOV is recorded. This is the background image, and it represents the intensity variations that occur without a specimen in the light path, only due to any inhomogeneity in illumination source, the system optics, or camera, and can then be used to correct all subsequent images. When the background image is subtracted from a given image, areas that are similar to the background will be replaced with values close to the mean background intensity. The process is called background subtraction and is applied to flatten or even out the background intensity variations in a microscope image. It should be noted that, if the camera is logarithmic with a gamma of 1.0, then the background image should be subtracted. However, if the camera is linear, then the acquired image should be divided by the background image. Background subtraction can be used to produce a flat background and compensate for nonuniform lighting, nonuniform camera response, or minor optic artifacts (such as dust specks that mar the background of images captured from a microscope). In the process of subtracting (or dividing) one image by another, some of the dynamic range of the original data will be lost.

27.4.2.2 *Surface Fitting*

The second approach is to use the process of surface fitting to estimate the background image. This approach is especially useful when a reference specimen or the imaging system is not available to experimentally acquire a background image [31]. Typically, a polynomial function can be used to estimate variations of background brightness as a function of location. The process involves an initial determination of an appropriate grid of background sampling points. By selecting a number of points in the image, a list of brightness values and locations can be acquired. In particular, it is critical that the points selected for surface fitting represent true background areas in the image and not foreground (or object) pixels. If a foreground pixel is mistaken for a background pixel, the surface fit will be biased, resulting in an overestimation of the background. In some cases, it is practical to locate the points automatically for background fitting. This is feasible when working with images, which have distinct objects that are well distributed throughout the image area and contain the darkest (or lightest) pixels present. The image can then be subdivided into a grid of smaller squares or rectangles, the darkest (or lightest) pixels in each subregion located, and these points used for the fitting [31]. Another issue is the spatial distribution and frequency of the sampled points. The greater the number of valid points which are uniformly spread over the entire image, the greater the accuracy of the estimated surface fit. A least-squares fitting approach may then be used to determine the coefficients of the polynomial function. For a third-order polynomial, the functional form of the fitted background is

$$B(x,y) = a_0 + a_1 \cdot x + a_2 \cdot y + a_3 \cdot xy + a_4 \cdot x^2 + a_5 \cdot y^2 + a_6 \cdot x^2 y + a_7 \cdot xy^2 + a_8 \cdot x^3 + a_9 \cdot y^3.$$
(27.3)

This polynomial has 10 $(a_0 - a_9)$ fitted constants. In order to get a good fit and diminish sensitivity to minor fluctuations in individual pixels, it is usual to require several times the minimum number of points. We have found that using approximately three times the total number of coefficients to be estimated is sufficient. Figure 27.3(A–E) demonstrates the process of background subtraction. Panel A shows the original image, panel B presents its 2D intensity distribution as a surface plot, panel C shows the background surface estimated via the surface fitting algorithm, panel D shows the background subtracted image, and panel E presents its 2D intensity distribution as a surface plot.

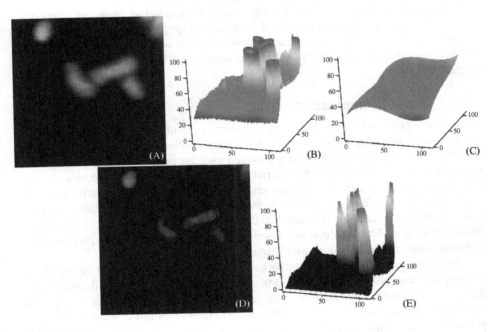

FIGURE 27.3

Background subtraction via surface fitting. Panel A shows the original image; panel B presents its 2D intensity distribution as a surface plot; panel C shows the background surface estimated via the surface fitting algorithm; panel D shows the background subtracted image; and panel E presents its 2D intensity distribution as a surface plot.

27.4.2.3 *Other Approaches*

Another approach used to remove the background is frequency domain filtering. It assumes that the background variation in the image is a low-frequency signal and can be separated in frequency space from the higher frequencies that define the foreground objects in the image. A highpass filter can then be used to remove the low-frequency background components [30].

Other techniques for removing the background include nonlinear filtering [32] and mathematical morphology [33]. Morphological filtering is used when the background variation is irregular and cannot be estimated by surface fitting. The assumption behind this method is that foreground objects are limited in size and smaller than the scale of background variations, and the intensity of the background differs from that of the features. The approach is to use an appropriate structuring element to describe the foreground objects. Neighborhood operations are used to compare each pixel to its neighbors. Regions larger than the structuring element are taken as background. This operation is performed for each pixel in the image, and a new image is produced as a result. The result of applying this operation to the entire image is to shrink the foreground objects by the radius of the structuring element and to extend the local background brightness values into the area previously covered by objects.

Reducing brightness variations by subtracting a background image, whether it is obtained by measurement, mathematical fitting, or image processing, is not a cost-free process. Subtraction reduces the dynamic range of the grayscale, and clipping must be avoided in the subtraction process or it might interfere with subsequent analysis of the image.

27.4.3 **Color Compensation**

Many of the problems encountered in the automatic identification of objects in color (RGB) images result from the fact that all three fluorophores appear in all three color channels due to the unavoidable overlap among fluorophore emission spectra and camera sensitivity spectra. The result is that the red dye shows up in the green and blue channel images, and the green and blue dyes are smeared across all three color channels as well. Castleman [34] describes a process that effectively isolates three fluorophores by separating them into three color channels (RGB) of the digitized color image. The method, which can account for black level and unequal integration times [34], is a preprocessing technique that can be applied to color images prior to segmentation.

The technique yields separate, quantitative maps of the distribution of each fluorophore in the specimen. The premise is that the imaging process linearly distributes the light emitted from each fluorophore among the different color channels. For example, for an N-color system, each $N \times 1$ pixel vector needs to be premultiplied by an $N \times N$ compensation matrix. Then for a three color RGB system, the following linear transformation may be applied:

$$y = ECx + b, \tag{27.4}$$

where y is the vector of RGB gray levels recorded at a given pixel, and x is the 3×1 vector of actual fluorophore brightness at that pixel. C is the 3×3 color smear matrix, which specifies how the fluorophore brightnesses are spread among the three color channels. Each element c_{ij} is the proportion of the brightness from fluorophore i that appears in the color channel j of the digitized image. The elements of this matrix are determined experimentally for a particular combination of camera, color filters, and fluorophores. E specifies the relative exposure time used in each channel, i.e., each element e_{ij} is the ratio of the current exposure time for color channel i, to the exposure time used for the color spread calibration image. The column vector b accounts for the black level offset of the digitizer, that is, b_i is the gray level that corresponds to zero brightness in channel i.

Then the true brightness values for each pixel can be determined by solving Eq. (27.4) as follows:

$$x = C^{-1}E^{-1}[y - b], \tag{27.5}$$

where C^{-1} is the color compensation matrix. This model assumes that the gray level in each channel is proportional to integration time, and that the black levels are constant with integration time. With CCD cameras both of these conditions are satisfied to a good approximation.

27.4.4 Image Enhancement

In microscopy, the diffraction phenomenon due to the wave nature of light introduces an artifact in the images obtained. The OTF, which is the Fourier transform of the point spread function (PSF) of the microscope, describes mathematically how the system treats periodic structures [35]. It is a function that shows how the image components at different frequencies are attenuated as they pass through the objective lens. Normally the OTF drops off at higher frequencies and goes to zero at the optical cutoff frequency and beyond. Frequencies above the cutoff are not recorded in the microscope image, whereas mid-frequencies are attenuated (i.e., mid-sized specimen structures lose contrast).

Image enhancement methods improve the quality of an image by increasing contrast and resolution, thereby making the image easier to interpret. Lowpass filtering operations are typically used to reduce random noise. In microscope images, the region of interest (specimen) dominates the low and middle frequencies, whereas random noise is often dominant at the high end of the frequency spectrum. Thus lowpass filters reduce noise but discriminate against the smallest structures in the image. Also, highpass filters are sometimes beneficial to restore partially the loss of contrast of mid-sized objects. Thus, for microscope images, a properly designed filter combination has not only to boost the midrange frequencies to compensate for the optics but also must attenuate the highest frequencies since they are dominated with noise. Image enhancement techniques for microscope images are reviewed in [36].

27.4.5 Segmentation for Object Identification

The ultimate goal of most computerized microscopy applications is to identify in images unique objects that are relevant to a specific application. Segmentation refers to the process of separating the desired object (or objects) of interest from the background in an image. A variety of techniques can be used to do this. They range from the simple (such as thresholding and masking) to the complex (such as edge/boundary detection, region growing, and clustering algorithms). The literature contains hundreds of segmentation techniques, but there is no single method that can be considered good for all images, nor are all methods equally good for a particular type of image. Segmentation methods vary depending on the imaging modality, application domain, method being automatic or semiautomatic, and other specific factors. While some methods employ pure intensity-based pattern recognition techniques such as thresholding followed by connected component analysis [37, 38], some other methods apply explicit models to extract information [39, 41]. Depending on the image quality and the general image artifacts such as noise, some segmentation methods may require image preprocessing prior to the segmentation algorithm [42, 43]. On the other hand, some methods apply postprocessing to overcome the problems arising from over-segmentation. Overall, segmentation methods can be broadly categorized into point-based, edge-based, and region-based methods.

27.4.5.1 *Point-based Methods*

In most biomedical applications, segmentation is a two-class problem, namely the objects, such as cells, nuclei, chromosomes, and the background. Thresholding is a point-based approach that is useful for segmenting objects from a contrasting background. Thus, it is commonly used when segmenting microscope images of cells. Thresholding consists of segmenting an image into two regions: a particle region and a background region. In its most simple form, this process works by setting to white all pixels that belong to a gray level interval, called the threshold interval, and setting all other pixels in the image to black. The resulting image is referred to as a binary image. For color images, three thresholds must be specified, one for each color component. Threshold values can be chosen manually or by using automated techniques. Automated thresholding techniques select a threshold, which optimizes a specified characteristic of the resulting images. These techniques include clustering, entropy, metric, moments, and interclass variance. Clustering is unique in that it is a multiclass thresholding method. In other words, instead of producing only binary images, it can specify multiple threshold levels, which result in images with three or more gray level values.

27.4.5.2 *Threshold Selection*

Threshold determination from the image histogram is probably one of the most widely used techniques. When the distributions of the background and the object pixels are known and unimodal, then the threshold value can be determined by applying the Bayes rule [44]. However, in most biological applications, both the foreground object and the background distributions are unknown. Moreover, most images have a dominant background peak present. In these cases, two approaches are commonly used to determine the threshold. The first approach assumes that the background peak shows a normal distribution, and the threshold is determined as an offset based on the mean and the width of the background peak. The second approach, known as the triangle method, determines the largest vertical distance from a line drawn from the background peak to the highest occurring gray level value [44].

There are many thresholding algorithms published in the literature, and selecting an appropriate one can be a difficult task. The selection of an appropriate algorithm depends upon the image content and type of information required post-segmentation. Some of the common thresholding algorithms are discussed. The Ridler and Calvard algorithm uses an iterative clustering approach [45]. The mean image intensity value is chosen as an initial estimate of the threshold is made. Pixels above and below the threshold are assigned to the object and background classes, respectively. The threshold is then iteratively estimated as the mean of the two class means. The Tsai algorithm determines the threshold so that the first three moments of the input image are preserved in the output image [46]. The Otsu algorithm is based on discriminant analysis and uses the zeroth- and the first-order cumulative moments of the histogram for calculating the threshold value [47]. The image content is classified into foreground and background classes. The threshold value is the one that maximizes between-class variance or equivalently minimizes within-class variance. The Kapur *et al.* algorithm uses the entropy of the image [48]. It also classifies the image content as two classes of events with each class characterized by a probability density function (pdf). The method then maximizes the sum of the entropy of the two pdfs to converge to a single threshold value.

Depending on the brightness values in the image, a global or adaptive approach for thresholding may be used. If the background gray level is constant throughout the image, and if the foreground objects also have an equal contrast that is above the background, then a global threshold value can be used to segment the entire image. However, if the background gray level is not constant, and the contrast of objects varies within the image, then an adaptive thresholding approach should be used to determine the threshold value as a slowly varying function of the position in the image. In this approach, the image is divided into rectangular subimages, and the threshold for each subimage is determined [44].

27.4.5.3 *Edge-based Methods*

Edge-based segmentation is achieved by searching for edge points in an image using an edge detection filter or by boundary tracking. The goal is to classify pixels as edge pixels or non-edge pixels, depending on whether they exhibit rapid intensity changes from their neighbors.

Typically, an edge-detection filter, such as the gradient operator, is first used to identify potential edge points. This is followed by a thresholding operation to label the edge points and then an operation to connect them together to form edges. Edges that are several pixels thick are often shrunk to single pixel width by using a thining operation, while algorithms such as boundary chain-coding and curve-fitting are used to connect edges with gaps to form continuous boundaries.

Boundary tracking algorithms typically begin by transforming an image into one that highlights edges as high gray level using, for example, a gradient magnitude operator. In the transformed image, each pixel has a value proportional to the slope in its neighborhood in the original image. A pixel presenting a local maximum gray level is chosen as the first edge point, and boundary tracking is initiated by searching its neighborhood (e.g., 3×3) for the second edge point with the maximum gray level. Further edge points are similarly found based on current and previous boundary points. This method is described in detail elsewhere [49].

Overall, edge-based segmentation is most useful for images with "good boundaries," that is, where the intensity varies sharply across object boundaries and is homogeneous along the edge. A major disadvantage of edge-based algorithms is that they can result in noisy, discontinuous edges that require complex postprocessing to generate closed boundaries. Typically, discontinuous boundaries are subsequently joined using morphological matching or energy optimization techniques. An advantage of edge detection is the relative simplicity of computational processing. This is due to the significant decrease in the number of pixels that must be classified and stored when considering only the pixels of the edge, as opposed to all the pixels in the object of interest.

27.4.5.4 *Region-based Methods*

In this approach, groups of adjacent pixels in a neighborhood wherein the value of a specific feature (intensity, texture, etc.) remains nearly the same are extracted as a region. Region growing, split and merge techniques, or a combination of these are commonly used for segmentation. Typically, in region growing a pixel or a small group of pixels is picked as the seed. These seeds can be either interactively marked or automatically picked. It is crucial to address this issue carefully, because too few or too many seeds can result in under- or over-segmented images, respectively. After this the neighboring seeds are grouped together or separated based on predefined measures of similarity or dissimilarity [50].

There are several other approaches to segmentation, such as model-based approaches [51], artificial intelligence-based approaches [52], and neural network-based approaches [53]. Model-based approaches are further divided into two categories: (1) deformable models and (2) parametric models. Although there is a wide range of segmentation methods in different categories, most often multiple techniques are used together to solve different segmentation problems.

27.4.6 Object Measurement

The ultimate goal of any image processing task is to obtain quantitative measurement of an area of interest extracted from an image or of the image as a whole. The basic objectives of object measurement are application dependent. It can be used simply to provide a measure of the object morphology or structure by defining its properties in terms of area, perimeter, intensity, color, shape, etc. It can also be used to discriminate between objects by measuring and comparing their properties.

Object measurements can be broadly classified as (1) geometric measures, (2) ones based on the histogram of the object image, and (3) those based on the intensity of the object. Geometric measures include those that quantify object structure, and these can be computed for both binary and grayscale objects. In contrast, histogram- and intensity-based measures are applicable to grayscale objects. Another category of measures, which are distance-based, can be used for computing the distance between objects, or between two or more components of objects. For a more detailed treatment of the subject matter, the reader should consult the broader image analysis literature [54–56]. In computing measurements of an object, it is important to keep in mind the specific application and its requirements. A critical factor in selecting an object measurement is its robustness. The robustness of a measurement is its ability to provide consistent results on different images and in different applications. Another important consideration is the invariance of the measurement under rotation, translation, and scale. When deciding on the set of object measures to use these considerations should guide one in identifying a suitable choice.

27.4.7 The User Interface

The final component of the software package for a computerized microscopy system is the graphical user interface. The software for peripheral device control, image capture, preprocessing, and image analysis has to be embedded in a user interface. Dialogue boxes are provided to control the automated microscope, to adjust parameters for tuning the object finding algorithm, to define the features of interest, and to specify the scan area of the slide and/or the maximum number of objects that have to be analyzed. Parameters such as object size and cluster size are dependent on magnification, specimen type, and quality of the slides. The operator can tune these parameters on a trial and error basis. Windows are available during screening to show the performance of the image analysis algorithms and the data generated. Also, images containing relevant information for each scan must be stored in a gallery for future viewing, and for relocation if required. The operator can scroll through this window and rank the images according to the features identified. This allows the operator to select for visual inspection those images containing critical biological information.

27.5 A COMPUTERIZED MICROSCOPY SYSTEM FOR CLINICAL CYTOGENETICS

Our group has developed a computerized microscopy system for the use in the field of clinical cytogenetics.

27.5.1 Hardware

The instrument is assembled around a Zeiss Axioskop or an Olympus BX-51 epi-illumination microscope, equipped with a 100 W mercury lamp for fluorescence imaging and a 30 W halogen source for conventional light microscopy. The microscope is fitted with a ProScan motorized scanning stage system (Prior Scientific Inc., Rockland), with three degrees of motion (X, Y, and Z), and a four-specimen slide holder. The system provides 9×3-inch travel, repeatability to $\pm 1.0\,\mu m$, and step size from 0.1 to $5.0\,\mu m$. The translation and focus motor drives can be remotely controlled via custom computer algorithms, and a high precision joystick is included for operator control. The spatial resolution of the scanning stage is $0.5\,\mu m$ in X and Y and $0.05\,\mu m$ in the Z direction, allowing precise coarse and fine control of stage position. A Dage 330T cooled triple chip color camera (Dage-MTI Inc., Michigan) capable of on-chip integration up to 8 seconds and 575-line resolution is used in conjunction with a Scion-CG7 (Scion Corporation, Frederick, ML) 24-bit frame grabber to allow simultaneous acquisition of all three color channels ($640 \times 480 \times 3$). Alternatively, the Photometrics SenSys^{TM} (Roper Scientific, Inc., Tucson, AZ) camera, which is a low light CCD having 768×512 pixels (9×9 mm) by 4096 gray levels and 1.4 MHz readout speed, is also available. For fluorescence imaging, a 6-position slider bar is available with filters typically used in multispectral three-color and four-color fluorescence in situ hybridization (FISH) sample. Several objectives are available, including the Zeiss (Carl Zeiss Microimaging Inc., Thornwood, NY) PlanApo 100X NA 1.4 objective, CP Achromat 10X NA 0.25, Plan-Neofluar 20X NA 0.5, Achroplan 63X NA 0.95, Meiji S-Plan 40X NA 0.65, Olympus UplanApo 100X NA 1.35, Olympus UplanApo 60X NA 0.9, and Olympus UplanApo 40X N.A. 0.5–1.0. The automated microscope system is controlled by proprietary software running on a PowerMac G4 computer (Apple Inc., Cupertino, CA).

27.5.2 Software

The software that controls the automated microscope includes functions for spatial and photometric calibration, automatic focus, image scanning and digitization, background subtraction, color compensation, nuclei segmentation, location, measurement, and FISH dot counting [31].

27.5.2.1 *Autofocus*

Autofocus is done by a two-pass algorithm designed to determine first whether the field in question is empty or not, and then to bring the image into sharp focus. The first pass of the algorithm examines images at three Z-axis positions to determine whether there is enough variation among the images to indicate the presence of objects in the field to focus on. The sum over the image of the squared second derivatives described by Groen *et al.* [18] is used as the focus function $f(x)$;

$$f(x) = \sum_i \sum_j \left(\frac{\partial^2 g(x,y)}{\partial x^2} \right)^2,$$

(27.6)

where $g(i,j)$ is the image intensity at pixel (i,j). A second-order difference is used to estimate the second-order derivative (Laplacian filter):

$$\frac{\partial^2 g(x,y)}{\partial x^2} \approx \frac{\Delta^2 g}{\Delta x^2} = g(i,j+1) - 2g(i,j) + g(i,j-1).$$

(27.7)

The Laplacian filter strongly enhances the higher spatial frequencies and proves to be ideal for our application. At the point of maximal focus value, the histogram is examined above a predetermined threshold to determine the presence of cells in the image.

Once the coarse focus step is complete, a different algorithm brings the image into sharp focus. The focus is considered to lie between the two Z-axis locations that bracket the location that gave the highest value in the course focus step. A hill-climbing algorithm is then used with a "fine focus" function based on gradients along 51 equispaced horizontal and vertical lines in the image. Images are acquired at various Z-locations, "splitting the difference" and moving toward locations with higher gradient values until the Z-location with the highest gradient value is found, to within the depth of focus of the optical system. To ensure that the background image of all the color channels is in sharp focus, the fine focus value is taken to be the sum of the fine focus function outputs for each of the three (or four) color channels.

The coarse focus routine determines the plane of focus (3 frames) and is followed by a fine focus algorithm that finds the optimal focus plane ($\sim 5-8$ frames). The total number of images analyzed during the fine focus routine depends upon how close the coarse focus algorithm got to the optimal focus plane. The closer the coarse focus comes to the optimal focus position, the fewer steps are required in the fine focus routine. The autofocus technique works with any objective by specifying its numerical aperture, which is needed to determine the depth of focus, and focus step size. It is conducted at the beginning of every scan, and it may be done for every scan position or at regular intervals as defined by the user. A default interval of 10 scan positions is programmed. We found that the images are "in-focus" over a relatively large area of the slide, and frequent refocusing is not required. For an integration time of 0.5 seconds we recorded an average autofocus time of 28 ± 4 seconds. The variability in the focusing time is due to the varying number of image frames captured during the fine focus routine. The total time for autofocus depends upon image content (which will affect processing time), and the integration time for image capture.

The autofocusing method described above is based on image analysis done only at the resolution of the captured images. This approach has a few shortcomings. First, the high-frequency noise inherent in microscope images can produce an unreliable autofocus function when processed at full image resolution. Second, the presence of multiple peaks (occurring due to noise) may result in a local maximum rather than the global maximum being identified as the optimal focus or at least warrant the use of exhaustive search techniques to find optimum focus. Third, computing the autofocus function values at full resolution involves a much larger number of pixels than computing them at a lower image resolution. To address these issues, a new approach based on multiresolution image analysis has been introduced for microscope autofocusing [14].

Unlike its single-resolution counterparts, the multiresolution approach seeks to exploit salient image features from image representations not just at one particular resolution but across multiple resolutions. Many well-known image transforms, such as the Laplacian pyramid, B-splines, and wavelet transforms, can be used to generate multiresolution representations of microscope images. Multiresolution analysis has the following characteristics: (1) salient image features are preserved and are correlated across multiple resolutions, whereas the noise is not, (2) it yields generally smoother autofocus function curves at lower resolutions than at full resolution, and (3) if the autofocus measurement and search are carried out at lower resolutions, the computational load is reduced exponentially. A wavelet-transform-based method to compute autofocus functions at multiple resolutions has been developed by our group and is described in detail elsewhere [14].

27.5.2.2 Slide Scanning

The algorithm to implement automated slide scanning moves the slide in a raster pattern. It goes vertically down the user-selected area and then retraces back to the top. It moves to a predetermined fixed distance across and then starts another scan vertically downward. This process is continued until the entire user-defined area has been scanned. The step size in the X- and Y-directions is adjusted (depending on the pixel spacing for the objective in use) such that there is no overlap between the sequentially scanned fields.

The system was designed to implement slide scanning in two modes depending on the slide preparation. A "spread" mode allows the entire slide to be scanned, whereas a "cytospin" mode may be used to scan slides prepared by centrifugal cytology. Both the spread and cytospin modes also have the capability to allow user-defined areas (via fixed area or lasso) to be scanned. The average slide-scanning rate recorded for the system is 12 images/min. This value represents the total scanning and processing (autofocusing and image analysis) rate. Image analysis algorithms are tailored for each specific application.

27.6 APPLICATIONS IN CLINICAL CYTOGENETICS

Cytogenetics is the study of chromosomes, especially in regard to their structure and relation to genetic disease. Clinical cytogenetics involves the microscopic analysis of chromosomal abnormalities such as an increase or reduction in the number of chromosomes or a translocation of part of one chromosome onto another. Advances in the use of DNA probes have allowed cytogeneticists to label chromosomes and determine if a specific DNA sequence is present on the target chromosome. This has been useful in detecting abnormalities beyond the resolution level of studying banded chromosomes in the microscope and also in determining the location of specific genes on chromosomes. Clinical tests are routinely performed on patients in order to screen for and identify genetic problems associated with chromosome morphology. Typical tests offered include karyotype analysis, prenatal and postnatal aneuploidy screening by PCR or FISH, microdeletion and duplication testing via FISH, telomere testing via FISH, MFISH (multiplex FISH), and chromosome breakage and translocation testing. The computerized microscopy system described above has been applied to the following cytogenetic screening tests.

27.6.1 Fetal Cell Screening in Maternal Blood

Scientists have documented the presence of a few fetal cells in maternal blood and have envisioned using them to enable noninvasive prenatal screening. Using fetal cells isolated from maternal peripheral blood samples eliminates the procedure-related risks associated with amniocentesis and chorionic villus sampling [57].

The minute proportion of fetal cells found in maternal blood can now be enriched to one per few thousand using magnetic activated cell sorting [58] or fluorescence activated cell sorting [59], or a combination of the two. Aneuploidies can then be detected with chromosome-specific DNA probes via FISH [60]. Microscopy-based approaches have been used to identify fetal cells in maternal blood, but the small number of fetal cells present in the maternal circulation limits accuracy and makes cell detection labor intensive. This creates the need for a computerized microscopy system to allow repeatable, unbiased, and practical detection of the small proportions of fetal cells in enriched maternal blood samples.

FISH is one of the methods currently under investigation for the automated detection of fetal cells. It is a quick, inexpensive, accurate, sensitive, and relatively specific method that allows detection of the autosomal trisomies 13, 18, and 21, X and Y abnormalities, and any other chromosome abnormality for which a specific probe is available.

We used the system to detect fetal cells in FISH-labeled maternal blood. The separated cells in enriched maternal blood were examined for gender and genetic aneuploidy using chromosome-specific DNA probes via FISH. The nucleus was counterstained with DAPI (4',6-Diamidino-2-phenylindole), and chromosomes X and Y were labeled with SpectrumGreen and SpectrumOrange, respectively (Vysis Inc., Downers Grove, IL).

If the fetus is male, FISH can be used directly, with one probe targeting the Y-chromosome, and different colored probes for other chromosomes, to detect aneuploidies. An automated system can examine enough cells to locate several fetal (Y-positive) cells and then make a determination about aneuploidy in the fetus. If the fetus is female, one must analyze a number of cells that is sufficient to rule out the possibility of aneuploid fetal cells.

Specific image analysis algorithms were employed to detect the cells and FISH dots, following background subtraction and color compensation. The digitized images were initially thresholded in the user-defined cell channel (generally, blue for the DAPI counterstain) to obtain binary images of cells. The cells were then uniquely identified using a region labeling procedure [61]. The 8-connected pixel neighborhood is used to determine the pixel belonging to a certain object. Each pixel in the connected neighborhood is then assigned a unique number so that finally all the pixels belonging to an object will have the same unique label. The number of pixels in each object is computed and used as a measure of cell size. Subsequently, shape analysis is used to discard large cell clusters and noncircular objects. Further, a morphological technique is used for automatically cutting touching cells apart. The morphological algorithm shrinks the objects until they separate and then thins the background to define cutting lines. An exclusive OR operation then separates cells. Cell boundaries are smoothed by a series of erosions and dilations, and the smoothed boundary is used to obtain an estimate of the cellular perimeter. ANDing this thresholded and morphologically processed mask with the other two red and green planes of the color compensated image yields grayscale images containing only dots that lie within the cells. Objects are then located by thresholding in the probe color channels, using smoothed boundaries as masks. A minimum size criterion is used to eliminate noise spikes, and shape analysis is used to flag noncompact dots. The remaining objects are counted. The locations of dots found are compared with the cell masks to associate each chromosomal dot with its corresponding cell. Finally, we implemented a statistical model to determine unbiased estimates of the proportion of cells having a given number of dots. The befuddlement theory provides guidelines for dot counting algorithm development by establishing the point at which further reduction of dot-counting errors will not materially improve the estimate [62]. This occurs when statistical sampling error outweighs dot-counting error. Isolated cells with dots are then evaluated to determine gender and/or aneuploidy and finally classified as fetal or maternal cells. Once the fetal cells have been identified by the automated image analysis algorithms, the stage and image coordinates of such cells are stored in a table along with the cell's morphological features, such as area, shape factor, and dot count. The detected cells can be automatically relocated at any subsequent time by centering upon the centroid of the cells using the previously stored stage and image coordinates. The results of automated image analysis are illustrated in Fig. 27.4. The software accurately (1) detects single cells, (2) separates touching cells, and (3) detects the green dots in the isolated cells. The fetal cell screening system evaluation is presented in a recent publication [63].

27.6.2 Subtelomeric FISH for Detection of Cryptic Translocations

Subtelomeric FISH (STFISH) uses a complete set of telomere region-specific FISH probes designed to hybridize to the unique subtelomeric regions of every human chromosome. Recently, a version of these probes became commercially available (ChromoProbe Multiprobe^{TM} T-System, Cytocell Ltd.). The assay allows for simultaneous analysis of the telomeric regions of every human chromosome on a single microscope slide, except the p-arms of the acrocentric chromosomes. It is anticipated that these probes will be

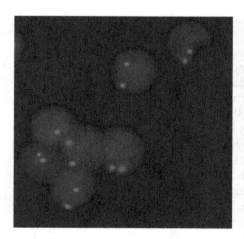 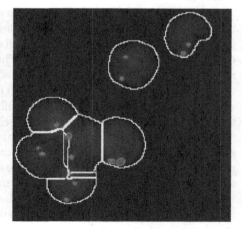

FIGURE 27.4

Fluorescence image of seven female (XX) cells. Adult female blood was processed via FISH. Cells are counterstained blue (DAPI); X chromosomes are labeled in green (FITC). Results of automated image analysis. As illustrated in the right panel, the software accurately detects single cells, separates touching cells, and detects the green dots in individual cells.

extremely valuable in the identification of submicroscopic telomeric aberrations. These are thought to account for a substantial, yet previously under-recognized, proportion of cases of mental retardation in the population. The utility of these probes is evident in that numerous recent reports describe cryptic telomere rearrangements or submicroscopic telomeric deletions [64].

27.6.2.1 *The STFISH assay*

STFISH uses a special 24-well slide template that permits visualization of the subtelomeric regions of every chromosome pair at fixed positions on the slide template (Fig. 27.5). Each well has telomeric-region-specific probes for a single chromosome; for example, well 1 has DNA probes specific to the telomeric regions of chromosome 1 and well 24 has DNA probes specific for the Y chromosome telomeres. At present, the assay requires a manual examination of all 24 wells. When screening anomalies, first each of the 24 regions on the slide must be viewed to find metaphases. The second step involves image acquisition, followed by appropriate image labeling (to indicate the region on the slide from which the image was captured), and saving the images. This is required to identify the chromosomes correctly. The third step involves an examination of the saved images of one or more metaphases from each of the 24 regions. This examination involves the identification of the (labeled green) p-regions and the (red labeled) q-regions for each pair of chromosomes in each of the 24 regions. Finally, the last step requires the correlation of any deleted or additional p- or q-arm telomeric material within the 24 regions to allow the interpretation of the telomeric translocation, if present. A trained cytogenecist takes approximately 3 hours to complete reading a slide for the STFISH assay, and an additional hour to complete data analysis. Furthermore, the procedure is not only labor intensive, but it requires trained cytogenecists for slide reading and data

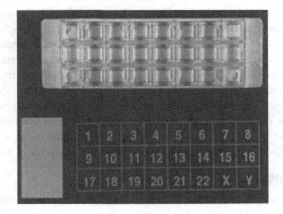

FIGURE 27.5

Illustration of the "Multiprobe™ coverslip device" (top) divided into 24 raised square platforms and the "template microscope slide" (bottom) demarcated into 24 squares.

interpretation. This procedure is even more tedious in cases without prior knowledge of the chromosomal anomaly.

It is apparent that computerized microscopy can be applied to produce labor and time savings for this procedure. Automated motorized stages, combined with computer controlled digital image capture, can implement slide scanning, metaphase finding, and image capture, labeling, and saving (steps 1 and 2). This removes the tedious and labor-intensive component of the procedure, allowing a cytogeneticist to examine a gallery of stored images rapidly for data interpretation. Image analysis algorithms can also be implemented to automatically flag images that have missing or additional telomeric material (steps 3 and 4). This would further increase the speed of data interpretation. Finally, automated relocation capability can be implemented, allowing the cytogeneticist to perform rapid visual examination of the slide for any of the previously recorded images.

We recorded a slide scanning time (including autofocusing, scanning, and image analysis) of 4 images/min ($\sim$0.04 mm^2/min) for an integration time of 0.5 seconds. The slide-scanning algorithm was designed to scan the special Cytocell, Inc. template slide that is used for the STFISH. As seen in Fig. 27.5, the template slide is divided into 24 squares (3 rows of 8) labeled from 1 to 22, X and Y. Each square in the grid is scanned, and the metaphases found in each square are associated with the corresponding chromosome label. This is accomplished by creating a lookup table that maps each square in the grid to fixed stage coordinates. The stage coordinates of the four vertices of each square are located and stored.

27.6.2.2 *User Interface*

The user interface for the newly designed slide-scanning algorithm is presented in Fig. 27.6. The 24 well regions of the Cytocell template slide are mapped to the corresponding stage coordinates as shown in Fig. 27.6. The crosshair (seen in region 12) indicates the current position of the objective. The user can select a particular slide region, or a range of slide regions, as desired for scanning. For each selected region, scanning begins at the center and continues in a circular scan outward, toward the periphery. This process is continued until either the entire selected region is scanned or a predefined number of metaphases have been found. The default is to scan the entire slide, starting at region 1 and ending at region 23 (for female specimens) or 24 (for male specimens), with a stop limit of 5 metaphases per region. For example, at the end of the default scan, the image gallery would have a total of 120 metaphase images for a male specimen. The step size in both the X- and Y-directions can be adjusted (depending on pixel size, as dictated by the objective in use) so that there is no overlap between sequential scan fields. This is controlled by the X- and Y-axis factors shown in the user interface in Fig. 27.6.

27.6.2.3 *Metaphase Finding*

Image analysis capability for this application includes locating the metaphases in the images. The software first uses gray-level thresholding and boundary tracking algorithms to find objects in the image. The isolated objects are then classified using a set of user-defined parameters to identify metaphases. The key classification parameters include the size and shape of the objects, clustering of similar objects in a group, and the number

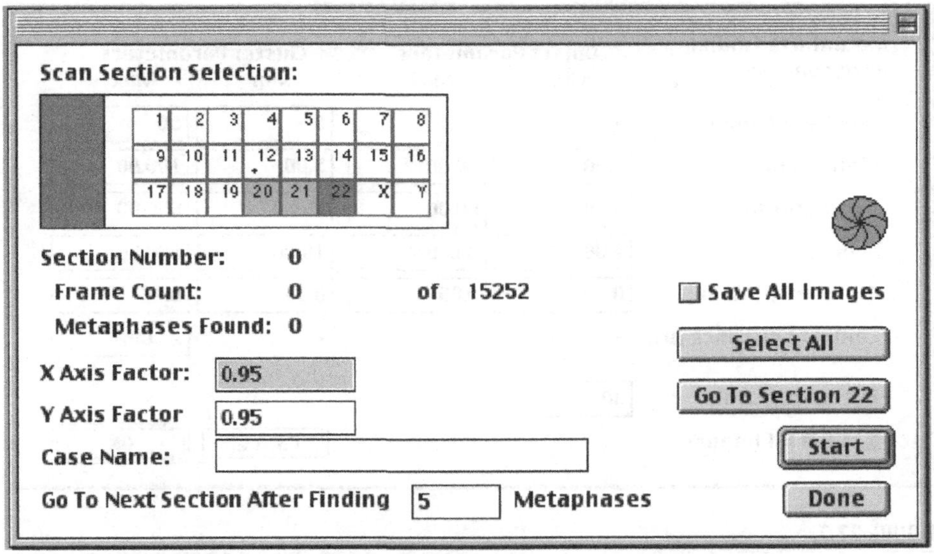

FIGURE 27.6

User interface for automated scanning of the Cytocell Multiprobe™ template microscope slide. The user can select a region to scan at the click of a mouse button (Ex: regions 20, 21, and 22 were selected above). Either the entire selected region can be scanned or the user can define the number of metaphases per region (Ex: 5 metaphases, as shown above). Scanning then continues with the next selected region. The X-axis and Y-axis factors adjust the scanning step size in X and Y, and may be used to capture overlapping regions to avoid the loss of cells that fall between adjacent image frames.

of objects in a group. This works because chromosomes in a metaphase are typically rod-like and are clustered together in groups of approximately 46.

Figure 27.7 shows the user interface for metaphase finding, with default object parameters for images captured with a 100X objective. These parameter values, when tested on more than ten images, accurately identified all the metaphases therein. The result for a representative metaphase image appears in Fig. 27.8. The objects shown in green were selected as members of a cluster, and the clustering algorithm rejected the objects shown in red. The green box encloses the cluster of objects identified as a metaphase. Red boxes show clusters that were rejected (see the lower left corner in Fig. 27.8). Metaphases located at a distance of 15 mm from the boundaries of the squares are also discarded to avoid attempting to analyze metaphases that overlap two neighboring squares. Every metaphase located in an individual square on the Cytocell slide is assigned to a group numbered like the square on the template slide. Images in which metaphases are found are labeled according to their slide region and are stored in an image gallery. These metaphases can be relocated automatically at a later time, using previously stored stage and image coordinates. The automatically identified metaphases are then visually examined for the

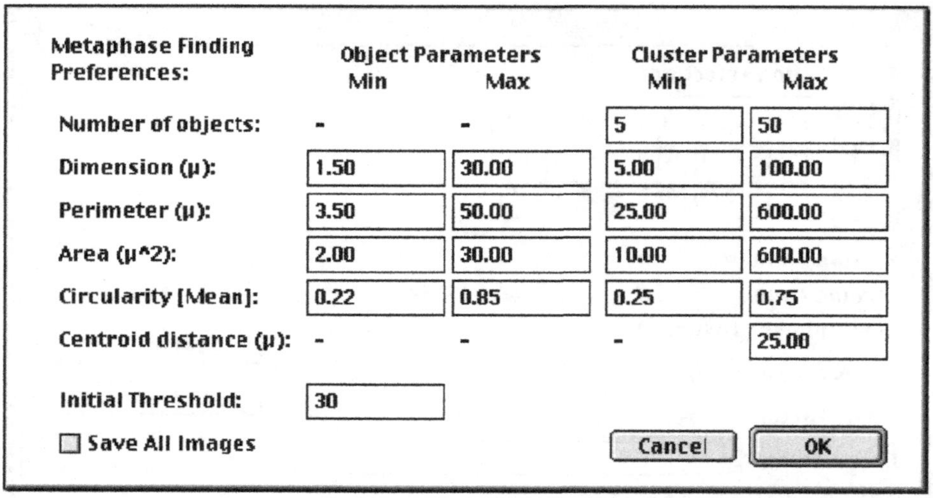

Metaphase Finding Preferences:	Object Parameters		Cluster Parameters	
	Min	Max	Min	Max
Number of objects:	-	-	5	50
Dimension (µ):	1.50	30.00	5.00	100.00
Perimeter (µ):	3.50	50.00	25.00	600.00
Area (µ^2):	2.00	30.00	10.00	600.00
Circularity [Mean]:	0.22	0.85	0.25	0.75
Centroid distance (µ):	-	-	-	25.00
Initial Threshold:	30			
☐ Save All Images			Cancel	OK

FIGURE 27.7

User interface for automated metaphase finding preferences. The object parameters were empirically determined to operate best on typical metaphase specimens captured using a 100X objective.

detection of subtelomeric rearrangements. Figures 27.9 and 27.10 show images of the subtelomeric assay. This specimen has a distal monosomic 2q deletion and is trisomic for distal 17q. The subtelomeric regions on the shorter arms (p) are labeled green with FITC, and the subtelomeric regions on the longer arms (q) are labeled red using Texas Red. As seen in Fig. 27.9, chromosome 2 is deleted for distal q. Figure 27.10 shows trisomy for distal 17q, with a cryptic translocation of distal 17q on chromosome 2.

27.6.3 Detection of Gene Duplications

Recent studies have shown that chromosomal deletions and duplications result in human diseases with complex phenotypic abnormalities [65, 66]. The current understanding is that duplications of segments of the human genome may eventually be shown to be responsible for many human traits [67]. Following the recent sequencing of the human genome, a future task of the human genome project is to delineate genome architectural features, such as low-copy and region-specific repeats (duplications). The eventual identification of these may enable prediction of several regions susceptible to rearrangements associated with genomic disorders. However, current-screening methods for genetic anomalies that use FISH, especially duplication analysis, have not advanced beyond manual screening of specimens. We used the system for computerized microscopy to support fast, accurate, and inexpensive screening of gene duplications. Our approach is to use readily available DNA probes for the specific disorders, such as (1) neuropathies: Charcot-Marie-Tooth Disease (CMT1A) and hereditary neuropathy with pressure palsies,

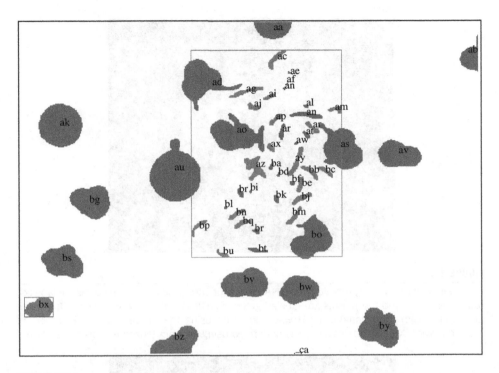

FIGURE 27.8

The output of the metaphase finding algorithm. The isolated objects are labeled aa-az, ba-bz, and ca (total 53 objects), and then classified using the parameters described in Fig. 27.7. The objects in green were classified as objects of a cluster, while red objects are rejected. A cluster of objects outlined by a green box is identified as a metaphase, while cluster objects outlined by a red box are rejected.

(2) neurological disorders: Pelizaeus-Merzbacher disease and X-linked spastic paraplegia, (3) muscular wasting disorders: Duchene and Becker muscular dystrophy, and (4) contiguous-gene syndromes: Smith-Magenis syndrome, for interphase FISH, followed by automated genetic screening to detect gene duplications.

27.6.3.1 *Dot-Finding*

Our system software was tailored for this particular application to perform the following tasks. After an image is acquired, it was to be analyzed to identify nuclei and to detect dots. This involves the following six steps: (1) find the nucleus objects and find the dot objects within the nucleus, (2) determine if each dot object represents a single FISH signal or multiple signals, (3) measure the separation distance between duplicated dots, (4) classify the isolated dots as single, double, split, or overlapping, (5) count the dots, and (6) generate a report. The algorithms for cell and dot finding are described earlier in

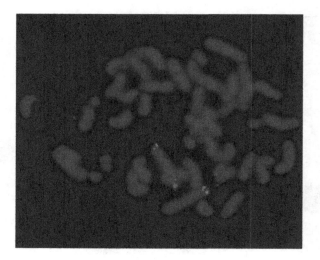

FIGURE 27.9

FISH was performed using subtelomeric DNA probes for chromosome 2. The q-arms are labeled red (Texas Red) and the p-arms are labeled green (FITC). This specimen is deleted for distal 2q (monosomic). Subtelomeric FISH was performed using the Chromoprobe Multiprobe™ T System from Cytocell Ltd. Imaging was performed using a computerized microscopy system.

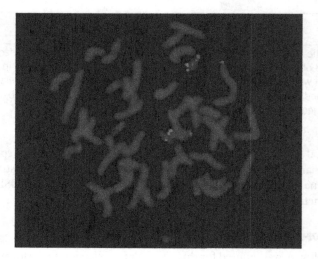

FIGURE 27.10

FISH was performed using subtelomeric DNA probes for chromosome 17. The q-arms are labeled red (Texas Red) and the p-arms are labeled green (FITC). The specimen is trisomic for distal 17q and carries a cryptic translocation (derivative 2). Subtelomeric FISH was performed using the Chromoprobe Multiprobe™ T System from Cytocell Ltd. Imaging was performed using a computerized microscopy system.

the fetal cell project. Specialized algorithms were developed to identify duplicated gene signals.

We implemented the following algorithm to separate individual neighboring dots and to measure the distance between the two dots. Following the dot-finding algorithm, we initially determined the number of dot objects for the target fluorophore (which labeled the gene of interest). In cells carrying a duplication, the nuclei would have a total of three FISH signals for the target gene, of which two signals would occur on the abnormal chromosome (indicating a duplication) and one signal would occur on the normal chromosome. The shape of each dot object was initially measured (using existing shape analysis algorithms) to determine if it is a single or double dot. If the boundary of the dot had low eccentricity, the dot was initially tagged as a single dot. If the eccentricity was relatively high, the object had a higher probability of being a double dot. If the cell nuclei had three dot objects for the target gene, we initially isolated the two objects that were tagged as potential single dots and were nearest to each other to represent FISH signals on the abnormal chromosome. This was achieved by determining the centroid of each dot object [61] as a measure of its spatial position in the cell nuclei. A subimage of the dot objects was then obtained by cropping a square region enclosing the FISH signals. Figure 27.11(a) shows a cropped imaged of two FISH signals.

27.6.3.2 *Surface Fitting*

We obtain morphological and image features for the dot objects by surface fitting, using the sum of two rotated Gaussian surfaces as a model. The sum of two rotated Gaussian surfaces was modeled as follows:

$$f\left(x_a, y_a, A_a, x_{a0}, y_{a0}, \sigma_{ax}, \sigma_{ay}, \theta_a, x_b, y_b, A_b, x_{b0}, y_{b0}, \sigma_{bx}, \sigma_{by}, \theta_b\right) = A + B, \qquad (27.8)$$

where

$$A = e^{\left[-\frac{[(x_a \cos\theta_a - y_a \sin\theta_a) - (x_{a0}\cos\theta_a - y_{a0}\sin\theta_a)]^2}{2\sigma_{ax}^2} - \frac{[(x_a \sin\theta_a + y_a \cos\theta_a) - (x_{a0}\sin\theta_a + y_{a0}\cos\theta_a)]^2}{2\sigma_{ay}^2} \right]}, \qquad (27.9)$$

and

$$B = e^{\left[-\frac{[(x_b \cos\theta_b - y_b \sin\theta_b) - (x_{b0}\cos\theta_b - y_{b0}\sin\theta_b)]^2}{2\sigma_{bx}^2} - \frac{[(x_b \sin\theta_b + y_b \cos\theta_b) - (x_{b0}\sin\theta_b + y_{b0}\cos\theta_b)]^2}{2\sigma_{by}^2} \right]}. \qquad (27.10)$$

In the equations above, A is the amplitude, x_o, y_o are the position, σ_x and σ_y are the standard deviations (radii) in the two directions, x, y, are surface points, and θ is the angle of rotation with respect to the X-axis. These parameters are used with the subscript a or b to represent the two Gaussian surfaces. A least-squared minimization of the mean-squared error was performed using the Quasi Newton Minimization technique [68]. To recover the surface, we estimated the following 12 parameters:

$$A_a, x_{a0}, y_{a0}, \sigma_{ax}, \sigma_{ay}, \theta_a, A_b, x_{b0}, y_{b0}, \sigma_{bx}, \sigma_{by}, \text{ and } \theta_b.$$

The image data points from the subimage containing the dot objects (Fig. 27.11(a)) were used as input points for the minimization routine. Initial estimates for the size parameters

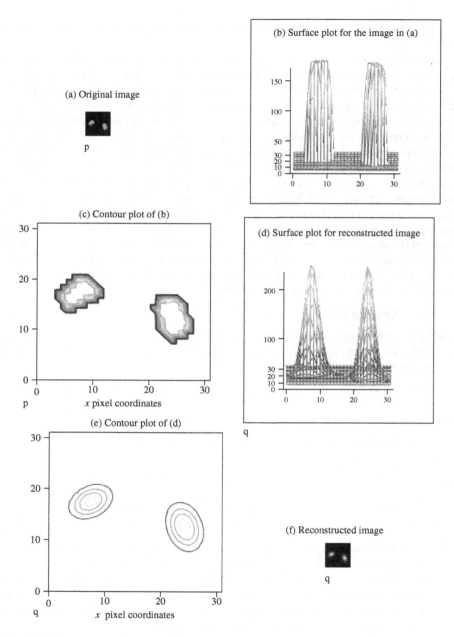

FIGURE 27.11

Surface fitting using the sum of two rotated Gaussians as a model. (a) original image; (b) surface plot of (a); (c) contour plot of (b); (d) surface plot of reconstructed image; (e) contour plot of (d); and (f) reconstructed image.

were obtained from the input data points, as follows. The centroid of the dot objects was used as an estimate for (x_0, y_0), the average image intensity was used to estimate A, the angle of rotation was set to an initial value of 45°, and the standard deviations (σ) in the x and y directions were set to a value of 1.0. The minimization was performed using a constraint tolerance (CTOL) of 0.001 and a convergence tolerance (TOL) of 0.001. The value of CTOL controls the precision of the solution. The larger the value, the less precise the solution may be. For smaller values of CTOL, a more precise solution may be found, but the processing time is increased. The value of TOL controls the duration of an iteration. Typically, we were able to estimate the parameters with negligible error values computed as the square root of the sum of squared residuals (computed value-expected value).

For double dots, the estimated parameters for A and B (Eq. 27.8) differ, and may then be used to represent two single dots that are each modeled as a 2D Gaussian surface. If surface-fitting procedure is actually performed on a single (elliptical) dot, then the estimated parameters from the two Gaussian surfaces in the model have equal σ_x and σ_y values, and their position (x, y) was nearly equal (i.e., within 2 or 3 pixels of each other).

The performance of the surface-fitting algorithm is illustrated in Fig. 27.11. An image of FISH signals (dots) and its corresponding surface and contour plots are illustrated in Figs. 27.11(a) and (b), respectively. A contour plot of the surface is presented in Fig. 27.11(c). Surface fitting was performed to obtain the model parameters, and Figs. 27.11(d)–(e) show the surface plot and contour of the estimated model. Figure 27.11(f) presents the image that was reconstructed using the estimated parameters from the surface fitting.

We tested the algorithm and it performed successfully in all the cases tested. Overall the performance of the algorithm was optimal, except for poor quality images. For images that had an extremely low signal-to-noise ratio, the iteration procedure took slightly longer to converge to the solution resulting in a 1–2% reduction in the processing speed. Similarly, the single dot (from the normal homologous chromosome) was modeled using a single 2D rotated Gaussian to compute its size and integrated intensity.

27.6.3.3 *Ellipse Fitting*

In order to compute the separation distance between double dots, the boundary for each dot was computed using the parameters estimated from the surface-fitting algorithm. This was achieved by modeling each dot as an ellipse. The following equation was used to model a single rotated ellipse:

$$f(x,y) = \frac{[(x\cos\theta - y\sin\theta) - (x_0\cos\theta - y_0\sin\theta)]^2}{2\sigma_x} + \frac{[(x\sin\theta + y\cos\theta) - (x_0\sin\theta + y_0\cos\theta)]^2}{2\sigma_y}.$$

(27.11)

The estimated values for x_0, y_0, σ_x, σ_y, and q obtained from the surface modeling were used in the equation above, and the equation was solved to compute the boundary points by setting $f(x,y)$ to the value of 1.0. Figure 27.12(a) and (b) illustrates the boundary points obtained for the sample image presented in Fig. 27.11. The next step was to compute the separation distance between two dots. The procedure is illustrated in Fig. 27.13. Briefly, the line segment joining the centroids of the two dots was computed as follows:

$$y = \frac{(x - x1)(y1 - y2)}{(x1 - x2)} + y1. \qquad (27.12)$$

This is called the peak to peak distance (PP). The point of intersection of PP with the boundary of each of the dots was then determined by simultaneously solving Eqs. (27.11) and (27.12). Segment PP intersects each dot boundary at two points (4 points total). The point of intersection closest to the midpoint of segment PP was chosen for each dot (shown as (ix1, iy1) and (ix2, iy2) in Fig. 27.12). Then the shortest distance between two dots was taken as the separation distance (SD) and computed as the length of the line segment joining (ix1, iy1) and (ix2, iy2) using the following equation:

$$SD = \sqrt{(ix1 - ix2)^2 + (iy1 - iy2)^2}. \qquad (27.13)$$

The separation distance was then normalized with respect to the size of the cell (cell radius) to obtain a relative measure of the distance. Finally, the total integrated fluorescence intensity and average intensity for each dot were computed using intensity values of all pixels with the boundary. The separation distance and the average fluorescence intensity were then used to classify the dots as described below.

27.6.3.4 *Multiple Dots*

In gene duplication studies, it is important to determine whether a gene is duplicated or single. Duplicated genes are represented in FISH images as two dots of the same color that are separated by a distance greater than or equal to the diameter of a single dot. We developed image analysis algorithms to classify dots and to determine the separation distance between the dots.

FISH dots can occur as touching dots, split dot signals, overlapping dots, or separated dots. The measured values of the separation distance (SD), average intensity (IS) and diameter (DS) for single signals, and average intensity I1 and I2 for duplicated signals were used to classify the dots. The single dot represents the unduplicated gene on the homologous chromosome, and double dots represent the target pair of dots to be classified. Typically, for touching dots the separation distance is zero. A FISH signal is sometimes smeared so that a single dot splits into (appears as) two dots. This is called a split signal. In this case one dot is usually smaller than the other. The separation distance for split dots is less than one-fourth the size of the single dot, and the intensity of both or at least one dot is less than the intensity of the single dot. During the "S" phase of the cell cycle during DNA synthesis, chromosomes are replicated and thus two dots are seen in FISH images. These are called replicated signals. The distance between replicated dots is typically small, because the separation distance is proportional to the width of the sister chromatids. However, since the gene locus is itself replicated, the intensity and

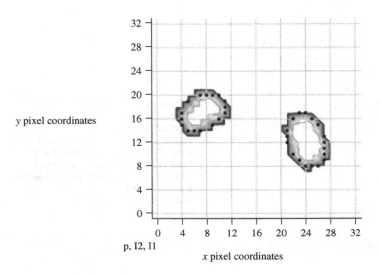

p, I2, I1

x pixel coordinates

(a) Contour plot of original image with dot boundaries superimposed

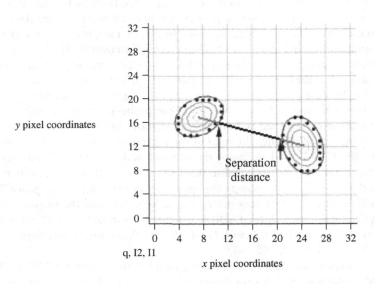

q, I2, I1

x pixel coordinates

(b) Contour plot of reconstructed image (obtained via surface fitting) with dot boundaries superimposed

FIGURE 27.12

Automated measurement of separation distance between duplicated dots.

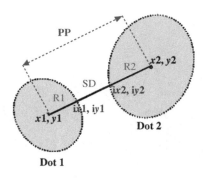

----------	:	Dot boundary estimated from equation (2)
$x1, y1$	:	Centroid of Dot 1
$x2, y2$	:	Centroid of Dot 2
PP	:	Peak to peak distance between Dot 1, and Dot 2
$ix1, iy1$	:	Point of intersection of PP with Dot 1
$ix2, iy2$	:	Point of intersection of PP with Dot 2
SD	:	Separation distance between Dot 1 and Dot 2
R1	:	Radius of Dot 1
R2	:	Radius of Dot 2

FIGURE 27.13

Schematic illustrating the computation of the separation distance.

size of each dot is equal to that of the single dot. Finally, duplicated signals are used to represent true gene duplication. These dots are well separated from each other such that the separation distance between the dots is greater than or equal to the half the size of the single dot, and each has a size and intensity equal to that of a single dot. These criteria are illustrated and outlined in Fig. 27.14. The diameter and intensity of the signal (on the normal chromosome) are chosen for the single dot parameters. Intensity values are considered significant only if the intensity values change by > 40% (for either an increase or a reduction). This is because several other factors such as background noise, homogeneity of the light source, and type and concentration of the probe affect the intensity value. Thus, small changes were neglected and only large variations in intensity are considered while classifying the signals. Finally, the ratio of the separation distance to the diameter of a single dot (SD/DS) was used to classify the signals based on the criteria outlined in Fig. 27.14. If this ratio takes values equal to $0.5, 1.0, 2.0, 3.0, \ldots$, this indicates that one half, one, two, three, $\ldots$, dots can occupy the space between the duplicated genes. The dots were classified based on two parameters: the ratio SD/DS and the intensity ratio $(I1 + I2)/IS$. Split signals have values of SD/DS $\approx 0.0 - 1.0$, and IS $\approx I1 + I2$, replicated signals have a SD/DS value <$0.0 - 0.5$, and $(I1 + I2)/IS \approx 2.0$, and duplicated signals have a SD/DS ratio ≥ 0.5, and $(I1 + I2)/IS \approx 2.0$.

Figure 27.15 presents an image of a cell showing a duplication pattern for CMT1A. The PMP22 cosmid contig was labeled with digoxigenin and detected with antidigoxigenin conjugated to rhodamine, which fluoresces red. The FL1 cosmid contig was labeled with biotin and detected with avidin conjugated to FITC, which fluoresces green. FL1 cosmid was used as an internal control to facilitate chromosome identification and to check hybridization efficiency. In each interphase nucleus, the normal chromosome 17 displays one green and one red signal. In cells carrying the duplication, the abnormal chromosome 17 shows one green signal and two red signals (Fig. 27.15).

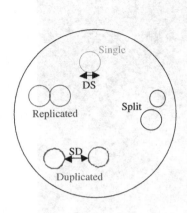

Let Single represent the unduplicated signal on the homologous chromosome, with diameter DS and integrated intensity I_s.

Double dots can be classified as Split, Replicated and Duplicated. Let D_1 and D_2 represent the diameter, and I_1 and I_2 represent the integrated intensities of the two dots in the pair.

Let SD be the shortest distance, and PP be the peak to peak distance (as defined in Fig. 27.5).

Criteria for identifying FISH signals

	$\frac{SD}{DS}$	$\frac{(I_1+I_2)}{IS}$	$D_1 = D_2 = DS$
Split	0.0−0.5	1.0	No
Replicated	0.0−0.5	2.0	Yes
Duplicated	>0.5	2.0	Yes

FIGURE 27.14

Criteria for classifying split, replicated, and duplicated signals.

27.6.3.5 *Performance*

The results of our study are presented in Table 27.1. A total of 10 patient samples were analyzed, with 3−15 cells per sample. An average value of 1.46 ± 0.43 and 2.21 ± 0.45 was measured for SD/DS and I1 + I2 / IS, respectively. The SD between duplicated dots is of a known size, i.e., 1.5 MB, whereas since the signal is duplicated, the intensity of each replicated signal should match that of a single gene giving an intensity ratio of 2.0. The within and across sample variability for SD/DS was found to be ~30% and ~15%, respectively. The within and across sample variability for I1 + I2/IS was found to be ~22% and ~20%, respectively. The automated analysis was able to correctly identify 10/10 (100%) of the samples as gene duplications.

27.6.4 **Four-Color FISH for Aneuploidy Screening**

The AneuVysion Assay (Vysis Inc., Downers Grove) is a prenatal test that provides rapid (24 to 48 hour) detection of trisomy 13, 18, and 21 (Down syndrome) and aneuploidy of sex chromosomes X and Y. The probe mixture for chromosomes 18 (SpectrumAquaTM D18Z1, alpha satellite DNA (18p11.1-q11.1)), X (SpectrumGreenTM DXZ1, alpha satellite DNA (Xp11.1-q11.1)), and Y (SpectrumOrangeTM DYZ3, alpha satellite DNA (Yp11.1-q11.1)) is a three-color FISH assay. This results in four-color images, using aqua, green, and orange for the dots, and blue for the cell nuclei. The system uses a cooled three-chip RGB camera and a 24-bit frame grabber board, allowing simultaneously capture of three color channels in a single video frame time. Use of the cooled three-chip camera facilitates the capture of high-resolution RGB images that are free

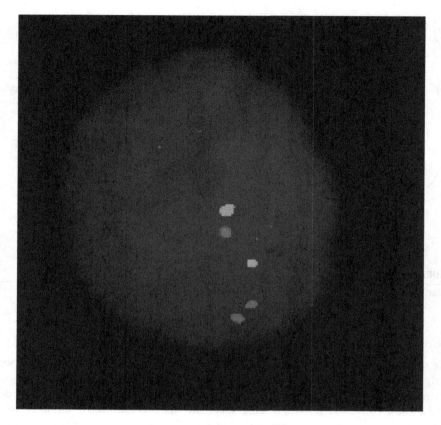

FIGURE 27.15

Representative cell for CMT1A showing a duplication pattern. The PMP22 cosmid contig was labeled with digoxigenin and detected with antidigoxigenin conjugated to rhodamine, which fluoresces red. The FL1 cosmid contig (internal control) was labeled with biotin and detected with avidin conjugated to FITC, which fluoresces green. In each interphase nucleus, the normal chromosome 17 displays one green and one red signal. A duplication is seen here with the abnormal chromosome 17 showing one green signal and two red signals.

of color aliasing. The system can capture four-color FISH images (blue cells with red, green, and aqua signals). Typically, when digitizing four-color images with an RGB camera, one of the three color channels captures the fourth color in a second scan. For FISH samples that use, for example, a blue counterstain, with SpectrumAqua, SpectrumGreen, and SpectrumOrange labeling chromosomes 18, X and Y, respectively, the SpectrumOrange fluorophore is imaged in the red channel, the SpectrumGreen fluorophore in the green channel, and both DAPI and SpectrumAqua are imaged in the blue channel.

TABLE 27.1   Results using CMT1A specimens with a known 1.5 MB duplication.

No.	Signal percentage (N ≈ 100)		# of cells	Automated analysis for separation distance (SD) of duplicated signals (Dot$_1$, Dot$_2$)							
				Radius of signals (μ)			Intensity of signals (μ)			SD/DS	I$_1$+I$_2$/I$_S$
	3 dots	2 dots		Single	Dot$_1$	Dot$_2$	I$_S$	I$_1$	I$_2$	$\mu \pm \sigma$	$\mu \pm \sigma$
1.	70	30	15	2.0	1.5	1.7	42.9	39.1	41.5	1.4 ± 0.4	2.0 ± 0.7
2.	80	20	5	1.3	1.3	0.9	18.7	30.2	16.0	1.5 ± 0.3	2.2 ± 0.4
3.	80	20	9	1.6	1.3	1.4	26.0	25.3	25.4	1.3 ± 0.4	2.0 ± 0.4
4.	80	20	10	1.5	1.1	1.2	27.3	29.4	31.7	1.4 ± 0.3	2.2 ± 1.0
5.	70	30	3	1.6	1.5	1.4	51.2	39.7	45.3	1.6 ± 0.3	1.6 ± 0.5
6.	70	30	3	2.9	2.9	2.5	100.4	79.5	93.7	1.1 ± 0.6	1.7 ± 0.1
7.	80	20	4	2.2	2.1	1.7	110.4	112.5	108.1	1.5 ± 0.5	2.0 ± 0.5
8.	74	26	3	2.0	1.8	2.3	107.7	118.6	111.8	1.0 ± 0.4	2.1 ± 0.2
9.	74	26	3	2.1	2.9	2.0	31.0	50.5	49.0	1.7 ± 0.5	3.2 ± 0.2
10.	70	30	5	3.2	2.5	2.1	134.7	144.8	138.4	1.7 ± 0.6	2.2 ± 0.5
									Average	**1.46 ± 0.43**	**2.12 ± 0.45**

27.6.4.1 *Image Acquisition*

The system employs a novel solution to allow four-color sample processing using an RGB camera. The system employs a filter wheel fitted with a 420 nm longpass (LP) filter in the excitation path, and it captures RGB images using a Quad-band (red, green, blue, aqua) optical filter set. In this setup, the Quad filter is stationary, and image capture is done in two frame-grab cycles. First, the LP filter is positioned in the light path to block DAPI excitation, and the red, green, and aqua channels are captured simultaneously. Next, the LP filter automatically moves out of the excitation path, and the DAPI image is acquired in the blue channel. Thus, we automatically capture four-color images with an RGB camera in two video frame times. The resulting digital image is stored with four channels, one for each color.

27.6.4.2 *Spectral Overlap*

Since each of the four fluorophores is imaged in a separate channel, the problem of analyzing different color dots occurring at the same X-Y location is eliminated. However, to achieve total discrimination between different color signals, the effects of spectral overlap should be minimized. As seen in Fig. 27.16 (top panel), the RGBA image clearly shows the blue nucleus, and the red, green, and aqua dots. The individual red, green, blue, and aqua components of the image are also shown. As seen in each component image, in addition to the true color for each channel (white arrows), there is color bleed-through that occurs from neighboring spectral regions (yellow arrows). This is due to the unavoidable overlap among fluorophore emission spectra and RGB camera sensitivity spectra. The RGBA image was corrected to remove the overlap and separate the fluorophores using color compensation [34]. Figure 27.16 (bottom panel) shows the results of color compensation. The spectral bleed-through is effectively removed, and the different color dots are clearly separated in the individual color component images. Thus, using a 3-CCD color camera, along with the background subtraction and color compensation algorithms discussed above, we obtain good spectral separation with rapid image capture. Similarly, appropriate filter optics, used in conjunction with image processing, allows the capture of multicolor images. Following image capture, the cell and dot finding algorithms described above are applied to implement automatic aneuploidy screening.

27.6.5 Thick Specimen Imaging

Automated microscope instruments almost uniformly do their analysis on 2D images, and their ability to handle thick specimens is severely limited. In thick specimen preparations, structures that fall above or below the focal plane are obscured or lost. The reliability of automated microscopy could be highly increased if the thick specimen limitation were resolved. We have developed a technique for enhancing the image content available in microscope images of thick sections by performing optical section deblurring, followed by image fusion using wavelet transforms.

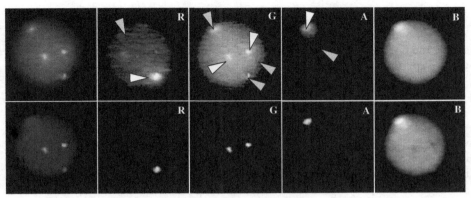

Color compensation applied to an image of a FISH labelled lymphocyte. The nucleus is
counterstained with DAPI, and dots are labelled with red (chromosome Y), green
(chromosome 21), and aqua (chromosome X). The original image and individual red (R),
green (G), aqua (A) and blue (B) channels are shown in the top panel. The compensated
image and its component channels are shown in the bottom panel. The spectral overlap
and its effective removal via color compensation is evident in the green channel image.

Intensity profile of a line segment (white) through aqua, green and red dots.
Graph A presents the intensity profile before processing, and Graph B
presents the intensity profile of the compensated image.

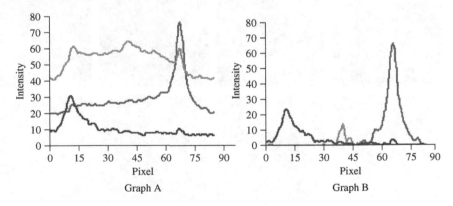

FIGURE 27.16

A four-color image captured using a RGB camera and color compensation.

27.6.5.1 *Deblurring*

Weinstein and Castleman pioneered the deblurring of optical section images using a simple method that involves subtracting adjacent plane images that have been blurred with an appropriate defocus psf [69], given by

$$f_j \approx g_j - \sum_{i=1}^{M} (g_{j-i} * h_{-i} + g_{j+i} * h_i) * k_0, \qquad (27.14)$$

where f_j is the specimen brightness distribution at focus level j, g_j is the optical section image obtained at level j, h_i is the blurring psf due to being out of focus by the amount i, k_0 is a heuristically designed highpass filter, and the $*$ represents the convolution operation.

Thus one can partially remove the defocused structures by subtracting 2M adjacent plane images that have been blurred with the appropriate defocus psf and convolved with a suitable highpass filter k_0. The filter, k_0, and the number, M, of adjacent planes must be selected to give good results. While this technique cannot recover the specimen function exactly, it does improve optical section images at reasonable computational expense. It is often necessary to use only a small number, M, of adjacent planes to remove most of the defocused information. Figure 27.17 shows images from transmitted light microscopy and fluorescence microscopy that have been deblurred using optical sections above and below at a Z-interval of 1 μm. While this technique cannot recover the specimen function exactly, it does improve optical section images at reasonable computational expense.

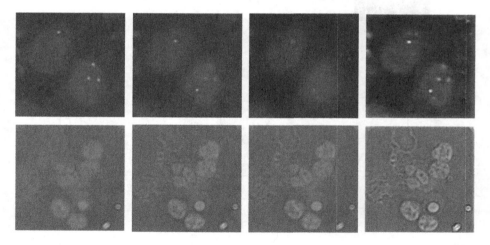

FIGURE 27.17

Deblurring. The top row shows images of FISH-labeled lymphocytes. The left three images are from an optical section stack taken one micron apart. The right image is the middle one deblurred. The in-plane dots are brighter, while the out-of-plane dots are removed. The bottom row shows transmitted images of May-Giemsa stained blood cells. The left three images are from an optical section stack taken one-half micron apart. The rightmost image is the middle one deblurred.

27.6.5.2 *Image Fusion*

One effective way to combine a set of deblurred optical section images into a single 2D image containing the detail from each involves the use of the wavelet transform [8]. A linear transformation is defined by a set of basis functions. It represents an image by a set of coefficients that specify what mix of basis functions is required to reconstruct that image. Reconstruction is effected by summing the basis functions in proportions specified by the coefficients. The coefficients thus reflect how much each of the basis functions resembles a component of the original image. If a few of the basis functions match the components of the image, then their coefficients will be large and the other coefficients will be negligible, yielding a very compact representation. The coefficients that correspond to the desired components of the image can be increased in magnitude, prior to reconstruction, to enhance those components.

27.6.5.3 *Wavelet Design*

A wavelet transform is a linear transformation in which the basis functions (except the first) are scaled and shifted versions of one function, called the "mother wavelet." If the wavelet can be selected to resemble components of the image, then a compact representation results. There is considerable flexibility in the design of basis functions. Thus it is often possible to design wavelet basis functions that are similar to the image components of interest. These components, then, are represented compactly in the transform by relatively few coefficients. These coefficients can be increased in amplitude, at the expense of the remaining components, to enhance the interesting content of the image. Fast algorithms exist for the computation of wavelet transforms.

Mallat's iterative algorithm for implementing the one-dimensional discrete wavelet transform (DWT) [70, 71] is shown in Fig. 27.18. In the design of an orthonormal DWT, one begins with a "scaling vector," $h_0(k)$, of even length. The elements of the scaling vector must satisfy certain constraints imposed by invertibility. For example, the elements must

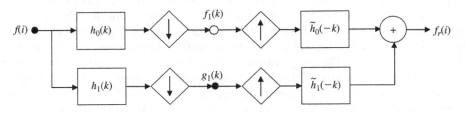

FIGURE 27.18

Mallat's (1D) DWT algorithm. The left half shows one step of decomposition, while the right half shows one step of reconstruction. The down and up arrows indicate downsampling and upsampling by a factor of two, respectively. For an orthonormal transform, the two filters on the right are the same as the two on the left. Further steps of decomposition and reconstruction are introduced at the open circle.

sum to $\sqrt{2}$, their squares must sum to unity, and the sum of the even-numbered elements must equal the sum of the odds [65]. From $h_0(k)$ is generated a "wavelet vector"

$$h_1(k) = \pm (-1)^k h_0(-k). \tag{27.15}$$

These two vectors are used as discrete convolution kernels in the system of Fig. 27.18 to implement the DWT. For example, all possible four-element orthonormal scaling vectors are specified by

$$h_0 = [c_1 c_3 \quad c_1 \quad c_2 \quad -c_2 c_3]^T, \tag{27.16}$$

where

$$c_3 = \frac{\sqrt{1 - (c_1^2 + c_2^2)}}{\sqrt{c_1^2 + c_2^2}}, \tag{27.17}$$

and c_1 and c_2 are free choice parameters. All even-length orthonormal scaling vectors, and biorthogonal scaling vectors of any length, can be similarly parameterized.

Mallat's algorithm leads to the "cascade" algorithm of Daubechies [70, 71], which is a simple method for constructing the basis functions that correspond to specified scaling and wavelet vectors. With these tools it is then simple to specify and design wavelet transforms with desired properties. Given parameterized scaling and wavelet vectors, first select the parameter values (e.g., c_1 and c_2, above) and then use the cascade algorithm to construct the corresponding scaling function and basic wavelet. These show the form of the basis functions of that wavelet transform. Repeat the process using different parameter values until the desired basis function shape is attained. Then use $h_0(k)$ and $h_1(k)$ in the 2D version of Mallat's algorithm to implement the wavelet transform and its inverse.

27.6.5.4 *Wavelet Fusion*

Image fusion is the technique of combining multiple images into one that preserves the interesting detail of each [72]. The wavelet transform affords a convenient way to fuse images. One simply takes, at each coefficient position, the coefficient value having maximum absolute amplitude and then reconstructs an image from all such maximum-amplitude coefficients. If the basis functions match the interesting components of the image, then the fused image will contain the interesting components collected up from all of the input images. The images can be combined in the transform domain by taking the maximum-amplitude coefficient at each coordinate. An inverse wavelet transform of the resulting coefficients then reconstructs the fused image. We found that deblurring prior to wavelet fusion significantly improves the measured sharpness of the processed images. An example of wavelet image fusion using transmitted light and fluorescence images is shown in Fig. 27.19. Optical section deblurring followed by image fusion produced an image in which all of the dots are visible for the fluorescence images. We use these techniques to improve the information content of images from thick samples. Specifically, this technique improves the dot information in acquired FISH images because it incorporates data from focal planes above and below.

27.7 COMMERCIALLY AVAILABLE SYSTEMS

Computer-assisted microscopy systems can vary in price, sensitivity, and capability. The selection of a system depends upon the experimental applications for which it will be used. Typically, the selection is based on requirements for image resolution, sensitivity, light conditions, image acquisition time, image storage requirements, and most importantly the postacquisition image processing and analysis required. Other considerations are the technical demands of assembling the component hardware and configuring software. Computerized imaging systems can be assembled from component parts or obtained from a supplier as a fully integrated system. Several companies offer fully integrated computerized microscopy systems and/or provide customized solutions for specialized systems.

A brief listing of some of the commercially available systems is provided here. Applied Precision Inc. (Issaquah, WA) provides a computerized imaging instrument, the DeltaVision™ Restoration Microscopy System for applications such as 3D time course studies with live cell material. Applied Precision also offers the softWoRx™ Imaging Workstation for post-acquisition image processing such as deconvolution, 3D segmentation, and rendering. Universal Imaging Corp. (West Chester, PA) provides software, including the MetaMorph™, MetaView™, and MetaFluor™ systems, which can be customized for computerized microscopy applications in transmitted light, time

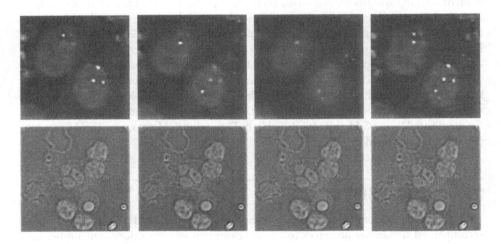

FIGURE 27.19

Image fusion using transmitted light and fluorescence images. The top row shows FISH-labeled lymphocytes. The left three images are from a deblurred optical section stack taken one micron apart. The right image is the fusion of the three using the biorthogonal 2,2 wavelet transform. Notice that the fused image has all of the dots in focus. The bottom rows demonstrate a similar effect in transmitted light images. The deblurring process, followed by image fusion, enhances image detail.

lapse studies, and fluorescence microscopy. VayTek Inc. (Fairfield, Iowa) provides an integrated microscopy imaging system, the Proteus^{TM} system, that can be custom configured to any microscopy system. VayTek's proprietary software for deconvolution and 3D reconstruction, including MicroTome^{TM}, VoxBlast^{TM}, HazeBuster^{TM}, Vtrace^{TM}, and Volume-Scan^{TM}, can also be custom configured for most current microscopy systems. ChromaVision Medical Systems Inc. (San Juan, CA) provides an Automated Cellular Imaging System that allows cell detection based on color-, size-, and shape-based morphometric features. MetaSystems GmbH (Altlussheim, Germany) provides a computerized microscopy system based on Zeiss optics for scanning and imaging pathology slides, cytogenetic slides for FISH, MFISH, and metaphase detection, oncology slides, and for rare cell detection, primarily from blood, bone marrow, or tissue section samples. Applied Imaging Corp. (Santa Clara, CA), now part of MetaSystems GmbH, Germany, provides fully automated scanning and image analysis systems. Their MDS^{TM} system provides automated slide scanning using brightfield or fluorescent illumination to allow standard karyotyping, FISH, and comparative genomic hybridization, as well as rare cell detection. They also have the Oncopath^{TM} and Ariol Sl-50^{TM} image analysis systems for oncology and clinical pathology applications.

The field of automated imaging is also of great interest to pharmaceutical and biotechnology companies. Many are now developing high-throughput and high-content screening platforms for automated analysis of intracellular localization and dynamics of proteins and to view the effects of a drug on living cells more quickly. High-content imaging systems for cell-based assays have proliferated in the past year, examples include Cellomic's ArrayScan system and KineticScan workstation (Cellomics, Inc., Pittsburgh, PA); Amersham's INCell Analyzer 1000 and 3000 (Amersham Biosciences Corp., Piscataway, NJ); Acumen Bioscience's Explorer system (Melbourn, United Kingdom); CompuCyte's iCyte imaging cytometer and LSC laser scanning cytometer (CompuCyte Corporation, Cambridge, MA); Atto Bioscience's Pathway HT kinetic cell imaging system (Atto Bioscience Inc., Rockville, MD); Universal Imaging's Discovery-1 system (Universal Imaging Corporation, Downingtown, PA); and Q3DM's (now part of Beckman Coulter, San Diego, CA), EIDAQ 100 High-Throughput Microscopy (HTM) system (recently discontinued).

27.8 CONCLUSIONS

The rapid development of microscopy techniques over the last few decades has been accompanied by similar advances in the development of new fluorescent probes and improvements in automated microscope systems and software. Advanced applications such as deconvolution, FRET, and ion ratio imaging require sophisticated routines for controlling automated microscopes and peripheral devices such as filter wheels, shutters, automated stages, and cameras. Computer-assisted microscopy provides the ability to enhance the speed of microscope data acquisition and data analysis, thus relieving humans of tedious tasks. Not only the cost efficiency is improved due to the corresponding reduction in labor costs and space but also errors associated with operator bias are eliminated. Researchers are not only relieved from tedious manual tasks but may also quickly examine thousands of cells, plates, and slides, as well as precisely determine some informative activity against a cell, and collect and mine massive amounts of data. The process is also repeatable and reproducible with a high degree of precision.

We have described a specific configuration of a computerized fluorescence microscope with applications in clinical cytogenetics. Fetal cell screening from maternal blood has the potential to revolutionize the future of prenatal genetic testing, making noninvasive testing available to all pregnant women. Its clinical realization will be practical only via an automated screening procedure because of the rare number of fetal cells available. Specialized slides, based on the grid template, such as the subtelomeric FISH assay, require automated scanning methods to increase accuracy and efficiency of the screening protocol. Similarly, automated techniques are necessary to allow the quantitative analysis for the measurement of the separation distance for detection of duplicated genes. Thick specimen imaging using deblurring methods allows the detection of cell structures that are distributed throughout the volume of the entire cell. Thus, there are sound reasons for pursuing the goal of automation in medical cytogenetics. Not only does automation increase laboratory throughput, it also decreases laboratories' costs for performing tests. And as tests become more objective, the liability of laboratories also decreases. The market for comprehensive automated tests is vast in terms of both size (whether measured in test volume or dollars) and potential impact on people's lives.

The effective commercial use of computer-assisted microscopy and quantitative image analysis requires the careful integration of automated microscopy, high-quality image acquisition, and powerful analytical algorithms that can rationally detect, count, and quantify areas of interest. Typically, the systems should provide walk-away scanning operation with automated slide loaders that can queue several (50 to 200) slides. Additionally, the automated microscopy systems should have the capability to integrate with viewing stations to create a network for reviewing images, analyzing data, and generating reports. There has been an increase in the commercialization of computerized microscopy and high-content imaging systems over that past five years. Clearly, future developments in this field will be of great interest to biotechnology. All signs indicate that superior optical instrumentation and software for cell research are on the development horizon.

ACKNOWLEDGMENTS

We would like to thank Vibeesh Bose, Hyohoon Choi, and Mehul Sampat for their assistance with the development and testing of the computerized microscopy system. The development of the automated microscopy system was partially supported by NIH SBIR Grant Nos. HD34719-02, HD38150-02, and GM60894-01.

REFERENCES

[1] M. Bravo-Zanoguera, B. Massenbach, A. Kellner, and J. H. Price. High-performance autofocus circuit for biological microscopy. *Rev. Sci. Instrum.*, 69:3966–3977, 1998.

[2] J. C. Oosterwijk, C. F. Knepfle, W. E. Mesker, H. Vrolijk, W. C. Sloos, H. Pattenier, I. Ravkin, G. J. van Ommen, H. H. Kanhai, and H. J. Tanke. Strategies for rare-event detection: an approach for automated fetal cell detection in maternal blood. *Am. J. Hum. Genet.*, 63:1783–1792, 1998.

[3] L. A. Kamenstsky, L. D. Kamenstsky, J. A. Fletcher, A. Kurose, and K. Sasaki. Methods for automatic multiparameter analysis of fluorescence in situ hybridized specimens with a laser scanning cytometer. *Cytometry*, 27:117–125, 1997.

[4] H. Netten, I. T. Young, L. J. van Vliet, H. J. Tanke, H. Vroljik, and W. R. Sloos. FISH and chips: automation of fluorescent dot counting in interphase cell nuclei. *Cytometry*, 28:1–10, 1997.

[5] I. Ravkin and V. Temov. Automated microscopy system for detection and genetic characterization of fetal nucleated red blood cells on slides. *Proc. Opt. Invest. Cells In Vitro In Vivo*, 3260:180–191, 1998.

[6] D. J. Stephens and V. J. Allan. Light microscopy techniques for live cell imaging. *Science*, 300:82–86, 2003.

[7] T. Lehmann, J. Brendo, V. Metzler, G. Brook, and W. Nacimlento. Computer assisted quantification of axo-somatic buttons at the cell membrane of motorneurons. *IEEE Trans. Biomed. Eng.*, 48:706–717, 2001.

[8] J. Lu, D. M. Healy, and J. B. Weaver. Contrast enhancement of medical images using multiscale edge representation. *Opt. Eng.*, 33:2151–2161, 1994.

[9] J. S. Ploem, A. M. van Driel-Kulker, L. Goyarts-Veldstra, J. J. Ploem-Zaaijer, N. P. Verwoerd, and M. van der Zwan. Image analysis combined with quantitative cytochemistry. *Histochemistry*, 84:549–555, 1986.

[10] E. M. Slayter and H. S. Slayter. *Light and Electron Microscopy*. Cambridge University Press, New York, NY, 1992.

[11] I. T. Young. Quantitative microscopy. *IEEE Eng. Med. Biol.*, 15:59–66, 1996.

[12] J. D. Cortese. Microscopy paraphernalia: accessories and peripherals boost performance. *The Scientist*, 14(24):26, 2000.

[13] E. Gratton and M. J. vandeVan. Laser sources for confocal microscopy. In J. B. Pawley, editor, *Handbook of Biological Confocal Microscopy*. Plenum, New York, 69–97, 1995.

[14] Q. Wu. Autofocusing. In Q. Wu, F. A. Merchant, and K. R. Castleman, editors, *Microscope Image Processing*. Academic Press, Boston, MA, 441–467, 2008.

[15] E. T. Johnson and L. J. Goforth. Metaphase spread detection and focus using closed circuit television. *J. Histochem. Cytochem.*, 22(7):536–545, 1974.

[16] B. Dew, T. King, and D. Mighdoll. An automatic microscope system for differential leucocyte counting. *J. Histochem. Cytochem.*, 22:685–696, 1974.

[17] H. Harms and H. M. Aus. Comparison of digital focus criteria for a TV microscope system. *Cytometry*, 5:236–243, 1984.

[18] F. C. Groen, I. T. Young, and G. Ligthart. A comparison of different focus functions for use in autofocus algorithms. *Cytometry*, 6(2):81–91, 1985.

[19] F. R. Boddeke, L. J. van Vliet, H. Netten, and I. T. Young. Autofocusing in microscopy based on the OTF and sampling. *Bioimaging*, 2:193–203, 1994.

[20] L. Firestone, K. Cook, K. Culp, N. Talsania, and K. Preston. Comparison of autofocus methods for automated microscopy. *Cytometry*, 12(3):195–206, 1991.

[21] D. Vollath. Automatic focusing by correlative methods. *J. Microsc.*, 147:279–288, 1987.

[22] D. Vollath. The influence of scene parameters and of noise on the behavior of automatic focusing algorithms. *J. Microsc.*, 152(2):133–146, 1988.

[23] J. F. Brenner, B. S. Dew, J. B. Horton, T. King, P. W. Neurath, and W. D. Selles. An automated microscope for cytologic research. *J. Histochem. Cytochem.*, 24:100–111, 1976.

[24] A. Erteza. Depth of convergence of a sharpness index autofocus system. *Appl. Opt.*, 15:877–881, 1976.

[25] A. Erteza. Sharpness index and its application to focus control. *Appl. Opt.*, 16:2273–2278, 1977.

[26] R. A. Muller and A. Buffington. Real time correction of atmospherically degraded telescope images through image sharpening. *J. Opt. Soc. Am.*, 64:1200–1210, 1974.

[27] J. H. Price and D. A. Gough. Comparison of phase-contrast and fluorescence digital autofocus for scanning microscopy. *Cytometry*, 16(4):283–297, 1994.

[28] J. M. Geusebroek, F. Cornelissen, A. W. Smeulders, and H. Geerts. Robust autofocusing in microscopy. *Cytometry*, 39(1):1–9, 2000.

[29] A. Santos, C. Ortiz de Solorzano, J. J. Vaquero, J. M. Pena, N. Malpica, and F. del Pozo. Evaluation of autofocus functions in molecular cytogenetic analysis. *J. Microsc.*, 188(3):264–272, 1997.

[30] J. C. Russ. *The Image Processing Handbook*. CRC Press, Boca Raton, FL, 1994.

[31] K. R. Castleman, T. P. Riopka, and Q. Wu. FISH image analysis. *IEEE Eng. Med. Biol.*, 15(1):67–75, 1996.

[32] E. R. Dougherty and J. Astola. *An Introduction to Nonlinear Image Processing*. SPIE, Bellingham, WA, 1994.

[33] J. Serra. *Image Analysis and Mathematical Morphology*. Academic Press, London, 1982.

[34] K. R. Castleman. Digital image color compensation with unequal integration periods. *Bioimaging*, 2:160–162, 1994.

[35] K. R. Castleman and I. T. Young. Fundamentals of microscopy. In Q. Wu, F. A. Merchant, and K. R. Castleman, editors, *Microscope Image Processing*. Academic Press, Boston, MA, 11–25, 2008.

[36] Y. Wang, Q. Wu, and K. R. Castleman. Image enhancement. In Q. Wu, F. A. Merchant, and K. R. Castleman, editors, *Microscope Image Processing*. Academic Press, Boston, MA, 59–78, 2008.

[37] W. E. Higgins, W. J. T. Spyra, E. L. Ritman, Y. Kim, and F. A. Spelman. Automatic extraction of the arterial tree from 3-D angiograms. *IEEE Conf. Eng. Med. Biol.*, 2:563–564, 1989.

[38] N. Niki, Y. Kawata, H. Satoh, and T. Kumazaki. 3D imaging of blood vessels using x-ray rotational angiographic system. *IEEE Med. Imaging Conf.*, 3:1873–1877, 1993.

[39] C. Molina, G. Prause, P. Radeva, and M. Sonka. 3-D catheter path reconstruction from biplane angiograms. *SPIE*, 3338:504–512, 1998.

[40] A. Klein, T. K. Egglin, J. S. Pollak, F. Lee, and A. Amini. Identifying vascular features with orientation specific filters and b-spline snakes. *IEEE Comput. Cardiol.*, 113–116, 1994.

[41] A. K. Klein, F. Lee, and A. A. Amini. Quantitative coronary angiography with deformable spline models. *IEEE Trans. Med. Imaging*, 16:468–482, 1997.

[42] D. Guo and P. Richardson. Automatic vessel extraction from angiogram images. *IEEE Comput. Cardiol.*, 25:441–444, 1998.

[43] Y. Sato, S. Nakajima, N. Shiraga, H. Atsumi, S. Yoshida, T. Koller, G. Gerig, and R. Kikinis. 3D multi-scale line filter for segmentation and visualization of curvilinear structures in medical images. *IEEE Med. Image Anal.*, 2:143–168, 1998.

[44] K. R. Castleman. *Digital Image Processing.* Prentice Hall, Englewood Cliffs, NJ, 1996.

[45] T. Ridler and S. Calvard. Picture thresholding using an iterative selection method. *IEEE Trans. Syst. Man Cybern.*, 8:629–632, 1978.

[46] W. Tsai. Moment-preserving thresholding. *Comput. Vis. Graph. Image Process.*, 29:377–393, 1985.

[47] N. Otsu. A threshold selection method from gray-level histograms. *IEEE Trans. Syst. Man Cybern.*, 9:62–66, 1979.

[48] J. Kapur, P. Sahoo, and A. Wong. A new method for gray-level picture thresholding using the entropy of the histogram. *Comput. Vis. Graph. Image Process.*, 29(3):273–285, 1985.

[49] Q. Wu, and K. R. Castleman. Image segmentation. In Q. Wu, F. A. Merchant, and K. R. Castleman, editors, *Microscope Image Processing.* Academic Press, Boston, MA, 159–194, 2008.

[50] A. C. Bovik, editor. *The Handbook of Image and Video Processing*, Chap. 4.8, 4.9, and 4.12. Elsevier Academic Press, 2005.

[51] T. McInerney and D. Terzopoulos. Deformable models in medical image analysis: a survey. *IEEE Med. Image Anal.*, 1:91–108, 1996.

[52] C. Smets, G. Verbeeck, P. Suetens, and A. Oosterlinck. A knowledge-based system for the delineation of blood vessels on subtraction angiograms. *Pattern Recognit. Lett.*, 8:113–121, 1988.

[53] R. Nekovei and Y. Sun. Back-propagation network and its configuration for blood vessel detection in angiograms. *IEEE Trans. Neural Netw.*, 6:64–72, 1995.

[54] L. Dorst. *Discrete Straight Lines: Parameters, Primitives and Properties.* Delft University Press, Delft, The Netherlands, 1986.

[55] T. Y. Young, and K. Fu. *Handbook of Pattern Recognition and Image Processing.* Academic Press, San Diego, CA, 1986.

[56] A. K. Jain. *Fundamentals of Digital Image Processing.* Prentice-Hall, Englewood Cliffs, NJ, 1989.

[57] H. Firth, P. A. Boyd, P. Chamberlain, I. Z. Mackenzie, R. H. Lindenbaum, and S. M. Hudson. Severe limb abnormalities after chorionic villus sampling at 56 to 66 days. *Lancet*, 1:762–763, 1991.

[58] D. Ganshirt-Ahlert, M. Burschyk, H. P. Garritsen, L. Helmer, P. Miny, J. Horst, H. P. Schneider, and W. Holzgreve. Magnetic cell sorting and the transferrin receptor as potential means of prenatal diagnosis from maternal blood. *Am. J. Obstet. Gynecol.*, 166:1350–1355, 1992.

[59] D. W. Bianchi, G. K. Zickwolf, M. C. Yih, A. F. Flint, O. H. Geifman, M. S. Erikson, and J. M. Williams. Erythroid-specific antibodies enhance detection of fetal nucleated erythrocytes in maternal blood. *Prenat. Diagn.*, 13:293–300, 1993.

[60] S. Elias, J. Price, M. Dockter, S. Wachtel, A. Tharapel, J. L. Simpson, and K. W. Klinger. First trimester prenatal diagnosis of trisomy 21 in fetal cells from maternal blood. *Lancet*, 340:1033, 1992.

[61] F. A. Merchant, S. J. Aggarwal, K. R. Diller, K. A. Bartels, and A. C. Bovik. Three-dimensional distribution of damaged cells in cryopreserved pancreatic islets as determined by laser scanning confocal microscopy. *J. Microsc.*, 169:329–338, 1993.

[62] K. R. Castleman and B. S. White. Dot-Count proportion estimation in FISH specimens. *Bioimaging*, 3:88–93, 1995.

[63] F. A. Merchant and K. R. Castleman. Strategies for automated fetal cell screening. *Hum. Reprod. Update*, 8(6):509–521, 2002.

[64] S. J. Knight and J. Flint. Perfect endings: a review of subtelomeric probes and their use in clinical diagnosis. *J. Med. Genet.*, 37(6):401–409, 2000.

[65] X. Hu, A. M. Burghes, P. N. Ray, M. W. Thompson, E. G. Murphy, and R. G. Worton. Partial gene duplication in Duchenne and Becker muscular dystrophies. *J. Med. Genet.*, 25:369–376, 1988.

[66] K. S. Chen, P. Manian, T. Koeuth, L. Potocki, Q. Zhao, A. C. Chinault, C. C. Lee, and J. R. Lupski. Homologous recombination of a flanking repeat gene cluster is a mechanism for a commom contiguous gene deletion syndrome. *Nat. Genet.*, 17:154–163, 1997.

[67] L. Potocki, K. Chen, S. Park, D. E. Osterholm, M. A. Withers, V. Kimonis, A. M. Summers, W. S. Meschino, K. Anyane-Yeboa, C. D. Kashork, L. G. Shaffer, and J. R. Lupski. Molecular mechanism for duplication 17p11.2 - the homologous recombination reciprocal of the Smith-Magenis microdeletion. *Nat. Genet.*, 24:84–87, 2000.

[68] W. H. Press, W. T. Flannery, S. A. Teukolsky, and B. P. Vetterling. *Numerical Recipes in C*. Cambridge University Press, New York, 1992.

[69] M. Weinstein and K. R. Castleman. Reconstructing 3-D specimens from 2-D section images. *Proc. SPIE*, 26:131–138, 1971.

[70] S. Mallat. A theory for multiresolution signal decomposition: the wavelength representation. *IEEE Trans. Pattern Anal. Mach. Intell.*, 11:674–693, 1989.

[71] I. Daubechies. Orthonormal bases of compactly supported wavelets. *Commun. Pure and Appl. Math.*, 41:909–996, 1988.

[72] J. Aggarwal. *Multisensor Fusion for Computer Vision*. Springer-Verlag, New York, NY, 1993.

Towards Video Processing

Alan C. Bovik

The University of Texas at Austin

28

Hopefully the reader has found the *Essential Guide to Image Processing* to be a valuable resource for understanding the principles of digital image processing, ranging from the very basic to the more advanced. The range of readers interested in the topic is quite broad, since image processing is vital to nearly every branch of science and engineering, and increasingly, in our daily lives.

Of course our experience of images is not limited to the still images that are considered in this *Guide*. Indeed, much of the richness of visual information is created by scene changes recorded as time-varying visual information. Devices for sensors and recording moving images have been evolving very rapidly in terms of speed, accuracy, and sensitivity, and for nearly every type of available radiation. These time-varying images, regardless of modality, are collectively referred to as *video*.

Of course the main application of digital video processing is to provide high-quality visible-light videos for human consumption. The ongoing explosion of digital and high-definition television, videos on the internet, and wireless video on handheld devices ensures that there will be significant interest in topics in digital video processing for a long time.

Video *analysis* is still a young field with considerable work left to be done. By mining the rich spatio-temporal information that is available in video, it is possible to analyze the growth or evolutionary properties of dynamic physical phenomena or of living specimens. More broadly, video streams may be analyzed to detect movement for security purposes, for vehicle guidance or navigation, and for tracking moving objects, including people.

Digital video processing encompasses many approaches that derive from the essential principles of digital image processing (of still images) found in this *Guide*. Indeed, it is best to become conversant in the techniques of digital image processing before embarking on the study of digital video processing. However, there is one important aspect of video processing that significantly distinguishes it from still image processing, makes necessary significant modifications of still image processing methods for adaptation to video, and also requires the development of entirely new processing philosophies. That aspect is *motion*.

833

Digital videos are taken from a real world containing 3D objects in motion. These objects in motion project to images that are in motion, meaning that the image intensities and/or colors are in motion at the image plane. Motion has attributes that are both simple and complex. Simple, because most visual motion is relatively smooth in the sense that the instantaneous velocities of 3D objects do not usually change very quickly. Yet object motion can also be complex, and includes deformations (when objects change shape), occlusions (when one object moves in front of another), acceleration (when objects change their direction), and so on.

It is largely the motion of these 3D objects and their 2D projections that determines our visual experience of the world. The way in which motion is handled in video processing largely determines how videos will be perceived or analyzed. Indeed, one of the first steps in a large percentage of video processing algorithms is *motion estimation*, whereby the movement of intensities or colors is estimated. These motion estimates can be used in a wide variety of ways for video processing and analysis.

Other ways wherein video presents special challenges relate to the significant increase in data volume. The extra (temporal) dimension of video implies significant increases in required storage, bandwidth, and processing resources. Naturally it is of high interest to find efficient algorithms that exploit some of the special characteristics of video, such as temporal redundancy, in video processing.

The companion book to this one, the *Essential Guide to Video Processing,* explains the significant problems encountered in video processing, beginning with the essentials of video sampling, through motion estimation and tracking, common processing steps such as enhancement and interpolation, the extremely important topic of video compression, and on to more advanced topics such as video quality assessment, video networking, video security, and wireless video. Like the current book, the companion video *Guide* finishes with a series of interesting and essential applications including video surveillance, video analysis of faces, medical video processing, and video-speech analysis.

It is our hope that the reader will embark on the second leg of their voyage of discovery into one of the most exciting and timely technological topics of our age. The first leg, digital image processing, while extremely fascinating and important on its own intellectual and practical merits, is in many ways a prelude to the broader, more sophisticated, and more challenging topic of digital video processing.

Index

Page numbers followed by "f" indicate figures and "t" indicate tables.

835